Isozymes

III

Developmental Biology

Organizing Committee and Editorial Board

CLEMENT L. MARKERT

Chairman and Editor

Department of Biology
Yale University
New Haven, Connecticut

BERNARD L. HORECKER
Roche Institute of
  Molecular Biology
Nutley, New Jersey

ELLIOT S. VESELL
Department of Pharmacology
Pennsylvania State University
College of Medicine
Hershey, Pennsylvania

JOHN G. SCANDALIOS
Department of Biology
University of South Carolina
Columbia, South Carolina

GREGORY S. WHITT
Provisional Department of
  Genetics and Development
University of Illinois
Urbana, Illinois

The Third International Conference on Isozymes
Held at Yale University, New Haven, Connecticut
April 18-20, 1974

Isozymes

III

Developmental Biology

EDITED BY

Clement L. Markert

Department of Biology
Yale University

ACADEMIC PRESS   New York   San Francisco   London   1975

A Subsidiary of Harcourt Brace Jovanovich, Publishers

ACADEMIC PRESS RAPID MANUSCRIPT REPRODUCTION

ACADEMIC PRESS, INC.
111 Fifth Avenue, New York, New York 10003

United Kingdom Edition published by
ACADEMIC PRESS, INC. (LONDON) LTD.
24/28 Oval Road, London NW1

Library of Congress Cataloging in Publication Data

International Conference on Isozymes, 3d, Yale Univer-
    sity, 1974
    Isozymes.

    Includes bibliographies and indexes.
    CONTENTS: v. 1.   Molecular structure.   v. 2.   Phys-
iological functions.   v. 3.   Developmental biology.
    1.   Isoenzymes–Congresses.   I.   Markert, Clement
Lawrence, 1917-   II.   Title.   [DNLM: 1.   Isoenzymes –
Congresses.   W3 IN182A1974i/QU135 I587 1974i]
QP601.I48   574.1'925   74-31288
ISBN 0–12–472703–4 (v. 3)

Contents

CONTENTS

CONTENTS

Contributors

Louisa M. Atherton, Department of Biology, The Pennsylvania State University, University Park, Pennsylvania 16802

Doris Balinsky, Enzyme Research Unit, The South African Institute for Medical Research, Post Office Box 1038, Johannesburg, South Africa

Hans Peter Bernhard, Department of Biology, Yale University, New Haven, Connecticut 06520

I. Bersohn, Enzyme Research Unit, The South African Institute for Medical Research, Post Office Box 1038, Johannesburg, South Africa

A. Blanco, Cátedra de Química Biológica, Facultad de Ciencias Médicas, Universidad Nacional de Córdoba, Córdoba, Argentina

G. M. Booth, Department of Zoology, Brigham Young University, Provo, Utah 84601

James L. Brewbaker, Department of Horticulture, University of Hawaii, 3190 Maile Way, Honolulu, Hawaii 96822

Efthihia Cayanis, Enzyme Research Unit, The South African Institute for Medical Research, Post Office Box 1038, Johannesburg, South Africa

Michael J. Champion, Department of Zoology, University of Illinois, Urbana, Illinois 61801

Su-En Chao, Division of Biochemical Genetics, School of Medicine, State University of New York, Buffalo, New York 14207

Verne M. Chapman, Department of Molecular Biology, Roswell Park Memorial Institute, Buffalo, New York 14203

S. Chou, Department of Biological Chemistry, Washington University, St. Louis, Missouri 63110

M. E. Conklin, Department of Biology, San Diego State University, San Diego, California 92182

D. W. Cooper, School of Biological Sciences, MacQuarie University, North Ryde, N. S. W. 2113, Australia

Wayne E. Criss, Departments of Obstetrics and Gynecology and Biochemistry, University of Florida College of Medicine, Gainesville, Florida 32610

Gretchen J. Darlington, Department of Medicine, Division of Human Genetics, The New York Hospital-Cornell Medical Center, New York, New York 10021

Marc E. De Broe, Department of Clinical Chemistry, University Hospital Gent, De Pintelaan 135, 9000 Gent, Belgium

Darrell Doyle, Department of Molecular Biology, Roswell Park Memorial Institute, Buffalo, New York 14203

Wolfgang Engel, Institut fur Humangenetik und Anthropologie, der Universitat, Albertstrasse 11, D-7800 Freiburg, West Germany

Charles J. Epstein, Department of Pediatrics, University of California, San Francisco, California 94143

Robert P. Erickson, Department of Pediatrics, University of California, San Francisco, California 94143

Robert N. Feinstein, Division of Biological and Medical Research, Argonne National Laboratory, Argonne, Illinois 60439

Michael R. Felder, Department of Biology, University of South Carolina, Columbia, South Carolina 29208

William H. Fishman, Tufts Cancer Research Center, Tufts University School of Medicine, 136 Harrison Avenue, Boston, Massachusetts 02111

Winifried Franke, Institut fur Humangenetik und Anthropologie, der Universitat, Albertstrasse 11, D-7800 Freiburg, West Germany

Paul J. Fritz, Department of Pharmacology, The Milton S. Hershey Medical Center, Pennsylvania State University, College of Medicine, Hershey, Pennsylvania 17033

Lawrence I. Gilbert, Department of Biological Sciences, Northwestern University, Evanston, Illinois 60201

Erwin Goldberg, Department of Biological Sciences, Northwestern University, Evanston, Illinois 60201

Robert W. Harrington, Florida Medical Entomology Laboratory, Florida Division of Health, Vero Beach, Florida 32960

Gary E. Hart, Genetics Section, Plant Sciences Department, Texas A & M University, College Station, Texas 77843

Yoichi Hasegawa, Department of Horticulture, University of Hawaii, 3190 Maile Way, Honolulu, Hawaii 96822

A. Hatzfeld, Institut de Pathologie Moleculaire, Centre Hospitalo-Universitaire Cochin, 24, rue du Faubourg St. Jacques, Paris 75014, France

J. Robert Heckman, Department of Biology, Elizabethtown College, Elizabethtown, Pennsylvania 17022

Mark A. Hermodson, Division of Medical Genetics, Department of Medicine, University of Washington, Seattle, Washington 98195

Roger R. Hewitt, Department of Biology, The University of Texas System Cancer Center, M. D. Anderson Hospital and Tumor Institute, Houston, Texas 77025

Kazuya Higashino, Third Department of Internal Medicine, Osaka University School of Medicine, Fukushima-Ku, Osaka, Japan

Akira Ichihara, Institute for Enzyme Research, School of Medicine, Tokushima University, Tokushima 770, Japan

P. I. Ittycheriah, Department of Biological Sciences, Northwestern University, Evanston, Illinois 60201

Samson T. Jacob, Department of Pharmacology, Milton S. Hershey Medical Center, Pennsylvania State University College of Medicine, Hershey, Pennsylvania 17033

J. A. Jaehning, Department of Biological Chemistry, Division of Biology and Biomedical Sciences, Washington University, St. Louis, Missouri 63110

Olli Jänne, Department of Pharmacology, Milton S. Hershey Medical Center, Pennsylvania State University College of Medicine, Hershey, Pennsylvania 17033

P. G. Johnston, School of Biological Sciences, MacQuarie University, North Ryde, N. S. W. 2113, Australia

Sylvia J. Kerr, Department of Surgery, University of Colorado Medical Center, 4200 East Ninth Avenue, Denver, Colorado 80220

Hyram Kitchen, Division of Comparative Medical Research, Department of Biochemistry, Michigan State University, East Lansing, Michigan 48824

Lester A. Klein, Department of Biological Chemistry, Harvard Medical School, Boston, Massachusetts 02115

Levy Kopelovich, Laboratory of Applied and Diagnostic Biochemistry, Memorial Sloan-Kettering Cancer Center, 410 East 68th Street, New York, New York 10021

Leonid Korochkin, Institute of Cytology and Genetics, Siberian Academy of Sciences, Novosibirsk 630090, U. S. S. R.

Shunjiro Kudo, Third Department of Internal Medicine, Osaka University School of Medicine, Fukushima-ku, Osaka, Japan

J. R. Larsen, Department of Entomology, University of Illinois, Urbana, Illinois 61801

Herbert G. Lebherz, Department of Molecular Biology, Roswell Park Memorial Institute, Buffalo, New York 14203

Ted C. Leung, Halifax Laboratory, Fisheries and Marine Service, Research and Development Directorate, Environment Canada, Halifax, Nova Scotia B3J 2R3, Canada

Jonathan J. Li, Department of Biological Chemistry, Harvard Medical School, Boston, Massachusetts 02115

Sara A. Li, Department of Biological Chemistry, Harvard Medical School, Boston, Massachusetts 02115

Thomas A. Linkhart, Department of Avian Sciences, University of California, Davis, California 95616

William F. Loomis, Department of Biology, University of California, San Diego, La Jolla, California 92037

Franco Mangia, University di Roma, Instituto di Istologia et Embriologia Generale, Via Alfonso Borelli, 50, 00161 Roma, Italy

George A. Marzluf, Department of Biochemistry, Ohio State University, Columbus, Ohio 43210

Edward J. Massaro, Department of Biochemistry, State University of New York, Buffalo, New York 14214

Janet C. Massaro, Department of Biochemistry, State University of New York, Buffalo, New York 14214

Colin J. Masters, Department of Biochemistry, University of Queensland, St. Lucia 4067, Australia

Hiroshi Masuji, Cancer Institute, Okayama University Medical School, Okayama, 700, Japan

Elizabeth T. Miller, Department of Zoology, University of Illinois, Urbana, Illinois 61801

Robert Mitchell, Department of Molecular Biology, Roswell Park Memorial Institute, Buffalo, New York 14203

Harold P. Morris, Department of Biochemistry, Howard University School of Medicine, Washington, D. C. 20001

Carolyn E. Murtagh, School of Biological Sciences, MacQuarie University, North Ryde, N. S. W., 2113, Australia

Jerome S. Nisselbaum, Laboratory of Applied and Diagnostic Biochemistry, Memorial Sloan-Kettering Cancer Center, 410 East 68th Street, New York, New York 10021

Paul H. Odense, Halifax Laboratory, Fisheries and Marine Service, Research and Development Directorate, Environment Canada, Halifax, Nova Scotia B3J 2R3, Canada

Gilbert S. Omenn, Division of Medical Genetics, Department of Medicine, University of Washington, Seattle, Washington 98195

Hironobu Ozaki, Department of Zoology, Michigan State University, East Lansing, Michigan 48824

Kenneth Paigen, Department of Molecular Biology, Roswell Park Memorial Institute, 666 Elm Street, Buffalo, New York 14203

Ulrich Petzoldt, Max-Planck Institut fur Immunobiologie, Stefan-Meierstrasse 8, D-7800 Freiburg, West Germany

W. E. Poole, C. S. I. R. O. Division of Wildlife Research, Lyneham, A. C. T., 2602, Australia

Tapas K. Pradhan, Departments of Obstetrics and Gynecology and Biochemistry, University of Florida College of Medicine, Gainesville, Florida 32610

Kenneth M. Pruitt, Laboratory of Molecular Biology, University of Alabama Medical Center, Birmingham, Alabama 35233

R. G. Roeder, Department of Biological Chemistry, Washington University, St. Louis, Missouri 63110

Frank H. Ruddle, Department of Biology, Yale University, New Haven, Connecticut 06520

Elizabeth M. Sajdel-Sulkowska, Rosentiel Basic Medical Sciences Research Center, Brandeis University, Waltham, Massachusetts 02154

Jiro Sato, Cancer Institute, Okayama University Medical School, Okayama 700, Japan

Kiyomi Sato, Fels Research Institute, Temple University School of Medicine, Philadelphia, Pennsylvania 19140

Tsuyoshi Sato, Department of Biochemistry, Hirosaki University School of Medicine, Hirosaki, Japan

John G. Scandalios, Department of Biology, University of South Carolina, Columbia, South Carolina 29208

F. Schapira, Institut de Pathologie Moleculaire, Centre Hôpital-Universitaire Cochin, 24, rue du Faubourg St. Jacques, Paris 75014, France

Robert Schiff, Hyland Division Travenol Laboratories, Costa Mesa, California 92626

L. B. Schwartz, Department of Biological Chemistry, Washington University, St. Louis, Missouri 63110

Harold L. Segal, Biology Department, State University of New York, Buffalo, New York 14214

James B. Shaklee, Department of Zoology, University of Illinois, Urbana, Illinois 61801

G. B. Sharman, School of Biological Sciences, MacQuarie University, North Ryde, N. S. W., 2113, Australia

Charles R. Shaw, M. D. Anderson Hospital and Tumor Institute, University of Texas Medical Center, Houston, Texas 77025

Michael I. Sherman, Department of Cell Biology, Roche Institute of Molecular Biology, Nutley, New Jersey 07110

Thomas B. Shows, Roswell Park Memorial Institute, Buffalo, New York 14203

Robert M. Singer, Tufts Cancer Research Center, Tufts University School of Medicine, Boston, Massachusetts 02111

V. E. F. Sklar, Department of Biological Chemistry, Washington University, St. Louis, Missouri 63110

H. H. Smith, Biology Department, Brookhaven National Laboratory, Upton, New York 11973

Horst Spielmann, Pharmakologisches Institut, Abt. Embryonal., Pharmakologie der Freien Universitat, 1 Berlin 33, Thielallee, Berlin, West Germany

C. L. Stratton, Medical School, University of Nevada, Reno, Nevada 89503

James E. Strong, M. D. Anderson Hospital and Tumor Institute, University of Texas Medical Center, Houston, Texas 77025

Richard T. Swank, Department of Molecular Biology, Roswell Park Memorial Institute, 666 Elm Street, Buffalo, New York 14203

Daniel Tennenbaum, Department of Pediatrics, University of California, San Francisco, California 94143

Peter E. Thompson, Department of Zoology, University of Georgia, Athens, Georgia 30602

Shiro Tomino, Department of Molecular Biology, Roswell Park Memorial Institute, 666 Elm Street, Buffalo, New York 14203

David C. Turner, Institute for Cell Biology, Swiss Federal Institute of Technology, CH-8006, Zurich, Switzerland

Tito Ureta, Departamento de Bioquimica, Facultad de Medicina Sede Norte, Universidad de Chile, Santiago-4, Chile

J. L. VandeBerg, School of Biological Sciences, MacQuarie University, North Ryde, N. S. W., 2113, Australia

Sharanjit S. VedBrat, Department of Zoology, University of Illinois, Urbana, Illinois 61801

Larry R. Versteegh, Department of Biochemistry and Biophysics, Iowa State University, Ames, Iowa 50010

Claude A. Villee, Department of Biological Chemistry, Harvard Medical School, Boston, Massachusetts 02115

Charles R. Walker, Department of Avian Sciences, University of California, Davis, California 95616

D. G. Walker, Department of Anatomy, Johns Hopkins University School of Medicine, Baltimore, Maryland 21205

Carol M. Warner, Department of Biochemistry and Biophysics, Iowa State University, Ames, Iowa 50010

A. Weber, Institut de Pathologie Moleculaire, Centre Hôpital-Universitaire Cochin, 24, rue du Faubourg St. Jacques, Paris 75014, France

Sidney Weinhouse, Department of Biochemistry, Temple University School of Medicine, Philadelphia, Pennsylvania 19140

R. Weinmann, Department of Biological Chemistry, Washington University, St. Louis, Missouri 63110

Thomas E. Wheat, Department of Biological Sciences, Northwestern University, Evanston, Illinois 60201

E. Lucile White, Department of Pharmacology, The Milton S. Hershey Medical Center, The Pennsylvania State University, Hershey, Pennsylvania 17033

Donald H. Whitmore, Jr., Biology Department, The University of Texas, Arlington, Texas 76019

Elaine Whitmore, Department of Biological Sciences, Northwestern University, Evanston, Illinois 60201

Gregory S. Whitt, Department of Zoology, University of Illinois, Urbana, Illinois 61801

Roger J. Wieme, Department of Clinical Chemistry, University Hospital Gent, De Pintelaan 135, 9000 Gent, Belgium

Barry W. Wilson, Department of Avian Sciences, University of California, Davis, California 95616

James E. Wright, Department of Biology, Pennsylvania State University, University Park, Pennsylvania 16802

Linda Wudl, Merck Institute, Rahway, New Jersey 07065

Yuichi Yamamura, Third Department of Internal Medicine, Osaka University School of Medicine, Fukushima-ku, Osaka, Japan

Yasuhide Yamasaki, Institute for Enzyme Research, School of Medicine, Tokushima University, Tokushima 770, Japan

T. Yamauchi, Department of Biology, M. D. Anderson Hospital and Tumor Institute, University of Texas Medical Center, Houston, Texas 77025

Ning-Sun Yang, Roche Institute of Molecular Biology, Nutley, New Jersey 07110

G. Wendel Yee, Department of Avian Sciences, University of California, Davis, California 95616

William H. Zinkham, Department of Pediatrics, Johns Hopkins University, School of Medicine, Baltimore, Maryland 21205

Preface

Isozymes are now recognized, investigated, and used throughout many areas of biological investigation. They have taken their place as an essential feature of the biochemical organization of living things. Like many developments in the biomedical sciences, the field of isozymes began with occasional, perplexing observations that generated questions which led to more investigation and, finally, with the application of new techniques, to a clear recognition and appreciation of a new dimension of enzymology.

The area of isozyme research is only about 15 years old but has been characterized by an exponential growth. Since the recognition in 1959 that isozymic systems are a fundamental and significant aspect of biological organization, many thousands of papers have been published on isozymes. Several hundred enzymes have already been resolved into multiple molecular forms, and many more will doubtless be added to the list. In any event, it is now the responsibility of enzymologists to examine every enzyme system for possible isozyme multiplicity.

Two previous international conferences have been held on the subject of isozymes, both under the sponsorship of the New York Academy of Sciences—the first in February 1961, and the second in December 1966. And now, after a somewhat longer interval, the Third International Conference was convened in April 1974, at Yale University. For many years, a small group of investigators has met annually to discuss recent advances in research on isozymes. They have published a bulletin each year and have generally helped to shape the field; in effect, they have been a standing committee for this area of research. From this group emerged the decision to hold a third international conference, and an organizing committee of five was appointed to carry out the mandate for convening a third conference. This Third International Conference was by far the largest of the three so far held with 224 speakers representing 21 countries, and organized into nine simultaneous sessions for three days on April 18, 19, and 20, 1974. Virtually every speaker submitted a manuscript for publication, and these total almost 4,000 pages. The manuscripts have been collected into four volumes entitled, *I. Molecular Structure; II. Physiological Function; III. Developmental Biology; and IV. Genetics and Evolution.* The oral reports at the Conference and the submitted manuscripts cover a vast area of biological research. Not every manuscript fits precisely into one or another of the four volumes, but the most appropriate assignment has been made wherever possible. The quality of the volumes and the success of the Conference must be credited to the participants and to the organizing committee. The scientific community owes much to them.

xix

Acknowledgments

I would like to acknowledge the help of my students and my laboratory staff in organizing the Conference and in preparing the volumes for publication. I am grateful to my wife, Margaret Markert, for volunteering her time and talent in helping to organize the Conference and in copy editing the manuscripts.

Financial help for the Conference was provided by the National Science Foundation, the National Institutes of Health, International Union of Biochemistry, Yale University, and a number of private contributors:

Private Contributors

American Instrument Company
Silver Spring, Maryland 20910

Canalco
Rockville, Maryland 20852

CIBA-GEIGY Corporation
Ardsley, New York

Gelman Instrument Company
Ann Arbor, Michigan 48106

Gilford Instrument Laboratories, Inc.
Oberlin, Ohio

Hamilton Company
Reno, Nevada 89502

Kontes Glass Company
Vineland, New Hampshire 08360

The Lilly Research Laboratories
Indianapolis, Indiana

Merck Sharp & Dohme
West Point, Pennsylvania

Miles Laboratories, Inc.
Elkhart, Indiana

New England Nuclear
Worcester, Massachusetts 01608

Schering Corporation
Bloomfield, New Jersey

Smith Kline & French Laboratories
Philadelphia, Pennsylvania

Warner-Lambert Company
Morris Plains, New Jersey

ISOZYMES AND DEVELOPMENTAL BIOLOGY

GREGORY S. WHITT
Provisional Department of Genetics and Development
University of Illinois, Urbana, Illinois   61801

The isozyme concept, one of the most fruitful concepts of
contemporary biology and medicine, was first introduced by
Markert and Møller in 1959. Developmental biology is one of
the many areas which has benefitted from the application of
this concept with its supporting technology.

INTRODUCTION AND HISTORICAL BACKGROUND

In the years following the first visualization of isozymes
on gels by histochemical procedures (Hunter and Markert, 1957)
the possibilities for utilizing these "zymograms" for monitor-
ing gene expression during development was quickly appreciated.
One enzyme system, lactate dehydrogenase, proved to be particularly amenable to developmental analysis (Markert and Møller,
1959; Flexner et al., 1960; Cahn et al., 1962; Markert and
Ursprung, 1962; Vesell et al., 1962). Studies of this and
other enzymes revealed that the complex isozyme patterns observed during embryogenesis and within differentiated tissues
could be attributed to the activities of relatively few genes.
Subsequent to this initial exploratory period, there has
been a tremendous proliferation of investigations employing
isozyme analyses of ontogeny. Although a large number of
enzyme systems and a wide spectrum of different organisms have
been utilized, space does not permit the detailed enumeration
of all pertinent reports either in the general literature or
in this specific volume. Comprehensive reviews of isozyme
analyses of developmental processes and disease states can be
found elsewhere (Markert and Ursprung, 1971; Criss, 1971;
Masters and Holmes, 1972; Weinhouse and Ono, 1972; Dreyfus,
1972; Kirkman, 1972; Vesell, 1973; Scandalios, 1974). This
general introduction to volume III (Isozymes: Developmental
Biology) will concentrate on some of the highlights of this
symposium and indicate those areas of "isozymology" that appear
particularly promising for future developmental analysis.
Many new areas of research have been developed which could
scarcely be imagined at the time of the First International
Isozyme Conference in 1961. At that time isozymic analyses
of gene expression during development were indeed leading to
various technical and theoretical breakthroughs. Concomitantly
with the growth of the new experimental approaches a redifini-
tion of developmental biology was also occurring. Indeed, the

1

concept of embryology was evolving into the much broader biological umbrella of "developmental biology", which has now
encompassed areas previously considered the exclusive domain
of cytology, genetics, chemistry, or medicine. It is, therefore, quite appropriate that this volume should cover such
diverse issues as somatic cell hybridization, subcellular localization of isozymes, reproductive biology, normal and abnormal embryogenesis, and gene expression in normal and abnormal
differentiated cells.

Many facets of developmental biology have been described
in this symposium and the various experimental interventions
described have been very creative and productive. In some
developmental systems the changes in isozyme patterns were due
to intrinsic genetic events. In other systems the researcher
was much more closely involved in perturbing genomic expression by the formation of allophenes, teratomas, hybrid organisms, or hybrid cells. In some instances the stimuli employed
to elicit gene function were specific hormones or metabolites.

ISOZYMES AND MEDICINE

In certain areas of medicine the utility of isozymes in
the diagnosis of the site of cellular and tissue damage has
been demonstrated for several diseases (myocardial infarction,
muscular dystrophy, etc.). Furthermore, it is now possible to
foresee a time when routine medical examination will employ
isozyme procedures to detect incipient disease states. In
the more distant future, isozyme analyses may even help assess
the predisposition of some individuals to develop specific
metabolic diseases.

Moreover, the isozymic analysis of certain disease states
has helped us to better understand basic mechanisms of cellular
differentiation. The isozyme repertories of some neoplasias
lend credence to the concept that cancer is best viewed as a
disease of cellular differentiation. In some neoplastic tissues the isozyme patterns are similar to those of embryonic
tissues; this appears to be due to a repression of genes normally expressed in the adult differentiated cell and to a re-
expression of "fetal genes". The isozymic analysis of the
"dedifferentiated" neoplastic cell may help to elucidate some
of the causes of cancer, and in addition these studies should
provide insights into the mechanisms of gene modulation during
normal cellular differentiation.

The examination of isozyme patterns of diseased or differentiating tissues enables us to infer the mechanisms underlying the changes in isozyme patterns. However, as a number
of the participants in this symposium have pointed out, caution

2

should be exercised in making these inferences since isozyme changes during most developmental processes are the result of a long chain of events distantly removed from initial cause(s).

MECHANISMS RESPONSIBLE FOR CHANGES IN ISOZYME PATTERNS DURING DEVELOPMENT

The simplest (but not necessarily most correct) interpretation of isozyme alterations during disease or differentiation is that a change has occurred at the level of gene transcription. Although differential gene regulation is probably the most important component contributing to developmental changes in isozyme patterns, the necessity of considering additional levels of cellular regulation has been stressed by many participants in this symposium.

The specific mRNA for an isozyme may have its access to the cytoplasm regulated, and if it does enter the cytoplasm, the translational process may be controlled. After translation the assembly of polypeptide subunits and the expression of enzymatic activity can be modified by epigenetic events. And, of course, it is important to recognize that the final gene product, the isozyme, can be drastically altered in its cellular levels simply by changing its rate of degradation, or in the cases of some diseases, the amount of isozyme leakage from the cell. Any one (or more) of these steps may be contributing to a change in the isozyme pattern during development.

Regardless of the level of regulation of isozyme synthesis, it is still difficult to determine from an isozyme pattern the extent to which the isozymes observed are functioning in the cell. It has been indicated in this conference that the activity of some isozymes can be regulated by their position at specific subcellular locales and by the presence or absence of various effector molecules within the cell.

These considerations are made more complex by the fact that developmental studies often involve the isozymic analyses of heterogeneous populations of cells or tissues. In these instances it is difficult to determine whether the change in an isozyme pattern reflects a change in the rate of isozyme synthesis per cell or whether it reflects a change in the numbers of a specific cell type within a tissue composed of heterogeneous cell populations. Procedures such as microelectrophoresis and light and electron microscopic histochemical localization of isozymes *in situ* will be utilized even more extensively in the future to help resolve this issue.

Furthermore, in cases where embryos are very small, embryos at the same morphological stage are sometimes pooled in order to conduct isozymic analyses. Problems of interpretation might

3

to conduct isozymic analyses. Problems of interpretation might arise because polymorphisms at structural and regulatory genes could result in quite different timings of gene expression for different embryos within the population. Assumptions of uniform genetic background should not be made lightly.

Changes in isozymes patterns are indicators of changes in gene expression and are reflections of genetic and epigenetic events inside and outside the cell. In addition to inferences which can be drawn about the mechanisms of gene expression in differentiating or in diseased tissue, the changing isozyme pattern tells us something about the rapidly changing metabolism of the embryo. Considerable insights into this important area of developmental physiology have been gained by analyzing developmental progressions of groups of enzymes and their isozymes within the same and different metabolic pathways. These studies permit the determination of the synchrony of gene functions involved in the isozymic, morphological, and functional differentiation of the cell.

No matter how informative the changing isozyme patterns are in providing a picture of both gene expression and changing emphases on different metabolisms, these changes are usually the consequence and not the cause of differentiation. However, in some instances, a change in the isozyme repertory of certain enzyme systems may be directly involved in determining subsequent developmental events or in the manifestation of a disease state. As a number of researchers have shown, metabolic deficiencies can be altered by mutational events at structural and regulatory genes controlling the levels of certain isozymes. Mutations of structural genes may also result in allelic isozymes with maladaptive kinetic or physical properties.

In addition to these more familiar causes of genetic alteration of isozyme patterns, some mutational events appear to affect the subcellular localization of the isozyme. Furthermore, mutations may affect the efficiency of enzymes responsible for the normal intracellular conversion of one isozymic form to another. Regardless of the site of the genetic lesion, the change in some isozyme patterns may be a primary or secondary expression of the defects responsible for causing the disease.

In a parallel manner, many isozymic changes during normal development are simply reflections of altered metabolism. However, a few classes of isozymes appear to do more than reflect changes in gene expression necessary for new metabolic emphases: they may also trigger profound developmental events by their presence or absence. For example, it has been suggested that an insect esterase isozyme, synthesized at a specific developmental period, may catabolize, and thus regulate the titer of, a hormone responsible for major developmental changes. Another class of isozymes which are important in directly regulating gene function during development and cellular differentiation is the RNA polymerases. Different RNA polymerases appear to occupy different cellular compartments, and in some instances unique isozymes exist at different developmental stages. Perhaps a portion of the specificity underlying gene activation is invested in these isozymes. It is not known yet whether additional isozymic forms of RNA polymerase, such as those formed in prokaryotes by the attachment of various "factors", are formed in eukaryotic cells. This should be a very promising area of future investigation. Still another class of enzymes, whose isozymes may play significant developmental roles, are those that participate in controlling levels of cyclic nucleotides.

MULTIPLE LOCUS ISOZYME ANALYSIS OF DEVELOPMENT

A particular class of isozymes -- the multi-locus isozyme systems -- has been especially informative in developmental analyses. Isozyme systems such as lactate dehydrogenase, aldolase, creatine kinase, and others, have been shown, at this symposium, to be particularly powerful probes for studying embryogenesis and cytodifferentiation.

In a number of developing systems the existence of embryo specific and stage specific isozymes has been reported. Such observations raise a number of questions about the uniqueness of embryonic metabolism. In addition to the isozymes synthesized in a temporally specific manner there are a number of multi-locus isozyme systems in which a specific isozyme is found primarily in one cell type, e.g., the lactate dehydrogenase isozymes of the primary spermatocyte of birds and mammals and of the teleost photoreceptor cell. These isozymes provide some of the more dramatic examples of cellular and temporal specificity of gene function. Investigations into the sperm specific isozyme should afford an excellent opportunity to elucidate mechanisms responsible for specificity of gene regulation and sperm metabolism. Moreover, it is conceivable that increased understanding of sperm specific isozymes will permit new approaches to contraception. The presence of unique isozymes in highly differentiated cells does suggest an unique metabolism and this in turn raises the question of whether entire constellations of enzymes will ultimately be found to exist in unique isozymal forms in such specialized cell types. The physiological significance of such multi-locus isozymes may reside not only in their kinetic properties but also in their physical properties, thus permitting them to

5

be positioned differentially within specific subcellular locales. These observations and conclusions necessitate new
approaches to investigating cellular developmental processes.
Future studies of isozyme ontogeny must also involve analyses
of the ontogeny of the subcellular structures to which these
isozymes specifically bind. Furthermore, the process of enzyme
degradation may be partially dependent on enzyme localization.
The amount of time the isozyme spends "bound" to the cell may
help to determine its sensitivity to catabolism. This may
explain observations on some isozymes which indicated that an
isozyme can be synthesized at similar rates within two different tissues while having vastly different rates of degradation.

Since some of the isozymes of multi-locus isozyme systems
are restricted in their synthesis to specific cells or tissues, it is possible to determine how tightly coupled these
specific isozyme syntheses are to certain specific morphological
and functional facets of differentiation.

The multi-locus isozymes also provide an excellent medium
for examining the evolution of specificity of gene regulation.
Since most multiple locus isozymes are probably derived from
a single ancestral gene, and since some of the isozymes are
still structurally and functionally very similar, one can
investigate the distance that initially identical loci must
diverge in order to come under separate developmental control.
This problem becomes particularly acute for those isozyme loci
which are tightly linked, yet have diverged in their structure
and developmental specificity.

ALLELIC ISOZYME ANALYSES OF DEVELOPMENT

The mechanisms of gene activation and the specificity of
gene regulation as a function of evolutionary distance can also
be assessed by examining allelic isozymes and their expression in developing systems. Interspecific hybrid embryos have
been formed from species in the same genus as well as from
those which are only very distantly related. There is an
increase in incidence of allelic asynchrony and allelic repression during development as the evolutionary distance between
the genomes increases. Also there appears to be a predominance
of repression of paternal isozyme synthesis which can be interpreted as a failure of gene activating molecules in the maternal
cytoplasm to recognize the paternal genome. However, pertur-
bations at other levels such as lesions in translation or in
the degradative process could produce the same effects. However, the absence of some maternal isozymes and the corresponding expression of the paternal isozymes in some interspecific

6

hybrids indicates that the mechanisms inovlved in allele expression are not yet fully elucidated.

Another way of using isozymes to analyze the perturbation of the genome is through the fusion of somatic cells. Somatic cell hybridization has proven to be a powerful procedure in studying gene expression of differentiated cells. Unlike the situation in the interspecies hybrid embryos, cell fusion is a mixing of both nuclei and cytoplasms and generally a fusion of cells whose state of differentiation is already largely determined. In addition, isozymic analyses indicate that for some enzymes the structural gene resides on a different chromosome from the gene regulating its function. Isozymic markers have permitted the analysis of positive and negative controls over gene function, as well as the extent to which regulatory molecules can recognize homologous, but different, structural genes.

Another way of using allelic isozymes to analyze the mechanisms of gene modulation in differentiated cells is to examine levels of these isozymes under conditions under which dosage compensation may occur, as, for example, during mammalian X-chromosome inactivation and during gene expression in polyploid cells. One of the newest and most exciting uses of allelic isozymes is their employment as markers for blastomeres making up chimeric embryos. In allophenic mice one can determine whether the origin of a syncytial tissue results from cell fusion or nuclear multiplication. In addition, there are various ways in which these allelic isozyme markers can be used to determine and trace the clonal origin of cells. In many of the studies of gene expression in mammalian embryos very sensitive microelectrophoretic and microspectrophotometric techniques have been developed to give entirely new insights into mammalian development. It is in this way that the time of first function of the embryo genome will be better determined.

This symposium has clearly shown the power of isozymes as tools to describe and analyze many facets of developmental biology. The large numbers of enzyme systems employed, the increasing sensitivity of detection, and the variety of developmental systems currently being studied suggest that our understanding of the genetic and epigenetic regulation of gene expression in both normal and diseased systems will be greatly enhanced by the continued application of isozyme analyses.

REFERENCES

Cahn, R.D., N.O. Kaplan, L.Levine, and E. Zwilling 1962. Nature and development of lactate dehydrogenases. *Science* 136:962-969.

Criss, W.E. 1971. A review of isozymes in cancer. *Cancer Res.* 31:1523-1542.

Dreyfus, J.C. 1972. Bases moléculaires des maladies enzymatiques génétiques. *Biochimie* 54:559-571.

Flexner, L.B., J.B. Flexner, R.B. Roberts, and G. de la Haba 1960. Lactic dehydrogenases of the developing cerebral cortex and liver of the mouse and guinea pig. *Dev. Biol.* 2:313-328.

Hunter, R.L. and C.L. Markert 1957. Histochemical demonstration of enzymes separated by zone electrophoresis in starch gels. *Science* 125:1294-1295.

Kirkman, H.N. 1972. Enzyme defects. In: *Progress in Medical Genetics*, Vol. VIII (A.G. Steinberg and A.G. Bearn, eds). Grune and Stratton, Inc., pp. 125-168.

Markert, C.L. and F. Møller 1959. Multiple forms of enzymes: Tissue, ontogenetic, and species specific patterns. *Proc. Nat. Acad. Sci. U.S.A.* 45:753-763.

Markert, C.L. and H. Ursprung 1962. The ontogeny of isozyme patterns of lactate dehydrogenase in the mouse. *Dev. Biol.* 5:363-381.

Markert, C.L. and H. Ursprung 1971. Developmental genetics. In: *Foundations of Developmental Biology Series* (C.L. Markert, ed.). Prentice-Hall, Inc., Englewood Cliffs, N.J., pp. 214.

Masters, C.J. and R.S. Holmes 1972. Isoenzymes and ontogeny. *Biol. Rev.* 47:309-361.

Scandalios, J.G. 1974. Isozymes in development and differentiation. *Ann. Rev. Plant. Physiol.* 25:225-258.

Vesell, E.S. 1973. Advances in pharmacogenetics. In: *Progress in Medical Genetics*, Vol. IX (A.G. Steinberg and A.G. Bearn, eds.). Grune and Stratton, Inc., pp. 291-367.

Vesell, E.S., J. Philip, and A.G. Bearn 1962. Comparative studies of the isozymes of lactic dehydrogenase in rabbit and man. Observations during development and tissue culture. *J. Exp. Med.* 116:797-800.

Weinhouse, S. and T. Ono 1972. Isozymes and enzyme regulation in cancer. In: *Gann Monograph on Cancer Research 13*. University of Tokyo Press, Tokyo, pp.322.

HORMONAL REGULATION OF RNA POLYMERASES IN LIVER AND KIDNEY

SAMSON T. JACOB, OLLI JÄNNE
and
ELIZABETH M. SAJDEL- SULKOWSKA *
Department of Pharmacology
Milton S. Hershey Medical Center
Pennsylvania State University College of Medicine
Hershey, Pennsylvania   17033,

Physiological Chemistry Laboratories, Massachusetts
Institute of Technology, Cambridge, Massachusetts   02139
and
*Rosentiel Basic Medical Sciences Research Center
Brandeis University, Waltham Massachusetts 02154

ABSTRACT.   Nucleolar RNA polymerases (forms I and IV) of rat liver were preferentially stimulated within 2-4 hrs of hydrocortisone treatment.   Inhibition of protein synthesis prior to hydrocortisone treatment did not prevent its stimulatory effect on polymerases.   Hormone administration neither caused preferential extraction of nucleolar RNA polymerases nor increased the number of new RNA chains initiated by the nucleolar enzymes.   These results suggest that hydrocortisone does not increase *de novo* synthesis of nucleolar polymerases, but rather activates pre-existing enzyme molecules.

Concurrently, the effect of testosterone on mouse kidney RNA polymerases was investigated.   Polymerases I and II were stimulated as early as 15 min with a peak at 1 hr after a single subcutaneous injection of testosterone propionate. A later peak of stimulation for the same polymerases was observed at 20 hrs after hormome administration.   In a series of experiments, the effect of testosterone-cytosolic receptor complex on chromatin and polymerases I and II from mouse kidney was investigated in vitro.   The hormome-receptor complex stimulated the chromatin template activity when assayed with polymerase I or II.   When native DNA was substituted for chromatin, the stimulation of RNA synthesis in vitro by the hormone-receptor complex was considerably reduced.   The results suggest that the increased transcription of chromatin by isolated polymerases I and II in presence of hormone-receptor complex is mainly due to activation of the template rather than a direct activation of polymerases in vitro.

INTRODUCTION

It is widely recognized that eukaryotic DNA-dependent RNA polymerase (EC 2.7.7.6) occurs in multiple forms (see Jacob, 1973). These forms are compartmentalized inside the cell nucleus. Thus forms I and IV (IA) are localized in the nucleolus whereas forms II and III are of nucleoplasmic origin (Jacob et al., 1970a; Blatti et al., 1970). The nucleolar forms are insensitive to the mushroom toxin α-amanitin (Jacob et al., 1970b; Kedinger et al., 1970), whereas form II is completely sensitive to low concentrations of the toxin (Jacob et al., 1970a,b; Lindell et al., 1970; Kedinger et al., 1970). When used in high concentrations, α-amanitin also inhibits form III (Weinmann and Roeder, 1974).

One of the significant functional differences between the nucleolar and nucleoplasmic RNA polymerases is their independent regulations by physiological factors such as hormones (summarized in Jacob, 1973; O'Malley and Means, 1974). We have previously demonstrated that the activity of nucleolar RNA polymerases IA and I is stimulated within 1-2 hrs of administration of hydrocortisone to adrenalectomized rats (Sajdel and Jacob, 1971). The stimulation of nucleolar polymerases is consistent with the observed increase in the rate of ribosomal RNA precursor (45 S RNA) synthesis (Jacob et al., 1969). Other steroid hormones have also been shown to stimulate preferentially nucleolar RNA polymerases, at least at the early stages of their action (Jacob, 1973). In this communication we show that the early action of hydrocortisone on rat liver is to increase the activity rather than *de novo* synthesis of nucleolar RNA polymerases.

In order to understand further the molecular mechanisms by which steroids regulate RNA synthesis, we have initiated a series of studies on an animal model (tfm/y mouse), where a sex-linked recessive mutation has resulted in either defective or deficient cytosolic androgen receptors (Lyon and Hawkes, 1970; Gehring et al., 1971; Bardin et al., 1973) Subsequently, animals are unable to accumulate testosterone in the nuclei of androgen-responsive tissues such as kidney (Bullock and Bardin, 1974; Goldstein and Wilson, 1972). If the androgen insensitivity of tfm/y mouse is characterized by an inherited decrease of androgen receptor activity in target organs, the chromatin from these tissues would still have the ability to bind the steroid receptor complex in vitro, if supplied with an exogenous viable receptor. RNA synthesizing system should, therefore, theoretically not be affected by androgens in tfm/y mouse kidney. Using tfm/y mice (no cytosolic androgen receptor), the carrier females of this disease (decreased receptor concentration) and their unaffected littermates with normal receptor

10

concentration (Bullock and Bardin, 1974) as our model system, we hope to elucidate the role of cytoplasmic steroid receptor in the biochemical expression of hormome action in the target organs. We now report the results of our initial studies on the early androgen action on normal mouse kidney.

MATERIALS AND METHODS

Treatment of animals. Hydrocortisone phosphate (5 mg/100 g) was given intravenously in 0.2 ml of 0.15 M saline to anesthetized adrenalectomized rats. Female mice were administered testosterone propionate (4 mg/100 g) in 0.2 ml of sesame oil subcutaneously. Animals were sacrificed after hormone treatments at various time points as indicated in the text and figure legends. Cycloheximide (2 mg/100 g) was given intraperitoneally, 15 min prior to hormone treatment.

Isolation of nuclei. Rat liver nuclei were isolated as previously described (Busch et al., 1967). Rat and mouse kidney nuclei were prepared essentially by the same procedure with the exception that, for initial homogenization, the sucrose concentration was decreased to 2.0 M. When nuclei were used for solubilization of RNA polymerases, the nuclear pellet was suspended directly in the solubilization buffer (see below). For chromatin isolation, the isolated nuclei were further washed with 0.5 volumes of buffer composed of 0.32 M sucrose, 2 mM $MgCl_2$, and 0.2% Triton X-100.

Isolation of chromatin. Kidney chromatin was prepared from Triton X-100 washed nuclei, using modifications of the procedures of Shaw and Huang (1970) and Cox et al., (1973). The method consisted of repeated washings of nuclei with Tris-HCl buffers (pH 8.0) of decreasing molarities and subsequent isolation of chromatin by centrifugation through 1.7 M sucrose. Chromatin preparation was finally taken up in 0.4 mM Tris-HCl (pH 8.0) and used within the following 20 hrs. Mouse spleen chromatin used in some control studies was prepared by similar procedure.

Isolation and partial purification of mouse kidney androgen. receptor. Mouse kidney tissue was homogenized in TGMED buffer (50 mM Tris-HCl, 25% glycerol (v/v), 5 mM $MgCl_2$, 0.1 mM EDTA, 0.5 mM dithiothreitol, pH 7.9). After incubation of 105,000 x g supernatant (referred to as cytosol) with $[1,2-^3H_2]$ testosterone (5×10^{-9}M) for 3-4 hrs at 4°C, the cystosolic testosterone receptor complex was isolated by ammonium sulfate

precipitation (33% fractional saturation). The precipitate
was spun down, dissolved in a small volume of TGMED buffer,
and dialyzed against 500 volumes of the same buffer to remove
ammonium sulfate. Occasionally, "androgen receptor fraction"
was prepared for control studies from non-target tissues
using similar procedures.

*Solubilization and purification of liver and kidney RNA poly-
merases.* Rat liver nuclear RNA polymerases were solubilized
and separated by DEAE Sephadex chromatography essentially as
previously described (Sajdel and Jacob, 1971; Jacob, 1973).
Rat and mouse kidney nuclei were homogenized in the solubilization buffer composed of 50 mM Tris-HCl (pH 8.5), 50 mM KCl,
1 mM $MgCl_2$, 0.5 mM dithiothreitol and 40% glycerol (v/v).
After sonication, the glycerol concentration was decreased
to 25%. The sample was incubated at 37° for 30 min, centrifuged, precipitated with ammonium sulfate, and dialyzed as
previously described for rat liver nuclear enzyme. Different
forms of kidney RNA polymerase were resolved by chromatography on DEAE Sephadex using a linear ammonium sulfate
gradient (0.03 M to 0.5 M) as described by Sajdel and Jacob
(1971).

Measurement of RNA polymerase activities. Activities of
different forms of RNA polymerase in whole nuclei were
assayed under three conditions (Blatti et al., 1970): (1)
in low ionic medium in the presence of Mg^{2+} and α-amanitin,
(2) in high ionic medium in the presence of Mn^{2+}, and (3)
in high ionic medium in the presence of Mn^{2+} and α-amanitin.
The activity under the condition (1) preferentially measures
nucleolar RNA polymerase (form I) activity. The activity
obtained after subtracting (1) from (2) measures the predominant nucleoplasmic RNA polymerase (form II) activity. Usually the activity under the condition (3) approximates the
nucleolar RNA polymerase activity. RNA polymerase activities
in chromatographically separated fractions were measured
using saturating amounts of calf thymus DNA as template.
When chromatin was used as template, unlimiting amounts of
various RNA polymerases were used.

Measurement of RNA-chain length. After incubation of whole
nucleoli, the labeled RNA synthesized was precipitated with
10% trichloroacetic acid and subjected to alkaline hydrolysis.
Uridine-3'-phosphate (UMP) and uridine (3' end) were separated
on poly(ethyleneimine)-cellulose thin-layer chromatography
(Randerath and Randerath, 1967).

RESULTS

Effect of hydrocortisone on the nucleolar RNA polymerase activities. Intravenous administration of hydrocortisone to adrenalectomized rats led to an increase in liver nucleolar RNA polymerase (forms I and IV) activities, whereas no significant changes were observed in nucleoplasmic polymerases at the same time point. These results are summarized in Table I. Adrenalectomy resulted in reduction of activities of nucleolar RNA polymerases (results not shown). Within 4 hrs of administration of a single dose of hydrocortisone, forms I and IV were increased in activity by 156% and 210%, respectively (Table 1).

Effect of cycloheximide on the hydrocortisone-induced stimulation of nucleolar RNA polymerases. Cycloheximide was given intraperitoneally 15 min before hormone treatment and the animals were sacrificed 3 1/2 hrs later. The hormone-induced increase in the activities of forms I and IV were not reduced by the inhibition of protein synthesis. The basal activities of liver nucleolar enzymes from adrenalectomized rats were not altered by cycloheximide (Table II). Since the dose of cycloheximide used in these experiments inhibits protein synthesis almost completely (Muramatsu et al., 1970), the persistent stimulation of nucleolar RNA polymerases in the presence of cycloheximide argues against *de novo* synthesis of forms I and IV in response to hydrocortisone administration.

Evidence against preferential extraction of nucleolar RNA polymerases after hormone treatment. The total amounts of protein extracted from the nuclei of control and hormone-treated livers were identical implying that hydrocortisone did not increase the efficiency of enzyme extraction from the nucleoli. Furthermore, RNA polymerase activity left in the residue after extraction of the enzyme was less than 5% of the activity originally present in the nuclei. In order to rule out preferential extraction of RNA polymerase I after hormone treatment the amount of enzyme protein from control and hydrocortisone-treated livers corresponding to polymerase I (pooled from DEAE-Sephadex column fractions) that is required to saturate the binding sites on relatively limiting template was determined. For these studies enzyme protein ranging from 10 to 75 µg of protein was used with 5 µg of calf thymus DNA in RNA polymerizing reaction. The enzyme activity was proportional to enzyme concentration up to about 50 µg of enzyme protein before it reached a plateau (Fig. 1). The difference in the nucleolar enzyme activity between the control and hormone-treated samples persisted at all enzyme concentrations.

TABLE I

EFFECT OF HYDROCORTISONE ON RAT LIVER
POLYMERASES I AND IV

The experimental details concerning the dose and injection
schedule of hydrocortisone, extraction and chromatographic
resolution of the polymerases are given in the text. The data
is the average of five experiments.

	RNA POLYMERASE ACTIVITY (nmoles UMP incorporated)	
SOURCE	POLYMERASE I	POLYMERASE IV
Adrenalectomized	10	7
+Hydrocortisone	31	18

TABLE II

EFFECT OF CYCLOHEXIMIDE ON THE STIMULATION OF
NUCLEOLAR RNA POLYMERASE I AND IV (IA) BY HYDROCORTISONE

Experimental details are given in the text. Polymerases
I and IV were pooled from the peak fractions and used in satur-
able quantities. The results are the average of three experiments. a = 9.5 nanomoles UMP incorporated; b=8.4 nanomoles
UMP incorporated.

	NUCLEOLAR RNA POLYMERASE ACTIVITY	
SOURCE	POLYMERASE I	POLYMERASE IV
Control	100[a]	100[b]
Control plus cycloheximide	95	97
Hydrocortisone	190	227
Hydrocortisine plus cycloheximide	185	213

[a]9.5 nanomoles UMP incorporated.

[b]8.4 nanomoles UMP incorporated.

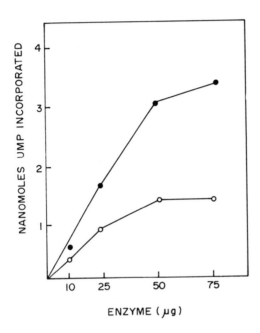

Fig. 1. *Activity of liver nucleolar RNA polymerase I as a function of enzyme concentration.* The peak activity fractions of polymerase I from DEAE Sephadex columns were pooled and the activity of the enzyme from control and hydrocortisone-treated livers of adrenalectomized rats was determined with increasing amounts of enzyme protein O——O control; ●——● hormone.

Further evidence for qualitative changes in nucleolar RNA polymerases. An alternate line of evidence for qualitative changes in RNA polymerase is to demonstrate that the number of polymerase molecules remains constant after hormone treatment. It can be assumed that the number of RNA chains is a reflection of the number of RNA polymerase molecules actively engaged in transcription (Bremer et al., 1965; Richardson, 1966) and that under conditions where chain re-initiation is minimal, each transcribing enzyme is responsible for the synthesis of only one nucleotide chain. Such an experiment can be performed in isolated nucleoli where RNA synthesis in vitro catalyzed by the endogenous nucleolar RNA polymerases is essentially a chain elongation reaction. The reaction was carried out using isolated nucleoli from adrenalectomized and hydrocortisone-treated livers. The enzyme activity was linear for 25 minutes. The reaction was terminated by addition of 10% TCA and the product was subjected to alkaline hydrolysis. Uridine 3'-phosphate (UMP) and uridine (3' end) were separated by PEI cellulose thin-layer chromatography as described under Materials and Methods. The radioactivity in UMP gives measurements of average chain length and that in uridine provides measurements of the number of chains, because only the 3'-ends would measure the number of chains already in progress in the isolated nucleoli.

The incorporation of [^{14}C] UTP into total nucleolar RNA, UMP, and uridine is given in Table III. Hydrocortisone stimulated the incorporation into UMP by 88% whereas no significant change in the extent of incorporation into uridine was detected. Thus the average chain length of RNA, as determined by the ratio CMP

TABLE III

EFFECT OF HYDROCORTISONE ON THE CHAIN LENGTH OF
RNA PRODUCED BY NUCLEOLAR RNA POLYMERASES

RADIOACTIVITY* IN

SOURCE	UMP	URIDINE	CHAIN LENGTH
Control	9,100	140	65
Hydrocortisone	17,100	143	116

*Counts/min.

---

in UMP/CPM in uridine was increased by 78% after hydrocortisone treatment, whereas the number of new chains initiated by the hormone was negligible. These experiments suggest that hydrocortisone stimulates nucleolar RNA polymerase activity by increasing the rate at which the existing enzyme molecules polymerize nucleotides into RNA rather than causing a *de novo* synthesis of enzyme. Similar experiments were performed by Barry

16

and Gorski (1971), who used isolated uterine nuclei (under conditions favorable for nucleolar RNA polymerase) to demonstrate
that estradiol did not increase nucleolar RNA polymerase molecules.

Effect of testosterone on mouse kidney nuclear RNA polymerase
activities. In order to determine the time required for the
maximal early stimulation of kidney nuclear RNA polymerases,
the enzyme activities were measured in isolated mouse kidney
nuclei at different time intervals ranging from 15 min to 28
hrs after a single injection of testosterone propionate (Fig.
2). Female mice were used in these initial studies for the
following reasons: (1) female and male kidney have approximately the same androgen receptor concentration (Bullock and
Bardin, 1974), (2) the saturation of this receptor by endogenous steroids is probably lower in female kidney, and subsequently (3) testosterone administration leads to comparable changes

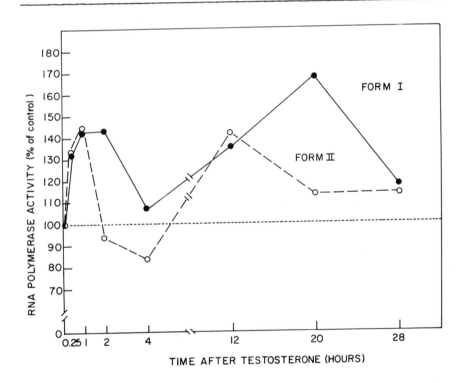

Fig. 2. *Time-course of mouse kidney RNA polymerase activation*
by testosterone. Mice (10 in each group) 8 weeks old, were
given subcutaneously 1 mg of testosterone propionate in sesame
oil at 0.25, 1, 2, 4, 12, 20, and 28 hrs prior to sacrifice.

Fig. 2 Cont'd. Control animals were given oil only. Kidney nuclei were isolated and RNA polymerase activities measured as described in Materials and Methods. The incorporation of UMP into acid-insoluble material was calculated per mg of nuclear DNA. The changes in RNA polymerase I and II activities are expressed as percentages of control values, which were 56 and 585 pmoles of UMP/mg DNA for polymerase I and II, respectively.

---

in some renal enzyme activities in female and castrated male mice (Bardin et al., 1973).

The activities of both polymerases I and II were stimulated as early as 15 min after a single subcutaneous injection of testosterone propionate. Polymerase II activity reached its maximum (140% compared to control) within 1 hr and declined to normal levels within 2 hrs. On the other hand, form I activity was at its maximum (140% compared to control) at 1 hr and remained at the increased level for at least an additional hour. Both forms had a second peak of stimulation at later time intervals. Thus the activities of forms I and II were augmented again at 20 hrs (165%) and at 12 hrs (140%), respectively, after testosterone treatment.

In subsequent studies RNA polymerase was extracted from mouse kidney nuclei at 1 1/2 and 20 hrs after a single injection of testosterone propionate. Forms IA (IV), I and II were resolved chromatographically (see Materials and Methods). The results obtained are summarized in Table IV. Forms I and II were present in the ratio of 1:6-7 and were eluted at 0.08 M and 0.22 M ammonium sulfate, respectively. At 1 1/2 hrs after testosterone administration, the activities of forms IA (IV) and I were increased approximately two-fold, whereas from II activity was enhanced by 60% (Table IV). Since the enzyme activities were assayed with exogenous DNA as template, the early stimulation of form II activity measured in whole nuclei (Fig. 1) should have resulted partly from alterations in RNA polymerase II itself. All forms of RNA polymerase were stimulated to a higher extent at 20 hrs after hormone administration. It is possible that this late increase of RNA polymerase activity is largely due to *de novo* synthesis of the enzyme.

In vitro activation of mouse kidney chromatin with testosterone receptor-complex. The effect of partially purified cytosolic testosterone receptor-complex on chromatin template activity and on kidney RNA polymerases I and II was investigated in vitro in order to find out whether the hormone receptor-complex directly activates the polymerases or whether the activation is an event subsequent to chromatin activation.

Chromatin was prepared from mouse kidney and RNA polymerases

TABLE IV

EFFECT OF TESTOSTERONE ADMINISTRATION ON MOUSE KIDNEY RNA POLYMERASE
A single dose of testosterone was given to mice, which were killed 1½
hrs (T-1½) and 20 hrs (T-20) after hormone administration. Control
group (C) was given sesame oil.

RNA POLYMERASE ACTIVITY

EXPERIMENTAL GROUP	POLYMERASE IV		POLYMERASE I		POLYMERASE II	
	P moles UMP incorporated[a]	±%[b]	P moles UMP incorporated	±%	P moles UMP incorporated	±%
C	150	--	330	--	1560	--
T-1 1/2	325	+115	625	+90	2530	+69
T-20	1330	+785	890	+170	3400	+118

[a] picomoles of UMP incorporated into TCA insoluble material/0.2 ml of
enzyme fraction.

[b] Percent change.

from both mouse and rat kidney as described under Materials and Methods. In most of the experiments rat kidney RNA polymerases were used, but similar results were obtained with mouse kidney enzymes. Testosterone receptor-complex was capable of stimulating in vitro the chromatin template activity when assayed with polymerase I or II (Fig. 3). The activity of polymerase

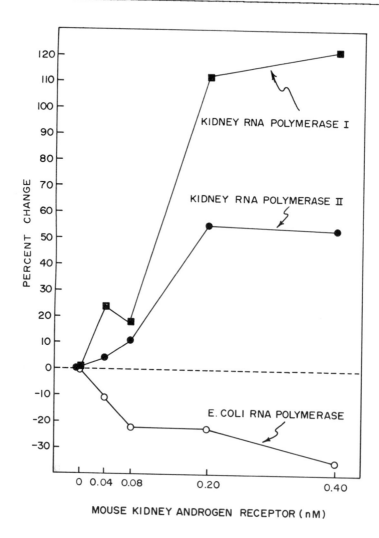

Fig. 3. *Effect of mouse kidney androgen receptor testosterone-complex on the in vitro transcription of mouse kidney chromatin with various RNA polymerases.* Rat kidney RNA polymerases, mouse kidney androgen receptor and mouse kidney chromatin were

Fig. 3. Cont'd. prepared as described in the text and bacterial RNA polymerase (*E. coli* K-12) was purchased from Miles Laboratories. The amount of androgen receptor added is expressed as the final bound [^{3}H] testosterone concentration in the incubation. The changes in activities are expressed as percentages of control values, which were 3.7, 17 and 140 pmoles of UMP incorporated using rat kidney form I, rat kidney form II and *E.coli* RNA polymerase, respectively.

I or II measured with exogenous DNA was not augmented to the same extent under these conditions. The results of these experiments suggest that the increased transcription of kidney chromatin in vitro by isolated polymerases I and II in presence of testosterone-cytosolic receptor complex is mainly due to activation of chromatin rather than an effect on the polymerases per se by the steroid-receptor complex. The stimulation of both polymerases from kidney was increased with increasing concentrations of hormone receptor-complex (Fig. 3). An important observation was the inability of bacterial RNA polymerase (*E.coli* K-12) to reveal the steroid receptor-complex induced increase in the chromatin template activity; in fact, testosterone receptor-complex in increasing concentration decreased the transcription of chromatin by bacterial RNA polymerase (Fig. 3). These studies add to the growing evidence that mammalian chromatin is transcribed inefficiently and non-specifically by prokaryotic RNA polymerase (Butterworth et al., 1971; Keshgegian and Furth, 1972; Jacob, 1973; Reeder, 1973; Maryanka and Gould, 1973). Further proof for the specificity in the interaction of kidney chromatin with cytoplasmic testosterone receptor-complex was obtained by the lack of significant stimulation of RNA synthesis when chromatin or "androgen receptor fraction" (see Materials and Methods) from non-target tissues such as spleen was used (results not shown).

DISCUSSION

It is now well accepted that almost all steroid hormones augment gene transcription in target tissues through the following series of reactions (Gorski et al., 1968; Jensen et al., 1968; O'Malley and Means, 1974): (a) formation of a hormone receptor complex in the cytoplasm; (b) activation of the complex and its translocation to the nucleus; (c) binding of the receptor to specific sites (acceptor) on the chromatin; (d) early stimulation of the chromatin activity; (e) stimulation of nucleolar RNA polymerase (Form I) activity, consequent increase in ribosomal RNA synthesis and in the production of ribosomes; (f) late stimulation of RNA polymerases I and II; (g) translation of specific mRNA.

We have focused our studies to elucidation of the molecular

mechanism by which steroid hormones exert their ultimate bio-chemical expression. The present studies clearly show that hydrocortisone and testosterone stimulate nucleolar RNA poly-merases in liver and kidney respectively, within the first few hours of hormone administration. The most striking finding is the enhanced transcription of kidney chromatin activity by homologous RNA polymerase on exposure to testosterone-cytoso-lic receptor complex in vitro. The tissue specificity of the hormonal effect on chromatin has been borne out by the inabil-ity of kidney androgen-receptor complex to stimulate the trans-cription of spleen chromatin. Davies and Griffiths (1973) have also reported activation of prostate chromatin by 5α-dihydro-testosterone-receptor complex. Raynaud-Jammet and Baulieu (1969) and Mohla et al. (1972) have demonstrated stimulation of RNA synthesis in uterine nuclei by estradiol-receptor com-plex. The lack of significant stimulation of RNA synthesis in vitro when native DNA was used instead of chromatin in the RNA synthesizing system strongly suggests that polymerases are not directly activated by hormones, although we cannot rule out such a possibility in vivo.

Almost all hormones studied to date stimulate nucleolar RNA polymerase activity in vitro within 1-3 hrs of their admin-istration. Thus, in addition to hydrocortisone, other steroids such as cortisone (Yu and Feigelson, 1971) triiodothyronine and growth hormone (Smuckler and Tata, 1971) stimulate form I in liver. Similarly, polymerase I is stimulated by testoster-one in prostate (Mainwaring et al., 1971) by ACTH and dibutyryl cyclic AMP in adrenals (Fuhrman and Gill, 1974), and by estradiol in uterus (Barry and Gorski, 1971). Our data suggest that at early stages of their action hormones do not stimulate synthesis of nucleolar RNA polymerases, but rather they activate the pre-existing enzyme molecules thereby enhancing their abil-ity to transcribe the template efficiently. The stimulation of polymerase I and II at later stages of hormone action appears to arise from increased synthesis of these enzymes. It is not yet known whether the early stimulation of nucleolar RNA poly-merase and chromatin activation are due to two independent processes.

The activation of the nucleolar enzyme may be induced by several factors. For example, the enzyme itself could have undergone conformational (allosteric) changes. Alternatively, the steroids could stimulate the synthesis of a "rapidly-turn-ing over" cytoplasmic protein factor(s) directly involved in the synthesis of ribosomal RNA. Evidence for such a factor has been presented by several groups (see Jacob, 1973). This factor(s) may be a subunit of nucleolar RNA polymerase, just as "sigma factor" is a part of *E.coli* holoenzyme. It is like-ly that the very early increase in the chromatin activity could

22

lead to *de novo* synthesis of mRNA for this protein factor(s). An examination of the subunit composition of nucleolar RNA polymerases from hormone-treated tissues may reveal whether the hormone has induced any structural alterations in these enzymes.

ACKNOWLEDGEMENT

This work was supported in part by Public Health Service Research Grant No. CA-15733 from the National Cancer Institute and by USPHS Contract No. NO1-HD-2-2730.

REFERENCES

Bardin, C. W., L. P. Bullock, R. J. Sherins, I. Mowszowicz, and W. R. Blackburn. 1973. Androgen metabolism and mechanism of action in male pseudohermaphroditism: A study of testicular feminization. *Recent Progr. Hormone Res.* 29: 65-109.

Barry, J., and J. Gorski. 1971. Uterine RNA polymerase: effect of estrogen on nucleotide incorporation into 3' chain termini. *Biochemistry* 10: 2384-2390.

Blatti, S. P., C. J. Ingles, T. J. Lindell, P. W. Morris, R. F. Weaver, F. Weinberg, and W. J. Rutter. 1970. Structure and regulatory properties of eukaryotic RNA polymerase. *Cold Spring Harb. Symp. Quant. Biol.* 35: 649-657.

Bremer, H., M. W. Konrad, K. Gaines, and G. S. Stent. 1965. Direction of chain growth in enzymic RNA synthesis. *J. Mol. Biol.* 13: 540-553.

Bullock, L. P., and C. W. Bardin. 1974. Androgen receptors in mouse kidney: a study of male, female and androgen-insensitive (tfm/y) mice. *Endocrinology* 94: 746-756.

Busch, H., K. S. Narayan, and J. Hamilton. 1967. Isolation of nucleoli in a medium containing spermine and magnesium acetate. *Exp. Cell Res.* 47: 329-336.

Butterworth, P.H.W., R. F. Cox, and C. J. Chesterton. 1971. Transcription of mammalian chromatin by mammalian DNA-dependent RNA polymerase. *Eur. J. Biochem.* 23: 229-241.

Cox, R. F., M. E. Haines, and N. H. Carey. 1973. Modification of the template capacity of chick-oviduct chromatin for form-B RNA polymerase by estradiol. *Eur. J. Biochem.* 32: 513-524.

Davies, P., and K. Griffiths. 1973. Stimulation of ribonucleic acid polymerase activity in vitro by prostatic steroid-protein receptor complexes. *Biochem. J.* 136: 611-622.

Fuhrman, S. A., and G. N. Gill. 1974. Hormonal control of adrenal RNA polymerase activities. *Endocrinology* 94: 691-700.

Gehring, U., G. M. Tomkins, and S. Ohno. 1971. Effect of the androgen-insensitivity mutation on a cytoplasmic receptor for dihydrotestosterone. *Nature New Biol.* 232:106-107.

Goldstein, J. L., and J. D. Wilson. 1972. Studies on the pathogenesis of the pseudohemaphroditism in the mouse with testicular feminization. *J. Clin. Invest.* 51: 1647-1657.

Gorski, J., D. Toft, G. Shyamala, D. Smith, and A. Notides. 1968. Hormone receptors: Studies on the interaction of estrogen with the uterus. *Recent Progr. Hormone Res.* 24: 45-80.

Jacob, S. T. 1973. Mammalian RNA polymerases. *Progr. Nuc. Acid Res. Mol. Biol.* 13: 93-126.

Jacob, S. T., E. M. Sajdel, and H. N. Munro. 1969. Regulation of nucleolar RNA metabolism by hydrocortisone. *Eur. J. Biochem.* 7:449-453.

Jacob, S. T., E. M. Sajdel, and H. N. Munro. 1970a. Different responses of soluble whole nuclear RNA polymerase to divalent cations and to inhibition by α-amanitin. *Biochem. Biophys. Res. Commun.* 38: 765-770.

Jacob, S. T., E. M. Sajdel, W. Muecke, and H. N. Munro. 1970b. Soluble RNA polymerases of rat liver nuclei: properties, template specificity, and amanitin responses in vitro and in vivo. *Cold Spring Harb. Symp. Quant. Biol.* 35: 681-691.

Jensen, E. V., T. Suzuki, T. Kawashima, W. E. Stumpf, P. W. Jungblut, and E. R. DeSombre. 1968. A two-step mechanism for the interaction of estradiol with rat uterus. *Proc. Nat. Acad. Sci. U.S.A.* 59: 632-638.

Kedinger, C., M. Gniazdowski, J. L. Mandel, Jr., F. Gissinger, and P. Chambon. 1970. α-amanitin: a specific inhibitor of one of two DNA-dependent RNA polymerase activities from calf thymus. *Biochem. Biophys. Res. Commun.* 38: 165-171.

Keshgegian, A. A., and J. J. Furth. 1972. Comparison of transcription of chromatin by calf thymus and *E.coli* RNA polymerases. *Biochem. Biophys. Res. Commun.* 48: 757-763.

Lindell, T. J., F. Weinberg, P. W. Morris, R. G. Roeder, and W. J. Rutter. 1970. Specific inhibition of nuclear RNA polymerase II by α-amanitin. *Science* 170: 447-448.

Lyon, M. F., and S. G. Hawkes. 1970. X-linked gene for testicular feminization in the mouse. *Nature* (London) 227: 1217-1219.

Mainwaring, W. I. P., F. R. Mangan, and B. M. Peterken. 1971. Studies on the solubilized RNA polymerase from rat-ventral prostate gland. *Biochem. J.* 123: 619-628.

Maryanka, D., and H. Gould. 1973. Transcription of rat liver chromatin with homologous enzyme. *Proc. Nat. Acad. Sci. U.S.A.* 70: 1161-1165.

Mohla, S., E. R. DeSombre, and E. V. Jensen. 1972. Tissue-specific stimulation of RNA synthesis by transformed estradiol-receptor complex. *Biochem. Biophys. Res. Commun.* 46: 661-667.

Muramatsu, M., N. Shimada, and T. Higashinakagawa. 1970. Effect of cycloheximide on the nucleolar RNA synthesis. *J. Mol. Biol.* 53: 91-106.

O'Malley, B. W., and A. R. Means. 1974. Female steroid hormones and target cell nuclei. *Science* 183: 610-620.

Randerath, K., and E. Randerath. 1967. Thin layer separation methods of nucleic acid derivatives. *Methods Enzymol.* 12: 323-347.

Raynaud-Jammet, C., and E. E. Baulieu. 1969. Action de l' estradiol In Vitro: Augmentation de la biosynthese d'acide ribonucleique dans les noyaux uterins. *C. R. Acad. Sci.* (Paris) 268:3211-3214.

Reeder, R. H. 1973. Transcription of chromatin by bacterial RNA polymerase. *J. Mol. Biol.* 80: 229-241.

Richardson, J. P. 1966. Enzymic synthesis of RNA from T7 DNA. *J. Mol. Biol.* 21: 115-127.

Sajdel, E. M., and S. T. Jacob. 1971. Mechanism of early effect of hydrocortisone on the transcriptional process: stimulation of the activities of purified rat liver nucleolar RNA polymerase. *Biochem. Biophys. Res. Commun.* 45: 707-715.

Shaw, L. M. J., and R. C. C. Huang. 1970. A description of two procedures which avoid the use of extreme pH conditions for the resolution of components isolated from chromatins prepared from pig cerebellar and pituitary nuclei. *Biochemistry* 9: 4530-4542.

Smuckler, E. A., and J. R. Tata. 1971. Changes in hepatic nuclear DNA-dependent RNA polymerase caused by growth hormone and triiodothyronine. *Nature* (London) 234: 37-39.

Weinmann, R., and R. G. Roeder. 1974. Role of DNA-dependent RNA polymerase III in the transcription of the tRNA and 5 S RNA genes. *Proc. Nat. Acad. Sci. U.S.A.* 71: 1790-1794.

Yu, F. L., and P. Feigelson. 1971. Cortisone stimulation of nucleolar RNA polymerase activity. *Proc. Nat. Acad. Sci. U.S.A.* 68: 2177-2180.

STRUCTURE, FUNCTION, AND REGULATION OF
RNA POLYMERASES IN ANIMAL CELLS

R. G. ROEDER, S. CHOU, J. A. JAEHNING,
L. B. SCHWARTZ, V. E. F. SKLAR, and R. WEINMANN
Department of Biological Chemistry
Division of Biology and Biomedical Sciences
Washington University
St. Louis, Missouri 63110

ABSTRACT. The mouse plasmacytoma class I, II, and III RNA polymerases have been purified and shown to have distinct structures and functions. The two large molecular weight subunits and some low molecular weight subunits differ between the class I, II, and III enzymes. However, some small subunits appear common to two or to three enzyme classes. α-Amanitin was used to distinguish the endogenous RNA polymerase activities of the class I, II, and III RNA polymerases in isolated nuclei and to show their functions, respectively, in transcription of the genes for the rRNAs, the HnRNAs, and the transfer and 5S RNAs. The cellular activity levels of solubilized RNA polymerases I and III vary with the physiological state of the cell in different mouse tissues, suggesting that the activities of the genes which they transcribe are regulated in part by specific enzyme levels in adult cell types. The increased RNA polymerase I levels in malignant mouse plasmacytoma cells reflect primarily increased enzyme concentrations. The levels of RNA polymerase II in mouse tissues show much less variability suggesting similar rates of HnRNA synthesis in these cell types or cellular excesses of this enzyme. During very early embryonic development in X. laevis all RNA polymerases are present in vast cellular excess and changes in specific gene function are not accompanied by changes in total enzyme levels, suggesting that factors other than enzyme levels are involved in regulating the transcription of all genes during this developmental period.

INTRODUCTION

Eukaryotic cells contain multiple forms of nuclear DNA-dependent RNA polymerases which may be grouped in three major classes (I, II, and III) on the basis of their distinctive catalytic and chromatographic properties (Roeder, 1969; Roeder and Rutter, 1969) and, in the case of animal cells, on the basis of their unique α-amanitin sensitivities (reviewed in Schwartz et al, 1974b). The enzymes are also partially dis-

tinguished on the basis of their subnuclear localizations
(Roeder and Rutter, 1970; Schwartz et al, 1974b). These
observations suggest that the class I, II, and III enzymes
might differ in structure and in function and that selective
gene transcription might be regulated in part via distinct
enzymes which are specific for different classes of genes.
More direct evidence supports these ideas. RNA polymerases
I and II have been shown to have distinct subunit patterns
(see, for example, Schwartz and Roeder 1974a,b; Kedinger et
al, 1974). Nucleoplasmic RNA polymerase II appears to synthesize heterogeneous nuclear (DNA-like) RNA whereas the nucleolar
RNA polymerase I presumably transcribes the rRNA cistrons
(Roeder and Rutter, 1970; Blatti et al, 1970; Zylber and
Penman, 1971; Reeder and Roeder, 1972). Furthermore, the
levels of the enzymes (especially RNA polymerase I) vary in
different cell types and fluctuate in response to various
stimuli (for references see Schwartz et al, 1974b). However,
the mechanisms by which the activities of the various enzymes
are altered have not been elucidated.

The present report describes more recent observations concerning the structure and function of the class III RNA polymerases which are compared with homologous class I and class II
RNA polymerases. The levels of activity of RNA polymerases I,
II, and III have also been determined in different mouse tissues, including malignant plasmacytomas, and during early
amphibian development to investigate possible relationships
between specific enzyme levels and specific gene function or
between specific enzyme levels and the physiological state of
the cell. In addition, at least one mechanism by which the
level of RNA polymerase I activity is increased in rapidly
proliferating cells is demonstrated.

EXPERIMENTAL PROCEDURES

Biological Materials

MOPC (mineral oil plasmacytoma) 315 solid plasmacytomas
were propagated subcutaneously in BALB/c mice (Schwartz et al,
1974b). Liver and spleen tissue were from normal or from
tumor-bearing 10-week old mice. Oocytes, unfertilized eggs,
embryos, and cultured kidney cells of *Xenopus laevis* were
obtained as described (Roeder, 1974a,b).

Methods

The procedures for RNA polymerase isolation and purification for sodium dodecyl sulfate-polyacrylamide gel electro-

phoresis, for RNA synthesis in nuclei and nucleoli, and for electrophoretic analyses of RNA products are detailed or referenced in the appropriate figure legends. One unit of RNA polymerase activity represents the incorporation of one picomole UMP in 20 min under standard assay conditions (Schwartz et al, 1974b; Roeder, 1974a).

RESULTS

Structural Features of the RNA Polymerases

Mouse plasmacytoma cells contain high levels of RNA polymerase activity (see below). When analyzed by DEAE-Sephadex chromatography (Fig. 1) two major peaks of activity are observed with native DNA as template and correspond, respectively, to the nucleolar RNA polymerase I and the nucleoplasmic RNA polymerase II reported in many other eukaryotes. Using low levels of α-amanitin to inhibit RNA polymerase II, two peaks of activity are observed in the upper region of the salt gradient, using as template either native DNA or poly(dAT). These latter peaks correspond to class III RNA polymerases described in other systems (Roeder and Rutter, 1969; Roeder, 1974a) based on their chromatographic and catalytic properties and their α-amanitin sensitivities (below). RNA polymerases I, II, III_A and III_B have been further purified by ion exchange chromatography and sucrose gradient sedimentation as reported elsewhere (Roeder and Schwartz, 1974a,b; Sklar and Roeder, manuscript in preparation).

To investigate the subunit composition of the murine plasmacytoma class I, II, and III RNA polymerases, purified enzyme preparations were subjected to electrophoresis on polyacrylamide gels in the presence of sodium dodecyl sulfate. The left panel of Fig. 2 shows the results obtained with RNA polymerase I (I_A and I_B, see below) on cylindrical gels. The right panel of Fig. 2 shows the results obtained with RNA polymerase II (II_0, II_A, and II_B; see below) and with RNA polymerases III_A and III_B which were run concomitantly on slab gels. In addition, RNA polymerase III purified from *Xenopus laevis* oocytes (Roeder, 1974b; Roeder and Chou, manuscript in preparation) is shown for comparison. Based on relative molar ratios and on the constancy of polypeptide mass to enzyme activity in different chromatographic fractions following gradient elution (data not shown), those bands indicated by arrows appear to represent enzyme subunits. The molecular weights of these polypeptide bands are summarized in Table I. This table also summarizes more detailed analyses of the different class I and different class II enzyme forms from plasmacytoma cells which are reported elsewhere (Schwartz and Roeder, 1974b).

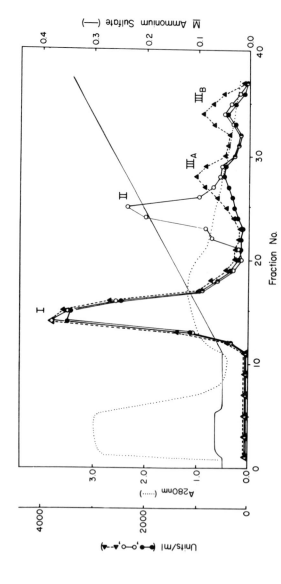

Fig. 1. DEAE-Sephadex chromatography of RNA polymerases from MOPC 315 tumors. The RNA polymerase activity in 10 gm of tumor was solubilized and subjected to chromatography on a column (325 ml) of DEAE-Sephadex. Fractions of 20 ml were collected and activity was measured with calf thymus DNA in the absence (O——O) or presence (●——●) of 0.5 μg α-amanitin per ml or with poly[d(A-T)] in the presence of 0.5 μg α-amanitin per ml (▲---▲). All procedures were as described by Schwartz et al (1974b).

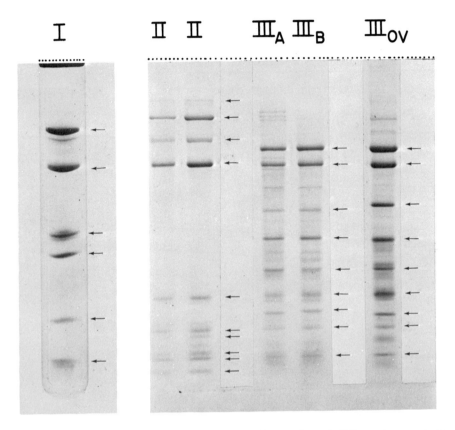

Fig. 2. Polyacrylamide Gel Electrophoresis of RNA polymerases in the presence of Sodium Dodecyl Sulfate.

Plasmacytoma RNA polymerase I was purified through the second phosphocellulose step as previously described (Schwartz and Roeder, 1974a).

Plasmacytoma RNA polymerase II was purified through the DEAE-Sephadex step (Schwartz and Roeder, 1974b). RNA polymerases III_A and III_B were separated by DEAE-Sephadex chromatography (Fig. 1) and were subsequently purified by ion exchange chromatography (CM-Sephadex, phosphocellulose) and sucrose density gradient centrifugation (Sklar and Roeder, manuscript in preparation). The soluble RNA polymerase III from mature *X. laevis* ovaries was purified by similar procedures (Roeder and Chou, manuscript in preparation). Electrophoresis of RNA polymerase I was on cylindrical gels containing 10% acrylamide in the presence of sodium dodecyl sulfate (Schwartz and Roeder, 1974a). Electrophoresis of RNA poly-

(*Fig. 2 legend continued*)
merases II and III was carried out on a slab gel (9.0 x 9.0 x 0.1 cm) containing a 5 to 15% polyacrylamide gradient and in the presence of sodium dodecyl sulfate. The arrows denote the bands whose approximate molecular weight values (determined from standards run simultaneously) are indicated in Table I.

TABLE I
PLASMACYTOMA ENZYME SUBUNITS (DALTONS X 10^{-3})

I_A	I_B	II_O	II_A	II_B	III_A	III_B	$III_{X.\ laevis}$
		240					
			205				
195	195						
				170			
					155	155	155
		140	140	140			
					138	138	137
117	117						
					89	89	92
					70	70	68
61							
52	52				52	52	52
		42	42	42	42	42	42
					34	34	33
29	29	29	29	29	29	29	29
		25	25	25			
		21	21	21			
19	19	19	19	19	19	19	19
		16	16	16			

These data summarize the apparent subunit compositions of the class I, II, and III RNA polymerases shown in Fig. 2. The compositions of the different forms of plasmacytoma RNA polymerase I (tentatively designated here as I_A and I_B) and plasmacytoma RNA polymerase II (II_O, II_A, and II_B) which are separable by polyacrylamide gel electrophoresis under non-denaturing conditions (Schwartz and Roeder, 1974a,b) are shown separately. The apparent molecular weight values for the small polypeptides are approximate and vary slightly in different electrophoresis systems. Thus the three smaller enzyme I subunits here correspond to subunits I_d, I_e, and I_f reported previously (Schwartz and Roeder, 1974a) and the five smaller enzyme II associated polypeptides here correspond, respectively, to subunits II_d, II_e, II_f, II_g, II_h, and II_i reported previously (Schwartz and Roeder, 1974b). See text for discussion of subunits common to different enzyme classes.

Thus, electrophoresis of RNA polymerase I under non-denaturing conditions resolves two forms of the enzyme (tentatively designated here as I_A and I_B). One form has six subunits as indicated in Table I. The other is identical except for the absence of the 61,000 dalton subunit. Similarly this technique resolves three forms of RNA polymerase II. Forms II_O, II_A, and II_B contain, respectively, the 240,000, the 205,000, and the 170,000 dalton subunits as well as all of the lower molecular weight species. The RNA polymerase II preparation used in Fig. 2 appears to contain little of the form II_O since the 240,000 subunit is barely detectable. No differences in the subunit compositions of the chromatographically separated plasmacytoma class III enzymes are apparent upon polyacrylamide gel electrophoresis in the presence of sodium dodecyl sulfate (Fig. 2), suggesting that their distinct chromatographic properties may reflect charge differences between otherwise similar subunits. The close structural similarity between the mouse plasmacytoma RNA polymerase III and the *X. laevis* oocyte RNA polymerase III suggests that these enzymes have similar functions in these distinct cell types and that the class III enzymes in other eukaryotic cells may have similar, if not identical, structures.

A comparison of the data in Table I reveals that the class I, II, and III enzymes are readily distinguished on the basis of their large subunits (> 100,000 daltons) which are distinct for each enzyme class. All the high molecular weight subunits indicated in Table I are readily separated when the various enzymes are mixed prior to electrophoresis in the presence of sodium dodecyl sulfate (Sklar, Schwartz and Roeder, manuscript in preparation).

Although the situation with respect to the low molecular weight polypeptides is not as clear, some of these appear to be unique to each enzyme class, e.g., the 61,000 dalton polypeptide in RNA polymerase I, and the 89,000 and 70,000 dalton polypeptides in RNA polymerase III. On the other hand, some subunits appear to be common, at least with respect to molecular weight, to two or more RNA polymerase classes as suggested from the data summarized in Table I and as revealed by more critical analyses in which the enzymes are mixed prior to denaturation and electrophoresis in the presence of sodium dodecyl sulfate (Sklar et al, manuscript in preparation). These subunits include the 52,000 dalton polypeptides (common to I and III) and the 29,000 dalton polypeptide (common to I, II, and III). It is not yet clear whether the 42,000 and the 19,000 dalton polypeptides are shared among different enzyme classes as the data in Table I suggest. It should be stressed that the indicated molecular weights in Table I are approxi-

mate since slightly different values are obtained in different
analytical gel systems. Furthermore, the conclusions with
respect to the number and identity of the low molecular weight
subunits in the enzyme III preparations must be regarded as
tentative until stringent criteria for enzyme homogeneity are
obtained.

Properties and Functions of the RNA Polymerases

The class I, II, and III RNA polymerases exhibit divalent
metal ion (Mg^{++} and Mn^{++}) and salt (ammonium sulfate) activation profiles which are distinct but which are similar to
those reported for the analogous enzymes from other systems
(Schwartz et al, 1974b). The enzymes also show differential
sensitivities to α-amanitin as shown in Fig. 3 by the dashed

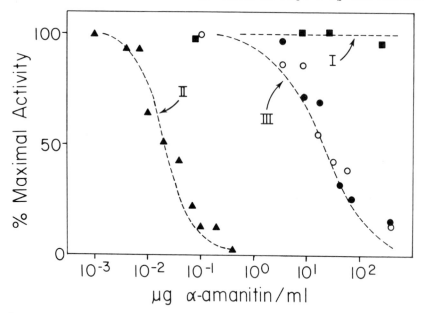

Fig. 3. Effect of α-amanitin concentration on purified RNA
polymerases and on endogenous nuclear and nucleolar RNA synthesis.
    Purified RNA polymerases I, II, and III from MOPC 315 cells
were assayed with exogenous calf thymus DNA as template at
varying concentrations of α-amanitin (dashed lines). Total
endogenous RNA synthesis was measured in isolated nucleoli
(squares) and isolated nuclei (triangles) in the absence of
exogenous templates. In the case of intact nuclei the α-
amanitin sensitive activity is normalized to 100% and corresponds to only 40% of the total activity. The remaining 60%

(*Fig. 3 legend continued*)   is completely insensitive to 400 µg
α-amanitin per ml. The closed and open circles represent,
respectively, synthesis of 5S RNA and pre-4S RNA in isolated
nuclei, monitored electrophoretically as in Fig. 4. For
details see Weinmann and Roeder (1974a).

lines. RNA polymerase I is uninhibited by α-amanitin ($\leq$ 400
µg per ml). RNA polymerases II and III can both be completely
inhibited although II (50% inhibition at 0.02 µg α-amanitin
per ml) is much more sensitive than is III (50% inhibition at
20 µg α-amanitin per ml). No differences in α-amanitin sensitivity are apparent for the different class II enzymes or for
the different class III enzymes (Schwartz et al, 1974b;
Weinmann and Roeder, 1974a; Roeder, 1974a). These distinct
α-amanitin sensitivities have been used to distinguish functions of the class I, II, and III enzymes in isolated plasmacytoma cell nuclei which continue to synthesize RNA in vitro.
As indicated in Fig. 3 (solid triangles) about 40% of the
endogenous activity in nuclei shows the same sensitivity to
α-amanitin as does purified RNA polymerase II, and previous
studies (see Introduction) indicate that this represents synthesis of DNA-like or heterogeneous nuclear RNA which presumably is precursor to mRNA. Previous studies (Reeder and
Roeder, 1972) showed that rRNA synthesis in isolated nuclei
was resistant to 1 µg α-amanitin/ml, ruling out the direct
involvement of RNA polymerase II in the synthesis of this RNA.
As shown in Fig. 3, endogenous RNA synthesis in isolated nucleoli (solid squares) is insensitive to 400 µg α-amanitin/ml,
thereby ruling out the involvement of RNA polymerase III in
rRNA synthesis and confirming the previous supposition that
RNA polymerase I, the predominant enzyme which is extracted
from nucleoli (Roeder and Rutter, 1970a; Schwartz and Roeder,
1974b), is responsible for rRNA synthesis. Furthermore, these
data show that non-specific inhibition of RNA synthesis by
high α-amanitin levels does not occur in these systems.
     To ascertain the enzyme(s) responsible for tRNA and 5S RNA
synthesis the α-amanitin sensitivities of the syntheses of
these species of RNA were monitored (Weinmann and Roeder,
1974a). Previous studies have shown that these RNAs are synthesized in isolated nuclei although the tRNAs are synthesized
as 4.5S RNA precursors which are not further processed in
isolated nuclei. These RNAs were defined by polyacrylamide
gel electrophoresis and in the case of the 4.5S tRNA precursors
by conversion to 4S RNA species with cytoplasmic enzymes
(McReynolds and Penman, 1974; references in Weinmann and
Roeder, 1974). The data in Fig. 4 show that these RNA species
are synthesized in MOPC 315 nuclei in the presence of 1 µg

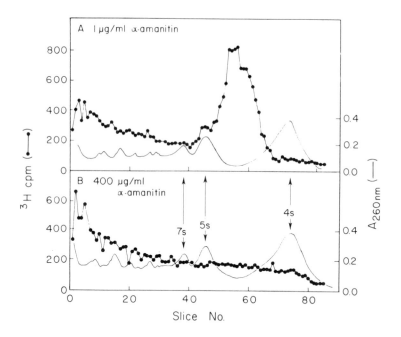

Fig. 4. Synthesis of 5S RNA and pre-4S RNA in isolated nuclei.
Nuclei were isolated from MOPC 315 cells and allowed to synthesize RNA (labeled with ^{3}H-UTP) in the presence of 1.0 or
400 μg α-amanitin per ml. RNA was extracted and analyzed by
polyacrylamide gel electrophoresis in the presence of nonradioactive marker 4S, 5S, and 7S RNAs (continuous solid line)
to detect newly synthesized (solid circles) 5S RNA and 4S RNA
precursors (major peak at 4.5S). For details see Weinmann
and Roeder (1974).

α-amanitin per ml (sufficient to inhibit II) but not in the
presence of 400 μg α-amanitin per ml (sufficient to inhibit
III). More detailed data are summarized in Fig. 3 which show
clearly that pre 4S RNA synthesis (open circles) and 5S RNA
synthesis (solid circles) have exactly the same sensitivity
to α-amanitin as RNA polymerase III. Since both III_A and III_B
have identical α-amanitin sensitivities it is not possible to
distinguish specific functions of these enzymes.

Levels of RNA Polymerases in Different Mouse Tissues

To determine possible relationships between the levels of specific enzymes and the physiological state of the cell, enzyme levels have been monitored in different mouse tissues from normal and plasmacytoma (tumor) bearing mice. The enzymes were solubilized and the absolute amounts of each enzyme measured with exogenous non-specific templates after resolution by chromatography on DEAE-Sephadex. Class I, II, and III RNA polymerases were detected in all tissues and the quantitative results are summarized in Table II. The cellular levels of RNA polymerase II were relatively invariant in the different tissues. In contrast, the cellular levels of RNA polymerases I and III (especially the former) showed large variations and were highest in the malignant plasmacytoma cells. However, the enzyme I and III levels were also higher in the liver and in the spleen (considerably enlarged, presumably in response to the invading tumor) from tumor bearing mice than in the corresponding tissues from normal mice. These increased levels presumably reflect increased rates of growth and proliferation (and gene transcription) in these cell types.

TABLE II
LEVELS OF RNA POLYMERASE ACTIVITIES
IN DIFFERENT MOUSE TISSUES

Tissue Fraction	Units/mg DNA		
	I	II	III
MOPC 315 Tumors	5,500	2,600	1,100
BALB/c Liver	764	1,350	186
MOPC 315 BALB/c Liver	2,330	1,780	292
BALB/c Spleen	210	1,138	137
MOPC 315 BALB/c Spleen	1,150	1,380	310

RNA polymerase levels were determined by chromatographic (DEAE-Sephadex) analyses following solubilization of the RNA polymerase activity in MOPC 315 tumors and in spleen and liver tissues from control and tumor-bearing BALB/c mice. Data taken from Schwartz et al (1974b).

The increased levels of RNA polymerase activity in rapidly proliferating tissues could be a result of increased concentrations of RNA polymerase or of alterations in the activity (e.g., by effector molecules) of a constant population of molecules. To distinguish these possibilities in the case of

RNA polymerase I the intrinsic specific activities were compared for several enzymes using non-specific templates. The
specific activity of the homogeneous plasmacytoma RNA polymerase I (400-600 units per µg protein) was found to be essentially the same as that of the purified mouse liver enzyme I
(500-600 units/µg protein) and similar to that reported for
the calf thymus enzyme I (Gissinger and Chambon, 1972). Thus
it appears that plasmacytoma cells contain increased concentrations of RNA polymerase I molecules relative to the concentrations in normal cells. This conclusion is also consistent
with the observation that the subunit composition of the
plasmacytoma enzyme I is analogous to that reported for the
calf thymus enzyme I (Schwartz and Roeder, 1974a).

Levels of RNA Polymerases during Embryonic Development

Because the absolute and relative rates of synthesis of the
major classes of RNA vary enormously during early embryonic
development in *X. laevis* (Brown and Dawid, 1969), this system
also seemed advantageous for investigating relationships between specific gene activation and the levels of specific RNA
polymerases. Oocytes, embryos, and adult somatic cells contain
class I, II, and III RNA polymerases analogous to those found in
in other eukaryotic cells (Roeder, 1974a,b). The relative
proportions and absolute amounts of these enzymes per embryo
at selected stages are summarized in Table III. It can be
seen that these parameters are constant for stages between the
unfertilized egg and the gastrula stages, despite the fact that
the unfertilized egg snythesizes no RNA and that the genes for
HnRNA, tRNA, and 5S and rRNA are sequentially activated at the
clevage (8,000 cells), blastula (15,000 cells), and the early
gastrula (30,000 cells) stages, respectively (Brown and Dawid,
1969).

The enzyme levels present in the unfertilized egg and in
the pre-gastrula embryos are similar to those present in mature
oocytes and thus appear to be synthesized during oogenesis
(Roeder, 1974b). However, as discussed previously, it is not
yet clear whether these enzymes function both in oocytes and
in embryos. The absolute amounts of each enzyme in the unfertilized egg are 5×10^4 to 5×10^5 higher on a per cell
basis than the enzyme levels in adult somatic cells (Table III).
The enzyme levels per cell subsequently decline with cell division, however, so that near the late gastrula stages (slightly
later for enzyme III) the enzyme levels approximate those found
in adult cells. Subsequent to these stages, however, the
embryos accumulate net amounts of each enzyme (Table III),
presumably in order to maintain these cellular enzyme levels
with continued cell division.

TABLE III

LEVELS OF RNA POLYMERASE ACTIVITIES DURING EMBRYONIC DEVELOPMENT IN *X. LAEVIS*

Stage	Average Cell No.	Units/10^4 Embryos			Units/10^4 Cells		
		I	II	III	I	II	III
Egg	1	10,500	24,900	9,500	10,500	24,900	9,500
Cleavage	8	10,900	25,100	11,800	1,700	3,137	1,500
Gastrula	62,000	11,500	33,700	11,600	0.19	0.54	0.19
Tadpole	420,000	90,800	140,000	20,000	0.19	0.33	0.04
Kidney Cell	---	---	---	---	0.18	0.17	0.02

The RNA polymerase levels were determined by DEAE-Sephadex analyses following solubilization of RNA polymerase activity in unfertilized eggs, in embryos at various stages, and in cultured kidney cells. The RNA polymerase II activity represents combined II_A (minor form) and II_B (major form) species. From Roeder (1974b).

39

DISCUSSION AND CONCLUSIONS

The class I, II, and III RNA polymerases differ not only in their catalytic and chromatographic properties (Roeder and Rutter, 1969; Schwartz et al, 1974a) but also in their molecular structures and in their respective functions. Thus the large molecular weight subunits and certain of the smaller polypeptides differ between the class I, class II, and class III enzymes suggesting that these enzymes are composed primarily of products of distinct genes and that they are not readily interconvertible by simple structural alterations. However, certain low molecular weight subunits appear to be common to two or to three enzyme classes suggesting that some features of catalytic function or of enzyme regulation may be common to functionally distinct enzymes. The enzymes within a class appear to differ in only a single polypeptide in the case of RNA polymerases I (Schwartz and Roeder, 1974a) and II (Kedinger et al, 1974; Weaver et al, 1971; Schwartz and Roeder, 1974b) and possibly only by charge differences in the case of RNA polymerase III (present data; Sklar, Schwartz, and Roeder, manuscript in preparation). No distinct catalytic or functional properties have yet been found for various heterogeneous forms of the RNA polymerases (Chesterton and Butterworth, 1971; Roeder, 1974a; Schwartz et al, 1974a,b).

Collectively the class I, II, and III RNA polymerases appear to be sufficient for synthesis of all major classes of RNA. The rRNA cistrons are transcribed by RNA polymerase I, the DNA sequences coding for HnRNA are transcribed by RNA polymerase II, and the tRNA and 5S RNA genes are transcribed by RNA polymerase III. This suggests that all of the RNA polymerases, including RNA polymerase III, serve essential functions and therefore should be ubiquitous in all cell types, as has recently been demonstrated (Schwartz et al, 1974b). Recent studies of cellular enzyme function in the transcription of viral genes (those coding for mRNAs and the non-translated 5.5S RNA) during adenovirus-2 infection of human cells provide additional evidence for the general function of the class II and III enzymes in the synthesis, respectively, of mRNAs (or mRNA precursors) and low molecular weight "structural" or non-messenger RNAs (Weinmann, Raskas, and Roeder, 1974).

In different adult mouse tissues the cellular levels of RNA polymerases I and III show considerable variability depending on the physiological state of the tissue. These observations are consistent with the belief that the activities of the rRNA genes and, presumably, the tRNA and 5S genes, are more active in more rapidly growing tissues (Mauck and Green, 1973) and suggest that these classes of reiterated and clustered genes may be regulated in part by alterations of the cellular levels of enzyme activity. In contrast, the cellular levels of RNA polymerase II activity show much less variability. This is consistent either with the suggestion that cellular rates of heterogeneous RNA synthesis do not show drastic quantitative differences in cells with different growth rates (Mauck and Green, 1973) or with the possibility that this enzyme is present in excess. Furthermore, because of the limited heterogeneity within the class II enzymes, additional factors (see below) are obviously required to effect selective gene transcription within the heterogeneous DNA sequences which this enzyme transcribes. Other investigators have also found that the cellular activity levels of solubilized RNA polymerases I and II are regulated independently in relation to various stimuli or growth conditions (for references see Schwartz et al, 1974b; Schwartz and Roeder, 1974a).

While variations in specific RNA polymerase levels may reflect physiological changes or changes in gene activity in certain cell types (e.g., in adult differentiated tissues) this correlation may not always hold. For example, during very early embryonic development and concomitant cell differentiation, all enzymes appear to be present in vast excess on a per cell basis and changes in RNA polymerase levels and patterns are not apparent during changes in specific gene activities. Only during later development do enzyme levels appear to become rate limiting. Thus either templates or other undefined transcription factors appear to be limiting during the early development period.

The molecular mechanisms by which cells regulate the activity and specificity (selectivity) of each of the enzymes is not yet clear. In the case of the mouse plasmacytoma RNA polymerase I, the increased level of enzyme activity per cell (relative to normal cells) appears to be due primarily to an increased concentration of catalytically active RNA polymerase molecules and not to an increased efficiency (e.g., by effector molecules) of a constant population of enzyme molecules. However, there are indications that other short-lived factors may be necessary for RNA polymerase I activity in vivo (Wu and Feigelson, 1972; Schwartz et al, 1974a) and such factors might in some instances be limiting and hence regulatory (e.g., during embryonic development). Although factors which alter the activity of RNA polymerase II specifically (Stein and Hausen, 1970; Seifert et al, 1973) or of all enzymes in general (diMauro et al, 1972) have been described, their specific functions remain to be defined.

Although the class I, II, and III enzymes have distinct subunit patterns, the overall complexities and structural organizations of all these enzymes (two high molecular weight subunits and several low molecular weight subunits in each case) resemble those of the prokaryotic enzymes. This suggests that some of the mechanisms which regulate the activity of the eukaryotic enzymes and their ability to transcribe selectively certain genes may be similar to those which regulate the pro-karyotic enzyme(s). These mechanisms may involve factors which interact with the enzyme or with the template (Chamberlin, 1974).

ACKNOWLEDGMENTS

This research was supported by Grants from the National Institutes of Health to R.G.R. (GM-19096-02-M4B) and to Washington University (5-S04-FR-06115). R.G.R. is a Research Career Development Awardee (1-K04-GM-70661-01), L.B.S. a Medical Scientist Trainee (5-T05-GM-02016) and J.A.J. and V.E.F.S. Predoctoral Trainees (5-T01-GM-1311) of the National Institutes of Health.

REFERENCES

Blatti, S. P., C. J. Ingles, T. J. Lindell, P. W. Morris, R. F. Weaver, R. Weinberg, and W. J. Rutter 1970. Structure and Regulatory Properties of Eukaryotic RNA Polymerase. *Cold Spring Harb. Symp.* 35: 649-657.

Brown, D. D., and I. B. Dawid 1969. Developmental Genetics. *Ann. Rev. Genet.* 3: 127-154.

Chamberlin, M. J. 1974. The Selectivity of Transcription. *Ann. Rev. Biochem.* 43: 721-775.

Chesterton, C. J., and P. H. W. Butterworth 1971. Selective Extraction of Form I DNA-Dependent RNA polymerase from Rat Liver Nuclei and its Separation into Two Species. *Eur. J. Biochem.* 19: 232-241.

DiMauro, E., C. P. Hollenberg, and D. B. Hall 1972. Transcription in Yeast: A Factor that Stimulates Yeast RNA Polymerases. *Proc. Natl. Acad. Sci.* 69: 2818-2822.

Gissinger, F., and P. Chambon 1972. Animal DNA-Dependent RNA Polymerases. 2. Purification of Calf-Thymus AI Enzyme. *Eur. J. Biochem.* 19: 232-241.

Kedinger, C., F. Gissinger, and P. Chambon 1974. Molecular Structures and Immunological Properties of Calf-Thymus Enzyme AI and of Calf-Thymus and Rat-Liver Enzymes B. *Eur. J. Biochem.* 44: 421-436.

Mauck, J. C., and H. Green 1973. Regulation of RNA Synthesis in Fibroblasts during Transition from Resting to Growing State. *Proc. Natl. Acad. Sci.* 70: 2819-2822.

McReynolds, L., and S. Penman 1974. A Polymerase Activity Forming 5S and pre-4S RNA in Isolated HeLa Cell Nuclei. *Cell* 1: 139-145.

Reeder, R. H., and R. G. Roeder 1972. Ribosomal RNA Synthesis
in Isolated Nuclei. *J. Mol. Biol.* 67: 433-441.

Roeder, R. G. 1969. Multiple RNA Polymerases and RNA Synthesis
in Eukaryotic Systems. Ph.D. Thesis, University of Washington.

Roeder, R. G., and W. J. Rutter 1969. Multiple Forms of DNA-
Dependent RNA Polymerase in Eukaryotic Organisms. *Nature*
224: 234-237.

Roeder, R. G., and W. J. Rutter 1970. Specific Nucleolar and
Nucleoplasmic RNA Polymerases. *Proc. Natl. Acad. Sci.*
65: 675-682.

Roeder, R. G. 1974. Multiple Forms of DNA-Dependent RNA Polymerase in *Xenopus laevis*, Isolation and Characterization
of Enzymatic Properties. *J. Biol. Chem.* 248: 241-248.

Roeder, R. G. 1974b. Multiple Forms of DNA-Dependent RNA
Polymerase in *Xenopus laevis*, Levels of Activity during
Oocyte and Embryonic Development. *J. Biol. Chem.* 248:
249-256.

Schwartz, L. B., C. Lawrence, R. E. Thach, and R. G. Roeder
1974a. Encephalomyocarditis Virus Infection of Mouse
Plasmacytoma Cells, II. Effect on Host RNA Synthesis
and RNA Polymerases. *J. Virology* in press.

Schwartz, L. B., and R. G. Roeder 1974a. Purification and Subunit Structure of Deoxyribonucleic Acid-Dependent Ribonucleic Acid Polymerase I from the Mouse Myeloma, MOPC
315. *J. Biol. Chem.* in press.

Schwartz, L. B., and R. G. Roeder 1974b. Purification and Subunit Structure of Deoxyribonucleic Acid-Dependent Ribonucleic Acid Polymerase II from the Mouse Plasmacytoma,
MOPC 315. *J. Biol. Chem.* in press.

Schwartz, L. B., V. E. F. Sklar, J. A. Jaehning, R. Weinmann,
and R. G. Roeder 1974b. Isolation and Partial Characterization of the Multiple Forms of Deoxyribonucleic Acid-
Dependent Ribonucleic Acid Polymerase in the Mouse
Myeloma, MOPC 315. *J. Biol. Chem.* in press.

Seifert, K. H., P. P. Juhasz, and B. J. Benecke 1973. A Protein Factor from Rat Liver Tissue Enhancing the Transcription of Native Templates by Homologous RNA Polymerase B. *Eur. J. Biochem.* 33: 181-191.

Stein, H., and P. Hausen 1970. A Factor from Calf Thymus Stimulating DNA-Dependent RNA Polymerase Isolated from this
Tissue. *Eur. J. Biochem.* 14: 270-277.

Weaver, R. F., S. P. Blatti, and W. J. Rutter 1971. Molecular
Structures of DNA-Dependent RNA Polymerases (II) from
Calf Thymus and Rat Liver. *Proc. Natl. Acad. Sci.* 68:
2994-2999.

Weinmann, R., and R. G. Roeder 1974. Role of DNA-Dependent RNA Polymerase III in Transcription of the tRNA and 5S RNA Genes. *Proc. Natl. Acad. Sci.* 71: 1790-1794.

Weinmann, R., H. Raskas, and R. G. Roeder 1974. Role of DNA-Dependent RNA Polymerases II and III in the Transcription of the Adenovirus Genome Late in Productive Infection. *Proc. Natl. Acad. Sci.* in press.

Wu, F.-Y., and P. Feigelson 1972. The Rapid Turnover of RNA Polymerase of Rat Liver Nucleolus, and of Its Messenger RNA. *Proc. Natl. Acad. Sci.* 69: 2833-2837.

Zylber, E. A., and S. Penman 1971. Products of RNA Polymerases in HeLa Cell Nuclei. *Proc. Natl. Acad. Sci.* 68: 2861-2865.

MULTIPLE FORMS OF RNA POLYMERASE
IN PREIMPLANTATION MOUSE EMBRYOS

CAROL M. WARNER and LARRY R. VERSTEEGH
Department of Biochemistry and Biophysics
Iowa State University
Ames, Iowa   50010

ABSTRACT. Multiple forms of DNA-dependent RNA polymerase have been characterized from adult mouse liver nuclei. The form I isozymes are most active at low ionic strength, have a low Mn^{2+}/Mg^{2+} activity ratio, and are insensitive to inhibition by 0.30 µg/ml of α-amanitin. Isozyme II is most active at high ionic strength, has a high Mn^{2+}/Mg^{2+} activity ratio, and is completely sensitive to inhibition by 0.30 µg/ml of α-amanitin.

Solubilized RNA polymerase from mouse blastocysts is inhibited 35% by 1.1 µg/ml of α-amanitin. In contrast, the total RNA polymerase activity solubilized from adult mouse liver nuclei is inhibited 82% at an α-amanitin concentration of 0.3 µg/ml. This suggests that there is more isozyme I than II in mouse blastocysts compared to adult mouse liver nuclei. The RNA polymerase activity in preimplantation mouse embryos is not detectable at the two-cell stage of development, but rises rapidly from the eight-cell stage on. Also, the relative amounts of the I and II isozymes change with the stage of development, as is shown by sensitivity of the preimplantation mouse embryo enzymes to α-amanitin.

INTRODUCTION

During the past several years it has become apparent that eukaryotic cells contain multiple forms of DNA-dependent RNA polymerase (Roeder and Rutter, 1969; Kedinger et al., 1972). We have isolated and characterized three isozymes of this enzyme from adult mouse liver nuclei (Versteegh and Warner, 1973). In addition, we have studied preimplantation mouse embryos for evidence of the multiplicity of enzyme forms of RNA polymerase (Warner and Versteegh, 1974).

The DNA, RNA, and protein content of preimplantation mouse embryos has been carefully measured as shown in Table I (Brinster, 1967; Olds et al., 1973). The mouse embryos used in these studies were collected after superovulation of the mice with 5 I.U. PMS (pregnant mare serum) followed 48 hours later by 5 I.U. HCG (human chorionic gonadotropin). The correlation of the stage of development with the number of hours after HCG administration is shown for reference to work to be described later (Figure 5). It is seen that early blastocysts contain only 24 ng of protein/embryo. Thus, if we make the very liberal estimate that 1% of the protein is the enzyme RNA polymerase, 100 embryos would be needed to obtain 24 ng of the enzyme. It becomes readily apparent that actual isolation of enzyme forms would prove to be a monumental task, so indirect methods for studying the enzyme forms were used in these studies.

TABLE I

PROTEIN[1], RNA[2], AND DNA[2] CONTENT OF
PREIMPLANTATION MOUSE EMBRYOS

Stage of Development	Approx. Hours Post HCG	Protein/ embryo (ng)	RNA/ embryo (ng)	DNA/ embryo (pg)
1-cell	18	28	0.55	29
2-cell	43	26	0.40	41
8-cell	71	24	0.46	155
morula	80	21	N.R.[3]	N.R.
early blastocyst	90	24	1.37	439

[1]Data from Brinster, R.L., *J. Reprod. Fert.* (1967) 13:413-420.
[2]Data from Olds, P.J., Stern, S. and Biggers, J.D., *J. Exp. Zool.* (1973) 186:39-46.
[3]N.R. = Not reported.

METHODS AND RESULTS

ADULT MOUSE LIVER RNA POLYMERASES

The scheme for the purification of adult mouse liver RNA polymerases is shown in Table II. One enzyme unit is defined as the amount of enzyme which catalyzes the incorporation of one picomole of ^{3}H-UMP into high molecular weight product per minute at 37°.

When the F4 enzyme is passed over a DEAE-Sephadex column, the elution pattern shown in Figure 1 is obtained. The three isozymes have been designated IA, IB and II. The properties of these isozymes are summarized in Table III. It is seen that the form I enzymes are most active at low ionic strength, have a low Mn^{2+}/Mg^{2+} activity ratio, and are not inhibited by α-amanitin. The form II enzyme is most active at high ionic strength, has a high Mn^{2+}/Mg^{2+} activity ratio, and is completely

TABLE II

PREPARATION OF RNA POLYMERASE FROM MOUSE LIVER

Purification Step	Total Activity (Units)[1]	Total Protein (mg)	Specific Activity (Units/mg)
F1. Liver homogenate	7,920	14,400	.55
F2. Nuclear suspension	36,000	359	100
F3. Lysis and sonication	40,000	314	127
F4. Centrifugation	49,400	166	298

[1]One enzyme unit is defined as that amount of enzyme which catalyzes the incorportion of one picomole of ^{3}H-UMP into high molecular weight product in one minute under standard assay conditions.

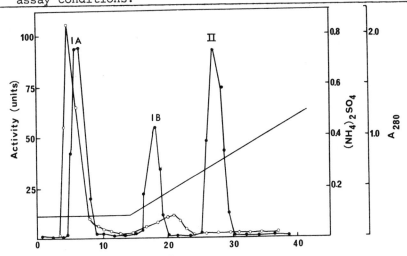

Fig. 1.   DEAE-Sephadex chromatography of F4 RNA polymerases. 3 mg of protein was placed on a 0.9 X 12 cm column and eluted with a linear gradient of $(NH_4)_2SO_4$.   (●——●) enzyme activity; (O——O) A_{280};   (——) $(NH_4)_2SO_4$ gradient.   The three peaks of activity are designated IA, IB, and II.

inhibited by 0.30 µg/ml α-amanitin.   All three isozymes are inhibited by exotoxin, but the inhibition of isozyme II is more readily reversed than that of either isozymes IA or IB by the addition of excess ATP.   None of the isozymes is significantly inhibited by rifampin, so that the presence of mitochondrial contamination can probably be excluded (Mukerjee and Goldfeder, 1973).

TABLE III

PROPERTIES OF RNA POLYMERASE FROM ADULT MOUSE LIVER

	Isozyme		
	IA	IB	II
$(NH_4)_2SO_4$ concentration (M) of elution from DEAE-Sephadex	0.10	0.14	0.28
Optimal $(NH_4)_2SO_4$ concentration (M)	0.03	0.03	0.09
Mn^{2+}/Mg^{2+} activity ratio	3.3	2.2	5.7
% Inhibition by 0.30 µg/ml α-amanitin	0	0	100
% Inhibition by 5 µM exotoxin at:			
.6 mM ATP	71	68	76
2.1 mM ATP	40	48	28
% Reversal of exotoxin inhibition by ATP	44	29	63
% Inhibition by 40 µg/ml rifampin	-9	-4	8

PREIMPLANTATION MOUSE EMBRYO RNA POLYMERASES

We have been able to assay for RNA polymerase activity in groups of 100 blastocysts. The embryos were collected 91 hours post-HCG injection, treated with pronase to remove the zona pellucida, transferred from Whitten and Bigger's (1968) medium to 0.05 M Tris, 0.5% BSA, pH 7.4 and frozen and thawed 3X in liquid nitrogen. The amount of ^{3}H-UTP used in the assay mixture was 200X higher than that used in studying the adult liver isozymes, and the incubation time at 37° was increased from 20 minutes to 40 minutes (Warner and Versteegh, 1974). The amount of incorporated ^{3}H-UMP was determined by the method of Litman (1968).

Figure 2 shows the activity of the blastocyst enzymes as a function of ammonium sulfate concentration. The adult liver F4 mixture of enzymes show a combined maximum activity at about 0.04 M $(NH_4)_2SO_4$ (unpublished) while the blastocyst enzymes are most active at an $(NH_4)_2SO_4$ concentration of about 0.03 M (Fig. 3). It was impossible for us to assay the blastocysts at less than 0.03 M $(NH_4)_2SO_4$ for technical reasons.

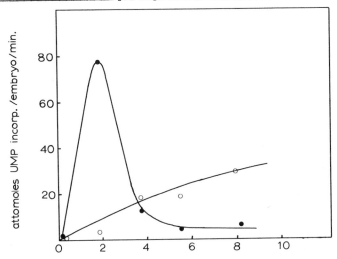

Fig. 2. Activity of blastocyst enzymes as a function of metal ion concentration. (●) Mn^{2+}; (O) Mg^{2+}.

The blastocysts show RNA polymerase activity in the presence of Mn^{2+} or Mg^{2+} very similar to that for the F4 and isolated enzyme forms. Mn^{2+} enhances the enzyme activity better than Mg^{2+} and shows a sharp optimum at 2mM Mn^{2+}. The curve for Mg^{2+} stimulation of activity is much broader than that for Mn^{2+}.

Inhibition of the blastocyst enzymes by varying concentrations of α-amanitin is shown in Figure 4. All these assays were performed at an $(NH_4)_2SO_4$ concentration of 0.09 M to favor detection of isozyme II. It is seen that maximal inhibition of enzyme activity occurs at an α-amanitin concentration of 0.8 μg/ml, and that the maximum observed inhibition is 35%. Table IV compares the effect of α-amanitin on blastocyst and F4 enzymes. Maximal inhibition of the F4 enzymes occurs at 0.30 μg/ml of α-amanitin, and the maximum observed inhibition is 82%. Thus, it seems probable that the mouse blastocysts have less RNA polymerase activity that is sensitive to inhibition by α-amanitin than do adult mouse liver nuclei. The

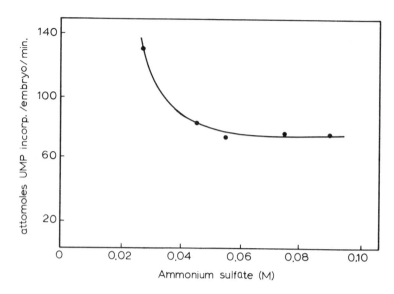

Fig. 3. Activity of blastocyst enzymes as a function of
$(NH_4)_2SO_4$ concentration.

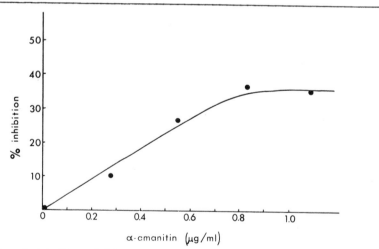

Fig. 4. Activity of blastocyst enzymes as a function of α-
amanitin concentration.

enzyme activity in the blastocysts which is sensitive to α-
amanitin needs almost three times more α-amanitin than the
adult liver enzyme for maximal inhibition. The decreased sensitivity is probably not due to protection of the enzyme by
endogenous DNA, since the blastocyst enzyme activity is completely dependent on exogenous DNA (see Table IV). Exotoxin has

an inhibitory effect on the F4 and blastocyst enzymes. Although

TABLE IV

A COMPARISON OF EFFECTS OF VARIOUS
INHIBITORS ON F4 AND BLASTOCYST ENZYMES

A. F4 Enzymes

Inhibitor (concentration)	$(NH_4)_2SO_4$ Concentration (M)	% Inhibition
α-amanitin (0.30 µg/ml)	0.04	73
α-amanitin (0.30 µg/ml)	0.09	82
exotoxin (300 µM)	0.01	92
rifampin (40 µg/ml)	0.04	0
no exogenous DNA	0.01	91

B. Blastocyst Enzymes

Inhibitor (concentration)	$(NH_4)_2SO_4$ Concentration (M)	% Inhibition
α-amanitin (1.10 µg/ml)	0.03	6
α-amanitin (1.10 µg/ml)	0.09	35
exotoxin (270 µM)	0.09	80
rifampin (40 µg/ml)	0.09	14
no exogenous DNA	0.09	100

the F4 enzymes are completely insensitive to rifampin, the blastocyst enzymes may show some inhibition by rifampin. One possible
explanation is that the F4 enzymes are isolated from nuclei,
while the mixture of blastocyst enzymes may contain some mitochondrial RNA polymerase which is sensitive to rifampin. It
should be noted, however, that the error in the F4 results is
< 1% whereas the error in the blastocyst results is ± 10%.

Our next experiment was to assay for RNA polymerase activity at different stages of preimplantation mouse development. The embryos were collected, as described previously
(Warner and Versteegh, 1974), at the two-cell, eight-cell,
morula, and blastocyst stages. Whereas 100 blastocysts contained sufficient enzyme activity for the assay, we had to assay
activity at the two-cell stage in 600 embryos, activity at the
eight-cell stage in 200 embryos, and activity in the morula
stage in 150 embryos to obtain sufficient activity for reproducible results. The RNA polymerase activity/embryo as a function of the stage of development is shown in Figure 5. It
is seen that a burst of RNA polymerase activity begins at the
eight-cell stage, with very little detectable activity before
that.

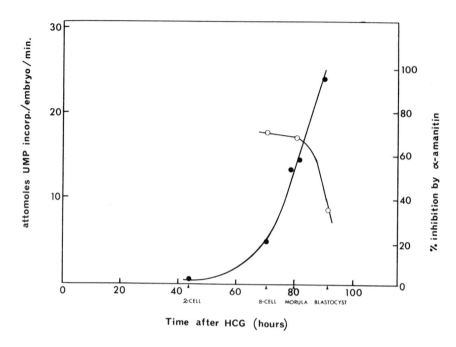

Fig. 5. Activity of embryonic enzymes as a function of time of development. (●) without α-amanitin; (O) with α-amanitin.

DISCUSSION AND CONCLUSIONS

RNA synthesis in preimplantation mouse embryos has been studied in a number of laboratories (Mintz, 1964; Thomson and Biggers, 1966; Monesi and Salfi, 1967; Ellem and Gwatkin, 1968; Woodland and Graham, 1969; Monesi et al., 1970; Piko, 1970; Tasca and Hillman, 1970; Daentl and Epstein, 1971; Epstein and Daentl, 1971; Knowland and Graham, 1972; Globus et al., 1973). It has recently been shown that even the first cleavage division probably requires RNA synthesis, since it is inhibited by actinomycin D and by α-amanitin (Globus et al., 1973). This differs from the sea urchin and amphibians where early cleavage divisions are not blocked by actinomycin D (Gross and Cousineau, 1963; Brachet et al., 1964). RNA synthesis cannot be directly demonstrated in one cell mouse embryos because labeled precursor cannot enter the cells at this stage (Daentl and Epstein, 1971). Knowland and Graham (1972) have demonstrated that all classes of RNA are synthesized at the two-cell stage of development. Earlier autoradiographic studies by Mintz (1964) had

suggested that incorporation of ^{3}H-uridine into the nucleus
was only into the non-nucleolar portion up to the four-cell
stage. At the eight-cell stage there is a burst of RNA synthetic activity and it seems to be almost exclusively in the
nucleoli (Mintz, 1964) and to be rRNA (Ellem and Gwatkin, 1968).
Thus, it seems that at least some RNA synthesis occurs very
early in mouse embryo development and that this RNA is mostly
heterogeneous or mRNA. Then, at about the eight-cell stage of
development there is a large burst of rRNA synthesis, although
small amounts of rRNA may have been synthesized earlier.

We sought to investigate the types and amounts of the enzyme responsible for RNA synthesis, RNA polymerase, in preimplantation mouse embryos. Since isozyme II is believed mainly to synthesize mRNA while isozyme I synthesizes rRNA (Blatti
et al., 1970), the ratio of these two isozymes might be expected to change during early mouse embryo development, just as
the type of RNA being synthesized changes.

As can be seen in Figure 5, we find no detectable RNA
polymerase activity until the eight-cell stage of development.
There is then a burst of RNA polymerase activity similar to the
burst of rRNA synthesis that is seen in these embryos. More
interesting, though, is that inhibition of enzyme activity by
α-amanitin (α-amanitin inhibits only isozyme II in vitro) shows
that the amount of isozyme II decreases from the eight-cell to
the blastocyst stage of development. This would correspond
with more type I isozyme being needed for rRNA synthesis. At
all stages of development there is less isozyme II than is
found in adult liver nuclei (see Table IV). We conclude that
the types and amounts of RNA being synthesized in preimplantation mouse embryos is directly correlated to the types of RNA
polymerases present.

ACKNOWLEDGEMENTS

We thank Professor T. Wieland for the α-amanitin, Dr. R.
P. M. Bond for the exotoxin, and Carla Tollefson for technical
assistance.
This work was supported by the Population Council, New
York, and by NIH grant 1R01 AI 11752-01 IMB.

REFERENCES

Blatti, S.P., C.J. Ingles, T.J. Lindell, M.W. Morris, R.F.
Weaver, F. Weinberg, and W.J. Rutter 1970. Structure and
Regulatory Properties of Eukaryotic RNA Polymerase. *Cold
Spring Harbor Symp. Quant. Biol.* 35:649-657.

Brachet, J., H. Denis, and F. de Vitry 1964. The Effects of
Actinomycin D and Puromycin on Morphogenesis in Amphibian
Eggs and *Acetabularia mediterranea*. *Develop. Biol.*
9:398-434.

Brinster, R.L. 1967. Protein Content of the Mouse Embryo
During the First Five Days of Development. *J. Reprod.
Fert.* 13:413-420.

Daentl, D.L. and C.J. Epstein 1971. Developmental Interrelationships of Uridine Uptake, Nucleotide Formation and Incorporation into RNA by Early Mammalian Embryos. *Develop.
Biol.* 24:428-442.

Ellem, K.A.O. and R.B.L. Gwatkin 1968. Patterns of Nucleic
Acid Synthesis in the Early Mouse Embryo. *Develop. Biol.*
18:311-330.

Epstein, C.J. and D.L. Daentl 1971. Precursor Pools and RNA
Synthesis in Preimplantation Mouse Embryos. *Develop.
Biol.* 26:517-524.

Globus, M.S., P.G. Calarco, and C.J. Epstein 1973. The Effects
of Inhibitors of RNA Synthesis (α-Amanitin and Actinomycin
D) on Preimplantation Mouse Embryogenesis. *J. Exp. Zool.*
186:207-216.

Gross, P. and G. Cousineau 1963. Effect of Actinomycin D on
Macromolecule Synthesis and Early Development in Sea Urchin
Eggs. *Biochem. Biophys. Res. Commun.* 10:321-326.

Kedinger, C., F. Gissinger, M. Gniazdowski, J. Mandel, and P.
Chambon 1972. Animal DNA-Dependent RNA Polymerases: 1.
Large-Scale Solubilization and Separation of A and B Calf-
Thymus RNA-Polymerase Activities. *Eur. J. Biochem.* 28:269-
276.

Knowland, J. and C. Graham 1972. RNA Synthesis at the two-cell
Stage of Mouse Development. *J. Embryol. Exp. Morph.* 27:
167-176.

Litman, R.M. 1968. A Deoxyribonucleic Acid Polymerase from
Micrococcus luteus (*Micrococcus lysodeikticus*) Isolated on
Dexoyribonucleic Acid-Cellulose. *J. Biol. Chem.* 243:6222-
6233.

Mintz, B. 1964. Synthetic Processes and Early Development in
the Mammalian Egg. *J. Exp. Zool.* 157:85-100.

Monesi, V. and V. Salfi 1967. Macromolecular Synthesis During
Early Development in the Mouse Embryo. *Exptl. Cell Res.*
46:632-635.

Monesi, V., M. Molinaro, E. Spalletta, and C. Davoli 1970.
Effect of Metabolic Inhibitors in Macromolecular Synthesis
and Early Development in the Mouse Embryo. *Exptl. Cell
Res.* 59:197-206.

Mukerjee, H. and A. Goldfeder 1973. Purification and Properties of Ribonucleic Acid Polymerase from Rat Liver

Mitochondria. *Biochem.* 12:5096-5101.

Olds, P.J., S. Stern, and J.D. Biggers 1973. Chemical Estimates of the RNA and DNA Contents of the Early Mouse Embryo. *J. Exp. Zool.* 186:39-46.

Piko, L. 1970. Synthesis of Macromolecules in Early Mouse Embryos Cultured in Vitro: RNA, DNA, and a Polysaccharide Component. *Develop. Biol.* 21:257-279.

Roeder, R.G. and W.J. Rutter 1969. Multiple Forms of DNA-dependent RNA Polymerase in Eukaryotic Organisms. *Nature* 224:234-237.

Tasca, R.J. and N. Hillman 1970. Effects of Actinomycin D and Cycloheximide on RNA and Protein Synthesis in Cleavage Stage Mouse Embryos. *Nature* 225:1022-1025.

Thomson, J.L. and J.D. Biggers 1966. Effect of Inhibitors of Protein Synthesis on the Development of Preimplantation Mouse Embryos. *Exptl. Cell Res.* 41:411-427.

Versteegh, L.R. and C. Warner 1973. DNA-Dependent RNA Polymerases from Normal Mouse Liver. *Biochem. Biophys. Res. Commun.* 53:838-844.

Warner, C.M. and L.R. Versteegh 1974. In Vivo and In Vitro Effects of α-Amanitin on Preimplantation Mouse Embryo RNA Polymerase. *Nature* 248: 678-680.

Whitten, W. and J. Biggers 1968. Complete Development In Vitro of the Preimplantation Stages of the Mouse in a Simple Chemically Defined Medium. *J. Reprod. Fert.* 17:399-401.

Woodland, H.R. and C.F. Graham 1969. RNA Synthesis during Early Development of the Mouse. *Nature* 221:327-332.

THE EXPRESSION OF β-GLUCURONIDASE DURING MOUSE EMBRYOGENESIS

VERNE M. CHAPMAN
Department of Molecular Biology
Roswell Park Memorial Institute
Buffalo, New York   14203

LINDA WUDL
Merck Institute
Rahway, New Jersey   07065

ABSTRACT. β-glucuronidase activity was measured in single mouse embryos on day 2 (2- to 4-cell stage) through day 4 (blastocyst stage) to determine when embryonic genes begin to function during embryogenesis. β-glucuronidase was chosen for this study because it is a biochemically well characterized gene product with genetic variants of the structural gene. Additionally, it is possible to assay activities of single embryos using a microfluorometric assay. We observed a pronounced increase in activity of 100-fold between day 3 (8-cell stage) and day 4. Activity changes between days 2 and 3 were not as pronounced but still significant. Activity changes were examined in embryos of the strains C3H/HeJ and C57BL/6J and the F_1 hybrid from C3H/HeJ females and C57BL/6J males. These strains differ in activity levels of glucuronidase in adult tissues and heat denaturation kinetics. On the basis of activity changes in embryos on days 2, 3, and 4, and a comparison of embryos with different glucuronidase genotypes, we have determined that embryonic genes are functioning on day 2 and that the rate of production of glucuronidase increases sharply on day 3.

An important question in the study of early mammalian development is when does the embryonic genome become functional and by extention capable of directing the processes of differentiation and development. Experimentally, this requires determining when individual structural genes become functional and how the onset of function is controlled. To determine when genes become functional it would be desirable to have a significant increase in a biochemically and genetically well-defined gene product in embryonic cells. Since maternally derived message has been demonstrated in embryos of other species (Raff, et al, 1972; Gross, 1968) it is necessary to distinguish between product derived from embryonic genes and from stored message.

β-glucuronidase, encoded by a single structural gene, *Gus*, on chromosome 5 of the mouse, fulfills the requirements of such an experimental system. In studies of activity changes during preimplantation development we found a dramatic increase in β-glucuronidase activity of 100-fold between day 3 (8-cells) and day 4 (morulae). Because genetic variants of the structural gene are present among normally developing inbred strains it was possible to establish embryonic gene functions in this 100-fold increase in activity by a qualitative determination of paternal gene product in hybrid embryos. The experimental utility and value of this system is additionally enhanced by the availability of a microfluorometric assay technique (Wudl and Paigen, 1974) which can detect β-glucuronidase in single embryos and potentially in single blastomeres.

MATERIALS AND METHODS

Inbred mice of the strains C3H/HeJ (C3) and C57BL/6J (B6) were obtained from the Jackson Laboratories, Bar Harbor, Maine. A heat labile β-glucuronidase is produced by the Gus^h structural allele from C3. When tissue homogenates of C3 mice are heated for 30 min at 70° C less than 10 per cent of the original activity remains. By contrast, enzyme from B6 with the Gus^b allele has more than 50 per cent of its activity remaining in similarly treated homogenates. Enzyme from adult tissues of (C3H/HeJ x C57BL/6)F_1 hybrids (C3B6F_1) has heat denaturation kinetics which are intermediate to the parental strains. β-glucuronidase activity in B6 is about 6- to 10-fold greater than C3. Activity levels in the C3B6F_1 hybrid are intermediate (Ganschow and Paigen, 1968). Whether the same mutation is responsible for both heat denaturation and activity level differences has not been established (Paigen, 1964).

Studies on the molecular basis of the activity differences between C3 and B6 have established that the catalytic efficiency of Gus^h and Gus^b gene products is similar (Swank, Ganschow, and Paigen, 1973). More recently, studies of the turnover of β-glucuronidase in C3 and B6 indicate that B6 has higher activity because it has a greater rate of synthesis than C3 (Ganschow, 1974).

Ovulation was induced in 8- to 16-week old females by injecting 4 I.U. pregnant mares serum, PMS (pregnyl, Ayerst) followed 48 hours later with 4 I.U. human chorionic gonadotropin, HCG (Sigma). The females were paired with stud males late in the afternoon following HCG injection and checked the following morning for vaginal plugs.

Embryos were flushed from the fallopian tubes or uterine horns with phosphate buffered saline (PBS), pH 7.2 containing 0.3% bovine serum albumin (BSA). Embryos were collected on

day 2 (2- to 4-cell), at three times on day 3 approximately
9 a.m., 12 noon, and 4 p.m. (8- to 16-cells) and day 4 (morulae
or early blastocyst). Embryos were separated from the cellular
debris with micropipets and transferred twice to fresh PBS
with BSA.

β-glucuronidase activity was assayed in single embryos using
a single cell microfluorometric assay technique (Wudl and
Paigen, 1974). For the enzyme assay intact embryos in PBS·BSA
solution were picked up with micropipets in a volume of less
than 2 μl and transferred to 0.5 ml of enzyme assay mixture
which contained 0.1 M acetate, pH 6.0, 0.9% NaCl, 0.1% BSA,
3×10^{-4} M 4-methyl umbelliferyl-β-D-glucuronic acid. The
embryos were picked up with finely drawn polyethylene tubing
(0.3- to 0.4 mm diameter) and deposited singly in microdroplets
of 1- to 10 nl under oil on a glass microscope slide. The
embryos were disrupted by alternately freezing and thawing the
slides in an atmosphere of N_2. The liberated enzyme was
allowed to react with substrate in the droplet by incubating
the slides in a moist chamber at 37°C for one or two hours.
The slide was then placed in a closed chamber with a reservoir
of triethylamine, which diffuses through the oil and raises
the pH of the droplet to 10. This stops the enzyme reaction
and develops the fluorescence.

The fluorescence of droplets with and without embryos was
measured using a Nikon fluorescence microscope equipped with
an Aminco photomultiplier microphotometer. Interference filters (Baird Atomic) were positioned to obtain monochromatic
light for excitation (365 nm) and emission (455 nm). The
fluorescence of a droplet is directly proportional to the number of moles of product in the droplet.

RESULTS

Mean specific activities obtained from single embryo measurements from day 2 through day 4 are shown in Table 1. The
most striking feature of these data is the 100-fold increase
in activities between day 3 and day 4 which occurred in each
of the genotypes examined. Even though the embryos are undergoing cleavage this increase in activity represents an estimated 20- to 40-fold increase in the level of β-glucuronidase
per cell among the different strains. On the other hand, the
difference in activities between day 2 and day 3 is also a
significant increase in activity per embryo but on a per cell
basis there is no significant activity change. Thus, from the
character of the activity change we conclude that the synthesis
of β-glucuronidase may begin as early as day 2 and that a pronounced increase in the rate of synthesis occurs on day 3.

TABLE 1

Mean β-Glucuronidase Activities (10^{-15} moles hr^{-1} embryo^{-1}) ± S.E. From Single Mouse Embryo Measurements

Time of Development	C3H/HeJ	C3B6F$_1$	C57BL/6J
Day 2	0.10 ± 0.01 (16)[a]	0.21 ± 0.03 (16)	0.85 ± 0.05 (32)
Day 3 (early)	0.20 ± 0.02 (10)	1.04 ± 0.18 (19)	1.75 ± 0.10 (49)
Day 3 (middle)	0.60 ± 0.07 (19)	1.80 ± 0.28 (18)	5.50 ± 2.70 (5)
Day 3 (late)	0.78 ± 0.09 (30)	3.13 ± 0.40 (42)	20.20 ± 3.50 (31)
Day 4	65.90 ± 6.35 (29)	98.30 ± 9.50 (34)	150.15 ±31.20 (20)

[a] Numbers of individual embryos measured.

To ascertain the timing of embryonic gene function in the synthesis of β-glucuronidase we compared the increase in activity of B6, C3, and the C3B6F$_1$ hybrid. We reasoned that the properties of the embryonic F$_1$ hybrid β-glucuronidase would be similar to the C3 mother if maternal or stored message were responsible for the production of β-glucuronidase during early embryogenesis. However, if embryonic genes were being transcribed and translated the properties of the hybrid would be intermediate to the parental strains.

A comparison of C3 and B6 activities (Table 1) during the preimplantation period shows two important features. First, B6 consistently has higher activity than C3 at each stage of development which is in accord with adult tissue activity differences. Second, while the fold increase in activity is similar for both strains the actual amount of enzyme produced is different. Thus, between day 2 and day 3 B6 makes nearly 10 times more enzyme than C3 and between day 3 and day 4 B6 makes 2.3 times more enzyme than C3. Enzyme turnover studies on these amounts of enzyme in embryos is presently difficult; however, these results are consistent with the finding that B6 has a higher rate of β-glucuronidase synthesis than C3.

Activity levels of C3B6F$_1$ embryos on day 2 are significantly higher than C3 levels. By early day 3 the hybrid is intermediate to the parental activities and they remain intermediate through day 4. The amount of enzyme produced by the F$_1$ between days 2 and 3 is nearly equal to that made by the B6 parent but

between days 3 and 4 the amount of enzyme produced is intermediate to the parental strains.

From the enzyme activity levels in C3B6F$_1$ hybrids we conclude that the gene product of the paternally derived Gus^b
allele is produced as early as day 2 and that the 100-fold
increase in activity between days 3 and 4 is directed by the
embryonic genome. At present, we cannot distinguish whether
the difference between C3 and C3B6F$_1$ embryos on day 2 is the
result of synthesis of β-glucuronidase by the embryo or whether
it simply represents B6 sperm enzyme incorporated at fertilization.

The production of the paternal gene product in C3B6F$_1$
hybrids was independently determined by examining the heat
denaturation kinetics of β-glucuronidase in day 4 embryos.
For these assays a modification of the microassay technique
was used. Mice 21-25 days of age were superovulated and
embryos collected at the 2-cell stage and grown in in vitro
culture (16) to the late morulae-early blastocyst stage (early
day 4). Embryos were then collected, counted, and washed in
phosphate buffered saline. Approximately 100 embryos of each
genotype (C3 homozygous, B6 homozygous, and C3B6F$_1$ heterozygous) were placed in 60 μl of buffer (.1M acetate pH 4.6)
in small (0.9 ml) test tubes. The embryos were frozen and
thawed five times by alternately placing the test tubes in dry
ice-acetone and warm water. The test tubes were then placed
in a controlled temperature circulating water bath preheated
to 70°C. At t=0 and t=30 minutes, 10 μl samples were removed
and added to a second micro test tube containing 10 μl of
assay buffer plus substrate (.1 M acetate, pH 6.0, 6 x 10^{-4} M
4-MU-glucuronic acid). These tubes were covered with parafilm
and placed in a closed chamber with a reservoir of buffer and
incubated at 37°C for 24 hours. β-glucuronidase is extremely
stable at 37°C and activity is linear for more than 100 hours.

The concentration of product in the micro test tube assay
was determined by the following procedure. The assay mixture
was deposited in nl (nanoliter) droplets under oil on microscope slides and made basic by incubating the slides in the
presence of triethylamine. The fluorescence and diameter of
several droplets from each assay mixture were measured. Fluorescence is directly proportional to the number of moles or
product present in a droplet and volume can be derived from the
droplet diameter. From these data the concentration of 4-MU
in the droplets is calculated.

The results of the heat inactivation experiment described
above are presented in Table 2, experiment 1, as moles of product per hour per embryo at 0 time and after incubating 30 minutes at 70°C. The percent survival of β-glucuronidase in B6

TABLE 2
Heat Inactivation of Embryonic β-Glucuronidase

Strain	Exp. Number	Activity t_0 10^{-14} moles/ hr/emb	Activity t_{30} 10^{-14} moles/ hr/emb	% Survival	% Survival Adult Liver Enzyme Paigen and Ganschow 1965
C57BL/6J	1	3.2	1.7	53.1	60
(C3H x C57BL/6)F_1	1	2.86	1.07	37.4	40
	2	3.3	1.2	36.4	
C3H/HeJ	1	0.75	0.13	17.3	10
	2	5.0	0.5	10.0	

homozygous, C3 homozygous, and F_1 hybrid embryos are compared.
If the increase in enzyme activity on day three was due to
translation of maternally derived messenger RNA or to transcription and translation of maternal genetic information only,
the enzyme in the F_1 hybrid, which was produced by crossing a
C3 mother with a B6 father, should have the same heat inactivation kinetics as the C3 homozygous embryos. This is not what
our results demonstrate.

The F_1 embryos contain β-glucuronidase which is similar in
heat sensitivity to that of the adult hybrid animal, and intermediate to that of the two parental strains.

Since the embryos used in experiment 1 were obtained from
superovulated 21-25 day old females and grown in vitro, it was
possible that we were not looking at the normal developmental
pattern. Therefore, we repeated the heat inactivation experiment using 6- to 8-week old females induced to ovulate and
mated according to the standard procedure. Embryos developed
in vivo and were collected from the uterus (blastocyst stage).
The procedure for testing heat inactivation of β-glucuronidase
described for experiment 1 was followed. The results are presented in Table 2 as experiment 2. The B6 homozygotes were
omitted in this experiment because it was difficult to obtain
sufficient numbers of day 4 embryos from these females.

Again, we find that the F_1 hybrid contains a heat-stable
form of β-glucuronidase which indicates that the paternal gene
for β-glucuronidase is expressed by day three of development.

DISCUSSION

On the basis of activity changes during preimplantation
development and a comparison of embryos with different genotypes for β-glucuronidase, we have established that 1) the
embryonic genes for β-glucuronidase are functioning on day 2
of development and 2) that the rate of production of β-glucuronidase increases sharply on day 3.

Embryonic gene expression during cleavage has been established for one other locus in the mouse, glucose phosphate
isomerase (*Gpi-1*). Using electrophoretic variants of *Gpi* it
has been possible to detect paternal gene product preimplantation possibly as early as the 8-cell stage (Chapman, Whitten,
and Ruddle, 1971; Brinster, 1973). While GPI has been useful
for establishing the timing of embryonic gene function it has
the limitation of being present in high activity from prefer-
tilization synthesis so that activity per embryo decreases
during cleavage when paternal gene product is detectable.
Also, large numbers of embryos are required to detect paternal
gene product at the 8-cell stage with currently available techniques.

Other enzymes which increase in activity during preimplantation development include uridine kinase (Daentl and Epstein, 1971), hypoxanthine-guanine phosphoribosyl transferase, HGPRT (Epstein, 1970) and adenine phosphoribosyl transferase, APRT (Epstein, 1970). These enzymes increase in activity from 6- to 10-fold between days 3 and 4; however, none of these enzymes have genetic variants to verify whether the increase in activities are embryonic or maternal gene products.

A second group of enzymes is characterized by relatively high activity at ovulation that remain unchanged until day 3 when they begin to decrease. These enzymes include lactate dehydrogenase (LDH), glucose-6-phosphate dehydrogenase (G6PD), and GPI. The most prominent example of these enzymes, LDH, comprises about 5 per cent of the total protein in the egg at fertilization (Erickson, 1974). No synthesis of LDH is detectable during early cleavage which suggests that the coordinate decreases in the activity of LDH, G6PD, and GPI on day 3 reflects the onset of degradation. In this regard, it is of interest that decrease in these activities is coincident with the increased β-glucuronidase production by the embryo.

Since the embryos are beginning to differentiate into compact morulae and blastocysts on days 3 and 4 it would be interesting to know if the increased activities of β-glucuronidase, uridine kinase, HGPRT, and APRT as well as the decreased activity of other enzymes represent a coordinate activation of genes associated with this morphogenetic change. Conversely, on the basis of the rates of incorporation of labelled precursors into RNA (Woodland and Graham, 1969; Daentl and Epstein, 1971) and protein (Epstein and Smith, 1973) in preimplantation embryos it may be possible that the changes in enzyme activity on day 3 result from a complex of recruitment of protein synthesizing machinery, an increase in the number of gene copies available for transcription, and an increased cell surface area available for transport of nucleotides and amino acids for supporting transcription and translation.

REFERENCES

Brinster, R. 1973. Parental glucose phosphate isomerase activity in three-day mouse embryos. *Biochem. Genet.* 9: 187-191.

Chapman, V. M., W. K. Whitten, and F. H. Ruddle 1971. Expression of paternal glucose phosphate isomerase-1 (*Gpi-1*) in preimplantation stages of mouse embryos. *Develop. Biol.* 26: 153-158.

Daentl, D. L. and C. J. Epstein 1971. Developmental interrelationships of uridine uptake, nucleotide formation and incorporation into RNA by early mammlian embryos. *Develop. Biol.* 24: 428-442.

Epstein, C. J. 1970. Phosphoribosyl transferase activity during early mammalian development. *J. Biol. Chem.* 245: 3289-3294.

Epstein, C. J. and S. A. Smith 1973. Amino acid uptake and protein synthesis in preimplantation mouse embryos. *Develop. Biol.* 33: 171-184.

Erickson, R. P. 1974. Studies on LDH isozymes in gametes and early development. These proceedings

Ganschow, Roger and Kenneth Paigen 1968. Glucuronidase phenotypes of inbred mouse strains. *Genet.* 59: 335-349.

Ganschow, R. E. 1974. Simultaneous genetic control of the structure and rate of synthesis of murine β-glucuronidase. These proceedings

Gross, P. R. 1968. Biochemistry of differentiation. *Ann. Rev. Biochem.* 37: 631-660.

Paigen, K. 1964. The genetic control of enzyme realization during differentiation. in *Congenital malformations-- 2nd International Conference on Congenital Malformations.* (Fishbein ed.) The International Medical Congress, New York, 181-190.

Raff, R. A., H. V. Colot, S. E. Selvig, and P. R. Gross 1972. Oogenetic origin of messenger RNA for embryonic synthesis of murotubule proteins. *Nature* 235: 211-214.

Swank, R. T., R. E. Ganschow, and K. Paigen 1973. Genetic control of glucuronidase induction in mice. *J. Mol. Biol.* 81: 225-243.

Woodland, H. R. and C. F. Graham 1969. RNA synthesis during early development of the mouse. *Nature* 221: 327-332.

Wudl, L. R. and K. Paigen 1974. Enzyme measurements on single cells. *Science* (in press).

ISOZYMES AS GENETIC MARKERS IN EARLY MAMMALIAN DEVELOPMENT

WOLFGANG ENGEL and WINFRIED FRANKE
Institut für Humangenetik
und Anthropologie der Universität
D-7800 Freiburg, Albertstrasse 11
and
ULRICH PETZOLDT
Max-Planck Institut für Immunbiologie
D-7800 Freiburg, Stefan-Meierstrasse 8

ABSTRACT. Early mammalian development might be dependent in part on maternally transmitted storage products. One of these products, an enzyme lactate dehydrogenase, shows species differences with respect to its isozyme pattern in mammalian oocytes. In species of the orders *Rodentia* and *Lagomorpha* only LDH-1 (B-subunits) is demonstrable in the oocyte, while in species of the orders *Carnivora* and *Artiodactyla* and in man LDH isozymes formed of A and B-subunits are present in the oocyte. In the rabbit the LDH isozyme pattern changes during oocyte maturation. Oocytes in the dictyotän stage exhibit only LDH-1, while those induced by HCG for proceeding further in the meiotic process change their LDH pattern: LDH isozymes formed of A- and B-subunits become visible after the HCG injection. The newly appearing LDH isozymes might be due to an inactive maternally transmitted storage product, not further characterized till now. Furthermore it its shown that in rodent species in general the total preimplantation LDH activity is based on the maternally transmitted B-subunits while the activation of the embryonic LDH genes starts only with implantation.

Since the developmental information stored in the mammalian egg is rather small, the embryonal genome should initiate transcription soon after fertilization. With respect to the onset of synthesis of definite proteins due to embryonic gene activity during preimplantation development only the results on the first appearance of the paternally coded PGI isozyme in the mouse are known, which do not contribute to the question of the time sequence of allele activation. On the other hand alleles at the gene loci for S-NADP-IDH, S-NADP-MDH and M-NADP-MDH expressed only after implantation in the mouse are reported to be activated synchronously.

It is assumed that maternally transmitted reserve substances, namely proteins, ribosomes, and long-lived m-RNA formed during the lampbrush stage of oogenesis may govern early development in lower vertebrates. Recently Gross et al (1973) and Skoultchi and Gross (1973) provided direct proof that histone m-RNA is stored in the unfertilized sea urchin egg, demonstrating for the first time a maternally inherited message with a definite coding sequence.

Since lampbrush chromosomes have been demonstrated in the oocyte of various vertebrate species belonging to different taxonomic groups including mammals (for review see Davidson and Hough, 1972), it might be concluded that lampbrush chromosomes are of ancient evolutionary origin. This suggests that early mammalian development in some part is guaranteed by maternally transmitted gene products (Wolf and Engel, 1972). RNA synthesized in the oocyte nucleus is presumably metabolized during growth and contributes as well to a storage of ribosomes and developmental information to be used after fertilization (Mintz, 1964). The investigations of Burkholder et al (1972) point to the formation of inactive maternal ribosome-mRNA complexes during oocyte maturation that disappear from the cytoplasm during the early cleavage stages after fertilization in the mouse.

Lactate dehydrogenase as a storage product in the mammalian embryo

One of the maternally derived gene products found to be present in oocytes of all mammalian species so far examined is an enzyme lactate dehydrogenase (LDH; E.C. 1.1.1.27). LDH activity in mature oocytes varies within a wide range between different species. The mouse has the highest value observed and the human and the rabbit show the lowest one (Brinster, 1968). Until recently, electrophoretic studies had been performed only on the mouse oocyte. It was shown that the entire LDH activity in the mouse oocyte is based only on the presence of B-subunits (LDH-1) (Engel, 1972). In an attempt to evaluate the situation in other mammals we have studied the LDH isozyme pattern in oocytes of 15 species of the orders *Rodentia, Lagomorpha, Carnivora, Artiodactyla,* and in man (Fig. 1). The LDH isozyme pattern was demonstrated by vertical micro disc electrophoresis in 5 µl capillaries as described by Grossbach (1965). The results obtained point to the existence of two types of oocytes in mammals: in species of the orders *Rodentia* and *Lagomorpha* only LDH-1 (B-subunits) is demonstrable, in species of the orders *Carnivora* and *Artiodactyla* and in man LDH isozymes formed of A and B-subunits are present in the oocyte. Especially in rodent species, the LDH isozyme pattern was also studied in ova and fertilized eggs. No change

ORDER	SPECIES
RODENTIA	Mus musculus
	Microtus arvalis
	Microtus agrestis
	Microtus ochrogaster
	Clethrionomys glareolus
	Apodemus sylvaticus
	Rat
	Guinea pig
	Syrian hamster
LAGOMORPHA	Rabbit
CARNIVORA	Cat
	Dog
ARTIODACTYLA	Pig
	Cow
	Sheep
PRIMATES	Man

Fig. 1. Mammalian species in which the LDH isozyme pattern
was investigated in oocytes.

in the isozyme pattern was seen during this developmental
period (Engel and Kreutz, 1973). The course of LDH activity
during the preimplantation phase has been followed in the
mouse (Brinster, 1965), in the rat (Brinster, 1967a), and in
the rabbit (Brinster, 1967b). During the first 3 days of
development the activity remains constant and is 20 times
higher in the mouse than in the rabbit. When the embryo moves
from the oviduct into the uterus the activity in the mouse
decreases and reaches values near zero shortly before implantation (Wolf and Engel, 1972) while in the rabbit the activity
rises sharply (Brinster, 1967b). Electrophoretic studies of
the LDH isozyme pattern during this developmental period have
been done exclusively with the mouse. It was shown that the
whole preimplantation development is characterized by the
presence of LDH-1, while shortly after implantation only LDH-
5 (A subunits) is present; additional isozymes appear subsequently (Auerbach and Brinster, 1967). A similar shift of
the LDH isozyme pattern has been postulated during rat development (Cornette et al, 1967). We studied the LDH isozyme
pattern in preimplantation and early postimplantation embryos
of the mouse, rat, guinea pig, and Syrian hamster as well as
in the middle and late preimplantation rabbit embryo. Our

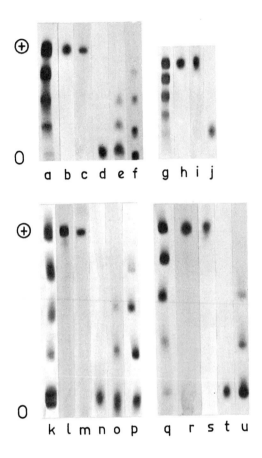

Fig. 2a–u. LDH isozymes of organ homogenates and pre- and postimplantation embryos of the mouse, guinea pig, rat, Syrian hamster. Mouse: (a) kidney, (b) 3 fertilized eggs, (c) 4 morulae, (d) 5 embryos day 7, (e) 3 embryos day 8, (f) 3 embryos day 9. Guinea pig: (g) kidney, (h) 8 fertilized eggs, (i) 7 morulae, (j) 3 embryos day 9. Rat: (k) mixture of heart and muscle, (l) 4 fertilized eggs, (m) 4 morulae, (n) 2 embryos day 8, (o) 2 embryos day 9, (p) 2 embryos day 10. Syrian hamster: (q) kidney, (r) fertilized eggs, (s) 6 morulae, (t) 3 embryos day 7, (u) 4 embryos day 8.

results indicate that in rodent species in general the total preimplantation LDH activity is based on the maternally trans-mitted B-subunits while the activation of the embryonal LDH genes starts only with implantation (Fig. 2). Moreover, the gene for A-subunits is activated first and the gene for

B-subunits follows later (Engel and Petzoldt, 1973).

As in rodents, LDH-1 is the only LDH isozyme present in the ovarian oocyte of the rabbit. However, in contrast to rodents the LDH isozyme pattern in the rabbit embryo changes long before implantation: in the 86 hr old rabbit embryo LDH isozymes formed of A- and B-subunits are already present. This suggests that the embryonal LDH genes in the rabbit start synthesizing long before implantation (Engel and Petzoldt, 1973). Meanwhile, Brinster (1973a) has demonstrated A- and B-subunits as early as in the fertilized egg. To elucidate further these differences in the LDH isozyme pattern between the oocyte and the fertilized egg we studied the electrophoretic pattern of LDH during oocyte maturation and in the unfertilized egg of the rabbit.

Materials and Methods

Randomly bred rabbits were treated with 4 daily subcutaneous injections of 0.33 mg follicle stimulating hormone (FSH) from Armour Pharmaceutical Company (Chicago) in 20% Kollidon[R] 25 and 100 i.u. human chorionic gonadotropin (HCG: Prolan[R] from Bayer AG, Leverkusen, Western Germany) intravenously at the 5th day. Unfertilized eggs were flushed from the oviduct with Krebs-Ringer-bicarbonate solution (KRB) containing 0.1% bovine serum albumin (BSA) 16 hr after the HCG injection. Fertilized eggs were prepared from mated females 16 hr after copulation. Oocytes were obtained from rabbits 24 hr after the last FSH injection as well as 2, 8, and 16 hr after the HCG treatment. The cumulus cells were removed by exposure to trypsin (0.025% trypsin in KRB with BSA; Difco Laboratories, Detroit) or by mechanical treatment only, drawing the oocytes back and forth in a narrow pipette. Oocytes, ova, and fertilized eggs were checked under the stereomicroscope to be free of surrounding cells. The denuded eggs were then washed two times in KRB with BSA, placed in small tubes and stored at -78°C. Generally 10-15 eggs were placed in each tube. Micro disc electrophoresis was used to demonstrate the LDH isozyme pattern in the oocytes, ova, and fertilized eggs, The method has been described recently by Engel and Kreutz (1973). As a modification a 10% separation gel and a 2.5% spacer gel in 10 µl capillaries were used; the pH of the electrode buffer tris-glycine (50 mM; 383 mM) was adjusted to pH 11.0 with 3 N NaOH. For the identification of the LDH isozymes in the eggs, the positions of the LDH bands were compared to those from organ homogenates of adult animals.

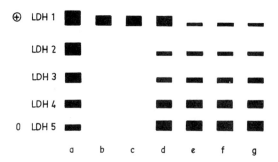

Fig. 3. Schematic presentation of the LDH isozyme pattern in (a) kidney, (b-e) oocytes, (f) unfertilized and (g) fertilized eggs of the rabbit. For further details see text.

RESULTS

As shown in Fig. 3, in the unfertilized egg as well as in the fertilized egg LDH isozymes formed of A- and B-subunits are present; the staining intensity of the bands decreases from LDH-5 to LDH-1 (Fig. 3f,g). In oocytes from rabbits treated only with FSH, LDH-1 is the only LDH isozyme (Fig. 3b); after HCG injection the LDH isozyme pattern in the oocyte changes. 2 hr after the HCG injection the LDH isozyme pattern corresponds to that of HCG untreated animals (Fig. 3c), while 8 hr after the HCG injection a complex LDH isozyme pattern is demonstrable. In addition to LDH-1 also LDH 5, 4, 3, and 2 are present. The staining intensity distribution over the five LDH bands is as follows: LDH-1 and LDH-5 are heavily stained, while the staining intensity decreases from LDH-5 to LDH-2 (Fig. 3d). Since we pooled always 10-15 oocytes for electrophoresis it might be suggested that the oocyte samples investigated consist of those oocytes exhibiting only LDH-1 and others exhibiting LDH-5, 4, 3, 2 and 1 with decreasing staining intensity from LDH-5 to LDH-1, thus corresponding to ovulated eggs. The following results point in this direction:

1. Ovulation in the rabbit occurs between 9.5 and 10 hr after copulation or HCG injection (Pincus and Enzmann, 1935). However, not all mature follicles are ovulated, which might be due to oocyte immaturity. Oocytes prepared from these non-ovulated follicles 16 hr after the HCG injection exhibit only LDH-1, while the ovulated eggs exhibit LDH-5, 4, 3, 2 and 1 in a decreasing staining intensity as mentioned above.

2. Of 15 capillary electrophoresis tests with 10 oocytes
each 8 hr after the HCG injection 1 gel exhibited only LDH-1
(Fig. 3c), while another exhibited LDH-5, 4, 3, 2 and 1 with
decreasing staining intensity from LDH-5 to LDH-1 (Fig. 3e).
The other gels showed an isozyme pattern with varying intensity for the five LDH isozymes, indicating that different
numbers of oocytes with only LDH-1 and those with LDH-5 to
1 had been pooled together.

From these results we conclude that the LDH isozyme pattern
in the rabbit oocyte changes during the gonadotropin induced
progression of the germ cell from late prophase (germinal
vesicle stage) of the first meiotic division to metaphase of
the second meiotic division when ovulation occurs. Furthermore, oocytes which become ovulated have undergone the shift
in the LDH isozyme pattern.

Since gene activity during oogenesis seems to be restricted
to the lampbrush stage of oocyte growth (Bloom and Mukherjee,
1972; Moore et al, 1974), the LDH isozymes newly appearing
during oocyte maturation in the rabbit should not be due to
transcriptional activity but rather to stored products which
are synthesized during the lampbrush stage but preserved in
an inactive form for later use. Preliminary results from
our laboratory have shown that in oocytes from FSH pretreated
rabbits, thus exhibiting only LDH-1, the LDH isozyme pattern
of the oocytes is changed after 5 min of hyaluronidase incubation. All 5 LDH isozymes with a decreasing staining activity
from LDH-5 to LDH-1 are demonstrable. This LDH isozyme pattern was described as characteristic for the freshly ovulated
egg. Furthermore the LDH activity in the hyaluronidase
treated oocytes seems to be higher than in untreated oocytes.
The question arises whether activation of stored products
during oocyte maturation in the rabbit starts on the level of
inactive mRNA for A- and B-subunits of LDH, on the level of
already existing A- and B-subunits prevented from associating,
or with inactive isozymes LDH-5, 4, 3, 2 and 1 in the oocyte.
Since the activation process has taken place after 5 min of
hyaluronidase incubation the storage should be rather due to
preexisting inactive LDH subunits of LDH isozymes than inactive mRNA. The mechanism by which hyaluronidase acts in the
oocytes is not known. Since the active isozyme LDH-1 in the
immature oocyte is destroyed during hyaluronidase incubation,
the activation process may take place by liberating protein
coated precursors. It might be speculated that in the ovary
hyaluronidase is set free from the abundant lysosomes in the
oocyte (Anderson, 1972) at a definite stage of oocyte maturation. Whether this is true or not, our results might repre-

73

sent the first example of the existence of a definite maternally transmitted gene product which is inactive, but activated
during oocyte maturation.

Gene activation during pre- and postimplantation development
of mammals

In contrast to the externally developing embryo of lower
vertebrates, the mammalian embryo has only small amounts of
available storage substances in the egg cytoplasm. Therefore
the embryonic genome in mammals should initiate transcription
very early during preimplantation development. However, the
onset of synthesis of a certain product does not prove that
further development is dependent on that particular product.
The rRNA synthesis, which reaches a noteworthy rate at the
late cleavage stages (mouse) or at the onset of blastulation
(rabbit), could lead to storage of rRNA for later use. It is
still unknown at what stage the first mRNA is synthesized.
However, even the demonstration of mRNA synthesis does not
prove that this messenger is used for protein synthesis, and
therefore that activation of the particular gene is a pre-
requisite for further development. Presumably, the embryonic
genome is activated unspecifically and transcription starts,
but no translation takes place (degeneration of messenger).
The crucial event in the course of gene activation during
development should be the translation of embryonically coded
messenger, especially if the specific regulation is a translational one (Tomkins and Martin, 1970; Ohno, 1971). This
question can be studied by the demonstration of the first
appearance of embryonically coded individual proteins. Numerous enzymes already present as storage substances in the
mouse oocyte have been quantified during the course of preimplantation development, and different patterns have been
found (for review see Wolf and Engel, 1972; Biggers, 1972).
Changes in enzyme activity allow only indirect conclusions,
if any, regarding the onset of embryonic genetic activity.
However, qualitative variation, e.g. differential mobility
of different allele products on electrophoresis, permits definite conclusions as to the onset of activity of embryonic
gene products. Through crossings of different homozygous
inbred strains it is possible in the F_1 generation to ascertain the time of appearance of the respective gene products
of the maternally and paternally derived alleles. If this
event occurs during the preimplantation phase, while maternally
transmitted storage products are still available, conclusions
are only possible regarding the onset of activity of the
paternal allele. In this case a further possibility is the

use of either a three allele system, which at present is not available in the mouse, or a back cross of heterozygous F_1 animals.

Until now only 4 polymorphic isozyme systems of the mouse have been used to investigate the activation of individual alleles: Phosphoglucose isomerase (PGI; E.C. 5.3.1.9), the supernatant form of NADP-dependent isocitrate dehydrogenase (S-NADP-IDH; E.C.1.1.1.42), and the mitochondrial as well as supernatant form of NADP-dependent malate dehydrogenase (M- and S-NADP-MDH; E.C. 1.1.1.40). Of these only PGI shows electrophoretic resolution before implantation.

a) <u>PGI</u>

Mouse inbred strains possess two electrophoretic variants of this enzyme (De Lorenzo and Ruddle, 1969). In heterozygotes obtained from crosses of the appropriate strains (SJL/J x F_1 of a C57BL/6J x C3H/HeJ cross). Chapman et al (1971) studied the stage at which the paternally derived phenotype appears for the first time. Until the 4th day (blastocyst) F_1 animals only possess the maternal phenotype; on the 5th day (late blastocyst) the heteromeric isozyme first appears, resulting from the presence of subunits coded for by the paternal allele. Recently Brinster (1973b) stated that the first paternal PGI activity occurred as early as in the 8-cell embryo.

These experiments give no information on the stage at which the maternal allele is activated, since the homozygous phenotype of the mother is indistinguishable from the hemizygous phenotype of the maternal embryonic allele. However, those investigations, especially on allelic systems which become activated during preimplantation development might be of far reaching significance for understanding the process of gene activation during early mammalian development. If maternally transmitted alleles in the heterozygous embryos are preferentially activated first, it might be suggested that the egg cytoplasm has a role in recognizing and activating individual gene loci during early mammalian development. Asynchronous activation of the parental alleles has been reported to some extent in lower vertebrates (Hitzeroth et al, 1968), birds (Castro-Sierra and Ohno, 1968), and plants (Davies, 1973). In all of these cases the maternally transmitted allele was the first to be activated.

To study the mode of activation of paternal PGI alleles in the mouse an electrophoretic system might be useful by which PGI activity of a single embryo can be demonstrated. Chapman et al (1971) as well as Brinster (1973b) used starch gel

electrophoresis, pooling 15-45 and 500 embryos respectively.
We succeeded recently in the demonstration of PGI isozymes of
a single mouse ovum by using micro disc electrophoresis.
Studies are now under way to elucidate the mode of activation
of parental PGI alleles in the mouse. The results which might
be expected in the case of synchronous and of asynchronous
activation, respectively, are shown in Fig. 4. The cross is
performed between a heterozygous F_1 mother AA'(SJL/J x C3H)
and a homozygous father AA (C3H). It is suggested that the
embryonic PGI genes are activated despite the presence of
maternally transmitted PGI storage products. From Fig. 4 it
can be inferred that in the case of synchronous activation
as well as asynchronous activation initially two different
phenotypes occur, differing in storage products with respect
to activity distribution.

b) S-NADP-IDH, S-NADP-MDH, M-NADP-MDH

These enzymes are demonstrable electrophoretically in the
mouse only during organogenesis. We determined the point of
initial electrophoretic appearance for S-NADP-IDH 248 hr
after vaginal plug, while S-NADP-MDH and M-NADP-MDH could be
demonstrated for the first time 228 and 270 hr after vaginal
plug, respectively (Wolf and Engel, 1972; Engel and Wolf,
1971; Engel, 1973). S-NADP-IDH has been demonstrated by
Donahue and Stern (1972) in the 12-day mouse embryo and by
Epstein et al (1972) in the 9-day mouse embryo.
Since the three enzymes are demonstrable electrophoretically rather late during mouse embryonic development, an effect
of maternally derived storage products is no longer to be
expected. Therefore these enzymes can easily be used to
differentiate between the synchronous and asynchronous appearance of the allelic products. In the case of synchronous
activation the heteromeric bands should occur first, while in
the case of asynchronous activation the product of one allele
appears first and the heteromeric band occurs with the onset
of synthesis of the other allelic gene product. Using such
a system, synchronous activation of the parental alleles at
the gene locus for S-NADP-IDH (Donahue and Stern, 1970; Wolf
and Engel, 1972) and S-NADP-MDH (Engel and Wolf, 1971) in
the mouse has been inferred. A further example for this mode
of activation in the mouse represents the gene for M-NADP-
MDH (Engel, 1973).
M-NADP-MDH behaves electrophoretically as a tetrameric
molecule. In the mouse there is one gene locus with two
different alleles for this enzyme (Shows et al, 1970). Homozygous animals exhibit a single band, heterozygotes 5 bands.

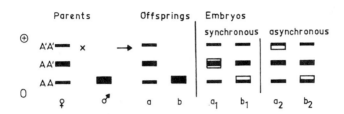

Fig. 4. Schematic presentation of the PGI phenotypes in offspring (a,b) and in preimplantation embryos after asynchronous (a_1,b_1) and synchronous (a_2, b_2) allele activation respectively of a heterozygous mother AA' and a homozygous father
AA.

The two strains SM/J and CBA differ from each other in the
electrophoretic position of this enzyme. Offspring in the F_1
generation uniformly exhibit 5 bands (Fig. 5). For the experiments we used embryos 265-300 hr after vaginal plug. The
material consisting of 10-15 embryos was subjected to starch
gel electrophoresis. The following electrophoretic conditions
were used: 14% starch in tris citrate buffer (tris: 0.015 M;
citrate: 0.0029 M) pH 8.4; bridge: tris-citrate buffer (tris:
0.22 M, citrate 0.043 M) pH 8.4; horizontal electrophoresis
at 12V/cm for 5 hr. Staining of the gel slices was performed
as described by Engel and Wolf (1971).

M-NADP-MDH in the mouse embryo can first be detected upon
electrophoresis 270 hr after the vaginal plug, thus exhibiting
a 3 banded pattern. These bands coincide in their position
with the 3 heteromeric bands of adult heterozygotes. The 2
homomeric bands first appear in embryos 295 hr after vaginal
plug. These findings indicate that both alleles contribute
to the initial M-NADP-MDH synthesis in embryos 270 hr after
vaginal plug. The conclusion is therefore justified that both
parental alleles for M-NADP-MDH start functioning synchronously. With these investigations on S-NADP-IDH and S-NADP-MDH
the function of different alleles at 3 gene loci have now been
studied. The results suggest synchronous activation of these
alleles.

Final remarks

Isozyme studies may provide an important tool for the
characterization of proteins stored in the oocyte of mammals.

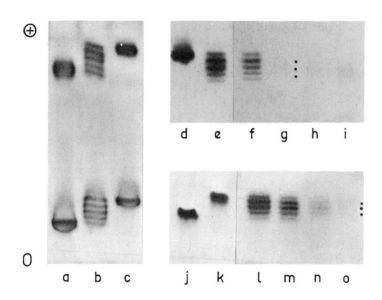

Fig. 5a–o. Isozyme pattern of the kidney of the mitochondrial (upper part of the gel) and the supernatant form (lower part of the gel) of NADP-dependent MDH of adult CBA (a), SM/J (c) and F_1-hybrid (b) mice. In d–o the M-NADP-MDH isozyme pattern of adult CBA (j) and SM/J mice (d,k), and of the F_1-hybrid at different ages after vaginal plug is shown: (e) adult hybrid, (l) newborn hybrid, (m) 295 hr, (n) 282 hr, (h,i) 280 hr, (g,o) 270 hr. The sample of 15 embryos per gel slot was subjected to electrophoresis.

As we have shown for an enzyme lactate dehydrogenase in the rabbit, proteins might be present not only as active enzyme molecules but also in an inactive form for later use. Furthermore, electrophoretic investigations in preimplantation embryos from suitable breedings can elucidate whether parental alleles at one gene locus are activated synchronously or a-asynchronously. The results may contribute to solving the problem of whether maternally derived gene products function in activating genes during early mammalian development. Such a hypothesis was presented as early as 1896 by E. B. Wilson.

ACKNOWLEDGMENTS

We thank Prof. U. Wolf for critical discussions and for reading the manuscript. Thanks are also due to Mrs. Margareta Lindig for skillful assistance in animal breeding. This work was supported by the Deutsche Forschungsgemeinschaft (SFB46).

REFERENCES

Anderson, E. 1972. The localisation of acid phosphatase and the uptake of horseradish peroxidase in the oocytes and follicle cells of mammals. In *Oogenesis* (J. D. Biggers and A. W. Schuetz, eds.), pp. 87-117. Baltimore: University Park Press; London: Butterworths.

Auerbach, S. and R. L. Brinster 1967. Lactate dehydrogenase isozymes in the early mouse embryo. *Exp. Cell. Res.* 46: 89-92.

Biggers, J. D. 1972. Metabolism of the oocyte. In *Oogenesis* (J. D. Biggers and A. W. Schuetz, eds.), pp. 241-251. Baltimore: University Park Press; London: Butterworths.

Bloom, A. M. and B. B. Mukherjee 1972. RNA synthesis in maturing mouse oocytes. *Exp. Cell Res.* 74: 577-582.

Brinster, R. L. 1965. Lactic dehydrogenase activity in the preimplanted mouse embryo. *Biochim. Biophys. Acta* (Amst.) 110: 439-441.

Brinster, R. L. 1968. Lactate dehydrogenase activity in the oocytes of mammals. *J. Reprod. Fert.* 17: 139-146.

Brinster, R. L. 1967a. Lactic dehydrogenase activity in preimplantation rat embryo. *Nature* (Lond.) 214: 1246-1247.

Brinster, R. L. 1967b. Lactate dehydrogenase activity in the preimplantation rabbit embryo. *Biochim. Biophys. Acta* (Amst.) 148: 298-300.

Brinster, R. L. 1973a. Lactate dehydrogenase isozymes in the preimplantation rabbit embryo. *Biochem. Genet.* 9: 229-234.

Brinster, R. L. 1973b. Parental glucose phosphate isomerase activity in three-day mouse embryos. *Biochem. Genet.* 9: 187-191.

Burkholder, G. D., D. E. Comings, and T. A. Okada 1972. A storage form of ribosomes in mouse oocytes. *Exp. Cell Res.* 69: 361-371.

Castro-Sierra, E. and S. Ohno 1968. Allelic inhibition at the autosomally inherited gene locus for liver alcohol dehydrogenase in chicken-quail hybrids. *Biochem. Genet.* 1: 323-335.

Chapman, V. M., W. K. Whitten, and F. H. Ruddle 1971. The expression of paternal glucose phosphate isomerase (Gpi-1) in preimplantation stages of mouse embryos. *Devel. Biol.* 26: 153-158.

Cornette, J. C., B. B. Pharriss, and G. W. Duncan 1967. Lactic dehydrogenase isozymes in the ovum and embryo of the rat. *Physiologist* 10: 146.

Davidson, E. H. and B. R. Hough 1972. Utilization of genetic information during oogenesis. In *Oogenesis* (J. D. Biggers and A. W. Schuetz, eds.), pp. 129-139. Baltimore: University Park Press; London: Butterworths.

Davies, D. R. 1973. Differential activation of maternal and paternal loci in seed development. *Nature New Biol.* 245: 30-32.

De Lorenzo, R. J. and F. H. Ruddle 1969. Genetic control of two electrophoretic variants of glucose phosphate isomerase in the mouse *(Mus musculus)*. *Biochem. Genet.* 3: 151-162.

Donahue, R. P. and S. Stern 1970. Isocitrate dehydrogenase in mouse embryos: activity and electrophoretic variation. *J. Reprod. Fert.* 22: 575-577.

Engel, W. and U. Wolf 1971. Synchronous activation of the alleles coding for the S-form of the NADP-dependent malate dehydrogenase during mouse embryogenesis. *Humangenetik* 12: 162-166.

Engel, W. 1972. Lactate dehydrogenase isoenzymes in oocytes and unfertilized eggs of mammals. *Humangenetik* 15: 355-356.

Engel, W. and R. Kreutz 1973. Lactate dehydrogenase isoenzymes in the mammalian egg: Investigations by micro disc electrophoresis in 15 species of the orders *Rodentia, Lagomorpha, Carnivora, Artiodactyla* and in man. *Humangenetik* 19: 253-260.

Engel, W. 1973. Onset of synthesis of mitochondrial enzymes during mouse development. *Humangenetik* 20: 133-140.

Engel, W. and U. Petzoldt 1973. Early developmental changes of the lactate dehydrogenase isoenzyme pattern in mouse, rat, guinea-pig, Syrian hamster and rabbit. *Humangenetik* 20: 125-131.

Epstein, C. J., J. A. Weston, W. K. Whitten, and E. S. Russel 1972. The expression of the isocitrate dehydrogenase locus (Id-1) during mouse embryogenesis. *Devel. Biol.* 27: 430-433.

Gross, K. W., M. Jacob-Lorena, C. Baglioni, and P. R. Gross 1973. Cell-free translation of maternal messenger RNA from sea urchin eggs. *Proc. Natl. Acad. Sci. USA* 70: 2614-2618.

Grossbach, U. 1965. Acrylamide gel electrophoresis in capillary columns. *Biochim. Biophys. Acta* (Amst.) 107: 180-182.

Hitzeroth, H., J. Klosse, S. Ohno, and U. Wolf 1968. Asynchronous activation of parental alleles at the tissue-specific gene loci observed on hybrid trout during early development. *Biochem. Genet.* 1: 287-300.

Mintz, B. 1964. Synthetic processes and early development in the mammalian egg. *J. Exp. Zool.* 157: 85-100.

Moore, G. P. M., S. Lintern-Moore, H. Peters, and M. Faber 1974. RNA synthesis in the mouse oocyte. *J. Cell. Biol.* 60: 416-422.

Ohno, S. 1971. Simplicity of mammalian regulatory systems inferred by single gene determination of sex phenotypes. *Nature* (Lond.) 234: 134-137.

Pincus, G. and E. V. Enzmann 1935. The comparative behavior of mammalian eggs in vivo and in vitro. I. The activation of ovarian eggs. *J. Exp. Med.* 62: 665-675.

Shows, T. B., V. M. Chapman, and F. H. Ruddle 1970. Mito-chondrial malate dehydrogenase and malic enzyme: Mendelian inherited electrophoretic variants in the mouse. *Biochem. Genet.* 4: 707-718.

Skoultchi, A. and P. R. Gross 1973. Maternal histone messenger RNA: Detection by molecular hybridization. *Proc. Natl. Acad. Sci. USA* 70: 2840-2844.

Tomkins, G. M. and D. W. Martin, Jr. 1970. Hormones and gene expression. *Ann. Rev. Genet.* 4: 91-106.

Wilson, E. B. 1896. On cleavage and mosaic-work. *Arch. Entwicklungsmech. Org.* 3: 19-35.

Wolf, U. and W. Engel 1972. Gene activation during early development of mammals. *Humangenetik* 15: 99-118.

ESTERASE ISOZYMES DURING MOUSE EMBRYONIC DEVELOPMENT *IN VIVO* AND *IN VITRO*

MICHAEL I. SHERMAN
Department of Cell Biology
Roche Institute of Molecular Biology
Nutley, New Jersey 07110

ABSTRACT. The electrophoretic profile of non-specific esterase isozymes in mouse trophoblast differs from that of embryo proper and yolk sac at midgestation. Of two bands of activity (EsA, EsF) prominent only in trophoblast, one (EsF), identical to the genetically defined Es-1, is of maternal origin. The enzyme profiles are the same in both diploid and polyploid trophoblast cells. Embryo and yolk sac possess a region of esterase activity (EsD) not present in trophoblast. Furthermore, yolk sac preparations have a band of activity (EsG) similar to the genetically defined isozyme Es-2. Two regions of enzyme activity, EsB and EsC, are observed in all tissues. These esterase isozymes are all present in one or more adult tissues.

The different cell types developing in blastocyst cultures have isozyme patterns consistent with their *in vivo* profiles. Uptake of EsF by trophoblast cells *in vitro* can also be demonstrated. Esterase isozymes have been used in efforts to characterize cells in long term blastocyst cultures.

Analysis of esterase isozymes from rat embryonic and extraembryonic tissues indicate that some analogies with mouse tissues exist. Uptake by rat trophoblast of a serum esterase may also be taking place, although probably not to the extent observed in mouse trophoblast.

INTRODUCTION

Since the early studies of Mintz (1964), Cole and Paul (1965), and Gwatkin (1966), substantial improvements have been realized in the culture of mouse blastocysts (Hsu, 1971; 1973; Sherman, 1972a; 1974b; Bell and Sherman, 1973). In this laboratory, derivatives of both cell types in the blastocyst, trophectoderm and inner cell mass, proliferate *in vitro*, and can be maintained in culture for long periods of time (Sherman, 1974b). It has been our aim to utilize this system to study various aspects of mammalian embryogenesis at the molecular level. For these purposes, we have characterized a number of biochemical properties unique to trophoblast (fetal part of the placenta), yolk sac or embryo proper (see Sherman, 1974a), and we are investigating ways in which the expression of these

83

biochemical markers may be controlled, both *in vivo* and *in vitro*. We have found the electrophoretic profiles of non-specific esterase isozymes to be particularly useful as indicators of the degree of differentiation of embryonic and extra-embryonic tissues (Sherman, 1972a; Sherman and Chew, 1972). The current status of these studies is described below.

METHODS

Unless otherwise indicated, the embryonic and extraembryonic mouse tissues used in this study were obtained by mating SWR/J females with SJL/J males (Jackson Laboratories, Bar Harbor, Me.). Rat embryos were of the Sprague-Dawley strain (Marland Farms, Hewitt, N.J.). The day on which copulation plugs were observed is considered the first day of pregnancy. Detailed descriptions of the preparation of animals and dissection of tissues (Sherman, 1972b) as well as culture of embryos (Sherman, 1972a; 1974b; Bell and Sherman, 1973) have appeared elsewhere.

For electrophoresis, in vivo tissues were homogenized at a protein concentration of 1-2 mg/ml either in water or in phosphate-buffered saline (Dulbecco and Vogt, 1954). Samples contained from 5 to 200 µg protein and were applied to the gel in 1-2% ethylene glycol and bromphenol blue. Cultured cells were suspended in small volumes of water or phosphate buffered saline. Cells were disrupted by homogenization (for larger samples) or by repeated freeze-thawing, sonication, treatment with 0.5% Nonident P-40, (Shell Chemicals, East Orange, N.J.), or combinations of the above, when only small amounts of tissue were available. Electrophoresis was carried out in a water-jacketed vertical polyacrylamide slab gel apparatus. Gels were 11.5 cm long, 15.5 cm wide, and 0.3 cm thick. Most often, an 8% separating gel and a 4.5% spacer gel containing 0.2% "Tween 80" (Fisher Scientific, Co.) were used (Sherman, 1972a). The gel buffer was 0.031 M boric acid - 0.0031 M Tris pH 7.2 and the reservoir buffer contained 0.05 M boric acid - 0.005 M Tris pH 7.2. Electrophoresis was carried out at 35-55 volts/cm across the gel, and the duration of the run (terminated when the bromphenol blue marker reached the end of the gel) in this buffer system was 2 hours or less. After the run, gels were stained for esterase activity at 37° with α-naphthyl acetate and/or α-naphthyl butyrate coupled with Fast Blue BB dye; the former procedure was according to Shaw and Koen (1968), except that twice the amount of substrate was used, the latter according to Ruddle and Roderick (1966).

RESULTS

Esterase Isozyme Profiles in Midgestation Tissue Homogenates

Homogenates of midgestation embryo, yolk sac, trophoblast and decidua (maternal moiety of the placenta) can all be distinguished by their esterase isozyme electrophoretic profiles (Figure 1). Distinctions can be drawn whether α-naphthyl acetate (Figure 1a) or α-naphthyl butyrate (Figure 1b) is used as substrate. While EsA and EsF are prominent in trophoblast, the two activities are present only in trace amounts in yolk sac, and absent in the embryo. EsF is not synthesized by the trophoblast or yolk sac, however; it is of maternal origin (Sherman and Chew, 1972; see below).

The trophoblast layer of the midgestation mouse placenta contains large numbers of diploid cells (ectoplacental cone) as well as a population of polyploid cells with DNA contents ranging from 8-512 times the haploid amount (Barlow and Sherman, 1972). Chew and Sherman (1974) have demonstrated that fractions enriched for either diploid, intermediate sized polyploid, or large polyploid trophoblast cells can be dissected apart on the tenth day of pregnancy. Enzyme analysis of these fractions revealed that Δ^5,3β-hydroxysteroid dehydrogenase activity was clearly associated with polyploid cells, and reached its highest levels in the largest cell fraction. Esterase isozyme analysis on these fractions (Figure 2) reveals that diploid and polyploid cells have very similar profiles (with the exception of a variable band of activity, just beyond EsA, which, when present, appears to be restricted to the small cell fractions). In three separate experiments, EsF activity was highest in the intermediate sized cell fraction.

EsG occurs only in yolk sac, and EsD is observed both in embryo and yolk sac, but is absent from trophoblast profiles (Figure 1). Decidua profiles also contain EsD activity, and EsE, a slightly faster migrating band (with some preparations, two bands of activity are present in the EsE region, and they are better resolved from EsD than in Figure 1). The absence of EsD and EsE in trophoblast profiles demonstrates that the maternal and foetal parts of the midgestation mouse placenta can be cleanly separated (though the separation cannot be achieved after the twelfth day). EsB and EsC are present in all tissues; the former appears to selectively utilize the acetyl ester as substrate. Occasionally, other minor bands of activity have been detected, e.g., the band migrating slightly faster than EsA in some trophoblast preparations (Figure 2, slot S).

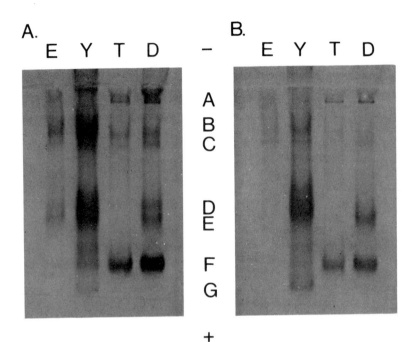

Fig. 1. Esterase isozyme electrophoretic profiles in midgestation embryonic and extraembryonic tissues. Whole homogenates of eleventh day tissues were electrophoresed and stained with (A) α-naphthyl acetate or (B) α-naphthyl butyrate. Direction of migration is from top to bottom. E: embryo; Y: yolk sac; T: trophoblast; D: decidua.

Time of Appearance of Esterase Isozymes

The characterization of esterase isozyme profiles is difficult in preimplantation and early postimplantation embryos due to the very small amounts of tissue available, and the relatively low levels of total esterase activity present in the embryo, even by the eighth gestation day (see Sherman, 1972a). Yet another complication arises from the fact that uterine fluid contains EsA, EsF, and a band of activity migrating between EsA and EsB; contamination with uterine fluid or uptake of esterase from the fluid by preimplantation embryos could therefore lead to faulty conclusions. In an effort to overcome these difficulties, large numbers of preimplantation embryos (generally between 500 and 1000) were washed several times in relatively large volumes of phosphate buffered saline before electrophoretic analysis. After removal of cumulus

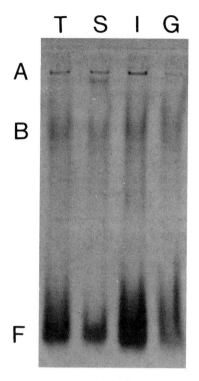

Fig. 2. Esterase isozyme profiles in different fractions of
the trophoblast layer. Trophoblast cells were crudely
fractionated according to size as described by Chew and
Sherman (1974). Samples of whole homogenates containing 50 μg
protein from total trophoblast (T) or from fractions enriched
for small (diploid) cells (S), intermediate sized polyploid
cells (I) or giant sized polyploid cells (G) were analyzed by
electrophoresis. Stain substrate was α-naphthyl acetate.

cells with hyaluronidase, recently fertilized one-cell embryos
(first day of gestation) have extremely low levels of esterase
activity. Traces of EsC and EsE activity (both observed in
ovary preparations) may be present. Two-cell (second day)
embryos possess EsC activity, with traces of EsB and EsE.
Blastocysts collected on the fourth day of gestation have been
analyzed in lots of 300 to 750 on five different occasions.
Twice, activity in the EsA, EsB, EsC and EsF regions were
observed; three times, only EsB and EsC activity were observed.
On two other occasions 265 and 295 fourth day blastocysts
after twenty-four hours in culture possessed only EsB and EsC
activity. Finally, in separate experiments, lots of 600 and

750 two-cell embryos cultured to the blastocyst stage in the medium of Whitten and Biggers (1968) contained only EsB and EsC activity. The most likely interpretation of these results is that EsB and EsC are present in, and possibly synthesized by, preimplantation embryos, but that EsA and EsF are taken up by blastocysts from the uterine fluid, presumably into the blastocoel (see below). It is not clear from these studies whether EsB and EsC are present in presumptive embryo and yolk sac (inner cell mass), or trophoblast (trophectoderm) cells. However, the presence of these isozymes in eighth day tropho-blast preparations and their absence in eighth and ninth day embryo-yolk sac homogenates (Sherman, 1972a), might suggest that the isozymes are restricted to trophectoderm cells.

EsA appears to be synthesized in trophoblast on or before the eighth day of gestation; the isozyme is present in tenth day yolk sac preparations, but does not appear in the embryo proper before the 14th day. EsD is detectable in both embryo and yolk sac homogenates by the tenth day of gestation.

Identity of Esterase Isozymes

Strain variation of the electrophoretic mobility of EsF has been observed (Figure 3; Sherman and Chew, 1972). On the basis of a comparison of the mobility of EsF from several mouse strains, it has been concluded that EsF is identical to the isozyme genetically defined as Es-1 by Popp and Popp (1962). EsG migrates identically with an isozyme in kidney preparations designated Es-2 (Ruddle et al., 1969); both iso-zymes also show a preference for the butyryl ester over the acetyl ester as substrate. The other isozymes described here have not been further identified.

Table 1 illustrates that each of the esterase isozymes in embryonic and extraembryonic tissues is also present in several adult tissues. In some cases, the isozymes of adult tissues show multiple banding where only single bands are observed in embryonic and extraembryonic tissues (see Sherman, 1972a). There are also activities present in some of the adult tissues which are not observed in any of the midgestation embryonic or extraembryonic tissues.

Maternal Esterase Uptake

By the use of isozyme variants, we have been able to demonstrate conclusively that EsF in trophoblast and yolk sac cells is of maternal origin (Sherman and Chew, 1972). In trophoblast and yolk sac taken from Fl crosses, only the maternal type of EsF is expressed. If blastocysts are trans-ferred to incubator mothers differing in EsF phenotype, and

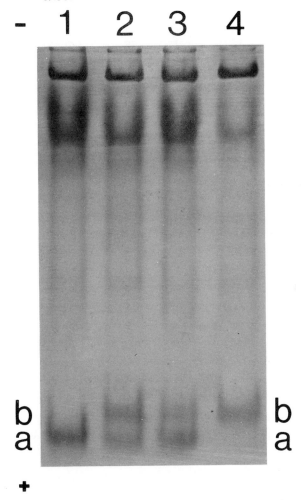

Fig. 3. Electrophoretic variation of EsF. Crude homogenates
of tenth and eleventh day decidua were from the following
strains: slot 1: C57Bl/6J; slot 2: C57Bl/6J x DBA/2J; slot 3:
a mixture of homogenates analyzed in slots 1 and 4; slot 4:
SWR/J. Samples were stained with α-naphthyl acetate as substrate. Similar variations were observed in trophoblast and
yolk sac preparations (Sherman and Chew, 1972).

---

fetal tissues dissected out after implantation, the phenotype
of EsF in trophoblast and yolk sac is always that of the incubator mother. When blastocysts are incubated under the
kidney capsule of a male host, a prominent band of EsF
activity is observed in the profiles of the trophoblastic

TABLE 1

OCCURRENCE OF ESTERASE ISOZYMES IN TISSUES OF ADULT MICE

TISSUE	ESTERASE ISOZYME						
	A	B	C	D	E	F	G
ERYTHROCYTES		X		X			
HEART	X					X	
KIDNEY	X		X	X		X	X
INTESTINE	X	X		X	X		X
LEUCOCYTES	X	X	X	X		X	
LIVER	X	X		X			
LUNG	X	X		X		X	
OVARY	X	X	X		X	X	
SERUM	X					X	X
SPLEEN	X	X	X		X	X	X
TESTIS	X	X	X	X		X	
UTERINE FLUID	X					X	
UTERUS	X	X	X	X		X	X

growths so obtained; the phenotype is that of the host. A
substantial amount of the enzyme activity remains with the
trophoblast and yolk sac pellets after several cycles of
washing and sedimenting the cells, but is released upon
breaking them. Therefore, at least some of the EsF is intracellular. The profiles in Figure 2 indicate that all parts of
the trophoblast layers are capable of accumulating EsF.
However, the relatively high levels of EsF activity in the
"intermediate" cell population suggest that the lateral
(rather than polar) portions of the trophoblast layer, wherein
the majority of these cells are localized, are most active in
EsF uptake. The most likely conclusion from all these observa-

tions is that trophoblast cells are capable of taking up EsF
from the maternal serum, with which it is in intimate contact,
and that some of this enzyme is passed on to yolk sac cells
(Sherman and Chew, 1972). The enzyme apparently does not
reach the embryo proper, since embryos do not have EsF
activity. Because some serum esterases are never detected
in trophoblast preparations, the process must be selective.

Without variants of EsA, we cannot exclude the possibility
that some of the EsA in trophoblast and later, yolk sac,
preparations is also of maternal origin. However, unlike EsF,
EsA can still be detected two weeks after eighth day trophoblast cells have been placed in culture, and the isozyme also
appears as trophoblast differentiates in blastocyst cultures
(see below). Also, the band of esterase activity in rat
trophoblast and yolk sac, which is probably analagous to EsA
in the mouse, is not present in rat serum (see Fig. 5). Mouse
serum also contains small amounts of EsG. It is unlikely that
yolk sac EsG is of maternal origin, though, since this enzyme
is also synthesized by cultures of cells resembling yolk sac
(Figure 4). Also, EsG is never observed in trophoblast preparations, and at midgestation, yolk sac cells are prevented
from direct contact with maternal serum by the trophoblast
layer.

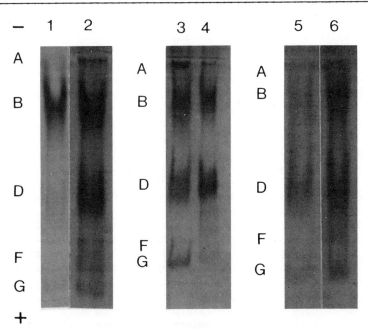

Fig. 4. Esterase profiles of blastocyst-derived vesicles in
short-term and long-term cultures. Blastocysts were cultured

(Fig. 4 legend continued)
in NCTC-109 medium supplemented with 10% heat-inactivated
fetal calf serum and antibiotics. Preparations of vesicles
separated from monolayers after 8 (slot 1) and 35 (slot 3)
days of culture were compared with eleventh day *in vivo* yolk
sac preparations (slots 2 and 4). The sample in slot 5 is a
homogenate of a blastocyst-derived cell line after eight months
in culture. Slot 6 is an eleventh day *in vivo* yolk sac profile. α-naphthyl acetate and α-naphthyl butyrate were used
together as stain substrates.

Esterase Isozymes in Blastocyst Cultures

In our early efforts at blastocyst culture, we succeeded
only in maintaining trophectoderm cells which eventually gave
rise to giant trophoblast cells (Sherman, 1972a; 1972b;
Barlow and Sherman, 1972); cells of the inner cell mass did
not survive much longer than four days *in vitro*. In these
cultures, polyploidization could take place (Barlow and
Sherman, 1972) and low levels of trophoblast alkaline phosphatase (Sherman, 1972b) and EsA (Sherman, 1972a) appeared after
four days *in vitro*. With improved culture conditions (Sherman,
1972a; 1974b), inner cell mass proliferates, often forming
large multicellular vesicles which resemble yolk sac both
morphologically (Sherman, 1974b) and in the content of two
lysosomal enzymes (Bell and Sherman, 1973). Cells with
various morphologies also grow as monolayers, intermingling
with giant trophoblast cells. After eight days in culture,
the observed esterase isozyme profiles of these monolayers is
consistent with that expected from a mixture of trophoblast
and embryo cells: small amounts of EsA, and larger amounts of
EsB, EsC, and EsD are present. EsF and EsG are never observed.
On the other hand, vesicle preparations have esterase isozyme
profiles identical to that of eleventh day yolk sac, except,
of course, that EsF is absent (Fig. 4, slot 1): EsA (trace
amounts) EsB, EsC, EsD and EsG are all present. These vesicles
can be maintained in culture for long periods of time; even
after 35 days in culture, the isozyme profile resembles that
of yolk sac (Fig. 4, slot 3). After about two months in culture, vesicles and monolayers from fourth day blastocysts
were disaggregated and subcultured. A number of different
lines were established, and after a total of eight months in
culture, one of these lines still possesses a profile similar
to that of yolk sac (Fig. 4, slot 5). This cell line also
shares other biochemical properties in common with midgestation yolk sac (Sherman, 1974b).

Esterase Uptake in Blastocyst Cultures

We reported earlier that trophoblast cells in culture medium supplemented with mouse serum showed no evidence of EsF uptake. It was also noted, however, that long periods of culture in mouse serum affected blastocysts adversely (Sherman and Chew, 1972). In more recent experiments, we have cultured blastocysts in the presence of 10% heat-activated fetal calf serum, and then exposed them for short periods (3-24 hours) to medium supplemented with 5% heat-inactivated fetal calf serum and 5% strain C57B1/6J mouse serum. Cultures were then washed several times with phosphate-buffered saline and collected. As a control, cultures were incubated for the same period of time in medium containing only heat-inactivated fetal calf serum, exposed briefly to chilled medium containing mouse serum, and washed in the same manner as the experimental cultures. In this way, the presence of the C57B1/6J variant of EsF could be demonstrated in the esterase profiles of experimental, but not control, cultures. The esterase profiles of blastocysts cultured for twenty-four hours and then exposed to mouse serum for six hours contain high levels of EsF, EsA and some of the other serum esterases. Even control cultures (exposed to mouse serum in the cold for less than a minute) demonstrate some uptake of serum esterases. It is likely that these isozymes are in fact taken up into the blastocoel fluid. Four or seven day blastocyst cultures are also capable of taking up added EsF from the medium, but there is no detectable uptake of EsA or any of the other serum esterases; therefore, the same specificity is exhibited in vitro as in vivo. Control cultures show no evidence of esterase uptake. Seven day cultures contain increasing amounts of EsF between 3 and 24 hours of incubation in mouse serum, but EsF content falls by 48 hours, presumably due to an adverse effect of mouse serum upon the cells. When vesicles and monolayers are separated before electrophoretic analysis, EsF uptake appears to be restricted to the latter fraction of the culture. In some cases, after washing, cultured cells were disrupted, and cell debris was sedimented. All the EsF activity was in the supernatant fraction, indicating that the enzyme is not merely bound to the outside of the cell.

Esterase Activity in Midgestation Rat Tissues

Esterase isozyme analysis was carried out on rat embryonic and extraembryonic tissues to determine the extent to which the resultant profiles resemble those of the mouse, as well as to follow isozyme patterns in trophoblast late in development. The latter studies could be carried out because, unlike

93

the mouse, rat trophoblast and decidua can be separated throughout pregnancy (Marcal et al., 1974). A number of analogies between mouse and rat esterases can be drawn (Fig. 5). For example, EsA', the slowest moving rat isozyme, is present in trophoblast and decidua at least as early as the eleventh day of gestation, but does not appear in the embryo proper before the fourteenth or fifteenth day (as noted earlier, rat serum does not contain EsA' activity, but does contain a band which migrates slightly more anodally). EsF' is the major band in rat serum and is also present in yolk sac, trophoblast and decidua. EsB' (a series of bands in the rat demonstrating marked polymorphism) and EsD' migrate between EsA' and EsF', while EsG' is the fastest migrating band.

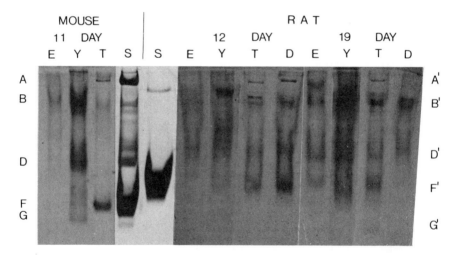

Fig. 5. Comparison of mouse and rat embryonic and extraembryonic esterase profiles. Electrophoretic profiles from twelfth and nineteenth day rat embryo (E), yolk sac (Y), trophoblast (T) and decidua (D) preparations were compared with equivalent tissues from eleventh day mouse preparations (the gross development of the rat embryo lags behind that of the mouse by about one day). Serum (S) profiles from pregnant adult females are also included. Both α-naphthyl acetate and α-naphthyl butyrate were used as stain substrates.

The analogies between mouse and rat esterases do, however, break down. For example, EsF' activity is prominent in embryo profiles as early as the fifteenth day of gestation. EsD' is present in trophoblast preparations; in fact, trophoblast contains all the isozyme bands observed in the other tissue profiles. Finally, EsG' is restricted to trophoblast, rather than yolk sac, preparations, and appears relatively late in gestation, between the fifteenth and seventeenth days.

DISCUSSION

We have shown in other studies that yolk sac and trophoblast are capable of developing *in vitro* from cultured blastocysts, and that their biochemical properties at least qualitatively reflect those of cells developing *in vivo* (Sherman, 1972b; Barlow and Sherman, 1972; Bell and Sherman, 1973; Chew and Sherman, 1974). Experiments with non-specific esterases described previously (Sherman, 1972a) are consistent with this view. With improved culture conditions and techniques, we have now extended these observations. When monolayers and vesicles developing from blastocyst cultures are assayed separately, each possesses an esterase profile consistent with that expected on the basis of morphologies: the monolayer profile has both trophoblast and embryo esterases, while that of the vesicles closely resembles the yolk sac profile. It is notable that the time required for expression of these isozymes *in vitro* is very similar to that *in vivo*. For example, vesicles after seven or eight days in culture (total gestation age eleven to twelve days) possess esterase profiles first observed in tenth day *in vivo* yolk sac preparations (Fig. 4). Finally, our results also suggest that trophoblast cells are capable of taking up EsF in culture. The maternal environment is therefore clearly unnecessary for the appearance of these properties. Indeed, when released from the uterine milieu, cells resembling yolk sac are capable of surpassing their normal limited lifespan and maintaining unique biochemical characteristics for long periods of time (Fig. 4; Sherman, 1974b).

Recently, experiments involving the transmission of maternal macromolecules into the preimplantation (Glass, 1970) and post-implantation (Brambell, 1969) embryo have been reviewed. It can be inferred from these reviews that some transfer of macromolecules from mother to fetus takes place throughout the gestation period; our studies on esterase uptake described here and elsewhere (Sherman and Chew, 1972) support this view. In their work Glass (1970) and Gitlin and Morphis (1969) have stressed the selectivity of the uptake process. With the apparent exception of non-specific uptake of esterases into the blastocoel (see also Brambell, 1958 and Sugiwara and Hafez, 1965 for studies on serum proteins in blastocoelic fluid), the process of esterase uptake is also selective. There are indications that at least some of the esterase activity in one- and two-cell embryos is of maternal origin: EsE, for example, is present in the ovary and in the early embryo, but not in the blastocyst. Also, immunozymograms have revealed that a number of the serum fractions detected in oocytes have non-specific esterase activity (Glass, 1970). Yet, we find only some of the esterase isozymes in ovary and serum preparations, not necessarily the most prominent species, to be present in early embryos.

The selective uptake of EsF by trophoblast has been described above. The inability of EsF to reach the embryo proper parallels a report by Anderson (1959) that γ-globulins could penetrate the trophoblast and yolk sac layers of the rat, but did not enter the fetus. The nature of the barrier is unknown, but its presence has obvious implications as a protective device for the fetus.

Without access to genetic variants, it cannot be conclusively stated that EsF' esterase in rat trophoblast and yolk sac is of maternal origin, although it would appear from the staining intensity on the gels that uptake, if it occurs in the rat, is not so marked as it is in the mouse. On the other hand, the presence of EsF' in the rat embryo at later stages of development might suggest that some of the enzyme is capable of finding its way through the placental barrier, though the possibility that this isozyme is actually synthesized by the embryo is at least as likely. These, and other, differences between mouse and rat esterase profiles reinforce earlier observations (Bell and Sherman, 1973) which indicate clear differences at the biochemical level during embryonic development of the mouse and the rat, even though gross morphology is quite similar.

A survey of both Fig. 1 and Table 1 reveals that remarkably few tissues, either embryonic or adult, share identical esterase electrophoretic profiles (see also Holmes and Masters, 1967). While this fact is useful in studies such as ours where each tissue can be defined by its peculiar combination of esterases, the *in vivo* role of these bewildering assortments of isozymes is unexplained. Further studies must be carried out to clarify the relationship between these enzymes and differentiation.

REFERENCES

Anderson, J.W. 1959. The placental barrier to gamma-globulins in the rat. *Amer. J. Anat.* 104: 403-430.
Barlow, P.W., and M.I. Sherman. 1972. The biochemistry of differentiation of mouse trophoblast: polyploidy. *J. Embryol. Exp. Morphol.* 27: 447-465.
Bell, K.E., and M.I. Sherman. 1973. Enzyme markers of yolk sac differentiation. *Develop. Biol.* 33: 38-47.

Brambell, F.W.R. 1958. The passive immunity of the young mammal. *Biol. Rev.* 33: 488-531.

Brambell, F.W.R. 1969. *The Transmission of Passive Immunity From Mother to Young.* North Holland Publishing Co., London.

Chew, N.J., and M.I. Sherman. 1974. Biochemistry of differentiation of mouse trophoblast: Δ^5,3β-hydroxysteroid dehydrogenase. Submitted for publication.

Cole, R.J., and J. Paul. 1965. Properties of cultured pre-implantation mouse and rabbit embryos, and cell strains derived from them. In *"Preimplantation Stages of Pregnancy"* (Wolstenholme, G.E.W., and M. O'Connor, eds.), pp. 82-112. Little-Brown, Boston, Massachusetts.

Dulbecco, R., and M. Vogt. 1954. Plaque formation and isolation of pure lines with poliomyelitis viruses. *J. Exp. Med.* 99: 167-182.

Gitlin, D., and L.G. Morphis. 1969. Systems of materno-foetal transport of γG immunoglobulin in the mouse. *Nature* 223: 195-196.

Glass, L.E. 1970. Transmission of maternal proteins into oocytes. *Advan. Biosci.* 6: 29-58.

Gwatkin, R.B.L. 1966. Defined media and development of mammalian eggs *in vitro*. *Ann. N.Y. Acad. Sci.* 137: 79-90.

Holmes, R.S., and C.J. Masters. 1967. The developmental multiplicity and isoenzyme status of cavian esterases. *Biochem. Biophys. Acta* 132: 379-399.

Hsu, Y. 1971. Post-blastocyst differentiation *in vitro*. *Nature,* London 231: 100-102.

Hsu, Y. 1973. Differentiation *in vitro* of mouse embryos to the stage of early somite. *Develop. Biol.* 33: 403-411.

Marcal, J., N.J. Chew, D.S. Salomon, and M.I. Sherman. 1974. Δ^5,3β-hydroxysteroid dehydrogenase activity in rat trophoblast and ovary during the gestation period. In preparation.

Mintz, B. 1964. Formation of genetically mosaic embryos and early development of 'lethal (t^{12}/t^{12})-normal' mosaics. *J. Exp. Zool.* 157: 273-292.

Popp, R.A., and D.M. Popp. 1962. Inheritance of serum esterases having different electrophoretic patterns. *J. Heredity* 53: 111-114.

Ruddle, F.H., and T.H. Roderick. 1966. The genetic control of two types of esterases in inbred strains of the mouse. *Genetics* 54: 191-202.

Ruddle, F.H., T.B. Shows, and T.H. Roderick. 1969. Esterase genetics in *Mus musculus*: expression, linkage, and polymorphism of locus Es-2. *Genetics* 62: 393-399.

Shaw, C.R., and A.L. Koen. 1968. Starch gel zone electrophoresis of enzymes. In *"Chromatographic and Electro-*

phoretic Techniques" (Smith, I., ed.), Vol. 2, pp. 325-364. Heinemann, London

Sherman, M.I. 1972a. Biochemistry of differentiation of mouse trophoblast: esterase. *Exp. Cell Res.* 75: 449-459.

Sherman, M.I. 1972b. The biochemistry of differentiation of mouse trophoblast: alkaline phosphatase. *Develop. Biol.* 2*r*: 337-350.

Sherman, M.I. 1974a. *In vivo* and *in vitro* differentiation during early mammalian embryogenesis. *Front. Rad. Therapy* 9: 122-134.

Sherman, M.I. 1974b. Long term culture of cells derived from mouse blastocysts. *Differentiation.* (in press.)

Sherman, M.I., and N.J. Chew. 1972. Detection of maternal esterase in mouse embryonic tissues. *Proc. Natl. Acad. Sci. U.S.A* 69: 2551-2555.

Sugiwara, S., and E.S.E. Hafez. 1965. Electrophoretic patterns of proteins in the blastocoelic fluid of the rabbit following ovariectomy. *Anat. Rec.* 158: 115-120.

Whitten, W.K., and J.D. Biggers. 1968. Complete development *in vitro* of the preimplantation stages of the mouse in a simple chemically defined medium. *J. Reprod. Fertil.* 17: 399-401.

GENETIC CONTROL AND DEVELOPMENTAL EXPRESSION OF ESTERASE ISOZYMES IN DROSOPHILA OF THE VIRILIS GROUP

LEONID KOROCHKIN
Institute of Cytology and Genetics
Siberian Academy of Sciences
Novosibirsk 630090, USSR

ABSTRACT. The appearance in development of esterases in different species of the virilis group of Drosophila has been examined by the zymogram technique. The patterns of esterases in different organs and tissues at different stages of differentiation were investigated with respect to genetic control, whether due to structural genes or regulatory genes. Hybrids between different members of the virilis group proved to be particularly instructive. The map location of esterase genes were determined and the influence of the general genetic background on gene expression ascertained at successive developmental stages.

INTRODUCTION

It is well known that genes must be differentially expressed in different cell types and in the same cell type at successive stages of development. Thus, this phenomenon is the basis of tissue and cell differentiation (Markert and Ursprung, 1971).

Some morphological regularities of gene action were described by analysis of the development of many mutants (Hadorn, 1961). But investigations of the phenotypic expression of biochemical traits is of great importance for our knowledge of the molecular genetic mechanisms of gene function and gene interaction during ontogenesis (Astauroff, 1968; Ohno, 1970; Whitt et al., 1973). Isozymes have proved to be especially convenient markers of gene activity and are now much used for genetic and developmental analysis of many experimental organisms such as mice (Markert and Hunter, 1959) and Drosophila (Ursprung et al., 1968).

We have used different species of Drosophila of the virilis group in our investigations. This group of Drosophila is suitable for biochemical experiments because: 1. They are quite large; 2. interspecific hybrids are fertile; 3. based on this fact M. Evgeniev (personal communication) worked out a rather simple method for localization of genes on a cytological map.

The esterase system was chosen as biochemical markers because: 1. Esterase activity is very high and it is possible to determine activity in different organs, tissues, and cells;

99

2. there are some data about the role of esterases in morphogenetic and metabolic processes. For instance, JH-esterases
(Whitmore et al., 1974) break down juvenile hormone; acetyl
choline esterases (ACHE) are active in specific nerve centers;
and some esterases break down ethers and amides of amino acids.

MATERIAL AND METHODS

The outbred stocks of Drosophila species (*D.virilis*, *D.
texana*, *D.littoralis*, *D.imeretensis*) used in this study were
kindly provided by Prof. N. N. Sokolov and M. Evgeniev
(Institute of Developmental Biology, Academy of Sciences,
Moscow, USSR). Different inbred strains were derived from
these stocks in the Laboratory of Developmental Genetics.

Homogenates of single flies and larvae (1 to 3 or 4) were
used to determine the activity of esterases and their electrophoretic mobilities. Each fly was homogenized in 0.04 ml
double distilled water on the rough surface of a slide. Esterase activity was assayed by the method of Bamford and Harris
(1964) and α-naphtyl propionate was used as substrate. Enzyme
activity was expressed in micromoles of α-naphthol formed per
mg of protein per minute. Lowry's method was used to estimate
protein concentration in the homogenate.

Starch gel electrophoresis was performed vertically using
14% starch and 10% sucrose in a medium containing 0.05M Tris-
EDTA-borate buffer at pH 8.1. The electrode buffer was 0.3M
sodium borate, pH 8.0. A droplet of the homogenate was
absorbed on rectangular strips of Watmann 3MM paper (5 x 7 mm
or 3 x 5 mm). These strips were inserted into slots cut in
the starch gel. Electrophoresis took 3-4 hr at 5-10 C° with
a voltage of 300 v and current intensity of 30-40 ma. After
electrophoresis the upper layer of the gels was sliced off
and the bottom slice stained using the standard technique
(Sims, 1965; Shaw and Prasad, 1970). The substrates were
α-naphthyl propionate (Sigma Chemical Co., USA, or Chemapol.,
Praha and Lachema N. P., Brno, Czechoslovakia) or a mixture of
of α- and β-naphthyl acetates (Chemapol, Praga or Lachema,
N. P., Brno Czechoslovakia) and fast blue RR salt (dye-coupler,
Sigma Chemical Co., USA, or Chemapol, Praha). The electrophoresis in 7.5% polyacrylamide gel was performed with flat
microcapillaries using a method which has been described
elsewhere (Korochkin, 1972; Korochkin et al., 1972). The gels
were preparated by the procedure of Raymond and Wang (1960)
using 1M Tris=EDTA buffer (pH 8.3). Standard chambers of
smaller size were used for electrophoresis. The electrode
buffer was 0.3M sodium borate, pH 8.0. The electrophoresis
took 25-30 min at a current intensity of 1-1.5 mA per capillary

tube. Densitometry of the electropherograms was carried out in a microphotometer IFO-451 (USSR) and in a microdensitometer MK-III CS (Joyce and Loebl). To determine the relation between esterase zones and inhibitors the slabs were treated with 10^{-3} M $CuSO_4$, 10^{-3} M $Pb(NO_3)_2$, 10^{-5} M E-600, 10^{-4} M physostigmine.

RESULTS

BRIEF CHARACTERIZATION OF ESTERASES IN DROSOPHILA OF THE VIRILIS GROUP

By starch and polyacrylamide gel electrophoresis, six fractions of esterase activity have been detected in Drosophila of the virilis group (Fig. 1). These esterases have been characterized in detail using a series of substrates, inhibitors, and thermal treatments (Korochkin et al., 1973).

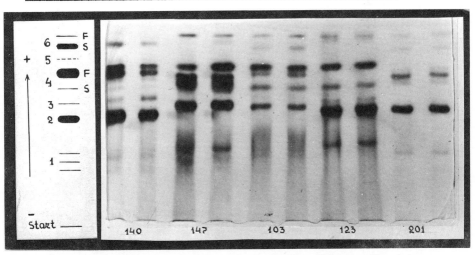

Fig. 1. Zymograms of esterases of single flies. Stocks: *D.virilis, D.texana, and D.littoralis.* The number at the bottom designates the number of the stock from the collection. 140, 147, 103-*D.virilis,* 123-*D.texana,* 201-*D.littoralis.* Left scheme of esterase pattern in *D.virilis* with designation of the number assigned to the fractions.

A Pakhomoff and I. Kiknadze in my laboratory found that subfractions of the slowest esterase-1 are ACHE molecular aggregates. Probably esterase-3 also belongs to ACHE. Esterase-2 stains very intensely and is a β-esterase according to the classification of Johnson et al. (1966) because it splits preferentially β-naphtyl acetate and stains red unlike all the other fractions of esterases of *D.virilis.* Esterases 4 and 6 are represented by two subfrations (Fig. 1), fast (F) and slow (S). Esterase-5 stains more faintly in most stocks and varies in activity. These esterases split mainly α-naphthyl acetate (α-esterases) and belong to the group of carboxyl esterases (Narise, 1973). Sometimes two special types of esterases (p-esterase and S-esterase) are expressed. We also found that interspecific and interstrain differences exist in the esterase patterns of different flies of the virilis group.

Interspecific differences. Different esterase patterns were demonstrated between *D. virilis* and *D. texana* and *D. littoralis*. Esterase-2 in *D. texana* and *D. littoralis* is intermediate between the A (fast) and B (slow) forms of *D. virilis*, and esterase-4 is usually very weakly expressed in both species (Fig. 1). *D. imeretensis* is characterized by faster mobility of all its esterase fractions in comparison with the other species (Fig. 2).

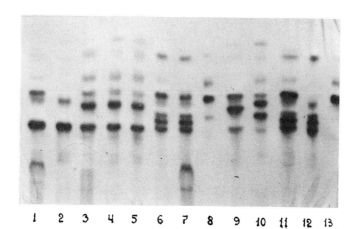

Fig. 2. Zymograms of *D. virilis* and *D. imeretensis*. 1-the mixture of homogenates of *D. virilis* and *D. imeretensis*, 2-*D. virilis*, 3-5-♀ *D. virilis* x ♂ *D. imeretensis*, 6-7-the mixture of homogenates of *D. virilis* and *D. imeretensis* (*D. virilis* was heterozygous for β-esterase), 8-*D. imeretensis*, 9-10- ♀ *D. virilis* x ♂ *D. imeretensis*, 11-12- the mixture of homogenates of *D. virilis* and *D. imeretensis* (*D. virilis* was heterozygous for β-esterase). 13-*D. imeretensis*. Starch gel electrophoresis. A mixture of α- and β-naphthyl acetate was used as substrate.

The differences between stocks. Two types of differences of esterase variants were observed among some mutant stocks of *D. virilis*: 1. differences in electrophoretic mobility (fast-A and slow-B variants of esterase-2), and 2. differences in the distribution of activities between two subfractions, fast (F) and slow (S) of esterase-4 and esterase-6.

LOCALIZATION OF GENES CONTROLLING ESTERASES. ORGANIZATION OF THE GENETIC SYSTEM REGULATING PHENOTYPIC EXPRESSION OF ESTERASES

For this investigation we made use of the fact that stock 147 of *D.virilis* (b bk dt - mutations which were localized in the 2nd chromosome) is homozygous for the fast variant (A) of esterase-2 and has high activity of the slow subfraction of esterase-4 (S-variant). In addition, we exploited the fact that stock 140 (va - mutation which marks the 2nd chromosome) is homozygous for the B-variant of esterase-2 and for the F-variant of esterase-4. Appropriate crosses were carried out. We found that these esterase variants were under monogenic control (Table 1, 2) and we located two kinds of genes: 1. genes controlling the differences in electrophoretic mobility, and 2. genes controlling the differences in the distribution of activities between the two subfractions. Among offspring from crosses between ♀ ♀ 147 x F_1 (147 x 140)♂ ♂ or between

$$♀ ♀ \quad \frac{b \ + \ bkdt \quad Est - 2A \quad Est - 4S}{b \ + \ bkdt \quad Est - 2A \quad Est - 4S}$$

X

$$\frac{b \quad + \ bkdt \quad Est - 2A \quad Est - 4S}{+va + \ + \quad \quad Est - 2B \quad Est - 4F} \ ♂ ♂$$

all the flies with bbkdt phenotypes have only the A variant of Est-2 and the S-variant of Est-4. We conclude that both esterase loci are linked with visible markers which are located on the second chromosome.

The esterase genes of *D.virilis* were localized by Ohba and Sasaki (1967, 1968) on the second chromosome also. For localization of the corresponding genes in *D.texana, D. littoralis,* and *D.imeretensis* we crossed all of these species to *D.virilis* marker stock 160 [2nd chromosome-broken (b), 3rd chromosome-gapped (gp), 4th - cardinal (cd), 5th-peach (pe), 6th-glossy (gl)]. Electrophoresis of the progeny of the backcross to the *virilis* marker stock showed that flies which were homozygous for broken (b), the second chromosome marker, were also homozygous for the virilis α- and β-esterases. Similar data are also known for *D.americana* (McReynolds, 1967) and *D.montana* (Roberts and Baker, 1973) of the virilis group.

The following crosses were performed for a more precise

TABLE 1

MODE OF INHERITANCE OF ESTERASE-2 IN *D.VIRILIS*

NN	Crosses		Est-2 phenotype in offsprings		
	Females	Males	AA	AB*	BB
1	147 AA	103 AA	19	0	0
2	103 AA	147 AA	5	0	0
3	147 AA	140 BB	0	10	0
4	140 BB	147 AA	0	10	0
5	F_1(147x140) AB	147 AA	26	26	0
6	F_1(147x140) AB	140 BB	0	9	7
7	F_1(147x140) AB	F_1(147x140) AB	5	11	4
8	103 AA	140 BB	0	33	0
9	140 BB	103 AA	0	27	0
10	F_1(103x140) AB	103 AA	6	6	0
11	F_1(103x140) AB	140 BB	0	7	5
12	F_1(103x140) AB	F_1(103x140) AB	7	17	8

*Esterase-2 is dimer and therefore represented by three fractions (A, AB, and B) in hybrids.

TABLE 2

MODE OF INHERITANCE OF ESTERASE-4 IN *D.VIRILIS*

	Crosses		Est-4 phenotype in offsprings		
NN	Females	Males	FF	FS	SS
1	147 BB	140 AA	0	23	0
2	140 AA	147 BB	0	14	0
3	F_1(147x140) AB	147 BB	0	12	11
4	F_1(147x140) AB	140 AA	6	6	0
5	F_1(147x140) AB	F_1(147x140) AB	4	10	6

localization of Est-2 and Est-4 loci (Table 3). From table 3 it is evident that 83% of crossovers occurred in the course of recombination between genes bk and Est-2, and 17% between genes Est-2 and dt. It is inferred that the position of the Est-2 locus is 209.3±. On the other hand, 79.7% crossovers occurred during the recombination between genes b – Est-4, and 20.3% between the genes Est-4 and bk. It is concluded that locus Est-4^m is located in the region of 192.0±.

Next we investigated the stock of *D.virilis* which was characterized by a fast (A) variant of esterase-4. The data concerning the localization of this gene (Est-4) are presented in table 4.

TABLE 3

LOCALIZATION OF GENES Est-2 AND Est-4^{m}*

Crosses	Phenotype and number of recombinants	Est-2 phenotype in recombinants		Crosses	Phenotype and number of recombinants	Est-4 phenotype in recombinants	
		A	AB			S	FS
♀♀ $\dfrac{\text{b+bk dt Est-2A}}{\text{+va+ } \text{Est-2B}}$x	b bk+27	3	24	♀♀ $\dfrac{\text{b+bk dt Est-4S}}{\text{+va+ } \text{Est-4F}}$x	b+ +90	14	76
	++dt 30	24	6		+bk dt90	69	21
♂♂ $\dfrac{\text{b+bk dt Est-2A}}{\text{b+bk dt Est-2A}}$	b+dt 17	15	2	♂♂ $\dfrac{\text{b+bk dt Est-4S}}{\text{b+bk dt Est-4S}}$	+bk +17	12	5
	+bk+ 20	15	5		b+dt 15	3	12

*The gene controlling the differences in the distribution of activities between two subfractions, fast and slow, was designated Est-4^{m}. The gene controlling the differences in electrophoretic mobility is Est-4.

TABLE 4

LOCALIZATION OF GENE Est-4

Crosses	Phenotype and number of recombinants		Est-4 phenotype in recombinants	
			AB	B
♀♀ $\frac{b\ bk\ dt\ Est\ 4B}{+\ +\ +\quad Est-4A}$ x	b + +	33	29	4
	+ bk dt	39	10	29
♂♂ $\frac{b\ bk\ dt\ Est-4B}{b\ bk\ dt\ Est-4B}$	+ bk +	20	3	17
	b + dt	10	6	4

The position of gene Est-4 is 191.5±, practically the same as for gene Est-4^m. Gene Est-6^m was localized at 190.6± m.u. (Korochkin et al., 1973). We also estimated the percentage of crossovers between different esterase genes and determined that the order of the loci is 2, 4, 6.

Roberts and Baker (1973) on the basis of population analysis concluded that in D.montana the total distance between loci of α-esterase genes is 8 recombinants out of 2,140 or 0.37%, demonstrating that the loci are very tightly linked. The authors supposed that the α-esterase loci were developed phylogenetically by tandem duplication.

To localize genes on a cytological map we (Korochkin and Evgeniev, in press) conducted two series of experiments. In one of these we used a stock of D.littoralis marked by the dominant, lethal (in the homozygous state) mutation Shorty. V. Mitrofanoff (personal communication) localized this mutation on the second chromosome inside the inversion, Ki-j-Tl-Va (according to Hsu map, 1952). The progeny of mating (D.virilis 147 B bk dt x D.littoralis Shorty) x D.virilis 140 (va) were analyzed electrophoretically. It was observed that esterase genes are localized outside the inversion. In the second experiment we produced stocks of D.virilis whose 2nd chromosome contained segments of the second chromosome of D.littoralis. The experimental matings were as follows: ♀ D.virilis stock 160 was crossed to D.littoralis. F_1 males were then crossed to ♀ stock 160 and, as a result, males heterozygous for the 2nd chromosome with $\frac{b\ gp\ cd\ pe\ gl}{+\ gp\ cd\ pe\ gl}$ constitution were produced. These males were crossed to D.virilis females of stock E_1 whose second chromosome was marked with the mutations Pu eb b dk dt. Females $\frac{Pu\ eb\ b\ dk\ dt\ (virilis)}{+\ +\ +\ +\ +\ (littoralis)}$ were crossed to males of the stock E_1. In F_1, these crossover males of stock E_2 - Pu eb b, E_3- b bk dt, E_4 - Pu eb and E_5^{-b} were analyzed cytogenetically and electrophoretically for (esterase pattern).

According to preliminary data esterase genes are localized between the IIF8-IIG2 regions (Griffin's designations, Fig.3). We plan to localize these loci more exactly (1-2 bands). It is now possible to construct an hypothesis of the organization of the genetic material controlling the synthesis of esterase-4 and esterase-6, which are characterized by the existence of F- and S-subfractions. The corresponding variations in F- and S-subfractions of esterases are attributable to the same cistron or to two closely linked (probably adjacent) cistrons (Fig. 4). According to the first hypothesis the corresponding esterase fractions can exist in two states, free and membrane-bound. In the process of protein extraction for electrophoretic separation, membrane-bound fractions are extracted still attached to membrane fragments. Due to this situation this esterase appears on electropherograms as a separate region of slower mobility than does a free esterase. The existence of such esterases was observed in different membrane-containing cell components (mitochondrial, microsomal, cellular membrane debris) of D.virilis and D.texana in our laboratory. The second hypothesis is based on the assumption that a single structural gene controls the synthesis of each subfraction. Some data are in accord with this assumption but are nevertheless not contradictory to the first hypothesis. These data are: 1. the existence of independent null-variants of each subfraction in some individuals; 2. the specificity of subcellular distribution of the F- and S-subfractions; 3. the organ specificity of the distribution of these subfractions 4. the existence of their independent electrophoretic variants; and 5. the ontogenetic coordination of their phenotypical expression (Korochkin et al., 1973). A similar model was presented for alkaline phosphatase by Wallis and Fox (1968). It is possible to explain the appearance of single individuals characterized by an equal expression of F- and S-subfractions (Korochkin and Matveeva, 1973) as resulting from a crossover between tightly linked genes (Fig. 4) like the amylase genes in D.melanogaster (Bahn, 1967).

THE INFLUENCE OF DIFFERENT CHROMOSOMES ON THE FORMATION OF ESTERASE PATTERNS

With respect to the interspecific and interstrain differences in the esterase patterns observed in the virilis group of Drosophila, two possibilities appear plausible. The question may be posed whether these differences are determined only by genes directly controlling esterases and located on the second chromosome or whether they are dependent on other chromosomes as well. In view of the fact that in D.virilis

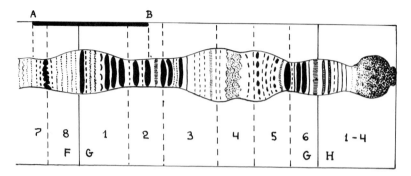

Fig. 3. The proximal part of the second chromosome of
D.virilis (designations according to J. Paterson, W. Stone and
A. Griffin, 1940, The University of Texas Publ. no. 4032,
Houston, Texas). A-B- the region of the position of esterase
genes.

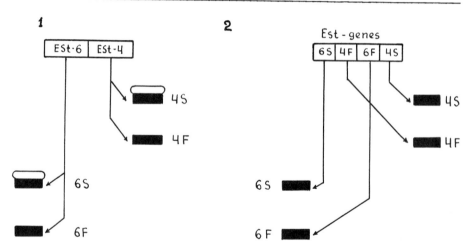

Fig. 4. Two hypotheses of the regulation of the formation of
F- and S-subfractions of esterases.
1- It is suggested that the S-subfraction differs from the
F-subfraction by binding to proteins of cellular membranes.
2- Each subfraction of esterase is controlled by a single
special gene.(Explanation in text).

the esterases are located on the second chromosome and mutations
occurring in this chromosome have marked phenotypic expression
(for instance, detached), we have introduced a second chromosome of *D.virilis* and a second chromosome of *D.texana* (in
heterozygous condition) into a "chromosomal milieu" formed by

the chromosomes of *D. texana*. In *D. texana* the proximal ends
of the second and third chromosomes are connected; for this
reason these two chromosomes could be introduced into an environment of *D. virilis* chromosomes. To achieve this, the following crosses were performed:

1) *D. texana* $\frac{+}{+}$ x *D. virilis* $\frac{dt}{dt}$
   (Stock 123)   (Stock 147)

2) From F_1 $\frac{dt\ (vir)}{+\ (tex)}$ only males were used (they

had a normal phenotype since $\underline{dt}$ is a recessive mutation) and
crossed with females from the 147 $\frac{dt}{dt}$ of *D. virilis*. From each
subsequent generation normal males were chosen and crossed with
147 $\frac{dt}{dt}$ to maintain the stock. Thus, all the texana chromosomes were substituted gradually by the chromosomes of *D. virilis*, except chromosomes 2 and 3, which were selected for in
each generation. The elimination of *D. texana* chromosomes was
also checked cytologically; stock P_1 was obtained this way. On
the contrary, stock P_2, which was produced in a similar manner,
had the second chromosome of *D. virilis* in a heterozygous
condition introduced into a *D. texana* genome. Crosses between
stock 119 of *D. texana* homozygous for the mutation "detached"
and stock 142 of *D. virilis* were made. Normal $\frac{dt}{dt}$ males were
chosen from each generation and crossed with $\frac{dt}{dt}$ females from
stock 119, insuring the progressive substitution of *D. virilis*
chromosomes by those of *D. texana*. The electropherogram of
stock P_1 corresponded to the "virilis" type (active esterase-4),
whereas the esterase pattern of the P_2 stock was identical to
the "texana" type (Fig. 5) slightly active esterase-4. The
conclusion may be drawn that the expression of these characters
is determined by the chromosomal surroundings.

Which of the chromosomes influences the expression of
esterase-4? Trying to answer this question, we have crossed
D. virilis stock 160 with *D. texana* and *D. littoralis* and then
crossed the hybrid to *D. virilis* 160. The offspring (as well
as interspecific hybrids) were analyzed electrophoretically
using markers to recognize the individual's genotype (Table 5).
The zymograms show that when an individual is homozygous for
the second chromosome, the heterozygous condition of the other
chromosome does not affect the activity of esterase-4. In the
case in which the second chromosome is in a "heterozygous"
condition to the controls, then 60% of individuals are "heterozygous" for the 4th chromosome.

It was observed that in the progeny of the mating (160 x
201) x 160 the activity of esterase-4 is 11.5±1.2 when the 2nd
and 4th chromosomes are in a "heterozygous" state (control -

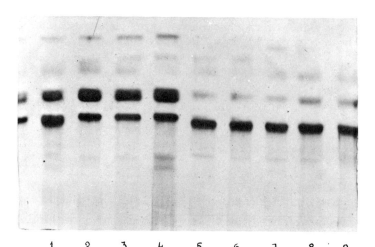

Fig. 5. Zymograms of esterases of single flies, P and P stocks, Starch gel electrophoresis. 1 and 4 - P_1. 2-3- stock 147, 5-9- P_2. Esterase-2 of the parental species has merged into one band in this zymogram because of the slight differences in mobility (see Fig. 1). Differences in mobilities of esterase-2 in stocks P_1 and P_2 are due to the absorbing crosses used in the different $D.virilis$ stocks (see text). To derive P_1 stock, 147 was used whose esterase-2 migrates faster than that of $D.texana$; to obtain P_2 stock, 142 was used whose esterase-2 migrates slower than that of $D.texana$

TABLE 5

THE ACTIVITY OF ESTERASE-4 (IN RELATIVE UNITS)
IN INTERSPECIFIC HYBRIDS

Species	$D.viri-lis$ Stock 147	$D.tex-ana$ Stock 123	123 x 147	$D.viri lis$ Stock 140	$D.littora-lis$ Stock 201	140 x 201
Activity	39.0±0.7	17.5±6.0	28.3±1.4	34.8±1.3	13.0±2.8	15.7±1.0
Species	$D.viri-lis$ Stock 140	$D.litto-ralis=$ Stock 201	201 x 140	$D.viri-lis$ Stock 160	$D.littora lis$ Stock 201	160 x 201
Activity	36.0±0.5	14.0±0.7	19.0±0.5	40.0±2.0	10.3±2.0	20.0±1.5

22.6±1.5). The "heterozygous" condition of the 5th chromosome affects the activity of esterase-6.

These results prompted us to identify the region of the 5th chromosome which is responsible for this effect. For this purpose, different lengths of segments of the 5th chromosome of *D. littoralis* were inserted into the 5th chromosome of *D. virilis*.. It was found that the region B2-G6 influences the esterase pattern (Fig. 6). There was also an indication that region A5-B2 perhaps exerts an influence on the activity of esterase-6, whereas region B2-G6 affects the expression of esterase-4. It was also found that when segment A5-G6 of the 5th chromosome is present, a slow esterase that is usually not observed appears on the zymograms (Fig. 6). This esterase is expressed only in males and localized in the bulbus ejaculatorius. This esterase is present in the control in *D. imeretensis* and *D. montana*. The nature of these esterases in now under investigation.

What are the plausible mechanisms controlling the relations between different esterase fractions in Drosophila of the virilis group during ontogenesis?

We propose that two mechanisms might operate to control esterase activities. One mechanism involves the binding of the esterase fractions to cellular membranes. In standard homogenates we examine only water-soluble esterases, whereas in cells, membrane-bound fractions are also present. Possibly, species and stocks differ in the capacity of their membranes to bind different esterase fractions. Partial support for this view has been obtained by experiments in which esterases were extracted with triton X-100, and also by analyses of esterases in different cell sub-fractions. It may be suggested that corresponding chromosome regions may influence the activities of esterase fractions by controlling their binding to intracellular membranes. This would constitute cellular level of regulation.

The second mechanism may be called the tissue level of regulation. The main point is that in organs of Drosophila the relationships of various esterase fractions are different. Thus, in the Malpighian tubules of *D. virilis* the activity of esterase-4 is higher than is true for *D. texana*, and vice versa. This fact determines the different sensitivities of the esterases of Malpighian tubules to thermal treatment, since esterase-2 is thermo-sensitive and esterase-4 is thermo-resistant.

As a result, thermal pretreatment sharply reduces the histochemical reaction of *D. texana* tubules to esterases and not so for *D. virilis*. An interesting phenomenon has been observed in hybrids. Thermal treatment and subsequent staining

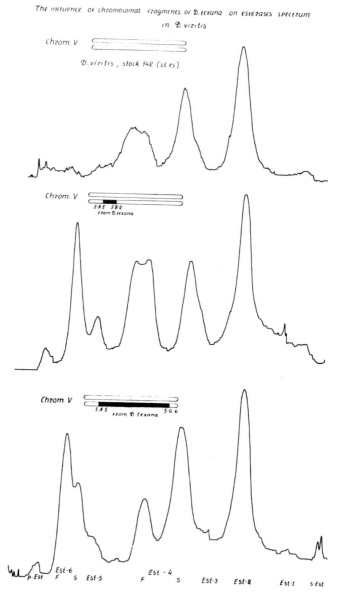

Fig. 6. Influence of presence of a *D.texana* 5th chromosome segment in the *D.virilis* 5th chromosome on esterase patterns in *D.virilis*. Segment of *D.texana* chromosome is black. Densitogram obtained with microdensitometer IFO-451.

for esterases reveal a peculiar mosaicism in cell distribution. In some cells, esterase-2 predominates and in other cells, esterase-4. The total activity of either of these esterases may be regulated by the ratio of these two cell types.

Thus, two forms of the regulation of the expression of allelic and non-allelic genes are conceivable: one is cellular, where two variants of the same protein (isozymes, for instance) are present; the other is tissue, where each protein variant is present only in cells of a definite type. Tissues, on the whole, would constitute a heterogeneous mosaic structure for a given character. Within each clone of cells a protein variant would be manifested as a dominant character, while at the tissue level it would be co-dominant. The complex interrelations between different cell types in the course of ontogenesis may play an important role in the mechanisms of dominance in the genetic basis of cellular differentiation, and possibly in heterosis.

Finally, the relation between different esterase fractions also depends upon the relative rates of their synthesis and degradation. That such a mechanism functions has been shown by Quail and Scandalios (1972) with respect to certain plant isozymes.

THE REGULARITIES OF THE ACTIVATION OF ESTERASE GENES DURING ONTOGENESIS

Studies of hybrids between stocks 140 and 147 of *D.virilis* have demonstrated that the embryonic period is characterized by low activity of the esterases and that it rises slightly at the end of embryogenesis (stage 16-20hr.) Esterase activity rises considerably just after hatching. It continues to rise till the middle of the 3rd instar and then gradually declines so that pupal stage is characterized by relatively low activity. After emergence from the pupa, enzyme activity begins to rise again and increases for 10 days (Fig. 7).

In view of the fact that esterase activity increases immediately after hatching, it should be interesting to investigate the sequence of appearance of the paternal and maternal isozymes in the course of ontogenesis in hybrids between different stocks characterized by different electrophoretic mobilities of corresponding esterases. With respect to such hybrids, starch and polyacrylamide gel procedures yielded identical results. Hybrids from the two types of crosses did not differ in the sequential expression of parental alleles in the course of ontogenesis. During the embryonic period the hybrid pattern of esterases corresponds to that of the maternal stock 140. This condition is maintained up to the stage of 16-20 hr.,

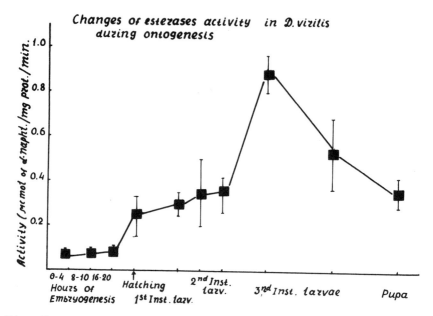

Fig. 7. Changes of esterase activity in *D.virilis* during ontogenesis.

when the paternal form appears. The paternal allele of ester-ase-2 is manifested only after hatching, judging by the very faint trace of the paternal and hybrid fractions at the first early larval instar. In the middle of the 1st instar larval period, all three forms of esterase-2 are distinct.

The paternal fraction of esterase-6 becomes clear-cut only in the middle of the 1st instar larval period.

At the larval stage, and during pupation, the faster α- and β-esterases (2-3) fractions) were observed; these probably break down juvenile hormone (Berger and Canter, 1973; Whitmore et al., 1974).

From the data obtained it is difficult to define exactly when the corresponding loci are activated, because some time may separate transcription and translation, as was shown re-garding the synthesis of some hemolymph proteins of *D.melano-gaster* during ontogenesis (Roberts, 1971).

Differential expression of some esterase isozymes is apparent during ontogenesis of some interspecific hybrids. We have observed maternal and paternal effects in interspecific hybrids. Two types of third instar larvae have been distin-guished in ♀ *D.texana* 123 x *D.virilis* 140 ♂ , and in ♀ *D.texana* 123 x *D.virilis* 103 ♂ hybrids: 1) larvae with equal expression of both parental forms of esterase-4; and 2) larvae

with stronger expression of the maternal form. In ♀ 123 x
103 ♂ hybrids a maternal fraction of esterase-4 may frequently
by the only one observed. In ♀ *D.texana* 123 x *D.virilis* 140
♂ hybrids only the paternal form of esterase-4 is displayed
at the late pupal stage; on the zymogram of imagos it usually
stains more intensely than the maternal form. In third instar
larvae of reciprocal hybrids the esterases are expressed co-
dominantly; at the stage of middle and late pupae, however,
only the maternal fraction of esterase-4 is present. Esterase-
2 exerts a maternal effect at the larval stage of ♀ *D.texana*
123 x *D.virilis* 147 ♂ hybrids; in imagos both parental forms
are rather intensely stained. In larvae and pupae of ♀ *D.*
littoralis x *D.virilis* ♂ hybrids, esterase-2 is sometimes
represented by both parental bands, though in this instance,
the "maternal esterase" is expressed more weakly (Korochkin
et al., 1973).

The activation of genetic systems that modify esterase
patterns takes place during pupation as a result of imaginal
disc differentiation. The activity of esterase-4 is identical
at the larval stage in *D.virilis* and *D.texana* (or *D.littoralis*)
and the corresponding interspecific differences appear only
during pupation. Probably messenger RNA for synthesizing
membrane proteins is transcribed at this time. These proteins
would be responsible for the binding of esterase isozymes, or
perhaps this regulative process occurs at the translational
level as a result of the activation of pre-existing mRNA.

In extending these data it should prove interesting to
analyze isozyme synthesis in single cells containing polytene
chromosomes. In this case we have the unique possibility to
study a correlation between the activity of chromosome segments
controlling esterase synthesis and the phenotypic expression
of this biochemical trait in the process of cell
differentiation.

ACKNOWLEDGEMENTS

The author is very grateful to Prof. N. Sokoloff and Dr.
M. Evgeniev for providing with Drosophila stocks. I wish to
thank the following workers for supplying some of the chemicals used in this study: Professors C. Shaw, M. Green, V.
Novak, H. Callan, J. E. Edström, and A. McLaren.

REFERENCES

Astauroff, B. 1968. The problems of the developmental viology.
   J. Com. Biol. 29: 139. (Russ.)
Bahn, E. 1967. Crossing over in the chromosomal region

determining amylase isozymes in *D.melanogaster*. *Hereditas*. 58: 1

Bamford, H., and H. Harris 1964. Studies on "usual" and "atypical" serum cholinesterases. *Am. J. Hum. Genet*. 27: 417-

Berger, E., and R. Canter 1973. The esterases of Drosophila. *J. Dev. Biol*. 33: 48-

Chino, M. 1937. The genetics of *Drosophila virilis*. *Japan. J. of Genet*. 12: 189-

Hadorn, E. 1961. *Developmental Genetics and Lethal Factors*. Academic Press, London-New York.

Hsu, T. 1952. Chromosomal variation and evolution in the virilis group of Drosophila. Univ. Texas Publ. 5204: 35-

Johnson, F., C. Kanapi, R. Richardson, M. Wheeler, and W. Stone 1966. An operational classification of Drosophila esterases for species comparisons. *Univ. Tex. Publ*. 6615, pp. 517

Korochkin, L. 1972. A simple ultramicromethod of electrophoretic separation of proteins and isozymes from isolated cells in polyacrylamide gel. *Tsitologiya* 14: 670-

Korochkin, L., N. Mertvetsov, N. Matveeva, and O. Serov 1972. Microelectrophoresis of proteins and isozymes in flat capilaries. *FEBS Letters* 22: 213-

Korochkin, L., and N. Matveeva 1973. Some regularities of phenotypical expression of esterases in *D.virilis*. *Ontogenez*. 5: 201- . (Russ.).

Korochkin, L., N. Matveeva, M. Golubovsky, and M. Evgeniev 1973. Genetics of esterases in Drosophila of the virilis group. 1. *Biochem. Genet*. 10: 363-

Markert, C. L. and R. L. Hunter 1959. The distribution of esterases in mouse tissues. *J. Histochem. and Cytochem*. 7: 42-49.

Markert, C. L. and H. Ursprung 1971. *Developmental Genetics*, Prentice-Hall, Englewood Cliffs, N. Y.

McReynolds, M. 1967. Homologous esterases in three species of the virilis group in *D.virilis*. *Genetics* 56: 527-

Narise, S. 1973. Esterase isozymes of *D.virilis*. *Japan. J. Genetics*. 48: 119.

Ohba, S., and F. Sasaki 1967. Electrophoretic variants of esterase in *D.virilis*. *Drosophila Inform. Serv*. 42: 75-

Ohba, S., and F. Sasaki 1968. Esterase isozyme polymorphisms in *D.virilis* populations. *Proc. 12th Internat. Congr. Genet*. (Tokyo) 2: 156-

Ohno, S. 1970. *Evolution by Gene Duplication*, Springer-Verlag, New York.

Quail, P., and J. Scandalios 1971. Turnover of genetically defined catalase isozymes in maize. *Proc. Nat. Acad. Sci. USA*, 68: 1402-

Raymond, S., and Y. Wang 1960. Preparation and properties of acrylamide gel for use in electrophoresis. *Anal. Biochem.* 1: 391-

Roberts, D. 1971. Antigens of developing *D.melanogaster*. *Nature* 233: 394-

Roberts, R., and W. Baker 1973. Frequency distribution and linkage disequilibrium of active and null esterase isozymes in natural populations of *D.montana*. *Amer. Naturalist.* 107: 709-

Sasaki, F. 1971. Immunological studies on esterase isozymes of Drosophila. *Japan. J. Genet.* 46: 439-

Shaw, C., and R. Prasad 1970. Starch gel electrophoresis of enzymes. *Biochem. Genet.* 4: 297-

Sims, M. 1965. Methods of detection of enzymatic activity after electrophoresis on polyacrylamide gel in Drosophila. *Nature* 207: 757-

Ursprung, H., K. Smith, W. Sofer, and D. Sullivan 1968. Assay systems for the study of gene function. *Science* 160: 1075-

Wallis, B., and A. Fox 1968. Genetic and developmental relationship between two alkaline phosphatases in *D.melanogaster*. *Biochem. Genet.* 2: 141

Whitmore, D., L. Gilbert, and P. Ittycherian 1974. The origin of hemolymph carboxylesterases "induced" by the insect juvenile hormone. *Molec. Cell. Endocrin.* 1: 37-

Whitt, G., E. Miller, and J. Shaklee 1973. Developmental and biochemical genetics of LDH isozymes in fishes. In: *Genetics and Mutagenesis of Fish.* Ed. J. Schröder. Springer-Verlag. Berlin-Heidelberg- N. Y. pp. 243-276.

SELECTIVE EXPRESSION OF MULTIPLE HEMOGLOBINS IN THE DEVELOPMENT OF *Chironomus* LARVAE

PETER E. THOMPSON
Department of Zoology
University of Georgia
Athens, Georgia 30602

ABSTRACT. Previous studies have shown that two of the ten monomeric hemoglobins present as a normal complement in larvae of *Chironomus tentans* are interrelated in their genetic regulation, showing competitive or compensatory shifts in synthesis from one genetic strain to another. These two hemoglobins, which are structurally quite similar and seem to represent the least divergent of a number of duplicate loci, both show a striking increase in quantity at the start of the fourth and last larval instar when they become the major hemoglobin components. On the supposition that a high post-moult titer of juvenile hormone is responsible for this synthesis, third-instar larvae were exposed to a synthetic juvenile hormone for eight days. Instead of stimulating early synthesis of the two hemoglobins, the synthetic hormone blocked production of both forms dramatically into the fourth instar. The findings not only confirm that the two components are involved in some common and highly selective scheme of regulation, but also indicate that synthesis of important proteins may be coupled with insect hormone titer independent of morphological change. It seems likely, however, that ecdysone is a factor in the stimulation of these late hemoglobins.

The hemoglobins of the genus *Chironomus* are striking from several points of view, some long recognized and others only recently delineated. The fact that this insect genus has an abundance of authentic hemoglobins has been recognized for over a hundred years. The sharply hyperbolic curve of oxygenation of hemoglobin in hemolymph extracts is well established (reviews by Buck, 1953, and Weber, 1965). Some few years ago, however, it was found that the hemoglobin of individual larvae actually consists of several hemoglobins in many if not all *Chironomus* species (Thompson and English, 1966; Manwell, 1966; Tichy, 1968). It has ultimately been shown that the physiological properties ascribed to "the hemoglobin" are actually the net effect in oxygen binding of a family of hemoglobins differentiated with respect to their individual properties (Sick and Gersonde, 1969).

In structure, these hemoglobins seem to be more or less closely related (Braunitzer and Braun, 1965; Braun et al, 1968; Buse et al, 1969), and in *C. tentans* consist almost entirely of monomers (Thompson et al, 1968; Tichy, 1968). Heterozygosity does not contribute allelic forms to the total number; in *C. tentans* and its sibling species *C. pallidivittatus* the number is relatively constant at 8-11 (some components are such minor ones that quantitative variation brings their assay to near zero). Their control is apparently by a series of duplicate loci, most if not all located on chromosome 3 (Tichy, 1968, 1970; Thompson et al, 1969), but in at least 3 regions (Tichy, 1970).

Our recent interest has focused upon the rate of synthesis of the several forms of hemoglobin in *C. tentans*, and upon the regulatory mechanisms which underlie differential synthesis. All strains of this species taken from North American localities have essentially the same complement of ten or so hemoglobins when electrophoresed on polyacrylamide gels. Strain differences commonly involve the relative quantities of these components, as determined by gel scanning. Most striking among such differences is an unusual pattern associated with collections from a limited locality in northwestern Iowa (the "Jemmerson" strain), which contrast with nearly all other strains by having a hemoglobin with R_f of 0.61 as the major last-instar component. All other strains have had as their major component another hemoglobin with R_f of 0.50 (Fig. 1).

Structural analysis has shown that this difference is not due to a qualitative or structural change in any hemoglobins, but rather to a regulatory shift in relative quantity (Thompson and Patel, 1972). The shift is due to a mutation which maps near the structural loci of one, and possibly both, of these hemoglobins. Its effect is thought to be transcriptional, since the decrease in Hb 0.50 acts only on a *cis* allele while the increase in Hb 0.61 occurs both in *cis* and in *trans* (Thompson and Horning, 1973). The two hemoglobins are structurally very similar, differing in only two peptides, and may be encoded by loci with common control elements. The present study strengthens our interpretation of selective regulation and expression of these two among the several hemoglobin components.

The pattern of relative quantity shown by electrophoresis and scanning differs not only between strains, but also through the larval instars or developmental stages of the same strain. It has seemed possible, although unlikely, that these progressive differences might be due to differential degradation of the various hemoglobins, rather than to

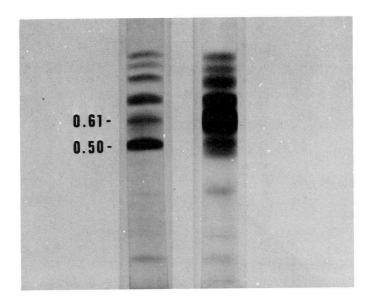

Fig. 1. Electrophoretic patterns of hemoglobins in the Jemmerson strain (right) and a normal strain of *C. tentans*.
Origin is at bottom, anode at top.

straightforward differences in the expression of loci. As a
test of this alternative, second-instar larvae of a normal
strain (from Lake Okoboji in Iowa) were divided into two groups,
both from the same egg mass. One group was placed in 25 ml
of our usual medium (a suspension of ground nettle-leaf in
0.15 percent saline) containing 25 micro-curies of tritium-
labeled δ-aminolevulinic acid (New England Nuclear). After
four days, this group was withdrawn, washed, and placed in
unlabeled medium for two weeks. The control group of small
second-instar larvae were kept in their usual medium, but
were bled at the end of the four-day period to establish electrophoretically what the pattern of relative concentrations
during the period of treatment must have been. After the two
weeks' lapse, many of the treated larvae had advanced to the
fourth instar. These were pooled for electrophoresis on a
series of gels, the gels were scanned, and twenty regions of
the gels were cut to assay how much radioactivity was associated with each peak or region in the electrophoretic pattern.
Substantial quantities of label ran with the hemoglobin components, indicating that labeled aminolevulinic acid is freely
incorporated as heme in the synthesis of *Chironomus* hemoglobin,
just as it is in vertebrates (Levere and Granick, 1965).

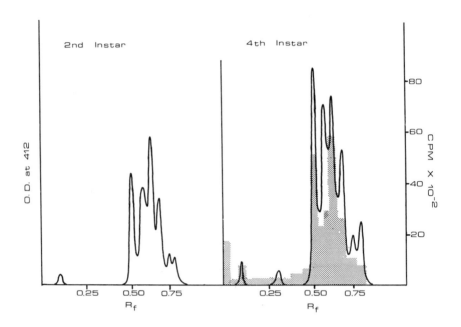

Fig. 2. Electrophoretic scans showing quantitative relations among hemoglobins in second instar (left) and fourth instar (right). Shading represents distribution of label incorporated in second and assayed in fourth instar.

The results of electrophoretic analysis, with distribution of label, are shown in Fig. 2. The left-hand figure is a scan of the base pattern of hemoglobin concentrations in second-instar larvae at the end of the four-day period. It was obtained from the unlabeled one-half of the larvae from the Okoboji egg mass. This hemoglobin pattern is strikingly different from that found in the experimental group after growth to the fourth instar (right). Nevertheless, the distribution in fourth-instar larvae of label incorporated during second-instar synthesis is still very close to the second-instar pattern observed directly. This indicates that hemoglobins synthesized in young larvae persist, and persist equally well, through later larval instars. Differential degradation does not appear to be a significant factor, the hemoglobins are long-lived, and the electrophoretic pattern at a any late stage probably represents an accumulation of hemoglobins over all previous stages rather than an approach to *de novo* synthesis.

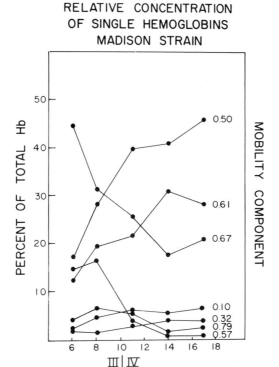

RELATIVE CONCENTRATION
OF SINGLE HEMOGLOBINS
MADISON STRAIN

Fig. 3. Shifts in relative quantities of hemoglobins of a
normal (Madison) strain, spanning the moult between third and
fourth instar.

Systematic estimates of how various individual hemoglobins
enter into a change in relative quantities through these larval stages have shown that a striking shift takes place in the
early fourth instar, and that the 0.50 and 0.61 hemoglobins
appear to express selectively at that time. Fig. 3 shows the
sudden increase in Hb 0.50 and a lesser increase in Hb 0.61
after the III-IV transition or moult in a normal Madison
strain. Fig. 4 shows that in the Jemmerson strain, where Hb
0.61 becomes the prevalent component, a dramatic burst of
synthesis of that hemoglobin takes place in early fourth instar. In earlier instars, these hemoglobins are minor ones.
Such findings closely parallel the observation of Tichy (1970)
that certain hemoglobins of *C. pallidivittatus* and German

123

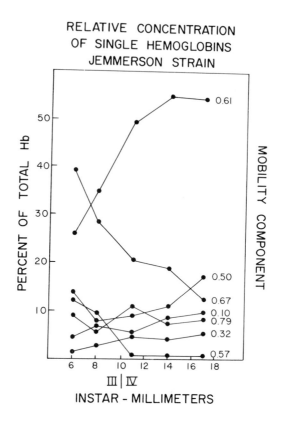

RELATIVE CONCENTRATION
OF SINGLE HEMOGLOBINS
JEMMERSON STRAIN

Fig. 4. Shifts in relative quantities of hemoglobins in the Jemmerson strain.

strains of *C. tentans* increase suddenly at the onset of the fourth instar.

This stage of sudden active synthesis is a suggestive one. In a number of insect systems, the titer of endogenous juvenile hormone (JH) increases at or after the completion of each larval-larval moult (summary in Doane, 1973). Although few cases of a highly specific effect of juvenile hormone on enzymes or other proteins have been described, Whitmore et al (1972) have demonstrated that juvenile hormone injected into pupae of the saturniid moth *Hyalophora gloveri* stimulates production of the carboxylesterases which are normally active in degrading JH itself during the last larval instar. While the carboxylesterase case might represent a uniquely direct interaction with the inducing hormone, it has seemed worthwhile to explore a possible role of juvenile hormone in stimulating the major hemoglobins of the last instar in *Chironomus*.

Our approach has been to apply juvenile hormone at earlier stages and to look for a shift to earlier increase in Hb 0.50 or Hb 0.61. A synthetic growth regulator expected to have juvenile hormone (JH) activity in dipterans was supplied by Dr. G. B. Staal of Zoecon Corporation. This material was dissolved in acetone at a one percent solution, of which 0.2 ml were added to a number of 500-ml cultures of the aqueous medium described previously, for a final concentration of four parts per million of the insect growth regulator (IGR). Third-instar larvae grown in such cultures were matched with other larvae from the same egg masses, grown in controls with 0.2 ml of acteone only added to their medium.

It was expected, and our earliest results quickly confirmed, that the synthetic juvenile hormone is rather quickly broken down in our standard cultures, which are not free of bacteria and protozoans. With renewal of the medium and its additives at 2-day intervals, however, a striking effect was seen.

The effect of the synthetic growth regulator was essentially the reverse of that postulated on the basis of a high titer of JH in normal larvae at early fourth instar. Fig. 5 summarizes the patterns observed in third-instar larvae (left) of a normal Madison strain, and fourth-instar patterns of the same group grown to 18 mm with synthetic JH (shaded profile at right) or with acetone only (dotted profile). The growth substance has dramatically blocked the synthesis of what should otherwise have been the major hemoglobin components at this stage of the last larval instar.

In keeping with the regulatory interactions and commonality in stage of synthesis shown by the Hb 0.50 and Hb 0.61, a similar inhibition in the Jemmerson strain is seen to apply to Hb 0.61, which normally represents more than 50 percent of the total hemoglobin in late fourth-instar larvae of that strain. This effect of the synthetic JH is represented in Fig. 6.

These effects of juvenile hormone are reversible, in that larvae withdrawn from treatment after 4-6 days have invariably reverted within a few days to a pattern of relative quantities near that normally encountered in fourth-instar larvae. This was also the case in our early experiments where the synthetic hormone was not replenished at short intervals, and it is reversion to normal pattern that has led us to conclude that the material is degraded rather quickly in our cultures.

Apart from this highly specific effect on two hemoglobins among the complement of ten or so, the synthetic JH has no outwardly detectable effect upon morphology or physiology, except to retard larval growth and moulting in an irregular

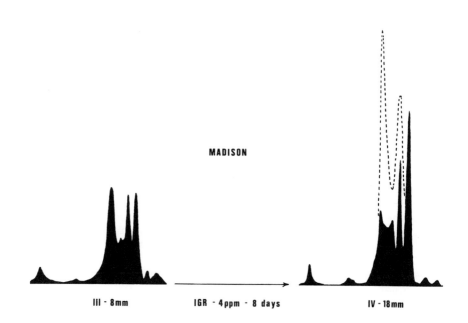

MADISON

III - 8mm   IGR - 4ppm - 8 days   IV - 18mm

Fig. 5. Effect of synthetic juvenile hormone (IGR) on expression of hemoglobins in a normal strain.

---

fashion. It is particularly relevant that the oenocytes which are the primary if not the sole site of hemoglobin synthesis (Thompson, unpublished) follow a normal course of development through all stages of third and fourth instar, as determined by total length, size of head capsule, and development of gonads (Wuelker and Goetz, 1968). These cells, occurring as paired clusters located in each segment in the lateral blood vessels, have completed all proliferation during early larval stages and always consist during the third and fourth instars of a cluster of four flat oenocytes compressed into a lozenge shape and interconnected by a slender filament to one spherical oenocyte. The only change normally observed in these cells from third to fourth instar is enlargement of the single, spherical oenocyte from a very small size to 80-100μ in diameter. This size change appears to be unaffected by synthetic JH.

It is worthwhile to note in this regard that the striking changes in hemoglobin complement or pattern induced by the growth regulator are associated strictly with post-mitotic activities of oenocytes and are not mediated by rounds of DNA

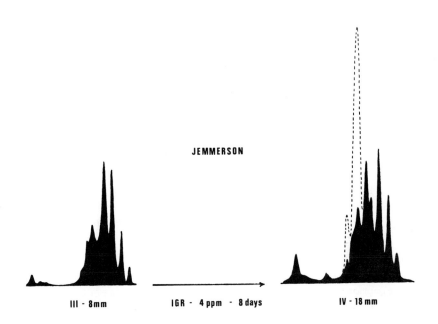

Fig. 6. Effect of IGR on hemoglobins of the Jemmerson strain.

synthesis. This adds to the accumulation of cases in which postmitotic regulatory shifts in gene activity can be demonstrated (Willis, 1974).

The directness of the effect of synthetic juvenile hormone in our present experiments is of course open to question. Since the generalized role of JH seems commonly to be the inhibition of late developmental events, and particularly events that require a high titer of ecdysone, a thorough examination of the effect of ecdysone and of antagonistic interactions between ecdysone and juvenile hormone is in order. Regardless of the actual basis, however, it is already clear that the genetic regulation of a specific protein requiring unusual rates of synthesis in a specific stage may become coupled with hormone levels in some fairly direct fashion.

REFERENCES

Braunitzer, G. and V. Braun 1965. Untersuchungen an Insekt-hamoglobinen *(Chironomus thummi)*. *Hoppe-Seyler's Z. Physiol. Chem.* 340: 88-91.

Braun, V., R. R. Crichton, and G. Braunitzer 1968. Über monomere und dimere Insektenhämoglobine (Chironomus thummi). Hoppe-Seyler's Z. Physiol. Chem. 349: 197-210.

Buck, J. B. 1953. The internal environment in regulation and metamorphosis. In: K. D. Roeder, ed., Insect Physiology, Wiley and Sons, New York, pp. 191-217.

Buse, G., S. Braig, and G. Braunitzer 1969. Die Konstitution einer Hämoglobin-Komponente (Erythrocruorin) der Insekten (Chironomus thummi thumii; Diptera). Hoppe-Seyler's Z. Physiol. Chem. 349: 1686-1690.

Doane, W. W. 1973. Role of hormones in insect development. In: S. J. Counce and C. H. Waddington, eds., Developmental Systems: Insects, Academic Press, New York, pp. 291-497.

Levere, R. D. and S. Granick 1965. Control of hemoglobin synthesis in the cultured chick blastoderm by delta-aminolevulivic acid synthetase: increase in the rate of hemoglobin formation with delta-aminolevulinic acid. Proc. Natl. Acad. Sci. USA 54: 134-137.

Manwell, C. 1966. Starch gel electrophoresis of the multiple hemoglobins of small and large larval Chironomus; a developmental hemoglobin sequence in an invertebrate. J. Embryol. Exp. Morph. 16: 259-270.

Sick, H. and K. Gersonde 1969. Die O_2-Bindungseigenschaften einiger Larvalhämoglobine von Chironomus th. thummi. Eur. J. Biochem. 7: 273-279.

Thompson, P. E. and D. S. English 1966. Multiplicity of hemoglobins in the genus Chironomus. Science 125: 75-76.

Thompson, P. E., W. Bleeker, and D. S. English 1968. Molecular size and subunit structure of the hemoglobins of Chironomus tentans. J. Biol. Chem. 243: 4463-4467.

Thompson, P. E., D. S. English, and W. Bleeker 1969. Genetic control of the hemoglobins of Chironomus tentans. Genetics 63: 183-192.

Thompson, P.E., and G. Patel 1972. Compensatory regulation of two closely related hemoglobin loci in Chironomus tentans. Genetics 70: 275-290.

Thompson, P. E. and M. J. Horning 1973. Regulatory interactions involving two hemoglobin loci of Chironomus. Biochem. Genet. 8: 309-319.

Tichy, H. 1968. Hemoglobins of Chironomus tentans and pallidivittatus, in Molecular Genetics (Springer-Verlag, Berlin), pp. 248-252.

Tichy, H. 1970. Biochemische und cytogenetische Untersuchungen zur Natur des Hämoglobin-Polymorphismus bei Chironomus tentans und Chironomus pallidivittatus. Chromosoma 29: 131-188.

Weber, R. E. 1965. On the haemoglobin and respiration of

Chironomus larvae with special reference to *Ch. plumosus plumosus* L. Ph.D. Dissertation, Rijks-universiteit, Leiden. 91 pp.

Whitmore, D., E. Whitmore, and L. I. Gilbert 1972. Juvenile hormone induction of esterases: a mechanism for the regulation of juvenile hormone titer. *Proc. Natl. Acad. Sci. USA* 69: 1592-1595.

Willis, J. H. 1974. Morphogenetic action of insect hormones. *Ann. Rev. Ent.* 19: 97-115.

Wuelker, W. and P. Goetz 1968. Die Verwendung der Imaginalscheiben zur Bestimmung des Entwicklungszustandes von *Chironomus*-Larven. *Z. Morph. Tiere* 62: 363-388.

ISOZYME ONTOGENY OF THE MOSQUITO, *ANOPHELES ALBIMANUS*

SHARANJIT S. VEDBRAT, and GREGORY S. WHITT
Department of Zoology
University of Illinois
Urbana, Illinois 61801

ABSTRACT: Changes in gene expression during mosquito development have been analyzed by investigating the ontogeny of isozymes resolved by starch gel electrophoresis. The genetic and molecular bases of nine esterase systems have been studied, as well as their developmental specificity. The use of specific inhibitors indicates that of the nine esterases investigated in this species, seven are carboxylesterases, one is an arylesterase, and one is an acetylesterase. The acetylesterase, Est-8, the arylesterase, Est-2, and two of the carboxylesterases, Est-4 and Est-6 are present during all stages of development. These are encoded in four separate loci located on the same chromosome. The linkage map is: Est-8 ⊢__12%__⊣ Est-4 ⊢__22%__⊣ Est-2 ⊢_9%_⊣ Est-6. Developmental analysis of the five remaining carboxylesterases reveals that Est-1 exhibits cyclic expression, Est-5 and Est-7 are larvae specific, and Est-3 and Est-9 are found only in the pupae. The developmental progression of other enzyme and isozyme systems was also investigated. L-lactate dehydrogenase and glycerol-3-phosphate dehydrogenase undergo a reciprocal change at pupation. Subsequently, there is a transitory increase in LDH activity in the young adults. Acid phosphatase is synthesized in the males significantly earlier than in females. The results of this study indicate that the isozyme systems of the mosquito are very useful for analyzing the mechanisms of insect development.

INTRODUCTION

Mosquitoes are excellent organisms for developmental studies (VedBrat and Whitt, 1974). Moreover, they are medically and economically important as vectors of malaria. Despite their importance, very little is known about their genetic makeup or their developmental biochemistry.

Isozymes (Markert and Møller, 1959) have been shown to be ideal gene products for investigating the patterns of gene expression during insect development. The esterases (EC 3.1. 1.-) and their isozymes have been useful gene markers for studying mechanisms of insect development (Whitmore et al., 1972; Laufer, 1961; Leibenguth 1973; and Cho et al., 1972;

131

Korochkin et al., 1973).

Esterase polymorphism and inheritance of the allelic isozymes have been studied for several different species of mosquitoes (Kitzmiller, 1974). Genetic analysis of the linkage
relationships of these esterases has not been carried out
with mosquitoes. Furthermore, only a few investigators have
employed esterases to study the developmental genetics of
mosquitoes (Briegel and Freyvogel, 1971; Simon, 1969). Our
present study incorporates a multifaceted approach to mosquito
developmental genetics and includes developmental analyses,
genetic studies, and biochemical characterization of the esterase isozymes in the mosquito species, *Anopheles albimanus*.

MATERIALS AND METHODS

Mosquitoes were reared according to the technique of
Rabbani and Kitzmiller (1972), except that rat chow served as
larval food.

Genetic crosses. A stock homozygous for at least four of the
esterase loci was established. Mass matings were carried out
by using males from the homozygous stock and females from a
stock polymorphic for esterase loci. Analysis of phenotypes
of individuals from single egg batches along with the enzyme
phenotype of the female parent permits the analysis of matings
equivalent to single pair crosses. Single pupae from the egg
batches of females heterozygous for at least two of the esterase loci were scored for their enzyme phenotypes in order to
analyze two-point and three-point crosses.

Electrophoresis. The preparation of the enzyme extract and the
electrophoretic conditions are described in VedBrat and Whitt
(1974). Staining procedures are those listed in Shaw and
Prasad (1970), with some slight modifications.

RESULTS

In *Anopheles albimanus*, at least nine zones of non-specific
esterase activity have been resolved by starch gel electrophoresis. These have been numbered Est-1 to Est-9 according to
their relative electrophoretic mobility (fig. 1). The esterases have been classified using various inhibitors according
to Holmes and Whitt (1970) and Booth et al. (1973). Fig. 2
summarizes the effect of various inhibitors on these esterases.
Est-2 appears to be an arylesterase (EC 3.1.1.2), Est-8 an
acetylesterase (EC 3.1.1.6) while the remaining seven zones
may be classified as carboxylesterases (EC 3.1.1.1).

132

Esterases in *Anopheles albimanus*

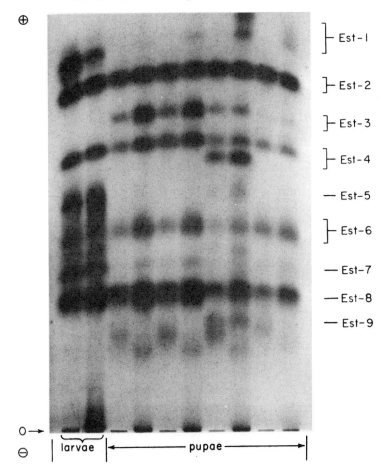

Fig. 1. The nine esterase systems resolved by gel electrophoresis in *Anopheles albimanus*.

A schematic representation of the ontogeny of these
esterases is shown in fig. 3. Four of the esterases, Est-2,
Est-4, Est-6, and Est-8 persist through all the stages of
development after their first appearance in the earlier stages.
These are the same four esterases which are later shown to be
on the same chromosome. The five other esterases are restricted to a particular developmental period. The stage specific
esterases are Est-5 and Est-7 (larvae specific), Est-1 (Larvae-
pupae), Est-3 and Est-9 (pupae). The time of onset of gene

INHIBITOR EFFECTS

INHIBITORS	ESTERASES								
	1	2	3	4	5	6	7	8	9
CONTROL	-	-	-	-	-	-	-	-	-
ESERINE	-	-	-	-	-	-	-	-	-
DFP	+++	-	+++	+++	+++	+++	+++	++	++
DDVP	+++	+++	+++	+++	+++	+++	+++	+++	+++
PARAOXAN	+++	+++	+++	+++	+++	+++	+++	-	++
P-HMB	-	+++	-	-	-	-	-	-	-
	CE	ArE	CE	CE	CE	CE	CE	AcE ?	CE

INHIBITION (+),   NO INHIBITION (-).

Fig. 2. Classification of anopheline esterases on the basis of inhibitor specificity, DFP = diisopropyl phosphorofluoridate, DDVP = 0,0-dimethyl 0-(2,2-dichlorovinyl) phosphate, and Para-oxon = 0,0-diethyl 0- (4-nitrophenyl) phosphate are organophos-phates. Eserine is a carbamate, and pHMB = parahydroxy mercur-ibenzoate is a sulfhydryl reagent. CE = carboxylesterase; ArE = arylesterase; and AcE = acetylesterase.

expression varies for different esterases. Est-2 and Est-4 are detected as early as the egg stage while Est-1, Est-5, Est-7, and Est-8 are not detected until the first instar larvae have hatched out. However, Est-1, Est-5 and Est-7 are not found in later pupal or adult stages. The activity of Est-6 appears for the first time in the second instar larvae. However, this kind of coarse-grained analysis which does not subdivide the various stages, can be somewhat misleading. This has been revealed by a more detailed analysis before and after metamorphosis. Est-1 disappears in the last phase of fourth instar larvae before pupation and is also absent 2-3 hours after pupation (fig. 4). However, the enzyme activity subsequently reappears and per-sists until 18 hours after pupation (figs. 5 and 6). Similarly, Est-3 and Est-9 are detectable for the first 12 to 18 hours of the pupal stage (figs. 5 and 6).

Significant changes in the activity of various esterase isozymes takes place twice during the postembryonic development

Ontogeny of Esterases in *A. albimanus*

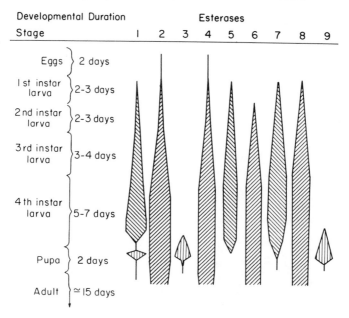

Fig. 3. Schematic representation of ontogeny of esterases in *A. albimanus*. The width of each strip represents a semi-quantitative estimate of enzyme activity judged from the amount of dye deposited on the gel.

of *Anopheles albimanus*, the first period at the time of pupation and the second some 12 to 15 hours after pupation. Before pupation, the loss of Est-1 activity occurs simultaneously with the abrupt appearance of Est-3 and Est-9. Then, three hours after pupation, Est-1 reappears at the time Est-5 and Est-7 are greatly diminished (figs. 4 and 5).

The genetic basis for four of the nine esterases, Est-2, Est-4, Est-6 and Est-8 has been determined. Each of these four esterase loci is polymorphic. Some have two different alleles, the others have three (fig. 7). Two-point and three-point crosses demonstrated that these four loci are linked and their tentative map distances are given in fig. 8. It is interesting to note that these linked loci code for isozymes with similar developmental behavior.

The ontogeny of other enzyme systems in this species has also been studied. There is a reciprocal change in enzyme

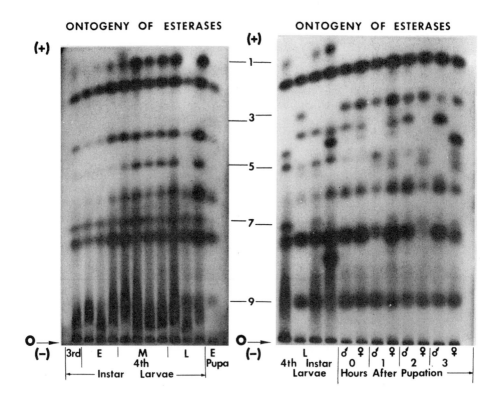

Fig. 4. Esterase ontogeny at about the time of metamorphosis from larva to pupa. E=early, M=mid, L=late.

activity for L-lactate dehydrogenase (LDH) (EC 1.1.1.27) and glycerol-3-phosphate dehydrogenase (G-3-PDH) (EC 1.1.1.8) in the pupal stage. G-3-PDH-3 activity was first detected in the eggs while that of LDH appears for the first time in the first instar larvae. In the early pupae the LDH activity is much higher than G-3-PDH activity. The situation is reversed in the late pupae with almost no LDH activity while that of G-3-PDH-1 reaches a higher level. However, there is in the young adult a transitory increase in the LDH activity (see VedBrat and Whitt, 1974 for a more detailed discussion).

Acid phosphatase (EC 3.1.3.2) is temporally preferentially expressed in the male. The males possess considerable activity at pupation. In fact, the activity first observed in some of the fourth instar larvae is probably due to the contribution of males (fig. 9). However, it is impossible to sex individuals at this stage. In females, this acid phosphatase activity is expressed for the first time three hours after

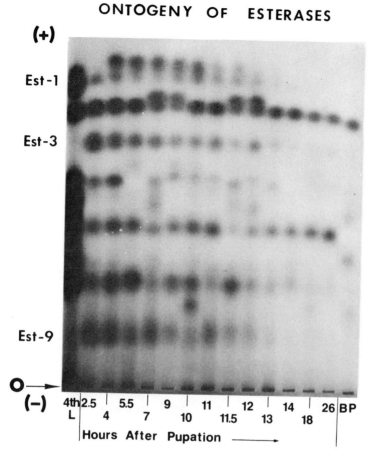

Fig. 5. Ontogeny of esterases during the early pupal stage. L = larva, BP = black pupa ready to emerge as adult.

pupation. By the mid-pupal stage there does not appear to be any difference in the enzyme activities between male and females (fig. 9).

DISCUSSION

The developmental and genetic analyses of various isozymic systems of anopheline mosquitoes indicate that these mosquito species are very useful model systems for studying dipteran biochemical and developmental genetics. Determination of the time and tissue specificity of gene expression as well as the linkage relationships of the loci encoding the same isozymes

Esterases – Ontogeny in Pupal Stage

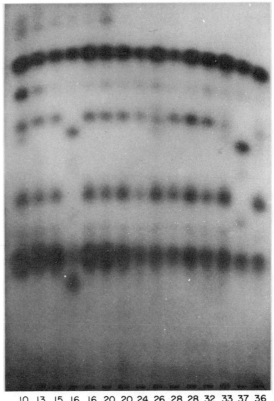

| 10 | 13 | 15 | 16 | 16 | 20 | 20 | 24 | 26 | 28 | 28 | 32 | 33 | 37 | 36 |
| ♂ | ♀ | ♀ | ♂ | ♀ | ♂ | ♀ | ♂ | ♀ | ♂ | ♀ | ♂ | ♀ | ♂ | ♀ |

Hours after Pupation

Fig. 6. Ontogeny of the esterases during the late pupal stage. The most anodal band is Est-1 and the most cathodal Est-8. Est-5 and Est-9 are absent.

should be quite useful for testing postulates regarding mechanisms of developmental regulation. Coupling of the morphological differentiation of a particular tissue or organ with the onset of the synthesis of a specific isozymic form has been demonstrated clearly for other enzymes e.g., LDH C4, Champion et al., (1974) and Miller and Whitt (1974). The degree to which isozyme synthesis is tightly coupled to morphological differentiation should help provide insight into the physiological role of the esterase.

ESTERASE   ALLELIC   VARIANTS

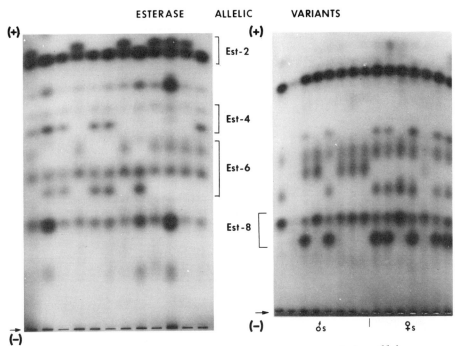

Fig. 7. Allelic variation in the esterases of *A. albimanus*.

A cyclic pattern is observed for the Est-1 activity during
development. Its greatly diminished activity before and immediately after pupation suggests that it might play some role
in the regulation of the levels of ecdysone, the moulting
hormone. This postulate would be similar to the one of Whitmore et al. (1972) for the regulation of juvenile hormone
levels. However, in their case the levels of the esterase
isozyme would be higher just before pupation to lower the titer
of juvenile hormone.

Those esterases (Est-5 and Est-7) synthesized only in
larval stages may be associated with the larval digestive system and its distinctive metabolism. Larvae feeding on microorganisms, powdered rat chow, or similar foods may require
these esterases whereas adults which are only fed on sugar
solution, would not. The esterases, (Est-3 and Est-9) which
are restricted to the early pupal stage may be involved in the
process of tissue and cellular breakdown which occurs at
this time.

The four esterases present in all stages probably possess
some very generalized function. Perhaps, they participate in
the breakdown of lipids with smaller fatty acid chain length

LINKAGE RELATIONSHIPS FOR FOUR ESTERASE LOCI: EST-2, 4, 6 AND 8.

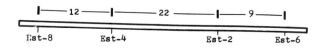

2 point crosses for	Number of progeny per cross	Map distances and gene sequence
		8 4 2 6
Est-2 & 6	43	⊢— 13.9 —⊣
Est-4 & 6	43	⊢——— 25.6 ———⊣
Est-4 & 2	12	⊢—— 25 ——⊣
Est-8 & 6	46	⊢———— 34.8 ————⊣
Est-8 & 6	29	⊢———— 24.1 ————⊣
3 point crosses for		
Est-8,4 & 6	48	⊢———— 37.5 ————⊣
		⊢— 4.2 —⊢——— 35.4 ———⊣
Est-8,4 & 2	43	⊢——— 48.8 ———⊣
		⊢— 20.9 —⊢—— 27.9 ——⊣
Est-4,2 & 6	74	⊢——— 33.7 ———⊣
		⊢—— 23 ——⊢— 10.8 —⊣
Est-4,2 & 6	86	⊢——— 21 ———⊣
		⊢—— 17.4 ——⊢— 3.5 —⊣
Est4,2 & 6	43	⊢——— 22.7 ———⊣
		⊢—— 15.9 ——⊢— 6.8 —⊣

TENTATIVE MAP

⊢— 12 —⊢——— 22 ———⊢— 9 —⊣

Est-8   Est-4   Est-2   Est-6

Fig. 8. Linkage relationships for the four esterase loci;
Est-2, Est-4, Est-6 and Est-8.

which may be used as an energy source. This kind of activity
has been reported by Castillon et al. (1973) in three insects,
Cerratitis capitata, *Dacus oleae* and *Musca domestica*. The
chain length of glycerides acted upon by esterases lies towards
the smaller side of the range of those acted upon by lipases.

Furthermore, it is also conceivable that some esterases
might play a role in insecticide resistance. One of the esterases is an arylesterase. The arylesterases have been found to
hydrolyze certain organophosphates (Augustinsson, 1961).
These esterases may provide resistance to some of the pesticides.

Acid Phosphatase Ontogeny

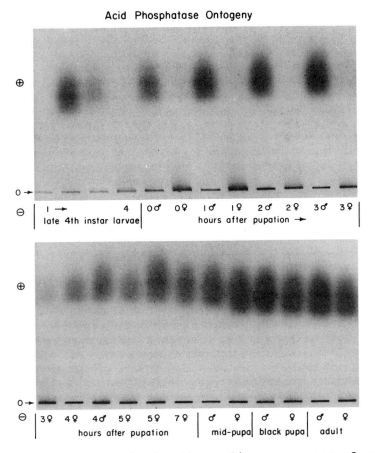

Fig. 9. Zymograms showing the earlier appearance of acid phosphatase in the males.

In order to test some of these postulates, future studies must determine the tissue and cellular specificities of synthesis for many of these esterases and their isozymes. In order to understand the control and coordination of enzyme activities at various stages, some of the following means of controlling activity should be rigorously excluded. The appearance and disappearance of a particular enzyme or isozyme at a particular stage may be due to 1) replacement of one type of cell line with another. A good example has been reported by Berger and Canter (1973) where an eclosion-specific enzyme remains with the pupal case after the adult emerges and the adult does not synthesize the enzyme ; 2) there may be an activation or inactivation of previously existing enzymes,

3) or as has been demonstrated for some enzyme systems, repression or derepression of a specific gene locus.

The linkage of four loci coding for the four esterases with similar temporal behavior of gene expression during development has been established. Because there are only three pairs of chromosomes in this species and these are the only four loci analyzed for their linkage relationships, it is possible that this linkage association is just a chance event. However, esterase loci have been shown to be linked in a number of unrelated species. This suggests that the mechanism by which esterases have evolved is by gene duplication and later divergence as Roberts and Baker (1973) have suggested for the closely linked loci in *Drosophila*. The linkage pattern in mosquitoes with the four loci spread over about 50 map units could have initially existed as four closely linked loci which were subsequently separated by inversions. These inversions occur in natural populations of a related Anopheline species (Kreutzer et al., 1972). Further studies will test this postulate by determining the exact position of esterase loci with respect to specific polytene chromosome bands.

ACKNOWLEDGEMENTS

This research was supported by NSF GB 16425 and GB 43995 to G. S. Whitt and USPHS grant AT-03486 to J. B. Kitzmiller.

REFERENCES

Augustinsson, K. B. 1961. Multiple forms of esterases in vertebrate blood plasma. *Ann. N. Y. Acad. Sci.* 94: 844-860.

Berger, E. and R. Canter 1973. The esterases of *Drosophila*. I. The anodal esterases and their possible role in eclosion. *Dev. Biol.* 33: 48-55.

Booth, G. M., J. Connor, R. A. Metcalf, and J. R. Larsen 1973. A comparative study of effects of selective inhibitors on esterase isozymes from the mosquito, *Anopheles punctipennis*. *Comp. Biochem. Physiol.* 44B: 1185-1195.

Briegel, H. and T. A. Freyvogel 1971. Non-specific esterases during development of culicine mosquitoes. *Acta Tropica* 28: 291-297.

Castillon, M. P., R. E. Catalan, S. Vega, and A. M. Municio 1973. Biochemistry of the insects *Dacus oleae* and *Ceratitis capitata*. Changes of cholinesterases and triglyceride hydrolyzing enzymes. *Comp. Biochem. Physiol.* 44B: 639-646.

Champion, M. J., J. B. Shaklee, and G. S. Whitt 1974. Developmental genetics of teleost isozymes. *III Isozymes:Developmental Biology*. C. L. Markert, editor. Academic Press, N.Y.

Cho, P., R. Davenport, G. S. Whitt, G. M. Booth and J. R.

Larsen 1972. Electrophoretic and histochemical analysis of esterase synthesis in the milkweed bug *Oncopeltus fasciatus* (Dall.) *Wilhelm Roux. Archiv.* 170: 209-220.

Holmes, R. S. and G. S. Whitt 1970. Developmental genetics of the esterase isozymes of *Fundulus heteroclitus. Biochem. Genet.* 4: 471-480.

Kitzmiller, J. B. 1974. Genetics of mosquitoes. *Adv. Genet.* (In preparation.)

Korochkin, L. I., N. M. Matveeva,and M. B. Evgeniev.1973. Expression of esterase isozymes in hybrids of *Drosophila virilis. Genetika* 9: 55-61.

Kreutzer, R. D., J. B. Kitzmiller,and E. Ferreira 1972. Inversion polymorphism in the salivary gland chromosomes of *Anopheles darlingi. Mosq. News* 32: 555-565.

Laufer, H. 1961. Forms of enzymes in insect development. *Ann. N. Y. Acad. Sci.* 94: 825-835.

Leibenguth, F. 1973. Esterase-2 in *Ephestia kuhniella.* II. Tissue specific patterns. *Biochem. Genet.* 10: 231-242.

Markert, C. L. and F. Møller,1959. Multiple forms of enzymes: Tissue, ontogenetic and species specific patterns. *Proc. Nat. Acad. Sci. U. S. A.* 45: 753-763.

Miller, E. T. and G. S. Whitt 1974. Lactate dehydrogenase isozyme synthesis and cellular differentiation in the teleost retina. *III Isozymes: Developmental Biology.* C. L. Markert, editor, Academic Press, N. Y.

Rabbani, M. G. and J. B. Kitzmiller 1972. Chromosomal translocations in *Anopheles albimanus.* Wiedmann. *Mosq. News.* 32: 421-432.

Robert, R. M. and W. K. Baker 1973. Frequency distribution and linkage disequilibrium of active and null esterase isozymes in natural populations of *Drosophila montana. Am. Nat.* 107: 709-726.

Shaw, C. R. and R. Prasad 1970. Starch gel electrophoresis of enzymes- A compilation of recipes. *Biochem. Genet.* 4: 297-320.

Simon, J. P. 1969. Esterase isozymes in mosquito *Culex pipiens fatigans.* Developmental and genetic variation. *Ann. Ent. Soc. Am.* 62: 1307-1311.

VedBrat, S. S. and G. S. Whitt 1974. Lactate dehydrogenase and glycerol-3-phosphate dehydrogenase gene expression during ontogeny of the mosquito (*Anopheles albimanus*). *J. Exp. Zool.* 187: 135-140.

Whitmore, D., Jr., E. Whitmore,and L. I. Gilbert 1972. Juvenile hormone induction of esterases: A mechanism for the regulation of juvenile hormone titer. *Proc. Nat. Acad. Sci. U. S. A.* 69: 1592-1595.

ISOZYME TRANSITIONS OF CREATINE KINASE AND ALDOLASE
DURING MUSCLE DIFFERENTIATION IN VITRO

DAVID C. TURNER
Institute for Cell Biology, Swiss Federal Institute
Of Technology, CH-8049 Zürich, Switzerland

ABSTRACT. Parallel transitions in the isozyme patterns of
creatine kinase (CPK) and fructose diphosphate aldolase
are known to occur during skeletal muscle development.
The B form of creatine kinase and the C form of aldolase
are replaced by M-CPK and aldolase A, respectively. These
proteins can be used as markers in studies of muscle cell
differentiation in vitro and particularly in studies of the
importance of the process in which mononucleated cells fuse
to form syncytial myotubes. Plates from a series of cultures in which the course of fusion was highly synchronous
were extracted at various times and assayed for total CPK
and aldolase enzyme activities. Although the data obtained suggest that the accumulation of M-CPK and aldolase
in the cultures as a whole begins prior to the main "burst"
of cell fusion, such experiments cannot provide conclusive
evidence that such an accumulation occurs in unfused cells
and not merely in the few myotubes also present from the
time of plating. The indirect fluorescent antibody technique was therefore used to detect marker proteins in individual cells. In cultures in which the fusion process
was blocked by addition of a calcium-specific chelator,
bipolar mononucleated cells showed the same pattern of
staining as myotubes: both fused and unfused cells reacted positively for CPKs M and B and for aldolases A and C.
Accumulation of marker enzymes typical of adult muscle can
thus occur in unfused cells.

INTRODUCTION

Cells derived from embryonic avian or mammalian skeletal
muscle, when cultured under appropriate conditions, give rise
to multinucleated, cross-striated, spontaneously contractile
muscle cells. Among the many morphological and biochemical
changes known to occur during the overall process of myogenesis, both in vitro and in vivo, are dramatic changes in the
isozyme patterns of several enzymes. This report will deal
with two such cases: the isozyme transitions of creatine
kinase (CPK) and fructose diphosphate aldolase in chicken
muscle cell cultures. Reports on two other enzymes which
undergo isozyme transitions during muscle differentiation ,

145

acetylcholinesterase and phosphoglyceric acid mutase, appear elsewhere in these volumes (Wilson, 1974; Omenn and Hermodson, 1974).

In most vertebrates there are three principal isozymes of CPK, occurring in characteristic tissue-specific distributions (Eppenberger, et al., 1964). These CPKs, designated MM, MB, and BB, are dimers of approximately 84,000 daltons, formed by the association of two types of subunits, M and B (Dawson, et al., 1965; 1967). A form of CPK with a higher molecular weight has been isolated from mitochondria (Farrell, et al., 1972; Jacobus and Lehninger, 1973); its relationship (if any) to the three principal isozymes is unclear. In this report the discussion will be restricted to the well-characterized three-membered M-B isozyme set.

The CPK isozyme transition in developing muscle was first described by Eppenberger, et al. (1964). While BB-CPK is the only CPK isozyme found in the breast muscle of the 7-day chick embryo, all three isozymes are prominent at 11 days, and by 15 days there has been an almost complete transition to the MM form, which is the only one present in adult muscle cells. Some of this MM-CPK in differentiated skeletal muscle cells appears to be bound to specific sites along the myofibrillar contractile apparatus (Turner, et al., 1973). The different- ial localization of CPK within different subcellular compart- ments, and the possible functional significance thereof, are the subject of another paper (Eppenberger, et al., 1974).

The transition from B to M CPK also occurs during myogene- sis in vitro (Fig. 1), although for reasons not yet understood the transition does not go to completion even after many days in culture. Both in vivo and in vitro the CPK transition is accompanied by a large (12 to 20-fold) increase in CPK specific activity (Eppenberger, et al., 1964; Turner, et al., 1974). This increase reflects an actual accumulation of CPK molecules and not merely the replacement of B by M subunits, since the specific activities of the purified BB and MM enzymes differ only by a factor of 2 under the assay conditions used (H.M. Eppenberger, unpublished observations).

The molecular basis of the occurrence of multiple mole- cular forms of aldolase is well understood. The active aldo- lase molecules are tetramers with a molecular weight of ca. 160,000 (Kawahara and Tanford, 1966; Penhoet, et al., 1967). Three different subunit types (A,B, and C) are known. The characteristic isozyme patterns of different tissues result from the association of these subunits into the various pos- sible types of homo-and heterotetramers (Penhoet, et al., 1966; Lebherz and Rutter, 1969). Further information on the aldo- lases is given in the papers presented at this conference by

CPK

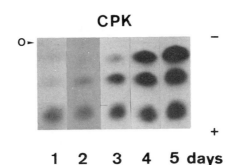

1 2 3 4 5 days

Fig. 1. The CPK isozyme transition in cell cultures derived from embryonic chicken muscle. Extracts were prepared at the indicated times after plating, electrophoresed on cellulose acetate strips and then stained for CPK activity (Turner, et al., 1974).

Lebherz (1974a) and Marquardt (1974).

During muscle development in vivo there is a transition from an "embryonic" pattern consisting primarily of aldolases C_4 and AC_3 to aldolase A_4, the classical adult muscle enzyme (Herskovits, et al., 1967; Lebherz and Rutter, 1969). At no stage of muscle development do aldolases containing B subunits appear. As with CPK, there is evidence that the "adult" form, in this case aldolase A_4, binds to distinct sites on the myofibrils (Arnold, et al., 1969). The aldolase transition also occurs in muscle cell cultures (Fig. 2), though here, too, the the transition does not go all the way to the "adult" form under the usual conditions. An increase in aldolase specific

Aldolase

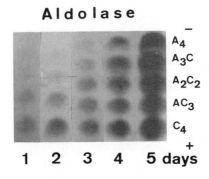

A_4
A_3C
A_2C_2
AC_3
C_4

1 2 3 4 5 days

Fig. 2. The aldolase isozyme transition in the same cultures as in Fig. 1. Cellulose acetate electrophoresis followed by specific enzymatic staining were performed according to Penhoet, et al. (1966).

activity occurs concomitant with the isozyme transition (Eppenberger, et al., 1962; Turner, et al., 1974).

The CPK and aldolase isozyme transitions are of interest to the developmental biologist for several reasons: 1) A smooth transition from one isozyme type to another presumably reflects coordinated changes in gene expression. The relative concentration of one gene product (e.g., M-CPK) is increased (presumably as a result of *de novo* synthesis, but possibly because of decreased degradation, activation of a proenzyme, etc.) at the same time that the relative concentration of another gene product (B-CPK) is diminished. Most developmental biologists believe that cell differentiation depends upon the selective expression of a limited number of genes within each cell type. In most cases studied so far, however, the mechanisms involved, and even the levels of control (e.g., transcription, messenger stability, translation, subunit exchange, degradation) are largely unknown. Because isozyme transitions represent coordinated changes affecting two gene products, investigation of isozyme systems seems an especially promising approach to the understanding of control mechanisms underlying cell differentiation. 2) A distinct, but related, point is that there is evidence (summarized in Turner and Eppenberger, 1974) that the CPK and aldolase transitions occur together in a temporally coupled, tissue-specific manner. This suggests that a joint analysis of these two transitions may provide an additional "handle" for getting at factors controlling differential expression of whole "teams" of genes. 3) The well-characterized CPK and aldolase isozymes are useful markers for cell differentiation. The marker protein most often used heretofore in studies of muscle differentiation, myosin, has been less satisfactory because of its poor activity as an immunogen and because the molecular basis of its multiple molecular forms is only now being clarified (Taylor, 1972).

Progress in clarifying the regulation of differential gene expression during myogenesis (points 1 and 2, above) has been largely hampered by insufficiencies in the muscle cell culture system itself. We have concentrated (Turner, et al.,1974,1975a, b) on obtaining homogeneous cell populations, synchronizing their development, and defining the stages of differentiation through which they pass -- all necessary before analysis of the underlying control mechanisms can begin. The isozymes of CPK and aldolase have, however, been very useful as markers for differentiation (point 3, above) throughout the work of developing and defining the experimental system (Turner, et al., 1974, 1975a,b). In this report I shall focus primarily on one such application: the detection by immunofluorescent methods of CPK and aldolase isozymes within individual cultured cells.

RESULTS AND DISCUSSION

Much attention has been given to the importance of the fusion event in the overall process of myogenesis. One favored approach has been to try to determine when, relative to the time at which fusion begins, a given biochemical process (e.g. CPK accumulation or myosin synthesis) is initiated (Coleman and Coleman, 1968; Morris and Cole, 1972; Yaffe and Dym, 1972; Zalin, 1973; Turner, et al., 1974,1975a). The success of this "relative timing" approach has often been limited by the poor synchronization of fusion achieved and in some cases by the fact that no time can be singled out as marking the "onset" of fusion (Turner, et al., 1974). Recently, however, procedures have been introduced which make it possible to manipulate the fusion process more or less at will. One such method (Paterson and Strohman, 1972) involves the addition of the calcium ion chelator, EGTA, to the cultures at the end of the first day. Under these conditions, fusion is blocked, and the post-mitotic cells that otherwise would have fused accumulate. Then, when sufficient calcium is added back to the culture medium, fusion proceeds in a single highly synchronous "burst". Using this method we have obtained evidence that increases in the specific activities of CPK and aldolase precede the onset of fusion (Fig. 3). Electrophoreses of extracts of EGTA-treated cultures show that there is a shift in the isozyme patterns toward M-CPK and aldolase A (Turner, et al., 1975a). The increases in enzyme activity and the isozyme transitions observed in such fusion-blocked cultures are, however, less pronounced than the corresponding changes seen in normally fusing cultures. Experiments to quantitate the accumulation, and ultimately the synthesis, of the "adult" subunits, M-CPK and aldolase A, are in progress.

Much as the biochemist might wish to work with extracts of mass cultures, as in the experiments just described, there is a problem with this approach when applied to primary muscle cell cultures. Even in cultures treated with EGTA there is invariably a small but significant number of myotubes present before the readdition of calcium. In the experiment shown in Fig. 3, approximately 2% of all nuclei scored at 69 hr. were found to be in multinucleated cells. Most of these myotubes are probably present from the time of plating (Dawkins and Lamont, 1971), and others may form during the first day in culture before the addition of EGTA. These "contaminant" myotubes, like the myotubes formed in the main fusion burst, presumably accumulate large amounts of M-CPK and aldolase A. One could, therefore, argue that the increases in the measured

levels of CPK and aldolase, and the accompanying isozyme transitions, observed with extracts of whole cultures occur entirely
within those myotubes present from the time of plating. Although this explanation is an unlikely one in view of the very
small number of such myotubes, it cannot easily be ruled out
with mass culture experiments.

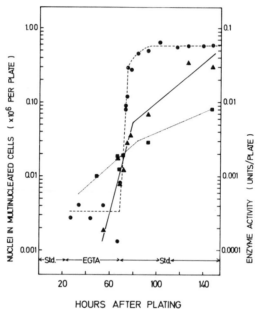

HOURS AFTER PLATING

Fig. 3. Relationship of the accumulation of total activity
per plate for CPK (▲———▲) and aldolase (■----■) to the
time course of myoblast fusion in vitro (●----●). (Adapted
from Turner, et al., 1975a).

As an alternative, one can turn to methods (e.g., histochemistry, fluorescent antibody staining) which permit the
detection of specific marker proteins within individual cells.
One can ask in what cell types and, possibly, in what subcellular locations a given marker can be found. Are there
any mononucleated cells which contain M-CPK and aldolase A?
Of course, as long as one continues to focus on the importance
of the fusion event, asking where these marker enzymes can be
detected is not fundamentally different from asking when the
markers first appear. The two approaches should be regarded
instead as different sides of the same coin. A multinucleated
myotube is by definition post-fusion. Similarly, a bipolar
myogenic cell with a single nucleus has to be pre-fusion.

In the remainder of this report I shall summarize experiments in which the indirect fluorescent antibody method was

used to detect the presence of specific gene products in the different cell types (Turner, et al., 1975b). Antisera were elicited in rabbits against homogeneous preparations of the following chicken enzymes: MM-CPK, BB-CPK, aldolase A_4, aldolase C_4. It has been shown (Eppenberger, et al., 1967; Lebherz, 1974b) that these antisera react in a subunit-specific manner. That is, anti-MM-CPK antiserum reacts with CPKs MM and MB, but shows no immunological reaction with BB-CPK. Similarly, anti-BB-CPK antiserum precipitates only the MB and BB isozymes. Anti-aldolase A_4 antiserum reacts well with aldolases A_4, A_3C, A_2C_2, very weakly with AC_3, and not at all with C_4. Antiserum against aldolase C_4 precipitates C_4, AC_3, A_2C_2, and to a much smaller extent A_3C. Fluorescent staining involved: 1) brief fixation to make cells permeable to antibodies, followed by washing with buffered salt solution; 2) a first incubation with one of the above mentioned antisera (or control serum), followed by thorough washing to remove non-specifically adsorbed antibody; 3) a second incubation with fluorescein-labelled horse antiserum directed against rabbit IgG antibodies, again followed by careful washing; and 4) mounting in a mixture of glycerin and glycine buffer. It was possible to stain different regions of the same culture with different antisera (or control sera) by carrying out steps 2) and 3) in small cylindrical plastic wells that had been pressed lightly onto the surface of the dish.

A typical result is shown in Fig. 4 (phase contrast on the left, fluorescence on the right). In this case the cells were stained for M-CPK. The myotubes have a brightly fluorescent cytoplasm, against which the many nuclei stand out as dark spots. Other cells visible under phase contrast, including a prominent binucleated fibroblast-like cell, are non-fluorescent. In standard medium cultures like these, no unfused cells showing clearcut positive staining were observed in any of the many fields examined. It is therefore concluded that the only cells that contain M-CPK under such conditions are the myotubes.

A similar analysis of the cell-specific staining patterns has been carried out for cultures maintained in 3 different culture media (Turner, et al., 1975b): 1) standard medium (as in Fig. 4); 2) standard medium containing EGTA to block fusion; and 3) standard medium containing 7×10^{-6} M 5-bromodeoxyuridine (BrdUrd), a thymidine analog. In cultures treated with BrdUrd, fusion is almost completely prevented (Stockdale, et al., 1964). This action of BrdUrd is reversible (Stockdale, et al., 1964) and appears to involve blockage of a step in myogenesis prior to the calcium-dependent fusion process itself (Turner, et al., 1975a). Electrophoresis

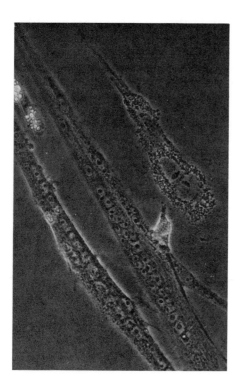 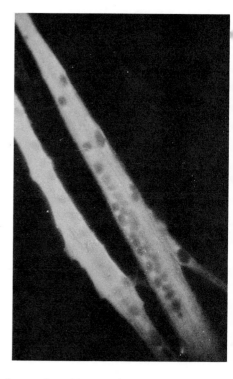

Fig. 4. Detection of M-CPK in cultured cells by the indirect fluorescent antibody method (see text). Left: phase contrast micrograph of a standard medium culture, showing several un-fused cells in addition to the multi-nucleated myotubes. Right: Fluorescence micrograph of the same field. Under these culture conditions myotubes fluoresce intensely while unfused cells are non-fluorescent (X 270).

of extracts of whole cultures treated with BrdUrd had indicated nearly complete suppression of the CPK and aldolase isozyme transitions (Turner, et al., 1974). Nevertheless, small amounts of CPKs MM and MB were detected. Although we suspected that these M-subunit-containing CPK isozymes were provided by the few "contaminating" myotubes present from the time the cultures were initiated, experiments with whole culture extracts could again not settle the point.

Although it is not possible to go through all of the evidence in detail here, the relatively low magnification photographs in Fig. 5 are intended to give some idea of the different patterns of staining obtained with the antisera against the MM and BB isozymes of CPK. The top row (Fig. 5 a-c) shows

Anti - M   Control   Anti - B

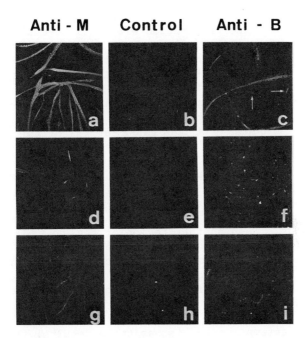

Fig. 5 a-i. Immunofluorescent detection of creatine kinases
M and B in standard (a-c), EGTA-treated (d-f) and BrdUrd-treated
(g-i) cultures. Fields in the left-hand column were reacted
with anti-MM-CPK antiserum, those in the middle with nonimmune
serum, and those in the right-hand column with anti-BB-CPK
antiserum. This figure is discussed in detail in the text
(X 36).

3 different fields of a 10-day culture maintained in standard
medium. Myotubes again fluoresced brightly (cf. Fig. 4) after
reaction with anti-MM-CPK antiserum (Fig. 5a). In another
region of the same dish reacted instead with anti-BB-CPK antiserum, the myotubes fluoresced less brightly and often had a
mottled appearance (Fig. 5c). Areas incubated with non-immune
serum showed no appreciable fluorescence (Fig. 5b). The standard medium culture shown in Figs. 5 a-c also contained numerous
mononucleated cells (both myogenic and fibroblastlike) between
the myotubes. Whereas essentially none of these cells stained
for M-CPK, a significant fraction did react positively for B-
CPK (arrows, Fig. 5c).

The middle row (Fig. 5 d-f) shows the pattern of staining
for CPKs M and B in a 10-day old EGTA-treated culture. In
Fig. 5d a few bipolar mononucleated cells are seen to have

reacted positively for M-CPK. Many other cells in this field, both spindle shaped and fibroblastlike, failed to react. By contrast, in another region of the same plate many fluorescent cells are seen after incubation with anti-BB-CPK antiserum (Fig. 5f). The strongly fluorescent cells comprise roughly half of the cells present in the field. Again, no fluorescent cells are seen in the area reacted with control serum (Fig. 5e).

The bottom row of Fig. 5 shows a 10-day old BrdUrd-treated culture reacted with anti-MM-CPK antiserum (g), non-immune serum (h), and anti-BB-CPK antiserum (i). The field in Fig. 5g was chosen because it contained several of the few myotubes present in the culture. These myotubes are clearly fluorescent and thus contain M-CPK. The rest of the field is covered by a monolayer of flattened cells, none of which reacted positively for M-CPK. It seems clear, therefore, that the traces of the MM and MB isozymes seen in electrophoreses of BrdUrd cultures are derived exclusively from those few myotubes invariably present. When another region of the same plate was reacted with anti-BB-CPK antiserum, not only the few myotubes, but also a number of mononucleated cells, were seen to fluoresce (Fig. 5i). Although many of the unfused cells reacted weakly, the intensity of the fluorescence was significantly greater than that due to nonspecific adsorption of control serum (Fig. 5h).

Before attempting to discuss these results, it is necessary to raise a technical point. The intensity of fluorescence observed in the experiments just described depends not only on the concentration of antigen-antibody complexes within the cytoplasm, but also on the thickness of the cytoplasm. Other things being equal, a tubular cell (such as a myotube or a bipolar myogenic cell in EGTA) will appear brighter than a flattened cell (such as a fibroblast or a myogenic cell in BrdUrd). This presumably explains why only a small percentage of the cells in the field shown in Fig. 5i, in which most of the cells are extremely flattened, actually can be said to show a positive fluorescent staining. In addition to differences in cell geometry, fixation artifacts (denaturation of antigen, failure of antibodies to penetrate) may also contribute to the observed differences in fluorescence intensity. Whatever the reasons, there are, in any of the fields reacted with anti-BB-CPK antiserum, and in some reacted with anti-MM-CPK antiserum as well, some cells which, although morphologically indistinguishable from the fluorescent cells, nevertheless fail to fluoresce. What this means is that we cannot claim, for example, that all bipolar cells prevented from fusing in EGTA will accumulate M-CPK. What we can say, however, is that at least some such cells do contain M-CPK.

Experiments with antisera directed against aldolases C_4 and A_4 gave results very similar to those obtained with antisera against BB-CPK and MM-CPK, respectively (Turner, et al., 1975b). I have summarized the results of the immunofluorescence experiments in Table 1.

TABLE I

Medium	Principal Cell Type	Marker Enzymes			
		B-CPK	M-CPK	aldolase C	aldolase A
Standard	Myotubes	+	+	+	+
EGTA	Bipolar mono-nucleated cells	+	+	+	+
BrdUrd	Flattened mononucleated cells	+	−	+	−

The major point that emerges from these studies is that unfused cells can accumulate marker enzymes typical of differentiated muscle. (One could ask why bipolar mononucleated cells in standard medium rarely if ever showed a positive reaction for M-CPK, while bipolar cells in EGTA often fluoresced brightly when reacted in the same way with anti-MM-CPK antiserum. The simplest explanation would be that the cells in standard medium fuse before they are able to accumulate detectable levels of the marker enzyme.) The results further suggest, when taken together with other evidence on the action of BrdUrd (Turner, et al., 1975a, b), that myogenic cells can synthesize M-CPK and aldolase A only if they have already passed through the step in myogenesis that is blocked by BrdUrd. Electrophoresis of whole-culture extracts had indicated that the CPK and aldolase isozyme transitions did not go to completion in culture (see above). The results presented here show that this is true for the individual cells as well: essentially all myotubes still contain the "embryonic" B-CPK and aldolase C along with the "adult" markers M-CPK and aldolase A.

In conclusion it is evident that the fluorescent antibody method for the detection of specific gene products, such as isozymes, has many potential applications to developmental problems. The method has particular promise as a means to study the early determinative steps in processes such as myogenesis, which,because they involve such small numbers of cells, cannot be investigated with conventional biochemical techniques.

ACKNOWLEDGEMENTS

I thank Miss Marianne Stäheli for excellent technical assistance and Drs. H. Lebherz and J.C. Perriard for reading the manuscript. I am indebted to Dr. H.M. Eppenberger for his support and advice. Supported by the Swiss National Science Foundation and by the Muscular Dystrophy Associations of America.

REFERENCES

Arnold, H., D. Nolte, and D. Pette 1969. Quantitative and histochemical studies on the desorption and readsorption of aldolase in cross-striated muscle. *J. Histochem. Cytochem.* 17:314-320 .

Coleman, J.R. and A.W. Coleman 1968. Muscle differentiation and macromolecular synthesis. *J. Cell Physiol.* 72, Sup. 1:19-34.

Dawkins, R.L. and M. Lamont 1971. Myogenesis in vitro as demonstrated by immunofluorescent staining with anti-muscle serum. *Exp. Cell Res.* 67:1-10.

Dawson, D.M., H.M. Eppenberger,and N.O. Kaplan 1965. Creatine kinase: evidence for a dimeric structure. *Biochem. Biophys. Res. Commun.* 21:346-353.

Dawson, D.M., H.M. Eppenberger, and N.O. Kaplan 1967. The comparative enzymology of creatine kinases. II. Physical and chemical properties. *J. Biol. Chem.* 242:210-217.

Eppenberger, H.M., D.M. Dawson, and N.O. Kaplan 1967. The comparative enzymology of creatine kinases. I. Isolation and characterization from chicken and rabbit tissue. *J. Biol. Chem.* 242:204-209.

Eppenberger, H.M., M. Eppenberger, R. Richterich,and H. Aebi 1964. The ontogeny of creatine kinase isozymes. *Develop. Biol.* 10:1-16.

Eppenberger, H.M., R. von Fellenberg, R. Richterich, and H. Aebi 1962. Die Ontogenese von zytoplasmatischen Enzymen beim Hühnerembryo. *Enzymol. Biol. Clin.* 2:139-174.

Eppenberger, H.M., T. Wallimann, H.J. Kuhn, and D.C. Turner 1974. Localization of creatine kinase isozymes in muscle cells: physiological significance. *II. Isozymes: Physiology and Function*, C.L. Markert, ed. Academic Press, N.Y.

Farrell, Jr., E.C., N. Baba, G.P. Brierley, and H.-D. Grumer 1972. On the creatine kinase of heart muscle mitochondria. *Laboratory Investigation* 27:209-213.

Herskovits, J.J., C.J. Masters, P.M. Wassarman, and N.O. Kaplan 1967. On the tissue specificity and biological significance of aldolase C in the chick. *Biochem. Biophys. Res. Commun.* 26:24-29.

Jacobus, W.E. and A.L. Lehninger 1973. Creatine kinase of rat heart mitochondria: coupling of creatine phosphorylation to electron transport. *J. Biol. Chem.* 248:4803-4810.

Kawahara, K. and C. Tanford 1966. The number of polypeptide chains in rabbit muscle aldolase. *Biochemistry* 5:1578-1584.

Lebherz, H.G. 1974a. On the regulation of fructose diphosphate aldolase isozyme concentrations in animal cells. *III. Isozymes: Developmental Biology*, C.L. Markert, ed., Academic Press, N.Y.

Lebherz, H.G. 1974b. Studies on the regulation of fructose diphosphate aldolase isoenzyme concentrations during chick skeletal muscle development. (In preparation).

Lebherz, H.G. and W.J. Rutter 1969. Distribution of fructose diphosphate aldolase variants in biological systems. *Biochemistry* 8:109-121.

Marquardt, R.R. 1974. Comparative physical, chemical and enzymatic properties of the isozymic forms of avian aldolases and fructose diphosphatase. *II. Isozymes: Physiology and Function*, C.L. Markert, ed. Academic Press, N.Y.

Morris, G.E. and R.J. Cole 1972. Cell fusion and differentiation in cultured chick muscle cells. *Exptl. Cell Res.* 75:191-199.

Omenn, G.S. and M.A. Hermodson 1974. Human phosphoglycerate mutase: isozyme marker for muscle differentiation and for neoplasia. *III. Isozymes: Developmental Biology*, C.L. Markert, ed., Academic Press, N.Y.

Paterson, B. and R.C. Strohman 1972. Myosin synthesis in cultures of differentiating chicken embryo skeletal muscle. *Develop. Biol.* 29:113-138.

Penhoet, E., M. Kochman, R. Valentine, and W.J. Rutter 1967. The subunit structure of mammalian fructose diphosphate aldolase. *Biochemistry* 6:2940-2949.

Penhoet, E., T. Rojkumar, and W.J. Rutter 1966. Multiple forms of fructose diphosphate aldolase in mammalian tissues. *Proc. Nat. Acad. Sci. U.S.A.* 56:1275-1282.

Stockdale, F., K. Okazaki, M. Nameroff, and H. Holtzer 1964. 5-Bromodeoxyuridine: effect on myogenesis in vitro. *Science* 146:533-535.

Taylor, E.W. 1972. Chemistry of muscle contraction. *Ann. Rev. Biochem.* 4: 577-616.

Turner, D.C., T. Wallimann, and H.M. Eppenberger 1973. A protein that binds specifically to the M-line of skeletal muscle is identified as the muscle form of creatine kinase. *Proc. Natl. Acad. Sci. U.S.A.* 70: 702-705.

Turner, D.C. and H.M. Eppenberger 1974. Developmental changes in creatine kinase and aldolase isoenzymes and their possible function in association with contractile elements. *Enzyme* 15: 224-238.

Turner, D.C., V. Maier, and H.M. Eppenberger 1974 . Creatine kinase and aldolase isoenzyme transitions in cultures of chick skeletal muscle cells. *Develop. Biol.* 36: 63-89.

Turner, D.C., R. Gmür, M. Stäheli, and H.M. Eppenberger 1975a. Differentiation in cultures derived from embryonic chicken muscle I. Comparison of the effects of EGTA and 5-bromo-deoxyuridine (in press).

Turner, D.C., R. Gmür, H.G. Lebherz, M. Stäheli, T. Wallimann, and H.M. Eppenberger 1975a. Differentiation in cultures derived from embryonic chicken muscle II. Phosphorylase histochemistry and fluorescent antibody staining for creatine kinase and aldolase (in press).

Wilson, B.W., T.A. Linkhart, C.R. Walker, and G.W. Yee 1974. Acetylcholinesterase isozymes and muscle development: Newly synthesized enzymes and cellular site of action of dystrophy of the chicken. *III. Isozymes: Developmental Biology*, C.L. Markert, editor, Academic Press, New York.

Yaffe, D. and H. Dym 1972. Gene expression during differentiation of contractile muscle fibers. *Cold Spring Harbor Symp. Quant. Biol.* 37: 543-548.

Zalin, R.J. 1973. The relationship of the level of cyclic AMP to differentiation in primary cultures of chick muscle cells. *Exptl. Cell Res.* 78: 152-158.

THE SYNTHESIS AND REGULATION
OF ASPARTATE AMINOTRANSFERASE ISOZYMES

JEROME S. NISSELBAUM AND LEVY KOPELOVICH
Laboratory of Applied and Diagnostic Biochemistry
and the Laboratory of Molecular Biology
Memorial Sloan-Kettering Cancer Center
New York, New York   10021

ABSTRACT: The rate constants of synthesis and degradation
of aspartate aminotransferase (AAT) isozymes in red blood
cells were studied during induction of and recovery from
phenylhydrazine-produced reticulocytosis. The increase
in total AAT activity during the induction of reticulocytosis was due primarily to the 47-fold increase in the
activity of the mitochondrial isozyme (K_s = 0.131; K_d =
0.015; $t_{1/2}$ = 45 hr). The activity of the cytosolic
isozyme increased to about 3 times its control value
(K_s = 0.036; K_d = 0.017; $t_{1/2}$ = 48 hr). The rate constants following recovery from phenylhydrazine treatment were: K_s = 0.005; K_d = 0.028; $t_{1/2}$ = 25 hr. for
the mitochondrial isozyme and K_s = 0.0043; K_d = 0.0044;
$t_{1/2}$ = 155 hr. for the cytosolic isozyme.
   This report also shows that glyceraldehyde-3-P and
erythrose-4-P, in low concentrations, are time-dependent
inhibitors of rat liver AAT isozymes and that inhibition
by glycolaldehyde-P is instantaneous. The mitochondrial
isozyme was more sensitive to inhibition than the cytosolic isozyme. Kinetic analysis of the types of inhibition and direct binding studies suggest the formation
of a Schiff base between an ε-amino lysyl residue and
the inhibitor at the enzymically active site. Amino
acid substrates potentiate the inhibition by converting
the isozyme to the pyridoxamine form, thereby exposing
a second ε-amino lysyl group which would react with the
inhibitor. It seems likely that the divalent phosphoryl
group on the inhibitor molecule is involved in the competition at the keto acid binding site. Our results
raise the possibility that glyceraldehyde-3-P and
erythrose-4-P may be implicated in the regulation of
gluconeogenesis as feed-forward inhibitors of AAT isozymes.

INTRODUCTION

   Changes in enzyme amount and the modulation of enzyme activity by effector molecules have long been recognized as pri-

159

mary mechanisms in the control of cell metabolism. The occurrence of isozymes that reside in different subcellular compartments provides yet another mechanism for the modulation of metabolic pathways. Aspartate aminotransferase (L-aspartate: 2-oxoglutarate aminotransferase, E.C. 2.6.1.1) is a key enzyme, shunting metabolites into several different metabolic pathways. It consists of two isozymes: a cationic, mitochondrial isozyme, and an anionic, cytosolic isozyme (Fleisher et al., 1960; Boyd, 1966). As part of a study of the regulation of aspartate aminotransferase (AAT), we demonstrated (Kopelovich et al., 1970, 1972) that glyceraldehyde 3-phosphate, erythrose 4-phosphate, and glycolaldehyde phosphate inhibit both the mitochondrial and cytosolic isozymes from rat liver. These compounds may be involved in the regulation of several metabolic pathways in vivo by modifying the activity of AAT isozymes. This short-term regulation would complement the long-term changes in the isozyme levels that occur as a result of dietary and hormonal alterations (Nakata et al., 1964; Sheid and Roth, 1965; Shrago and Lardy, 1966).

This report describes the kinetics of synthesis and degradation of AAT isozymes in rat peripheral red blood cells during the induction of and recovery from phenylhydrazine-produced reticulocytosis. The study of red blood cells during cytodifferentiation provides a unique system which, for the first time, permitted a comparison of the turnover rates of the isozymes in vivo. The report also summarizes additional work from this laboratory on the mechanism of inhibition of AAT isozymes by glyceraldehyde 3-phosphate, erythrose 4-phosphate, and glycolaldehyde phosphate. The physiological implications of the control of AAT isozymes by these compounds are discussed.

SYNTHESIS AND DEGRADATION OF ASPARTATE AMINOTRANSFERASE ISOZYMES

MATERIALS AND METHODS

Female buffalo rats (Simonsen), weighing 150-200 g, were made reticulocytotic by subcutaneous injections with phenylhydrazine on days 0, 1 and 3. At intervals, blood samples were taken by cardiac puncture (Kopelovich and Nisselbaum, 1974), and the activity of the AAT isozymes in the red blood cell (RBC) lysates was determined as described previously (Kopelovich et al., 1970) except that 0.1 M Tris buffer, pH 7.4, was used instead of barbital buffer. All enzyme units were expressed as μ moles/ml RBC/min.

For the preparation of specific antiserum, 1 mg of each purified rat liver isozyme was emulsified in 2.5 ml of complete

Freund's adjuvant (Difco) and injected subcutaneously into
the hind foot pads and neck of female (New Zealand) rabbits.
At 4, 5 and 6 weeks following the single treatment, blood was
collected, allowed to clot, and the antiserums were pooled.
Ouchterlony double-diffusion precipitin analysis showed that
each antiserum formed a single precipitin line with its homologous isozyme and did not cross-react with the heterologous
isozyme. The activity of each isozyme was measured in appropriately diluted lysates, which were incubated with each antiserum in amounts shown to result in complete inactivation of
the homologous isozyme (Kopelovich and Nisselbaum, 1974).

The rate of constants for synthesis (K_s) and degradation
(K_d) were determined as described by Segal et al. (1965) and
Schimke and Doyle (1970).

RESULTS AND DISCUSSION

Fig. 1 shows the time course of changes in the activity
of AAT isozymes and of the hematocrit and reticulocyte count
in phenylhydrazine-treated rats. The hematocrit fell to its
lowest value, about 26%, on day 3 and attained a value equal
to that found in untreated animals, about 50%, on day 7 (Fig.
1A). The reticulocyte count increased to a value of about 50%
on day 5, and attained a value of essentially 100% on day 7.
Thereafter, there was a decline in the reticulocyte count to
a value of about 3%, the level observed in untreated animals
(Fig. 1A). The activity pattern of AAT isozymes changed from
predominantly cytosolic in untreated rats to predominantly
mitochondrial on the 5th day following phenylhydrazine treatment. Total AAT activity was maximal on day 5, about 10 μ
moles/ml RBC/min, which is about 10-fold higher than the
control value (Fig. 1B). The increase in total AAT activity
on day 5 was due primarily to a 47-fold increase in the activity of the mitochondrial isozyme. The activity of the cytosolic isozyme became maximal on the 8th day about 3 times its
control value (Fig. 1B). This was followed by a decline of
the activities of both isozymes from their respective maxima
to control levels.

On the assumption that reticulocytes in untreated rats
contained the same amount of mitochondrial isozyme as that
present on the 5th day after initiation of phenylhydrazine
treatment, we inferred that the value of 0.18 μ moles/ml
RBC/min for the mitochondrial isozyme in untreated rats (Fig.
1) was due to the presence of 2.5% reticulocytes. This value
is in good agreement with the microscopic determination of
reticulocytes present in untreated rats (Fig. 1A). Mature
erythrocytes, therefore, unlike all other tissues thus far

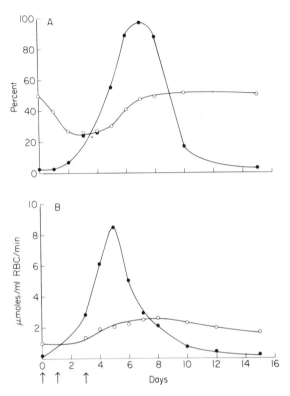

Fig. 1. Effect of phenylhydrazine administration in rats.
(A) Hematocrit open circles; reticulocyte count, closed
circles. (B) Activity of AAT cytosolic isozyme, open circles;
activity of mitochondrial isozyme, closed circles. Arrows
denote days on which phenylhydrazine was administered.

investigated contain only the cytosolic isozyme. It should be
noted that the isozymes in the RBC lysates from normal and
phenylhydrazine-treated rats reacted identically with their
homologous antiserums. Thus, the newly synthesized isozymes
were indistinguishable from the isozymes in rat RBC prior to
treatment.

The rate constants for synthesis (K_s) and degradation (K_d)
of both isozymes in untreated rats and rats treated with
phenylhydrazine were determined and are summarized in Table 1.
It is apparent that the rates of synthesis of both isozymes
were similar during recovery from phenylhydrazine-produced
reticulocytosis, but that the rate of degradation of the mitochondrial isozyme was about 7 times greater than that for the
cytosolic isozyme. Consequent to treatment with phenylhydrazine

TABLE 1

THE RATES OF SYNTHESIS AND DEGRADATION OF AAT ISOZYMES IN RAT
RBC DURING INDUCTION OF AND RECOVERY FROM PHENYLHYDRAZINE-
PRODUCED RETICULOCYTOSIS

Isozyme	Status	E_o or E_o'[a] Units	K_d[b] hr^{-1}	$t_{1/2}$[c] hr	Calculated K_s Units/hr
Mitochondrial	R^d	0.18	0.028	25	0.0050
"	I^e	8.50	0.015	45	0.131
Cytosolic	R^d	0.96	0.0044	155	0.0043
"	I^e	2.53	0.017	48	0.036

[a]Basal level of isozyme, E_o, or peak value E_o'

[b]Experimental values derived from plots of ln $(E-E_o)$ or ln $(E_o'-E)$ against time

[c]$t_{1/2} = \dfrac{0.693}{K_d}$

[d]R: Values calculated from enzyme levels during the recovery
period after discontinuation of phenylhydrazine treatment,
$K_s = K_d(E_o)$ (Kopelovich and Nisselbaum, 1974).

[e]I: Values calculated from enzyme levels during the period of
induction of reticulocytosis, $K_s = K_d(E_o')$.

K_d for the cytosolic isozyme increased about 3-fold, while K_d
for the mitochondrial isozyme declined to about half the value
that was found after discontinuation of treatment. At the
same time K_s for the mitochondrial isozyme increased about
26-fold and that for the cytosolic isozyme increased about
8-fold. Thus, the 47-fold increase in the activity of the
mitochondrial isozyme during phenylhydrazine treatment was the
result of both an increase in the rate of synthesis and a
decline in the rate of degradation. The 3-fold increase in
the activity of the cytosolic isozyme was the result of a
smaller increase in its rate of synthesis concomitant with a
moderate increase in its rate of degradation. The changes in
turnover rates of the isozymes could be due either to alter-
tions in stability of the protein molecules or to changes in
the activities of degradative enzymes.

REGULATION OF ASPARTATE AMINOTRANSFERASE ISOZYMES BY
GLYCERALDEHYDE 3-PHOSPHATE AND ITS HOMOLOGUES
ERYTHROSE 4-PHOSPHATE AND GLYCOLALDEHYDE PHOSPHATE

MATERIALS AND METHODS

^{14}C-labeled D-glyceraldehyde-3-P (specific activity of
0.34 Ci/mole) was prepared from labeled ^{14}C-(U)D-fructose-6-P
(New England Nuclear) by periodate oxidation according to the
procedure of Szewczuk et al., (1961) as described elsewhere
(Kopelovich et al., 1970). All materials were of the highest
purity commercially available. AAT isozymes were purified
from rat liver essentially as described by Nisselbaum and
Bodansky (1969). Cross contamination of the isozyme preparations was less than 0.1% as determined by starch gel electrophoresis. The activity of AAT isozymes in the presence of
aspartate and 2-oxoglutarate was measured by a modification
(Kopelovich et al., 1970) of the coupled reaction described
by Karmen (1955). The activity of AAT isozymes in the presence
of glutamate and oxaloacetate was measured in a coupled assay
with 2-oxoglutarate dehydrogenase (2-oxoglutarate:lipoate
oxidoreductase)(acceptor-acylating)(E.C.1.2.4.2.)(Nisselbaum
et al., 1971). Kinetic analysis of AAT activity in the
coupled assay with 2-oxoglutarate dehydrogenase showed a
family of parallel lines consistent with a Ping Pong Bi Bi
type mechanism on double reciprocal plots at varying concentrations of the alternate substrate. Data were processed as
described previously using programs for the Olivetti-Underwood
Programma 101 (Kopelovich et al., 1971).

RESULTS AND DISCUSSION

Table 2 shows the effect of various glycolytic intermediates and related compounds on the activity of AAT isozymes.
Of the compounds tested only glyceraldehyde-3-P, erythrose-4-
P, and glycolaldehyde-P substantially inhibited both isozymes.
The mitochondrial isozyme was more sensitive than the cytosolic isozyme to inhibition by these compounds. The D- and L-
isomers of glycerladehyde-3-P were equally effective inhibitors. This indicates that there is no stereo specific
requirement with regard to the second carbon of glycerladehyde-
3-P. The data also suggest that the conjoint presence of a
phosphate group and a free aldehyde are necessary for inhibition. Those compounds in which the aldehyde group exists
largely as the internal hemiacetal were ineffective (Kopelovich
et al., 1970).

The extent of inhibition of both isozymes was time

dependent in the case of glyceraldehyde-3-P and erythrose-4-P, but was instantaneous in the case of glycolaldehyde-P (Kopelovich et al., 1972). Maximal inhibition by glyceralde-hyde-3-P and erythrose-4-P was attained within 30 min. The presence of aspartate or glutamate during preincubation of the isozymes with inhibitor increased the extent of inhibition. In contrast, 2-oxoglutarate or oxaloacetate protected both isozymes against inhibition.

The addition of keto acid substrates to isozymes that had been preincubated with glyceraldehyde-3-P or erythrose-4-P resulted in a time-dependent release of inhibition. Addition of amino acid substrates resulted in a further time-dependent loss of activity (Fig. 2). In the case of the cytosolic iso-zyme in the presence of glyceraldehyde-3-P there appeared to be a second, irreversible phase of inhibition that was not affected by the presence of substrate. This second phase of inhibition was not apparent with erythrose-4-P (Kopelovich, et al., 1970, 1972). Turano et al., (1964) have noted that the extent of interaction of sulfhydryl groups in cytosolic AAT with p-chloromercuribenzoate parallels the degree of inhibition. The possibility exists that the second phase of inhibition of the cytosolic isozyme by glyceraldehyde-3-P involved such interaction, and that this reaction may play a role in the regulation in vivo of this isozyme.

The inhibition of both isozymes could be reversed sub-stantially by dialysis for 3 hr at 37° after they had been initially incubated for 30 min with glyceraldehyde-3-P. For the cytosolic isozyme, dialysis against buffer or buffer that contained 2-oxoglutarate, aspartate, or pyridoxal-5-P was equally effective. The second phase of inhibition of the cytosolic isozyme was not reversible by dialysis. In the case of the mitochondrial isozyme, dialysis against buffer or buffer that contained pyridoxal-5-P or aspartate resulted in only a slight increase in activity, whereas dialysis against buffer that contained 2-oxoglutarate gave almost complete recovery of enzyme activity. It is unlikely that pyridoxal-5-P was dissociated from either isozyme as a result of inhi-bition by glyceraldehyde-3-P since dialysis against pyridoxal-5-P was no more effective in restoring activity than was buffer alone (Kopelovich et al., 1974). The multipoint attachment of pyridoxal-5-P to AAT and its significance in catalysis have been discussed by Ivanov and Karpeisky (1969).

At pH 5.4 where about 95% of glyceraldehyde-3-P and its homologues exist as a monovalent ion, there was no inhibition of either isozyme (Fig. 3). Half-maximal inhibition of each isozyme occurred between pH 6.5 and 6.7, suggesting that the divalent anion of glyceraldehyde-3-P, pK=6.45 (Bergmeyer, 1965) and its homologues was the inhibitory species. As the pH was

TABLE 2

EFFECT OF GLYCOLYTIC INTERMEDIATES AND RELATED COMPOUNDS ON
AAT ISOZYME ACTIVITY

Compound	Concentration in Preliminary Incubation Mixture	Inhibition of Mitochondrial Isozyme	Inhibition of Cytosolic Isozyme
	mM	%	%
D-Glucose-1-P	2.0	1	1
D-Glucose-6-P	2.0	2	0
D-Fructose-1-P	2.0	0	1
D-Fructose-6-P	2.0	-1	1
D-Fructose-1, 6-P	2.0	2	0
DL-Glycerol-1-P	4.0	0	1
Dihydroxyacetone-P[a]	2.0	12	3
D-Glyceraldehyde	4.0	5	0
DL-Glyceraldehyde-3-P	1.0	81	31
D-3-Phosphoglycerate	2.0	0	1
2-Phosphoglycerate	2.0	0	0
Phosphoenolpyruvate	2.0	1	1
Erythrose-4-P	3.0	-	25
"	1.5	55	-
Glycolaldehyde-P	4.0	36	-

Isozymes were diluted in 2.5 ml of 0.04 M barbital-HCl, pH
7.4, that contained 0.15% human serum albumin and incubated
for 30 min at 37° with the indicated concentration of each
compound. Reactions were initiated with 0.5 ml of complete
substrate mixture.

[a]Enzymic analysis showed this preparation to contain 0.24%
 D-glyceraldehyde 3-phosphate, corresponding to 0.005 mM.

increased above 7.4 where glyceraldehyde-3-P exists completely
as the divalent anion, maximal inhibition of the cytosolic
isozyme was attained at approximately pH 8.4 and remained
constant up to pH 10.3. The inhibition of the mitochondrial
isozyme was optimal at pH 7.4 and decreased substantially at
higher pH values. Above pH 7.4 there is decreased ionization
of the positively charged groups that are responsible for
the electrophoretic behavior of the mitochondrial isozyme,
as indicated by the finding that this isozyme which is a cation
near neutrality has no net charge at pH 8.8 (Schwartz et al.,
1963). It would appear, therefore, that the change in net
charge at high pH values renders the mitochondrial isozyme
less sensitive to inhibition by glyceraldehyde-3-P. It
should be noted that the lower levels of inhibition at pH

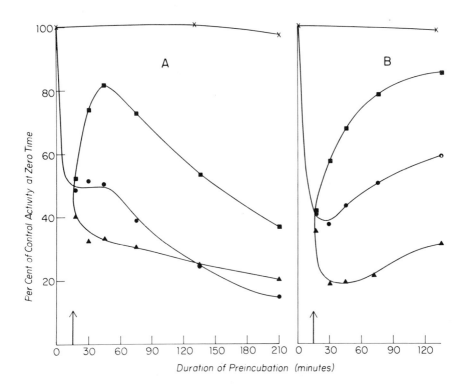

Fig. 2. Effect of time on inhibition of AAT isozymes by glyceraldehyde-3-P after addition of substrates. Cytosolic isozyme (A) was incubated at 37° with 3.6 mM, and mitochondrial isozyme (B) with 0.36 mM glyceraldehyde-3-P for 15 min. Each mixture was then divided into three portions, either buffer or substrate was added, and the incubation was continued. Control enzyme activity (X), activity with glyceraldehyde-3-P and buffer (●), 16.7 mM aspartate (▲), or 6.7 mM 2-oxoglutarate (■). All activities are expressed as percentage of the control activity at zero time.

values other than the optima were not due to alterations in the rates of interaction between the isozymes and inhibitors since maximal extent of inhibition was attained within 30 min at pH 6.4 and 9.2 as well as at pH 7.4.

The kinetics of inhibition of both AAT isozymes by each of the inhibitors was determined from double reciprocal plots of experiments in which the isozyme was incubated for 30 min at 37° in the presence of varying concentrations of inhibitor and one substrate. The reactions were initiated by addition

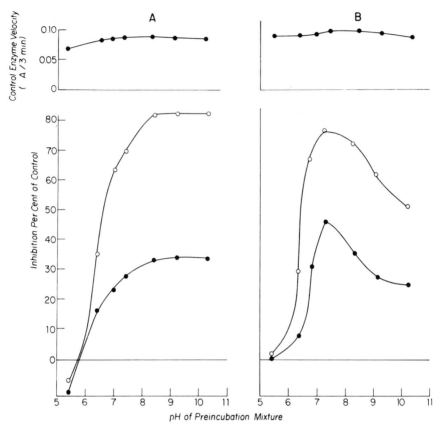

Fig. 3.   Effect of pH on inhibition of AAT isozymes by
glyceraldehyde-3-P.   Each isozyme was incubated with inhibitor
for 30 min at 37° at the pH values indicated.   The samples
were adjusted to pH 7.4 and substrate mixture was added to
start the reaction.   Controls incubated without inhibitor
(top curves) show that the isozymes were stable over the pH
range studied.   (A) Cytosolic isozyme with 4.0 mM (o) and
0.4 mM (●) glyceraldehyde-3-P.   (B) Mitochondrial isozyme
with 0.6 mM (o) and 0.16 mM (●) glyceraldehyde-3-P.   At each
pH value, percentage inhibition was calculated from the
controls.

of the alternate substrate, the appropriate coupling enzyme,
and cofactors.   The assumptions and the equations that were
used in processing the data and generating the kinetic
constants were delineated previously (Kopelovich et al., 1971).
Table 3 summarizes the types of inhibition and the inhibitor
dissociation constant for the isozymes with the three inhibi-

168

TABLE 3

INHIBITOR DISSOCIATION CONSTANTS AND TYPES OF INHIBITION OF AAT ISOZYMES BY GLYCERALDEHYDE-3-P, ERYTHROSE-4-P AND GLYCOLALDEHYDE-P

Variable Substrate	Inhibitor	Mitochondrial Isozyme K_i (mM)	Type of Inhibition	Cytosolic Isozyme K_i (mM)	Type of Inhibition
Oxaloacetate	Glyceraldehyde-3-P	---	Completely Competitive	1.5	Mixed Partially
2-Oxoglutarate	"	0.98	Competitive	1.9	Competitive Partially Noncompetitive
"	Erythrose-4-P	1.44	Completely Competitive	3.07	Completely Competitive
"	Glycolaldehyde-P	1.14	Completely Competitive	0.70	Completely Competitive
Glutamate	Glyceraldehyde-3-P	0.11	Completely Noncompetitive	0.57	Partially Noncompetitive
Aspartate	"	0.084	Noncompetitive	0.39	Noncompetitive
"	Erythrose-4-P	0.135	Completely Noncompetitive	0.33	Completely Noncompetitive
"	Glycolaldehyde-P	1.00	Completely Uncompetitive	0.88	Completely Uncompetitive

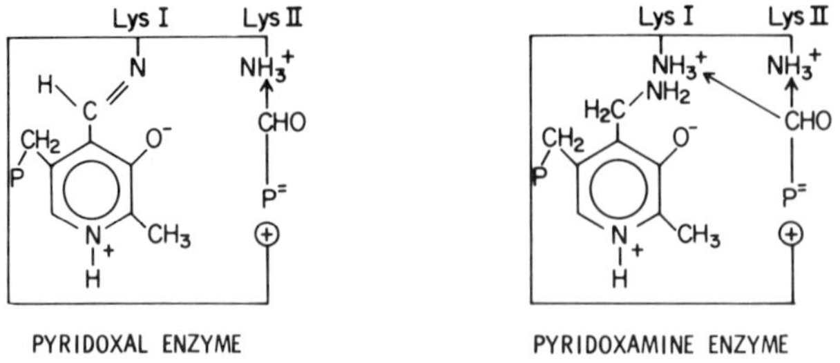

PYRIDOXAL ENZYME   PYRIDOXAMINE ENZYME

Fig. 4. Proposed mechanism for inhibition of AAT isozymes by glyceraldehyde-3-P, erythrose-4-P and glycolaldehyde-P. For the pyridoxamine enzyme the two arrows denote two molecules of inhibitors in the active center.

tors. The K_i values show that the pyridoxamine forms (incubation in the presence of an amino acid substrate) of both isozymes are more sensitive to inhibition by glyceraldehyde-3-P and erythrose-4-P than the pyridoxal forms (incubation in the presence of a keto acid substrate), and that the mitochondrial isozyme is more sensitive to these inhibitors than the cytosolic isozyme. Glycolaldehyde-P was equally effective with both forms of the isozymes.

The similarity between the types of inhibition obtained with both substrate pairs indicates that the keto acids bind to the isozymes at the same site and compete with glyceraldehyde-3-P, erythrose-4-P and glycolaldehyde-P for that site. It has been suggested (Ivanov and Karpeisky, 1969) that the α-carboxyl groups of all AAT substrates bind to a common site on the enzyme, but that the sites of binding of the terminal carboxyl group of C_4 and C_5 substrates may be different. The non-competitive nature of the inhibition in the presence of either amino acid substrate suggests that at least one point of attachment of the amino acids is outside of the site to which the keto acids and inhibitor bind. It seems likely that the divalent phosphoryl group on the inhibitor molecule is involved in the competition at the keto acid binding site.

Our proposed mechanism for the inhibition of AAT isozymes is summarized in Fig. 4. The phosphorylated aldehyde binds initially to the enzyme via the divalent phosphoryl group at a site to which the keto acid substrate binds and competes with with the keto acid for that site. The aldehyde group forms a Schiff base with the ε-amino group of a lysine at the active center of the enzyme. Amino acid substrates potentiate the inhibition by opening the internal aldimine linkage with pyridoxal-5-P, thereby exposing an ε-amino group of another

lysine. This would provide a second site for a Schiff base
formation with the inhibitor. It is unlikely that the
inhibitor, in the absence of amino acid substrates, can open
the internal aldimine linkage, since we were unable to detect
the spectral changes that are characteristic for this reaction
(Jenkins and Sizer, 1959).

In order to test the proposal that glyceraldehyde-3-P
forms a Schiff base with the enzyme, each isozyme was resolved
into its respective apoisozyme, in which both of the ε-amino
lysyl groups in the active center are free. Treatment of the
apoisozymes with sodium borohydride in the presence of glycer-
aldehyde-3-P essentially abolished the ability to restore
activity by addition of pyridoxal-5-P (Table 4). In contrast,
the addition of pyridoxal-5-P to apoisozymes that had been
treated with either inhibitor alone or sodium borohydride
alone resulted in substantial reactivation. Similar results
were obtained with glycolaldehyde-P.

TABLE 4

RECOVERY OF AAT ACTIVITY OF TREATED APOISOZYMES UPON INCUBATION
WITH PYRIDOXAL-5-P

| Treatment | Apoisozyme | |
	Cytosolic	Mitochondrial
	%	%
Control	100	100
Na BH$_4$ Reduction	49 ± 6	64 ± 5
Glyceraldehyde-3-P	75 ± 11	64 ± 8
Glyceraldehyde-3-P followed by Na BH$_4$	13 ± 1	15 ± 3

The reaction of ^{14}C-glyceraldehyde-3-P at the active center
of the holoisozyme was demonstrated in the presence of aspartate and in the presence of 2-oxoglutarate. Following treatment with sodium borohydride each preparation was denatured
by treatment with 10 M urea and carboxymethylated with iodo-
acetate. Chymotryptic digests of these samples were subjected
to peptide mapping by chromatography, butanol:acetic acid:
water, (4:1:5) and electrophoresis in pyridine:acetic acid:
water, (1:10:289) pH 3.6. Autoradiography of the peptide
maps showed that ^{14}C-glyceraldehyde-3-P and coenzyme, detected
by its fluorescence under untraviolet light, were bound to
the same peptide. Assuming two active sites per enzyme
molecule and that our proposed scheme is correct, the theoretical values for the binding of ^{14}C-glyceraldehyde-3-P would
be 4 and 2 moles per mole of enzyme in the presence of

aspartate and 2-oxogultarate respectively. The experimental values were 3.2 and 1.3 moles per mole of isozyme.

METABOLIC SIGNIFICANCE

The present studies have shown that both isozymes of AAT are substantially inhibited by glyceraldehyde-3-P, erythrose-4-P, and glycolaldehyde-P. The question arises as to whether glyceraldehyde-3-P concentrations in tissues can reach levels that may affect the activity of AAT isozymes in vivo. Trentham et al., (1969) have reported that glyceraldehyde-3-P exists as the geminal diol and the free aldehyde in a molar ratio of 29:1 in aqueous solution. They have also stated that only the free aldehyde form of glyceraldehyde-3-P is produced and utilized in vivo. Thus, under transient conditions, the concentration of the free aldehyde may be considerably higher than would be predicted from the equilibrium constant. Our K_i values for inhibition of AAT are related to the total concentration of both species of glyceraldehyde-3-P in aqueous solution. The K_i values for the free aldehyde form of the inhibitor would, therefore, be 1/30 of the reported values. This would be in the range of the in vivo concentration of glyceraldehyde-3-P.

The reported values for glyceraldehyde-3-P in normal rat liver range around 6 µ M (Rawat, 1968; Veech et al., 1970; Woods et al., 1970). Rat livers perfused with fructose contain up to 20 µ M glyceraldehyde-3-P (Woods et al., 1970). In red blood cells that were incubated at pH 8.0 in a medium that contained 20 mM phosphate, the glyceraldehyde-3-P concentration was 257 nmoles/ml of cells (Minakami et al., 1965). It has also been shown that the endogenous level of glyceraldehyde-3-P in Krebs ascites cells was 107 nmoles/g and rose to 137 nmoles/g when cells were incubated with 12.5 mM glucose (Gumaa and McLean, 1969). These data suggest that, under certain metabolic conditions, the concentration of glyceraldehyde-3-P is high enough to inhibit both isozymes of AAT. Erythrose-4-P and glycolaldehyde-P concentrations in mammalian tissues have not been reported.

Under conditions of energy excess when glycolosis and the hexose monophosphate shunt are highly active glyceraldehyde-3-P and erythrose-P could function as feed-forward inhibitors of AAT. The mitochondrial isozyme is considerably more sensitive to inhibition than the cytosolic isozyme. This would impede the diversion of oxaloacetate from the tricarboxylic acid cycle. Production of oxaloacetate from amino acids in the cytosol would also be impeded. The degree to which each isozyme might be inhibited by glyceraldehyde-3-P, erythrose-4-P and glycolaldehyde-P will depend on the tissue concentration of the keto acid and amino acid substrates relative to their K_m values, and the distribution of inhibitors and substrates of AAT between the mitochondria and the cytosol. Glyceraldehyde-3-P, in low concentrations, has also been found to be a time dependent inhibitor of malate dehydrogenase (E.C.1.1.1.37) (Kopelovich et al., 1972). It would appear, therefore, that these inhibitors may be implicated in the regulation of gluconeogenesis by modifying the activity of aspartate aminotransferase and malate dehydrogenase.

ACKNOWLEDGEMENT

This work was supported in part by grants from the National Cancer Institute (CA-08748) and The American Cancer Society (T-431L).

REFERENCES

Bergmeyer, H. V. (Editor) 1965. *Methods of Enzymatic Analysis*. Academic Press, New York, p. 1019.

Boyd, J. W. 1965. The extraction and purification of two iso-enzymes of L-aspartate: 2-oxoglutarate aminotransferase. *Biochim. Biophys. Acta.* 113: 302-311.

Fleisher, G. A., C. S. Potter, and K. G. Wakim 1960. Separation of 2 glutamic-oxaloacetic transaminases by paper electro-phoresis. *Proc. Soc. Exptl. Biol. Med.* 103: 229-231.

Gumaa, K. A. and P. McLean 1969. The pentose phosphate pathway of glucose metabolism. Enzyme profiles and transient and steady-state content of intermediates of alternative pathway of glucose metabolism in Krebs ascites cells. *Biochem. J.* 115: 1009-1029.

Ivanov, I. V. and M. Y. Karpeisky 1969. Dynamic three-dimensional model for enzymic transamination. *Advan. in Enzymol.* 32: 21-53.

Jenkins, W. T. and I. Sizer 1959. Glutamic aspartic transaminase II. The influence of pH on absorption spectrum and enzymatic activity. *J. Biol. Chem.* 234: 1179-1181.

Karmen, A. 1955. A note on the spectrophotometric assay of glutamic-oxaloacetic transaminase in human blood serum. *J. Clin. Investig.* 34: 131-133.

Kopelovich, L. and J. S. Nisselbaum 1974. The kinetics of synthesis and degradation of aspartate aminotransferase isozymes in rat peripheral blood cells during cytodiffer-entiation. *Proc. Soc. Exptl. Biol. Med.* 145: 504-507.

Kopelovich, L., L. Sweetman, and J. S. Nisselbaum 1970. Time-dependent inhibition of aspartate aminotransferase isozymes by DL-Glyceraldehyde 3-phosphate. *J. Biol. Chem.*

245: 2011-2017.

Kopelvich, L., L. Sweetman, and J. S. Nisselbaum 1971. Kinetics of inhibition of aspartate aminotransferase isozymes by DL-glyceraldehyde 3-phosphate. *Eur. J. Biochem.* 20: 351-362.

Kopelvich, L., L. Sweetman, and J. S. Nisselbaum 1972. Regulation of aspartate aminotransferase isozymes by D-Erythrose 4-Phosphate and glycoloaldehyde phosphate, the naturally occurring homologues of D-glyceraldehyde 3-phosphate. *J. Biol. Chem.* 247: 3262-3268.

Nakata, Y., T. Suematsu, K. Nakata, K. Matsumoto, and Y. Somomoto 1964. Activities of various aminotransferases in tumor-bearing rats. *Cancer Res.* 24: 1689-1699.

Nisselbaum, J. S. and O. Bodansky 1969. Quantitative immunochemical determination of the isozymes of aspartate aminotransferase in rat livers and transplantable rat hepatomas. *Cancer Res.* 29: 360-365.

Nisselbaum, J. S., L. Sweetman, and L. Kopelovich 1971. Effects of oxaloacetate and L-glutamate on glyceraldehyde 3-phosphate inhibition of aspartate aminotransferase isozymes as measured by a 2-oxoglutarate dehydrogenase coupled assay. *Eur. J. Biochem.* 23: 314-320.

Schwartz, M. K., J. S. Nisselbaum, and O. Bodansky 1963. Procedure for staining zones of activity of glutamic oxaloacetic transaminase following electrophoresis in starch gels. *Am. J. Clin. Pathol.* 40: 103-106.

Segal, H. L., Y. S. Kim, and S. Hooper 1965. Glucocorticoid control of rat liver glutamic-alanine transaminase biosynthesis. *Advan. Enzyme Reg.* 3: 29-42.

Sheid, B. and J. S. Roth 1965. Some effects of hormones and L-aspartate on the activity and distribution of aspartate aminotransferase on rat liver. *Advan. Enzyme Reg.* 3: 335-350.

Shimke, R. T. and D. Doyle 1972. Control of enzyme levels in animal tissues. *Ann. Rev. Biochem.* 39: 929-976.

Shrago, E. and H. A. Lardy 1966. Paths of carbon in gluconeogenesis and lipogenesis. II. Conversion of precursors to phosphoenolpyruvate in liver cytosol. *J. Biol. Chem.* 241: 663-668.

Szewczuk, A., E. Wolny, M. Wolny, and T. Baranowski 1961. A new method for the preparation of D-glyceraldehyde 3-phosphate. *Acta Biochim. Pol.* 8: 201-208.

Trentham, D. R., C. H. McMurray, and C. I. Pogson 1969. The active chemical state of D-glyceraldehyde 3-phosphate in its reaction with D-glyceraldehyde 3-phosphate dehydrogenase, aldolase, and triose phosphate isomerase. *Biochem. J.* 114: 19-24.

Turano, C., A. Giartosio, and P. Fasella 1964. Sulfhydryl groups and coenzyme binding in aspartic aminotransferase. *Arch. Biochem. Biophys.* 104: 524-526.

Veech, R. L., L. Raijman, and H. A. Krebs 1970. Equilibrium relations between the cytoplasmic adenine nucleotide system and nicotinamide-adenine nucleotide system in rat liver. *Biochem. J.* 117: 499-505.

Woods, H. F., L. V. Eggleston, and H. A. Krebs 1970. The cause of hepatic accumulation of fructose 1-phosphate on fructose loading. *Biochem. J.* 119: 501-510.

STAGE SPECIFIC ISOZYMES OF *DICTYOSTELIUM DISCOIDEUM*

WILLIAM F. LOOMIS
Department of Biology
University of California, San Diego
La Jolla, California 92037

ABSTRACT. A series of enzymes have been found to accumulate during discrete stages of the multicellular development of the cellular slime mold, *Dictyostelium discoideum*. Isozymes of several of these enzymes have been separated and analyzed. The developmental kinetics of accumulation of the isozymes of acetylglucosaminidase, threonine deaminase, β-glucosidase, and alkaline phosphate are described and related to their probable physiological functions. The pattern of accumulation of the stage specific enzymes was determined in a series of morphological mutant strains blocked at various points in the developmental pathway. A linear dependent sequence of 8 steps and a single independent step could be delineated by the mutant strains. The order of the steps does not correspond in all cases with the temporal order of the stage specific enzymes. Thus, accumulation of several enzymes appears to be a delayed response to earlier events. In at least three cases, the accumulation of a specific enzyme is not required for subsequent biochemical differentiation and so does not appear to be a component of the dependent pathway.

INTRODUCTION

Development in multicellular organisms has long been recognized to proceed through a temporal progression of discrete stages. Most often these have been characterized extensively only at the morphological level. In order to be able to resolve the molecular events which underlie and direct morphogenesis, investigations have been initiated in several systems to characterize stage specific proteins and determine the relationships of these proteins to the developmental program. Recent work in *Dictyostelium discoideum* has focused on some of the enzymes which accumulate during development (Table 1). These enzymes increase in specific activity several fold during discrete stages, apparently as a result of *de novo* synthesis. They are not synthesized in morphological mutants blocked in the early stages and many of them have been shown to be subject to the general temporal derangement which occurs in a "fast" mutant and in a "slow" mutant (Loomis, 1970b). By these criteria, the enzymes are referred to as developmentally regulated proteins.

177

TABLE 1

STAGE SPECIFIC ENZYMES

1.	Leucine amino peptidase	Firtel and Brackenbury 1972
2.	Alanine transaminase	Firtel and Brackenbury 1972
3.	N-acetylglucosaminidase	Loomis 1969a
4.	α-mannosidase	Loomis 1970a
5.	Trehalose-phosphate synthetase	Roth and Sussman 1968
6.	Threonine deaminase-2	Pong and Loomis 1973
7.	Tyrosine transaminase	Pong and Loomis 1971
8.	UDPG pyrophosphorylase	Ashworth and Sussman 1967
9.	UDP galactose poly-saccharide transferase	Sussman and Osborn 1964
10.	UDPG epimerase	Telser and Sussman 1971
11.	Glycogen phosphorylase	Firtel and Bonner 1970; Jones and Wright 1970
12.	Alkaline phosphatase	Loomis 1969c
13.	β-glucosidase-2	Coston and Loomis 1969

So as to focus on unique proteins, it has been essential to recognize isozymes which occur among the stage specific enzymes of this organism. In several cases, the various characteristics of the isolated isozymes have given us valuable hints as to the physiological role of the enzymes. In my laboratory we have concentrated on four sets of isozymes which change during development. The kinetics of accumulation of these isozymal pairs are shown in Fig. 1. A schematic representation of morphogenesis in *D. discoideum* is shown with the time scale.

RESULTS AND DISCUSSION

ACETYLGLUCOSAMINIDASE

Two discrete enzymes catalyze the hydrolysis of nitrophenyl acetylglucosaminidase (Loomis, 1969a, Dimond and Loomis, 1974). The major isozyme accumulates during aggregation and accounts for about 99% of the total activity. The presence of the minor isozyme was only recognized after we isolated a series of mutations inactivating the major enzyme (Dimond, Brenner and Loomis 1973). In these mutant strains the major enzyme was undetectable, while the minor isozyme remained active (Dimond and Loomis 1974). The minor isozyme, acetylglucosaminidase-1, was purified from a mutant strain by chromatography on hydroxylapatite and found to differ from the major isozyme in pH optimum, being maximally active at pH 3.5 as compared to the

178

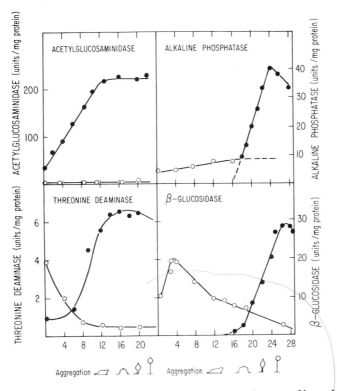

Fig. 1. Stage Specific Isozymes. Amoebae of *D. discoideum* NC-4 were grown in association with *Klebsiella aerogenes*, washed free of bacteria and allowed to develop at 22° on Millipore filters supported on pads soaked with buffered salt solution (Sussman 1966). At various times samples were taken and assayed for acetylglucosaminidase, threonine deaminase, alkaline phosphatase and β-glucosidase by the methods of Loomis 1969a, Pong and Loomis 1973, Loomis 1969b, and Coston and Loomis 1969, respectively. The minor isozyme of acetylglucosaminidase (o———o) was determined in a mutant strain (DBL 211) in which the major isozyme is inactive (Dimond and Loomis 1974). The early isozyme of threonine deaminase (o———o) was estimated as the activity sensitive to feed-back inhibition by 10^{-2}M L-leucine (Pong and Loomis 1973). The isozymes of β-glucosidase were assayed after separation by gel electrophoresis (Coston and Loomis 1969). A schematic representation of morphogenesis is given with the time scale.

pH optimum of 5 for the major isozyme (Dimond and Loomis 1974). Once the characteristics of the minor isozyme were recognized we were able to purify it from wild type cells as well, where it accounts for about 1% of the maximal activity. Both the major and minor isozymes were purified more than a thousand-fold by sequential column chromatography on Sephadex G-100, G-200, and hydroxylapatite. The most highly purified preparation of the minor isozyme contained a small proportion of the major isozyme but showed strikingly different thermostability when compared to purified preparations of the major isozyme (Fig. 2). The specific activity, thermostability, and substrate

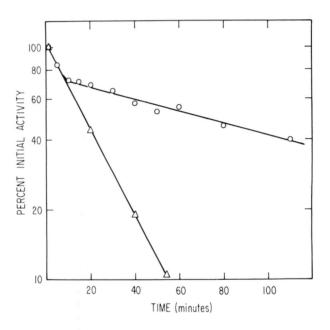

Fig. 2. Thermal inactivation of purified acetylglucosaminidase. The isozymes of strain A3 were purified over 1500 fold by gel filtration and chromatography on hydroxylapatite (Dimond and Loomis 1974). After incubation at 55° in 0.1 M sodium acetate buffer pH 5, the isozymes were cooled and assayed at 18°.

affinity (Km) of the minor isozyme was found to be unaffected in mutant strains in which the major isozyme was altered in these properties (Dimond and Loomis 1974). Thus, it is likely that the two isozymes are coded for by distinct structural genes and serve distinct physiological functions.

Although the minor isozyme makes up little of the total activity on the substrate, p-nitrophenyl-N-acetylglucosaminide, when assayed with bacterial murein as substrate it is far more active than the other isozyme (Dimond and Loomis 1974). In

fact the minor isozyme makes up the majority of the activity on
this substrate in cells growing on bacteria. We suggest that
the minor isozyme functions during growth on bacteria, while the
the major isozyme may function during development. Several
lines of evidence support this suggestion. Mutant strains
lacking the major isozyme grow as well as the wild type on bacteria and leave little or no murein undigested. Thus this isozyme does not appear to be required for bacterial assimilation.
Furthermore since the enzyme accumulates only after removal of
bacteria, the developmental kinetics indicate the major isozyme
plays a role in differentiation rather than growth. We also
have direct evidence that the major isozyme of acetylglucosaminidase is essential for maintenance of normal pseudoplasmodial
size during migration. Each of the five independent mutant
strains lacking the enzyme and four strains which synthesize
thermolabile enzyme are unable to form normal pseudoplasmodia
under non-permissive conditions (Dimond, Brenner and Loomis
1973). We are now studying the role of the major enzyme on
the components critical to migration.

THREONINE DEAMINASE

The second pair of isozymes we have studied catalyzes the
deamination of L-threonine (Pong and Loomis 1973). This reaction is the first step in the catabolism of threonine and is
also the first step in the biosynthesis of isoleucine. Separate
enzymes catalyze the reaction in the two pathways in all organisms studied to date. We have been able to isolate two distinct
isozymes of threonine deaminase from *Dictyostelium discoideum*
by gel filtration. Threonine deaminase-1 is associated with a
particle excluded by agarose 6B while threonine-deaminase-2
elutes as a molecule of about 120,000 daltons. The first isozyme can be dissociated from the particle by high salt treatment after which it sediments at 8S. After partial purification, the isozymes differ in thermostability and thus appear
to be distinct enzymes.

Threonine deaminase-1 of *D. discoideum* has been found to
be sensitive to feed-back inhibition by L-isoleucine and L-
leucine. This is a characteristic held in common by all biosynthetic threonine deaminases. Thus we surmise that threonine deaminase-1 plays a role in isoleucine biosynthesis.

Threonine deaminase-2, on the other hand, is insensitive
to feed-back inhibition by L-isoleucine or L-leucine and thus
appears to be a catabolic enzyme. Since development of *D.
discoideum* involves no net increase in protein and is fueled
in part by amino acid breakdown, we would expect the catabolic
function to predominate during development. In fact, we have

been able to show that developing cells contain almost exclusively the catabolic threonine deaminase, while growing cells
synthesize the biosynthetic enzyme (Fig. 1). The accumulation
of the catabolic enzyme is consistent with a role of this
enzyme in turnover of amino acids during development.

β-GLUCOSIDASE

The isozymes of β-glucosidase can be separated by electrophoresis on polyacrylamide gels   (Coston and Loomis 1969).
β-glucosidase-1 migrates at an R_f of 0.4 and has been shown to
be a molecule sedimenting at 8S while β-glucosidase-2 migrates
electrophoretically at an R_f of 0.3 and sediments as a molecule
of 10S. The Km of β-glucosidase-1 for p-nitrophenyl β-glucoside
is 1.28 x 10^{-3}M while the Km for β-glucosidase-2 is 0.83 x
10^{-3}M for this substrate. Growing cells contain exclusively
β-glucosidase-1 but do not continue to synthesize the isozyme
once development is initiated. β-glucosidase-2, on the other
hand, accumulates only late in development as the pseudoplasmodia culminate to form the terminal structure (Fig. 1). Both
isozymes are found associated with lysosomal-like structures
and probably play catabolic roles.

Since the isozymes differ in size, electrophoretic mobility
and substrate affinity, we would expect them to be distinct
proteins coded for by independent genes. However, we have recently isolated three independent mutant strains which lack β-
glucosidase-1 activity and have found that none of these
strains form β-glucosidase-2 (Dimond and Loomis, in manuscript).
The mutant strains develop almost normally and accumulate other
late stage specific enzymes, but accumulate less than 1% of the
normal amount of either β-glucosidase isozyme. Since none of
the strains selected for impaired β-glucosidase-1 form the
second enzyme, we surmise that the isozymes share a common component. This could be a common subunit or a regulatory element.
It is possible that a component of both isozymes is synthesized
from the same gene which is active during growth, repressed
during most of the developmental phase, and then reactivated
during culmination.

ALKALINE PHOSPHATASE

Two isozymes of alkaline phosphatase have been recognized
on starch gel electrophoresis (Krivanek 1956). The first is
found at all stages of development of D. *discoideum* while the
second is seen only late in development (Loomis 1969) (Fig. 1).
Although the developmental kinetics suggest that alkaline
phosphatase-2 might be necessary for the terminal stages in

development, this now seems unlikely since culmination has been observed to proceed in an apparently normal fashion under several conditions where alkaline phosphatase-2 does not accumulate. For instance, if cells of D. *discoideum* are incubated in the presence of 10^{-2}M L-cysteine, morphogenesis proceeds normally and the cells differentiate into normal spore and stalk cells, however, alkaline phosphatase-2 does not accumulate (Fig. 3).

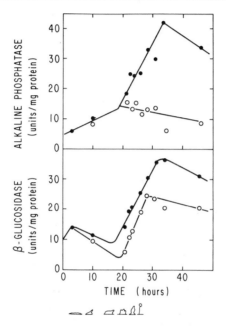

Fig. 3. Effect of cysteine on accumulation of alkaline phosphatase and β-glucosidase. D. *discoideum* was allowed to develop on filter supports with (o———o) or without (●———●) the addition of 10^{-2}M L-cysteine to the buffered solution.

Under similar conditions β-glucosidase-2 accumulates normally, indicating that these two stage specific isozymes are not coordinately controlled under all conditions.

ACCUMULATION IN MORPHOLOGICAL MUTANTS

A large number of mutant strains have been isolated which are blocked in morphogenesis at different stages. Biochemical analysis of some of these strains has indicated that the accumulation of the individual isozymes are separately controlled. For instance, several aggregateless strains accumulate the major isozyme of acetylglucosaminidase but do not accumulate threonine deaminase-2, β-glucosidase-2 or alkaline phosphatase-2. In other mutant strains threonine deaminase-2 accumulates normally but neither β-glucosidase-2 nor alkaline phosphatase-2 accumulate. The developmental pattern of a variety of stage specific enzymes including the four pairs of isozymes have been followed in a series of morphological mutants. The patterns in an aggregateless strain, Agg 206, and a fruitless strain, KY3, are shown in Fig. 4. The pattern seen in other strains suggests

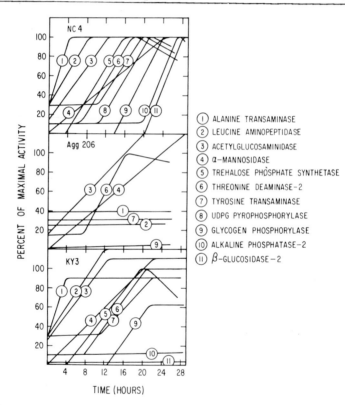

Fig. 4. Patterns of Accumulation of Stage Specific Enzymes. Cells of strain NC-4 (wild-type) Agg206 (aggregateless) and KY3 (fruitless) were incubated on filters and samples taken for enzymatic analysis at various times. The numbers refer to the specific enzyme activities given at the right. The kinetics of accumulation were determined in the wild-type and mutant strains in the studies referred to in Table I.

that the pleiotropic effects of the morphological mutants have a simple linear polarity (Fig. 5).

MUTANT STRAINS

ENZYMES	VA-5	DA 2	Agg 206	TS 2or VA 4	DTS 6	KY 3	MIN 2
3 NAG	−	+	+	+	+	+	
4 MAN		+	+	+	+	+	
6 TD			−	+	+	+	
I AT			−			+	
2 LAP			−	+	+	+	
5 TPS			−			+	
7 TT		−	−	+	+	+	
II GP		−	−	−		+	
12 AlKP				−	−	−	+
13 βG-2				−	−	−	+

Fig. 5. Pattern of Accumulation in Morphological Mutants. The accumulation of the enzymes referred to in Figure 4 was determined in a series of mutant strains. Strain VA-5, DA-2 and Agg 206 fail to aggregate. Strain TS 2 aggregates at 22° but not at 27° and was analyzed under non-permissive conditions. Strain DTS 6 develops normally at 22° but fails to form pseudoplasmodia or grex at 27° (Loomis 1969b). This strain was also analyzed at non-permissive temperatures. Strain KY 3 fails to culminate (Yanagisawa, Loomis and Sussman 1967). Strain Min 2 fails to form spores (Loomis 1968). Many of the determinations have been reported in the studies referred to in Table I. Others were performed for this study. Accumulation of specific enzymes was considered positive if at least 80% of peak specific activity was reached. Accumulation was considered negative if less than 20% of the normal accumulation occurred.

When considered together with the morphological aberrations in these mutant strains, these studies define 8 unequivocal steps in development which form a linear progression (Figure 6). The intervening stages are referred to as stages A through I. At present we have no idea what processes or functions comprise these steps. The stage specific enzymes themselves are probably not components of the main causal progression since structural gene mutations in at least three of them, acetylglucosaminidase-1, α-mannosidase, β-glucosidase do not modify subsequent patterns of biochemical differentiations (Dimond, Free ,and Loomis, unpublished).

185

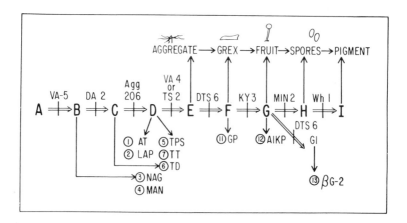

Fig. 6. Genetic Dissection of Developmental Steps in *Dictyostelium*. The data presented in Figures 4 and 5 was used to define 9 discrete steps (A through I) on which the accumulation of stage specific enzymes or morphological structures depend. The arrows indicate dependent relationships. Mutant strain Wh 1 forms normal fruiting bodies but does not synthesize the yellow carotenoid pigment. The numbers refer to the enzymes listed in Table I. In a few places there is an uncertainty of one step in assigning the dependence for a specific enzyme (see Fig. 5).

The sequence of stages defined by the mutant strains differs somewhat from the temporal sequence of accumulation of the various enzymes. This is most striking in the aggregate-less strain Agg 206 in which acetylglucosaminidase-2, α-mannosidase and threonine deaminase-2 accumulate almost normally but aminopeptidase and alanine transaminase do not accumulate (Fig. 7). These latter two enzymes are the first to accumulate in wild type cells. The most likely explanation is that accumulation of acetylglucosaminidase-2, α-mannosidase and threonine deaminase-2 is a delayed response to an event which occurs prior to the signal for accumulation of aminopeptidase and alanine transaminase.

A branch point appears to occur at stage G leading to the step necessary for accumulation of β-glucosidase-2. When strain DTS6 is incubated at the non-permissive temperature pseudoplasmodial formation and further development is blocked and neither alkaline phosphatase nor β-glucosidase-2 accumulate. However, if cells of this strain are shifted down to the permissive temperature after 26 hours, morphogenesis proceeds again and normal fruiting bodies are formed. Under these

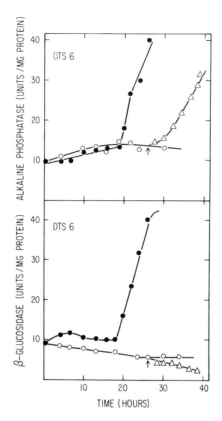

Fig. 7. Accumulation of Alkaline Phosphatase and β-glucosidase in Strain DTS 6. The cells were incubated either at 22° (●————●) or at 27° (o————o). After 28 hours at 27° some cells were shifted to 22° and incubated further (Δ————Δ) (Dimond and Loomis, unpublished).

conditions alkaline phosphatase accumulates normally but β-glucosidase-2 is not synthesized. These results indicate that a step specific to β-glucosidase-2 accumulation is irreversibly affected by non-permissive conditions in strain DTS6 while other steps are fully reversible in this strain.

CONCLUSION

Since we have chosen to focus on single gene products which are expressed during discrete stages of development it has been essential to separate and characterize isozymes. Undoubtedly these isozymes serve distinct functions in vivo and in several cases it has been possible to recognize their specific roles. The relationship of the changing enzymological spectrum to the visual stages of morphogenesis remains one of the most challenging aspects of developmental biology. It is hoped that the recognition of stage specific isozymes coupled with genetic analysis of pertinent mutations will shed light on this problem.

Moreover, the stage specific isozymes have provided convenient markers to delineate a few of the steps in the process of temporal differentiation of *Dictyostelium*. So far they have allowed us to genetically define 8 discrete steps. As the number of enzymes analyzed increases other mutant strains should be able to further subdivide the process of development and define other branch points and parallel pathways.

ACKNOWLEDGEMENT

These studies were carried out in collaboration with my colleagues Bruce Coston, Randall Dimond, Sheng-Shung Pong and Sally Shure. The work has been supported by the National Science Foundation.

REFERENCES

Ashworth, J. M. and M. Sussman 1967. The appearance and disappearance of uridine diphosphate glucose pyrophosphorylase activity during differentiation of the cellular slime mold *Dictyostelium discoideum. J. Biol. Chem.* 242: 1696-1700.

Coston, M. Bruce and W. F. Loomis 1969. Isozymes of β-glucosidase in *Dictyostelium discoideum. J. Bact.* 100: 1208-1217.

Dimond, R., M. Brenner,and W. F. Loomis 1973. Mutations affecting N-acetylglucosaminidase in *Dictyostelium discoideum. Proc. Natl. Acad. Sci. U.S.A.* 70:3356-3360.

Dimond, R. and W. F. Loomis 1974. Vegetative isozyme of N-acetylglucosaminidase in *Dictyostelium discoideum. J. Biol. Chem.,* 249: 5628-5632.

Firtel, R. A. and J. Bonner 1970. Developmental control of alpha 1-4 glucan phosphorylase in cellular slime mold *Dictyostelium discoideum. Fedn. Proc. Fedn. Am. Socs. Exp. Biol.* 29: 669.

Firtel, R. A. and Robert W. Brackenbury 1972. Partial characterization of several protein and amino acid metabolizing enzymes in the cellular slime mold *Dictyostelium discoideum. Develop. Biol.* 27: 307-321.

Jones, Theodore H. D. and Barbara E. Wright 1970. Partial

purification and characterization of glycogen phosphorylase from *Dictyostelium discoideum*. *J. Bact.* 104: 754-
761.

Krivanek, J. O. 1956. Alkaline phosphatase activity in the
developing slime mold, *Dictyostelium discoideum*. *Raper.*
J. Exp. Zool. 133: 459-480.

Loomis, W. F. 1968. The relation between cytodifferentiation
and inactivation of a developmentally-controlled enzyme
in *Dictyostelium discoideum*. *Exp. Cell Res.* 53: 282-287.

Loomis, W. F. 1969a. Acetylglucosaminidase, an early enzyme in
the development of *Dictyostelium discoideum*. *J. Bact.*
97: 1149-1154.

Loomis, W. F. 1969b. Temperature-sensitive mutants of *Dictyostelium discoideum*. *J. Bact.* 99: 65-69.

Loomis, W. F. 1969c. Developmental regulation of alkaline phosphatase in *Dictyostelium discoideum*. *J. Bact.* 100: 417-
422.

Loomis, W. F. 1970a. Developmental regulation of α-mannosidase
in *Dictyostelium discoideum*. *J. Bact.* 103: 375-381.

Loomis, W. F. 1970b. Temporal control of differentiation in
the slime mold, *Dictyostelium discoideum*. *Exp. Cell Res.*
60: 285-289.

Pong, S. S. and W. F. Loomis 1971. Enzymes of amino acid metabolism in *Dictyostelium discoideum*. *J. Biol. Chem.*
246: 4412-4416.

Pong, S. S. and W. F. Loomis 1973. Replacement of an anabolic
threonine deaminase by a catabolic threonine deaminase
during development of *Dictyostelium discoideum*. *J. Biol.*
Chem. 248: 4867-4873.

Roth, R. and M. Sussman 1968. Trehalose 6-phosphate synthetase
(uridine diphosphate glucose: D-glucose 6-phosphate 1-
glucosyltransferase) and its regulation during slime mold
development. *J. Biol. Chem.* 243: 5081-5087.

Sussman, M. 1966. Biochemical and genetic methods in the study
of cellular slime mold development. *Methods in Cell
Physiology*. Vol. 2 : 397-410 (ed. D. Prescott) Academic
Press, New York.

Telser, Alvin and M. Sussman 1971. Uridine diphosphate galac-
tose-4-epimerase, a developmentally regulated enzyme in
the cellular slime mold *Dictyostelium discoideum*. *J. Biol.*
Chem. 246: 2252-2257.

Yanagisawa, K., W. F. Loomis, and M. Sussman 1967. Developmental
regulation of the enzyme UDP-galactose polysaccharide
transferase. *Exp. Cell Res.* 46: 328-334.

BIOCHEMICAL PROPERTIES AND DEVELOPMENT EXPRESSION OF GENETICALLY DETERMINED MALATE DEHYDROGENASE ISOZYMES IN MAIZE

NING-SUN YANG[1]

MSU/AEC Plant Research Laboratory,
Michigan State University, East Lansing, Michigan 48824

ABSTRACT. Genetic and biochemical studies suggest that both soluble and mitochondrial malate dehydrogenase (MDH) isozymes in maize are genetically determined, and are not different conformational forms derived from the same primary structure. Results of the genetic analysis also show that the five commonly observed mitchondrial malate dehydrogenase (m-MDH) isozymes are under the control of two groups of loci residing on two different chromosomes. Biochemical properties for each of the soluble and mitochondrial MDH isozymes in the highly inbred strain W64A were examined. Not only the soluble malate dehydrogenases (s-MDHs) and m-MDHs differ in most of their physical and kinetic properties, the MDH isozymes within the same subcellular location may also differ significantly in several of these properties examined. Developmental expression of the multiple genes controlling MDH isozymes has been studied in the young maize seedlings. It is found that both s-MDHs and m-MDHs in the scutellum of developing maize seedlings are *de novo* synthesized. The increase of both classes of MDHs are inhibited by cycloheximide, but not by chloramphenicol. These results indicate that the nuclear gene controlled mitochondrial MDH isozymes are synthesized in the cytoplasm and then become associated with the mitochondria.

INTRODUCTION

The occurrence of multiple molecular forms of enzymes (isozymes) is now known to be a common characteristic in most organisms. Malate dehydrogenases (L-Malate: nicotinamide adenine dinucleotide (NAD) oxidoreductase; E.C.1.1.1.37), in a wide variety of eukaryotic organisms (Kitto et al., 1970; Meizel and Markert 1967; Shows et al., 1970; Whitt 1970; Zee et al., 1970; Bailey et al., 1970; Karig and Wilson 1971; Weimberg 1968; Shannon 1968; Longo and Scandalios

---

[1]Present address:
Roche Institute of Molecular Biology, Nutley, New Jersey 07110

1969) have been shown to commonly exist in isozymic forms. There exist at least two major classes of malate dehydrogenase. One class is restricted in occurrence to the mitochondria (m-MDHs) where it functions as a component of the Krebs cycle, while the other class (s-MDHs) occurs in the soluble fraction of the cell, where it may participate in the malate shuttle (Lehninger 1970), in crassulacean acid metabolism of plant tissues (Ting 1970) and in other metabolic pathways (Ting et al., 1966). In plant tissues, malate dehydrogenases have also been found in glyoxysomes (Breidenbach and Beevers 1967; Longo and Scandalios 1969) and peroxisomes (Yamazaki and Tolbert 1969).

Within each of the two major classes, multiple electrophoretic forms are usually observed, even within a single tissue. Both genetic variants (Shows et al., 1970; Whitt 1970; Zee et al., 1970; Bailey et al., 1970; Karig and Wilson 1971) and post-translational modifications (Kitto et al., 1970; Meizel and Markert 1967) have been reported to account for such heterogeneity of malate dehydrogenase. The significance of MDH isozymes does not lie in the multiplicity of the enzyme *per se*, but in its role in cellular physiology and in evolutionary adaptation. Therefore, the MDH isozymes once genetically defined may serve as useful intracellular markers for both biochemical and developmental studies.

In maize, isozymic forms of both soluble and mitochondrial malate dehydrogenase are observed. In this report, the following aspects of maize malate dehydrogenase isozymes are reported: 1) How the multiplicity of maize MDH isozymes is determined. 2) How they differ in their biochemical properties. 3) How the various maize MDH isozymes are expressed during early sporophytic development.

MATERIALS AND METHODS

GENETIC ANALYSIS OF THE MULTIPLE FORMS OF MAIZE MDH ISOZYMES

Maize strains that were inbred for at least ten generations were used in all experiments. MDH isozyme patterns in various inbred lines or genetic crosses were checked by using the triploid liquid endosperm from individual kernels.

Soluble and mitochondrial fractions of maize tissues are separated by modifying the method of Longo and Longo (1970). The MDH isozymes in the two different subcellular locations are identified by starch gel electrophoresis and zymogram techniques according to the method of Scandalios (1969).

PURIFICATION AND BIOCHEMICAL STUDIES
OF THE MAIZE MDH ISOZYMES

Seven MDH isozymes in inbred strain W64A were separated
and highly purified by the following 6 steps: 1) Preparation of crude extract. 2) Treatment at pH 5. 3) Ammonium
sulfate fractionation. 4) Gel filtration through Sephadex
G-150. 5) DEAE-cellulose column chromatography. 6) Starch
gel electrophoresis and high speed centrifugation. Using
the highly purified enzyme preparations, comparative biochemical properties of the maize MDH isozymes were studied.

Five methods were used to study the possible inter--
conversions of maize MDH isozymes. These included second
run gel electrophoresis, mercaptoethanol treatment, pH 2
treatment, 7.5 M guanidine hydrochloride treatment, and
freezing and thawing in 1M NaCl. Details of the above
experimental procedures have been reported elsewhere by
Yang and Scandalios (1974a).

DEVELOPMENTAL STUDIES OF MAIZE MDH ISOZYMES

Isolation and quantitative assay of the individual MDH
isozymes were performed as described below. After crude
extraction, the MDH isozymes were separated by starch gel
electrophoresis. One horizontal slice was taken from the gel
and stained for MDH activity. This stained slice was then
used as template for excising single isozyme bands from the
unstained portion of the gel. Each excised band was then
placed in a syringe and squeezed into a centrifuge tube and
centrifuged at 45,000 xg for one hour. The supernatant,
containing a single MDH isozyme was used for quantitative
assay of the MDH activity. Malate dehydrogenase was assayed
spectrophotometrically according to the method of
Ochoa (1955).

Density labeling of the newly synthesized proteins in
maize scutella was conducted according to the method of
Quail and Scandalios (1971). The procedures for density
gradient centrifugation were essentially those of Filner and
Varner (1967).

In order to infuse antibiotics into maize scutella,
scutella in intact 4 day-old etiolated seedlings were excised
and vacuum infiltrated with nutrient medium with or without
antibiotics against protein synthesis, cycloheximide (CH) or
chloramphenicol (CAP). The excised scutella were then
incubated in the corresponding media for 15 hours. After
incubation, MDH activities were extracted and measured.

RESULTS AND DISCUSSION

THE GENETIC BASIS FOR MAIZE MDH ISOZYMES

In maize, there are two major classes of malate dehydrogenase isozymes (Fig. 1).

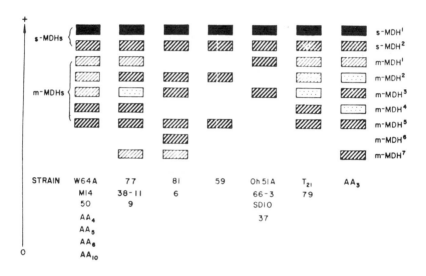

Fig. 1. Phenotypes of MDH isozymes observed in various inbred maize lines.

The two soluble forms (s-MDHs) appear in all the inbred lines
examined. Additional s-MDH isozyme variants have been
observed occasionally in inbred lines 59, 37, and T_{21}. These
variants are more anodal compared to the two s-MDHs
shown in Fig. 1. Seven variants of the mitochondrial MDH
isozymes (m-MDHs) were observed in the 20 highly inbred
lines examined (Fig.1). The m-MDHs are named from the anode
toward the cathode as m-MDH^1, m-MDH^2......m-MDH^6 and m-MDH^7.

Studies of open pollinated corn, in which the s-MDH
variants appear in a fairly high frequency, showed the
frequencies of the appearance of the s-MDH variants are the
same in plants with different m-MDH phenotypes (data not
shown). This result indicates that s-MDHs and m-MDHs in
maize are under the control of different genes.

As seen in Fig. 1, in the seven phenotypes of m-MDH

isozymes, each phenotype may consist of two to five major MDH isozymes. Since all of these lines have been inbred for at least 10 generations and the MDH isozyme patterns were observed consistently, therefore the isozymes in each of the specific inbred lines should not be the products of allelic genes. For example, in strain 59, m-MDH^2, and m-MDH^5 are not allelic isozymes, because they never segregate in the inbreds. This is also true for all the isozymes found in the other six MDH phenotypes.

The absolutely independent expression of each individual isozyme argues strongly against the possibility that any of the isozymes result from the modification of another. Since the isozymes also appear not to be allelic, it is suggested that each m-MDH isozyme, except the possible hybrid molecules, is coded by a separate structural locus.

Genetic control of the multiplicity of maize m-MDH isozymes has been studied by analyzing the back crosses and F_2 progenies made from several inbred lines seen in Fig.1. Segregation and differential gene dosage effects (in the triploid endosperm) were observed for the m-MDH isozymes. An example is given in Fig. 2.

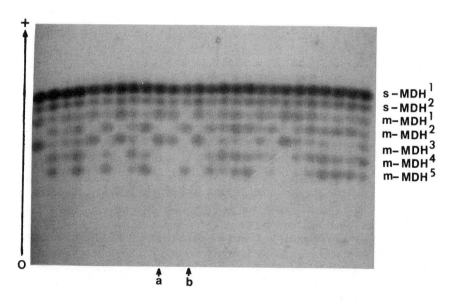

Fig. 2. MDH phenotypes of the F_2 progenies of (Oh51A x 59) x (Oh51A x 59). Note that m-MDH^1 and m-MDH^3 are always subjected to the same dosing effect, m-MDH^2 and m-MDH^5 are dosed simultaneously. However, these two types of MDHs were

195

Fig. 2. (cont.) never dosed simultaneously in the triploid endosperm. Letter chanels a and b represent the two parental MDH phenotypes of strains Oh51A and 59 respectively.

The results of some 15 different crosses suggest that m-MDH isozymes in maize are regulated by multiple structural loci. The m-MDH^1 and m-MDH^3 are encoded by two loci closely linked on one chromosome, while m-MDH^2 and m-MDH^5 are controlled by another two loci closely linked on another chromosome. The m-MDH^4 isozyme is derived from random association of the subunits of m-MDH^3 and m-MDH^5. The fact that two linkage groups residing on two nonhomologous chromosomes are involved in the expression of maize m-MDH isozymes adds further support to the findings that maize m-MDHs are controlled by nuclear genes (Longo and Scandalios 1969). Results of the genetic analysis also suggest that a third linkage group encoding for the less anodal m-MDHs (m-MDH^6, m-MDH^7) may also exist in some inbred strains.

Zee et al., (1970) suggested that s-MDHs in *Ascaris suum* are under the control of two separate genetic loci. Wheat et al., (1972) have demonstrated that s-MDHs in fish are controlled by two unlinked loci. In salmon (Bailey et al., 1970; Aspinwall 1974), duplicate loci for each of the two unlinked loci have been suggested. It is interesting that there is a great similarity between these results observed in aminal s-MDHs and the present results observed in maize m-MDHs.

Results of the genetic analysis strongly suggest that the polymorphism of maize MDH isozymes is genetically determined. Due to the limitation of the space in this report, details of the genetic control of maize m-MDH isozymes have not been described and will be demonstrated elsewhere by Yang and Scandalios (1974b).

BIOCHEMICAL PROPERTIES OF THE MAIZE MDH ISOZYMES

As shown in Fig. 1, strain W64A had the two s-MDHs and the five commonly observed m-MDHs and was therefore chosen as a source of isozymes for biochemical studies. In the etiolated seedlings, all of the organs examined had the same MDH isozyme patterns for both soluble and mitochondrial forms (Fig. 3). The s-MDH^2 in scutellum appears to be abundant compared to the same isozyme in other organs. Since purified or partially purified enzyme preparations are desired for studying biochemical properties, purification and separ-

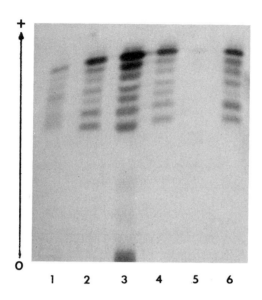

Fig. 3. Zmogram showing the MDH isozymes in the crude
extracts of different organs in etiolated maize seedlings
(strain W64A). (1) endosperm (2) shoot (3) scutellum (4)
root (5) pericarp (6) pericarp of the immature kernel.

ation of the seven MDH isozymes in strain W64A were performed.

After 4 steps of purification, the maize MDH isozymes
were eluted from a DEAE-cellulose column and were separated
into three distinct MDH peaks (Fig.4). The three DEAE-MDH
preparations were used for studies of the possible interconversion of maize MDH isozymes. For studies of the
kinetic and catalytic properties, the DEAE-MDH preparations
were further separated and purified by starch gel electrophoresis and high speed centrifugation.

As seen in Fig. 5, after second run electrophoresis, the
isozymes retain their original mobilities relative to each
other. No interconversion of the isozymes was observed.
Meizel and Markert (1967) observed that all of the supernatant MDH isozymes of Ilyanassa were apparently of the same
molecular weight and were all convertible to a single form
by prolonged exposure to mercaptoethanol. Results shown
in Fig. 6 indicate that both s-MDH and m-MDH isozymes of
maize are not mercaptoethanol convertible, conformational
isozymes. Kitto et al., (1970) showed that studies on

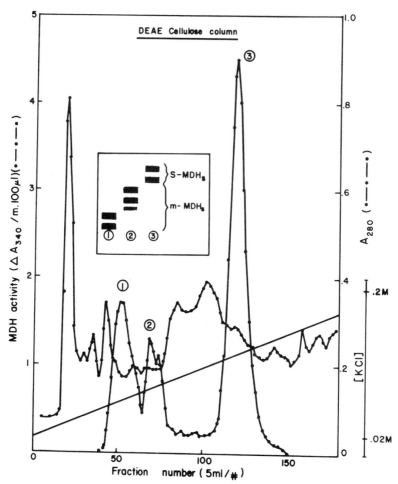

Fig. 4. Elution profile of malate dehydrogenase activites
from DEAE-cellulose column chromatography. Three peaks of
MDH activites were observed and the corresponding isozyme
patterns are shown in the inset.

reversible denaturation provides a useful test of the
conformer hypothesis of multiple electrophoretic forms of
isozymes. Using both acid and guanidine hydrochloride as
denaturants, reversible denaturation studies on maize MDH
isozymes were conducted. The results, which have been published elsewhere (Yang and Scandalios 1974a), showed that the
various maize MDH isozymes were not interconvertible from one
form to another. These results along with the genetic data
clearly indicate that the maize MDH isozymes (both s-MDHs and
m-MDHs) are not conformational isozymes; instead they are
genetically determined.

Using the highly purified enzyme preparations, biochemical

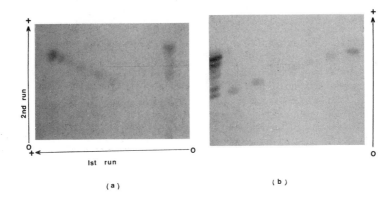

Fig. 5. MDH zymograms showing a second run electrophoresis.
All isozymes retained their original mobilities relative to
each other.
(a) After the first run electrophoresis of a crude MDH
preparation, the gel was turned to 90° and a second sample
serving as a control was inserted. The gel was then subjected to electrophoresis under the same condition as the
first gel.
(b) The pieces of starch gel containing particular MDH
isozymes were cut from the gel after first run electrophoresis
of a crude MDH preparation, placed in the slots of a second
gel and subjected to a second run electrophoresis under the
same condition as in the first.

properties for each of the seven MDH isozymes were examined.
Molecular weight, pI, pH optimum, thermolability, and
Michaelis-Menten constants (Kms) for oxaloacetic acid (OAA),
malate, nicotinamide adenine dinucleotide and NADH at different pHs were determined. Different kinetics of substrate
(OAA) inhibition and coenzyme (NAD) inhibition were observed
for the different isozymes. Effects of NAD analogs, chelating agents, reducing agents, metal ions, and TCA cycle acids
on the enzymatic activity of the isozymes were tested.
Details of the biochemical studies of maize MDH isozymes
have been reported previously by Yang and Scandalios (1974a).
In this paper, some of the biochemical properties will be
briefly described.
    Figure 7 shows that the various maize MDH isozymes have

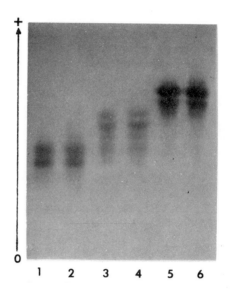

Fig. 6. Starch gel electrophoresis of maize MDH preparations after a 22-hour exposure to 100 mM 2-mercaptoethanol.
(1) Untreated DEAE-I MDHs (m-MDH^4, m-MDH^5).
(2) DEAE-I MDHs treated with 100 mM 2-mercaptoethanol.
(3) Untreated DEAE-II MDHs (m-MDH^1, m-MDH^2, m-MDH^3).
(4) DEAE-II MDHs treated with 100 mM 2-mercaptoethanol.
(5) Untreated DEAE-III MDHs (s-MDH^1, s-MDH^2).
(6) DEAE-III MDHs treated with 100 mM 2-mercaptoethanol.

different rates of thermal inactivation. Generally speaking, the s-MDHs are more heat labile than are the m-MDHs. According to their heat stabilities, m-MDH isozymes can be classified into two groups: m-MDH^1 and m-MDH^2 belong to one group; m-MDH^3, m-MDH^4, and m-MDH^5 belong to another. Michaelis constants (Kms) of the four substrates for maize MDH isozymes change as a function of pH (Fig. 8). According to the Km dependencies on pH, the seven MDH isozymes can be classified into three groups; the two soluble MDHs; the two most anodal m-MDHs; and the three most cathodal m-MDHs. Therefore in Fig. 8, Kms of three isozymes were chosen to represent those of the seven MDH isozymes. Inhibition of MDHs by OAA has been widely reported, but inhibition of MDHs by NAD has not, at least to my knowledge. In this investigation it is found that both s-MDHs and m-MDHs in maize are inhibited by their coenzyme, NAD. Inhibition of

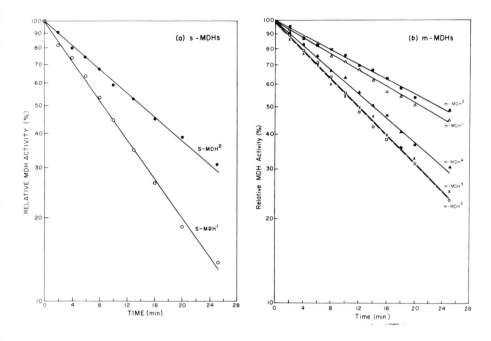

Fig. 7. Rate of heat-inactivation for s-MDHs (a) and m-MDHs (b).

maize MDHs by NAD are also pH dependent with greater inhibition at higher pH values. Kinetics of NAD inhibition for the various maize MDH isozymes may also differ as shown in Fig. 9.

Results of the biochemical studies showed that s-MDHs and m-MDHs are apparently different in most of the physical and kinetic properties examined. This may further indicate that these two classes of maize MDH isozymes are controlled by two different groups of structural genes. In comparing the biochemical properties of the mitochondrial MDH isozymes, it is found that the maize m-MDHs can be classified into two groups. The m-MDH[1] and m-MDH[2] belong to one group, while m-MDH[3], m-MDH[4] and m-MDH[5] belong to another. However, the genetic analysis shows that genes encoding for m-MDH[1] and MDH[3] are closely linked on one chromosome, those encoding for for m-MDH[2] and m-MDH[5] are closely linked and located on another chromosome. The m-MDH[4] isozyme is a hybrid molecule of m-MDH[3] and m-MDH[5]. These results lead me to suggest that the multiple structural loci coding for the above five m-MDHs in maize are possibly derived through evolution by gene dup-

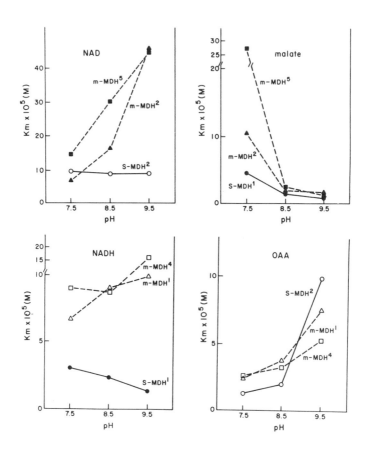

Fig. 8. Michaelis constants of maize MDH isozymes as a function of pH. For each substrate, Kms of three isozymes were chosen to represent those of the seven MDH isozymes.

lications, mutations and chromosome translocation. By these mechanisms, the genes coding for MDH isozymes having very similar biochemical properties (as those for m-MDH[1] and m-MDH[2]) may therefore locate on two nonhomologous chromosomes.

DEVELOPMENT OF MAIZE MDH ISOZYMES

The occurrence of soluble and mitochondrial MDH isozymes offers a good system for studying developmental processes within the cell. Not only does the expression of these isozymes represent a mechanism by which cells may regulate a group of

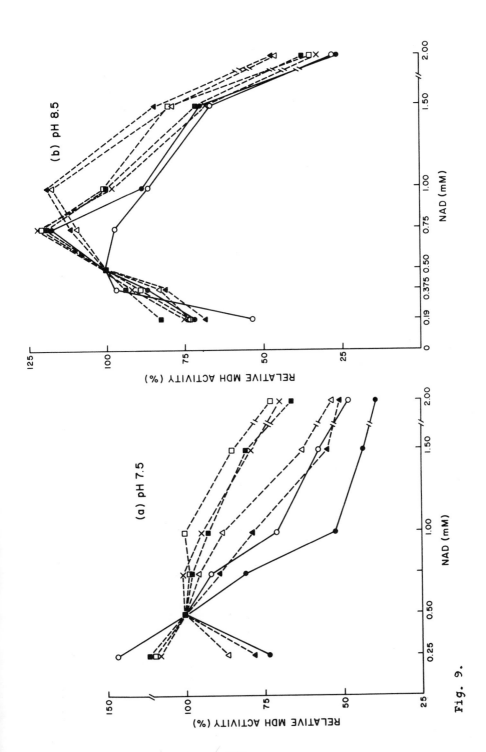

Fig. 9.

203

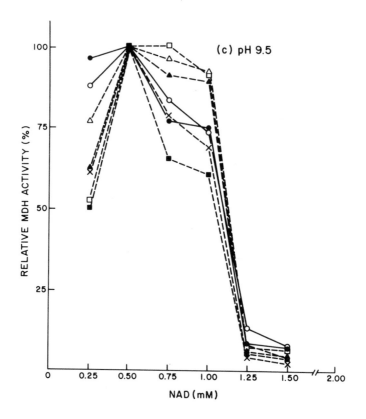

Fig. 9. Coenzyme inhibition (NAD)of soluble malate dehydrogenases (s-MDHs) and mitochondrial dehydrogenases (m-MDHs)
as a function of pH (25 mM sodium glycylglycine buffer, pH
7.5 to pH 9.5).

(a) pH 7.5
———●——— s-MDH[1]
- - -▲- - - m-MDH[2]
- - -■- - - m-MDH[5]

(b) pH 8.5
——— ○——s-MDH[2]
- - -□- - -m-MDH[3]

(c) pH 9.5
- - -△- - m-MDH[1]
- - -✗- - m-MDH[4]

proteins (enzymes) with similar characteristics during development, but the subcellular distribution of MDH isozymes
may serve as a model to study, both genetically and biochemically, the cooporation between organelles. Since the
maize MDH isozymes are genetically determined, expression
of these isozymes may reflect directly the activities of
their encoding genes.

Since the specific activity of MDH in scutellum is about 2 to 7 fold higher than those in other organs in the etiolated seedlings (see Fig. 3 for reference), development of maize MDH isozymes coded by different genes has been studied in the scutella of young seedlings. Inbred strains W64A and 59 have been used in the present study.

In etiolated W64A seedlings, the total MDH activity increases through the first five days and peaks about the 6th day. This is followed by a gradual decrease in activity, and by the 10th day the level is the same as that of the 4th day. Zymograms of scutellar MDH at various stages of seedling development, showed that the number and positions of the isozymes under study remained constant (data not shown). The developmental changes of MDH in scutella appear to be cor-related to the growth of the young maize seedlings. During the first 5 days, the etiolated seedlings grow at a fairly constant rate and reach a state in which the shoots are about 5 to 6 cm long. Between the 5th and 7th day, the shoots protrude the coleoptiles, the scutella and endosperms become highly liquified. Then the leaves and the stems start to elongate. The high levels of MDH activities observed in the scutella of the 5th - 7th day old seedlings may indicate that the scutella have reached a state for maximal supply of nutrient and energy to the etiolated seedlings and are ready to be degraded thereafter.

Time courses of development of the two s-MDHs and the five m-MDHs are shown in Fig. 10 and Fig. 11 respectively. The various s-MDH and m-MDH isozymes exhibit similar de-velopmental patterns. For the soluble isozymes, s-MDH[1] has always a higher activity than the s-MDH[2], especially during the late developmental stages. The activity of all mito-chondrial forms is less than that of the soluble forms. At any given point in scutellar development, the total MDH activity in mitochondria is only 60% of that in the soluble cytoplasm. Even though no apparent differential expression of m-MDH isozymes was observed, all of the m-MDHs do not seem to follow the same kinetics of accumulation. The observed differences are likely not due to the variability of the method, and some subtle regulation controlling the expression of each isozyme may be involved; this aspect has not been meaningfully examined at this point.

The density labeling technique was used to determine whether the development of s-MDHs and m-MDHs in scutella of developing maize seedlings is due to *de novo* synthesis of the enzyme moieties or to activation of the pre-existing enzymes. Figure 12 shows that both s-MDHs and m-MDHs become labeled after 5 days of germination indicating that both

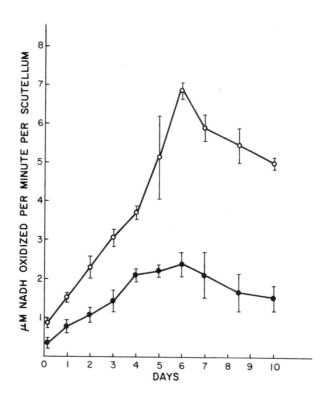

Fig. 10. Time course of development of the two s-MDHs in
scutella of germinating maize seeds.
——— O ——— s-MDH[1];   ——— ● ———   s-MDH[2].

classes of MDH isozymes are synthesized *de novo*. Therefore,
the newly appearing MDH isozymes in maize scutellum are
accumulated during the *de novo* synthesis of the enzyme
moieties. However, do they turn over as they accumulate?
The density labeling data indicate that the turn over of these
MDH isozymes must be rapid, since pre-existing, unlabeled
molecules are not detected in CsCl gradients after 5 days.
Therefore, it is suggested that the developmental changes of
the MDH isozyme content in germinating maize scutella result
from regulation of synthesis as well as degradation of the
enzyme moieties.

   After this result, it was of interest to determine the
intracellular site of synthesis of the two classes of MDH
isozymes. Chloramphenicol (CAP) and cycloheximide (CH), two
known inhibitors of protein synthesis were employed for such

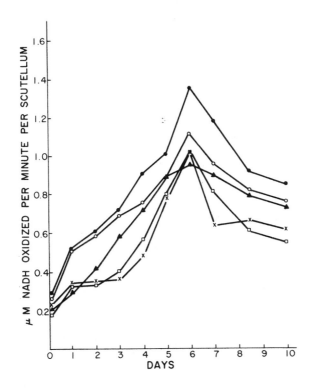

Fig. 11. Time course of development of the five m-MDHs in
scutella of germinating maize seedlings.
———●——— m-MDH[1] ——— ✕ ——— m-MDH[3] ——— □ ——— m-MDH[5]
———△——— m-MDH[2] ——— ○ ——— m-MDH[4]

studies. Effects of CAP and CH on protein synthesis in
maize scutella were studied by measuring their effects on
incorporation of radioactive leucine into TCA insoluble
materials. Cycloheximide (5 mg/ml) inhibits almost completely
(about 97%) the incorporation of leucine into proteins found
in soluble fractions; whereas, chloramphenicol (1 mg/ml)
inhibits less than 5% of the synthesis of proteins found
in soluble fractions. For the synthesis of proteins found
in crude particulate fractions (12,000 xg pellet), only
about 80%-85% was inhibited by cycloheximide, while about
30%-35% was also inhibited by chloramphenicol. Detailed
analysis by SDS gel electrophoresis indicates that
some proteins found in a dense particulate fraction

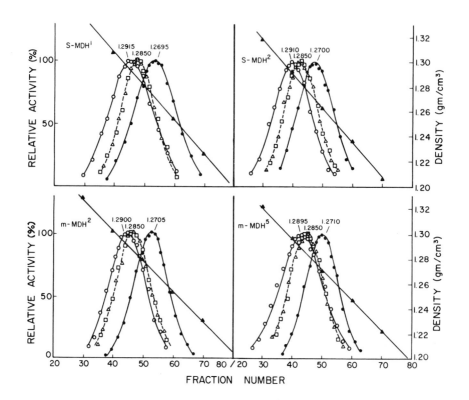

Fig.]2. Equilibrium distribution in CsCl gradients of
scutellar s-MDHs and m-MDHs from seeds (strain 59) grown
on either $^{14}NH_4Cl$ in H_2O for 5 days (●————●) or
$^{15}NH_4Cl$ in 70% D_2O for 7 days (O————O). The activity
of the LDH marker in the labeled (△————△) and unlabeled
(□————□) gradients have been superimposed and drawn as
one. Relative activity means that all points on these curves
are expressed as percentage of the highest point on each of
the individual curves. Density of CsCl gradient (▲————▲).

(likely the proteins of mitochondrial inner membrane)
are specifically inhibited by chloramphenicol and cyclohexamide.

TABLE I

Effects of Cycloheximide (CH) and Chloramphenicol (CAP) on
MDH Activities in Scutella Excised from 4-Day-Old Seedlings

Treatments	Total MDH Activity (μM NADH oxidized/min/scutella)
Intact scutella (4 day):	10.90 ± 0.11
Excised scutella (4.0 day – 4.62 day): control	10.96 ± 0.13
+ CH (5 μg/ml)	7.68 ± 0.06
+ CAP (1 mg/ml)	11.01 ± 0.07
Intact scutella (4.62 day	13.05 ± 0.06

Each fifteen scutella isolated from 4-day-old seedlings were
transferred to nutrient media containing cycloheximide (5 μg/
ml), or chloramphenicol (1 mg/ml), or no antibiotic against
protein synthesis. After vacuum infiltration for two min, the
samples were incubated in a water bath shaker at 25°C for 15
hr. After incubation, MDH activities in these three sets of
scutella and in scutella just isolated from 4.0 day and 4.62
day-old (4 day and 15 hr) seedlings were determined.

TABLE 2

Effects of Cycloheximide (CH)[a] and Chloramphenicol (CAP)[b]
on the Development of the Individual in Maize Scutella
Excised from 4-Day-Old Seedlings

MDH Isozymes	Comparative MDH Activity[c] (%)		
	15 hrs in nutrient medium only (control)	15 hrs in CH	15 hrs in CAP
s-MDH[1]	100	65	101
s-MDH[2]	100	69	97
m-MDH[1]	100	72	96
m-MDH[2]	100	62	102
m-MDH[3]	100	74	95
m-MDH[4]	100	67	106
m-MDH[5]	100	70	98

[a]CH: in a concentration of 5 μg/ml

[b]CAP: in a concentration of 1 mg/ml

[c]Comparative MDH activity is calculated on a per scutellum
basis. Isolation of each individual MDH isozyme is described
under "Materials and Methods".

The results indicate that under the present experimental conditions, chloramphenicol has penetrated into scutella and inhibited the synthesis of some specific mitochondrial proteins; on the other hand, cycloheximide does not inhibit nonspecifically the synthesis of all proteins in scutella; some specific proteins (likely mitochondrial membrane proteins) are inhibited to a much lesser extent. Details of the effects of CAP and CH will be reported elsewhere (Yang and Scandalios 1974b).

Effects of CAP and CH on the development of maize MDH isozymes have been studied under the same experimental conditions as were used for studying the effects of antibiotics on protein synthesis. Results in Table 1 show that the total MDH activities observed in the control and the CAP treated samples are the same. No inhibitory effect of CAP was observed. However, 15 hours after the addition of cycloheximide, about 30% of the MDH activity found in the control was lost. Table 2 shows that neither s-MDH isozymes nor m-MDH isozymes are affected by the treatment of chloramphenicol for 15 hours. However, when scutella are incubated with cycloheximide for the same period of time, both soluble and mitochondrial MDH isozymes decrease to a similar extent (Table 2). The results of the experiments dealing with the effects of chloramphenicol and cycloheximide on protein synthesis and on MDH development in maize scutella indicate that protein synthesis in the cytoplasm is necessary for the increase of both soluble and mitochondrial MDH activities which are observed in the course of sporophytic development. Protein synthesis in the mitochondira is not responsible for the increase of mitochondria MDH activities.

This result is consistent with the earlier findings that mitochondrial MDHs in maize are controlled by nuclear genes (Longo and Scandalios 1969). Since both s-MDHs and m-MDHs in the scutella of developing maize seedlings are synthesized *de novo* , the above observations may indicate that maize m-MDHs, which are controlled by nuclear genes, are synthesized in the cytoplasm and then become associated with the mitochondria.

ACKNOWLEDGEMENTS

It is a great pleasure to thank Dr. J.G. Scandalios for his generous advice and help during the course of these investigations at Michigan State University. My thanks go also to the U.S. Atomic Energy Commission for financial support of this investigation under Contract No. AT (11-1)-

1338. The contents presented in this paper are part of the author's Ph.D. thesis (1974) submitted to Michigan State University.

REFERENCES

Aspinwall, N. 1974. Genetic analysis of duplicate malate dehydrogenase loci in the pink salmon, *Onchorhynchus gorbuscha*. *Genetics* 76: 65-72.

Bailey, G.S., Wilson, A.C., Halver, J.E., and Johnson, C.L. 1970. Multiple forms of supernatant malate dehydrogenase in salmonid fishes. *J. Biol. Chem.* 245: 5927-5940.

Breindenbach, R.W. and Beevers, H. 1967. Association of the glyoxylate cycle enzymes in a novel subcellular particle from castor bean endosperm. *Biochem. Biophys. Res. Commun.* 27: 462-469.

Filner, P. and Varner, J.E. 1967. A simple and unquivocal test for *de novo* synthesis of enzymes: density labeling of barley α-amylase with H_2O^{18}. *Proc. Nat. Acad. Sci. U.S.A.* 58: 1520-1526.

Karig, L.M. and Wilson, A.C. 1971. Genetic variation in supernatant malate dehydrogenase of birds and reptiles. *Biochem. Genetics* 5: 211-221.

Kitto, G.B., Stolzenback, F.E., and Kaplan, N.O. 1970. Mitochondrial malate dehydrogenase: further studies on multiple electrophoretic forms. *Biochem. Biophys. Res. Commun.* 38: 31-39.

Lehninger, A.L. 1970. In *"Biochemistry"*, Ed. 1, Chapter 18, Mitochondrial compartmentation. p. 408. Worth publishers, Inc.

Longo, C.P. and Longo, G.P. 1970. The development of glyoxysomes in peanut cotyledons and maize scutella. *Plant Physiol.* 45: 249-254.

Longo, G.P. and Scandalios, J.G. 1969. Nuclear gene control of mitochondrial malic dehydrogenase in maize. *Proc. Nat. Acad. Sci. U.S.A.* 62:104-111.

Meizel, S. and Markert, C.L. 1967. Malate dehydrogenase isozymes of the marine snail, *Ilyanassa obsoleta*. *Arch. Biochem. Biophys.* 122: 753-765.

Ochoa, S. 1955. In *"Methods in Enzymology"* (Colowick, S.P. and Kaplain, N.O. eds) vol. 1. pp. 735-739. Academic Press, New York.

Quail, P.H. and Scandalios, J.G. 1971. Turn over of genetically defined catalase isozymes in maize. *Proc. Nat. Acad. Sci. U.S.A.* 68:1402-1406.

Scandalios, J.G. 1969. The genetic control of multiple

molecular forms of enzymes in plants: a review. *Biochem. Genetics*, 3: 73-79.

Shannon, L.M. 1968. Plant isoenzymes. *Ann. Rev. Plant Physiol.* 19: 187-210.

Shows, T.B., Chapman V.M., and Ruddle, F.H. 1970. Mitochondrial malate dehydrogenase and malic enzyme: Mendelian inherited electrophoretic variants in the mouse. *Biochem. Genetics*, 4: 707-718.

Ting, I.P. 1970. Nonautotrophic CO_2 fixation and crassulacean acid metabolism. In: *"Photosynthesis and photorespiration"*. (Hatch, M.D., Osmond, C.B., and Slayter, R.O. eds.,) pp.109-185. Interscience, New York.

Ting, I.P., Sherman, I.W., and Dugger, W.M., Jr. 1966. Intercellular location and possible function of malate dehydrogenase isozymes from young maize root tissue. *Plant Physiol.* 41: 1083-1084.

Weimberg, R. 1968. An electrophoretic analysis of the isozymes of malate dehydrogenase in several different plants. *Plant Physiol.* 43: 622-628.

Wheat, T.E., Whitt, G.S., and Childers, W.F. 1972. Linkage relationships between the homologous malate dehydrogenase loci in teleosts. *Genetics* 70: 337-340.

Whitt, G.S. 1970. Genetic variation of supernatant and mitochondrial malate dehydrogenase isozymes in the teleost *Fundulus heteroclitus*. *Separatum Experienttia* 26: 734-736.

Yamazaki, R.K. and Tolbert, N.E. 1969. Malate dehydrogenase in leaf peroxisomes. *Biochem. Biophy. Acta.* 178: 11-20.

Yang, N.S. and Scandalios, J.G. 1974a. Purification and biochemical properties of genetically defined malate dehydrogenases in maize. *Arch. Biochem. Biophys.* 161:335-353.

Zee, D.S., Isensee,H., and Zinkham, W.H. 1970. Polymorphism of malate dehydrogenase in *Ascaris suum*. *Biochem. Genetics* 4: 253-257.

DIFFERENTIAL GENE EXPRESSION AND BIOCHEMICAL PROPERTIES OF CATALASE ISOZYMES IN MAIZE

JOHN G. SCANDALIOS

Genetics Laboratory, Department of Biology
University of South Carolina
Columbia, South Carolina 29208

ABSTRACT. In *Zea mays* the regulation of the enzyme catalase ($H_2O_2:H_2O_2$ oxidoreductase, E.C.1.11.1.6) is being investigated. To date, a number of isozyme variants of catalase have been detected and shown to be under genetic control. The catalase system appears to be an excellent one for studying the control of gene expression in a eukaryotic organism, as a number of regulatory mechanisms appear to contribute to the overall pattern of catalase expression in the cell. In immature kernels, a single electrophoretic form of catalase coded by the Ct_1 gene is expressed. In mature seeds and soon after germination a second catalase form coded by the Ct_2 gene locus is expressed. Both the Ct_1 and Ct_2 catalases are tetramers composed of distinct subunit polypeptides. The subunits interact to generate hybrid isozymes in expected binomial proportions. The physicochemical properties of the catalases have been examined and are discussed. During early sporophytic development, the pattern of catalase expression appears to be regulated by (a) differential gene expression, (b) differential rates of synthesis and degradation of the gene products, (c) changing patterns in compartmentation, and (d) by a catalase-specific inhibitor.

INTRODUCTION

Cellular differentiation involves, at least in part, differential gene activity and control of synthesis and degradation of given proteins (enzymes) at specific times and places in the various cells and tissues of the developing organism. One approach to studying genetic control mechanisms during development is to select a specific protein or enzyme as an indicator of the state of differentiation of a tissue or organ and to analyze the factors that regulate its synthesis, properties, location, and ultimate fate. The elucidation of mechanisms responsible for selective gene expression seems to be the fundamental problem in attempts to understand developmental processes. Although at present our understanding of gene regulation in eukaryotes is meager, it is apparent that gene expression is controlled at several levels involving not only transcriptional controls but also translational and post-translational mechanisms (Rechcigl, 1971).

In *Zea mays* L., the regulation of the enzyme catalase (H_2O_2:H_2O_2 oxidoreductase, E.C.1.11.1.6) has been the subject of intense investigations in my laboratory for the past several years. The catalase system of maize appears to be an excellent one for studying the control of gene expression in an eukaryotic organism, as a number of regulatory mechanisms appear to contribute to the overall pattern of catalase expression during maize development.

I. *Genetic Control of the Ct_1 Catalases*

To date we have screened over one thousand inbred strains of maize from a very wide range of sources; in all cases, a single electrophoretically distinct form of catalase is found in the developing endosperm (16 days after pollination) of any one strain. Of the total inbred strains screened, six electrophoretic variants of the basic molecular form have been found (Scandalios, 1968); these six variants have also been independently recovered recently by Maletsky et al (1971). It was further demonstrated that the endosperm catalase variants are controlled by six allelic genes (i.e., Ct^F_1, Ct^K_1, Ct^M_1, Ct^S_1, Ct^V_1, CtV'_1) at the Ct_1 locus (Fig. 1). The frequency of the six alleles differs significantly in the population examined with the Ct^S_1 allele being most common and the Ct^V_1 the least common (Scandalios, 1968). In F_1 heterozygotes between any two allelic variants (e.g. F x K, F x V, etc.), catalase is resolvable into five distinct isozymes when liquid endosperm is subjected to electrophoresis - the two parental types and three new hybrid molecules with intermediate mobility to the parental molecules (Scandalios, 1965; 1968). The hybrid catalases can also be generated in vitro by dissociation-reassociation experiments in 1M NaCl (Scandalios, 1965). The genetic and chemical data support a tetrameric structure for the catalase of maize. Since endosperm is a triploid tissue with twice the maternal genomic contribution (Fig. 2), two kinds of individuals are observed in heterozygotes depending on the direction of the genetic cross. The apparent gene dosing is useful in studying its effects on enzyme structure and function and as an internal marker to study the expression of maternal versus paternal genome. The same genes are expressed also in the scutellum (a diploid tissue) at the time of kernel development, but the dosage is that

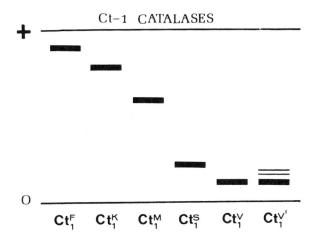

Fig. 1. Composite schematic zymogram of the six catalase variants found in immature endosperm of inbred strains of *Zea mays*. Horizontal axis indicates alleles at the Ct_1 locus. O = point of sample insertion. Migration is anodal at pH 7.4.

expected of diploid tissues (1:4:6:4:1) where the maternal and paternal genetic contributions are equal (Fig. 3).

II. *Developmental Control of Gene Expression*

The obvious question is whether the same catalase phenotypes are expressed throughout maize development. The time-course of catalase activity (Fig. 4) reflects a significant increase in early sporophytic development with a peak on the 4-5th days. It was soon determined that as seeds begin to mature, or during early germination (depending on the particular inbred strain), there are definite pattern shifts in catalase isozymes. In a given inbred strain homozygous for any allele at the Ct_1 gene, a series of "step-wise" shifts become apparent in zymogram patterns of catalase from scutellar extracts during early germination. The original Ct_1 isozyme, characteristic of the inbred strain, is totally excluded by the tenth day of sporophytic development (Fig. 5). The intermediate forms, seen at developmental stages when the Z and V isozymes overlap in their expression (e.g. 4th day) are not conformational isozymes of the Ct^V_1 allelic product; they are hybrid isozymes generated by the interaction of two kinds of polypeptide subunits (i.e., V and Z polypeptides). This has been verified by in vitro hybridization of the two extreme forms (Fig. 6). In fact, all evidence, some of which

215

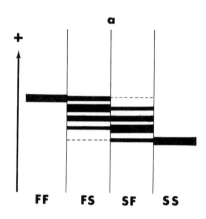

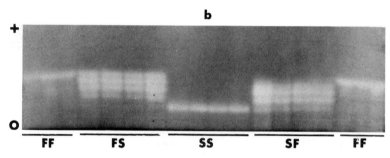

Fig. 2. Schematic (a) and zymogram (b) showing two allelic variants (F and S) and the two kinds of heterozygote patterns obtained from reciprocal F_1 crosses (F x S and S x F). The apparent gene dosage effects are due to the triploid nature of the endosperm. Migration is anodal.

will be discussed in detail in the ensuing pages, supports the hypothesis that the Z^4 isozyme is the product of a second catalase gene, Ct_2. The latter conclusion is strongly supported by a number of facts among which are: 1) Mutations affecting the Ct_1 locus (i.e., six alleles) do not simultaneously affect Ct_2; 2) Antisera prepared against Ct_1 absorb all Ct_1 allelic catalases, but do not do so for Ct_2 catalase; 3) Ct_1 and Ct_2 catalases have distinctly different turnover rates; and 4) Ct_1 and Ct_2 catalases have very distinct physicochemical properties. Thus, for any inbred maize strain homozygous for any one of the Ct_1 alleles which is expressed during kernel development, a second gene Ct_2 is also expressed, but at a different stage of development. The product of the Ct_2 gene is first detected in late seed matura-

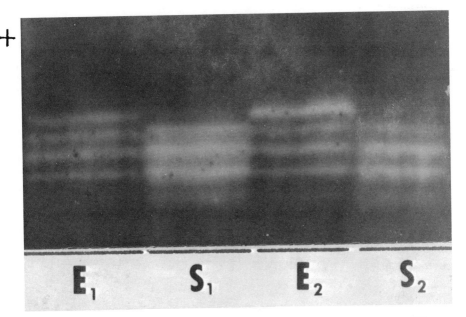

Fig. 3. Zymogram showing catalase pattern in endosperm (E) and scutellum (S) of two individual kernels (1 and 2). This experiment demonstrates that the same genes are operative in the endosperm and scutellum during kernel development. Note the characteristic gene dosage for the triploid endosperm and the diploid scutellum. F_1 kernels (F/V) were used. Migration is anodal.

tion or after seed imbibition (depending on the strain) and increases dramatically during the first four days of sporophytic development, while the Ct_1 gene product gradually disappears. At the developmental stage when both genes are simultaneously expressed and both classes of subunits are available five species of active tetramers are detected, two homotetramers (z^4 and v^4) and three heterotetramers (z^3v, z^2v^2, zv^3). Thus in the catalase system of maize, hybrid isozymes are generated by both intragenic and intergenic complementation, a fact which renders it an excellent system to examine the possible advantages and/or disadvantages of heteromultimers over the two parental homomultimers. Such information might be critical in our overall efforts to understand the mechanisms involved in heterosis.

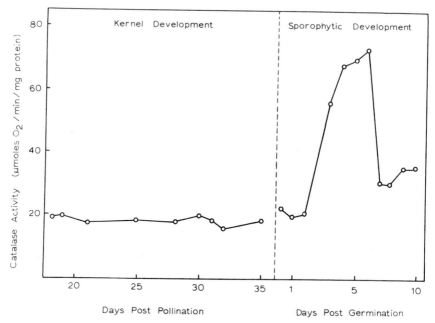

Fig. 4. Time course of catalase activity in developing kernels (days post pollination) and in scutella of germinating
maize seeds (sporophytic development). Compare to isozyme
pattern-shifts shown in Fig. 5.

III. *Factors Effecting the Differential Expression of Ct_1 and Ct_2*

The differential expression of the two catalase genes during maize development raises a number of questions with respect to the points of control. Are the major controls exerted at the transcriptional, translational, or post-translational levels? Is the newly appearing product of the Ct_2
locus synthesized *de novo* and does it turn over as it accumulates? Does the disappearance of the product of the Ct_1
locus represent degradation in the absence of further synthesis or are the subunits turning over as they disappear? Do
the rates of synthesis and degradation of the two gene products differ from one another?

a. *Differential Turnover of the Ct_1 and Ct_2 Catalases*

Some of the above questions were answered by employing the

218

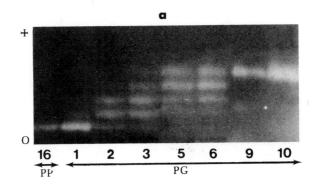

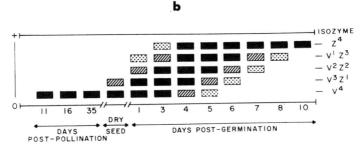

Fig. 5. Zymogram (a) and scheme (b) showing the time dependent pattern shift in catalase isozymes during early sporophytic development. (PP) = days post-pollination; (PG) = days
post-germination. Numbers indicate days. Scutellar extracts
were used since endosperm is degraded during germination. The
V^4 homotetramer is composed of subunits from the Ct^V_1 locus;
this isozyme is present in immature kernels as well as the
early seedling stages. The Z^4 homotetramer is composed of
subunits from the Ct_2 locus. This locus is expressed at the
onset of germination. The intermediate isozymes are heteromultimers composed of subunits from both genetic loci.
Migration is anodal.

techniques of gel electrophoresis and density labeling in conjunction with isopycnic equilibrium sedimentation (Quail and
Scandalios, 1971). This technique allows detection and quantitation of *de novo* synthesis without recourse to enzyme isolation and purification. Results from such experiments show
that both catalase isozymes V^4 (product of the Ct_1 gene) and
Z^4 (product of the Ct_2 gene) become density labeled during
the first 36 hours of germination (Fig. 7), indicating that
both classes of subunits are synthesized *de novo*. Since
there is a net loss of V^4 during germination (see Fig. 5),

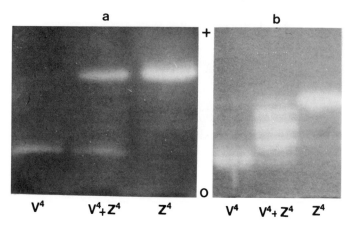

Fig. 6. Zymograms showing the effect of mixing the V^4 (Ct_1) and Z^4 (Ct_2) isozymes in phosphate buffer (pH 7.0) and subjecting to electrophoresis (a) and the pattern obtained by first freezing and thawing the mixture in phosphate buffer (pH 7.0) containing 1M NaCl as previously described (Scandalios, 1965). Note generation of hybrid molecules in vitro (b).

this isozyme is obviously being degraded as well as synthesized and is therefore turning over as it disappears. To determine whether or not Z^4 is also degraded, a pulse-chase density labeling experiment was designed to investigate this possibility and to compare the rates at which the V^4 and Z^4 isozymes decrease in density during the chase period. Fig. 8 shows the density gradient activity profiles obtained for the unlabeled and fully labeled catalase isozymes, along with those for chase periods of 6, 12, 18, and 24 hours. Both isozymes shift in density back towards the density of the unlabeled protein during the chase period. This is further evidence that V^4 is turning over, and on considering some other parameters (Quail and Scandalios, 1971), it also supports the idea that the Z^4 isozyme is turning over. Thus both catalase homotetramers (V^4 and Z^4) turn over in the scutella of maize seeds during germination--Z^4 as it accumulates; V^4 as it disappears. Z^4 accumulates because the rate of synthesis exceeds the rate of degradation, whereas V^4 slowly disappears because the rate of degradation exceeds the rate of synthesis. The results further indicate that the rate of synthesis of Z^4 exceeds that of V^4, suggesting that this may be a major factor in the differential expression of the two catalase genes.

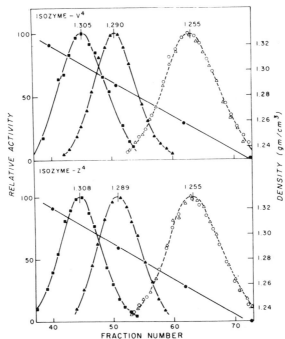

Fig. 7. Equilibrium distribution in CsCl gradients of scutellar catalase isozymes V^4 and Z^4 from seeds grown for 36 hours on either $K^{14}NO_3$ in H_2O (▲——▲) or $K^{15}NO_3$ in 70% D_2O (■——■). The activity of the lactate dehydrogenase (LDH) marker in the labeled (O--O) and unlabeled (△·-△) gradients have been superimposed and drawn as one. Relative activity means that all points on these curves are expressed as a percentage of the highest point on each of the individual curves. Density of CsCl gradient (●——●).

b. *Regulation by an Endogenous Catalase-Specific Inhibitor*

In addition to the above, we have recently found evidence of a catalase-specific inhibitor which appears to be differentially active during the first several days after germination. When crude scutellar homogenates from 1 and 4-day seeds (after imbibition) are mixed, the activity of the mixture is less than the sum of the activities added (Fig. 9). Experiments by J. C. Sorenson in my laboratory, clearly show that the inhibitor effect follows linear dose-response kinetics (Fig. 10). The inhibitor has been shown to be present in the 1-day extract but absent in the 4-day extract. Initial attempts at characterization show that the inhibitor is heat labile and non-dialyzable, suggesting that it may be a protein. Experiments are presently underway to purify the

221

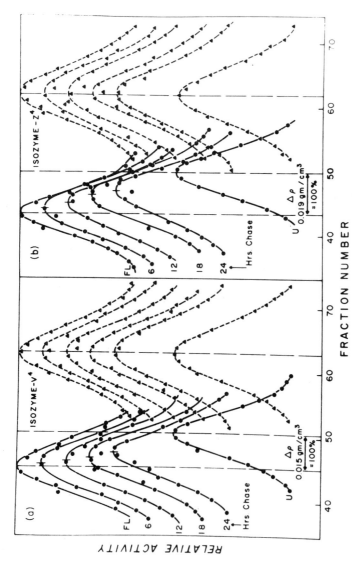

Fig. 8. Equilibrium distribution in CsCl gradients of scutellar catalase isozymes V^4 and Z^4 (●─●). Seeds were grown for 36 hours on Kl^4NO_3 in H_2O (U = unlabeled); 36 hours on Kl^5NO_3 in 70% D_2O (FL = "fully labeled"); or 36 hours on Kl^5NO_3 in 70% D_2O and then transferred to Kl^4NO_3 in H_2O for a further 6, 12, 18, or 24 hours. Peak centers of the LDH markers (◀─◀) from the gradients have been aligned so that their densities coincide. Density decreases from left to right.

222

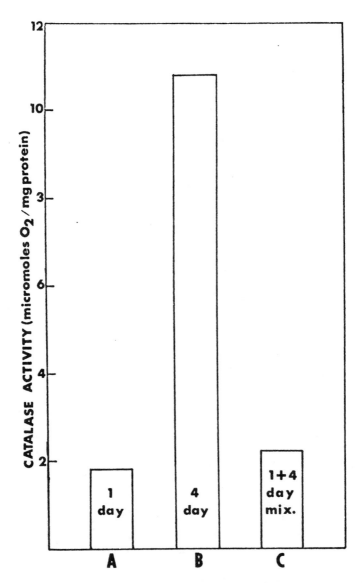

Fig. 9. Mixing experiments demonstrating the presence of a catalase-inhibitor in the 1-day extract. (A) Activity of scutellar extracts from 1-day imbibed kernels; (B) Activity of scutellar extracts from 4-day old seedlings; (C) Activity observed in a 1:1 mixture of (A) + (B). The reduction in catalase activity of the mixture is much greater than expected from a simple mixture. Expected activity = (A) + (B) x 2^{-1}.

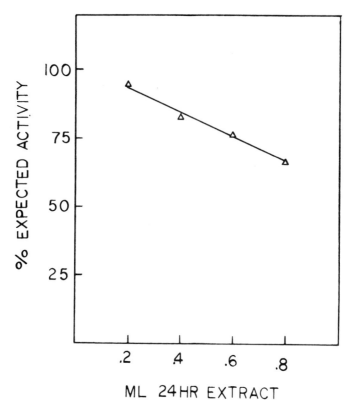

Fig. 10. Inhibitor dose-response curve. Various amounts of crude 24 hour scutellar extract were added to a constant amount of crude 96 hour extract. All assay volumes adjusted to 1 ml with buffer.

inhibitor in an effort to understand its physicochemical nature and its mode of action. A high degree of catalase specificity is suggested by the fact that the inhibitor from maize is fully active on beef liver catalase, but has no effect on maize peroxidases, a group of catalytically related hemoproteins (Table I). The possibility that the inhibitor is a protease has been ruled out by the high stability of the unmixed extracts, the extremely rapid inhibition kinetics (the reaction is essentially complete within 30 seconds), and the stability of the mixture after the initial inhibition reaction.

Table I. Specificity of the maize catalase inhibitor.

Sample	Catalase Activity[1] (U/ml)	Peroxidase Activity[2] (U/ml)
A) Day-1 Extract (Maize)	144	12.6
B) Day-4 Extract (Maize)	73	68.6
C) Beef Liver Catalase	77	-
D) Expected Activity of (A+B) Mix. (1:1)	109	40.6
E) Observed Activity of (A+B) Mix.	62	41.0
F) Inhibition (%)	43%	0.0
G) Expected Activity of (A+C) Mix. (1:1)	111	-
H) Observed Activity of (A+C) Mix.	69	-
I) Inhibition (%)	38%	-

[1]Catalase activity = micro-moles O_2 evolved/min.
[2]Peroxidase activity = chart units per min using o-dianisidine.

c. *Glyoxysomal Localization of Catalase Isozymes*

During kernel development, catalase is not associated with any distinct subcellular organelles. Whatever catalase is present is always associated with the soluble cytoplasm. However, as sporophytic development proceeds from the dry seed stage, glyoxysome biogenesis in scutella becomes apparent, reaching a peak at the fourth day of germination (Longo and Longo, 1970). It is at this time in scutellar development when approximately 25-30% of the total catalase present is associated with the glyoxysomes. None is associated with mitochondria, and the bulk is still associated with the soluble cytoplasmic fraction (Scandalios, 1974). There is no particular catalase isozyme specific to glyoxysomes; the isozyme complement of the organelles parallels that of scutellar development as a whole (Fig. 11; see also Fig. 5).

Since the Ct_2 gene is expressed at a time in development when cellular catalase is becoming increasingly compartmentalized, it is possible that the need for the second gene is to produce a more efficient catalase. The fact that both Ct_1 and Ct_2 catalases are found associated with glyoxysomes suggests non-specificity of the compartmentation process. Furthermore, the degree of compartmentation along with the presence of a catalase-specific inhibitor may be another factor controlling the Ct_1/Ct_2 catalase ratio in early sporophytic development.

IV. *Biochemical Properties of the Ct_1 and Ct_2 Catalases*

The fact that hybrid catalase isozymes can be generated in vivo by both intragenic (allelic) and intergenic (non-allelic) complementation affords an opportunity to study the possible functional advantages and/or disadvantages of heteromultimers over the parental homomultimers. Such information is essential in our efforts to understand the molecular mechanisms underlying such concepts as single gene heterosis, fitness, codominance, etc. Most important, knowledge of the physico-chemical properties of the various isozymes will afford us an opportunity to understand further the need for enzyme multiplicity in relation to the developmental and physiological needs of the organism, and perhaps shed some light on the basic question - Why isozymes?

We have demonstrated (Scandalios et al, 1972), that the homotetrameric and heterotetrameric catalases of maize differ significantly with respect to their biochemical properties (i.e., specific activity, K_m, photosensitivity, thermosta-

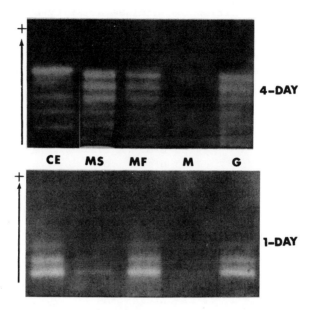

Fig. 11. Zymograms showing the distribution of catalase isozymes in different subcellular fractions isolated from maize
scutella of 1-day (bottom) and 4-days (top) after imbibition.
CE = crude extract; MS = mitochondrial supernatant (14,800g
supernatant); MF = mitochondrial fraction (14,800g pellet);
M = mitochondria; G = glyoxysomes. Latter two fractions were
isolated by sucrose gradient centrifugation. Purity of organelles was established by electron microscopy and by marker
enzymes (isocitratase and malate synthetase for glyoxysomes,
and cytochrome oxidase for mitochondria). Migration is
anodal.

---

bility, inhibitor sensitivity, etc.). In most instances the
heterotetramers generated by either intragenic or intergenic
complementation exhibit improved physicochemical properties
over the least efficient parental molecules (Tables II, III,
and IV); this suggests that hybrid enzymes may be advantageous to the organism. None of the catalases differed with
respect to molecular weight which was determined to be on the
order of 280,000 ± 10,000. In vitro hybridization experiments provide further evidence that interaction of non-ident-
ical subunits generates hybrid molecules with enzymatic activity higher than that of the parental homomultimers and independent of the total protein concentration (Fig. 12). The
pH optimum for all catalases was determined to be between pH
5-11 with a broad maximum from pH 6-8.

Table II. Inhibition of catalase activity by incubation in 10^{-4} M 3-amino-1, 2, 4-triazole for 1 hour at 25°C. Source of the isozymes was the F_1 heterozygote F X V. Protein concentration in each sample was adjusted to 0.2 mg per ml. Results are average of 3 independent experiments.

Catalase isozyme	Control	10^{-4} M AT*	% Inhibition
F^4	2.10 U/ml	0.84 U/ml	60.0
F^3V^1	3.23 U/ml	2.02 U/ml	37.5
F^2V^2	4.60 U/ml	2.11 U/ml	54.0
F^1V^3	2.25 U/ml	0.85 U/ml	62.3
V^4	0.65 U/ml	0.20 U/ml	69.7

*3-amino-1, 2, 4-triazole (AT) is a potent inhibitor of catalase by binding to the protein moiety of the enzyme molecule.

The above data suggest that the geometry of the subunits in catalase is important in the activity of the enzyme and make plausible the notion that arrangements of catalase subunits in the enzyme have some effect on the expression of enzyme activity. It is interesting to note that the differences between allelic gene products are less dramatic than those between the catalases coded by nonallelic genes. The Ct_1 and Ct_2 catalases differ significantly in most properties examined but they differ most dramatically with respect to two parameters: 1) The stability of the enzyme as measured by the heat denaturation of catalase at 50°C where all the Ct_1 allelic isozymes have a half-life of less than 10 minutes while Ct_2 catalase has a $t_{\frac{1}{2}}$ = 141 minutes - a more than 20-fold difference in thermostability; and 2) The photosensitivity of the enzyme where the Ct_2 catalase was shown to be at least five-times less sensitive ($t_{\frac{1}{2}}$ = 25 minutes) to light than the Ct_1 catalases ($t_{\frac{1}{2}}$ = 6 minutes). The half-lives of the hybrid isozymes reflect their subunit composition with intermediate sensitivities (Figs. 13 and 14). In all cases, the heteromultimers are catalytically more efficient than the least efficient parental homotetramer. Such differences in the physicochemical properties between V^4 and Z^4 catalases may be physiologically significant, since the Ct_2 gene, producing a more efficient catalase, is expressed at a time in development when there is a turning on of major metabolic processes essential for development and differentiation of the maize plant; it is at this stage that the root and shoot emerge.

Tissue (Organ) Specific Catalase Isozymes

In addition to the temporal changes in catalases discussed

Table III. Comparative kinetic properties of the isozymes formed from the random association of subunits from two alleles, Ct^F_1 and Ct^V_1, of the Ct_1 locus.

Electrophoretic Band	Isozyme	$K_{app}(H_2O_2)$	Thermostability ($t_{\frac{1}{2}}$ at 50°C)	Photosensitivity ($t_{\frac{1}{2}}$)
1	F^4	.047 M	3.7 min	7.0 min
2	F^3V^1	.047 M	5.8 min	6.8 min
3	F^2V^2	.035 M	5.0 min	7.0 min
4	F^1V^3	.045 M	6.0 min	7.8 min
5	V^4	.143 M	6.5 min	6.8 min

Table IV. Comparative kinetic properties of the isozymes formed from the random association of subunits from two genetic loci, Ct^V_1 and Ct^V_2.

Electrophoretic Band	Isozyme	$K_{app}(H_2O_2)$	Thermostability ($t_{\frac{1}{2}}$ at 50°C)	Photosensitivity ($t_{\frac{1}{2}}$)
1	Z^4	0.057 M	141 min	25.0 min
2	Z^3V^1	0.051 M	46 min	20.5 min
3	Z^2V^2	0.038 M	20 min	15.0 min
4	Z^1V^3	0.083 M	10 min	8.0 min
5	V^4	0.143 M	6.5 min	6.8 min

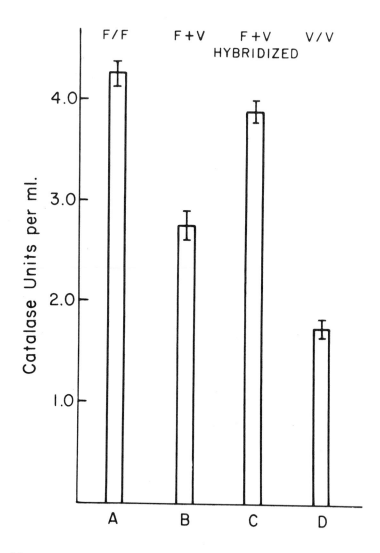

Fig. 12. Results of in vitro hybridization of the catalase homotetramers V^4 and F^4; extracts were obtained from the respective homozygous inbred strains. Catalase activity in the sample containing the F^4 = 13.25 ± 0.10 U/ml; activity of V^4 = 5.41 ± 0.08 U/ml. A mixture containing 40% of the F^4 extract and 60% of the V^4 extract had an activity of 8.49 ± 0.08 U/ml. This compares well with the theoretical catalase activity of such a mixture which is: (0.4) (13.25) + (0.6) (5.41) = 8.54 U/ml. On subjecting the F^4 and V^4 extracts,

as well as the mixture, to freezing and thawing in 1M NaCl about 70% of the catalase activity is lost. After thawing, the catalase activity of the F^4 = 4.24 ± 0.09 U/ml (A); V^4 = 1.73 U/ml (D). If there were no interaction between the F and V subunits, the expected catalase activity of the $40F^4$: $60V^4$ mixture would be: (0.4) (4.24) + (0.6) (1.73) = 2.73 ± 0.09 U/ml. However, the actual catalase activity of this mixture after hybridization is 3.89 ± 0.09 U/ml (C).

above and spatial differences reported earlier (Scandalios 1964; 1969), we have recently encountered a number of situations where a particular catalase appears in a given tissue at a specific period in development. In the pericarp, the situation exists where a unique catalase isozyme appears at the twenty-fifth day of kernel development (tentatively named Ct_3) and another at the seventh day of sporophytic development (tentatively named Ct_4). At the respective stage, either Ct_3 or Ct_4 is expressed in the pericarp along with the Ct_1 allele characteristic of the particular inbred strain (Figs. 15 and 16). Both Ct_3 and Ct_4 catalases seem to be independent of mutations affecting Ct_1 catalases and are probably products of distinct genes. The particular catalase complements detected thus far in pericarp are summarized in Fig. 17. In one inbred strain there are two isozymes at the Ct_4 position. The Ct_4 catalase is very precise in its time of expression at the seventh day with only minor traces of activity at the sixth and eighth days. The phenotypes detected thus far with respect to Ct_3 are shown in Fig. 18. In two rare cases there appeared to be subunit interactions between Ct^V_1 and Ct_3 subunits to generate hybrid molecules resulting in five isozymes in the pericarp (i.e., last frame Fig. 18). If indeed the subunits are different, why are such interactions not common whenever Ct_1 and Ct_3 are present? What factors control the interactions of catalases in pericarp? These are some questions we are presently attempting to answer.

DISCUSSION

The catalase isozyme system presents an opportunity to study gene regulation in eukaryotic cells because of the multiplicity of regulatory mechanisms operative during a period of intense developmental activity, the relative ease with which these mechanisms can be studied, the solid genetic background of maize, and the high suitability of the organism for experimental manipulation.

The results dealing with the interaction of genes, both

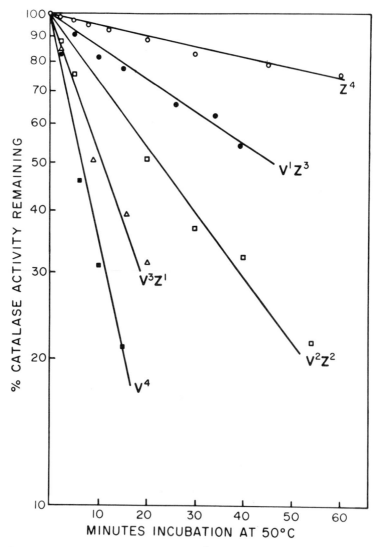

Fig. 13. Heat stability of the Z^4 (coded by Ct_2) and V^4 (coded by Ct_1) homotetramers and their interaction products (V^3Z^1, V^2Z^2 and V^1Z^3). Isozymes were isolated from starch gels and extracted by centrifugation. Individual isozymes were then placed in a 50.0°C water bath and catalase activity measured at timed intervals. The percent catalase remaining at any given time is calculated by comparison to the original activity and the results plotted on log scale.

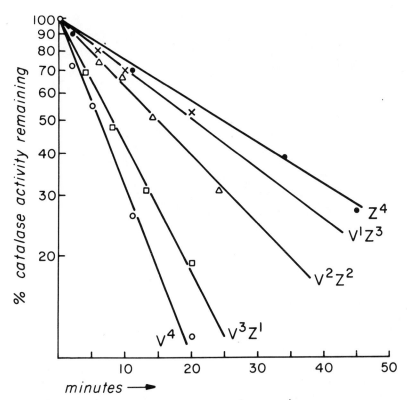

Fig. 14. Photosensitivity of the Z^4 and V^4 catalases and the hybrid isozymes generated by their subunit interactions. Extracts of individual isozymes are subjected to identical illumination conditions and activity of catalase was measured at definite time intervals. The percent catalase remaining at any given time is calculated by comparison to the original activity and plotted on a log scale.

allelic and nonallelic, to generate hybrid molecules render the catalase system an excellent one for understanding the mechanisms involved in single gene heterosis and the role of isozymes in terms of the physiology and metabolic efficiency of the organisms carrying them.

The presence of a catalase-specific inhibitor presents an opportunity to examine it not only for its intrinsic role in the overall scheme of catalase regulation, but as a regulatory mechanism in and of itself. The well documented protease inhibitors in several plant systems (Ryan, 1973) and the amylase inhibitors in wheat (Shainkin and Birk, 1970) appear

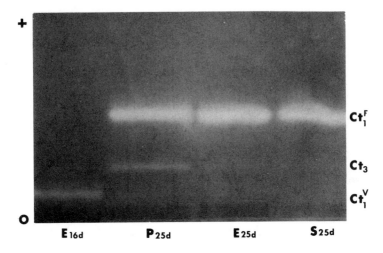

Fig. 15. Zymogram showing the Ct_3 catalase characteristic of pericarp tissue at 25-days after pollination. E_{16d} = endosperm from 16-day old Ct^V_1 kernels (marker); P_{25d} = pericarp E_{25d} = endosperm, S_{25d} = scutellum, all from 25-day old kernels of the genotype Ct^F_1. Migration is anodal.

not to inhibit endogenous enzymes. Conversely, the catalase inhibitor appears to be directly involved in regulating endogenous cellular catalase activity.

The subcellular distribution of catalase in the glyoxysomes may shed some light on a possible physiological role for the catalatic reaction of catalytic. Since glyoxysomes are known to contain oxidases (DeDuve and Baudhuin, 1966) which reduce oxygen to hydrogen peroxide, catalase which may act normally peroxidatically (with a rate constant k = $10^3 M^{-1} sec^{-1}$) may switch to its catalatic reaction (which has a rate constant k = $10^7 M^{-1} sec^{-1}$) as a safety mechanism to purge the cells of noxious amounts of H_2O_2 in the absence of a sufficient supply of hydrogen donors.

The results pertaining to tissue specific isozyme patterns pose questions as to the mechanisms involved in developing and maintaining such differences by organisms whose somatic cells have an identical genome. Some possible mechanisms are: differential gene action, differential availability of subunits, presence or absence of cofactors required to activate specific isozymes, differential rates of synthesis and degradation, differential subcellular localization, etc. In any event, the enzyme activity of an individual isozyme in a specific tissue is dependent on and reflects the maturity and cellular environment of that tissue.

PERICARP

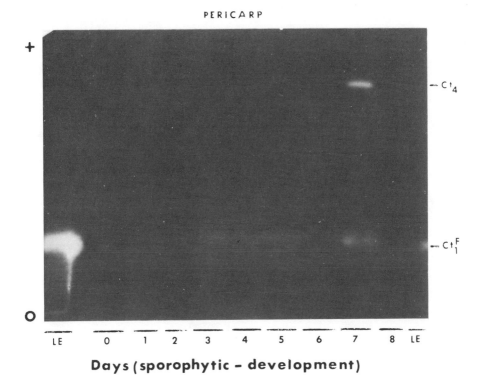

Days (sporophytic - development)

Fig. 16. Zymogram showing the time-dependent expression of the Ct_4 catalase in pericarp from kernels 8-days after germination. LE = endosperm from 16-day old kernels as markers. Migration is anodal.

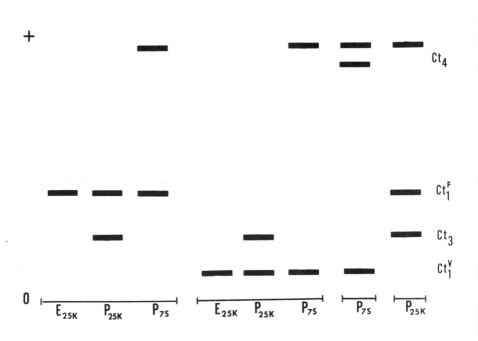

Fig. 17. Composite schematic zymogram showing the various tissue specific catalase patterns observed in different inbred strains. E = endosperm; P = pericarp; S = seedling; K = kernel; numbers on horizontal axis refer to day of development.

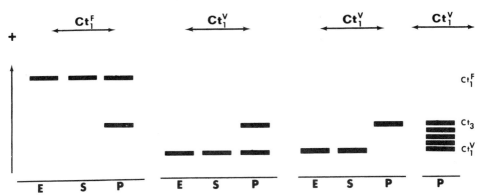

Fig. 18. Schematic zymogram showing kernel tissues from different inbreds of the genotype indicated on top. E = endosperm; S = scutellum; P = pericarp; all are from 25-day old kernels. In several inbred strains interaction between Ct_1 and Ct_3, resulting in hybrid isozymes, has been observed.

ACKNOWLEDGEMENT

Work supported by the U.S. Atomic Energy Commission under Contracts AT(11-1)-1338 and AT(38-1)-770.

REFERENCES

DeDuve, C. and P. Baudhuin 1966. Peroxisomes (microbodies and related particles). *Physiol. Rev.* 46: 323-357.

Longo, C.P. and G.P. Longo 1970. The development of glyoxysomes in peanut cotyledons and maize scutella. *Plant Physiol.* 45: 249-254.

Maletsky, S.I., E.V. Polyakova, E.V. Levites, and A.V. Aksenovich 1971. Molecular variants of catalase in inbred maize strains. *Isozyme Bull.* 4: 40.

Quail, P.H. and J.G. Scandalios 1971. Turnover of genetically defined catalase isozymes in maize. *Proc. Natl. Acad. Sci. U.S.A.* 68: 1402-1406.

Rechcigl, M. (Ed.) 1971. *Enzyme Synthesis and Degradation in Mammalian Systems.* University Park Press, Baltimore.

Ryan, C.A. 1973. Proteolytic enzymes and their inhibitors in plants. *Ann. Rev. Plant Physiol.* 24: 173-196.

Scandalios, J.G. 1965. Subunit dissociation and recombination of catalase isozymes. *Proc. Natl. Acad. Sci. U.S.A.* 53: 1035-1040.

Scandalios, J.G. 1968. Genetic control of multiple molecular forms of catalase in maize. *Ann. N.Y. Acad. Sci.* 151: 274-293.

Scandalios, J.G., E. Liu, and M.A. Campeua 1972. The effects of intragenic and intergenic complementation on catalase structure and function in maize: A molecular approach to heterosis. *Arch. Biochem. Biophys.* 153: 695-705.

Scandalios, J.G. 1974. Subcellular localization of catalase variants coded by two genetic loci during maize development. *J. Heredity* 65: 28-32.

GENETIC AND DEVELOPMENTAL REGULATION OF DISTINCT SULFATE PERMEASE SPECIES IN *NEUROSPORA CRASSA*

GEORGE A. MARZLUF
Department of Biochemistry
Ohio State University
Columbus, Ohio 43210

ABSTRACT. Sulfate uptake in *Neurospora crassa* occurs by an active transport system which is subject to an intricate and complex set of regulatory signals. This organism possesses two distinct sulfate permeases, one of which predominates in the mycelial stage while a second form is the only one present in conidiospores. The synthesis of both of these sulfate permeases is repressed by a metabolite derived from methionine. The synthesis of both permease species, as well as a number of related enzymes of sulfur metabolism, is also controlled by at least two unlinked regulatory genes, which involve both positive and negative signals. Superimposed upon the genetic and metabolic controls, the biosynthesis of the distinct permease isozymes is also subject to developmental regulation during spore germination and outgrowth.

The activity of the sulfate transport systems is regulated by transinhibition, which is not due to an intracellular pool of inorganic sulfate, but is instead exerted by an early intermediate of the assimilatory pathway. Finally, the sulfate permeases are subject to dynamic turnover, such that once new synthesis has terminated, the remaining activity is inactivated and declines to a low level.

SULFUR-CONTROLLED ENZYMES

In *Neurospora crassa* a family of related enzymes which function in the acquisition of sulfur from the environment are subject to an intricate set of genetic and metabolic controls. A metabolite derived from methionine, probably cysteine, acts to repress the synthesis of this entire group of enzymes which include aryl sulfatase, choline sulfatase, choline-0-sulfate permease, an aromatic sulfate permease, a specific methionine permease, an extracellular protease, and two permeases for inorganic sulfate (Metzenberg and Parson, 1966; Marzluf, 1970a; Marzluf, 1972a; Pall, 1971; Hanson and Marzluf, 1973). This same group of enzymes is controlled by at least two unlinked regulatory genes designated *cys-3* and *scon*, which cause a pleiotropic loss and constitutive presence

239

of the enzymes, respectively (Marzluf and Metzenberg, 1968; Burton and Metzenberg, 1972). The presence of dual transport systems, or isozymes, for the uptake of inorganic sulfate within this family of sulfur-controlled enzymes is of particular interest. Their properties and regulation will be the main subject of the remainder of this paper.

SULFATE TRANSPORT IN NEUROSPORA

Sulfate uptake in Neurospora occurs by an active transport process. It can be calculated that a single conidiospore can take up approximately 8 billion molecules of inorganic sulfate per min. The ion is transported into the cell against a concentration gradient and common inhibitors of energy coupling such as azide or dinitrophenol cause an immediate and complete inhibition of sulfate uptake. The observation that sulfate transport is also quite sensitive to inhibition by p-chloromercuribenzoate indicates that the permease possesses an essential and exposed sulfhydryl group (Marzluf, 1974). The toxic ion, chromate, is also transported by the sulfate transport system, which provided a basis for the isolation of mutants lacking the uptake system (Roberts and Marzluf, 1971).

Neurospora possesses two distinct sulfate permeases. Conidia possess only sulfate permease I while a second form predominates in the mycelial stage (Marzluf, 1970a). These two transport systems can easily be distinguished because the conidial system (sulfate permease I) has a dramatically higher K_m for sulfate than does sulfate permease II of mycelia (Fig. 1). The two permease species are also encoded by separate structural genes which are unlinked to one another. The *cys-13* gene, located on chromosome 1, encodes sulfate permease I while the gene for permease II, designated *cys-14*, is carried on chromosome 4 (Marzluf, 1970b).

Table 1 shows that conidia of the *cys-13* mutant strain completely lack the capacity for sulfate transport while mycelia of this same strain have a normal level of sulfate permease. The *cys-14* mutant lacks permease II but continues to synthesize the high K_m form characteristic of conidia during the mycelial phase, when wild type has switched to the second sulfate permease species. As would be expected, *cys-14* conidia have a normal amount of sulfate permease activity. The double mutant, *cys-13 cys-14*, is completely deficient in sulfate transport and cannot use inorganic sulfate for growth.

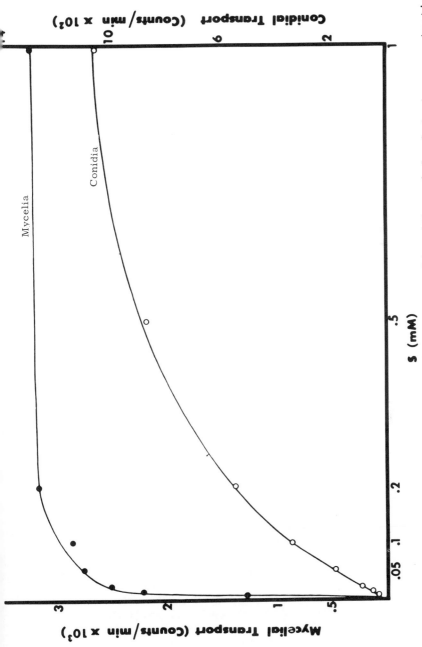

Fig. 1. Sulfate transport by wild-type conidia and mycelia with varied substrate concentrations. The details of the assays are reported elsewhere (Marzluf, 1970). Sulfate permease I, which has a low affinity for sulfate is found in conidia while mycelia possess primarily sulfate permease II.

TABLE 1

Transport of inorganic sulfate by conidia and mycelia
of wild-type and mutant strains[a]

Strain	Rate of sulfate transport	
	Conidia	Mycelia
	%	%
Wild-type	100	100
cys-3	0	0
cys-13 (Permease II)	0	100
cys-14 (Permease I)	100	25
cys-13; cys-14	0	0

[a] The uptake by the transport mutants is compared with the
wild-type strain which is arbitrarily assigned a value of 100%
for both mycelial and conidial transport. Details of the
assays and uptake values are reported elsewhere (Marzluf,
1970a).

GENETIC AND METABOLIC CONTROL
OF THE SULFATE PERMEASES

A complex set of genetic and metabolic controls, which
include both positive and negative signals, regulates in parallel the synthesis of both sulfate permease species. The
synthesis of these permeases is strongly repressed by a metabolite derived from methionine, an end product of sulfate assimilation. Sulfate permease II is also partially repressed by
adequate amounts of inorganic sulfate but the conidial uptake
system (Permease I) is not detectably repressed by sulfate.
This difference may be at least partially explained if an altered metabolism occurs during spore development which results
in a lower concentration of a corepressor derived from sulfate.
The synthesis of both sulfate permeases is also regulated
by *cys-3*, a positive control gene which is unlinked to either
cys-13 or *cys-14*. A functional product of the *cys-3* locus,
probably a regulatory protein, is required to activate the
synthesis of both permeases and several other related enzymes
as well, *cys-3* mutants having a pleiotropic loss of all of
these activities (Marzluf and Metzenberg, 1968; Marzluf,
1970a). Nonetheless, the structural genes for the two sulfate
permeases are not equally sensitive to the *cys-3* regulatory
product. This is apparent because different partial revertants

of the *cys-3* mutant have completely disproportionate levels of sulfate permease I and permease II (Table 2). This differential response is presumably due to receptor sites, perhaps a promoter region, adjacent to each of the two structural genes for the˙permeases, being somewhat different in their affinity for the common regulatory signal.

TABLE 2
Fractional restoration of enzyme activities
in selected *cys-3* revertants[a]

Revertant	Percent of wild type activity		
	mycelial sulfate permease (m)	conidial sulfate permease (c)	Ratio m/c
BPL − 30	2.1	66.5	0.03
BPL − 4	5.2	79.0	0.07
I CR − 8	7.9	57.5	0.14
DMS − 25	16.0	92.5	0.17
UV − 91	17.5	82.4	0.21
UV − 65t	18.5	23.0	0.80
UV − 112	19.4	19.6	0.99
XR − 51	39	11.0	3.55
EMS − 8	68.5	44.5	1.54
Wild-type	100	100	1.0
cys-3 mutant	0	0	−

[a] Transport assays were conducted as described before (Marzluf, 1970a) and the values reported are the average from at least two completely independent determinations. Equally disproportionate restoration of aryl sulfatase and choline sulfatase was also found with these revertants.

Mutation in another control gene, *scon*, which is unlinked to *cys-3*, *cys-13*, or *cys-14* eliminates the repression by both methionine and sulfate so that the synthesis of both permeases is constitutive (Burton and Metzenberg, 1972). The regulatory activity of this most interesting gene was shown to be restricted to its own nucleus whereas the *cys-3* signal can be transmitted from one nucleus to another (Burton and Metzenberg, 1972).

GEORGE A. MARZLUF

CONTROL OF SULFATE PERMEASE
ACTIVITY BY TRANSINHIBITION

Regulation of biosynthetic pathways is often achieved by feedback inhibition in which an end product inhibits the activity of the first enzymatic step. In the case of transport systems, such feedback inhibition is also referred to as transinhibition because the effect is exerted across the cell membrane and is presumably due to the interaction of the effector with sensitive sites on the inner membrane surface. The activity of both sulfate permease I and sulfate permease II is subject to control by transinhibition. When Neurospora cells were preloaded to provide an intracellular pool of nonradioactive sulfate, subsequent sulfate transport activity was strongly inhibited, suggesting that intracellular sulfate or a metabolite derived from it exerts feedback inhibition (Marzluf, 1973). Fig. 2 displays the results of transinhibition of sulfate permease I in conidia; similar transinhibition is found when mycelia are studied in the same manner. A possible alternative explanation for these results is that sulfate which enters during the preincubation period acts instead to very rapidly repress further synthesis of sulfate permease, which would mimic the effects of true feedback inhibition.

To distinguish between these possibilities, mutants with specific blocks in the sulfate assimilatory pathway which prevent the metabolism of sulfate to the true corepressor, e.g., mutants lacking sulfite reductase, were studied. With such mutants it was possible to demonstrate that feedback inhibition did take place and at its usual level in strains where repression could not occur. Most interestingly, however, a mutant strain which lacks ATP-sulfurylase, the very first enzyme in the sulfate assimilatory pathway, accumulated a large intracellular pool of inorganic sulfate but did not display feedback inhibition of sulfate permease activity. By contrast, dramatic feedback inhibition did occur in mutants which block at a point a few steps later in the pathway and lack sulfite reductase. Thus, it is clear that sulfate itself is not the feedback inhibitor, nor is either cysteine or methionine, but rather an early intermediate of the assimilatory pathway has this role. Intracellular 3'-phosphoadenosyl-5'-phosphosulfate (PAPS) is suspected to be the actual effector; PAPS is not only an intermediate in the pathway for cysteine and methionine biosynthesis, but is also the sulfate donor for a number of other biological reactions. This is a point where one might predict control would be exerted since when the cell has sufficient endogenous PAPS to supply its various synthetic requirements, it would be advantageous to limit further sulfate uptake.

244

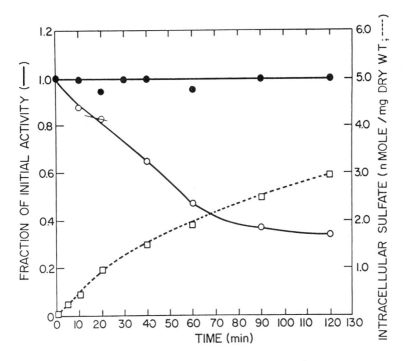

Fig. 2. Transinhibition of sulfate transport in wild type conidia. Conidia were preloaded with nonradioactive sulfate for various time intervals (0-120 min) when radioactive sulfate was added and its transport rate was determined with a 5 min assay. Control samples were incubated in a parallel manner except that the unlabeled sulfate was omitted. (●), control permease activity; (O), permease activity of samples preloaded with sulfate; the dashed line displays the intracellular pool of sulfate present after the various preincubation periods.

TURNOVER OF THE SULFATE PERMEASES

In addition to controlling the rate of enzyme synthesis at either the transcriptional or translational level, control of the quantity of cellular enzymes can be achieved by turnover. Both sulfate permease I and permease II are subject to dynamic turnover so that their immediate level represents the steady-state balance between their rates of synthesis and decay (Marzluf, 1972b). When new sulfate permease synthesis was prevented by the specific repression caused by methionine or by the inhibition of protein synthesis with cycloheximide, the permease originally present was not stable but disappeared with a half-life of approximately 2 hr (Fig. 3). The use of a *cys-3* temperature sensitive mutant made possible a study of turnover without the uncertainties usually present when inhibitors are employed. Upon transferring this mutant to the restrictive temperature, which prevents additional synthesis of the sulfate transport systems but which otherwise permits normal growth to continue, both permeases underwent turnover and again with a half life of about 2 hr.

Turnover of both sulfate permease I and sulfate permease II occurs rapidly enough to contribute to the control of the pathway for sulfate metabolism; inactivation of at least one essential element of each permease eliminates transport activity once *de novo* synthesis of the permease has been stopped. Another valuable point of control could be achieved if an end product accelerated the rate of turnover of an early enzyme of the relevant biosynthetic pathway; however, in this case methionine does not alter the turnover rate of the sulfate permeases.

One interesting difference does exist between conidia and mycelia in the turnover of the sulfate permease systems. In the presence of cycloheximide, a general inhibitor of cytoplasmic protein biosynthesis, sulfate permease I of conidia is fully stable and does not turnover, while the transport system of mycelia displays the usual rate of inactivation when this inhibitor is present. This outcome implies that the sulfate permeases do not have an innate lability but are subject to the activity of a turnover system. It appears that dormant conidia lack the enzyme system responsible for sulfate permease turnover, but that it is rapidly synthesized during germination and then serves to inactivate the permease.

Although the synthesis of the family of related enzymes of sulfur metabolism, including aryl sulfatase and the dual sulfate permeases, is controlled in parallel, interesting differences are found in the susceptibility of these same enzymes to turnover. Aryl sulfatase is a completely stable enzyme under all conditions which have been studied, while both permeases show significant turnover as just described. Since aryl sulfatase is an internal enzyme while the permeases are in contact with the extracellular environment by virtue of their membrane location, more immediate control over permease activity may be valuable. On the other hand, it is plausible that membrane-bound proteins, including many transport systems, are generally more susceptible to turnover than are most other classes of cellular proteins.

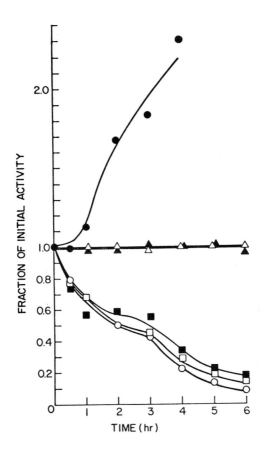

Fig. 3. Turnover of sulfate permease in wild type mycelia. After outgrowth of conidia into mycelia under conditions which permitted full derepression of sulfate permease activity, additional permease synthesis was abolished by addition of cycloheximide or repressing concentrations of methionine, or both. Mycelia in control flasks continued to synthesize the permease activity. Sulfate permease assays were conducted immediately following the additions and at intervals for 6 hr. Sulfate permease activities: (●), control; (O), high methionine; (■), cycloheximide; (□), high methionine plus cycloheximide. Possible turnover of aryl sulfatase was also followed for 6 hr. after the addition of cycloheximide (Δ) or repressing concentration of methionine (▲).

DEVELOPMENTAL REGULATION OF THE SULFATE
PERMEASE ISOZYMES DURING SPORE GERMINATION

The observation that conidia possess only sulfate permease I, the high K_m form encoded by *cys-13*[+], while permease II is the predominant form in actively growing mycelia indicates that these isozymes are developmentally regulated enzymes. Fig. 4 shows the development of sulfate transport activity during germination and the subsequent outgrowth of conidia into the mycelial stage. During germination of wild-type conidia, turnover of the pre-existing permease I present in the resting spores occurs but it is being continuously replaced by *de novo* synthesis of both permease I and permease II so that an increasing level of sulfate permease develops (Marzluf, 1972b). That permease I is the only form present in dormant conidia and that it continues to be synthesized during spore

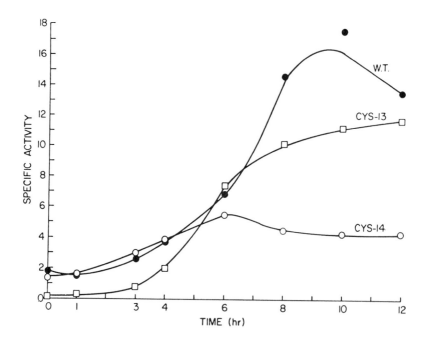

Fig. 4. Development of sulfate transport activity during conidial germination and outgrowth. Flasks containing low methionine medium were inoculated with conidia of wild type (●), cys-13 (□), and cys-14 (O) and incubated at 25° with rapid shaking. Sulfate permease activity of the germinating spores was assayed at various times for the subsequent 12 hr. (Marzluf, 1972).

germination is particularly evident in the case of the *cys-14* mutant, which lacks permease II, so that only the low affinity system can be synthesized. The specific activity of the sulfate permease of *cys-14* increases during conidial germination and reaches a maximum at about 6 hr before establishing the level found in mycelia of this strain. That only the sulfate permease I species is synthesized by *cys-14* at all developmental stages can be demonstrated by its characteristic low affinity for the sulfate ion.

Conidia of the *cys-13* mutant completely lack sulfate transport capacity, but during their germination a rapid synthesis of just permease II occurs, which results in a full complement of mycelial activity by 8 hr. In wild type it appears that both permease I and II are initially synthesized during conidial germination and outgrowth; by 12 hr of development, however, the amount of sulfate permease has already peaked and declined to a level only slightly greater than that possessed by the *cys-13* mutant. This outcome indicates that only a small fraction of the total sulfate uptake capacity of wild-type mycelia is via the low affinity system. Kinetic studies of the permease activity support this view.

It is interesting that during conidial germination a temporary peak of permease activity, above the usual derepressed level found in mycelia, is consistently observed at about 8-10hr. Nevertheless, it seems unlikely that the sulfate permeases themselves play any significant role in the germination process since these spores contain a sufficient endogenous sulfur supply to complete germination even in medium which is devoid of all sulfur (McGuire and Marzluf, 1974a). The pattern of differential synthesis of the specific sulfate permease isozymes during particular developmental stages indicates that an additional set of developmental controls is superimposed upon the genetic and metabolic controls described earlier which regulate both permeases in a parallel manner. Recent studies have also demonstrated that the synthesis of both choline sulfatase and aryl sulfatase during spore germination is subject to developmental controls (McGuire and Marzluf, 1974b). We do not as yet understand the mechanism responsible for the developmental regulation of the permeases or these related enzymes but this is clearly an intriguing area for future research.

SULFATE PERMEASE AS A MEMBRANE-ASSOCIATED COMPLEX

Since transport is a complex process, multiple elements may be required for the development of the complete permease system and its proper integration into the cell membrane. The basic components of a transport system would be expected to include a binding protein which is capable of binding the specific substrate for transport and an energy coupling system that permits significant accumulation to occur even against a concentration gradient. Components whose main function is integration of the transport system into the membrane may also exist and could even include a particular phospholipid (Scarborough, 1971). Such considerations suggest that additional genetic loci, besides the *cys-13* and *cys-14* genes described earlier, may encode elements of one or both of the sulfate permease systems. In a related study, we have demonstrated that both the development and maintenance of functional sulfate permease activity depends upon metabolic changes which affect the cell membrane. The synthesis of sulfate permease activity in an inositoless mutant shows an absolute requirement for an exogenous supply of inositol, and a drastic reduction in the synthesis of the sulfate transport system occurs when a limited amount of inositol is available (Marzluf, 1973). The maintenance of pre-existing functional sulfate transport activity apparently also requires a continuous renewal of membrane components since withdrawal of inositol from the medium of *inos* mutants results in a rapid inactivation of the transport activity. It appears that changes in membrane composition can cause a rapid loss of function of previously active sulfate transport systems. The high degree of dependence of the synthesis and activity of sulfate permease upon the continuous presence of inositol has been interpreted to mean that a lipid derivative of inositol is a component of the sulfate transport system (Marzluf, 1973).

Since sulfate permease I and sulfate permease II have significantly different K_m values for sulfate transport, it can be anticipated that they differ in the possession of distinct binding proteins with corresponding differences in their affinity for the sulfate ion. Pardee (1967) has shown that Salmonella possesses a specific membrane protein with a molecular weight of 32,000 which binds the sulfate ion and appears to be an integral part of the single sulfate transport system found in this organism. Current work in our laboratory is directed towards the isolation and study of the two different sulfate binding proteins predicted to occur in Neurospora; it will be of obvious interest to determine whether it is the synthesis and turnover of the binding proteins which regulates the level of activity of the sulfate permeases in the different developmental stages.

It is not yet clear why two permease systems for sulfate have evolved in Neurospora. No selective advantage which would explain the presence of the dual systems is apparent.

Mutants lacking one or the other of the two systems have excellent viability, germination, and grow well even on minimal medium. Furthermore, other organisms which have been studied have only a single transport system for inorganic sulfate; thus the reason for the presence of the dual systems in Neurospora remains a mystery.

ACKNOWLEDGMENT

The research was supported by Public Health Service Grant 5 RO1 GM-18642 from the National Institute of General Medical Sciences.

LITERATURE CITED

Burton, E. R. and R. L. Metzenberg 1972. Novel mutation causing derepression of several enzymes of sulfur metabolism in *Neurospora crassa*. *J. Bacteriol*. 109: 140-151.

Hanson, M. A. and G. A. Marzluf 1973. Regulation of a sulfur-controlled protease in *Neurospora crassa*. *J. Bacteriol*. 116: 785-789.

McGuire, W. G. and G. A. Marzluf 1974a. Sulfur storage in Neurospora: soluble pools of several developmental stages. *Arch. Biochem. Biophys*. in press.

McGuire, W. G. and G. A. Marzluf 1974b. Developmental regulation of choline sulfatase and aryl sulfatase in *Neurospora crassa*. *Arch. Biochem. Biophys*. in press.

Marzluf, G. A. 1970a. Genetic and biochemical studies of distinct sulfate permease species in different developmental stages of *Neurospora crassa*. *Arch. Biochem. Biophys*. 138: 254-263.

Marzluf, G. A. 1970b. Genetic and metabolic controls for sulfate metabolism in *Neurospora crassa*: Isolation and study of chromate-resistant and sulfate transport-negative mutants. *J. Bacteriol*. 102: 716-721.

Marzluf, G. A. 1972a. Genetic and metabolic control of sulfate metabolism in *Neurospora crassa*: A specific permease for choline-O-sulfate. *Biochem. Genet*. 7: 219-233.

Marzluf, G. A. 1972b. Control of the synthesis, activity, and turnover of enzymes of sulfur metabolism in *Neurospora crassa*. *Arch. Biochem. Biophys*. 150: 714-724.

Marzluf, G. A. 1973. Regulation of sulfate transport in Neurospora by transinhibition and by inositol depletion. *Arch. Biochem. Biophys*. 156: 244-254.

Marzluf, G. A. 1974. Uptake and efflux of sulfate in *Neurospora crassa*. *Biochem. Biophys. Acta* in press.

Marzluf, G. A. and R. L. Metzenberg 1968. Positive control by the cys-3 locus in regulation of sulfur metabolism in Neurospora. *J. Mol. Biol.* 33: 423-437.

Metzenberg, R. L. and J. W. Parson 1966. Altered repression of some enzymes of sulfur utilization in a temperature-conditional lethal mutant of Neurospora. *Proc. Natl. Acad. Sci.* 55: 629-635.

Pall, M. L. 1971. Amino acid transport in *Neurospora crassa*. IV. Properties and regulation of a methionine transport system. *Biochim. Biophys. Acta* 233: 201-214.

Pardee, A. B. 1967. Crystallization of a sulfate-binding protein from *Salmonella typhimurium*. *Science* 156: 1627-1628.

Roberts, K. R. and G. A. Marzluf 1971. The specific interaction of chromate with the dual sulfate permease systems of *Neurospora crassa*. *Arch. Biochem. Biophys.* 142: 651-659.

Scarborough, G. A. 1971. Sugar transport in *Neurospora crassa*. III. An inositol requirement for the function of the glucose active transport system. *Biochem. Biophys. Res. Commun.* 43: 968-975.

ON THE REGULATION OF FRUCTOSE DIPHOSPHATE ALDOLASE ISOZYME CONCENTRATIONS IN ANIMAL CELLS

HERBERT G. LEBHERZ[*]
Institute for Cell Biology
Swiss Federal Institute of Technology
CH-8006 Zurich, SWITZERLAND

ABSTRACT. The intracellular concentrations of the three aldolase subunits A, B, and C are independently regulated as shown by the tissue-specific and stage-specific patterns of aldolase isozymes. Some recent investigations concerning the regulation of intracellular aldolase isozyme concentrations are described here.
1) The aldolase of skeletal muscle is distributed between soluble (free) and particulate (bound) cell fractions.
a) The bound enzyme can be released by salt, substrate, and ATP. b) Free and bound aldolase A_4 were judged to be identical on the basis of several criteria. c) Bound A_4 readily exchanges with ^3H leucine labeled free A_4 in vitro.
d) Of the aldolases belonging to the A-C hybrid set, tetramers containing increasing numbers of A subunits showed increasing tendencies to bind to particulate muscle fractions.
2) Studies on chick brain showed that: a) The relative concentrations of aldolase tetramers are the same as those expected for the random combination of 5% A and 95% C subunits from a single subunit pool, suggesting that aldolase subunits associate in a random fashion. b) Aldolases AC_3 and C_4 of chick brain turnover at similar rates, suggesting that regulation at the level of subunit synthesis is primarily involved in determining the relative concentrations of isozymes in this tissue.
3) Studies on chick skeletal muscle showed that: a) The transition from predominantly C subunits to A subunits associated with development is accompanied by a large accumulation of A subunit protein. b) Breast ("white") muscle contains approximately four fold higher levels of aldolase A_4 than leg ("red") muscle. c) Breast muscle has a considerably higher rate constant for aldolase A subunit synthesis than does leg muscle.
4) These and other observations are discussed in relation to the mechanisms by which intracellular concentration of isozymes are regulated.

[*]Current address: Department of Molecular Biology, Roswell Park Memorial Institute, Buffalo, New York 14203

INTRODUCTION

Many enzymes have been shown to exist as multiple molecular forms (isozymes) within the tissues of a single organism. Several isozyme systems have been isolated in pure form and the molecular-genetic basis and properties of the individual isozymes defined. Considerably less is known about the possible specific physiological functions of the isozymes or about the mechanisms involved in regulating their concentrations within the cell. The in vitro catalytic properties of the mlltiple enzymes and observed correlations between tissue function ana localization of these proteins have suggested specific metabolic roles for some isozymes, but not for others. The in vivo regulation of isozyme concentrations must surely involve mechanisms operating at epigenetic levels in addition to those that regulate gene expression.

The isozymes of fructose diphosphate aldolase are now well characterized and represent a useful system for investigating the in vivo regulation of tissue-specific and stage-specific isozyme concentrations. The purpose of the present paper is to summarize briefly the molecular-genetic basis and possible physiological functions of vertebrate aldolases and to then describe some recent investigations concerned with the mechanisms involved in their in vivo regulation.

MOLECULAR-GENETIC BASIS OF ALDOLASE ISOZYMES

Three homologous aldolase subunit types, A, B, and C, have been found in every vertebrate organism so far tested (Penhoet et al., 1966; Rutter et al., 1968; Lebherz and Rutter, 1969). A forth subunit, aldolase "D", has been observed in salmonoid fish (Lebherz and Rutter, 1969). Detailed studies on the structure and properties of vertebrate aldolases, most notably those of the rabbit, allow the following conclusions to be made concerning the relationships between aldolase A, B, and C subunits. 1) The subunits have the same size (MW$\approx$40,000) and exist in tetrameric combinations (Kawahara and Tanford, 1966; Penhoet et al., 1967). 2) The subunits must have similar primary structures and conformations at the "active site" region since they catalyze the reversible cleavage. of fructose diphosphate by identical mechanisms via the formation of a Schiff-base between substrate and the E-NH$_2$ group of a lysyl residue at the "active site" (Horecker et al., 1963; Penhoet et al., 1969a). Indeed, the "active site" tryptic peptides of rabbit aldolases A and B have been shown to have very similar amino acid sequences.(Morse and Horecker, 1968). 3) The three subunits must have similar subunit association sites since they produce both homomeric and heteromeric tetramers in vitro and in vivo (Penhoet et al., 1966; 1967; Lebherz and Rutter, 1969). 4) The three subunits do have distinct antibody recognition sites since antisera directed towards any one subunit type do not cross react with the other two subunits (Penhoet et al., 1969a; Penhoet and Rutter, 1971). 5) The primary structures of A, B, and C subunits are encoded by homologous structural genes which presumably arose by divergent evolution from a common gene ancestry (Rutter et al., 1968). The uniquely different primary structures of the three subunits are reflected in the characteristic amino acid compositions and tryptic peptide "finger print" patterns of aldolases A, B, and C (Penhoet et al., 1969a). The differences in certain catalytic properties of the aldolases (Penhoet et al., 1969a) are also presumably determined by the differences in the structural genes of the three subunits. 6) Aldolase subunits appear to function as independent catalytic entities within the tetramer (Meigen and Schackman, 1970; Penhoet and Rutter, 1971) and, therefore, whether or not the tetrameric nature of the aldolases serves a specific function is not known.

TISSUE DISTRIBUTION AND PHYSIOLOGICAL FUNCTION OF ALDOLASES A, B, AND C

The aldolase isozymes present in extracts of vertebrate tissues are easily spearated and identified by electrophoresis followed by specific staining for fructose diphosphate cleavage activity (Penhoet et al., 1966). The restricted distribution of the aldolases in rat tissues is shown in figure 1. Aldolase A_4 is the only isozyme found in skeletal muscle, heart, and spleen. Aldolase B_4 is the predominant activity of liver and is presumably derived from liver parenchymal cells. The low levels of aldolase A_4 in this tissue are apparently derived from another liver cell type. The five members of the A-B hybrid set (A_4, A_3B, A_2B_2, AB_3, B_4) are present in kidney while the five activities of the A-C set (A_4, A_3C, A_2C_2, AC_3, C_4) are found in brain. The presence of heteromeric tetramers in kidney (A-B) and in brain (A-C) demonstrates that the two subunits are synthesized within the same cells. Isozyme distribution studies on a wide variety of vertebrates have been performed and showed the following general tissue distribution of the aldolases (Lebherz and Rutter, 1969): Aldolase A_4 was the only tetramer detected in skeletal muscles; Aldolase B subunits were restricted to liver and kidney and, within rabbit kidney, B subunits were localized in the cortex; Aldolase C subunits in C_4 tetramers and in combination with A subunits were found in all brains tested. The distribution

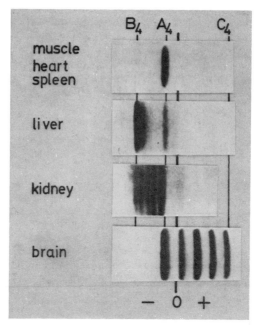

Fig. 1. Electrophoretic resolution of aldolase activities in rat tissue extracts. Methods for tissue extraction electrophoresis, and aldolase activity staining were as previously described (Penhoet et al., 1966; Lebherz and Rutter, 1969).

of C subunits in other tissues was quite variable.

The observed restricted distribution of aldolases A and B in vertebrate tissues is consistent with the proposed specialized physiological functions of these two proteins suggested by their in vitro catalytic properties. The catalytic properties of aldolase A_4 suggest that this enzyme functions in a glycolytic capacity while those of aldolase B_4 suggest that this enzyme functions mainly in gluconeogenesis and in the metabolism of fructose via fructose-1-phosphate (Rutter et al., 1963). Skeletal muscle is rich in glycolytic activity. Gluconeogenesis occurs in liver and kidney cortex while the metabolism of fructose via fructose-1-phosphate occurs primarily in liver (Benoy and Elliott, 1937; Krebs et al., 1963). Conclusive evidence for a specialized function of aldolase B in the metabolism of fructose was obtained with identification of the molecular basis of the metabolic lesion known as fructose intolerance. Patients with clinical symptoms of fructose intolerance have only slightly active, mutant forms of aldolase B (Nordmann et al., 1968; Gitzelmann et al., 1974).

There is as yet no known catalytic basis to suggsst
specialized functions for aldolases A and C and, except in
brain, the distribution of aldolase C varies considerably in
homologous tissues of different vertebrates (Lebherz and
Rutter, 1969). However, although the known catalytic properties of the two enzymes are similar (Penhoet et al., 1969a),
they do have opposite net electrostatic charges at physiological pH (see Penhoet et al., 1969a) and there is some evidence
that these isozymes are associated with different cellular
structures (see below; Clarke and Masters, 1972; 1973).
Consequently, the two enzyme subunits may be functionally distinguished in vivo, not on the basis of their physiological
function, but by their performance of the same function in
different regions of the cell.

REGULATION OF INTRACELLULAR ALDOLASE ISOZYME CONCENTRATIONS

The observed restricted tissue distribution of aldolases
A, B, and C (see above) as well as the observed transitions
from predominantly one subunit type to another associated with
the ontogenesis of certain vertebrate tissues (Weber and
Rutter, 1964; Rensing et al., 1967; Herskovitz et al., 1967;
Lebherz and Rutter, 1969; Lebherz, 1972a) demonstrate that the
intracellular concentrations of different aldolases are independently regulated. This regulation may involve a variety
of mechanisms operating at different levels within the cell
(see Discussion). The following experiments on the chick
were initiated in the hope of clarifying the intracellular
regulation of aldolase isozyme concentrations. Studies using
the isozymes of the A-C hybrid set are conveniently made since
these molecules are readily isolated by "affinity chromatography". The electrophoretic resolution of the five isozymes
is shown in figure 2 together with a schematic representation
of the subunit structure of each aldolase.

A. Relationship between "free" and "bound" aldolase.
Although the glycolytic enzymes are generally referred to as
"soluble" cell components, many reports of their being associated with "particulate" cell fractions have appeared in the
literature. It is now known that a significant portion of
the aldolase activity (up to 40%) in skeletal muscles is not
solubilized by repeated extraction of the tissue with buffers
of low ionic strength (Arnold and Pette, 1968). The remaining
activity (referred to here as "bound" enzyme) appears to be
associated with contractile muscle proteins, namely actin, and
can be solubilized by extraction of muscle particulate fractions
with buffers of high ionic strength or with low ionic strength

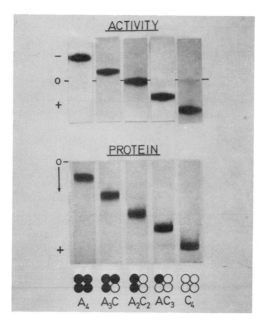

Fig. 2. Electrophoretic separation of purified aldolase isozymes of the A-C hybrid set. The enzymes were purified from
chicken tissues (Lebherz, 1972b). (top): Cellulose polyacetate electrophoresis followed by activity staining. (bottom):
Polyacrylamide gel electrophoresis followed by protein
staining with 1% amido Schwarz (Lebherz and Ursprung, 1974).

buffers containing certain metabolites (see below; Arnold and
Pette, 1968; 1970). These observations raise questions concerning the relationship between "free" and "bound" aldolase;
namely, 1) are the two forms structurally identical and 2)
are the two forms restricted to distinct cellular compartments
or do they readily exchange between the "free" and "bound"
condition. Since answers to these questions have implications
concerning the function and regulation of muscle aldolases,
the following studies were made.

The "bound" aldolase of muscle particulate fractions can
be released by low concentrations of fructose diphosphate and
ATP. Figure 3 shows the relationship between enzyme release
and concentration of these metabolites. As shown, considerably
lower concentrations of fructose diphosphate as compared to
ATP were required to release the "bound" enzyme. The sigmoidal
shapes of the two extraction curves indicate cooperative
mechanisms for the extractions as was previously reported by
Arnold and Pette (1968) for the desorption of aldolase A_4 from
F-actin effected by KCl. Fifty percent release of activity

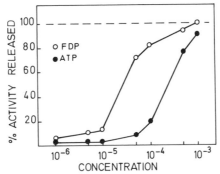

Fig. 3. Release of aldolase A₄ from particulate fractions of
chicken breast muscle. Breast muscle was minced and homogenized in 15 volumes of 0.01 M Tris-HCl, 0.001 M EDTA, 0.001 M
2-mercaptoethanol, pH 7.5 and the homogenate centrifuged at
10,000xg for 20 min. The pellet was washed by suspension in
the above buffer followed by centrifugation. The washing procedure was repeated until the A_{280} of the wash was less than
0.2. The final pellet contained approximately 25% of the total aldolase activity and was suspended in buffer. Aliquots
were extracted with fructose diphosphate or ATP at the concentrations indicated. After centrifugation, the supernatent
was assayed for aldolase activity (Blostein and Rutter, 1963).
100% release refers to the activity released by 0.2 M KCl.

was observed with approximately 2×10^{-5} M fructose diphosphate
which agrees well with the reported Michaelis constant, 4×10^{-5}
M, for chicken aldolase A₄ (Marquardt, 1969). Complete extraction of activity was observed with 1×10^{-3} M substrate.

As shown in figure 4, the purity of the aldolase preparation obtained by fructose diphosphate extraction of particulate
muscle fractions was considerably higher than that of soluble
preparations (S_1 and S_2) or of that prepared by extraction of
particulate fractions with 0.2 M KCl. A single protein zone
with the mobility of aldolase A₄ was detected upon electrophoresis in polyacrylamide gels. However, the specific catalytic activity of the preparation was only 50 to 60% as high
as pure aldolase A₄ and the preparation contained considerable
levels of glyceraldehyde-3-phosphate dehydrogenase activity;
this dehydrogenase and aldolase A₄ have similar mobilities in
the particular electrophoretic system used in figure 4. Even
though considerable purification of muscle aldolase can be
obtained by extraction of washed particulate fractions with
substrate, the relatively low amounts of "bound" enzyme compared to total muscle aldolase and the ease of aldolase

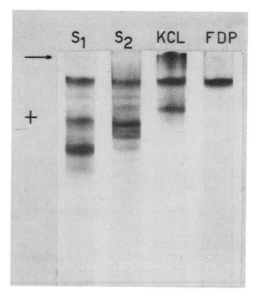

Fig. 4. Polyacrylamide gel electrophoresis of chicken breast
muscle fractions. Extractions of breast muscle were made as
described in the legend of figure 3. S_1 and S_2 refer to the
supernatants of the initial homogenate and first wash of the
particulate fractions, respectively. KCl and FDP refer to the
supernatants of particulate fractions after extraction with
fructose diphosphate or ATP, respectively (1×10^{-3} M). The
fractions were dialyzed, subjected to electrophoresis, and the
gels stained for protein (Lebherz and Ursprung, 1974).

isolation by "affinity chromatography" (Penhoet et al., 1969b)
argues against using substrate extraction in the purification
of muscle aldolase.

"Free" and "bound" aldolase were isolated from breast
muscle by a modification of the method described by Penhoet
et al. (1969b) for rabbit aldolase A_4 and were judged to be
identical by the following criteria: (figure 5) 1) Immunological identity in Ouchterlony immunodiffusion tests 2)
identical migration in cellulose polyacetate and polyacrylamide gel electrophoresis and 3) identical specific catalytic
activities for fructose diphosphate cleavage. The two forms
of the enzyme are also equivalent in that ^{3}H labeled "free" A_4
readily exchanges with unlabeled "bound" A_4 in vitro. As
shown in table I, the specific radioactivity of the "free"
enzyme rapidly decreased to that value expected assuming complete exchangeability between "free" and "bound" aldolase A_4.
From these observations, it is concluded that aldolase A_4

TABLE I

EXCHANGE BETWEEN "FREE" AND "BOUND" ALDOLASE A_4 OF CHICK SKELETAL MUSCLE

Sample[a]	CPM/unit	Observed ratio B/A	Predicted ratios of B/A complete exch.	No exch.
A	71			
B	150	2.1	2.3	1.0

Tritiated "free" aldolase A_4 was isolated from the breast muscle of a 3 week old chick 1.5 hours after injection with ^3H leucine (174 μCi/100 gm body weight). The purified enzyme contained 2400 CPM/mg. Unlabeled "bound" (particulate) aldolase A_4 was prepared by extensive washing of breast muscle particulate fractions with low ionic strength buffers as described in the legend of figure 3.

[a]Sample A: 0.28 mg particulate A_4 plus 0.21 mg ^3H labeled A_4 (final volume = 3 ml). After 3 min., the sample was centrifuged and the supernatant assayed for activity and radioactivity content. Sample B: Same as A except that the labeled aldolase was added after centrifugation of the particulate enzyme solution. Exchange between "free" and "bound" aldolase is apparent from the decrease in specific radioactivity of sample A as compared to sample B. The ratio of CPM/unit for the two samples (B/A) is close to that expected assuming complete exchange between the two forms of the enzyme (see predicted ratios).

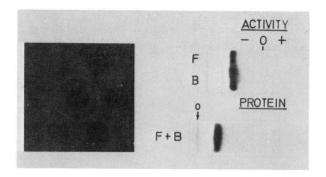

Fig. 5. Immunological and electrophoretic identity of "free" and "bound" aldolase A_4. "Free" aldolase was extracted from chicken breast muscle in 0.01 M Tris-HCl, 0.001 M EDTA, 0.001 M 2-mercaptoethanol, pH 7.5. "Bound" enzyme was extracted from washed particulate fractions with buffer containing 0.2 M KCl. Left: Ouchterlony double diffusion test (Lebherz, 1974a). The center well contains rabbit antisera directed against purified "free" aldolase A_4. The outer wells alternate between "free" and "bound" enzyme preparations containing approximately 0.1 mg/ml aldolase A_4. Right: (top) Cellulose polyacetate electrophoresis of "free" and "bound" enzyme fractions. The two samples were electrophoresed on the same strip. Electrophoresis and activity staining were performed as previously described (Penhoet et al., 1966). (bottom) Polyacrylamide gel electrophoresis (Lebherz and Ursprung, 1974) of purified "free" and "bound" aldolase A_4. The two enzymes were purified to homogeneity (Lebherz, 1972b) and a mixture of approximately 15 µgs of each preparation was electrophoresed and the gel stained for protein.

readily exchanges between the free and bound condition, at least after cell breakage.

Different aldolase isozymes show different tendencies to bind to muscle particulate fractions. During the C to A transition associated with chick skeletal muscle development, aldolase C_4 was never detected in the "bound" fraction while tetramers containing increasing numbers of A subunits showed increasing tendencies to be found in the particulate fraction (Lebherz, unpublished observation). Comparison of the electrophoretic patterns of "free" and "bound" aldolase fractions obtained from leg musculature of a newly hatched chick (figure 6) vividly demonstrates the preferential binding of aldolase tetramers to particulate fractions. The decreased binding of tetramers containing increasing numbers of negatively charged C subunits is probably due to electrostatic repulsion between the C subunits and actin; the binding of

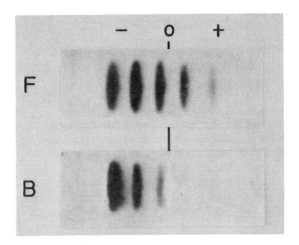

Fig. 6. Aldolase isozymes in the "free" and "bound" fractions of newly hatched chick leg musculature. "Free" and "bound" aldolases were extracted as described in the legend of figure 5 and were subjected to electrophoresis and activity staining (Penhoet et al., 1966).

aldolase A_4 to purified actin involves electrostatic attraction between positively charged aldolase A subunits and negatively charged actin molecules.(Arnold and Pette, 1968).

The above observations emphasize the need to consider extraction procedures when attempting to define qualitative as well as quantitive levels of aldolase isozymes in animal tissues.

B. *Regulation of aldolase isozyme concentrations in avian brain.* It is now clear that the intracellular proteins of animal cells are continually being catabolized and resynthesized, a process known as turnover (see reviews. Schimke and Doyle, 1970; Rechcigl, 1971). Consequently, the concentration of a protein is determined by the rate at which it is synthesized <u>and</u> rate at which it is degraded. Although rates of protein synthesis are influenced by a number of factors (availability of amino acids and ribosomes, rates of transcription and translation, etc.), the rate of protein synthesis (V_s) can normally be expressed in terms of a zero-order rate constant (K_s) for protein synthesis (Schimke and Doyle, 1970). Degradation in animal cells follows first-order kinetics and, therefore, the rate of protein degradation (V_d) is expressed in terms of a first-order rate constant (K_d) and the concentration of the protein present (P) (Schimke and Doyle, 1970). Under steady-state conditions, the concentration of a protein does not change with time and can be expressed in terms of the constants K_s and K_d as follows:

$$dP/dt = V_s - V_d = K_s - K_dP = 0, \text{ and } P = K_s/K_d$$

The time required to replace one half of the protein molecules with newly synthesized ones, or half-life, is related to the rate constant for degradation as follows: $t_{1/2} = \dfrac{\ln 2}{K_d}$.

Experiments designed to determine the relative roles of synthesis and degradation in maintaining steady-state concentrations of aldolase isozymes were performed on chick brain. This tissue was chosen since it is a major tissue containing multiple aldolases and also because the steady-state concentrations of A and C subunits are quite different. Steady-state levels (adult levels) of brain aldolases are reached shortly after hatching and the concentrations of aldolase tetramers in different brain regions are similar (Lebherz and Ursprung, 1974). After separation of the brain aldolases by ion exchange chromatography and quantitation of each enzyme, it was found that brain contains approximately 700 µg of aldolase tetramers per gm wet weight. (Lebherz and Ursprung 1974). The relative proportions of the isozymes in the total aldolase population are given in table II. Aldolase C_4 predominates with lesser amounts of AC_3 and A_2C_2; little or no A_3C or A_4 were ever detected. The tetramer distribution in chicken brain corresponds to a subunit population of 5% A and 95% C subunits. The last column shows the percentages of tetramers expected for the random combination of A (5%) and C (95%) subunits from a single subunit pool. The fact that the "observed" and "expected" levels for C_4 and AC_3 are similar suggests that aldolase subunit associations within brain cells are random: i.e., there do not appear to be favored tetrameric combinations of aldolase subunits.

A common method for investigating the rates of protein turnover is to determine the rate of decline in specific radioactivity in the protein after a single administration of a labeled amino acid. In order to make these measurements, the protein under investigation must be isolated in pure form. For this purpose, a rapid purification procedure, modified from that used by Penhoet et al. (1969b) for isolation of the rabbit aldolases was developed. The detailed procedure involves "affinity chromatography" on phosphocellulose columns and will be published elsewhere (Lebherz and Ursprung, 1974). Aldolase AC_3 and C_4 preparations of high purity are obtained with this method; as shown in figure 7, the aldolase C_4 preparations are homogeneous by the criteria of polyacrylamide gel electrophoresis while the AC_3 preparations contain only

TABLE II

RELATIVE CONCENTRATIONS OF ALDOLASE ISOZYMES
IN CHICK BRAIN[a]

Aldolase	% observed	% "expected"
C_4	81	81.4
AC_3	17	17.1
A_2C_2	2	1.4
A_3C	-	0.05
A_4	-	0

[a]Aldolase isozymes in fresh chicken brain extracts were
separated by ion exchange chromatography and the percentages
of each tetramer in the aldolase population determined after
corrections were made for the different specific catalytic
activities of A and C subunits. The tetramer distribution
observed corresponds to a subunit population of 5% A and 95%
C subunits. The % "expected" refers to the relative amounts
of each tetramer expected assuming random combination of A
and C subunits from a single subunit pool (see Lebherz and
Ursprung, 1974 for further details).

small amounts of contaminating aldolases (A_2C_2 and C_4).

Turnover of the brain aldolases was first investigated by
measuring the decline in specific radioactivity of the enzymes
as a function of time after ^3H leucine administration. Three
week old chicks were injected with ^3H leucine and at times
thereafter, 1 to 10 days, chicks were sacrificed and the brain
aldolases isolated. The decline in specific radioactivity of
AC_3 with time followed first-order kinetics and an apparent
half-life of 4.2 days was calculated (Lebherz and Ursprung,
1974). Unfortunately, difficulties in isolation of aldolase
C_4 were encountered at some time points and reliable data for
this enzyme were not obtained in this experiment.

The relative rates of turnover of the AC_3 and C_4 tetramers
were next investigated using the double isotope method
described by Arias et al.(1969). Using this method, relative
rates of turnover of different proteins within the same
tissue of a single animal may be determined. The animal re-
ceives an amino acid labeled with one isotope (in this case
^3H leucine) followed some time later with the administration
of the same amino acid labeled with another isotope (^{14}C

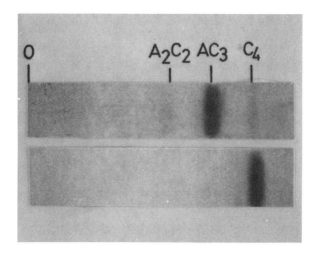

Fig. 7. Polyacrylamide gel electrophoresis of aldolase AC_3 and C_4 preparations isolated from chick brain. The two enzymes were purified by affinity chromatography (Lebherz and Ursprung, 1974).

leucine). After a short period of time, the animal is sacrificed, the proteins isolated, and their ^3H and ^{14}C contents determined. The ratio of $^{14}C/^3H$ is a measure of the relative rates of turnover of the proteins. The ^{14}C counts represent an early point on the degradation curve and the ^3H counts represent a later point on the same curve. Proteins with a high rate of turnover (short half life) will have higher $^{14}C/^3H$ ratios than proteins which turnover more slowly. Also, the difference in ratios for proteins which turnover at different rates will be increased as the time between injection of the two isotopes is increased; no differences in ratios would be observed if the two isotopes are given at the same time.

Chicks received ^3H leucine followed 2, 5, or 9 days later by ^{14}C leucine. One day after the ^{14}C leucine injection, the chicks were sacrificed, the brain aldolases isolated, and their $^{14}C/^3H$ ratios determined. Table III shows the normalized ratios for the aldolases from each group of chicks (normalized ratio for C_4 in each case is 1.00). As shown, the normalized ratios for the two enzymes are very similar. For comparison, notice the normalized ratios of the two aldolases that would be expected if AC_3 with an apparent half life of 4.2 days turns over twice as fast as C_4 (Table III, last column).

TABLE III

RELATIVE RATES OF TURNOVER OF ALDOLASES AC_3 AND C_4 IN CHICK BRAIN

Time between injections	Observed normalized $^{14}C/^3H$ ratios		Predicted normalized $^{14}C/^3H$ ratios			
			$t_{1/2}AC_3=t_{1/2}C_4$		$t_{1/2}AC_3=1/2t_{1/2}C_4$	
	AC_3	C_4	AC_3	C_4	AC_3	C_4
2 days	1.14	1.00	1.00	1.00	1.17	1.00
5 days	0.98	1.00	1.00	1.00	1.50	1.00
9 days	1.00	1.00	1.00	1.00	2.08	1.00

Three week old chicks received ^3H leucine followed 2, 5, or 9 days later by ^{14}C leucine injections. One day after the ^{14}C leucine injection, chicks were sacrificed, the brain aldolases isolated, and their $^{14}C/^3H$ ratios determined. The data were normalized so that the ratio for C_4 at each time was equal to 1.00. The predicted normalized ratios refer to those expected assuming that the two aldolases turnover at identical rates or, assuming that AC_3 with an apparent half life of 4.2 days turns over twice as fast as C_4. The detailed experiment is being published elsewhere (Lebherz and Ursprung, 1974).

Measurements of protein turnover using the methods employed here are influenced by the extent of isotope reutilization during the experiments. When labeled protein is degraded, some of the labeled amino acids liberated are reutilized as precursors for protein synthesis. The net result of this re- utilization is the calculation of apparent half lives which are longer than the true half lives (Poole, 1971). The ap- parent half life of 4.2 days calculated for aldolase AC_3 is, therefore, maximal due to reutilization. In addition, differ- ences in relative half lives as measured by $^{14}C/^3H$ ratios will be decreased as a result of reutilization (Poole, 1971). In fact, extensive reutilization of ^3H leucine in the present experiments could have masked differences in the half lives of AC_3 and C_4 if the enzymes truly turnover at different rates. However, two observations suggest that extensive isotope re- utilization is not responsible for the similar $^{14}C/^3H$ ratios calculated for the two aldolases. 1) The tritium counts in the total cellular amino acid pool of chick brain rapidly decreased after administration of ^3H leucine (Lebherz and Ursprung 1974). 2) Large differences in $^{14}C/^3H$ ratios of soluble brain proteins were detected after dissociation of the proteins in sodium dodecyl sulfate and separation of the sub- units by gel filtration in the presence of this detergent (Lebherz and Ursprung, 1974). Thus differences in relative half lives of soluble brain proteins are readily detected by this method. It is therefore concluded that aldolases AC_3 and C_4 and, hence, their subunits have similar rate constants for degradation. It follows, therefore, that the considerably higher steady-state concentration of C vs. A subunits in chick brain results primarily from the action of regulatory mechanisms operating at the level of subunit synthesis.

C. Regulation of aldolase isozyme concentrations during chick skeletal muscle development. Transitions from one aldolase subunit population to another are associated with the onto- genesis of certain vertebrate tissues. Developmental tran- sitions from A to B, A to C, C to A, and C to B subunits have been reported (Weber and Rutter, 1964; Rensing et al., 1967; Herskovitz et al., 1967; Lebherz and Rutter, 1969; Lebherz, 1972a). In all of these instances, heteromeric aldolase tetramers are produced during the transition. These obser- vations demonstrate that the transitions occur within indi- vidual cells and are not due to the proliferation of cells containing "differentiated" aldolase populations at the expense of cells containing "embryonic" aldolases. The onto- geny of aldolase in chick skeletal muscle was chosen as a system for investigating the regulation of aldolase

268

concentrations during development.

Early embryonic chick breast and leg muscle contain aldolase tetramers with a predominance of C subunits (Lebherz and
Rutter, 1969). As development proceeds, there is a transition
to tetramers containing only A subunits (A_4) (Lebherz and
Rutter, 1969; Herskovitz et al., 1967). A portion of the
breast muscle transition is shown in figure 8. In both
muscles, the transition is completed shortly after hatching.
A large accumulation of aldolase activity, expressed both as
enzyme activity per gram of tissue and as activity per mg of
soluble protein also occurs during development. As shown in
table IV, the amount of aldolase activity accumulated in
breast muscle is about 4 fold greater (Units/gm muscle) than
in leg muscle and the accumulation is completed approximately
3 weeks after hatching. The difference between aldolase
activity in breast and leg muscle is not due to the presence
of activators in breast muscle and/or inhibitors in leg
muscle. Immunotitration studies with rabbit antisera directed
towards chicken aldolase A_4 showed a direct correlation between
amount of aldolase activity and immunological cross reacting
material (CRM) in both muscle types (Lebherz, 1974a). Furthermore, approximately four times more aldolase protein per gram
of tissue can be isolated from breast as compared to leg
muscle (Lebherz, 1974a). It can be calculated that about 15%

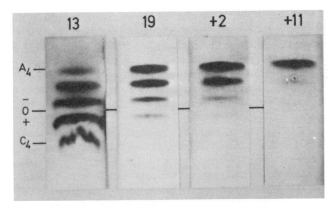

Fig. 8. Aldolase C to A subunit transition during chick
skeletal muscle development. Extracts of chick breast muscle
were prepared by homogenization in buffer containing 0.2 M
KCl (to release "bound" enzyme) and, after centrifugation,
the supernatant fractions were subjected to electrophoresis
and activity staining (Penhoët et al., 1966). Numbers refer
to days after fertilization or days after hatching (with
plus signs).

TABLE IV

ACCUMULATION OF ALDOLASE
DURING CHICK SKELETAL MUSCLE DEVELOPMENT

Days after fertilization	Aldolase activity (units/gm muscle)		Specific activity (units/mg sol. prot.)	
	Breast	Leg	Breast	Leg
11	0.39	0.60	0.04	0.04
15	0.70	0.85	0.04	0.04
19	3.3	7.1	0.20	0.30
Days after hatching				
2	6.9	17.2	0.30	0.68
11	72	21	1.63	0.68
20	143	38	2.1	0.76
38	156	43	2.3	0.89

Breast and leg skeletal muscles from pre and post hatching chicks were extracted in buffer containing 0.2 M KCl and the total activity/gm tissue and activity/mg soluble protein determined. More extensive data on the accumulation of aldolase during development is being collected (Lebherz, 1974a).

of the total soluble protein of breast muscle is aldolase A_4 while this enzyme comprises about 6% of the soluble leg muscle protein (Lebherz, 1974a).

The higher concentration of aldolase in breast vs leg muscle is apparently due to a higher synthetic rate for the production of A subunits in breast vs leg muscle cells. This was determined by measuring the relative incorporation of tritium into aldolase A_4 as compared to incorporation into total soluble protein after a short pulse (one hour) with ^3H leucine. The aldolase of breast and leg muscle were isolated by affinity chromatography and soluble protein was prepared by precipitation with trichloroacetic acid. Figure 9 shows a typical elution profile from phosphocellulose columns after application of the supernatant fraction of a muscle homogenate after centrifugation. Only that fraction which is eluted by fructose diphosphate (peak III) contained aldolase activity. Moreover, only peak III showed the presence of aldolase protein as judged by immunological cross reaction with antisera directed against aldolase A_4; no observable cross reactivity or precipitation of tritium counts occured when fractions from peaks I, II, or IV were incubated with antisera. This observation suggests that newly synthesized aldolase protein

270

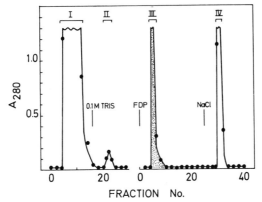

Fig. 9. Single-step isolation of muscle aldolase A_4 by substrate elution from phosphocellulose columns. Details of the procedure will be published elsewhere (Lebherz, 1974a). Only that fraction specifically eluted with fructose diphosphate (peak III) contained aldolase activity or aldolase protein (immunological cross reacting material). Cellulose polyacetate and polyacrylamide gel electrophoresis showed that the aldolase of peak III was homogeneous.

rapidly appears as active enzyme (protein capable of binding substrate). Thus, there is no evidence for the presence of inactive, but immunologically competent precursor forms of aldolase in skeletal muscle. For both breast and leg muscle preparations, the aldolase specifically eluted by substrate (peak III) was judged to be homogeneous by electrophoresis on cellulose polyacetate membranes or in polyacrylamide gels.

Table V shows the relative rate of aldolase synthesis in breast and leg muscle of a six week old chick as measured by the incorporation of tritium into aldolase or soluble protein per gram of muscle after a one hour pulse with ^3H leucine. About 21% of the total counts incorporated into soluble protein in breast muscle are found in aldolase while about 4.5% of the incorporation is found in aldolase for the leg muscle preparation. Further studies are in progress to determine the relative rates of degradation of aldolase in these two muscle types as well as to determine the relative roles of synthesis and degradation in regulating aldolase levels during development.

TABLE V

RELATIVE RATES OF SYNTHESIS OF ALDOLASE A_4 BY BREAST AND LEG SKELETAL MUSCLES

	% A_4 of total soluble protein	DPM ^{3}H in A_4 per gm muscle	DPM ^{3}H in sol. protein/gm muscle	% total incorp. in A_4
Breast	15.1	10.3×10^3	48.3×10^3	21.3
Leg	5.9	3.3×10^3	74.1×10^3	4.5

A 38 day old chick received 120 uCi ^{3}H leucine. One hour later, the chick was sacrificed. Aldolase was purified to homogeneity by affinity chromatography and soluble proteins collected by precipitation with 10% trichloroacetic acid. The mgs of A_4/gm muscle was calculated from the activity/gm using a specific catalytic activity of 15 units/mg. The mgs of soluble protein/ gm muscle was determined by the Lowry method (see Lebherz, 1974a for details and further data).

272

DISCUSSION

The in vivo regulation of protein concentrations may involve a variety of control mechanisms: 1) Mechanisms acting on the genome undoubtedly influence the transcriptional activities of the structural genes of protein subunits. 2) Factors may influence the rate of translation of mRNA into protein as well as affecting the stability and readability of the messages themselves. 3) Newly synthesized protein subunits may be released from the ribosome as inactive, precursor molecules; conversion of such precursors to functional protein subunits would require further processing which may be susceptible to regulatory constraints. 4) The cellular environment, including the concentrations of metabolites, may influence the state of association-dissociation of the subunits of oligomeric proteins. Such association-disassociation processes would allow for the exchange of subunits between oligomeric proteins and for the "uncoupling" of synthesis and degradation of subunits comprising the same oligomeric structure. 5) At the degradative level, interactions between a protein and the cellular environment as well as the inherent structural characteristics of the protein itself may be involved in determining the susceptibility of a protein to destruction.

The present work was initiated in the hope of gaining a general understanding of the regulation of tissue-specific and stage-specific concentrations of isozymes. The isozymes of fructose diphosphate aldolase were chosen for these studies.

Experiments on chick brain suggest that, under steady state conditions; 1) Aldolase subunits within the same cellular compartment are randomly distributed in tetrameric combinations. Therefore, the relative levels of the isozymes of chick brain appear to be governed solely by the relative amounts of the different subunits (A and C) present. 2) Aldolase tetramers and their subunits turnover at similar rates; That is, there does not appear to be discriminatory selection of the different aldolase tetramers and/or subunits for degradation. These data suggest that, although degradation is important in determining total aldolase concentrations, the relative levels of aldolase isozymes are controlled primarily by regulatory mechanisms operating at the level of subunit synthesis (transcription-translation). Whether or not these implications, based on the chick brain experiments, can be generalized to include other tissue or tissues of other organisms must await the results of further investigations.

In contrast to the emphasis on regulation at the level of subunit synthesis suggested above for the aldolases, a considerably different model for the regulation of steady state concentrations of lactate dehydrogenase isozymes has been proposed by Fritz and associates (1971). They suggest that epigenetic (post-translational) mechanisms are of considerable importance in regulating the relative levels of these isozymes within mammalian cells. Their observations on rat skeletal muscle and heart have led them to propose that lactate dehydrogenase subunit associations within cells of these tissues are not random (Fritz et al., 1970) and that different lactate dehydrogenase isozymes turn over at different rates within the same cell (Fritz et al., 1971; 1973). However, recent considerations of their data (Lebherz, 1974b) have shown that these observations may simply be a reflection of cellular heterogeneity (with respect to lactate dehydrogenase isozymes) of the tissues studied. The observed non-random tetrameric distributions of the subunits in extracts of heterogeneous tissues may be expected since more than one subunit pool (cell type) would be present in the intact tissue. Moreover, differential turnover of isozymes in different cell types of the same tissue would not be totally unexpected in view of the different metabolic characteristics of differentiated mammalian cells. Therefore, further experiments must be performed before concluding that regulatory mechanisms other than those that control subunit synthesis contribute significantly to the regulation of _relative_ concentrations of lactate dehydrogenase isozymes within the _same_ cell.

Studies on the aldolases of chicken skeletal muscle were undertaken in order to investigate the regulation of isozyme concentrations during development. The aldolase C to A subunit transition associated with development is accompanied by a large increase in aldolase activity (Table IV). Electrophoretic (Figure 8) and immunological criteria show that the increase in activity is due to the accumulation of aldolase A subunit protein. The observation that accumulation of aldolase in breast muscle is considerably greater than that in leg muscle was expected in view of the biochemical distinctions of different vertebrate skeletal muscle fiber types, including the concentrations of glycolytic enzymes (Peter et al., 1972). The "white" fibers of which chicken breast muscle is composed are particularly rich in glycolytic activity. It is proposed that the different concentrations of aldolase in the two muscles are predominantly due to a higher zero-order rate constant for aldolase A subunit synthesis in breast vs leg muscle; short term amino acid incorporation studies showed that the relative rate of aldolase subunit synthesis vs total soluble protein synthesis in breast muscle is approximately four-fold higher than in leg muscle. (Table V). It should now be possible to determine where in the synthetic process transcription, and/or translation, mechanisms act to determine the different aldolase synthetic activities of these muscle types.

Investigations on chick muscle aldolases are being performed in order to determine the mode of aldolase subunit assembly and disassembly in vivo. Since no evidence was found for the presence of inactive but immunologically competent forms of aldolase protein in chick muscle, (Fig. 9) it is suggested that newly synthesized aldolase polypeptide chains are released from the ribosome as functional subunits or are converted to functional subunits shortly thereafter. The newly formed subunits would then rapidly associate to form active aldolase tetramers.

Subunit exchange reactions, occurring via dissociation-association processes would allow for the production of "new" tetramers from pre-existing ones; for example, heteromeric tetramers could be produced by subunit exchange between homomeric enzymes. Subunit exchange reactions would allow for the continuous randomization of subunit associations; for example, during the developmental transitions from one subunit type to another. Finally, these reactions would allow for the "uncoupling" of turnover of different subunit types of the same tetrameric structure. The fact that subunits of certain cellular proteins have been found to turn over at heterogeneous rates; for example, subunits of the mammalian ribosome (Dice and Schimke, 1972), subunits of the enzyme fatty acid synthetase (Tweto et al., 1972), the "light" and "heavy" chains of myosin (Wikman-Coffelt et al., 1973), demonstrates that subunit exchange reactions do occur in vivo, since these reactions are required for the independent turnover of subunits of multi-subunit proteins. Furthermore, the observed general correlation between subunit size and rate of turnover in vivo (Dice et al., 1973; Glass and Doyle, 1972) further suggests that the subunit, not the oligomeric structure, is the unit of protein degradation and, therefore, that subunit exchange between oligomeric proteins commonly occurs in vivo.

One approach to investigate the possibility of whether or not subunit-exchange between isozymes is likely to occur in vivo, is to test for subunit-exchange reactions under simulated physiological conditions, in vitro. The observed subunit-exchange between lactate dehydrogenase isozymes under very mild in vitro conditions (Fritz et al., 1971; Millar et al., 1971) and the exchange between isozymes of glyceraldehyde-3-phosphate dehydrogenase observed in the presence of physiological concentrations of adenine nucleotides (Lebherz et al., 1973) do suggest that similar exchange processes occur within the cell. In contrast, previous studies on mammalian and avian aldolases suggest that the quaternary structure of these isozymes are of such stability as to preclude subunit-exchange between the aldolases under physiological conditions, at least in vitro (Lebherz, 1972b). No subunit exchange between aldolase isozymes occurred at high dilution, after freezing and thawing in high concentrations of salt, or by prolonged incubation of the isozymes in crude extracts in the presence or absence of a variety of cellular metabolites (Lebherz, 1972b; Lebherz and Abacherli, unpublished).

Recent studies on the in vivo stability of aldolase quaternary structure suggest that subunit exchange between aldolase tetramers does not occur within the cell (Lebherz, 1975). These observations are of considerable interest since they suggest a restrictive mechanism for the construction and destruction of aldolase tetramers in vivo; the lack of subunit exchange reactions would necessitate that all aldolase tetramers be produced at the time of the initial assembly of newly synthesized subunits and that the subunit associations established at this time be conserved until the tetramer is acted upon by the degradative mechanisms of the cell. Thus, in contrast to many, if not most, proteins (see above), the tetramer, not the subunit, may be the unit of aldolase protein turnover (Lebherz, 1975).

The present work which will be published in detail elsewhere (Lebherz and Ursprung, 1974; Lebherz, 1974a), has established the importance of regulation at the level of subunit synthesis in controlling tissue-specific and stage-specific patterns of aldolase isozymes. The transcriptional and translational regulation of aldolase A, B, and C subunit synthesis is now being investigated.

ACKNOWLEDGEMENTS

This work was supported by the Swiss National Science Foundation, project number 3.8640.72. I thank Miss E. Abacherli for her competent technical assistance and Drs. H. Ursprung, D. Turner, and H. Eppenberger for helpful discussions.

REFERENCES

Arias, I. M., D. Doyle, and R. T. Schimke 1969. Studies on the synthesis and degradation of proteins of the endoplasmic reticulum of rat liver. *J. Biol. Chem.* 244: 3303-3315.

Arnold, H. and D. Pette 1968. Binding of glycolytic enzymes to structural proteins of the muscle. *Eur. J. Biochem.* 6: 163-171.

Arnold, H. and D. Pette 1970. Binding of aldolase and trio-
    sephosphate dehydrogenase to F-actin and modification of
    catalytic properties of aldolase. *Eur. J. Biochem.* 15:
    360-366.
Benoy, M. P. and K. A. C. Elliott 1937. The metabolism of
    lactic and pyruvic acids in normal and tumor tissues.
    V. Synthesis of carbohydrates. *Biochem. J.* 31: 1268-1275.
Blostein, R. and W. J. Rutter 1963. Comparative studies of
    liver and muscle aldolase. II. Immunochemical and chroma-
    tographic differences. *J. Biol. Chem.* 238: 3280-3285.
Clarke, F. M. and C. J. Masters 1972. On the reversible and
    selective absorption of aldolase isoenzymes in rat brain.
    Arch. Biochem. Biophys. 153: 258-265.
Clarke, F. M. and C. J. Masters 1973. On the distribution of
    aldolase isoenzymes in subcellular fractions from rat
    brain. *Arch. Biochem. Biophys.* 156: 673-683.
Dice, J. F., P. J. Dehlinger, and R. T. Schimke 1973. Studies
    on the correlation between size and relative degradation
    rate of soluble proteins. *J. Biol. Chem.* 248: 4220-4228.
Dice, J. F. and R. T. Schimke 1972. Turnover and exchange of
    ribosomal proteins from rat liver. *J. Biol. Chem.*
    247: 98-111.
Fritz, P. J., E. L. White, K. M. Pruitt, and E. S. Vesell 1973.
    Lactate dehydrogenase isozymes. Turnover in rat heart,
    skeletal muscle and liver. *Biochemistry* 12: 4034-4039.
Fritz, P. J., E. L. White, and E. S. Vesell 1970. Biosynthesis
    and degradation of lactate dehydrogenases in rats. *Fed.
    Proc.* 29: 735.
Fritz, P. J., E. L. White, E. S. Vesell, and K. M. Pruitt 1971.
    New Theory of the control of protein concentrations in
    animal cells. *Nature New Biol.* 230: 119-122.
Gitzelmann, R., B. Steinmann, C. Bally, and H. G. Lebherz
    1974. Antibody activation of mutant human liver aldolase
    B from patients with hereditary fructose intolerance.
    Biochem. Biophys. Res. Commun. 59: 1270-1277.
Glass, R. D. and D. Doyle 1972. On the measurement of protein
    turnover in animal cells. *J. Biol. Chem.* 247: 5234-5242.
Herskovitz, J. J., C. J. Masters, P. M. Wasserman, and N. O.
    Kaplan 1967. On the tissue specificity and biological
    significance of aldolase C in the chick. *Biochem. Biophys.
    Res. Commun.* 26: 24-29.
Horecker, B. L., P. T. Rowley, E. Grazi, T. Cheng, and O.
    Tchola 1963. The mechanism of action of aldolases IV.
    Lysine as the substrate-binding site. *Biochem. Z.* 338: 36-
    51.
Kawahara, K., and C. Tanford 1966. The number of polypeptide
    chains in rabbit muscle aldolase. *Biochemistry* 5: 1578-
    1584.

Krebs, H. A., D. A. H. Bennett, P. de Gasquet, T. Gascoyne, and T. Yoshida 1963. Renal glyconeogenesis. The effect of diet on the gluconeogenic capacity of rat-kidney-cortex slices. *Biochem. J.* 86: 22-27.

Lebherz, H. G. 1972a. The ontogeny of fructose diphosphate aldolase B in the chick. *Develop. Biol.* 27: 143-149.

Lebherz, H. G. 1972b. Stability of quaternary structure of mammalian and avian fructose diphosphate aldolases. *Biochemistry* 11: 2243-2250.

Lebherz, H. G. 1974a. Studies on the regulation of fructose diphosphate aldolase isozyme concentrations in chick skeletal muscle (in preparation).

Lebherz, H. G. 1974b. On the regulation of lactate dehydrogenase isoenzyme concentrations in mammalian cells. *Experientia* 30: 655-658.

Lebherz, H. G. 1975. In vivo stability of Aldolase Quaternary Structure. *Proc. Nat. Acad. Sci. U.S.A.* (submitted).

Lebherz, H. G. and W. J. Rutter 1969. Distribution of fructose diphosphate aldolase variants in biological systems. *Biochemistry* 8: 109-121.

Lebherz, H. G., B. Savage, and E. Abächerli 1973. Adenine nucleotide mediated subunit exchange between isoenzymes of glyceraldehyde-3-phosphate dehydrogenase. *Nature New Biol.* 245: 269-271.

Lebherz, H. G. and H. Ursprung 1974. Synthesis and degradation of fructose diphosphate aldolase isoenzymes in avian brain (in preparation).

Meighen, E. A. and H. K. Schachman 1970. Hybridization of native and chemically modified enzymes. I. Development of a general method and its application to the study of the subunit structure of aldolase. *Biochemistry* 9: 1163-1176.

Millar, D. B., M. R. Summers, and J. A. Niziolek 1971. Spontaneous in vitro hybridization of lactate dehydrogenase homopolymers in the undenatured state. *Nature New Biol.* 230: 117-119.

Morse, D. E. and B. L. Horecker 1968. The mechanism of action of aldolases. *Adv. Enzymol.* 31: 125-181.

Nordmann, Y., F. Schapira, and J. Dreyfus 1968. A structural modified liver aldolase in fructose intolerance. Immunological and kinetic evidence. *Biochem. Biophys. Res. Commun.* 31: 884-889.

Penhoet, E., M. Kochman, and W. J. Rutter 1969a. Molecular and catalytic properties of aldolase C. *Biochemistry* 8: 4396-4402.

Penhoet, E., M. Kochman, and W. J. Rutter 1969b. Isolation of fructose diphosphate aldolases A, B, and C. *Biochemistry* 8: 4391-4395.

Penhoet, E., M. Kochman, R. Valentine, and W. J. Rutter 1967. The subunit structure of mammalian fructose diphosphate aldolase. *Biochemistry* 6: 2940-2949.

Penhoet, E., T. Rajkumar, and W. J. Rutter 1966. Multiple forms of fructose diphosphate aldolase in mammalian tissues. *Proc. Nat. Acad. Sci. U.S.A.* 56: 1275-1282.

Penhoet, E. and W. J. Rutter 1971. Catalytic and immuno-chemical properties of homomeric and heteromeric combinations of aldolase subunits. *J. Biol. Chem.* 246: 318-323.

Peter, J. B., R. J. Barnard, V. R. Edgeston, C. A. Gillespie, and K. E. Stempel 1972. Metabolic profiles of three fiber types of skeletal muscle in guinea pigs and rabbits. *Biochemistry* 11: 2627-2633.

Pool, B. 1971. The kinetics of disappearance of labeled leucine from the free leucine pool of rat liver and its effect on the apparent turnover of catalase and other hepatic proteins. *J. Biol. Chem.* 246: 6587-6591.

Rechcigl, M. Jr. 1971. Intracellular protein turnover and the roles of synthesis and degradation in regulation of enzyme levels. In: *Enzyme Synthesis and Degradation in Mammalian Systems*. (ed. by M. Rechcigl). S. Karger, New York pp 236-310.

Rensing, U., A. Schmid, and F. Leuthardt 1967. Veränderungen im Isoenzymmuster der Aldolasen aus Ratten-Organen während der Entwicklung. *Hoppe-Seyler's Z. Physiol. Chem.* 348: 921-928.

Rutter, W. J., R. E. Blostein, B. M. Woodfin, and C. S. Weber 1963. Enzyme variants and metabolic diversification. *Advan. Enzyme Regul.* 1: 39-56.

Rutter, W. J., T. Rajkumar, E. Penhoet, M. Kochman, and R. Valentine 1968. Aldolase variants: Structure and physiological significance. *Ann. N. Y. Acad. Sci.* 151: 102-117.

Schimke, R. T. and D. Doyle 1970. Control of enzyme levels in animal tissues. *Ann. Rev. Biochem.* 39: 929-976.

Tweto, J., P. Dehlinger, and A. R. Larrabee 1972. Relative turnover rates of subunits of rat liver fatty acid synthetase. *Biochem. Biophys. Res. Commun.* 48: 1371-1377.

Weber, C. S. and W. J. Rutter 1964. Differential synthesis of aldolases A and B during embryological development. *Fed. Proc. Fed. Amer. Soc. Exp. Biol.*, 23: 487.

Wikman-Coffeld, J., R. Zelis, C. Fenner, and D. T. Mason 1973. Studies on the synthesis and degradation of light and heavy chains of cardiac myosin. *J. Biol. Chem.* 248: 5206-5207.

ISOZYME REALIZATION AND ONTOGENY

COLIN MASTERS
Department of Biochemistry
University of Queensland
St. Lucia. 4067
AUSTRALIA

ABSTRACT. In interpreting the developmental progressions of isozymes and the influence of cellular factors on these phenotypic expressions, increasing attention is being directed to the contribution of particulate interactions. In this paper, the aldolase isozyme system has been examined, and evidence of a differential binding of the isozymes in rat brain provided. The adsorption process was shown to be reversible, of an electrostatic type, and to involve the active site region of the enzyme. In pursuing the microlocalization of this isozyme binding, a particular membrane type within the microsomal fraction was identified as the main site, and the characteristics of the adsorption were detailed.

The interaction between aldolase and the structural proteins of muscle was also studied by means of the electron microscope. In particular, aldolase was shown to form a regular lattice with reconstituted thin filaments, which allowed accurate dimensional analysis, the construction of a three-dimensional model, and a proposal for the mode of action of troponin in muscular contraction. Additionally, the first definitive evidence of a multi-component association between aldolase isozymes and other glycolytic components under physiological conditions has been advanced, and the probable contribution of this complex in ontogeny and different cell types discussed.

INTRODUCTION

The development patterns of isozymes have provided a rich source of information on many aspects of enzyme biology. Not only are these progressions of value to the developmental biologist in assessing the time-scale of differential gene function, but also in relation to such diverse and significant aspects as the structure and compositional inter-relationships of the proteins, the control of enzyme synthesis, the correlation between organ-specific patterns, and the distinctive physiological roles of isozymes, the etiology of diseases, and many other significant aspects of cell and tissue function (Masters and Holmes, 1972).

If we hark back to the early investigations on lactate dehydrogenase ontogeny, for example, it may be recalled that this information played a large part in establishing the nature of the subunit composition and gene control in this isozyme system (Markert and Ursprung, 1962; Cahn et al., 1962; and Lindsay, 1963). By my reference to enzyme realization in the title of this paper, however, I mean to lay emphasis on the fact that it is not only the structural characteristics and differential gene function which determine the final phenotype

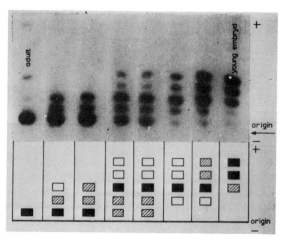

Fig. 1. Electrophoretic pattern of lactate dehydrogenase in ovine skeletal muscle during progressive stages of development.

of the isozymes. Other factors need to be considered as well, and of special significance during ontogeny is the possibility of interaction between the isozymes and the other components of the cell (Masters and Holmes, 1974).

In developing this theme, I have chosen to use as my main example, the aldolase isozyme system. With this enzyme, too, developmental progressions of isozymes are evident and have played a major part in defining structure and gene control of the enzyme (Masters and Holmes, 1972).

There is a further point which is evident from Figure 2, though, and that is the significant presence of aldolase C in the early stages of embryonic development in this animal. This point may be broadened and emphasized by reference to the early embryonic patterns of aldolase isozymes in other species (Figure 3). These observations establish the broad ontogenetic significance of aldolase C, but raise the further question of the nature of the physiological advantage

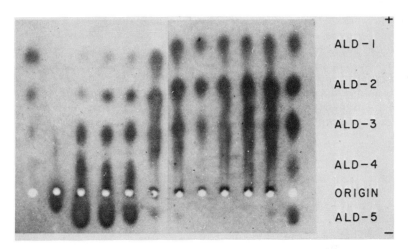

Fig. 2. Electrophoretic pattern of aldolase isozymes in chicken skeletal muscle during progressive stages of development.

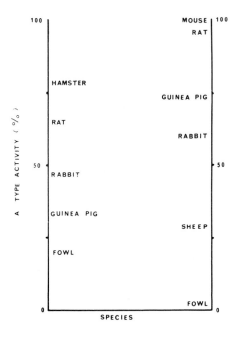

Fig. 3. The contribution of A-type activity to the aldolase (L.H.S.) and lactate dehydrogenase (R.H.S.) in early embryos of vertebrate species.

conferred on the animal by the developmental progressions from C- to A-type activity which occur in most tissues.

In relation to the latter question, I would like to examine some of the lines of evidence which may be considered to support the proposition that much of the physiological distinctiveness of these isozymes may relate, not so much to differences in kinetic properties in the soluble state, but rather to a differential interaction with cellular structure.

RESULTS AND DISCUSSION

THE ADSORPTION OF ALDOLASE ISOZYMES IN BRAIN

Firstly, then, it is necessary to establish whether or not a difference in binding properties exists between the isozymes, and a useful tissue for this purpose is rat brain which contains the full complement of aldolase AC isozymes. When sucrose homogenates of this tissue were centrifuged, it was found that a major portion of the aldolase was adsorbed to the particulate material, that the enzyme was desorbed in the presence of increasing concentrations of potassium chloride, and that quite distinct binding properties were observable between the isozymes (Figure 4). Other salts, and variation of pH, also exerted a marked influence on the binding of aldolase and produced similar patterns of desorption, with aldolase A always showing a far greater propensity for binding than aldolase C (Clarke and Masters, 1972).

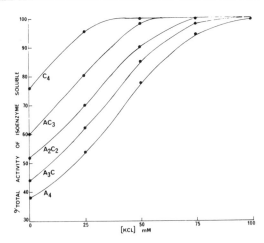

Fig. 4. The elution of aldolase isozymes from rat brain particulates by increasing concentrations of potassium chloride.

The picture of binding which emerges from these data, then, is one consistent with the existence of an electrostatic interaction occurring between the enzyme and the associated structural components. Aldolase A carries the more anodic charge at physiological pH than does aldolase C or the AC isozymes, and this order of charge diminution parallels the individual binding propensities of the isozymes. In regard to the membrane contribution, it would be anticipated that binding of the isozyme was occurring with negatively charged groups in the membranes, such as the phosphates of the phospholipids.

Another feature of particular interest in this situation is the property of specificity in the degree of desorption brought about by certain function dependent metabolites. The substrate of this enzyme, fructose-1:6-diphosphate, for example, exerts a pronounced effect on the desorption process, causing a considerable enhancement of release at very low concentration((Figure 5). Such data would suggest the participation of further, more subtle interactions in the binding of the enzyme, involving the active site either directly or indirectly.

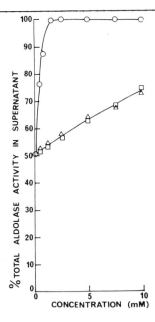

Fig. 5. The effect of fructose-1:6-diphosphate (O), fructose-1-phosphate (Δ) and fructose-6-phosphate (□) on the desorption of particulate bound aldolase in rat brain.

The binding data to this stage, then, may be summarized as indicating a marked tendency for aldolase to bind to membranes, a differential degree of adsorption among the isozymes, and an ability specifically to desorb the enzyme by certain metabolites.

In the sense of physiological correlation, one large remaining question is what happens to the activity of the aldolase when it binds to the membrane, and in order to provide some insight to this query, the effect of increasing FDP concentration on the activity of the soluble enzyme, and by comparison, on the aldolase activity of a rat brain homogenate system have been studied and are shown in Figure 6.

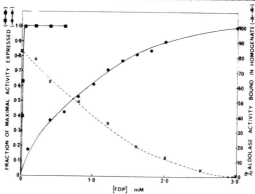

Fig. 6. The influence of fructose-diphosphate concentration on the expressed activity of rat brain aldolase. Cytosol (■), Whole Homogenate (●), Adsorbed Enzyme (X).

It may be seen that the activity response curves are markedly different in these two situations. At low FDP levels the catalytic activity of the bound enzyme was markedly reduced as compared with the soluble systems, and in the biphasic system, more than fifty times the level of FDP was required for maximal expression of catalytic activity than in the soluble system. As in the case of LDH-5 in skeletal muscle (Ehmann and Hultin, 1973) and mitochondrial bound hexokinase (Newsholme et al., 1968), we are again presented with the phenomenon of an apparent alteration of the kinetic properties of an isozyme system by adsorption to cellular material.

In relation to the implications of these data in respect to regulatory possibilities, it may be noted that previous studies on the binding of aldolase to cellulose phosphate (Masters et al., 1969) have pointed to a masking of the

active site of this enzyme by adsorption, with the desorption
of the enzyme from the model particulate phase by FDP re-es-
tablishing the ready accessibility and catalytic activity of
the aldolase. A similar mechanism in the cellular environment
would provide a ready basis of explanation for the differences
in activity which are observed in nervous tissue homogenates
between aldolase in the free and bound states. Additionally,
the existence of an equilibrium between free and bound forms
of aldolase, with each form having intrinsically different
catalytic properties, and with the equilibrium being dependent
on the level of function-dependent metabolites, provides all
the necessary ingredients of a control system for glycolysis;
and indeed several studies have noted high FDP levels in brain,
consistent with a limitation of flux at this step.

MICROLOCALIZATION OF ALDOLASE ISOZYMES IN BRAIN

So far then we have established the differential adsorption
characteristics of the aldolase isozymes, and noted the
possibilities of regulatory significance inherent in this
phenomenon. A logical continuation of these investigations
would be to establish the siting of these bound isozymes within
the cellular architecture, which is, of course, quite complicated in the case of brain. In pursuing these objectives by
differential centrifugation it may be noted from Figure 7,
that each of the primary fractions exhibited appreciable
aldolase activity, but the microsomal fraction was of special
interest in this connection since it exhibited a specific
activity which was more than half as great again as that in
the original homogenate (Clarke and Masters, 1973).
Electron microscope examination of this fraction revealed
a heterogeneous profile of smooth and rough vesicles, so
that subsequent subfractionation of this primary fraction by
gradient centrifugation was indicated and this identified the
source of the bound enzyme as one particular microsomal
membrane fraction.
Having identified and purified a particular membrane type
tn this way, more detailed information may be derived on the
nature of the adsorption phenomena contributing to this microlocalization, and the kinetic parameters of the enzyme in the
membrane-bound state. In the present instance, for example,
the membranes have been isolated and treated with specific
reagents such as phospholipases and ribonucleases in order to
clarify the role of specific groups in the adsorption process,
and the numbers of interactions of the involved binding sites
have been deduced by analysis of kinetic properties (Clarke
and Masters, 1974a). Thus, even in such a complex tissue as

287

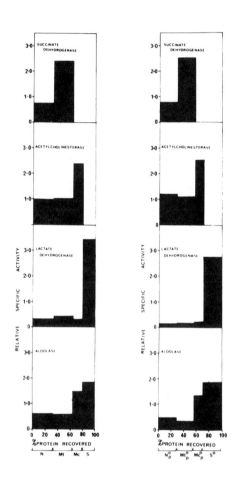

Fig. 7. Subcellular Distribution histograms of enzyme activities in primary fractions of rat brain homogenates. The right hand side shows the patterns remaining after osmotic shock treatment of these fractions.

brain, it has been possible to reach a reasonably sophisticated degree of understanding in regard to the molecular interpretation of microlocalization of the aldolase isozymes.

There are, however, limitations on the level of interpretation which may be reached in relation to a membrane system such as this - limitations which relate in the main to the difficulties of membrane purification and their complex molecular constitution. At this stage, then, rather than going into the detail of the microsomal membrane system, I would prefer to mention another type of approach to the question of

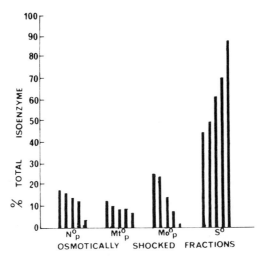

Fig. 8. Isozyme composition of aldolase in the primary fractions of rat brain homogenates; in each fraction the histograms read left to right, A_4, A_3C, A_2C_2, AC_3, C_4.

aldolase binding - one that is capable of leading to a very fine degree of resolution of the interaction.

ELECTRON MICROSCOPE STUDIES OF ALDOLASE INTERACTIONS IN MUSCLE

If we switch to the example of muscle, then we can say that the particulate involvement of aldolase binding has been studied in some detail in this tissue and that the thin filament has been identified as the principal site of location of this enzyme (Sigel and Pette, 1969). Furthermore, a comparison of the binding properties of the main structure proteins from this tissue has confirmed that F-actin is the main single particulate component in this interaction (Arnold and Pette, 1968). Since actin is a single defined protein, normally a part of the particulate material in muscle, but not subject to the difficulties of purification which attach to many membranous entities, the interaction of this substance with other purified components, such as aldolase, can be studied in some detail in in vitro systems. It is possible, for example, to extend our understanding of interactions in such a system through the medium of the electron microscope (Morton, Clarke, and Masters, 1974).

It may be observed that the characteristic appearance of pure F-actin is markedly altered by the addition of purified aldolase - A_4 (Figure 9). The enzyme addition appears to

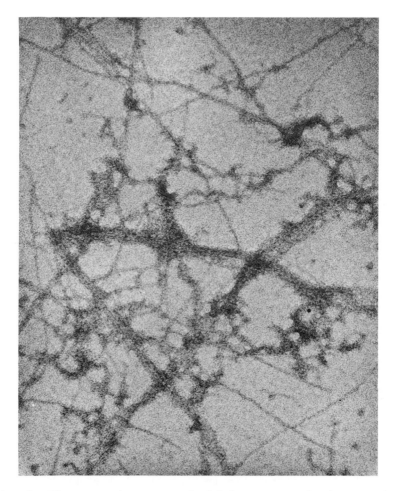

Fig. 9. Electron microgram of aldolase A_4-actin interaction.

broaden the individual filaments, to promote their aggregation, and to serve as a crosslinking agent.

A somewhat more physiological comparison may be drawn in the case of reconstituted thin filaments. We were led to this particular examination by the knowledge that troponin and tropomyosin are also intimately involved in the contractile process and that these components also interact wihh aldolase (Clarke et al., 1974). In the case of the thin filaments (actin - tropomyosin - troponin) a striking change also

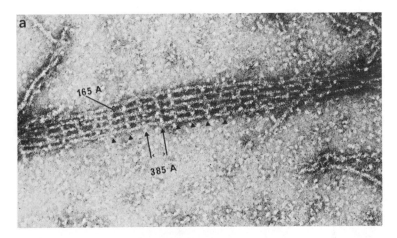

Fig. 10. Electron microgram of interaction between aldolase
A_4 and reconstituted thin filaments.

eventuated with the addition of aldolase-A_4 (Figure 10). A
well defined lattice structure was observed, with parallel
arrangement of the filaments and regular crosslinking. Once
again, this picture is distinct from that derived from actin -
aldolase alone, and testifies to the importance of tropomyosin - troponin in this complexing.

With such a regularity of interaction, it is possible to
measure the dimensions of the structural interactions with
some confidence, and in the present case a mean value of
385 Å for the center to center spacing of the crossbridge may
be deduced, and 165 Å for the center to center spacing be
between adjacent filaments.

It is of interest that this spacing of the crossbridges
corresponds to the established repeat distance for the troponin
molecule and leads to the assumption that the crossbridges may
be formed by the attachment of aldolase to troponin.

291

Confirmatory evidence for this conclusion also exists in the form of the stoichiometry of the reaction, and the essentiality of troponin in forming these structures.

On the basis of this dimensional information, it has been possible to contruct a model which describes the fit of aldolase to the component molecules of the thin filament (Figure 11), (i.e. the relative positions of aldolase to actin, tropomyosin, and troponins I, T, and C) and satisfies the established data on tropomyosin-troponin subunit interactions (Weber and Murray, 1973).

Fig. 11. Model of Aldolase - Thin Filament Interaction.

In this manner, then, a satisfying knowledge of the structural factors involved in the binding of aldolase to muscle structure has been reached, and this method appears to offer the promise of further rewarding insight into the mole- cular mechanisms of tissue function. It is of interest, for example, that the model suggests a mechanism for the mode of troponin implication in the regulation of muscle contraction.

In the activated state (i.e. when tropomyosin binds calcium), the tropomyosin moves to a more central position in the groove of the actin filament, and allows better access to the myosin action site. Consideration of the present model reveals that movement of the troponin molecule in response to variation in the Ca^{++} levels may account for this movement of the tropomyosin molecule in the groove of the actin helix, and thus provide a basis for a molecular latch mechanism during contraction.

EVIDENCE FOR AN ASSOCIATION OF ALDOLASE AND OTHER GLYCOLYTICAL ENZYMES

The fact that aldolase isozymes associate with particulate structure raises a number of important and novel issues in regard to the significance of this phenomenon in cell function. Two of the most commonly posed questions are:
1. Is the association real or merely artifactual?
2. Does the binding of aldolase hold significance in relation to the existence of the long-postulated "glycolytic complex"?

I would like now to consider some data bearing on both of these questions. A specific prescription for the presence of an association of cytosol components, has, of course, always been extremely difficult to accomplish, though a number of very ingenious approaches have been made to this problem (Green et al., 1965, Amberson and Bauer, 1971, De Duve, 1972).

The first definitive proof of the existence of a multi-enzyme aggregate of glycolytic enzymes occurring in tissue extracts under physiological conditions of pH and ionic strength was based upon the reasoning that cellular conditions involved both high saline and high protein concentrations, (the saline and protein concentrations in the cytosol have been calculated to be of the order of 0.9 and 15% respectively) and that both of these factors would heavily influence the binding of cytosol components to one another and to membranes. We have studied the differential centrifugation of glycolytic enzymes in concentrated myogen preparations of skeletal muscle, which were free from particulate matter, and maintained at pH 7.4, and an ionic strength of 0.15 M (Clarke and Masters, 1973).

Typical sedimentation profiles from these experiments are illustrated in the accompanying diagram, (Figure 12), the main point of interest being the significant second boundary, occurring in parallel for each of the individual enzymes studied, and hence establishing the presence of a complex of glycolytic enzymes in this soluble system.

Subsequent research has indicated that the size of the

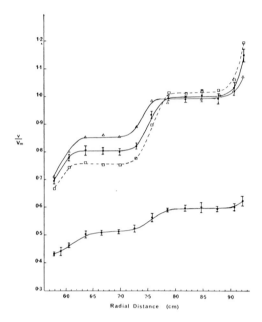

Fig. 12. Enzyme sedimentation in myogen preparation of sheep semitendinosus muscle. ● Aldolase, Δ Pyruvate Kinase, Lactate Dehydrogenase, o Phosphofructokinase.

observed complex is of sufficient magnitude to encompass the inclusion of all the glycolytic enzymes, and that a solubilized membranous component is necessary for complex formation under these experimental conditions. Also, the stability of the complex has proved to be very sensitive to factors such as pH, ionic strength, the concentration of proteins and metabolites, and the isozyme composition of the individual enzymic components, a behavior which goes far in explaining the previous lack of success in efforts aimed at demonstrating the existence of such an entity.

Overall, then, the picture which emerges in regard to the microlocalization of glycolytic enzymes is that these components appear to be associated in the cell in the form of a complex which exists not in free solution, but is plated on to the structural components containing actin or actin-like protein (Clarke and Masters, 1974b). This association would appear to act toward an increased efficiency in the functioning of the glycolytic sequence, while at the same time positioning this major source of cellular energy in a most apposite position in relation to the contractile processes requiring that energy. Developmental advantages conferred upon individual tissues during ontogeny by the maturational progressions of isozymes described at the beginning of this paper, may well be viewed in the same light.

SUMMARY

In conclusion, then, the various lines of evidence which have been presented in this communication are considered to provide supporting evidence for the proposition that the phenotypic expression of the aldolase isozymes is modified by interactions with the particulate components of the cell.

It is hoped that these data have also served to illustrate the means by which these biphasic interactions may be studied, and the level of detail which may derive from such studies.

As the study of isozymes and ontogeny advances, an increasing responsibility and need for further investigations into the biphasic interactions of isozymes would seem to be indicated.

ACKNOWLEDGEMENTS

It is a pleasure to acknowledge the major contributions to the recent research reported in this paper, by Drs. F. M. Clarke and D. Morton.

REFERENCES

Amberson, W. R. and A. C. Bauer 1971. Electrophoretic studies of muscle proteins. II. Complex formation between delta protein, myogen and myosin. *J. Cell Physiol*. 77: 281-299.

Arnold, H. and D. Pette 1968. Binding of glycolytic enzymes to structure proteins of muscle. *Eur. J. Biochem*. 6: 163-170.

Cahn, R. D., N. O. Kaplan, L. Levine, and E. Zwilling 1962. Nature and development of lactic dehydrogenases. *Science, N. Y*. 136: 962-969.

Clarke, F. M. and C. J. Masters 1972. On the reversible and selective adsorption of aldolase isoenzymes in rat brain. *Arch. Biochem. Biophys*. 153: 258-263.

Clarke, F. M. and C. J. Masters 1973. On the distribution of aldolase isoenzymes in subcellular fractions from rat brain. *Arch. Biochem. Biophys*. 156: 673-683.

Clarke, F. M., C. J. Masters, and D. J. Winzor 1974. Interaction of aldolase with the troponin-tropomyosin complex of beef muscle. *Biochem. J*. (in press).

Clarke, F. M. and C. J. Masters 1974a. On the reversible adsorption of aldolase to a microsomal membrane fraction from rat brain. *Int. J. Biochem*. (in press).

Clarke, F. M. and C. J. Masters 1974b. On the properties of the multi-enzyme aggregate of glycolytic components in muscle extracts. *Biochim. Biophys. Acta* (in press).

De Duve, C. 1972. In A. Akeson, and A. Ehrenberg. *Structure and Function of Oxidation Reduction Enzymes*. Pergamon Press, Oxford and New York, pp.715-728.

Ehmann, J. D. and H. O. Hultin 1973. Substrate inhibition of soluble and bound lactate dehydrogenase (Isoenzyme 5). *Arch. Biochem. Biophys*. 154: 471-475.

Green, D. E., E. Murer, H. O. Hultin, S. H. Richardson, B. Salmon, G. P. Brierley, and H. Baum 1965. Association of integrated metabolic pathways with membranes. I. Glycolytic enzymes of the red blood corpuscle and yeast. *Arch. Biochem. Biophys*. 112: 635-647.

Lindsay, D. T. 1963. Isozymic patterns and properties of lactate dehydrogenase from developing tissues of the chicken. *J. Exp. Zool*. 152: 75-89.

Markert, C. L. and H. Ursprung 1962. The ontogeny of isozymes patterns of lactate dehydrogenase in the mouse. *Dev. Biol*. 5: 363-381.

Masters, C. J. and R. S. Holmes 1972. Isoenzymes and ontogeny. *Biol. Rev*. 47: 309-361.

Masters, C. J. and R.S. Holmes 1974. In:Haemoglobin isoenzymes and tissue differentiation. ASP Biological and Medical Press, Amsterdam.

Morton, D., F. M. Clarke, and C. J. Masters 1974. A proposed model of the action of troponin in regulating muscle contraction. *Proc. Austral. Biochem. Soc*. (in press).

Newsholme, E. A., F. S. Rolleston, and K. Taylor 1968. Factors affecting the glucose-6-phosphate inhibition of hexokinase from cerebral cortex tissue of the guinea pig. *Biochem. J*. 106: 193-201.

Sigel, P. and D. Pette 1969. Intracellular localization of glycogenolytic and glycolytic enzymes in white and red rabbit skeletal muscle. *J. Histochem. Cytochem*. 17: 225-237.

Weber, A. and J. M. Murray 1973. Molecular control mechanisms in muscle contraction. *Physiol. Rev*. 53: 612-673.

LDH-X: CELLULAR LOCALIZATION, CATALYTIC PROPERTIES,
AND GENETIC CONTROL OF SYNTHESIS

A. BLANCO, WM. H. ZINKHAM, and D. G. WALKER
Cátedra de Química Biológica,
Facultad de Ciencias Médicas,
Universidad Nacional de Córdoba
Córdoba, Agrentina, and
Department of Pediatrics
The John Hopkins University School of Medicine, and
the Harriet Lane Service of the Children's
Medical and Surgical Center,
The Johns Hopkins Hospital,
Baltimore, Maryland 21205   and
Department of Anatomy,
The Johns Hopkins University School of Medicine,
Baltimore, Maryland 21205

ABSTRACT.   LDH-X is a unique form of lactate dehydrogenase
(LDH) in mature testes from many mammalian and avian
species. LDH-X activity accounts for 80-90 percent of
the LDH activity in human and rabbit sperm, and its
appearance in testicular homogenates of the rabbit coincides with proliferation and differentiation of spermatogonial stem cells.

LDH-X activity is confined to a subfraction of the
"heavy" mitochondrial fraction of rat testis. This fraction contains a unique type of mitochondrion, and only
when these peculiar mitochondria are present does one
detect LDH-X activity.

Kinetic studies have shown that values of Km's for
pyruvate and lactate are identical to or lower than those
for LDH-1, and much lower than those for LDH-5. Also,
LDH-X activity is most affected by high concentrations
of pyruvate, but is not inhibited by high levels of substrate when catalyzing the reverse reaction. LDH-X,
like LDH-1, may function as an "aerobic" dehydrogenase.

Genetic studies in pigeons have revealed that LDH-X
snythesis is controlled by a third genetic locus, LDH-C,
and that the LDH-C locus is linked to the LDH-B locus.
The most probable recombination fraction was zero, and
contiguity could not be excluded, thereby suggesting that
the LDH-B and LDH-C loci in pigeons acquired their separate identities by a duplication event.

Questions to be answered in the future concern the
functional significance of LDH-X , the nature of the mechanisms controlling the activity of the LDH-C locus, and

the nature of the evolutionary events responsible for the remarkable degree of polymorphism at the LDH-B and LDH-C loci in pigeons.

Mature testes from many mammalian and avian species contain a unique molecular form of lactate dehydrogenase (L-lactate: NAD oxidoreductase, EC 1.1.1.27). The discovery of this isozymic form of lactate dehydrogenase (LDH) was accomplished by a combination of electrophoretic and histochemical techniques (Blanco and Zinkham, 1963), a method by which the LDH isozymes are defined as bands on the surface of the electrophoresis medium; the isozyme of LDH unique to testis was designated "band X". In subsequent reports, "band X" has been referred to as LDH-X, a term which hopefully will be changed when a uniform system of nomenclature becomes available for classifying the isozymic forms of LDH.

LDH-X is a tetrameric molecule with a molecular size similar to that of the other isozymic forms of LDH (Zinkham, Holtzman, and Isensee, 1972; Stambaugh and Buckley, 1967). Although the C subunits of LDH-X differ from the A and B subunits present in the commonly occurring five isozymic forms of LDH (Goldberg, 1972; Goldberg, 1971), the C polypeptides can hybridize with the A or B polypeptides in vitro and in vivo to form heteropolymeric molecules with enzymatic activity (Zinkham, Blanco, and Kupchyk, 1963; Battellino and Blanco, 1970a; Goldberg, 1973). These findings indicate a high degree of structural similarity between the A, B, and C subunits.

The remarkable tissue specificity of LDH-X posed a number of questions, some of which have been answered during the past ten years. This report will summarize findings concerning 1) Tissue, Cellular, and Sub-cellular Localization of LDH-X, 2) Catalytic Properties of LDH-X, and 3) Genetic Control of LDH-X Synthesis.

TISSUE, CELLULAR, AND SUB-CELLULAR LOCALIZATION OF LDH-X

An extensive survey of many tissues from males and females of a variety of mammalian and avian species revealed that LDH-X activity is restricted to the testis. It is possible, of course, that the method of starch-gel electrophoresis lacks sufficient sensitivity to define the small amounts of LDH-X that might be present in only a few cells, e.g., the mature ovary in which only one cell matures and differentiates during the sexual cycle.

LDH-X activity accounts for 80-90 percent of the LDH activity in human and rabbit sperm (Zinkham, Blanco, and

Clowry, 1964). Whether LDH-X exists in other cellular components of the testis or male reproductive system is unknown.
Allen (1961) demonstrated LDH-X in the epididymis of the
mouse, a finding which may have been due to the presence of
intact or degenerating sperm. The small amount of LDH-X occasionally observed in seminal plasma may represent loss of
enzyme from sperm which have been altered during or after
the collection of the ejaculate.

Other observations which indicate that LDH-X is an enzymatic component of spermatogonial elements include the following. Ontogenetically, the appearance of LDH-X in testicular
homogenates of the mouse and the rabbit coincides with proliferation and differentiation of spermatogonial stem cells
(Zinkham, Blanco, and Clowry, 1964; Goldberg and Hawtrey,
1967; and Battellino and Blanco, 1970b). Also LDH-X activity
is present in post-pubertal, but not pre-pubertal human testes
(Blanco and Zinkham, 1963). And finally, induction of aspermatogenesis in guinea pigs by the method of Freund resulted in
disappearance of spermatogonial elements and a complete loss
of LDH-X activity without quantitative or qualitative changes
in the other LDH isozymes (Zinkham, Blanco, and Clowry, 1964).
The only histologic abnormality in the testes of these animals was a total lack of spermatogenesis.

Unlike the A and B subunit containing isozymes that are
distributed primarily in the cytosol, LDH-X activity is confined to the mitochondrial fraction of testicular homogenates
(Clausen, 1969; Domenech, Domenech, and Blanco, 1970) and
to the middle piece of spermatozoa (Clausen, 1969). A subfraction of the "heavy" mitochondrial fraction of rat testis
contains a unique type of mitochondrion. As described by
André (1962), these organelles have a very dense matrix and
dilated cristae, first appear in primary spermatocytes, and
become more numerous as differentiation progresses. During
spermiogenesis, these mitochondria migrate to the upper portion
of the axial filament to form the mitochondrial sheath of the
middle piece of spermatozoa (Fig. 1). Domenech, Domenech, and
Blanco (1970) have demonstrated that a preparation of "heavy"
mitochondria from rat testes (Fig. 1) contains most of the
LDH-X activity (Fig. 2). Furthermore, the appearance and
distribution of LDH-X coincides with the presence of these
peculiar mitochondria, i.e., only when they are observed does
one detect LDH-X activity (Domenech, Domenech, Aoki, and
Blanco, 1972). These observations indicate that LDH-X is
compartmentalized in the organelles unique to the differentiated germ cells of the mature male.

CATALYTIC PROPERTIES OF LHD-X

The unique biological features of LDH-X, especially

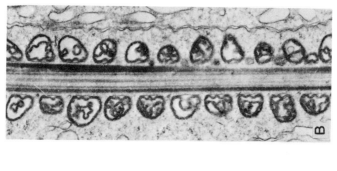

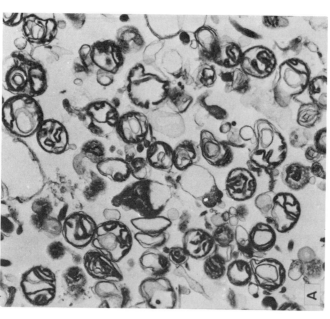

Fig. 1: A. Electron micrograph of a thin section of a mitochondrial subfraction from mature rat testis homogenate. This subfraction shows a predominance of a special type of mitochondria of dense matrix and dilated cristae (x 38,500). B. Electron micrograph of a longitudinal section through the middle piece of a maturing rat spermatid (x 38,500). Note the similarity of organelles of the mitochondrial sheath with those isolated in the preparation shown in A. For technical details, see Domenech, et al., (1972).

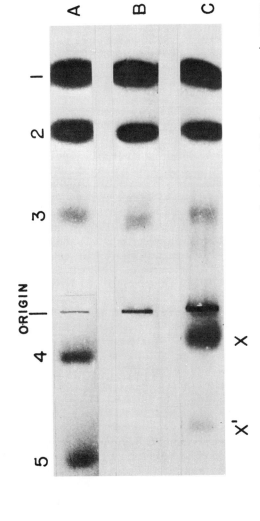

Fig. 2: Starch gel electrophoretic patterns of rat lactate dehydrogenase isozymes.
A: Extract of a reference tissue. B: Extract from a mitochondrial subfraction not containing the organelles shown in Fig. 1. C: Extract from a mitochondrial subfraction corresponding to the preparation shown in Fig. 1 A.-X and x^1 indicate the additional bands of testis.
Patterns of the soluble phase and extracts of the microsomal fraction from rat testis did
not exhibit the additional isozymes.

ontogeny and cellular distribution, led several investigators to study the catalytic properties of the enzyme. Allen (1961) first reported that LDH-X from mouse testis has a high affinity for α-hydroxy acids other than lactate. Later Zinkham, Blanco, and Clowry (1964) and also Blanco, Zinkham, and Kupchyk (1964) showed that LDH-X from several species exhibits a high rate of activity with α-keto and α-hydroxy acids of 4 and 5 carbon chains. Also rat sperm contains an LDH-X which has substantial activity for the pyridine nucleotide-linked interconversion of α-hydroxy glutarate (Schatz and Segal, 1969), a property not shared by the other isozymes of LDH. In view of the wide specificity of LDH-X for substrates other than pyruvate or lactate, some investigators have questioned whether LDH-X is an isozymic form of LDH. Current knowledge of sperm metabolism, however, suggests that LDH-X serves no significant function other than catalyzing the interconversion of lactate and pyruvate.

Comparative kinetic studies on LDH-1 (B_4), LDH-5 (A_4), and LDH-X (C_4) carried out by different investigators are in general agreement (Battellino and Blanco, 1970a: Schatz and Segal, 1969; Battellino, Ramos Jaime, and Blanco, 1968; Wilkinson and Withycombe, 1965; Kolb, Fleisher, and Larner, 1970; Hawtrey and Goldberg, 1970). Values of Km for pyruvate and lactate determined in several species are shown in Table I. In general, these values indicate that affinity of LDH-X for pyruvate and lactate is identical to or higher than that of LDH-1, and much higher than that of LDH-5. In most of the species studied, LDH-X is strongly inhibited by high concentrations of pyruvate, and in some species, the degree of inhibition is significantly higher than that determined for LDH-1 (Battellino and Blanco, 1970a; Schatz and Segal, 1969; Battellino, Ramos Jaime, and Blanco, 1968).

Fig. 3 presents plots of enzymatic activity as percentage of the maximum versus substrate concentration for the direct (pyruvate → lactate) and the reverse reactions catalyzed by LDH-1 (B_4), LDH-5 (A_4), and LDH-X (C_4) purified from rabbit testis. Activity of LDH-X is most affected by high concentrations of pyruvate, but is not inhibited by high levels of substrate when catalyzing the reverse reaction. A similar behavior at high concentrations of lactate has been demonstrated for the LDH-X isozymes of guinea pig (Battellino and Blanco, 1970a), man (Wilkinson and Withycombe, 1965), and bull (Kolb, Fleisher, and Larner, 1970).

GENETIC CONTROL OF LDH-X SYNTHESIS

Other investigators (Shaw and Barto, 1963; Boyer, Fainer, and Watson-Williams, 1963; Nance, Claflin, and Smithies, 1963)

TABLE I

K_m VALUES OF LACTATE DEHYDROGENASE FROM DIFFERNET SPECIES

K_m for Pyruvate (mM)	LDH-5 (A_4)	LDH-1(B_4)	LDH-X (C_4)
Rabbit			
3*	0.25	0.12	0.12
33	0.35	0.1	0.1
Guinea Pig			
4	0.13	0.05	0.04
Mouse			
5	0.2	0.05	0.05
19	0.12	0.042	0.026
Bat			
21	0.22	0.08	0.03
Human			
35	0.83	0.08	0.05
33	----	0.012	0.017

K_m for Lactate (mM)	LDH-5 (A_4)	LDH-1(B_4)	LDH-X (C_4)
Rabbit			
3	20.0	5.5	4.9
33	23.0	1.5	1.6
Guinea Pig			
4	25.0	4.3	5.1
Mouse			
5	11.0	7.5	4.3
Human			
33	----	0.95	0.66

* Refers to Reference Number in Bibliography

---

have shown that the synthesis of the A and B polypeptides of the usual five isozymes is under the control of two non-allelic structural genes, LDH-A and LDH-B. The presence of a unique form of LDH in mature testis suggested that a third genetic locus, LDH-C, was responsible for the synthesis of another LDH subunit, designated C. In order to substantiate this hypothesis, allelic forms of the LDH-C locus were searched for in a variety of mammalian and avian species. As shown in Fig. 4, three types of LDH isozyme patterns were found in testicular homogenates of pigeons. The frequency distribution of the

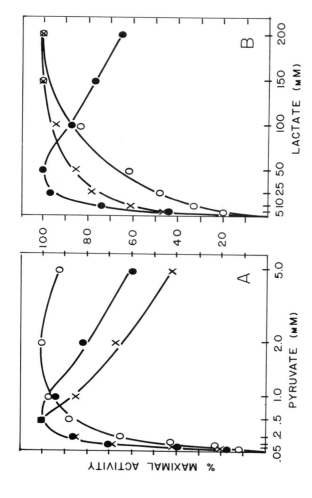

Fig 3: Effect of substrate concentration upon activity of rabbit LDH-5 (A$_4$), LDH-1 (B$_4$), and LDH-X (C$_4$) (● = LDH-1; o = LDH-5; x = LDH-X). Initial reaction velocity, expressed as percentage of maximal activity, is plotted against concentration of substrate. A: Direct reaction (pyruvate → lactate). Final concentrations in the reaction mixture were: NADH, 0.115 mM, phosphate buffer pH 7.4, 80 mM, and pyruvate, 0.05, 0.1, 0.2, 0.5, 1.0, 2.0, and 5.0 mM. B: Reverse reaction. Final concentrations in the reagent mixture were: NAD 1.0 mM, Tris pH 9.0, 10 mM and L-lactate, 5, 10, 25, 50, 100, 150, and 200 mM.

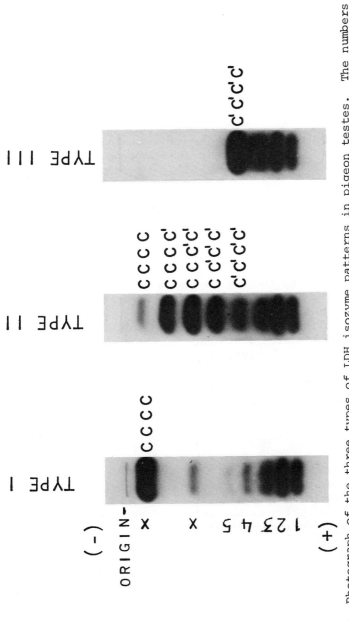

Fig. 4: Photograph of the three types of LDH isozyme patterns in pigeon testes. The numbers indicate the position of the usual isozymes, and the polypeptide composition of the LDH-X complex is listed by the X-bands. The minor X-band and the band in the position of LDH-5 in the Type I pattern probably represent tetramers of B and C subunits: B_1C_3 and B_2C_2 respectively. (Reproduced by permission of the Annals of the New York Academy of Science).

three phenotypes in approximately 1,000 wild and pure-bred pigeons showed that the distribution of the phenotypes follow-ed the expected distribution calculated according to the Hardy-Weinberg law for a single pair of alleles (Zinkham, Blanco, and Kupchyk, 1964). Frequencies of the C^1 and C^2 alleles were similar for wild pigeons, racing homers, and White Kings, the ratios of C^1/C^2 ranging from 4/1 to 5/1. In White Carneau pigeons the ratios were reversed, and in Silver Kings, approach-ed 1/1.

Further studies on pigeon tissues revealed three alleles at the LDH-B locus, B^1, B^2, B3, the frequencies of B^1/B^1, $B^1/B2$, $B2/B2$, $B^1/B3$, $B2/B3$, and $B3/B3$ genotypes being 0.649, 0.049, 0.002, 0.269, 0.005, and 0.026 respectively (Zinkham and Isensee, 1972). The presence of polymorphisms at the LDH-B and LDH-C loci made it possible to establish matings of appropriate genotypes to determine whether the B and C loci are linked. The results of breeding experiments conclusively demonstrated linkage between the B and C structural loci in the pigeon (Zinkham, Isensee, and Renwick, 1969). The most probable recombination fraction was zero, and contiguity could not be excluded,thereby suggesting that the LDH-B and LDH-C loci in the pigeon acquired their separate identities by a duplication event.

Despite the close linkage of the LDH-B and LDH-C loci in pigeons, there is a wide difference in the biological behavior of the two loci. The activity of the B locus is expressed in both sexes and in all tissues throughout life, whereas the activity of the C locus is restricted to a particular sex, a particular tissue and cell-line, and a particular stage of maturation. Another important example of the temporal speci-ficity of the LDH-C locus is the seasonal variation of LDH-X activity in the bat (*Tadarida brasiliensis*). Blanco, Gutierrez, Henquin, and Burgos (1969) have shown that between the end of Spring (November and December in Argentina) and early Autum (March to April), there is a cessation of sperma-togonial proliferation and differentiation, and LDH-X activity disappears. Resumption of maturation of germ cells during Winter and early Spring correlates with the reappearance of LDH-X activity in testicular homogenates.

As yet, the nature of the gene-activating mechanisms for LDH-X synthesis have not been defined. In one series of experiments, various procedures were employed to activate the LDH-C locus in the female pigeon. Sinistral ovariectomy during the first few days of life in some avian species results in the development of an ovo-testis on the right side (Domm, 1927; Miller, 1937; Kornfeld and Nalbandov, 1954; Taber and Solley, 1954), a structure which occasionally con-

tains mature sperm. Removal of the left ovary from 20 White
Carneau pigeons between the first and third days of life
resulted in the development of a small ovo-testis (20 to 60
mg wet weight) in 12 of the birds when sacrificed between
10 and 12 months of age. None of these glands contained
mature sperm, and none exhibited LDH-X activity (Zinkham
and Walker, 1974). Since administration of anti-estrogens
(Taber and Solley, 1954) has been shown to enhance the development of an ovo-testis in sinistral ovariectomized chickens,
another group of 16 White Carneau pigeons was given daily
intramuscular injections of 17-d-ethyl-19-nortestosterone
following removal of the left ovary before the third day of
life. Although 10 of the birds had a small ovo-testis when
sacrificed at one year of age, mature sperm and/or LDH-X
activity were not observed. Transplantation of a testis from
a one-day-old male pigeon into the peritoneal cavity of an
ovariectomized nest mate resulted in the development of a
mature testis at 8 months of age. The complement of LDH
isozymes in this testis, including the LDH-X phenotype, was
identical to that of the testis remaining in the donor bird.
Thus, it appears that the immature testis can develop into a
mature organ in the peritoneal cavity of the ovariectomized
female.

DISCUSSION

The presence of LDH-X activity in only one type of cell
suggests that the LDH-C locus functions to synthesize an
enzyme with a highly specific metabolic function. Certain
kinetic properties of the enzyme, namely, high affinities
for pyruvate and lactate and inhibition by large concentrations of the first substrate, are even more pronounced than
the corresponding values observed for LDH-1 (B_4). The
phenomenon of substrate inhibition to which Cahn et al (1962)
assigned functional significance has been extensively analyzed.
Recently Everse and Kaplan (1973) proposed that LDH-1 is a
regulatory enzyme, modulated by its oxidized substrate through
the formation of an abortive ternary complex (Enzyme-NAD-Pyru-
vate). In aerobic tissues most of the B_4 isozyme would be
inactive and unable to reduce pyruvate. An increase of lactate in the medium promotes the dissociation of the inactive
complex, followed by the oxidation of lactate.

If one extends this hypothesis to the metabolic characteristics of LDH-X, it would appear that this enzyme is especially
adapted to function in the direction of lactate oxidation.
Striking inhibition by pyruvate indicates that LDH-X readily
forms a stable abortive ternary complex. For the enzyme to

307

function as a pyruvate reductase would require extreme anaerobic conditions in which the NAD/NADH ratio would have to attain very low values.

On the other hand, the high affinity of LDH-X for lactate and the insensitivity of the enzyme to high concentrations of substrate or product (Battellino, Ramos Jaime, and Blanco, 1968) when catalyzing the reverse reaction, permit the enzyme to function "unidirectionally" under aerobic conditions. In fact, LDH-X appears to be more efficiently adapted for this metabolic role than does the B_4 isozyme, LDH-1.

High concentrations of lactate are present in the seminal fluid (Mann, 1964) and in the fluids of the female genital tract (Lutwak-Mann, 1962). During aerobiosis, spermatozoa can oxidize lactate from the surrounding medium as a source of energy (Salisbury and Lodge, 1962). Furthermore, oxygen tensions in the female genital tract are sufficiently high to support an aerobic type of metabolism (Bishop, 1956). Thus it is possible that LDH-X preferentially catalyzes the oxidation of lactate, and thereby generates NADH which in turn can be funneled into the respiratory chain. The mitochondrial localization of LDH-X also supports the concept that the enzyme functions primarily as an "aerobic" lactate dehydrogenase (Skilleter and Kun, 1972).

Many investigators have shown that isozymic forms of a variety of enzymes may be useful markers for studying gene expression during growth and cellular differentiation. In this respect, studies on LDH-X, because of its extraordinary temporal and cellular specificity, have been especially informative. The differentiation of spermatogonia into mature germ cells is accomplished by changes in morphology, function, and chemical composition, including enzyme systems. During the later stages of development, a genetic locus (LDH-C) is activated to synthesize an enzyme (LDH-X) which appears to be essential for normal sperm function. What is the nature of the factors that activate this locus? Further studies to answer this question should provide important information concerning the relationship between gene expression and cellular differentiation.

Also unanswered is the nature of the selective forces that have resulted in the remarkable degree of polymorphism involving the LDH-B and LDH-C loci in pigeons. The kinetic and physiochemical properties of the B^1, B^2, and B^3 polypeptides are identical, as are those for the C^1 and C^2 polypeptides (Zinkham, Kupchyk, Blanco, and Isensee, 1966; Zinkham and Isensee, 1972). Extensive observations on pigeons breeding in captivity indicate no selective advantage to any particular genotype as reflected by ability to reproduce. Perhaps this type of environment is selectively neutral, so that observations on wild-type pigeons would be required to define the effect of genotype on biological fitness. In this respect, the incidence of iso-alleles at the B and C loci is remarkably constant in wild pigeons collected from widely separated geographical areas. Such a constancy of allelic frequencies suggests that the B and C polymorphisms are biologically important. Other types of information are necessary, however, to define the relative contributions of random genetic drift or natural selection to the evaluation of the LDH polymorphisms in pigeons.

ACKNOWLEDGEMENTS

Supported in part by a Grant from the Consejo Nacional de Investigaciones Cientificas y Técnicas (CONICET). A. Blanco is a Career Investigator of the CONICET.

REFERENCES

Allen, J. M. 1961. Multiple forms of lactate dehydrogenase in tissues of mouse: Their specificity, cellular localization and response to altered physiological conditions. *Ann. N.Y. Acad. Sci.* 94: 937-951.

Andre, J. 1962. Contribution a la connaisance du chondriome. Etude de ses modifications ultrastructurales pendant la spermatogenese. *J. Ultrastruct. Res.* Suppl. 3: 1-185.

Battellino, L. J., F. Ramos Jaime, and A. Blanco 1968. Kinetic properties of rabbit testicular lactate dehydrogenase isozyme. *J. Biol. Chem.* 243: 5185-5192.

Battellino, L. J. and A. Blanco 1970a. Testicular lactate dehydrogenase isozyme: Nature of multiple forms in guinea pigs. *Biochim. Biophys. Acta* 212: 205-212.

Battellino, L. J. and A. Blanco 1970b. Catalytic properties of the lactate dehydrogenase isozyme "X" from mouse testis. *J. Exp. Zool.* 174: 309-316.

Bishop, D. W. 1956. Proceedings of the 3rd, International Congress on Animal Reproduction Physiology. Brown, Knight and Truscott, Ltd., London, p. 53.

Blanco, A. and W. H. Zinkham 1963. Lactate dehydrogenases in human testes. *Science* 139: 601-602.

Blanco, A., W. H. Zinkham, and L. Kupchyk 1964. Genetic control and ontogeny of lactate dehydrogenase in pigeon testes. *J. Exp. Zool.* 156: 137-152.

Blanco, A., M. Gutierrez, C. G. DeHenquin, and N. M. G. DeBurgos 1969. Testicular lactate dehydrogenase isozyme: Cyclic appearance in bats. *Science* 164: 835-836.

Boyer, S. H., D. C. Fainer, and E. J. Watson-Williams 1963. Lactate dehydrogenase variant from human blood: Evidence for molecular subunits. *Science* 141: 642-643.

Cahn, R. O., N. O. Kaplan, L. Levine, and E. Zwilling 1962. Nature and development of lactic dehydrogenases. *Science* 136: 962-969.

Clausen, J. 1969. Lactate dehydrogenase isoenzymes of sperm cells and testes. *Biochem. J.* 111: 207-218.

Domenech, E. M. de, C. E. Domenech, and A. Blanco 1970. Distribution of lactate dehydrogenase isozymes in subcellular fractions of rat tissues. *Arch. Biochem. Biophys.* 141: 147-154.

Domenech, E. M. de, C E. Domenech, A. Aoki, and A. Blanco 1972. Association of the testicular lactate dehydrogenase isozyme with a special type of mitochondria. *Biol. Reprod.* 6: 136-147.

Domm, L. V. 1928. Spermatogenesis following early ovariotomy in the brown Leghorn fowl. *Proc. Soc. Exp. Biol. and Med.* 26: 338-341.

Everse, J. and N O. Kaplan 1973. Lactate dehydrogenases: Structure and function. *Adv. Enzymol.* 37: 61-133.

Goldberg, E. and C. Hawtrey 1967. The ontogeny of sperm specific lactate dehydrogenase in mice. *J. Exp. Zool.* 164: 309-316.

Goldberg, E. 1971. Immunochemical specificity of lactate dehydrogenase X. *Proc. Nat. Acad. Sci. USA* 68: 349-352.

Goldberg, E. 1972. Amino acid composition and properties of crystalline lactate dehydrogenase X from mouse testes. *J. Biol. Chem.* 247: 2044-2048.

Goldberg, E. 1973. Molecular basis for multiple forms of LDH-X. *J. Exp. Zool.* 186: 273-278.

Gutierrez, M., N. M. G. DeBurgos, C. Burgos, and A. Blanco 1972. The sexual cycle of male bats: Changes of testicular lactate dehydrogenase isozymes. *Comp. Biochem. Physiol.* 43A: 47-52.

Hawtrey, C. O. and E. Goldberg 1970. Some kinetic aspects of sperm specific lactate dehydrogenase in mice. *J. Exp. Zool.* 174: 451-462.

Kornfeld, W. and A. V. Nalbandov 1954. Endocrine influences on the development of the rudimentary gonad of fowl. *Endocrinology* 55: 751-761.

Kolb, E., G. A. Fleisher, and J. Larner 1970. Isolation and characterization of bovine lactate dehydrogenase X. *Biochem.* 9: 4372-4380.

Lutwak-Mann, C. 1962. Some properties of uterine and cervical fluid in the rabbit. *Biochim. Biophys. Acta* 58: 637-639.

Mann, T. 1964. The biochemistry of semen and of the male reproductive tract. Methuen and Co., Ltd., London.

Miller, R. A. 1937. Spermatogenesis in a sex-reversed female and in normal males of the domestic fowl, *Gallus domesticus*. *The Anatom. Rec.* 70: 155-189.

Nance, W. E., A. Claflin, and O. Smithies 1963. Lactic dehydrogenase: Genetic control in man. *Science* 142: 1075-1077.

Salisbury, G. W. and J. R. Lodge 1962. Metabolism of spermatozoa. *Adv. Enzymol* 24: 35-104.

Schatz, L. and H. L. Segal 1969. Reduction of α-ketoglutarate by homogenous lactic dehydrogenase X of testicular tissue. *J. Biol. Chem.* 244: 4393-4397.

Shaw, C. R. and E. Barto 1963. Genetic evidence for the subunit structure of lactate dehydrogenase isozymes. *Proc. Nat. Acad. Sci.* 50: 211-214.

Skilleter, D. N. and E. Kun 1972. The oxidation of L-lactate by liver mitochondria. *Arch. Biochem. Biophys.* 152: 92-104.

Stambaugh, R. and J. Buckley 1967. The enzymic and molecular nature of the lactic dehydrogenase subbands and X_4 isozyme. *J. Biol. Chem.* 242: 4053-4059.

Tabor, E. and K. W. Solley 1954. The effects of sex hormones on the development of the right gonad in female fowl. *Endocrinology* 54: 415-424.

Wilkinson, J. H. and W. A. Withycombe 1965. Organ specificity and lactate dehydrogenase activity. Some properties of human spermatozoal lactate dehydrogenase. *Biochem. J.* 97: 663-668.

Zinkham, W. H., A. Blanco, and L. Kupchyk 1963. Lactate dehydrogenase in testis: Dissociation and recombination of subunits. *Science* 142: 1303-1304.

Zinkham, W. H., A. Blanco, and J. Clowry 1964. An unusual isozyme of lactate dehydrogenase in mature testes: localization, ontogeny, and kinetic properties. *Ann. N. Y. Acad. Sci.* 121: 571-588.

Zinkham, W. H., A. Blanco, and L. Kupchyk 1964. Lactate dehydrogenase in pigeon testes: Genetic control by three loci. *Science* 144: 1353-1354.

Zinkham, W. H., L. Kupchyk, A. Blanco, and H. Isensee 1966. A variant of lactate dehydrogenase in somatic tissues of pigeons: Physicochemical properties and genetic control. *J. Exp. Zool.* 162: 45-56.

Zinkham, W. H., N. A. Holtzman, and H. Isensee 1968. The molecular size of lactate dehydrogenase isozymes in mature testes. *Biochim. Biophys. Acta* 160: 172-177.

Zinkham, W. H., H. Isensee, and J. H. Renwick 1969. Linkage

of lactate dehydrogenase B and C loci in pigeons. *Science* 164: 185-187.

Zinkham, W. H. and H. Isensee 1972. Genetic control of lactate dehydrogenase synthesis in the somatic and gametic tissues of pigeons. *The John Hopkins Med. J.* 130: 11-25.

Zinkham, W. H. and D. G. Walker 1974. Personal observations.

STUDIES ON LACTATE DEHYDROGENASE ISOZYMES
IN GAMETES AND EARLY DEVELOPMENT OF MICE

ROBERT P. ERICKSON[1], HORST SPIELMANN[2], FRANCO MANGIA[3],
DANIEL TENNENBAUM, and CHARLES J. EPSTEIN
Department of Pediatrics
University of California
San Francisco, California 94143

ABSTRACT. The developing mouse oocyte shows a rate of
increase in lactate dehydrogenase activity which is greater
than the growth in volume. This appears to be LDH-1 since
only LDH-1 is found electrophoretically in early mouse
embryos and antibody to LDH-1 shows a precipitin line with
embryo extracts. The high level of LDH activity (48 nmoles/
hr/embryo) remains constant in mature oocytes and preimplantation embryos until day 2, whereafter it rapidly declines
with a half-time of 19.2 hr. Immunoinactivation studies
with anti-LDH-1 showed a loss of immunoreactive protein in
parallel with the loss of enzymatic activity, suggesting
degradation of the enzyme. Histochemical staining for LDH
with hydroxyvaleric acid as substrate of zymograms of extracts of early embryos did not disclose any LDH-X, and
antibody to LDH-X had no embryotoxic effect on preimplantation embryos in vitro or in vivo.

It has long been known that lactate or pyruvate, but not
glucose, are adequate energy sources for the mouse embryo to
the 8-cell stage (Whitten, 1957; Brinster, 1965). It is not
altogether surprising, then, that lactate dehydrogenase (LDH)
represents nearly 5% of the protein of the two-cell mouse
embryos (Lowenstein and Cohen, 1964; Brinster, 1965). Such
high levels of LDH are not typical of mammalian embryos; the
rabbit (Brinster, 1967a) and primates (Brinster, 1967b) have
levels of activity nearly two orders of magnitude lower.
There are major differences in LDH isozyme patterns between
mouse and rabbit embryos as well. Pre-implantation mouse embryos contain LDH-1 (B4) (Auerbach and Brinster, 1967; Rapola
and Koskimies, 1967; Epstein et al, 1971; Brinster, 1971),
while LDH-5 predominates in this stage of the rabbit embryo
(Brinster, 1973). We have studied the developmental pattern
of LDH in mouse oocytes and early embryos as a model for biochemical regulation in early development. In addition to
LDH-1, we have also examined the role of LDH-X during the
preimplantation period. Although we find some evidence for
LDH-X on the surface of spermatozoa, we find no evidence for
LDH-X in the embryo. This observation does not explain the

embryotoxic effect of rabbit anti-mouse LDH-X reported by
Goldberg and Lerum (1972) after passive immunization of impregnated female mice.

Biochemical Development of LDH in Oocytes and Preimplantation Embryos

Oocytes were obtained from the ovaries of immature Swiss
female mice by digestion with hyaluronidase, collagenase, and
lysozyme and sorted by size with a dissecting microscope fitted
with an ocular micrometer (Mangia and Epstein, 1974). Embryos
were obtained from superovulated, mated, 16-18 g, Swiss-Webster
females. The females were sacrificed on the appropriate day
(day 0 is referred to as the day on which a copulation plug
is found) and embryos flushed from the fallopian tube or
uterus. Assays for LDH were as previously described (Epstein
et al, 1969). The LDH activity is 5 nmoles/hr/oocyte in the
earliest recoverable oocytes and increases ten-fold with maturation (Fig. 1). This increase in LDH activity is greater than
the increase in volume of the oocyte during maturation and,
assuming constant specific activity, the proportion of oocyte
protein (which increases in parallel with volume) represented
by LDH increases with maturation (Mangia and Epstein, 1974).
The quantity of LDH per ovum or embryo remains unchanged with
ovulation, fertilization, and during the first two and one-
half days of development. There is then a rapid loss of LDH
to less than 20% of the plateau level in late blastocysts, the
decrease occurring with a $t_{\frac{1}{2}}$ of 19.2 hr (Epstein et al, 1969).
Assay of day 1 and day 4 embryos separately and in combination
showed summation of the separate values (Epstein et al, 1969).
This indicated that soluble activators or inhibitors were unlikely to be the cause of the changes in LDH activity.

Immunological Studies on LDH in Oocytes and Preimplantation Embryos

Previous immunochemical studies have established the immunological distinctness of LDH-1 and LDH-5 and the poor immuno-
genicity of LDH-1 (Plagemann et al, 1960; Nisselbaum and
Bodansky, 1961; Markert and Appella, 1963; Rajewsky et al,
1964). Chemical modification of the LDH-1 (acetylation) or
the use of the greater antigenicity of the A chain in LDH-3
(A_2B_2) for a carrier effect are maneuvers which have been
successfully used to produce anti-LDH-1 antibodies (Rajewsky,
1966; Rajewsky and Rottlander, 1967). We have used affinity
chromatography of LDH (O'Carra and Barry, 1972; Spielmann et
al, 1973), subsequent separation of the LDH isozymes by elec-

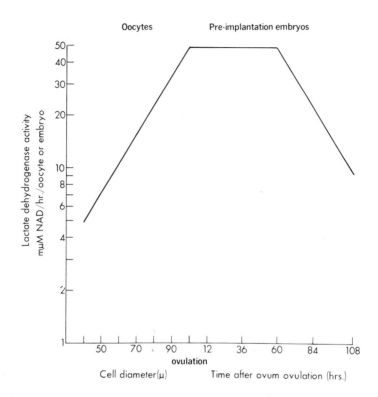

Fig. 1. Activity of LDH in developing oocytes and preimplantation embryos. Single mature oocytes and day 0-2 embryos were
assayed but immature oocytes and day 3 and 4 embryos were
pooled. (After Mangia and Epstein, 1974 and Epstein et al,
1969).

trophoresis on acrylamide, and immunization with the enzyme-
containing disc gel slice to provide a simple, rapid, and reproducible method for producing specific inactivating antibodies to LDH-1 with very small amounts of material (Spielmann
et al, 1974a).

When extracts of day 1 and day 4 embryos were titrated with
anti-LDH-1, inhibition curves identical to those obtained with
LDH-1 of heart were obtained (Fig. 2). These results indicate
that the ratio of enzymatic activities to immunoreactive protein for heart and the two embryonic stages are the same
(Spielmann et al, 1974b). Therefore, the decline in enzyme
activity between days 2 and 4 of embryonic development repre-

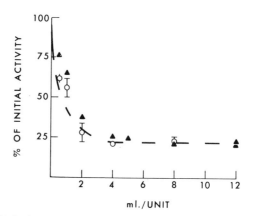

Fig. 2. Inhibition of LDH from day 1 and day 4 embryos compared to that of LDH from mouse heart. The 33% saturation $(NH_4)_2SO_4$ fraction (to remove serum LDH) of anti-LDH-1 was added in a series of increasing amounts to aliquots of extracts, incubated for 30 min at room temperature and the residual activity assayed. Curve: mouse heart extract: (O) day 1 embryos, average or average ± standard deviation; (▲) day 4 embryos. (After Spielmann et al, 1974b).

sents a concomitant loss of immunologically detectable protein, and embryonic LDH behaves immunologically as LDH-1. Double diffusion analyses also showed a precipitin reaction of anti-LDH-1 with day 2 embryos while a much higher titer anti-LDH-X (see below) showed no reaction (Fig. 3). Preliminary results of ^{3}H-tyrosine pulse-labeling and anti-LDH-1 immunoprecipitation studies suggest that the active LDH synthesis detectable by this method in the maturing oocyte ceases at the time of maturation. Hence, it is tentatively concluded that the plateau of enzyme activity between the mature oocyte and the 8-16 cell stage represents a period in which neither synthesis nor degradation is occurring while the decline from day 2 to day 4 must involve a semi-specific degradation process. We use the term semi-specific because glucose-6-phosphate dehydrogenase and guanidine deaminase activities also decrease in parallel while aldolase and malate dehydrogenase activities, for example, remain stable during this time period (Epstein et al, 1969).

Histochemical Staining of LDH from Embryos with Hydroxyvaleric Acid

Allen (1961) demonstrated the activity of LDH from testes

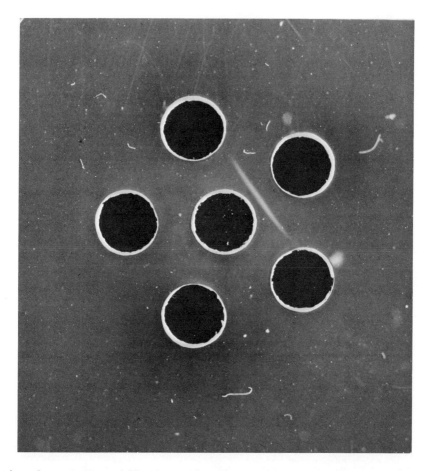

Fig. 3. Double diffusion analysis of 70 day 1 embryos (about 1.7 µg protein; lysed in 25 µl H_2O by freeze-thawing one time) against a) normal saline (2 o'clock), b) anti-LDH-1 (5 o'clock) and c) anti-LDH-X (7 o'clock).

on α-hydroxyvaleric acid and Goldberg (1965) demonstrated that this result was due to the LDH-X present in testes. We used this substrate to examine further the presence of LDH isozymes in preimplantation embryos. As is shown in Fig. 4, no LDH-X activity could be detected with even very large numbers of embryos when lysed and electrophoresed on acrylamide gels and stained histochemically with either substrate. This was confirmed by activity measurements. The activity measured using NAD and hydroxyvalerate (16.4 mM) was 0.09 nmoles/hr/embryo

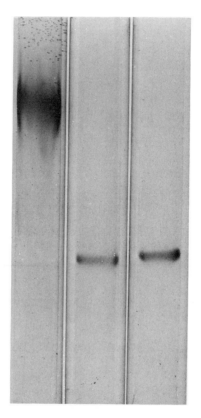

Fig. 4. Zymograms of LDH-X from testes (separated from the other LDH isozymes by affinity chromatography) compared to LDH of day 0 embryos. Left: LDH-X; center: 70, day 0 embryos; right: 12, day 0 embryos. Left and center: 0.05 M hydroxy-valerate as substrate; right 0.05 M lactate as substrate (acrylamide gel electrophoresis and staining conditions as in Spielmann et al, 1973). Top, cathode; bottom, anode.

compared to 21.2 nmoles/hr/embryo using 16.4 mM lactate.

Preparation of Rabbit Antisera to Mouse LDH-X

LDH was extracted from mouse testes and the 40-70% satura-tion $(NH_4)_2SO_4$ precipitate prepared according to Goldberg (1972). This precipitate was taken up in the buffer for affinity chromatography and LDH-X separated from LDH 1-5 by affinity chromatography (Spielmann et al, 1973). The excluded

fraction containing LDH-X was then heated at 55°C for 5 min.
Following concentration by vacuum dialysis, the material was
electrophoresed on 7.5% polyacrylamide gels and stained histochemically (Spielmann et al, 1973). The portion of the gel
with LDH-X activity was sliced out and frozen for use in immunization. Protein staining of such gels showed only one band
in the LDH-X activity position. The acrylamide slices containing an estimated 75 µg of LDH were pulverized in saline and
homogenized with equal volumes of complete Freund's adjuvant
for the first injection and with incomplete Freund's adjuvant
for subsequent injections. The animals were bled at weekly
intervals after the fourth weekly injection, and booster injections were given monthly. Fig. 5 compares the LDH inactivation titers of two anti-LDH-X sera with those of the anti-LDH-1
sera. The 50% inhibition for the anti-LDH-1 was 0.8 ml/unit
while for one of the anti-LDH-X sera it was 0.03 ml/unit and
for the other it was 0.012 ml/unit.

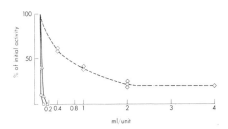

Fig. 5. Inhibition of LDH-X, separated from other testicular
isozymes by affinity chromatography (Spielmann et al, 1973),
by anti-LDH-X compared to inactivation of heart LDH by anti-
LDH-1:-O- anti-LDH-X, 473A and -△- anti-LDH-X, 273D, both with
testicular LDH-X,-◇- anti-LDH-1 with heart extracts. Assayed
as described in Fig. 2.

Effects of Anti-LDH-X on Embryos in vitro and in vivo

Embryos were obtained as described earlier (biochemical
development section) and incubated in Donahue's medium (Biggers
et al, 1967) for the 1-cell to 2-cell incubation or in
Whitten's medium (Whitten, 1971) with minor modifications
(Golbus et al, 1973) for the morula to blastocyst incubation.
The embryos were incubated with various additions of heat inactivated sera (Table I). There was no deleterious effect of
anti-LDH-X on either the 1-cell or morula embryos; if anything,
the anti-LDH-X was beneficial. The levels of anti-LDH-X titers
used in these in vitro experiments were very much greater than

319

TABLE I

Effects of antisera on development
of preimplantation embryos in vitro

	1-cell to 2-cell	morula to blastocyst		
	(# embryos) % 2-cell	(# embryos) % blasto-cysts		
% antiserum in medium:	~25% (2% by dry wt. of lyophilized serum)	0.5%	1%	5%
Control (no serum or 2% BSA)	(70) 71.5	------ (37) 54 ------		
Normal serum	-	(20) 70	(20) 85	(20) 0
Anti-LDH-1	-	(20) 60	(20) 45	(20) 0
Anti-LDH-X	(70) 88.5	(20) 90	(20) 65	(20) 20

would have been present in Goldberg and Lerum's (1972) in vivo experiments. Menge et al (1974) have recently shown embryopathic effects of uterine secretory IgA but not serum IgG. However, the passive immunization of Goldberg and Lerum (1972) would not affect secretory IgA but could result in the presence of immune IgG in the uterus.

Passive immunization of pregnant female mice was performed following Goldberg and Lerum (1972). Our anti-LDH-X had a lower inactivation titer than they reported (an average 100% inactivation titer of 0.15 ml/unit compared to 0.0059 ml/unit used by Goldberg, 1971). We injected 0.1 ml subcutaneously daily. Goldberg and Lerum's (1972) positive result was in terms of the number of treated mice which had litters. We would expect a direct action of the antiserum on embryos to affect the size of litters and not simply the presence or absence of litters. We have, therefore, plotted the average number of pups born per treated female in Fig. 6. We found no effect of anti-LDH-X on the presence or size of litters. Although differences in the specificity and titer of the antisera used may explain the difference between their and our results, our inability to detect LDH-X in embryos and to find an embryotoxic effect of anti-LDH-X in vitro are in accord with our failure to reduce litter size and number in vivo.

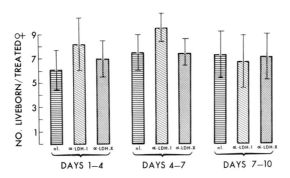

Fig. 6. Average number of pups born to impregnated females treated daily with 0.1 ml of the indicated sera injected subcutaneously on the indicated days.

One would not expect embryonic LDH to be on the surface of intact viable embryos and to be accessible to antibody. However, we have been able to detect LDH-X on the surface of maturing spermatozoa (Erickson et al, 1974) and this may provide a basis for the blocking of fertilization with anti-LDH-X (Goldberg, 1973). Since the plasma membrane of sperm fuses with the plasma membrane of the egg, some LDH-X might be brought to the egg membrane by the sperm. However, it is the one-cell embryo that was exposed to the highest concentrations of anti-LDH-X (Table I), and there still was no effect on development.

CONCLUSIONS

The increase of LDH-1 in the oocyte and subsequent decrease in the preimplantation embryo provide an excellent system for the study of gene control in development. There is no evidence for LDH-X in the embryo and we could find no embryotoxic effect with anti-LDH-X.

ACKNOWLEDGMENTS

These studies were supported by grants from the Population Council, New York, N. Y. and the National Institutes of Health, Bethesda, Md. (HD-03132).

FOOTNOTES

[1] Recipient of a Research Career Development Award from the National Institutes of Child Health and Human Development.

[2] Present address: Pharmakologisches Institut, Abt. Embryonal, Pharmakologie der Freien Universität, 1 Berlin 33, Thielallee, Berlin, West Germany.

[3] Present address: University di Roma, Instituto di Istologia ed Embriologia Generale, Via Alfonso Borelli, 50, 00161 Roma, Italy.

REFERENCES

Allen, J. M. 1961. Multiple forms of lactic dehydrogenase in tissues of the mouse: their specificity, cellular localization, and response to altered physiological conditions. *Ann. N. Y. Acad. Sci.* 94: 937-951.

Auerbach, S. and R. L. Brinster 1967. Lactate dehydrogenase isozymes in the early mouse embryo. *Exp. Cell Res.* 46: 89-92.

Biggers, J. D., D. G. Whittingham, and R. P. Donahue 1967. The pattern of energy metabolism in the mouse oocyte and zygote. *Proc. Natl. Acad. Sci.* 58: 560-567.

Brinster, R. L. 1965. Studies on the development of mouse embryos in vitro II. The effect of energy source. *J. Exp. Zool.* 158: 59-68.

Brinster, R. L. 1965. Lactic dehydrogenase activity in the pre-implanted mouse embryo. *Biochim. Biophys. Acta* 110: 439-441.

Brinster, R. L. 1967a. Lactate dehydrogenase activity in the preimplantation rabbit embryo. *Biochim. Biophys. Acta* 148: 298-300.

Brinster, R. L. 1967b. Lactate dehydrogenase activity in human, squirrel, monkey, and rhesus monkey oocytes. *Exp. Cell Res.* 48: 643-646.

Brinster, R. L. 1971. The lactate dehydrogenase in the pre-implantation embryos of Quackenbush and Swiss mice. *FEBS Letters* 17: 41-44.

Brinster, R. L. 1973. Lactate dehydrogenase isozymes in the preimplantation rabbit embryo. *Biochem. Genet.* 9: 229-234.

Epstein, C. J., E. A. Wegienka, and C. W. Smith 1969. Biochemical development of preimplantation mouse embryos: in vivo activities of fructose 1,6-diphosphate aldolase, glucose 6-phosphate dehydrogenase, malate dehydrogenase, and lactate dehydrogenase. *Biochem. Genet.* 3: 271-281.

Epstein, C. J., L. Kwok, and S. Smith 1971. The source of lactate dehydrogenase in preimplantation mouse embryos.
FEBS Letters 13: 45-48.

Erickson, R. P., D. Friend, and D. Tennenbaum 1974. in preparation.

Golbus, M. S., P. G. Calarco, and C. J. Epstein 1973. The
effects of inhibitors of RNA synthesis (α-amanitin and
actinomycin D) on preimplantation mouse embryos. *J. Exp.
Zool.* 186: 207-216.

Goldberg, E. 1965. Lactate dehydrogenase in spermatozoa: Subunit interactions in vitro. *Arch. Biochem. Biophys.* 109:
134-141.

Goldberg, E. 1971. Immunochemical specificity of lactate dehydrogenase-X. *Proc. Natl. Acad. Sci.* 68: 349-352.

Goldberg, E. 1972. Amino acid composition and properties of
crystalline lactate dehydrogenase X from mouse testes.
J. Biol. Chem. 247: 2044-2048.

Goldberg, E., and J. Lerum 1972. Pregnancy suppression by an
antiserum to the sperm specific lactate dehydrogenase.
Science 176: 686-687.

Goldberg, E. 1973. Infertility in female rabbits immunized with
lactate dehydrogenase X. *Science* 181: 458-459.

Lowenstein, J. E., and A. I. Cohen 1964. Dry mass, lipid content and protein content of the intact and zona-free mouse
ovum. *J. Embryo. Exp. Morph.* 12: 113-121.

Mangia, F. and C. J. Epstein 1974. Glucose-6-phosphate dehydrogenase and lactate dehydrogenase activities in growing
mouse oocytes. in preparation.

Menge, A. C., A. Rosenberg, and D. M. Burkons 1974. Effects of
uterine fluids and immunoglobulins from semen-immunized
rabbits on rabbit embryos cultured in vitro. *Proc. Soc.
Exp. Biol. Med.* 145: 371-378.

Markert, C. L., and E. Appella 1963. Immunochemical properties
of lactate dehydrogenase isozymes. *Ann. N. Y. Acad. Sci.*
103: 915-929.

Nisselbaum, J. S. and O. Bodansky 1961. Reactions of human
tissue lactic dehydrogenases with antisera to human heart
and liver lactic dehydrogenases. *J. Biol. Chem.* 236: 401-
404.

O'Carra, P., and S. Barry 1972. Affinity chromatography of
lactate dehydrogenase. Model studies demonstrating the
potential of the technique in the mechanistic investigation as well as in the purification of multi-substrate
enzymes. *FEBS Letters* 21: 281-285.

Plagemann, P. G. W., K. F. Gregory, and F. Wroblewski 1960.
The electrophoretically distinct forms of mammalian lactic
dehydrogenase. II. Properties and interrelationships of
rabbits and human lactic dehydrogenase isozymes. *J. Biol.
Chem.* 235: 2288-2293.

Rajewsky, K., S. Avrameas, S. P. Grabar, G. Pfleiderer and
E. D. Wachsmuth 1964. Immunologische Spezifität Von
lactatdehydrogenase isozymen Dreier Säugetier-Organismen.
Biochim. Biophys. Acta 92: 248-259.

Rajewsky, K. 1966. Kreuzreagierende antigene Determinanten
auf lactatdehydrogenasen I und V Durch acetylierung.
Biochim. Biophys. Acta 121: 51-68.

Rajewsky, D., and E. Rottländer 1967. Tolerance specificity
and the immune response to lactic dehydrogenase isoenzymes. *Cold Spring Harbor Symp. Quant. Biol.* 32: 547-554.

Rapola, J., and O. Koskimies 1967. Embryonic enzyme patterns:
characterization of the single lactate dehydrogenase isozyme in preimplanted mouse ova. *Science* 157: 1311-1312.

Spielmann, H., R. P. Erickson, and C. J. Epstein 1973. The
separation of lactate dehydrogenase X from other lactate
dehydrogenase isozymes of mouse testes by affinity chromatography. *FEBS Letters* 35: 19-23.

Spielmann, H., R. P. Erickson, and C. J. Epstein 1974a. The
production of antibodies against mammalian LDH-1. *Anal.
Biochem.* in press.

Spielmann, H., R. P. Erickson, and C. J. Epstein 1974b. Immunochemical studies of lactate dehydrogenase and glucose-6-
phosphate dehydrogenase in preimplantation mouse embryos.
J. Reprod. Fert. in press.

Whitten, W. K. 1957. Culture of tubal ova. *Nature* 179: 1081-
1082.

Whitten, W. K. 1971. Nutrient requirements for the culture of
preimplantation embryos. *Adv. Biosciences* 6: 129-141.

LDH-X: THE SPERM-SPECIFIC C_4 ISOZYME OF LACTATE DEHYDROGENASE

THOMAS E. WHEAT and ERWIN GOLDBERG
Department of Biological Sciences
Northwestern University
Evanston, Illinois 60201

ABSTRACT. Mature testes and spermatozoa of most mammalian and at least one avian species contain the C_4 isozyme of lactate dehydrogenase (LDH-X). This isozyme is unique in terms of the remarkable temporal and cellular specificity of its synthesis. It appears first in the primary spermatocyte, is found in no other tissue, and is completely absent from the female. Certain biochemical and enzymological properties of LDH-X correlate well with the requirements of sperm metabolism but the physiological rationale for a third subunit type remains unclear.

Procedures are now available for the isolation of large amounts of unequivocally pure, crystalline mouse LDH-X. This facilitates X-ray crystallographic studies and sequence analysis now in progress. Preliminary data suggest many structural similarities between LDH-X, LDH-1, and LDH-5.

The strict specificity of LDH-X to sperm provides a unique opportunity to study the immunosuppression of fertility. Both active and passive immunization of females reduces fertility. Immunochemical techniques clearly demonstrate LDH-X within the zona pellucida of mouse blastocysts. It is probably derived from supernumerary sperm and provides the antibody combining sites necessary for post-fertilization suppression of pregnancy.

It is now well-known that the molecular heterogeneity of vertebrate lactate dehydrogenase is based on a pattern of five isozymes resulting from the combination of A and B subunits into all possible tetramers. The number of molecular forms actually observed in a particular organ or individual may be decreased by restrictions on subunit assembly or increased by such mechanisms as conformational alternatives, multiple alleles, or additional gene loci (Markert and Whitt, 1968). One such case is the lactate dehydrogenase X (LDH-X) isozyme detected in the testes and sperm of birds and mammals.

LDH-X was first observed as a sixth electrophoretically resolvable form of LDH in human spermatozoa (Goldberg, 1963) and testes (Blanco and Zinkham, 1963). One or more additional LDH-X isozymes have been reported associated with the male reproductive tract of many other animals as shown in Table 1.

TABLE 1
Animals with LDH-X

Animal	Reference
Human	Goldberg, 1963 Blanco and Zinkham, 1963
Mouse	Goldberg, 1965
Rabbit	Battellino et al, 1968
Rat*	Blackshaw and Elkington, 1970a Goldberg, 1973a
Bull	Kolb et al, 1970
Guinea Pig*	Blackshaw and Samisoni, 1967 Battellino and Blanco, 1970 Goldberg, 1973a
Bat*	Valdivieso et al, 1968 Gutierrez et al, 1972
Ram	Blackshaw and Samisoni, 1967
Various Marsupials	Holmes, 1972 Holmes et al, 1973
Mongolian Gerbil*	Lindahl and Mayeda, 1972, 1973
Pig*	Valenta et al, 1967
Hamster	Goldberg, 1971
Dog	Zinkham et al, 1964b
Pigeon*	Blanco et al, 1964 Zinkham, 1968

* These animals contain more than one form of LDH-X. Except for the allelic isozymes of the pigeon, these additional forms are usually heterotetramers of C subunits with A or B subunits.

In every case, this unique isozyme activity was restricted to the sperm and testis, occurring in no other male tissue and in no female organ. It has now been demonstrated that multiple X-bands are usually heterotetramers of C subunits with A or B subunits (Goldberg, 1973a) except in the case of the pigeon allelic isozymes (Blanco et al, 1964).

The close association of LDH-X with the sperm has also been shown by a variety of developmental studies. It is known that LDH-X is absent in the prepubertal male human (Blanco and Zinkham, 1963; Goldberg, 1963), bat (Valdivieso et al, 1968), pigeon (Blanco et al, 1964), gerbil (Lindahl and Mayeda, 1972), and rat (Blackshaw and Elkington, 1970a). Detailed analysis of testicular maturation in the mouse reveals that the first appearance of LDH-X closely correlates with the onset of active spermatogenesis (Goldberg and Hawtrey, 1967; Hawtrey and Goldberg, 1968; Blackshaw and Elkington, 1970b).

LDH-X synthesis does not occur in aspermatogenic individuals. For example, in pathologic male infertility linked to azoospermia, LDH-X is absent (Szeinberg et al, 1966; Ishibe et al, 1971). In addition, the lack of LDH-X synthesis correlates to the absence of spermatogenesis in the cryptorchid guinea pig (Blackshaw and Samisoni, 1966), ram (Blackshaw and Samisoni, 1967), and mouse (Goldberg and Hawtrey, 1968). Finally when aspermatogenesis is induced by hypophysectomy, LDH-X synthesis declines with the reduction in sperm production (Blackshaw and Elkington, 1970b; Elkington et al, 1973). Both spermatogenesis and LDH-X synthesis can be maintained in hypophysectomized rats by hormonal injections (Winer and Nikitovich-Winer, 1971).

The close association between LDH-X synthesis and spermatogenesis is further exemplified by animals showing seasonal reproduction. In South American bats, the cyclic activation of LDH-X and the appearance of sperm correlate exactly (Gutierrez et al, 1972).

Most recently, immunofluorescence experiments have shown that LDH-X in the mouse testis is localized in the primary and secondary spermatocytes as well as in the spermatids around the lumen of the seminiferous tubules. This isozyme is absent from spermatogonia and non-spermatogenic elements. Furthermore, LDH-X is first detected during prophase of the first meiotic division (Hintz and Goldberg, in preparation).

Recent studies have clearly established that LDH-X is composed of a third, distinct subunit type, designated C, and that LDH-X is a true lactate dehydrogenase, rather than a non-specific α-hydroxy acid dehydrogenase. The distinctness of LDH-X has been demonstrated by genetic analyses and by immunochemical criteria. Independent electrophoretic variants of B and C subunits were discovered in the pigeon (Blanco et al, 1964;

Zinkham et al, 1964, 1966). Genetic analyses showed that these variants are inherited as codominant alleles at autosomal, nuclear loci and that the B and C loci are very closely linked (Zinkham et al, 1969; Zinkham and Isensee, 1972). These independent allelic variants conclusively demonstrate that LDH-X is encoded by a third gene locus distinct from the A and B loci.

In addition to genetic evidence of uniqueness, the structure of the three subunit types may be studied directly by immunochemical methods. Antisera to crude preparations of human A_4, B_4, or C_4 isozymes do not inhibit the activity of the heterologous isozymes while strongly inhibiting the homologous isozyme (Ressler et al, 1967). More recently, antibodies to highly purified, crystalline mouse LDH-X have been elicited in rabbits. This serum does not cross-react with A or B subunits as judged by inhibition of enzyme activity, precipitin reaction, Ouchterlony double diffusion method, immunoelectrophoresis, and microcomplement fixation. However, the antiserum to mouse C_4 cross-reacts with LDH-X from various mammalian sources and also from pigeon (Goldberg, 1971). This high degree of immunochemical specificity directly confirms the structural uniqueness of the C subunit which must, therefore, be encoded in a third locus.

Although there can be no doubt that LDH-X is distinct from the five isozymes of somatic tissues, it remains to be demonstrated that the C subunit is homologous to the A and B subunits. A great deal of direct and indirect biochemical evidence discussed below supports the conclusion, but one series of observations is particularly striking. Enzymically active hybrid tetramers combining C subunits with A and B subunits have been observed in vivo in a number of organisms as noted in Table 1, and can easily be formed by in vitro molecular hybridization (Zinkham et al, 1963; Goldberg, 1965; Goldberg, 1973). The formation of enzymically active heterotetramers indicates substantial structural similarity among the intersubunit binding sites and confirms the homology of A, B, and C subunits.

The existence of a third LDH subunit type showing such extraordinary tissue specificity has prompted much speculation on the physiological rationale for an additional isozyme. A large amount of enzymological data has been accumulated toward an understanding of the role of LDH-X, and it would be useful to briefly review several of these hypotheses.

Perhaps the most striking enzymic property of LDH-X is its ability to utilize a wide range of substrate and coenzyme analogues. For example, human LDH-X utilizes acetyl-pyridine adenine dinucleotide to a much greater extent than do the other

LDH isozymes (Goldberg, 1964). In addition, LDH-X from various species catalyzes the oxidation of a wide range of α-hydroxy acids. However, this property is somewhat variable from species to species. For example, mouse LDH-X catalyzes the reduction of α-ketovalerate and α-ketobutyrate much more efficiently than α-ketoglutarate (Wong et al, 1971) while the reverse is true for rat LDH-X (Schatz and Segal, 1969). However, these broad specificities are probably not physiologically relevant since the metabolic role of LDH-X in the sperm is the interconversion of lactate and pyruvate. The coenzyme analogues do not occur in vivo and the various substrate analogues are not important in sperm metabolism.

The response of LDH-X to varying concentrations of both pyruvate and lactate has been measured. These data show that LDH-X has a high affinity for pyruvate and is markedly inhibited by the substrate. The enzyme has a lower affinity for lactate and is not inhibited by excess lactate as substrate (Battelino et al, 1968; Battelino and Blanco, 1970; Hawtrey and Goldberg, 1970; Wong et al, 1971). Perhaps those parameters are suggestive of the suitability of LDH-X to function in an aerobic environment such as the female reproductive tract. However, it is most difficult to correlate in vitro substrate inhibition data with physiologically relevant parameters.

Another striking kinetic property of LDH-X is the low turnover number of 2,980 moles NADH/mole LDH-X/min (Goldberg, 1972). This contrasts with values from 41,500 to 160,000 for several other LDH isozymes (Pesce et al, 1967). The low rate of enzyme activity may prevent the accumulation of large quantities of lactate which would inhibit sperm motility.

LDH-X from various sources also shows a remarkable resistance to heat denaturation (Hawtrey and Goldberg, 1970; Goldberg, 1972; Kolb et al, 1970; Stambaugh and Buckley, 1967; Battelino et al, 1968; McKee et al, 1972). While this stability extends well above physiological temperatures, the resistance to heat denaturation probably reflects the overall stability of this isozyme. It would seem advantageous for a cell to have maximally stable components when it must function for a prolonged period without additional synthesis.

A most interesting difference between LDH-X and the other LDH isozymes is the relation of K_m to temperature shown in Fig. 1 (Goldberg, 1972). This striking change in substrate affinity between 32° and 37° may be very important to sperm metabolism since sperm pass phases of their functional life at different temperatures in the scrotum (32°) and in the female tract (37°). LDH-X could serve to differentially regulate pyruvate levels or the ratio of NAD to NADH.

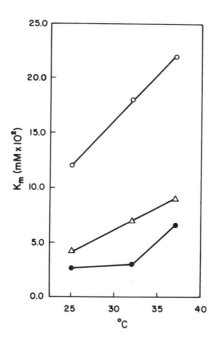

Fig. 1. K_m for pyruvate as a function of temperature. Note the sharp discontinuity for LDH-X (●—●) as compared to LDH-1 (△—△) and LDH-5 (O—O).

Further progress in understanding the structural and functional role of LDH-X will require the availability of protocols for reproducibly purifying large quantities of the enzyme. LDH-X has been purified to some extent from the testes or sperm of a wide variety of animals including mouse (Hawtrey and Goldberg, 1970; Wong et al, 1971; Goldberg, 1972, 1974), rat (Schatz and Segal, 1969), bull (Kolb et al, 1970), rabbit (Stambaugh and Buckley, 1967; Battellino et al, 1968), and human (Clausen and Øvlisen, 1965; Svasti, 1974). This discussion will focus on our preparation of mouse LDH-X. These published procedures (Goldberg, 1972; Goldberg, 1974) routinely give a high yield of crystalline enzyme whose purity has been conclusively demonstrated by all available enzymological, physical, and immunochemical criteria (Goldberg 1971, 1972). The starting material is testes from Swiss-Webster mice. After homogenization and centrifugation, the extract is heated to 55° and maintained at this temperature for 15 min. The heat treatment destroys none of the highly stable LDH-X while completely precipitating all of the other LDH isozymes as well as

many other proteins. After a series of ammonium sulfate frac-
tionations, the concentrated extract is chromatographed on
DEAE-Sephadex under such conditions that LDH-X elutes shortly
after the void volume. The pooled pure fractions are crystal-
lized in the presence of ammonium sulfate. A shower of small
crystals routinely forms within 24 hours when the preparation
is seeded with previously grown crystals. The two salient features of this scheme are the heat treatment, which greatly
simplifies the purification, and the ease of crystallization,
ensuring reproducible homogeneity. In addition, large crystals, suitable for X-ray diffraction studies, often form or
can be easily generated by recrystallization (Adams et al,
1973).

LDH-X may also be purified by affinity chromatography
(Goldberg and Whitmore, 1974; Goldberg, 1974). This has proven
particularly useful in the isolation of small quantities of
rabbit LDH-X from semen. Like the other LDH isozymes (O'Carra
and Barry, 1972) LDH-X in the presence of NADH binds to oxamate
immobilized on agarose gel. When NADH is omitted from the
elution buffer, LDH is released immediately and quantitatively.
However, in contrast to the other isozymes the binding of LDH-X
is exquisitely sensitive to the concentration of salt and of
NADH. It is bound only over a very narrow range of NADH concentration (Goldberg and Whitmore, 1974; Goldberg, 1974) which
is much lower than that commonly used for the other LDH isozymes (O'Carra and Barry, 1972). This sensitivity explains
the failure of other workers to achieve binding (Spielman et
al, 1973).

The role of coenzyme concentration in the affinity chromatography is shown in Fig. 2 and 3. Fig. 2 shows a crude
homogenate of mouse testis chromatographed in the presence of
200 μM NADH. The first peak, which is not retarded, contains
all of the LDH-X, and the second peak, which is eluted when
NADH is omitted from the column irrigants contains all of the
other isozymes. Fig. 3 shows a sample containing LDH-X free
of the other isozymes and chromatographed in 40 μM NADH. All
of the LDH-X is bound to the column and is quantitatively
released in the absence of NADH.

The availability of reasonably large amounts of absolutely
pure mouse LDH-X has enabled us to undertake a variety of
structural studies. The molecular size of LDH-X has been
determined in various ways. Crystalline mouse material migrates as a single symmetrical peak, 7.0 S, in sedimentation
velocity experiments, with no evidence of aggregation or dissociation (Goldberg, 1972). Zone velocity centrifugation in a
sucrose gradient yields a molecular weight of 140,000 and shows
the LDH-X peak superimposed on beef heart LDH-1 standard

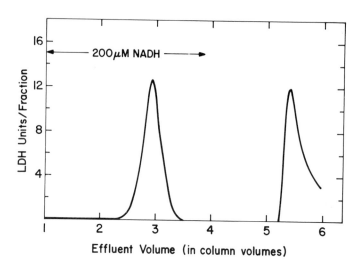

Fig. 2. Affinity chromatography of mouse LDH-X in the presence of 200 µM NADH. Note that the LDH-X is not retarded, eluting in the first peak. All the other isozymes are completely retained, eluting in the second peak when NADH is omitted from the column irrigant.

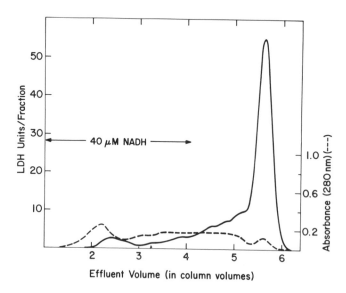

Fig. 3. Affinity chromatography of mouse LDH-X in the presence of 40 µM NADH. Note that the LDH-X binds to the oxamate and is released when NADH is omitted from the elution buffer.

(Goldberg, 1972). In addition, gel filtration analyses of LDH-X from pigeon, mouse, rabbit, and human reveals this isozyme to be identical in Stokes radii and molecular size to the isozymes of somatic tissues and to crystalline rabbit muscle LDH (Zinkham et al, 1968). Rat LDH-X is estimated to have a molecular weight of 125,000 by sedimentation equilibrium (Schatz and Segal, 1969). These values are in good agreement with the molecular size of other LDH isozymes.

Although the enzymically active form of LDH-X has a molecular weight of about 140,000, substantial evidence shows that it is composed of 4 subunits of 35,000 molecular weight. In vitro recombination experiments involving C subunits with A and B subunits yield the number of recombinants and binomial distribution of activity expected for a system of tetrameric isozymes (Goldberg, 1965; Zinkham et al, 1963; Goldberg, 1973a). In addition, various enzymic digests yield very nearly one-fourth the number of peptides expected from the amino acid composition calculated for a molecular weight of 140,000 (Goldberg, 1972). For example, tryptic digestion yields 35-40 peptides (104 lysines and 48 arginines) while restricting tryptic cleavage to arginine generates 11-12 fragments. Limited chymotryptic cleavage yields about 20 peptides which corresponds fairly well to the number of aromatic residues. These data are consistent only with a composition of four identical subunits for LDH-X.

Most recently, a preliminary X-ray diffraction examination of crystalline mouse LDH-X has yielded precise dimensions for the tetrameric enzyme. The C_4 homotetramer is P1 triclinic cell with dimensions a = 84.8 Å, b = 76.6 Å, c = 63.9 Å, α = 109.7°, β = 89.5°, γ = 96.5°. The triclinic cell therefore contains one tetramer (Adams et al, 1973). Similar results have been obtained for all other known crystalline forms of lactate dehydrogenase (Adams et al, 1972). Therefore, the enzymically active form of LDH-X is the same size and shape as other LDH isozymes.

Additional chemical studies of mouse LDH-X are in progress with a goal of determining the complete amino acid sequence. The published amino acid composition (Goldberg, 1972) has been refined as shown in Table 2 for a subunit molecular weight of 35,000. Cysteine was determined as the S-carboxymethyl derivative. The values are consistent with published data (Goldberg, 1972) with addition of cysteine and a higher, more accurate value for methionine. As noted previously, mouse LDH-X is quite similar to other mammalian LDH-1 and LDH-5 isozymes in amino acid composition, other than a reduced proportion of isoleucine and increased leucine. Glycine and threonine also appear to be more abundant in the C subunit.

TABLE 2

Amino acid composition of crystalline mouse LDH-X

CYS	4.23	ILE	21.4
ASP	35.8	LEU	38.9
THR	18.0	TYR	5.85
SER	23.0	PHE	7.75
GLU	25.4	HIS	6.81
GLY	33.7	LYS	24.7
ALA	20.1	ARG	10.5
VAL	36.2	PRO	11.5
MET	6.24		

Both dansylation and a single Edman cycle reveal no free amino terminal. This indicates that the N-terminal of LDH-X is blocked, probably by N-acetylation, as are other LDH isozymes (Brummel and Stegink, 1972). Sequence studies at this time are focused on the 11-12 peptides of crystalline mouse LDH-X generated by tryptic digestion restricted to arginine residues by reversible blocking of lysine with citraconic anhydride (Dixon and Perham, 1968). Fractionation of these peptides by Sephadex gel filtration shows that the number of fragments and their size distribution is remarkably similar to that derived from Dogfish LDH-A (Taylor et al, 1973).

With a detailed knowledge of the amino acid sequence and three-dimensional structure of LDH-X, it will be possible to undertake an analysis of the evolutionary relationship of the LDH-C subunit to the homologous A and B subunits. In addition, such complete characterization will make LDH-X a most useful model system for analyzing the immunosuppression of fertility.

There is a vast literature on the immunogenic impairment of fertility (reviewed by Edwards, 1970; Menge, 1970; Metz, 1972). Most of the immunological work has involved the antigenicity of spermatozoa and testes in homologous and heterologous species, and it has been observed that immunization of males with testicular antigen, usually in conjunction with some adjuvant, can result in testicular lesions, generally of the seminiferous epithelium. Similarly, there is evidence that fertility reduction occurs in females in response to testicular and sperm extracts, or semen as immunogens. However, the effect is never total and the mechanisms of action are unclear, undoubtedly because of the antigenic complexity of the preparations.

The major problem that has plagued this area of research has been the lack of antigenic specificity. Progress is unlikely until at least one antigen causing infertility has been identified chemically and immunologically. It appears that this major obstacle has been overcome by the purification and chemical characterization of LDH-X (Goldberg, 1972). Because of its localization, LDH-X must be sequestered from the organism's immune system. In the male, this is accomplished by the so-called blood-testis barrier (Dym and Fawcett, 1970; Vitale et al, 1973). In the female LDH-X is never synthesized. Thus it is reasonable to predict that this isozyme would be antigenic in both sexes. That LDH-X may function as an immunogen to impair reproductive processes in females has been demonstrated by both passive and active immunization studies (Goldberg and Lerum, 1972; Goldberg, 1973b).

When rabbits of either sex are injected with as little as 5 μg purified LDH-X emulsified in Freund's complete adjuvant, antibodies specific to the antigen are produced. Similarly, mouse LDH-X can provoke antibody formation in mice (Lerum and Goldberg, 1974). Immunization of rabbits with mouse testes extract (MTE) yields antiserum giving multiple precipitin lines on Ouchterlony plates against MTE, but only a single line against our purified LDH-X.

Immunization of female rabbits with mouse LDH-X caused a highly significant reduction of pregnancies. In order to determine whether the immunological lesion involved fertilization, embryonic development, or both, animals were sacrificed on day six post-coitus (p.c.) for blastocyst recovery, or laparotomized on day 10 p.c. for implant examination. The results presented in Table 3 clearly demonstrate that there is a marked reduction in blastocyst formation compared to control animals; however, all of the blastocysts which do develop, become implanted. Furthermore, the ratio of corpora lutea represented by atretic, unfertilized ova, on day 6 p.c., is 44.3 in the immune animals and only 14.5 in the controls. A somewhat surprising result, in view of the specificity of LDH-X to sperm, was obtained when development was allowed to proceed to term in those immunized females which did have implants. As shown in Table 4, embryo mortality was exceedingly high. While a post-fertilization effect of immunization was not predicted, this observation, nevertheless was consistent with previous findings (Goldberg and Lerum, 1972; Lerum and Goldberg, 1974) that injections of rabbit antisera to LDH-X into pregnant mice terminated pregnancies.

We reported initially (Goldberg and Lerum, 1972) that pregnancy suppression in mice occurs most frequently when antiserum is injected during the pre-implantational stages of embryonic development. An extensive analysis of this phenomenon on days 1-4 after mating is presented in Table 5. Apparently the effect of antiserum to LDH-X in the female is "all or none".

TABLE 3
Fertility of female rabbits immunized with mouse LDH-X

Treatment Group	Day 6 p.c. Blastocysts/C.L. x 100	Day 10 p.c. Implants/C.L. x 100
Control	74.0	75.4
Immunized	40.7	43.9

For details of methodology see Goldberg (1973b).

C.L. = corpora lutea

TABLE 4
Embryo survival in control and LDH-X immune rabbits

Embryos/Rabbit	Treatment Group[1]	
	Control	Immunized
day		
10	7.3 (13)	5.2 (23)
28	6.6 (9)	2.4 (14)

New Zealand White rabbits were laparotomized on day 10 p.c., and viable implants counted. They were then sacrificed on day 28. The same males were used for matings with immunized and control females.

1. No. animals in ( ).

A comparison of embryos per pregnancy reveals no significant difference between control and experimental groups (Table 5).
    As noted above, it is difficult to reconcile the findings on embryo mortality with the known specificity of the antibody to LDH-X. We have been unable to demonstrate LDH-X in whole embryos or in any embryonic tissues by the most sensitive methods available, and in our view its restriction to mature spermatogenic elements of the testes has been well-established. Thus, antibody-antigen interaction must occur either before

TABLE 5

Comparison (T test) of percentage of pregnancies and mean number of fetuses per pregnancy in antiserum and control serum. Injected female mice on days 1 - 4 after mating.

Treatment	Number of Matings*	% Pregnancies	Mean Number of Fetuses ± S.E.
Control Serum	401	71.0	11.0 ± 0.9
Antiserum	548	52.9	10.6 ± 0.7
P value	-	< 0.0001	> 0.2

* The mating was included only if corpora lutea could be observed. (Data from Lerum and Goldberg, 1974)

fertilization, or else prior to implantation when LDH-X combining sites would be accessible on the sperm within the female reproductive tract or associated with the ovum. Inhibition of fertilization, suggested by the data in Table 3, can occur by combination of sperm with antibodies released locally within the female reproductive tract so that spermatozoa are actually prevented from reaching the site of fertilization. Alternatively, interaction between antibody and the LDH-X of the sperm may block ovum penetration.

Post-fertilization effects of antiserum may be explained by a model in which antigenic determinants (i.e. LDH-X) are introduced into the ovum by the fertilizing sperm or more likely, by supernumerary sperm which accumulate within the perivitelline space and are phagocytized by the blastomeres (Thompson and Zamboni, 1974). These determinants would represent antibody receptors on or within the embryo. Thus, the degeneration of supernumerary sperm could lead to release and diffusion of LDH-X throughout the perivitelline space and into the zona pellucida where interaction with antibody would be possible.

Direct evidence has been obtained that specific antibody combining sites are contained within the mouse preimplantation embryo. Mouse blastocysts were flushed from uteri at 80 hr p.c. and incubated first with rabbit anti-LDH-X serum, and then after thorough washing, with fluorescein-conjugated goat anti-rabbit IgG. The intense fluorescence of the zona pellucida but not the embryo, shown in Fig. 4, confirms the contention that blastocysts can bind antibody to LDH-X under these conditions. Specific antibody localization within the zona

Fig. 4. Immunofluorescence of 80 hr mouse embryo. Blastocysts were incubated with the IgG fraction of antiserum to
mouse LDH-X, washed 3X with Brinster's medium for ovum culture
(BMOC), incubated with fluorescein-conjugated goat anti-rabbit
IgG, washed 3X with BMOC, placed under a coverslip and viewed

(*Fig. 4 legend continued*)   by fluorescence microscopy. The
blastocysts were ruptured to show the intense fluorescence of
the zona pellucida and fluorescence on the embryo proper.
There was no fluorescence of appropriately prepared controls.

---

pellucida and on the surface of the embryo proper was demonstrated with peroxidase-conjugated goat anti-rabbit IgG (Bené
and Goldberg, 1974) while the finding that no antibody binding
occurred with unfertilized ova provided persuasive evidence
that antigenic determinants were conferred by spermatozoa.

Embryo mortality in LDH-X immunized rabbits was investigated
further by transferring blastocysts from non-immune donors into
the uterine horns of control and LDH-X immune females. From
the data presented in Table 6 it can be seen that both implan-

TABLE 6
Survival rates of blastocysts transferred into the
uterine horns of control and LDH-X-immune rabbits

Treatment Group[1]	No. Embryos Transferred	Embryo Survival %	
		Day 10[2]	Day 28[3]
Control (11)	54	35.2	42.9
Immune (15)	95	30.5	41.7

Hank's balanced salt solution, pH 7.7, containing 200 IU/ml
pencillin and 10% normal rabbit serum, was used to flush
blastocysts from donors at 138 hr p.c. Recipients received
150 IU HCG, i.v., at the same time the donors were mated.

1. No. of recipients in ( ).
2. 10 day survival based on number of blastocysts transferred.
3. 28 day survival based on number of day 10 implants.

---

tation rate and embryo survival are similar in all recipients.
Conversely, pre-implantation development is markedly decreased
when embryos are transferred from the oviduct of immune donors
to non-immune recipients (Table 7). Therefore, when fertilization takes place in an immune environment, there is a high rate
of embryo mortality. The nature of the embryotoxic effect is
certainly of considerable interest. Perhaps, the interaction
between antibody and sperm LDH-X, while permitting ovum pene-

TABLE 7

Survival and development of 19 hr rabbit embryos
transferred from LDH-X immune and control recipients

Donors	Recipients	No. Embryos Transferred	% Blastocysts Recovered
Control	Control (2)	20	30
Control	Immune (2)	31	33
Immune	Control (7)	59	1.7

See Table 6 for description of experimental protocol.

tration, nevertheless causes a defect in the spermatozoan such that expression of the paternal genome in later embryonic development is adversely affected. This hypothesis is amenable to test with appropriate genetic markers as well as by karyo-type analysis.

From the results of these experiments on active immunization of female rabbits and passive immunization of pregnant mice, it is clear that LDH-X possesses antigenic properties which lead to an immunogenic impairment of fertility in females. This is the only work so far reported in which a monospecific immune system has been used to impair fertility. Completion of biochemical studies on the amino acid sequence of LDH-X will allow mapping of the antigenic determinants responsible for inducing fertility. It will then be possible to construct synthetic peptides of the appropriate sequence which may have similar biological properties.

ACKNOWLEDGMENTS

We wish to acknowledge the support of the National Insti-tutes of Health (Contract No. NIH 69-2162; Grant No. R01 HD05863; Biomedical Sciences Support Grant FR 7-28-05; Postdoctoral Fellowship 5 F02 GM 55188) and the Population Council (Grant No. M 71.0137C).

LITERATURE CITED

Adams, A. D., M. J. Adams, M. G. Rossmann, and E. Goldberg 1973. A crystalline form of testes-specific lactate dehydrogen-ase. *J. Mol. Biol.* 78: 721-722.

Adams, M. J., M. Buehner, K. Chandrasekhar, G. C. Ford, M. L. Hackert, A. Liljas, P. J. Lentz, S. T. Rao, M. G. Rossmann, I. E. Smiley, and J. L White 1972. Subunit interactions in lactate dehydrogenase. in *Protein-Protein Interactions* (Jaenicke and Helmreich, eds). Springer-Verlag, Heidelberg, 139-156.

Battelino, L. J. and A. Blanco 1970. Catalytic properties of the lactate dehydrogenase isozyme "X" from mouse testis. *J. Exp. Zool.* 174: 173-186.

Battellino, L. J., F. R. Jaime, and A. Blanco 1968. Kinetic properties of rabbit testicular lactate dehydrogenase isozyme. *J. Biol. Chem.* 243: 5185-5192.

Bené, M. and E. Goldberg 1974. Binding of antibody to LDH-X by the mouse blastocyst. *J. Exp. Zool.* 189: 261-266.

Blackshaw, A. W. and J. S. H. Elkington 1970a. Developmental changes in lactate dehydrogenase isoenzymes in the testis of the immature rat. *J. Reprod. Fert.* 22: 69-75.

Blackshaw, A. W., and J. S. H. Elkington 1970b. The effect of age and hypophysectomy on growth and the isoenzymes of lactate dehydrogenase in the mouse testis. *Biol. Reprod.* 2: 268-274.

Blackshaw, A. W., and J. I. Samisoni 1966. The effects of cryptorchism in the guinea pig on the isoenzymes of testicular lactate dehydrogenase. *Aust. J. Biol. Sci.* 19: 841-848.

Blackshaw, A. W., and J. I. Samisoni 1967. The testes of the cryptorchid ram. *Res. Vet. Sci.* 8: 187-194.

Blanco, A., and W. H. Zinkham 1963. Lactate dehydrogenases in human testes. *Science* 139: 601-602.

Blanco, A., W. H. Zinkham, and L. Kupchyk 1964. Genetic control and ontogeny of lactate dehydrogenase in pigeon testes. *J. Exp. Zool.* 156: 137-152.

Brummel, M. C., and L. D. Stegink 1972. Amino-terminal and carboxyl-terminal residues of rabbit muscle M_4 lactate dehydrogenase: Comparison with other vertebrate species. *Comp. Biochem. Physiol.* 41B: 487-492.

Clausen, J., and B. Øvlisen 1965. Lactate dehydrogenase isoenzymes of human semen. *Biochem. J.* 97: 513-517.

Dixon, H. B. F., and R. N. Perham 1968. Reversible blocking of amino groups with citraconic anhydride. *Biochem. J.* 109: 312-314.

Dym, M., and D. W. Fawcett 1970. The blood-testes barrier in the rat and the physiological compartmentation of the seminiferous epithelium. *Biol. Reprod.* 3: 308-326.

Edwards, R. G. 1970. Immunology of conception in pregnancy. *Br. Med. Bull.* 26: 72-78.

Elkington, J. S. H., A. W. Blackshaw, and B. DeJong 1973. The effect of hypophysectomy on testicular hydrolases, lactate dehydrogenase, and spermatogenesis in the rat. *Aust. J. Biol. Sci.* 26: 491-503.

Goldberg, E. 1963. Lactic and malic dehydrogenases in human spermatozoa. *Science* 139: 602-603.

Goldberg, E. 1964. Lactate dehydrogenases and malate dehydrogenases in sperm: studied by polyacrylamide gel electrophoresis. *Ann. N. Y. Acad. Sci.* 127: 560-570.

Goldberg, E. 1965. Lactate dehydrogenases in spermatozoa: subunit interactions in vitro. *Arch. Biochem. Biophys.* 109: 134-141.

Goldberg, E. 1971. Immunochemical specificity of lactate dehydrogenase-X. *Proc. Natl. Acad. Sci.* 68: 349-352.

Goldberg, E. 1972. Amino acid composition and properties of crystalline lactate dehydrogenase X from mouse testes. *J. Biol. Chem.* 247: 2044-2048.

Goldberg, E. 1973a. Molecular basis for multiple forms of LDH-X. *J. Exp. Zool.* 186: 273-278.

Goldberg, E. 1973b. Infertility in female rabbits immunized with lactate dehydrogenase X. *Science* 181: 458-459.

Goldberg, E. 1974. Lactate dehydrogenase-X (crystalline) from mouse testes. in *Methods in Enzymology*. In press.

Goldberg, E., and C. Hawtrey 1967. The ontogeny of sperm-specific lactate dehydrogenase in mice. *J. Exp. Zool.* 164: 309-316.

Goldberg, E., and C. Hawtrey 1968. The effect of experimental cryptorchism on the isozymes of lactate dehydrogenase in mouse testes. *J. Exp. Zool.* 167: 411-418.

Goldberg, E., and J. Lerum 1972. Pregnancy suppression by an antiserum to the sperm specific lactate dehydrogenase. *Science* 176: 686-687.

Goldberg, E., and E. Whitmore 1974. Purification of LDH-X by affinity chromatography. *Isozyme Bull.* 7: 10

Gutierrez, M., N. M. G. DeBurgos, C. Burgos, and A. Blanco 1972. The sexual cycle of male bats: Changes of testicular lactate dehydrogenase isoenzymes. *Comp. Biochem. Physiol.* 43A: 47-52.

Hawtrey, C., and E. Goldberg 1968. Differential synthesis of LDH in mouse testes. *Ann. N. Y. Acad. Sci.* 151: 611-615.

Hawtrey, C. O., and E. Goldberg 1970. Some kinetic aspects of sperm specific lactate dehydrogenase in mice. *J. Exp. Zool.* 174: 451-462.

Holmes, R. S. 1972. Evolution of lactate dehydrogenase genes. *FEBS Letters* 28(1): 51-55.

Holmes, R. S., D. W. Cooper, and J. L. Vandeberg 1973. Marsupial monotreme lactate dehydrogenase isozymes: phylogeny, ontogeny, and homology with eutherian mammals. *J. Exp. Zool.* 184: 127-147.

Ishibe, T., S. Matsuki, and H. Nihira 1971. Relationship between lactic dehydrogenase activity and other enzyme
activities of human seminal plasma. *Hiroshima J. Med. Sci.*
20: 215-222.

Kolb, E., G. A. Gleisher, and J. Larner 1970. Isolation and
characterization of bovine lactate dehydrogenase X. *Biochem.* 9: 4372-4380.

Lerum, J., and E. Goldberg 1974. Immunological impairment of
pregnancy in mice by lactate dehydrogenase X. *Biol.
Reprod.* 11: 108-115.

Lindahl, R., and K. Mayeda 1972. Lactate dehydrogenase isozymes
of the Mongolian gerbil, *Meriones unguiculatus* - I. Evidence for X subunits. *Comp. Biochem. Physiol.* 43B: 425-
433.

Lindahl, R., and K. Mayeda 1973. Lactate dehydrogenase isozymes
of the Mongolian gerbil, *Meriones unguiculatus* - II.
Ontogeny and relative isozyme composition. *Comp. Biochem.
Physiol.* 45B: 265-273.

Markert, C. L., and G. S. Whitt 1968. Molecular varieties of
isozymes. *Experientia* 24: 977-991.

McKee, R. W., E. Longstaff, and A. L. Latner 1972. Lactate
dehydrogenase in human testes. *Clin. Chim. Acta* 39: 221-
227.

Menge, A. C. 1970. Immune reactions and infertility. *J. Reprod.
Fert. Suppl.* 10: 171-186.

Metz, C. B. 1972. Effects of antibodies on gametes and fertilization. *Biol. Reprod.* 6: 358-383.

O'Carra, P., and S. Barry 1972. Affinity chromatography of
lactate dehydrogenase - Model studies demonstrating the
potential of the technique in the mechanistic investigation as well as in the purification of multi-substrate
enzymes. *FEBS Letters* 21: 281-285.

Pesce, A., T. P. Fondy, F. Stolzenbach, F. Castillo, and N. O.
Kaplan 1967. The comparative enzymology of lactic dehydrogenases. *J. Biol. Chem.* 242: 2151-2167.

Ressler, N., K. L. Stitzer, and D. R. Ackerman 1967. Reactions
of the lactate dehydrogenase X-band in human sperm with
homologous and heterologous antisera. *Biochim. Biophys.
Acta* 139: 507-510.

Schatz, L., H. L. Segal 1969. Reduction of α-ketoglutarate by
homogeneous lactic dehydrogenase X of testicular tissue.
J. Biol. Chem. 244: 4393-4397.

Spielmann, H., R. P. Erickson, and C. J. Epstein 1973. The separation of lactate dehydrogenase X from other lactate
dehydrogenase isozymes of mouse testes by affinity chromatography. *FEBS Letters* 35: 19-23.

Stambaugh, R., and J. Buckley 1967. The enzymic and molecular nature of the lactic dehydrogenase subbands and X_4 isozyme. *J. Biol. Chem.* 242: 4053-4059.

Svasti, J. and S. Viriyachai 1974. The properties of purified LDH-X from human testis. *II. Isozymes: Physiology and Function* C. L. Markert, editor. Academic Press. New York.

Szeinberg, A., A. Mor, H. Vernia, and S. Reischer 1966. "Band X" isozyme of lactic dehydrogenase in pathological spermatogenesis. *Life Sciences* 5: 1233-1238.

Taylor, S. S., S. S. Oxley, W. S. Allison, and N. O. Kaplan 1973. Amino acid sequence of dogfish M_4 lactate dehydrogenase. *Proc. Natl. Acad. Sci.* 70: 1790-1794.

Thompson, R. S., and L. Zamboni 1974. Phagocytosis of supernumerary spermatozoa by two-cell mouse embryos. *Anat. Rec.* 178: 3-13.

Valdivieso, D., E. Conde, J. R. Tamsitt 1968. Lactate dehydrogenase studies in Puerto Rican bats. *Comp. Biochem. Physiol.* 27: 133-138.

Valenta, M., J. Hyldgaard-Jensen, and J. Moustgaard 1967. Three lactic dehydrogenase isoenzyme systems in pig spermatozoa and the polymorphism of sub-units controlled by third locus C. *Nature* 216: 506-507.

Vitale, R., D. W. Fawcett, and M. Dym 1973. The normal development of the blood-testes barrier and the effects of clomiphene and estrogen treatment. *Anat. Rec.* 176: 333-344.

Winer, A. D., and M. B. Nikitovitch-Winer 1971. Hormonal effects on the rat gonadal lactate dehydrogenases. *FEBS Letters* 16: 21-24.

Wong, C., R. Yanez, D. M. Brown, A. Dickey, M. E. Parks, R. W. McKee 1971. Isolation and properties of lactate dehydrogenase isozyme X from Swiss mice. *Arch. Biochem. Biophys.* 146: 454-460.

Zinkham, W. H. 1968. Lactate dehydrogenase isozymes of testis and sperm: Biological and biochemical properties and genetic control. *Ann. N. Y. Acad. Sci.* 151: 598-610.

Zinkham, W. H., and H. Isensee 1972. Genetic control of lactate dehydrogenase synthesis in the somatic and gametic tissues of pigeons. *Johns Hopkins Med. J.* 130: 11-25.

Zinkham, W. H., A. Blanco, and L. Kupchyk 1963. Lactate dehydrogenase in testis: Dissociation and recombination of subunits. *Science* 142: 1303-1304.

Zinkham, W. H., A. Blanco, and L. Clowry 1964a. An unusual isozyme of lactate dehydrogenase in mature testes: Localization, ontogeny and kinetic properties. *Ann. N. Y. Acad. Sci.* 121: 571-588.

Zinkham, W. H., A. Blanco, and L. Kupchyk 1964b. Lactate dehydrogenase in pigeon testes: Genetic control by three loci. *Science* 144: 1353-1354.

Zinkham, W. H., L. Kupchyk, A. Blanco, and H. Isensee 1966. A variant of lactate dehydrogenase in somatic tissues of pigeons: Physicochemical properties and genetic control. *J. Exp. Zool.* 162: 45-56.

Zinkham, W. H., N. A. Holtzman, and H. Isensee 1968. The molecular size of lactate dehydrogenase isozymes in mature testes. *Biochim. Biophys. Acta* 160: 172-177.

Zinkham, W. H., H. Isensee, J. H. Renwick 1969. Linkage of lactate dehydrogenase B and C loci in pigeons. *Science* 164: 185-187.

INTRACELLULAR TURNOVER OF LACTATE DEHYDROGENASE ISOZYMES

PAUL J. FRITZ[1] , E. LUCILE WHITE[1] , AND KENNETH M. PRUITT[2]
[1] Department of Pharmacology
The Milton S. Hershey Medical Center
The Pennsylvania State University
Hershey, Pennsylvania 17033
[2] Laboratory of Molecular Biology
University of Alabama Medical Center
Birmingham, Alabama 35233

ABSTRACT. Data are presented showing how the levels of
five LDH isozymes in rat heart and two LDH isozymes in rat
liver and skeletal muscle change during postnatal development. The data are analyzed taking into account the
potential variability of synthetic and degradative rates;
curves representing these rates during the developmental
period are presented. In addition to the obvious conclusion that protein levels increase or decrease over a period
of time depending upon whether rates of synthesis are larger
or smaller than rates of degradation, it was also concluded
that the rate of change of the protein level depends upon
the magnitude of the difference in synthetic and degradative
rates. A higher steady state level of one enzyme compared
to another is not due to its higher steady state rate of
synthesis or lower steady state rate of degradation. Rather,
the steady state enzyme level as well as the level at any
other time is due to the difference between the areas under
the curves representing rates of synthesis and degradation
during the developmental period and to the level at the
beginning of the developmental period. Specific mechanisms
for regulation of LDH isozyme levels by post translational
assembly and disposition of subunits are also discussed.

The characteristic tissue distribution of mammalian lactate
dehydrogenase isozymes stimulated studies undertaken several
years ago to investigate the reasons for these tissue differences. The widely accepted explanation that differential synthetic rates were solely responsible seemed insufficient.
Intracellular amounts of isozymes, like other proteins, are
determined by a balance between how fast they are made and how
fast they are degraded. However, when isozymes are composed
of combinations of different subunits, the study of their
turnover takes on a dimension not present in the consideration
of single polypeptide chain proteins. This paper will be divided into two parts: (1) a general discussion of how rates of
synthesis and degradation of enzymes under conditions where
the enzyme level is changing, particularly during development,

contribute to the establishment of the adult or steady state level of the enzyme; and (2) a discussion of a specific mechanism of regulation of the lactate dehydrogenase isozyme levels involving *in vivo* exchange of subunits.

Enzyme Turnover at Times Other Than at Steady State

For most proteins in healthy tissues, intracellular turn-over involves primarily *de novo* synthesis and degradation. Thus, if the amount of an intracellular protein is decreasing, one can conclude that degradation is occurring, because the only way the amount can decrease is by the rate of degradation exceeding the rate of synthesis. Synthesis may or may not be occurring. The opposite conclusions are true if the amount of protein is increasing. However, these conclusions must be modified if there is extracellular input or output of the particular protein. There is no evidence for such contributions to lactate dehydrogenase turnover, and they have not been considered in the following analysis.

During postnatal development, particularly in the suckling period, the levels of rat lactate dehydrogenase isozymes fluctuate. In Figures 1-3 are presented the values for 5 isozymes in heart and 2 each in skeletal muscle and liver as a function of days after birth. The isozyme amounts were determined using DEAE-Sephadex chromatography (Fritz et al., 1970). Results obtained by mixing and assaying tissue extracts from different ages did not reveal the presence of activators or inhibitors, indicating that the changes were due to differences in the amount of enzyme protein. This conclusion was supported by quantitative precipitin studies of the type previously described (Osterman et al., 1973; Fritz and White, 1974). From birth to maturity the five lactate dehydrogenase isozymes in rat heart, as in mouse heart (Markert and Ursprung, 1962), undergo dramatic shifts in amounts (Figure 1). As a result of these time dependent alterations in pattern, the amount of A subunits remains virtually the same while the number of B subunits increases more than 3 fold (Table 1). Similarly, the changes in rat skeletal muscle lactate dehydrogenase isozymes during development result in a 10 fold increase in A subunits while the amount of B subunits decreases slightly from birth to maturity (Figure 2, Table 1). From the data in Table 1 it might appear that the amounts of lactate dehydrogenase A and B subunits in rat liver remain virtually unchanged during development. However, as seen in Figure 3, in the first 10 days after birth rat liver lactate dehydrogenase levels more than double and then decrease to steady state amounts.

Based on evidence from several laboratories (Murison, 1961;

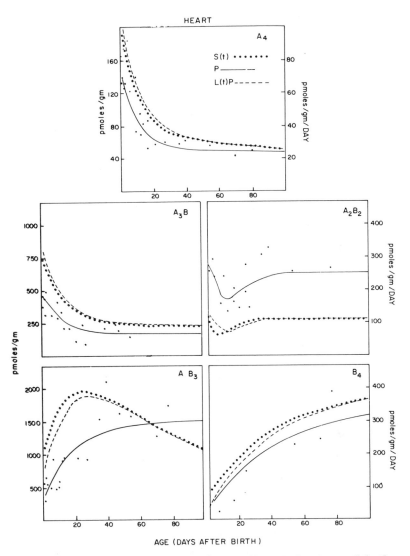

Fig. 1. Developmental changes in rat heart lactate dehydrogenase isozymes. The closed circles are the values for individual determinations after separation of the isozymes by DEAE-
Sephadex chromatography. The dotted line, S(t), and the
dashed line, L(t)P, represent the estimated values for the rates
(pmoles/gm/day) of input and output, respectively. The solid
line, P(pmoles/gm) was obtained by numerical integration of the
equation representing the overall rate of change of P with
respect to time. Details of this method are published elsewhere

(Fig. 1 legend continued)
(Fritz et al., 1974). Conversion of enzyme units to picomoles
was accomplished as previously described (Fritz et al., 1969).

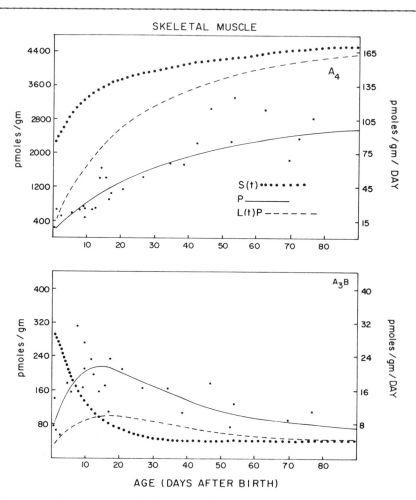

Fig. 2. Developmental changes in rat skeletal muscle lactate
dehydrogenase isozymes. For details see legend to figure 1.

Haining et al., 1970; Ove et al., 1972; Phillipidis et al.,
1972), there is reason to believe that changes in levels of
particular proteins during postnatal development are accompanied
by corresponding changes in rates of protein synthesis. Thus,
we have analyzed the data on protein concentration versus time
for rat heart (Figure 1), rat skeletal muscle (Figure 2), and

350

TABLE 1

LACTATE DEHYDROGENASE SUBUNITS: DISTRIBUTION AMONG ISOZYMES
IN THREE RAT TISSUES AT BIRTH AND 100 DAYS AFTER BIRTH

TISSUES AND ISOZYMES	BIRTH	pmoles/g		DAY 100	pmoles/g	
	Tetramer	Subunits A	Subunits B	Tetramer	Subunits A	Subunits B
HEART						
A4	136	544	---	50	200	--
A3B	463	1389	463	162	486	162
A2B2	676	1352	1352	632	1264	1264
AB3	396	396	1188	1502	1502	4506
B4	145	--	580	1580	--	6320
		3681	3583		3452	12,252
SKELETAL MUSCLE						
A4	176	704	--	2592	10,368	--
A3B	84	252	84	70	210	70
A2B2	29	58	58	42	84	84
AB3	82	82	246	27	27	81
B4	15	--	60	94	--	94
		1096	448		10,689	329
LIVER						
A4	2248	8992	--	2271	9084	--
A3B	119	357	119	26	79	26
A2B2		9349	119		9163	26
AB3						
B4						

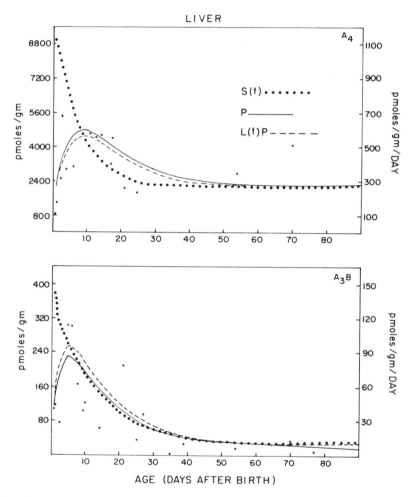

Fig. 3. Developmental changes in rat liver lactate dehydrogenase isozymes. For details see legend to Figure 1.

rat liver (Figure 3) lactate dehydrogenase isozymes, taking
into account the potential variability of synthetic as well
as degradative rates (Fritz et al., 1974).   Rates of synthesis
and degradation of the isozymes during development, as estimated by this method, are also shown in the Figures. Not only is
it true that protein levels increase or decrease over a period
of time depending upon whether rates of synthesis are larger
or smaller than rates of degradation but it is also true that
the rate of change in the protein level depends upon the

magnitude of the difference in synthetic and degradative rates. An example can be seen by comparing the skeletal muscle A_4 and heart B_4 isozymes. At birth both isozymes occur in roughly equal amounts but by 20 days after birth the amount of skeletal muscle A_4 is 1200 picomoles per gram of tissue, whereas the amount of heart B_4 at this time is only 600 picomoles per gram of tissue, in spite of the fact that by 20 days the heart isozyme is being synthesized at a faster rate than the skeletal muscle isozyme (200 versus 135 picomoles per gram of tissue per day). The explanation lies in the difference between rates of synthesis and degradation over this time period, as seen in Figures 1 and 2. It may seem paradoxical that in 3 cases (liver A_4 and A_3B, Figure 3, and skeletal muscle A_3B, Figure 2) the enzyme level is increasing while the rate of synthesis is decreasing and the rate of degradation is increasing. However, these observations merely emphasize the point that whether the enzyme level increases or decreases depends upon whether rates of synthesis or rates of degradation are greater. It should be kept in mind when comparisons are made between the steady state levels of two enzymes that for each of the enzymes the rate of synthesis must be equal to the rate of degradation because the enzyme levels in each case are constant. The steady state level is equal to the ratio of the rate of synthesis to the rate constant for degradation. Thus, there is no unique relationship between steady state level and either synthesis or degradation considered independently. Therefore, it is misleading to say that a higher steady state level of one enzyme compared to another is due to its higher rate of synthesis or lower rate of degradation.

The enzyme level at any time during development, including the steady state level, is related to (a) the level at the beginning of the developmental period and (b) the difference between the areas under the curves representing rates of synthesis and degradation during the developmental period up to the time of measurement. Steady state rate constants for synthesis and degradation of the lactate dehydrogenase isozymes derived from data on development up to 100 days after birth are shown in Table 2. As noted, at steady state the picomoles of enzyme synthesized per gram of tissue per day (k_s) is the same as the picomoles of enzyme degraded per gram of tissue per day; thus, k_s is a measure of absolute turnover. On the other hand, k_d is the ratio of absolute turnover to the total amount of enzyme present in a gram of tissue and thus is a measure of relative turnover. Expressed another way, k_d represents the per cent of the total amount of enzyme that is replaced or turned over in a given period of time, in this case 24 hours. Turnover is sometimes expressed as half-life,

TABLE 2

LACTATE DEHYDROGENASE ISOZYME TURNOVER IN RAT TISSUES

TISSUE AND ISOZYME	pmoles/gm at day 100	k_s pmoles/gm/day	k_d day^{-1}
HEART			
A_4	50	26	.52
A_3B	162	87	.54
A_2B_2	632	112	.18
AB_3	1502	208	.14
B_4	1580	367	.23
SKELETAL MUSCLE			
A_4	2592	169	.06
A_3B	70	4	.06
LIVER			
A_4	2271	267	.12
A_3B	26	11	.42

analogous to the first order decay of radioisotopes, and can be calculated by dividing ln2 by k_d. The data in Table 2 show that comparison of half-lives may give a false impression of fast or slow turnover. For example, the half-life of rat heart lactate dehydrogenase A_4 is 1.3 days (.693/.52 day^{-1}) whereas the same isozyme from rat skeletal muscle has a half-life of 11.5 days (.693/.06 day^{-1}). However, the absolute turnover of the skeletal muscle enzyme is more than 6 times greater than that of the heart enzyme.

Estimation of lactate dehydrogenase isozyme turnover in adult rat tissues by means of ^{14}C amino acid incorporation studies (Fritz et al., 1973) yielded results considerably different from those presented in Table 2. The major difference between the analysis of the radioisotope data and the developmental data presented here is the assumption, in the former case, that the isozymes are made and broken down independently. For reasons discussed below, we believe this is a questionable assumption. However, both methods of analysis yield results that are internally consistent; for example, the absolute turnover of A_4 as estimated by either method is liver > skeletal muscle > heart. Furthermore, from either method it is clear that a category of proteins compiled on the basis of their turnover shows lactate dehydrogenase isozyme turnover to be slow compared to serine dehydratase (1,900 pm/g/day), aldolase (6,500 pm/g/day), or tyrosine amino transferase (33,000 pm/g/day) (Fritz et al., 1974).

On Specific Mechanisms for Regulation of Lactate Dehydrogenase Isozyme Levels

The general treatment of isozyme turnover given above, although useful in understanding how changes in protein levels are related to rates of synthesis and degradation, nevertheless gives little insight into the mechanisms involved in regulating these levels. As mentioned previously, the subunit nature of isozymes complicates the understanding of regulatory mechanisms. Clearly, synthesis of the subunits proceeds in the usual transcription-translation manner, and most likely, degradation to amino acids is not unique for isozyme subunits. A major difference between single and multiple subunit proteins lies in the assembly and subsequent disposition of subunits. We have presented both *in vivo* and *in vitro* evidence in support of the idea that the distribution of lactate dehydrogenase isozymes in rat heart is established and maintained as a result of exchange of subunits among tetramers (Fritz et al., 1971).

In that study the data behaved kinetically as if (a) the homopolymers A_4 and B_4 were produced via the usual transcription-translation route, and (b) the heteropolymers were produced as a result of the bimolecular reactions:

$$A_4 + B_4 \rightleftarrows 2A_2B_2$$

$$A_4 + A_2B_2 \rightleftarrows 2A_3B$$

$$B_4 + A_2B_2 \rightleftarrows 2AB_3$$

and (c) the heteropolymers A_3B and AB_3 were degraded by first order reactions to amino acids. The subunit exchange model accounted entirely for the changes in rat heart lactate dehydrogenase isozyme patterns observed during development. Further evidence in support of the model was the fact that calculated and observed ^{14}C amino acid incorporation curves in the adult animal were in good agreement. We emphasized that intermediate reactions between monomers and dimers were not excluded by the model even though the kinetics were consistent with the simpler reactions. Thus, the model was compatible with the *in vitro* findings of Millar (Millar et al., 1971; Millar, 1974). No consideration was given to the effect of substrate, coenzyme, or phosphate ion on the proposed *in vivo* reactions in spite of their well-known effects on *in vitro* hybridization (Markert, 1963; Chilson et al., 1965; Anderson and Weber,1966; Trausch, 1972). It was also pointed out that other bimolecular reactions between isozymes were possible and that other models might be consistent with the data.

Studies on the *in vivo* synthesis of the subunit proteins hemoglobin and immunoglobulin revealed, in the case of hemo-globin, that completed α chains combine with incomplete β chains on polyribosomes and that αβ dimers are released from polyribosomes upon completion of the β chain (Baglioni and Campana, 1967). For immunoglobulins it was shown that com-pleted light chains combine with incomplete heavy chains still on the polyribosomes (Shapiro et al., 1966; Williamson and Askonas, 1967). No such direct studies have been reported for isozymes but similar mechanisms may be operative for isozymes composed of subunits. Thus, it seems that the release from polysomes of single polypeptide chains of lactate dehydrogenase A or B subunits is unlikely. The consequence of such events would be the random, three dimensional, diffusion controlled interaction of subunits to form tetramers. A more plausible explanation, in particular by analogy with hemoglobin and im-munoglobulin, is that the heteropolymers A_2B_2, A_3B, and AB_3 are produced by interaction of the homopolymers A_4 and B_4 either while one homopolymer is bound to a polyribosome or shortly after its release. Identifying the correct model is important and remains as a challenge in future research.

ACKNOWLEDGMENTS

Aided by Grant BC-50A from the American Cancer Society and by NIH Research Grant CA-12808 from the National Cancer Institute.

REFERENCES

Anderson, S., and G. Weber. 1966. The reversible acid dissociation and hybridization of lactic dehydrogenase. *Arch. Biochem. Biophys.* 116: 207-223.

Baglioni, C., and T. Campana. 1967. α-Chain and globin: intermediates in the synthesis of rabbit hemoglobin. *Europ. J. Biochem.* 2: 480-492.

Chilson, O.P., G.B. Kitto, and N.O. Kaplan. 1965. Factors affecting the reversible dissociation of dehydrogenases. *Proc. Natl. Acad. Sci. U.S.* 53: 1006-1014.

Fritz, P.J., E.L. White, J. Osterman, and K.M. Pruitt. 1974. Protein turnover during development. *Proc. Joint U.S.-Japan Symposium on Protein Turnover.* Academic Press, New York. In press.

Fritz, P.J., W.J. Morrison, E.L. White, and E.S. Vesell. 1970. Comparative study of methods for the quantitation of lactate dehydrogenase isozymes. *Anal. Biochem.* 36: 443-453.

Fritz, P.J., K.M. Pruitt, E.L. White, and E.S. Vesell. 1971. Control of protein levels in animal cells - a new theory. *Nature New Biology* 230: 119-122.

Fritz, P.J., and E.L. White. 1974. 3-Phosphoglycerate kinase from rat tissues. Further characterization and developmental studies. *Biochemistry* 13: 444-449.

Fritz, P.J., E.L. White, K.M. Pruitt, and E.S. Vesell. 1973. Lactate dehydrogenase isozymes: turnover in rat heart, skeletal muscle, and liver. *Biochemistry* 12: 4034-4039.

Fritz, P.J., E.S. Vesell, E.L. White, and K.M. Pruitt. 1969. The roles of synthesis and degradation in determining tissue concentrations of lactate dehydrogenase-5. *Proc. Natl. Acad. Sci. U.S.* 62: 558-565.

Haining, J.L., J.S. Legan, and W.J. Lovell. 1970. Synthesis and degradation of rat liver xanthine oxidase as a function of age and protein deprivation. *J. Geront.* 25: 205-209.

Markert, C.L. 1963. Lactate dehydrogenase isozymes: dissociation and recombination of subunits. *Science* 140: 1329-1330.

Markert, C.L., and H. Ursprung. 1962. The ontogeny of isozyme patterns of lactate dehydrogenase in the mouse. *Develop. Biol.* 5: 363-381.

Millar, D.B. 1974. The quaternary structure of LDH. 2. The mechanisms, kinetics and thermodynamics of dissociation, denaturation, and hybridization in ethylene glycol. *Biochim. Biophys. Acta.* In press.

Millar, D.B., M.R. Summers, and J.A. Niziolek. 1971. Spontaneous *in vitro* hybridization of LDH homopolymers in the undenatured state. *Nature New Biology* 230: 117-119.

Murison, G. 1969. Synthesis and degradation of xanthine dehydrogenase during chick liver development. *Develop. Biol.* 20: 518-543.

Osterman, J., P.J. Fritz, and T. Wuntch. 1973. Pyruvate kinase isozymes from rat tissues: developmental studies. *J. Biol. Chem.* 248: 1011-1018.

Ove, P., M. Obenrader, and A. Lansing. 1972. Synthesis and degradation of liver proteins in young and old rats. *Biochem. Biophys. Acta* 277: 211-221.

Phillipidis, H., R.W. Hanson, L. Reshef, M.F. Hopgood, and F.J. Ballard. 1972. The initial synthesis of proteins during development. *Biochem. J.* 126: 1127-1134.

Shapiro, A.L., M.D. Scharff, J.V. Maizel, Jr., and J.W. Uhr. 1966. Polyribosomal synthesis and assembly of the H and L chains of gamma globulin. *Proc. Natl. Acad. Sci. U.S.* 56: 216-221.

Trausch, G. 1972. On the determination of the molecular weight of lactate dehydrogenase by gel filtration on Sephadex G-200. *Biochimie* 54: 531-533.

Williamson, A.R., and B.A. Askonas. 1967. Biosynthesis of immunoglobulin: the separate classes of polyribosomes synthesizing heavy and light chains. *J. Mol. Biol.* 23: 201-216.

LACTATE DEHYDROGENASE ISOZYME
SYNTHESIS AND CELLULAR DIFFERENTIATION
IN THE TELEOST RETINA

E. T. MILLER and G. S. WHITT
Department of Zoology
University of Illinois
Urbana, Illinois 61801

ABSTRACT. Many orders of teleosts possess a unique lactate dehydrogenase isozyme (LDH C_4) predominantly synthesized in the retina. Because of its kinetic properties and its localization in the photoreceptor cells, this isozyme may play a role in visual metabolism. If the LDH C_4 isozyme is important to retinal metabolism, its appearance during embryogenesis should occur concomitantly with morphological and functional differentiation of the retina. Therefore, the timing of the synthesis of this retinal-specific isozyme with respect to the timing of the morphological differentiation was determined. Only LDH A and B subunits are present in the green sunfish (*Lepomis cyanellus*, Centrarchidae, Perciformes) from fertilization until several days after hatching. The C_4 isozyme is first detected on the fifth day post-hatch. Light and electron micrographs of the 3 day post-hatch retina reveal that the neural retina is immature at this time. The receptor cell nuclei are very large and the ellipsoids contain small numbers of mitochondria. Few photoreceptor cells have outer segments, indicating incomplete differentiation at 3 days post-hatch. However, by 5 days post-hatch, most photoreceptor cells have developed outer segments, and the ultrastructure of the 5 day post-hatch photoreceptor cells appears almost identical to that of the adult. On the basis of these studies, it appears that the synthesis of the C_4 LDH isozyme is tightly coupled with the morphological and functional differentiation of the retina.

Introduction

The first division of the fertilized egg begins a process of activation and modulation of gene function which leads to the formation of a complex, multicellular adult organism. One of the most powerful tools used in investigating gene expression during development is isozymes. The lactate dehydrogenase isozymes are a particularly useful model system (Markert and Ursprung, 1971).

The enzyme lactate dehydrogenase, which is found in all

359

vertebrates, is responsible for the conversion of pyruvate
to lactate, with the concomitant oxidation of the coenzyme
NADH to NAD+. The NAD+ produced can be utilized in the glycolytic pathway to facilitate the production of ATP for energy
under periods of relative anaerobiosis.

The two types of LDH subunits common to all vertebrates,
A and B, combine to form the tetrameric enzymes. While 5
isozymes are frequently seen in mammals, many fish contain
4, 3 or 2 isozymes composed of these subunits, (Markert and
Faulhaber, 1965), due to an apparent restriction on subunit
assembly (Whitt, 1970a,b) and differential instability
(Shaklee, 1972).

A third LDH locus has been found in many orders of fish.
The product of this third locus is referred to as the C_4 isozyme (Shaklee et al, 1973). This C_4 isozyme is like the B_4
isozyme in possessing a lower substrate concentration optimum
and greater susceptibility to substrate inhibition than the
A_4 isozyme in vitro (Whitt, 1970a; Whitt et al, 1973). It
is probably these kinetic properties which make the B_4 and C_4
isozymes suited to tissues which are relatively more aerobic.

The LDH isozymes in the tissues of a typical teleost, the
green sunfish (*Lepomis cyanellus*, Centrarchidae, Perciformes)
is shown in Fig. 1. The A_4 and B_4 isozymes are homologous to
those of other vertebrates yet show reversed electrophoretic
mobility in this species. The most anodal band is the C_4
isozyme synthesized predominantly in the eye of most teleosts.
Teleosts are noted for retinal oxygen tensions as high as
twenty times that of arterial blood (Fairbanks et al, 1969).
The presence of the C_4 isozyme in such a highly oxygenated
tissue as the eye is consistent with its kinetic properties.
The B_3C_1 heterotetramer is detected in other aerobic tissues
such as brain, kidney, and gill.

It has been postulated that the C_4 isozyme plays a role in
the retinal metabolism of teleosts (Whitt, 1970a). The major
biochemical process occurring in the photoreceptor cells is
the regeneration of the visual pigment. NAD serves as the
cofactor for retinol dehydrogenase in the regeneration pathway in fishes (Wald, 1968). The C4 isozyme may regulate the
NAD^+/NADH ratio in the photoreceptor cells, and thus play a
key role in the vision of these teleosts.

Whitt and Booth (1970) localized LDH activity in the neural
retina of the swordtail (*Xiphophorous*). Urea inactivation of
the isozymes containing A and B subunits permitted the selective staining of the C_4 isozyme. They found the C_4 activity
in the photoreceptor cells, primarily in the region referred
to as the ellipsoid.

The presence of three homologous LDH loci in teleosts

Tissue Pattern of LDH
Isozymes in Green Sunfish

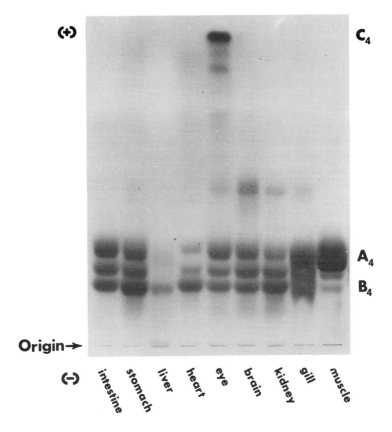

Fig. 1. Lactate dehydrogenase isozymes of the tissues of the green sunfish. The A_4, A_2B_2 and B_4 isozymes are present in most tissues. The highly anodal C_4 is found only in the eye. B_3C_1 heterotetramer is seen in the eye, brain, kidney, and gill.

facilitates the investigation of specific gene activation during differentation. Specifically, the time of gene activation with respect to cellular differentiation and the tissue specificity of gene function can be readily examined.

The retinal LDH isozyme has been the subject of several developmental studies. Nakano and Whiteley (1965) reported that the eye band appeared abruptly at hatching in the medaka (*Oryzias latipes*). They concluded this was due to the conversion from anaerobic to an aerobic metabolism at hatching. Whitt (1970a) observed the eye band at least 24 hours before the time of hatching in another teleost, *Fundulus heteroclitus,* and suggested that hatching was not the necessary stimulus to C-gene activation in this fish. Hitzeroth and co-workers (1968) showed that the retinal-specific LDH did not appear until considerably after hatching in the trout, and this has been confirmed by Goldberg et al, (1969).

Whitt (1968; 1970a) has postulated a correlation of the first expression of C-gene function in teleosts with differentiation of the neural retina. Recent support for this hypothesis has come from Kunz (1971) who observed C_4 appearance at the time of photoreceptor cell differentiation at the light microscope level in the guppy (*Lebistes*). Nakano and Hasegawa (1971) have shown on the basis of light microscopy that morphological differentiation may precede the appearance of the eye-band in the medaka by 2 days. Because the C_4 isozyme is primarily restricted to the photoreceptor cells, and because the sequence of events leading to LDH C-gene activation is unknown, it is the purpose of this investigation to determine how tightly coupled the isozymic differentiation of the retina is to its morphological differentiation in the green sunfish.

LDH isozyme synthesis during embryogenesis and cellular
    differentiation

The subunit composition of the green sunfish LDH isozymes was verified by immunochemical means. Fig. 2 shows the effect of anti-A_4 and anti-B_4 sera on the LDH isozymes of the sunfish eye. The specific anti-serum is mixed with enzyme extract and allowed to react for one-half hour. The mixture is centrifuged to precipitate some of the LDH-antibody complexes, and the supernatant is subjected to electrophoresis. The molecular sieving effect of the starch gel greatly retards the electrophoretic mobility of any LDH molecules bound to antibodies. The absence of certain bands on the starch gel is thus an indication of cross-reaction. In Fig. 2, the untreated eye homogenate and rabbit serum serve as controls. When the eye homogenate is reacted with anti-A_4 serum, low concentrations of serum precipitate the A_4 homotetramer and higher concentrations precipitate the A_2B_2 heterotetramer. Even at high concentrations of anti-A_4 serum, there is no effect on the B_4 isozyme. The eye extract was reacted with increasing amounts of anti-B_4 serum. At low concentrations, only isozymes containing B subunits are precipitated by anti-B_4 and anti-C_4 sera (Fig. 2). At high concentrations of anti-B_4 there is cross-reaction with the C_4 isozyme. The anti-C_4 serum reacts primarily with isozymes containing C subunits.

After determining the subunit compositions of the LDH isozymes, it was then possible to analyze their ontogeny in order to determine at what stage of development the C_4 isozyme is first synthesized. Zygotes were formed in the laboratory by stripping eggs, then fertilizing them with sperm expressed from the male. The eggs remain completely clear and all stages of development can be easily observed. Hatching occurs after 50 hours (2 days) at 24° Celsius. Development is very synchronous within the population (Childers, 1967).

A detailed examination of the initial stages of embryogenesis from unfertilized egg to hatching revealed no changes in the isozyme pattern. The unfertilized egg shows approximately equal amounts of A and B subunits. This A:B subunit ratio remains relatively constant during embryogenesis. Even at hatching, presumably a dramatic event for the larval fish, there is no sudden change in the isozyme pattern.

Developmental changes in gross morphology of the larval fish from hatching to 5 days post-hatch are portrayed schematically in Fig. 3. At the time of hatching, the sunfish is very immature. It has a large yolk sac and is entirely colorless, except for the red blood cells moving through the circulatory system. The eyes are poorly developed with no pigmentation. By one day post-hatch very small spots of pigmentation become visible in the eye. By 3 days post-hatch the eye is heavily pigmented, indicating development of the pigment epithelium. During the next two days, the eye becomes gold and then copper in color. It is this copper color which remains through the next weeks of larval development. The pigmentation appears complete by 5 days post-hatch. If a correlation exists between C-gene activity and retinal differentiation, then one would expect no C_4 appearance until after complete eye pigmentation because the photoreceptor cells are the last cells of the retina to differentiate. Not until 5 days post-hatch does the eye appear fully developed on the basis of external morphology.

The results of the electrophoretic analysis of the green sunfish LDH ontogeny are shown in Fig. 4. The whole embryo samples were collected from 3 to 8 days post-hatch and stored frozen until electrophoresis. There is a faint B_3C_1 band apparent on day 4. However, the C_4 isozyme synthesis is not detected until day 5, the time when the eye appears completely differentiated by external morphological criteria.

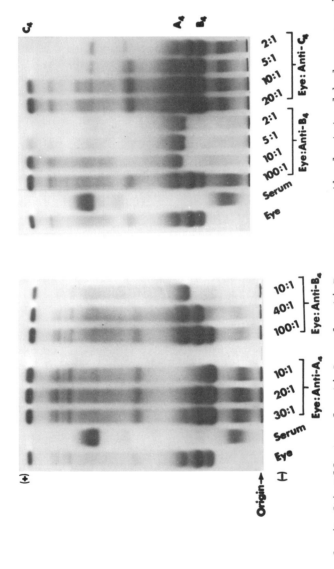

Fig. 2. (Left) Effects of anti-A_4 and anti-B_4 sera upon the lactate dehydrogenase isozymes of the green sunfish eye. Antisera were allowed to react with the eye extract for one-half hour prior to electrophoresis. Anti-A_4 serum precipitates only those isozymes containing A subunits (A_4 and A_2B_2); and anti-B_4 serum at low concentrations precipitates those containing B subunits.

(Right) Effects of anti-B_4 and anti-C_4 sera upon the lactate dehydrogenase isozymes of the green sunfish eye. Anti-B_4 serum precipitates the C_4 isozymes at high concentrations

Fig. 2 legend continued
showing immunochemical similarities between the B and C subunits. The anti-C_4 serum precipitates primarily those isozymes containing C subunits.

In addition to this cross, we were able to examine interspecific crosses in the genus *Lepomis*. A number of crosses
were made between the green sunfish ♀ and the bluegill ♂
(*Lepomis macrochirus*) (Fig. 4). The LDH isozymes of the adult
eye homogenate from each parent show nearly identical electrophoretic patterns. The LDH isozymes of the different developmental stages of hybrid progeny are very similar to those
of the green-green cross. On day 5 there is a faint B_3C_1
band, and the following day the C_4 was fully apparent. The
same results were obtained in the reciprocal cross.

Hybrids were also made between the green sunfish and the
warmouth, the red-ear, and the longear sunfish. In each cross,
development followed the same time course as in the green-green
cross, and the C_4 LDH appeared at 5 days post-hatch. Because
the C_4 isozymes of all these species have identical electrophoretic mobilities, nothing can be said about the activation
of one parental allele as compared to the other. However, no
gross perturbation occurs in the timing of the synthesis of
the retinal C_4 isozyme in the F_1 hybrid embryos. There is the
same apparent coupling of morphological and enzymatic differentiation as in the green sunfish embryos.

Morphological changes during retinal differentiation

After the time of C_4 appearance during development was
established, we investigated the morphology of the retina in
greater detail, first in the adult and then in the larval
fish. A schematic diagram of the retina shows the relationships between the innermost layer of ganglion cells and the
outermost layer of pigment epithelial cells (Fig. 5). Light
must pass through all the layers of the neural retina before
being detected by the photoreceptor cells. A diagram of a
single receptor cell (Fig. 5) demonstrates the relationships
of the ultrastructural components. The mitochondria-filled
ellipsoid region is of particular interest because this is
where the C_4 LDH was localized using light microscopy (Whitt
and Booth, 1970; Kunz, 1971). The highly folded membranes
of the outer segment contain the visual pigment (Hall et al,
1969) and must be present before retinal function is possible.

Both light and electron microscopic analyses of the adult
retina were carried out in order to determine the structure
of the fully differentiated retina. Dissected adult retinas

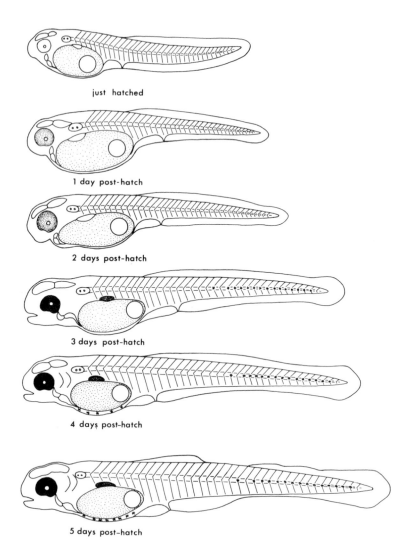

just hatched

1 day post-hatch

2 days post-hatch

3 days post-hatch

4 days post-hatch

5 days post-hatch

Fig. 3. Diagram of hatching and larval stages of green sunfish through 5 days post-hatch. At hatching, the fish is immature and no eye pigmentation is visible. Not until 5 days post-hatch does the eye gain the copper color maintained through the remaining stages of larval development.

were fixed in 2.5% glutaraldehyde, dehydrated through an ethanol series, and embedded in Epon. Thin sections were used for electron microscopic studies and thick sections of 1-2 μ were made for light microscopy.

The photoreceptor cells of the retina are of particular interest for several reasons. First, when the retina is undergoing cytodifferentiation, the innermost cell types (such as the ganglion cells) are the first to differentiate and the outermost cell types (such as the photoreceptors) are the last (Coulombre, 1965). Secondly, the outer segment is the last structure to be laid down; thus it serves as an index of the state of differentiation of the photoreceptor cell and consequently of the retina as a whole. And thirdly, it is in the inner segments, (predominantly the ellipsoid region), where the C_4 activity is localized. It is for these reasons that we focused the electron microscopic studies on the photoreceptor cell.

The ultrastructure of the outer segments and ellipsoid regions of twin cones is seen in an electron micrograph of the adult sunfish retina (Fig. 6). Twin cones are numerous in the sunfish retina and common to many species of teleosts (Lyall, 1957; Anctil, 1969; Wagner, 1972). Higher magnification (Fig. 6) reveals the tightly clustered mitochondria of the ellipsoid region in a radial pattern around the centriole. This concentration of mitochondria, in the same region where the retinal-specific LDH is localized (Whitt and Booth, 1970), is quite interesting because the sperm-specific LDH of birds and mammals has also been proposed to be in association with the mitochondria (Machado de Domenech et al, 1972).

After the morphology of the adult retina was determined, its structure was compared with those of the larval stages. Two stages of development were chosen, 5 days post-hatch (the time when C_4 isozyme synthesis is first detected) and 3 days post-hatch (the time prior to both C_4 synthesis and completion of external morphological differentiation).

There are a number of important differences between the structure of the retina at 3 days post-hatch and in the adult. The micrograph of the 3 day post-hatch retina reveals that the ellipsoid regions of the photoreceptor cells have reduced numbers of mitochondria (Fig. 7). Most photoreceptor cells do not possess outer segments at this stage. The outer segments which are present appear as small triangles. The nuclei are very much longer at this stage than those seen in the adult. Another feature characteristic of the 3 day post-hatch retina, and not seen in the adult, is the presence of lipid-like droplets in the pigment epithelium.

A number of changes have occurred by the fifth day post-hatch, that time when the C_4 isozyme is first detected (Fig. 7). Since the previous stage, there has been a dramatic increase in the number of photoreceptor cells with outer

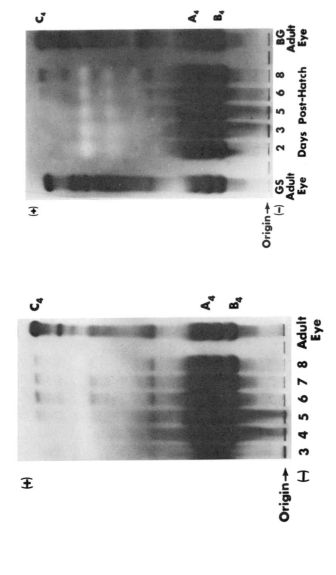

Fig. 4. (Left) Synthesis of the C_4 isozyme of lactate dehydrogenase during the development of the larval green sunfish. Whole larvae from days 3 to 8 were homogenized for electrophoresis. The right-hand slot shows the isozyme pattern of the adult eye. The B_3C_1 heterotetramer can be seen on day 4 post-hatch. On the fifth day post-hatch, the C_4 isozyme is fully apparent.

(Right) Synthesis of the C_4 isozyme of lactate dehydrogenase during the development of hybrid progeny from a BG ♂ x GS ♀ cross. Isozyme patterns of the eye extracts from each parent

Fig. 4 legend continued
are in the outside slots. On day 5 post-hatch there is a
faint B_3C_1 heteropolymer, and the C_4 isozyme is first detected
on day 6 post-hatch.

---

segments. Electron micrographs of the 5 day post-hatch retina
reveal a much closer arrangement of cells with larger outer
segments. The lipid droplets which were so prominent around
the outer segments of the earlier stage are almost entirely
absent by 5 days post-hatch. Furthermore, the receptor cells
now possess well-differentiated ellipsoids with tightly clustered mitochondria. In fact, cells of the central region of
the 5 day post-hatch retina appear very similar to those of
the adult retina.

Conclusion

The research described in this paper deals with the problem
of correlating morphological differentiation with specific
isozyme synthesis. It is shown that the retinal-specific LDH
appears on the fifth day after hatching, at the time when the
central retina is morphologically differentiated. Thus, it
appears that the first appearance of the C_4 LDH isozyme is
tightly coupled with the morphological (and probably functional) differentiation of the green sunfish retina.

Future research employing ultrastructural enzyme localization techniques will determine whether the C_4 LDH is specific
to certain receptor cell types and how closely it is associated
with the mitochondria of the ellipsoid. Furthermore, if the
retinal-specific LDH does play a role in the regeneration of
the visual pigment, there may be coordinate control of the
synthesis of C_4 LDH with other retinal-specific enzymes (e.g.,
retinol dehydrogenase) during differentiation.

Tissue-specific isozyme patterns in the adult organism
reflect the developmental changes which occur in the control
of gene expression. The presence of three homologous LDH
loci in the teleost provides a system for examining this
differential expression. The extent to which physical differences in the structure of the LDH-C locus play a role in
determining the stage and tissue specificity of its expression
will also be the subject of future investigations.

ACKNOWLEDGMENT

This research was supported by NSF grant GB 43995.

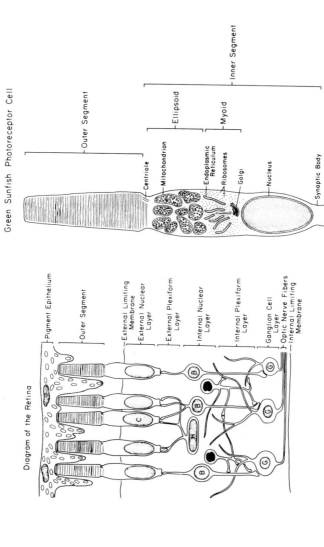

Green Sunfish Photoreceptor Cell

Diagram of the Retina

Fig. 5. (Left) Diagram of the retina of the adult green sunfish. Modified after Whitt et al. (1973). The cell layers are indicated on the right. Cell nuclei are labeled as follows: C: cone; H: horizontal; B: bipolar; A: amacrine; and G: ganglion. (Right) Diagram of a photoreceptor cell (cone) of the adult green sunfish. The ellipsoid region is densely packed with mitochondria.

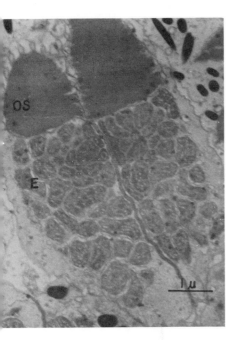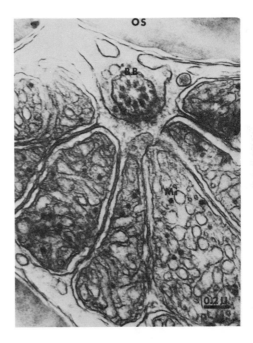

Fig. 6. (Left) Electron micrograph of a pair of twin cones from the adult green sunfish. Below the outer segment (OS) can be seen the dense concentration of mitochondria in the ellipsoid (E).
(Right) High magnification electron micrograph of a cone cell ellipsoid. The outer segment (OS) is at the top of the micrograph. The mitochondria (Mi) are arranged in a radial pattern around the basal body (BB) or centriole. LDH C_4 was localized in this region in a previous study (Whitt and Booth, 1970).

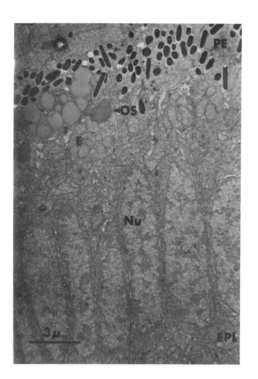 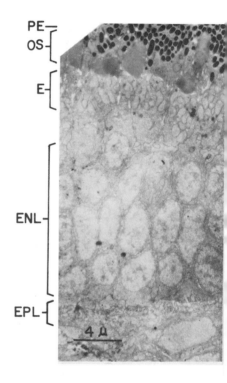

Fig. 7. (Left) Electron micrograph of the larval green sunfish at 3 days post-hatch. The structures are labeled as follows: PE, pigment epithelium; OS, outer segment; E, ellipsoid, Nu, nucleus; and EPL, external plexiform layer. Note the small number of outer segments, the presence of lipid-like droplets and the long nuclei.
(Right) Electron micrograph of the larval green sunfish at 5 days post-hatch. ENL refers to the external nuclear layer. Note the increased number of outer segments, the increased concentration of mitochondria, the lack of lipid-like droplets, and the smaller nuclear size.

REFERENCES

Anctil, M. 1969. Structure de la retine chez quelques
téléostéens marins du plateau continental. *J. Fish. Res.
B. Canada* 26: 597-628.

Childers, W. F. 1967. Hybridization of four species of sunfish. (Centrarchidae). *Ill. Nat'l Hist. Survey Bulletin*
29: 159-214.

Coulombre, A. J. 1965. The eye. In *Organogenesis* (R. L.
DeHaan and H. Ursprung, eds.). Holt, Rinehart, and
Winston, New York. pp. 219-251.

Fairbanks, M. B., J. R. Hoffert, and P. O. Fromm 1969. The
dependence of the oxygen-concentrating mechanism of the
teleost eye *(Salmo gairdneri)* on the enzyme carbonic
anhydrase. *J. Gen. Physiol.* 54: 203-211.

Goldberg, E., J. P. Cuerrier, and J. C. Ward 1969. Lactate
dehydrogenase ontogeny, paternal gene activation and
tetramer assembly in embryos of brook trout, lake trout
and their hybrids. *Biochem. Genet.* 2: 335-350.

Hall, M. O., D. Bok, and A. D. E. Bacharach 1969. Biosynthesis and assembly of the rod outer segment membrane system. Formation and fate of visual pigment in the frog
retina. *J. Mol. Biol.* 45: 397-406.

Hizeroth, H., J. Klose, S. Ohno, and U. Wolf 1968. Asynchronous activation of parental alleles at the tissue-specific
gene loci observed in the hybrid trout during early
development. *Biochem. Genet.* 1: 287-300.

Horowitz, J. J. and G. S. Whitt 1972. Evolution of a nervous
system specific lactate dehydrogenase isozyme in fish.
J. Exp. Zool. 180: 13-32.

Kunz, Y. 1971. Distribution of lactate dehydrogenase (and its
E-isozymes) in the developing and adult retina of the
guppy *(Lebistes reticulatus)*. *Rev. Suisse de Zool.*
78: 761-776.

Lyall, A. H. 1957. Cone arrangements in teleost retinae.
Q. J. Microsc. Sci. 98: 189-201.

Machado de Domenech, E., C. E. Domenech, A. Aoki, and A. Blanco
1972. Association of the testicular lactate dehydrogenase
isozyme with a special type of mitochondria. *Biol.
Reprod.* 6: 136-147.

Markert, C. L. and H. Ursprung 1971. *Developmental Genetics.*
Prentice-Hall, Englewood Cliffs, New Jersey. 214 pp.

Markert, C. L. and I. Faulhaber 1965. Lactate dehydrogenase
isozyme patterns of fish. *J. Exp. Zool.* 159: 319-332.

Nakano, E. and M Hasegawa 1971. Differentiation of the retina
and retinal lactate dehydrogenase isoenzymes in the teleost, *Oryzias latipes*. *Develop. Growth Differn.* 13:

337-351.

Nakano, E. and A. H. Whiteley 1965. Differentiation of multiple molecular forms of four dehydrogenases in the teleost, *Oryzias latipes*, studied by disc electrophoresis. *J. Exp. Zool.* 159: 167-180.

Salthe, S. M., O. P. Chilson, and N. O. Kaplan 1965. Hybridization of lactic dehydrogenases *in vivo* and *in vitro*. *Nature*, Lond. 207: 723-726.

Shaklee, J. B. 1972. A genetic, biochemical and evolutionary characterization of LDH isozyme structure in fish. Ph.D. Thesis, Yale University, New Haven, Connecticut.

Shaklee, J. B., K. L. Kepes, and G. S. Whitt 1973. Specialized lactate dehydrogenase isozymes: the molecular and genetic basis for the unique eye-and liver LDHs of teleost fish. *J. Exp. Zool.* 185: 217-240.

Wagner, H. J. 1972. Vergleichende Untersuchungen über das Muster der Sehzellen und Horizontalen in der Teleostier-Retina (Pisces). *Z. Morph. Tiere* 72: 77-130.

Wald, G. 1968. Molecular basis of visual excitation. *Science* 162: 230-239.

Whitt, G. S. 1968. Developmental genetics of lactate dehydrogenase isozymes unique to the eye and brain of teleosts. *Genetics* 60: 237.

Whitt, G. S. 1969. Homology of lactate dehydrogenase genes: E gene function in the teleost nervous system. *Science* 166: 1156-1158.

Whitt, G. S. 1970a. Developmental genetics of the lactate dehydrogenase isozymes of fish. *J. Exp. Zool.* 175: 1-36.

Whitt, G. S. 1970b. Directed assembly of polypeptides of the isozymes of lactate dehydrogenase. *Arch. Biochem. Biophys.* 138: 352-354.

Whitt, G. S. and G. M. Booth 1970. Localization of lactate dehydrogenase activity in the cells of fish *(Xiphophorus helleri)* eye. *J. Exp. Zool.* 174: 215-224.

Whitt, G. S., E. T. Miller, and J. B. Shaklee 1973. Developmental and biochemical genetics of lactate dehydrogenase isozymes in fish. In *Genetics and Mutagenesis of Fish* (J. H. Schroder, ed.). Springer-Verlag, New York. pp. 243-276.

GENETIC AND DEVELOPMENTAL ANALYSES OF LDH ISOZYMES IN TROUT

JAMES E. WRIGHT, J. ROBERT HECKMAN[1], AND LOUISA M. ATHERTON
Department of Biology
The Pennsylvania State University
University Park, Pennsylvania 16802

ABSTRACT. Tissue distributions and genetic and developmental aspects of three LDH isozyme systems specified by five loci in salmonids are described and a unifying system of their nomenclature is proposed. A fourth series of isozymes was shown to be an epigenetic modification of one autotetramer from the above.

Mutant alleles at the $Ldh-B_1$ and $-B_2$ loci found in different populations of rainbow trout (*Salmo gairdneri*) were used to determine that pseudolinkage of these duplicated loci occurs in males and females. Centric fusions under genetic control are proposed to account for the pseudolinkage. Independent assortment of the $Ldh-C$ from the $-B_1$ or $-B_2$ loci was found.

Synchronous activation of maternal and paternal alleles occurred at all LDH loci, but asynchronous activation of those loci associated in ontogeny with different developmental stages was found in rainbow trout. Practically synchronous activation of the B_1 and B_2 loci, which control the ubiquitous system of isozymes, normally occurred in species hybrids of rainbow with brown (*Salmo trutta*) or brook *Salvelinus fontinalis*) trout. Other loci exhibited both allelic and locus asynchrony in these hybrids.

Dual developmental patterns involving a mutant B'_2 allele from one strain of brook trout were shown to be genetically controlled and can be attributed to heterozygosity of a regulator locus.

INTRODUCTION

Genetically determined electrophoretic variants for protein types, including such enzymes as lactate dehydrogenase (LDH) which exhibit multiple molecular forms or isozymes, have proven useful in determining the nature of such proteins and the number of loci controlling them (Shaw, 1964). In the trout and salmon breeding program at the Pennsylvania Fish Commission's Benner Spring Fish Research Station, variant alleles at three of the five loci controlling

---

[1] Present address: Department of Biology, Elizabethtown College, Elizabethtown, Pennsylvania 17022.

three LDH systems in salmonids have been useful in differentiating populations of hatchery or wild accessions. They
have been particulary useful in maintaining the purity of
inbred lines which have become, or been made, homozygous for
these variants.

Availability of lines or strains of known genotypes
and phenotypes coming from controlled individual matings has
permitted use of the LDH variants in extensive family genetic
studies, particularly with *Salvelinus* species. Such studies
have been used in defining the complex LDH isozyme systems in
the salmonids. They have also been used in determining
normal and aberrant inheritance patterns, including linkage
relations and their cytogenetic basis, and in determining
that disomic rather than tetrasomic inheritance is operative
for these duplicated loci of this group of fishes believed to
have evolved through tetraploidization (Ohno, 1970).

In the present report results of studies of the above
genetic phenomena in rainbow trout (*Salmo gairdneri*) are
presented. In addition, ontogeny of LDH isozymes in developing embroyos of rainbows, of brook (*Salvelinus fontinalis*),
and brown (*Salmo trutta*) trout, and of hybrids of these
species with rainbow females was studied. Differential activation of loci and alleles and differentiation of isozyme
expression in organs and tissues are shown. Based on these,
and studies of others, we have adopted a unifying system of
nomenclature for LDH of salmonids.

MATERIALS AND METHODS

Populations of inbred lines and random-bred strains of
rainbow, brook, and brown trout, and accessions of lake trout
(*Salvelinus namaycush*), Atlantic salmon (*Salmo salar*), Ohrid
trout (*Salmo letnica*), coho salmon (*Oncorhynchus kisutch*), and
kokanee salmon (lacustrine form of sockeye, *O. nerka*) were
maintained at the Benner Spring hatchery. Wild type and
mutant LDH phenotypes and genotypes of the brook trout used
have been described by Morrison (1970) and by Wright and
Atherton (1970). Only preliminary accounts of the phenotypes
of the rainbow trout have been presented (Wright and Morrison,
1968 and Wright and Atherton, 1973), and these will be
fully described below. Crosses among lines or strains of
known LDH genotypes were made and F_1, F_2, and backcross generations studied. Letter designation of families used represent year of spawning, for example, O = fall of 1968, P'
= spring of 1970, T = fall of 1973. Incubation and growth
were in units receiving water at 10 ± 1 degrees C.

Electrophoresis was carried out on horizontal starch gels using EBT buffer of pH 8.6 according to procedures adapted from those of Markert and Faulhaber (1965) and described by Morrison (1970). Collection of eye humor and preparation of homogenates of other tissues have been described by Morrison (1970) and by Wright and Atherton (1970). Individual eggs or embryos were homogenized in 0.01 M tris-HCl buffer pH 7.5, centrifuged at 12,000 x g under refrigeration, and the supernatant subjected to electrophoresis.

NOMENCLATURE OF LDH SYSTEMS, SUBUNITS, AND LOCI IN SALMONIDS

Numerous investigators have studied the LDH isozymes of trout and other members of the family Salmonidae and have adopted various systems of nomenclature in their descriptions of the relatively complex zymogram patterns found in tissues of species of varied genotypes. In Table 1 we have compiled in relatively chronological order the reports of different groups to show that there has evolved general agreement on the nature of the isozymes but not on what to call them. The compilation reflects both electrophoretic mobility and the tissues in which the groups of isozymes are expressed. It reflects only indirectly the genetic, physico-chemical, molecular hybridization, and immunochemical studies from the cited work which show relationships and homologies of the isozyme systems to each other and to those of other vertebrates.

Recently Shaklee et al., (1973) have proposed a unifying system of nomenclature for fish LDH isozymes based on the extensive surveys of fish groups by Whitt and his colleagues (see Horowitz and Whitt, 1972) and by Markert and others of his associates. It is based on the finding that virtually all teleost fishes possess in common three loci specifying subunits A, B, and C. They point out that eye-band LDH of many groups of teleosts and the liver-band LDH of some gadoids and cyprinids show homologies to each other and possibly to the C-bands of sperm of birds and mammals. The authors further suggest that the duplicated A and B loci of salmonids and cyprinids be reflected by designating their products with superscript numbers. Basically, this leads to the subunit nomenclature system proposed on evolutionary grounds by the Ohno group, that is, C, B^1, B^2, A^1 and A^2 (Klose et al., 1968). Utter and Hodgins (1972) have adopted a variation of the latter scheme and have designated variant subunits with additional prime superscripts.

Taking the above for a framework plus the information contained in Figures 1, 2, 3, and 4 C and D as well as tissue distributions and homologies of LDH isozymes in salmonid

TABLE I

Comparison of Salmonid LDH Subunit and Isozyme System
Nomenclatures of Various Investigators

Eye (Anodal)	Digestive Tract	All Tissues	Skeletal Muscle (Most cathodal)	Investigator
n.d.	-	A B	- -	Goldberg, 1966
-	-	A-B-C (ABC system)	D E (DE system)	Hochachka, 1966
C (ABC system)	-	A B (AB system)	D E (DE system)	Morrison and Wright, 1966
C	-	B^1 B^2	A^1 A^2	Klose et al. 1968
-	system 1	system 2	system 3	Bouck and Ball, 1968
E',E (e group)	D',D (d group)	B' B (b group)	A' A (a group)	Massaro and Markert, 1968
E	Epigenetic (?)	B' B	A' A	Holmes and Markert, 1969
n.d.	-	A B (heart group)	n.d. n.d. (muscle group)	Bailey and Wilson, 1968
D^2,D^1	n.d.	B^2 B^1 C'(liver) C	A^2 A^1 E^2(gill)E^1	Williscroft and Tsuyuki, 1970
-	-	B'(liver)B''	-	Tsuyuki and Williscroft, 1973

TABLE 1, continued

n.d (e group)	n.d (d group)	B^2	B^1 (b group)	n.d	Utter & Hodgins, 1972

- = not reported

n.d = not designated

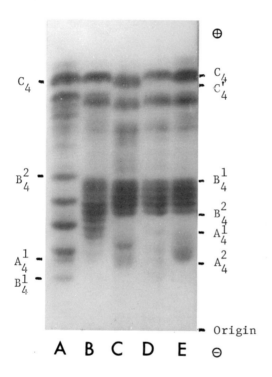

Fig. 1. Starch gel LDH zymograms of eye extracts of Salmonidae types. A = brook trout *(Salvelinus fontinalis)*; B = F_1 hybrid of rainbow x brook trout; C = rainbow trout *(Salmo gairdneri)*; D = kokanee salmon *(Oncorhynchus nerka)*; E = coho salmon *(O. kisutch)*.

species and hybrids studied by us and others, we propose the nomenclature system shown in Table 2. This scheme takes into account the genetic convention of designating duplicate gene loci with numbered subscripts; mutant alleles at those loci

TABLE 2

A unifying system of LDH nomenclature in Salmonids

LDH Systems	Tissues of greatest prominence	Homotetrameric Isozymes in order of decreasing Anodal mobility		Structural Gene Loci		Reported Mutant Alleles	
		Salmo	Salvelinus	Salmo	Salvelinus	Salmo	Salvelinus
Eye (group c)	Eye, nervous tissues	C_4	C_4	Ldh-C	-C	C'	C'
Ubiquitous (group b)	Liver, lens, red blood cells	B_4^1	B_4^2	Ldh-B_1	-B_2	B_1', B_1''	B_2', B_2''
"	Heart, active muscles	B_4^2	B_4^1	Ldh-B_2	-B_1	b_2	B_1', B_1'', B_1'''
Muscle (group a)	Red (active) muscles	A_4^1	A_4^1	Ldh-A_1	-A_1	–	–
"	White muscle	A_4^2	A_4^2	Ldh-A_2	-A_2	–	–

are then designated with superscript symbols. By using superscript numbers together with superscript symbols to designate
variant subunits it permits ready reference to what locus is
responsible for a given genetic variant zymotype. At the same
time conventional use of subscript numbers to represent combinations of subunits in homo- or heterotetramers is retained.

Thus there are three primary systems of LDH isozymes
formed from five subunits specified by five loci and forming
five homotetramers found in varying concentrations in the
different tissues. The three systems can be designated by
the indicated group designations of Massaro and Markert
(1968) or operationally on the basis of their primary tissue
distributions as shown.

Detailed descriptions of the nomenclature can be illustrated better when the rainbow trout zymograms in Figure 2 are
described below.

Reversed electrophoretic mobility of the homotetramers
involving B^1 and B^2 subunits in members of the genus *Salmo*
from those of *Salvelinus* are shown. All members of the
genus *Oncorhynchus* studied exhibit this *Salmo* pattern as
does *Coregonus clupeaformis*. *Thymallus articus*, and
probably *T. thymallus*, exhibits the mobility patterns of
Salvelinus. Assignment of the same electrophoretic mobility
patterns to homotetramers involving the A^1 and A^2 subunits
in *Salmo* and *Salvelinus* is somewhat equivocal. On developmental bases the two seem the same but our heat inactivation
and urea or guanidine hydrochloride inhibition experiments
indicate that they are reversed. No mutants for the latter
two loci have been found nor reported and the one fertile
species hybrid, splake, is of little use for a genetic
analysis since the muscle bands of the parental lake and
brook trout are of identical electrophoretic mobility. It
is of interest that *Thymallus articus* exhibits only one
muscle band *in vivo*.

The group of extra bands prominent in tissues of the
digestive tract of *Salmo* species, designated the d group by
Massaro and Markert (1968) and system 1 by Bouck and Ball
(1968), are an epigenetic modification of the B^1 homotetramer.
This was suggested by Holmes and Markert (1969)[4] on the basis
of immunological reaction to antiserum prepared against
weakfish B_4. More recently these bands have been designated
β by Massaro (1973) on these grounds. We have found
several pieces of evidence which lead to this conclusion.
In Figure 3 these bands in the pyloric caeca and intestine
of rainbow trout are seen to shift precisely with the electrophoretic shift in the variant B_4^1 tetramer. In Figure 4D
it is shown that by 90 days post fertilization the homo-

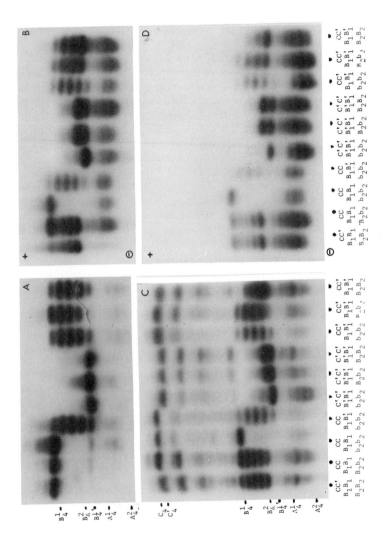

Legend for Figure 2: Comparative starch gel zymograms of
four tissues from representatives of 27 discernible genotypes
among the F_2 from rainbow trout parents heterozygous at the
Ldh-C, $-B_1$ and $-B_2$ loci (description in text). A = liver;
B = heart; C = eyes; D = pectoral muscle. Note the presence
of A^1-A^2 bands from muscle attachments of these tissues
in the yearling trout, and note contamination in slots 2
and 3 in A.

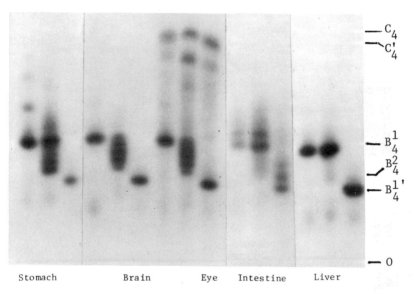

Fig. 3. Zymograms of three LDH types (slot 1 = $B_1B_1b_2b_2$;
slot 2 = $B_1B_1B_2B_2$; slot 3 = $B_1'B_1'b_2b_2$) in each of the labeled
tissues which permit distinction of heterotetramers involving
C with B^1 and B^2 subunits (eye, brain, stomach) and of
bands in digestive tissues and liver which represent epigenetic modification of the B_4^1 homotetramer.

tetramers and heterotetramers involving B^2 subunits begin to
disappear in the digestive tract; moreover, the bands anodal
to the B_4^1 tetramer begin to appear. The same phenomenon is
apparent with development of the liver shown in Figure 4C.
Long frozen storage of liver and other tissues, and particularly purified extracts of the liver B_4^1 tetramer, leads
to appearance of this epigenetic system. These observations
suggest digestive enzymatic modification of the B_4^1 tetramer.

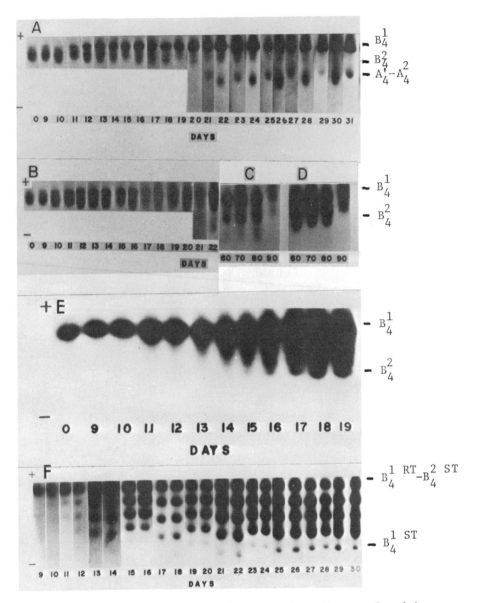

Fig. 4. Development of LDH isozymes in embryos of rainbow trout (RT) and of rainbow x brook trout (ST) hybrids at 10°C, in days after fertilization. A, B, C, and D = progeny from a rainbow female of genotype $B_1B_1B_2b_2$ x male rainbow of genotype $B_1B_1b_2b_2$. A = embryos of b_2b_2 genotype; B =

Legend for Fig. 4, continued: embryos of B_2b_2 genotype;
C = liver and D = digestive tract of B_2b_2 progeny exhibiting
disappearance of B^2 subunits and appearance of epigenetic
modification of B_4^1 by time of yolk sac absorption. E =
embryos from mating rainbow female genotype $B_1B_1b_2b_2$ x male
$B_1B_1B_2B_2$; F = embryos from same female rainbow x brook
trout $B_1B_1B_2B_2$ male.

Odense et al., (1969) were able to generate d group-type
bands by treatment of A_4 tetramers of the cod with pyloric
caeca extracts or with pepsin. A similar experiment has
been done in our laboratory. Plastic sponge bits were fed
to fasting rainbow trout and allowed to be trapped in the
duodenum for 24 hours. They were removed and squeezed into
rainbow heart and liver homogenates. After holding the mixture for two hours at 4°C no LDH bands were found after
electrophoresis of the treated heart extracts. In the
treated liver extracts a prominent B_4^1 band remained and the
atypical anodal bands (d group) had formed.

LDH PHENOTYPES AND GENOTYPES IN RAINBOW TROUT

Single genetic variants at three of the five LDH loci
have been uncovered among our rainbow trout populations.
Since gene dosage effects as well as mobility differences
are shown, it is possible to distinguish heterozygotes from
homozygotes. Therefore, 27 genotypes are distinguishable
from the zymotypes. Representative examples from fish of
an F_2 population which illustrate these are shown in zymograms
of four tissue extracts in Figure 2.
        The three genotypes involving the alleles at the B_2 locus
are shown in the first three slots in Figure 2. The
recessive null mutant (b_2) is found in relatively high frequency in numbers of fall spawning rainbow trout strains
maintained in eastern hatcheries as shown in Table 3. It
has been fixed in two of our inbred lines with no apparent
adverse effects. In the liver, Figure 2A, the genotypes cannot be distinguished since only the B_4^1 tetramer expresses
itself in this tissue. (The extra more cathodal bands are
contamination since they were not present in other runs).
In most tissues, including the eye, Figure 2C, the heterozygote exhibits a skewed distribution of 5 isozymes due
to dosage effects. In all tissues including the heart in
which normally the B^2 subunit predominates, the homozygote
appears as a single band in the ubiquitous (b group) system.
In Figure 2B the B_4^1 concentration in hearts of heterozygotes,

heart and pectoral muscle. However, as seen in Figure 2C, heterotetramers with C subunits, particularly dosage effects of $C_2B_2^2$ and $C_1B_3^2$ permit distinction.

A mutant C' allele occurred in relatively low frequency in all of the fall spawning eastern strains examined except a small sample from West Virginia (Table 3). As shown in Figure 2C the C_4' homotetramer shows such slightly less anodal mobility that the presumed 5 isozymes of the heterozygote appear as one wide band in the zymogram. The eye system (group c; formerly e) involves the tetramers of C or C' subunits and the heterotetramers formed between these and the B^2 and, with less affinity, the B^1 subunits. Because of the latter, the wild type eye pattern does not generally display *in vivo* the nine heterotetramers expected. However, examination shows that when the b_2 allele is homozygous all heterotetramers of C and B^1 or $B^{1'}$ subunits appear. It is possible to visualize the placement of all 15 tetramers in slot 2 into the expected groupings of 1-2-3-4- and 5, counting those of the ubiquitous B^1-B^2 system.

As shown in Figure 3, the C subunits also appear in the brain and stomach, as well as in a few other tissues containing nerves. Since fewer C subunits occur in these tissues, mostly heterotetramers involving C with B^1 or B^2 subunits are expected, and are found.

EXPERIMENTAL RESULTS

GENETIC ANALYSES

Numerous F_2 families from monohybrids for CC' or B_1B_1' have been classified and the three phenotypes fit well the 1:2:1 ratio expected for these autosomal codominant allelic pairs. Although b_2 acts as an autosomal recessive, apparent lack of complete dominance for the B_2 allele (dosage effect) permits progeny of crosses between heterozygotes to exhibit and to fit well the expected 1:2:1 ratios. Backcrosses of the heterozygote for any of the three loci to either mutant or wild type homozygotes produce progeny in expected 1:1 ratios.

Tests for joint segregation of the three allelic pairs required finding a late fall spawning strain 15 male of genotype $C'C'B_1B_1b_2b_2$ which was crossed to an early spawning Kamloops female of genotype $CCB_1'B_1'B_2B_2$ to produce a trihybird F_1. The resulting F_1 progeny showed such a spread in time of spawning that it has taken several years to find enough males or females ripe at the same time as those of homozygous

parental types to perform testcross matings. F_2 families
were obtained and the progeny proved useful in classification
of recombination types and particularly in determining that
the b_2 mutant was a null allele and not one involving electrophoretic mobility.

Testcross results for the B_1 and B_2 loci are shown in
Table 4. Three of the six males tested produced progeny in
non-random combinations with non-parental classes in signif-
cant excess - pseudolinkage. One of these males, used as a
precocious yearling (family P132), was testcrossed again at
four years of age and produced progeny (family S56) of
strikingly similar types, proportions, and parental class
percentage. The other three males produced progeny in frequencies exhibiting random assortment. Three of four families
from testcross matings of heterozygous females showed
random assortment. However, in family T174 the types and
frequencies of progeny fit the precepts of classical linkage;
over 98 percent of the female parent's gametes were of
parental types. Family T179 was made to test not only for
the joint segregation pattern but also to detect any intragenic
complementation in specifying the multimeric LDH protein. No
complementation occurred since no wild type LDH isozyme
appeared from this cross involving b_2 from stain 15 and the
b_2 from the Paradise strain male used. Thus the two mutants
from different sources are functionally allelic, if not
identical.

Independent assortment of the B_1 and C loci was shown
in all males and females tested. Among 1783 testcross
progeny from eight families, 889 or 49.9% were parental
types. Similarly, independent assortment was shown by the
B_2 and C loci; 493 of 938 testcross progeny (52.5%) were
parental types. The latter results are consistent with those
in the preliminary report of Wright and Morrison (1968).

DEVELOPMENTAL STUDIES

The persistence and the disappearance of maternal LDH
isozymes stored in the egg at oogenesis and the time of
activation of the five loci, at least in terms of tetramer
formation, was determined from the cross of rainbow female
genotype B_2b_2CC x male genotype $b_2b_2C'C'$(both were $A_1A_1A_2A_2B_1$
B_1). The results are shown in Figures 4A and B. From before
fertilization through day 9 of development the most concentrated isozymes were $B_1^1B_3^2$ and $B_2^1B_2^2$ heterotetramers. At day
10 intensification of the B_4^1 band and the heterotetramers

389

TABLE 4

Variation for pseudolinkage of Ldh -B_1 and -B_2 loci among

testcross families in rainbow trout

Family	Parental Genotypes ♀	♂	Progeny Genotypes $B_1B_1B_2B_2$	$B_1B_1B_2b_2$	$B_1B_1B_2B_2$	$B_1B_1B_2b_2$	Percent Parentals
P132 †	$B_1B_1B_2B_2$	B'_1B_2/B_1b_2	102	82	80	106	43.8*
P129	"	"	21	21	23	30	46.3
P'2	"	"	28	23	22	20	46.2
P'7	"	"	77	14	21	67	19.1*
Q114	"	"	32	35	34	28	53.5
Q116	"	"	58	28	32	59	33.9*
S56 †	"	"	84	62	66	81	43.7*
R22	B'_1B_2/B_1b_2	$B_1B_1B_2B_2$	29	30	26	19	53.8
S53	"	"	82	62	66	67	46.2
T174	"	"	--	80	84	3	98.2*

TABLE 4, continued

Family	Parental	Genotypes	$\begin{array}{c}B_1B_1B_2B_2\\B_1B_1B_2b_2\end{array}$	$\begin{array}{c}B_1B_1B_2b_2\\B_1B_1b_2b_2\end{array}$	$\begin{array}{c}B_1B_1B_2B_2\\B_1'B_1B_2b_2\end{array}$	$\begin{array}{c}B_1B_1B_2b_2\\B_1'B_1b_2b_2\end{array}$	Percent Parentals
T179	$B_1'B_2/B_1b_2$	$B_1B_1b_2b_2$	17	18	12	21	44.1

†   Same male parent

*   Significant deviations from independent assortment

involving B^2 subunits was obvious in the half of the embryos of genotype B_2b_2. In these the intensification continued until by day 15 the mature reversed pyramid pattern of this genotype was formed (Figure 4B). In the embryos of b_2b_2 genotype intensification of the B_4^1 tetramer continued; however, heterotetramers involving maternal B^2 subunits began disappearing shortly after day 10 but persisted until day 30, the time of hatching, before completely disappearing (Fig. 4A).

By day 18 tetramers involving A^1 subunits, and probably A^2 subunits, appeared and these reached a pronounced level at day 22, correlated closely with onset of melanophore formation. By day 60 the adult muscle pattern was obvious. At day 31, shortly after hatching, both the C and C' subunits of maternal and paternal origin, respectively, were equally activated.

Reciprocal crosses involving all possible genotypes at the B_1, B_2, and C loci were made to confirm and to extend the above observations. The redistribution of isozyme bands during development of different tissues, particularly of wild type embryos, was determined. In eye tissue at day 60 the B_4^1 isozyme was predominant in the ubiquitous (b) system and thus C-B^1 heterotetramers predominated in the eye system as they do through maturity. At day 90 the ubiquitous system showed the adult or binomial pattern occurring in wild type. The liver and digestive tract exhibited the most drastic change, as stated earlier - from a substantial amount of the B_4^2 tetramer at day 80 to none by day 90 (Figures 4C and D). This differentiation is correlated with the complete absorbance of the yolk sac and the begining of feeding by the fry. The gill showed relatively the same amount of B_4^2 isozyme at all stages examined but the B_4^1 isozyme continuously increased through day 90. Conversely, in heart and muscle tissue the B_4^2 isozyme increased, in comparison to B_4^1, through day 90.

A typical developmental pattern in B_2b_2 embryos from female genotype b_2b_2 x male genotype B_2B_2 is shown in Figure 4E. Only B_4^1 isozyme had been stored in the egg and at day 10 heterotetramers involving B^2 subunits appeared, becoming quite obvious by day 11.

In Table 5 are recorded the time of appearance of tetramers formed from subunits produced by each of the loci of each of the species rainbow, brown, and brook trout. They were roughly the same in embryos of intra- and interspecific crosses, at least for the products of the paternal allele at each locus. This information was obtained from that recorded above for rainbows, from intrastrain crosses of brown trout (where no mutants are known), from interstrain crosses among brook trout possessing different alleles at the B_2 locus, and from crosses of each of these to rainbow females of genotype b_2b_2.

TABLE 5

Days (after fertilization) of appearance of newly synthesized isozymes involving LDH subunits in rainbow, brown and brook trout embryos incubated at $10°C$

Species	B^1	B^2	C	A^1-A^2	Initiation of hatching
Rainbow	10	10	31	18	30
Brown	<12	<12	40	26	39
Brook	10	≤13	45	28	42

Identical, or practically identical, electrophoretic mobilities of various homo- and heterotetramers, the kinds and concentrations of tetramers stored in the eggs, and the somewhat smudged patterns sometimes obtained when homogenizing whole eggs made early developmental patterns extremely difficult to evaluate. Death of embryos at varying stages in development in the limited number of eggs collected daily compounded the difficulty. Thus, the time of appearance of B^1 tetramers and of heterotetramers involving brown B^2 subunits with B^1 subunits were not discernible unequivocally until day 12 although it is possible that they appeared earlier. Time of first involvement of subunits specified by alleles at the B_2 locus of brook trout was found to vary from possibly as early as day 10 to as late as day 22.

In Figure 4F is shown the developmental pattern in eggs and embryos of the cross of rainbow female genotype b_2b_2 with a brook of the wild type genotype B_2B_2. At day 10 the heterotetramers combining three B^1 subunits from the rainbow with one B^1 subunit of brook origin was visible. Although a double-banded pattern at the most anodal position seemed to be present at 10 days, it was not until day 13 that this was clearly apparent. This would indicate activity of the B^2 subunit from the brook trout. Other crosses of the rainbow b_2b_2 female with brook trout of geno-

type $B_2'B_2'$ provided information about the temporal activation
of the different brook mutant alleles. The $B^{2''}$ subunit
is more negatively charged and the $B^{2'}$ still more negatively
charged so that heterotetramers involving them with rainbow
B^1 subunits are distinguishable. Again, these heterotetramers
became clearly visible by day 13 in both hybrid genotypes.

Embryos of interstrain crosses of brook trout females of
wild type B_2B_2 (Inbreds 2 or 9) with males of genotype
B_2B_2' , B_2B_2' and $B_2'B_2'$ (Inbred 0) produced tetramers that
were different in mobility and/or concentration from those
stored in the egg. These were the tetramers expected from
involvement of various subunits produced by the B_2 locus,
but they were not unequivocally discernible until day 13.

Three brook trout females of genotype $B_2'B_2'$ from strain
22, of West Virginia origin, were used in crosses to Inbred
2 males of genotype B_2B_2. Dual patterns of LDH ontogeny,
in approximately 1:1 ratios, appeared among the embryos of
two crosses. These are shown under A and B in Figure 5.

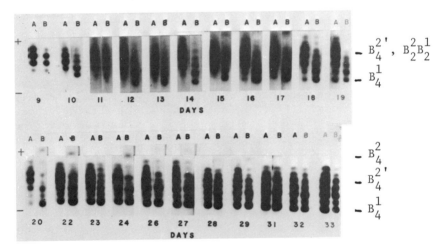

Fig. 5. Development of normal (A) and exceptional (B) isozyme
patterns in embryos from a brook trout female genotype
$B_1B_1B_2'B_2'$ x male of genotype $B_1B_1B_2B_2$ family. Late activation of the B' allele accounts for the abnormal patterns
shown under B (explanation in text).

A third cross produced embryos with the developmental
pattern typical of that found in other brook trout of genotype B_2B_2', and all like that under A of Figure 5. The
abnormal pattern shown (B in Figure 5) occurred suddenly at

day 10 when directed assembly of B_4^1 tetramers and those
tetramers involving B^1 and $B^{2'}$ subunits was apparent. At
the same time $B_4^{2'}$ tetramers stored in the eggs had begun
disappearing. Tetramers involving subunits produced by
the B_2 wild type allele appeared possibly as early as day
10 and clearly by day 12 in both types; the B_4^2 tetramer
was evident by day 20. In those embryos showing more
normal patterns the tetramers involving the three different
subunits were gradually added to form the fusiform, nine-
banded pattern shown.

Crosses of strain 22 males of genotype $B_2'B_2'$ to wild type
strain 2 females and to rainbow females of genotype b_2b_2
confirmed that two different activation patterns of B_2' alleles
were involved and that maternal effects were not present in
the abnormal pattern of early selective assembly of isozymes.
In 46 of 84 embryos examined from days 14 to 22 heterotetramers involving $B^{2'}$ subunits were present (normal); these
were absent in the other 38. Begining with day 23 zymograms of all embryos exhibited the heterotetramers involving
$B^{2'}$ subunits; however, half (22) were very intensely stained
while the remainder (20) gradually developed the same
formazan banding intensity.

DISCUSSION OF RESULTS

The finding that non-random assortment of Ldh-B_1 and
Ldh-B_2 loci occurred from half the males tested, while
independent assortment of these loci was exhibited by other
males of identical dihybrid parental genotypes, shows that
a form of pseudolinkage is involved. Particularly is this
so since non-parental classes were in excess in that half
of the testcross families. Similar reasoning leads to
the conclusion that pseudolinkage was exhibited by the testcross progeny of one female (practically all parental types)
while three other females exhibited independent assortment.
Pseudolinkage implies that a structural aberration of
chromosomes is involved and Davisson et al., (1973) have
provided evidence in *Salvelinus* that these are Robertsonian
translocations, or centric fusions of acro- or telocentric
chromosomes. From cytological studies by these and other
investigators it was pointed out that this form of chromosomal
aberration is a widespread feature of evolving karyotypes
in various species of salmonids.

Thus centric fusions would account for the data in rainbow trout. Ohno et al., (1965) reported intra- and inter-
individual polymorphism for the number of metacentric chromo-

somes in various tissues of male and female rainbow trout. They showed the metacentric number to range from 39 to 46 giving karyotypes of 2n = 58 to 65 with a constant arm number of 104. Meiotic multivalents were also found. Similarly, 2n numbers of 58 and 62 were found in cultured leucocytes of rainbow trout from the same basic Benner Spring stock as those used in the present study (Heckman et al., 1971). Pseudolinkage was found only among males in the *Salvelinus* material previously studied. Also Aspinwall (1974) found non-random joint segregation for duplicated MDH loci from pink salmon males which could be interpreted as pseudolink- age. Therefore, the pseudolinkage found in family T179 from a female dihybrid was surprising. The genetic data suggest that centric fusion does go on in females of rain- bow trout and this agrees with the cytological findings of Ohno et al., (1965). That centric fusion may be influenced by a genetic factor is suggested by the 1:1 ratio among males for patterns of joint transmission.

The absence of pseudolinkage between the Ldh-C and either the B_1 or B_2 loci is of interest. Studies by Massaro and Markert (1968) and the immunochemical homologies established by Holmes and Markert (1969) led these authors to conclude that the C gene was derived from the same ancestral B locus as the B_1 and B_2 duplicate loci. The same conclusion was made for other fishes by Whitt (1969). Indeed affinity of their subunits for forming eye system heterotetramers plus similarity for physico-chemical properties found in inhibit- ion studies (Morrison, unpublished) show that the C gene is more closely related to the B_2 than to the B_1. Thus, while one might have expected the chromosomes carrying them to be similar enough to undergo centric fusions, apparently they are not. Structural changes or rearrangement of centromeres on acrocentrics carrying them, or location of the C locus on a chromosome already of metacentric derivation could account for this. Alternatively, perhaps the dihybrid for these loci has not been constructed in an appropriate residual genotype for the particular centric fusions to occur.

Tissue ontogenesis of LDH isozymes has been studied in a wide range of vertebrates (see Markert and Ursprung, 1971) and our study provides a further illustration of changes that occur in isozyme distribution during cellular and tissue differentiation. Different tissues have particular patterns of isozymes which changed during embryonic development until the configuration for particular genotypes in adults was reached. The duplications that have led from three loci in most vertebrates to five in the salmonids provide a multiplicity of isozymes which can be synthesized or degraded depending on the molecular environment of the cells and tissues where they appear to function best.

After arising from duplication of a single B gene the genes B_1 and B_2 and their subunit products have themselves evolved to produce the b group ubiquitous isozyme system. Its isozyme pattern changes from that in the embryo to give differential tissue distributions which are strikingly similar to those involving the A-B system in many other vertebrates. Yet a homozygote for a null allele at the B_2 locus can exist because the subunits specified by the B_1 locus can be synthesized even in those tissues where the isozymes involving B^2 subunits predominate and seem to function best. However, one might suspect that a homozygote for a null allele at the B_1 locus might be lethal in view of the degradation of isozymes containing B^2 subunits in the liver or digestive tract during development. Or, would the C locus which also evolved from the ancestral B locus become active as it is in liver of some other fishes? Mutants, and particularly null mutants, at the A_1 and A_2 loci would be of real interest in studying the highly restricted muscle system.

Unique genotypes, particularly the b_2b_2 of rainbow trout which store only the B_4^1 tetramer in the egg, were valuable in the developmental studies. They permitted reexamination and extension of the work of others as to time of locus and allele activation in terms of the appearance of isozymes of the three different systems. Questions were raised about the basis for observed asynchronous activation of maternal and paternal alleles of some LDH loci in species and their hybrids of *Salmo* (Hitzeroth et al., 1968 and Klose et al., 1969). Goldberg et al., (1969) examined similar aspects of development in species and hybrids of *Salvelinus* and suggested that an odd case of non-random assembly of tetramers may be under genetic control. In the meantime Ballard (1973) has described the normal embryonic stages of rainbow and of brook trout when developing at 10°C, the incubation temperature used in our studies. Comparison of degree days required for development to given stages in the much colder temperatures used by the other investigators reveals no major disagreements between their and our work with some of the same species.

There is little real evidence from our data for asynchronous activation of maternal and paternal alleles within the three species for any of the five loci except for different genetic sources of the B_2' allele in brook trout. Other variation was found for activation of the brook trout B_2

Hitzeroth, H., J. Klose, S. Ohno, and U. Wolf 1968. Asynchronous activation of parental alleles at the tissue-specific gene loci observed on hybrid trout during early development. *Biochem. Genet.* 1: 287-300.

Hochachka, P.W. 1966. Lactate dehydrogenases in poikilotherms: definition of a complex isozyme system. *Comp. Biochem. Physiol.* 18: 261-269.

Holmes, R.S. and C.L. Markert 1969. Immunochemical homologies among subunits of trout lactate dehydrogenase isozymes. *Proc. Natl. Acad. Sci. U.S.A.* 64: 205-210.

Horowitz, J.J. and G.S. Whitt 1972. Evolution of a nervous system specific lactate dehydrogenase isozyme in fish. *J. Exp. Zool.* 180: 13-31.

Kaplan, N.O. 1964. Lactate dehydrogenase-structure and function. *Brookhaven Symp. Biol.* 17: 131-153.

Klose, J., U. Wolf, H. Hitzeroth, H. Ritter, N.B. Atkin, and S. Ohno 1968. Duplication of the LDH gene loci by polyploidization in the fish order *Clupeiformes*. *Humangenetik* 5: 190-196.

Klose, J., J. Hitzeroth, H. Ritter, E. Schmidt, and U. Wolf 1969. Persistence of maternal isoenzyme patterns of the lactate dehydrogenase and phosphoglucomutase system during early development of hybrid trout. *Biochem. Genet.* 3: 91-97.

Markert, C.L. and I. Faulhaber 1965. Lactate dehydrogenase isozyme patterns of fish. *J. Exp. Zool.* 159: 319-332.

Markert, C.L. and H. Ursprung 1971. *Developmental Genetics*. Prentice-Hall, Englewood Cliffs, New Jersey.

Massaro, E.J. 1973. Tissue distribution and properties of the lactate and supernatant malate dehydrogenase isozymes of the grayling, *Thymallus arcticus*. *J. Exp. Zool.* 186: 151-158.

Massaro, E.J. and C.L. Markert 1968. Isozyme patterns of salmonid fishes: evidence for multiple cistrons for lactate dehydrogenase polypeptides. *J. Exp. Zool.* 168: 223-238.

Morrison, W.J. 1970. Non-random segregation of two lactate dehydrogenase subunit loci in trout. *Trans. Amer. Fish. Soc.* 99: 193-206.

Morrison, W.J. and J.E. Wright 1966. Genetic analysis of three lactate dehydrogenase isozyme systems in trout: evidence for linkage of genes coding subunits A and B. *J. Exp. Zool.* 163: 259-270.

Odense, P.H., T.C. Leung, T.M. Allen, and E. Parker 1969. Multiple forms of lactate dehydrogenase in the cod, *Gadus morhua*. *Biochem. Genet.* 3: 317-334.

Ohno, S. 1970. *Evolution by Gene Duplication.* Springer-Verlag. New York.

Ohno, S., S. Stenius, E. Faisst, and M.T. Zenzes 1965. Post-zygotic chromosomal rearrangements in rainbow trout. *Cytogenetics* 4: 117-129.

Shaklee, J.B. K.L. Kepes, and G.S. Whitt 1973. Specialized lactate dehydrogenase isozymes: the molecular and genetic basis of the unique eye and liver LDHs of teleost fishes. *J. Exp. Zool.* 185: 217-240.

Shaw, C.R. 1964. The use of genetic variation in the analysis of isozyme structure. *Brookhaven Symp. Biol.* 17: 117-129.

Tsuyuki, H. and S.N. Williscroft 1973. The pH activity relations of two LDH homotetramers from trout liver and their physiological significance. *J. Fish. Res. Bd. Canada* 30: 1023-1026.

Utter, F.M. and H.O. Hodgins 1972. Biochemical genetic variation at six loci in four stocks of rainbow trout. *Trans. Amer. Fish. Soc.* 101: 494-502.

Whitt, G.S. 1969. Homology of lactate dehydrogenase genes. E gene function in the teleost nervous system. *Science* 166: 1156-1158.

Williscroft, S.N. and H. Tsuyuki 1970. Lactate dehydrogenase systems of rainbow trout - evidence for polymorphism in liver and additional subunits in gills. *J. Fish. Res. Bd. Canada* 27: 1563-1567.

Wright, J.E. and W.J. Morrison 1968. Genetic analysis of LDH isozymes in trout. *Isozyme Bull.* 1: 13.

Wright, J.E. and L.M Atherton 1970. Polymorphisms for LDH and transferrin loci in brook trout populations. *Trans. Amer. Fish. Soc.* 99: 179-192.

Wright, J.E. and L.M. Atherton 1973. *Genetics 74* (supplement): s300 (Abstract).

GLUCOSE-6-PHOSPHATE DEHYDROGENASE IN
BROOK, LAKE AND SPLAKE TROUT:
AN ISOZYMIC AND DEVELOPMENTAL STUDY[1]

T. YAMAUCHI[2] and E. GOLDBERG
Department of Biological Sciences
Northwestern University
Evanston, Illinois 60201

ABSTRACT. There are five G-6-P-specific G-6-PD isozymes in
both brook and lake trout. The most anodal isozyme in both
species has the same electrophoretic mobility, but the lake
trout pattern is more widely spaced on polyacrylamide gels.
In hybrids, i.e., splake trout, nine forms of G-6-PD can
be resolved. These results can best be explained by tetrameric molecules and two different subunit types. G-6-PD
isozymes in these trout are the products of two codominant
autosomal gene loci. G-6-PD activity profiles and isozyme
patterns were examined in developing splake trout. In all
crosses examined, the G-6-PD level of trout embryos dropped
immediately upon fertilization and then steadily increased
after gastrulation. In the F_1 cross Bk ♀ x Lk ♂ the maternal
isozyme pattern persisted until after completion of yolk
resorption. In F_2 crosses, all embryos at hatching revealed
either a brook or lake G-6-PD pattern. Thus, stable products from the oocyte (splake pattern) were no longer present, and the preferential expression of one parental gene
(probably maternal) was demonstrated. This asynchrony in
gene expression reflects incompatibility between the maternal protein synthesizing machinery and the paternal genome.
Perhaps compatibility requires recognizing more of the
paternal genome than just the structural gene loci.

INTRODUCTION

In most vertebrate systems there are two isozymic forms of
glucose-6-phosphate dehydrogenase (G-6-PD) which are encoded
in two distinct gene loci. One is highly though not absolutely
specific for glucose-6-phosphate (G-6-P) and NADP. This

[1] This research was supported in part by United States Public
Health Predoctoral Fellowship 4 FO1 GM 43657-03.

[2] Present address: Department of Biology, The University of
Texas System Cancer Center, M. D. Anderson Hospital and
Tumor Institute, 6723 Bertner, Houston, Texas 77025.

G-6-P-specific isozyme is localized primarily in the cytoplasmic fraction of the cell (Newburgh and Cheldelin, 1956). The
other isozyme is a microsomal enzyme and is characterized by
a broad substrate specificity, utilizing galactose-6-phosphate
(Gal-6-P) and 2-deoxyglucose-6-phosphate, as well as G-6-P.
Either NADP or NAD can serve as coenzyme for this isozyme,
which has been designated by Shaw (1966) and by Ohno et al,
(1966) as hexose-6-phosphate dehydrogenase (H-6-PD) to distinguish it from G-6-PD.

In the salmonids, studies on glucose-6-phosphate dehydrogenase have been limited primarily to the rainbow trout *(Salmo
gairdneri)* (Ohno et al, 1966; Kamada and Hori, 1970; Shatton
et al, 1971). Although both G-6-PD and H-6-PD have been detected in trout, the number of isozymes remains uncertain.
In brook trout *(Salvelinus fontinalis)*, lake trout *(Salvelinus
namaycush)*, and their viable hybrid, the splake trout, we
have an excellent system for studying the heterogeneity of
glucose-6-phosphate dehydrogenase and the relationship of
gene products. Stegeman et al, (1974) discuss the heterogeneity of H-6-PD and the immunological relatedness of H-6-PD
and G-6-PD. This report will examine the molecular basis for
the G-6-PD isozymes and the expression of G-6-PD during
development.

The developmental progression of isozymes provides a sensitive method for observing the expression of genes. If a
mechanism is operating in the early embryo which prevents gene
expression, only the maternal form, present in the egg cytoplasm, will be detected. When gene expression is no longer
inhibited, heterozygous individuals should express an enzyme
phenotype having both parental forms plus possible hybrid
forms of the enzyme(Shaw, 1964).

In the brook-lake-splake trout system, the parental types
have distinguishable G-6-PD patterns, and the hybrid pattern
appears to be the result of recombination of parental gene
products. An ontogenetic examination of the G-6-PD isozyme
patterns of splake trout from reciprocal crosses provides
an excellent opportunity to study the time of gene expression.

MATERIALS AND METHODS

Trout were obtained from hatcheries and lakes in Banff and
Jasper National Parks, Alberta, Canada. For the developmental
studies, samples of embryos were collected each week and
stored frozen. Specimens were shipped by air to Evanston,
Illinois, where live fish were normally maintained at 5°C.

The preparation of tissue samples and embryos for enzyme
assay was described by Yamauchi and Goldberg (1973, 1974).

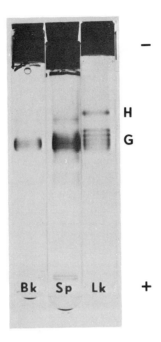

Fig. 1. Comparative zymograms of glucose-6-phosphate dehydrogenase in brook (Bk), lake (Lk), and splake (Sp) trout livers.
The sample was applied to the top of the gel, and the proteins
migrated toward the anode. The slower migrating region of
activity represents H-6-PD activity (H) and the faster region
represents G-6-P-specific G-6-PD activity (G).

G-6-PD isozyme patterns were resolved by electrophoresis on
5% polyacrylamide disc gels. The staining procedures used
were reported by Stegeman and Goldberg (1971).

RESULTS

G-6-PD Isozyme Patterns

The typical G-6-PD zymograms for brook, lake, and splake
trout are shown in Fig. 1. There were five G-6-P-specific
G-6-PD isozymes in both brook and lake trout (G region). The
most anodal band in each species had the same electrophoretic
mobility but the five isozymes of lake trout were more widely
spaced. In splake trout nine forms of G-6-PD could be resolved.
Although no variant pattern was observed for brook trout, two

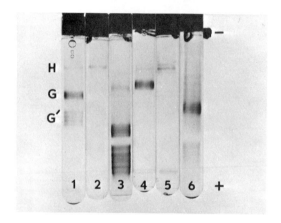

Fig. 2. Zymograms of glucose-6-phosphate dehydrogenase in brook trout liver. Homogenates were prepared in two buffers, one with NADP,gels (1-3) and the other with glycerol,gels (4-6). Gels were subjected to the standard electrophoretic run (50 min) and were stained with (1) G-6-P, (2) Gal-6-P, (3) G-6-P, (4) G-6-P, (5) Gal-6-P, (6) G-6-P. Gels 3 and 6 were subjected to electrophoresis for an additional 30 min.

additional patterns were observed in less than 10% of the lake trout examined. One variant pattern resembled that described for brook trout and the other resembled the F_1 splake trout pattern.

For brook and splake trout G-6-PD zymograms, an additional G-6-P-specific G-6-PD region (G') was occasionally resolved (Fig. 2). The number of bands in G' was usually four or five, but varied from zero to six. In the absence of NADP in the homogenizing buffer G' was barely detectable at best. The substitution of glycerol for NADP greatly stabilized G-6-PD activity but not G' stability. When G and G' regions were cut out of a gel, homogenized, and rerun on separate polyacrylamide gels, all the bands appeared in both gels (Fig. 3).

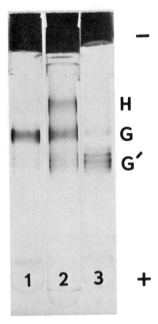

Fig. 3. Electrophoretic comparison of G and G' regions. A
brook trout liver homogenate was subjected to electrophoresis.
The G and G' regions were excised, and each was homogenized
and run separately on polyacrylamide gels. The patterns are:
(1) G region, (2) original homogenate, and (3) G' region.

Developmental Studies

The G-6-PD activity profiles are presented in Fig. 4
(Bk ♀ x Bk ♂ , Bk ♀ x Sp ♂ , and Sp ♀ x Bk ♂) and Fig. 5
(Bk ♀ x Lk ♂ and Lk ♀ x Bk ♂). Several trends were common
to most of the crosses examined. From fertilization to the
end of gastrulation (week 4), the level of G-6-PD activity
decreased slightly. By the completion of eye pigmentation
(week 8), G-6-PD activity increased to a level slightly
higher than that at fertilization. The level of enzyme
gradually increased until a week or two before hatching,when
G-6-PD activity virtually doubled. For all the crosses studied,
hatching occurred between weeks 12 and 13 of development. After
hatching the level of enzyme remained unchanged until commence-

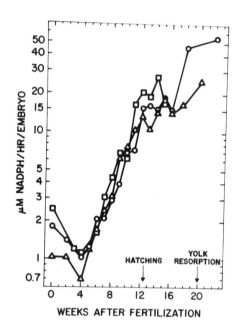

Fig. 4. Amount of G-6-PD in Bk ♀ x Bk ♂ (●-●), Bk ♀ x Sp ♂ (▲-▲), and Sp ♀ x Bk ♂ (■-■) crosses during development at 5°C. Each point before week 16 represents the mean of assays on 20-24 specimens. Each point after week 16 represents the mean of assays on three specimens.

ment of feeding (week 15) when the level again increased.

Examination of Bk ♀ x Lk ♂ embryos revealed that the maternal G-6-PD isozyme pattern persisted beyond hatching. In fact, the hybrid pattern did not appear until week 21 of development, a week after apparent completion of yolk resorption (Fig. 5). Unfortunately, due to a shortage of embryos the onset of paternal gene expression could not be identified in the reciprocal cross, Lk ♀ x Bk ♂.

From F_1 splake trout breeders two Sp ♀ x Sp ♂ crosses were obtained. After hatching, the alevins were maintained at 10° rather than 5°C. The higher temperature led to an increased

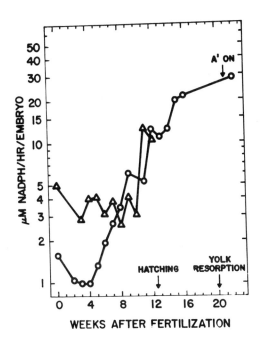

WEEKS AFTER FERTILIZATION

Fig. 5. Amount of G-6-PD in reciprocal hybrids of splake trout during development at 5°C. Each point before week 16 represents the mean of assays on 20-24 specimens. After week 16, each point represents the mean of assays on three specimens. Arrow indicates time at which expression of the paternal gene for G-6-PD can be detected electrophoretically. Symbols represent: Bk ♀ x Lk ♂ ,●—● ; Lk ♀ x Bk ♂ ,▲—▲.

rate of development such that completion of yolk resorption occurred by week 18. At hatching, only brook and lake trout isozyme patterns were observed in a 1:1 ratio. The hybrid pattern was not detected until week 16. Of the 24 fry examined during week 20 of development the observed frequency for the three phenotypes was 6 Bk: 14 Sp: 4 Lk, which closely approximates the 1 Bk: 2 Sp: 1 Lk ratio expected from random combination of gene products (Fig. 6).

DISCUSSION

G-6-PD Model

The proposed model for G-6-PD in the brook-lake-splake trout system is presented in Fig. 7. As there were no sex-

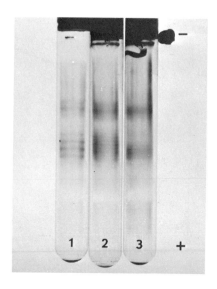

Fig. 6. Comparative zymograms of the G-6-PD phenotypes present in an F_2 splake trout population. Gels were subjected to an extended electrophoretic run of 80 min. The types of patterns are: (1) lake trout, (2) splake trout, and (3) brook trout.

related differences in isozyme patterns, G-6-PD isozymes in these trout are most likely the products of two codominant autosomal gene loci. The five-band pattern observed in the parental stocks, brook and lake trout, can best be explained by tetrameric molecules and two different subunit types. We have designated the least anodally migrating isozyme as the A_4 and the most anodally migrating isozyme as the B_4 isozyme. There is a difference in the net charge of the A subunit in brook and lake trout, as expressed by the difference in electrophoretic mobility of A_4 homotetramers. We arbitrarily assigned brook trout the A subunit and lake trout the A' subunit. The B subunits from brook and lake trout seem to have the same net charge; therefore, with respect to electrophoretic techniques the B_4 isozymes appear to be the same.

Consequently, in the splake trout four subunits are involved--A, A', and two B's--that cannot be distinguished electrophoretically and that yield 15 isozymes. That only nine are detected is a consequence of coincident electrophor-

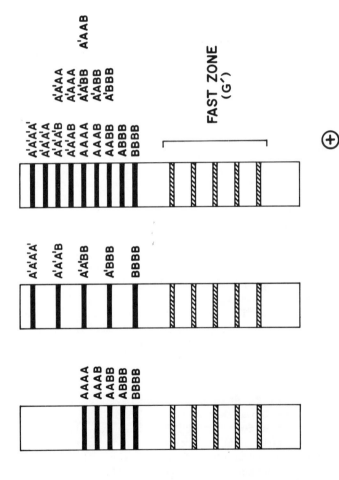

Fig. 7. Diagrammatic representation of G-6-PD isozymes showing the proposed subunit composition of (left to right) brook, lake, and splake trout. Note that the slowest anodally migrating isozyme for splake trout has the same mobility as the slowest isozyme in lake trout. Also, the most anodally migrating isozymes of all three species have the same electrophoretic mobility.

etic mobility of some of the possible subunit combinations as diagrammed in Fig. 7.

This genetic model is supported by two lines of evidence. The estimated molecular weights for the A and B subunits were approximately one fourth the molecular weight of the active molecule, thus suggesting an active tetrameric enzyme (Yamauchi, 1972). Also, an examination of several F_2 splake trout populations yielded a 1 Bk: 2 Sp: 1 Lk ratio of G-6-PD phenotypes, as would be expected from a one gene difference between the two parental types.

As noted earlier, a close relationship between the G and G' regions was evident. Further experiments were designed to examine the differences and similarities of these regions (Yamauchi, 1972). Electrophoresis on a polyacrylamide gel having a linear gradient pore size revealed that G-6-PD isozymes had a higher molecular weight than G' bands. Furthermore, during purification all attempts to separate and isolate these regions proved unsuccessful. But G6PD and G' regions were found to have subunits of similar molecular weight by SDS-gel electrophoresis. And finally, antibodies prepared against G-6-PD isozymes completely eliminated both G-6-PD and G' patterns.

The data therefore suggest that G' consists of active dissociation products of the G-6-PD isozymes. The inconsistency in the number of resolvable bands in G' indicates that a tetramer-dimer relationship is complicated by conformational and possibly ligand-induced (NADP) products. NADP plays an essential role in maintaining the quaternary structure of human erythrocyte G-6-PD (Bonsignore et al, 1969 and 1970; Yoshida and Hoagland, 1970). The apparent requirement for NADP in the expression of G' bands may indicate an equilibrium between the active "dimer" and the inactive monomer.

Asynchronous Expression of G-6-PD

At hatching, all F_2 splake embryos revealed either a brook or lake trout G-6-PD pattern. One-half of the F_2 embryos should be homozygous for one or the other allele and would be expected to have only either brook or lake patterns. The other half of these embryos should be heterozygotes, which is confirmed by the eventual expression of the hybrid pattern in one half of the F_2 population. Consequently at hatching, stable products from the oocyte (splake pattern) were no longer present, and the preferential expression of one parental gene (probably maternal) was evident.

In some interspecific crosses, the paternal gene contributes simultaneously with the maternal allele (Ohno et al,

1968) or may never be expressed (Castro-Sierra and Ohno, 1968; Whitt et al, 1972). More often though, expression of the paternally derived allele is delayed, as in the splake trout for LDH (Goldberg et al, 1969) and G-6-PD. In the F_2 hybrids, appearance of the splake trout pattern was observed several weeks after hatching, and in the F_1 hybrids, the pattern was further delayed until after yolk resorption. It would be tempting to postulate the occurrence of asynchronous gene activation. However, the appearance of a hybrid pattern merely reflects the end product of gene activation, and a delay in gene expression may occur at any one or more steps in the enzyme synthesizing process - from transcription to enzyme activation. Thus, although our data strongly suggest delayed expression of the paternal gene, asynchronous activation cannot be conclusively demonstrated.

The level of G-6-PD activity correlates well with the subordinate role of carbohydrate as an energy source (Yamauchi and Goldberg, 1974). Perhaps delayed gene expression may be related to either the commencement of feeding or the depletion of yolk. The former may be discounted since the splake pattern appeared in F_1 fry at the same time, whether fry had been fed or starved (Yamauchi, 1972). However, egg cytoplasm may play a role in gene output (Hitzeroth et al, 1968). In studying the brook-lake-splake trout system, Goldberg et al, (1969) proposed a cytoplasmic-mediated modulation of LDH which persisted until completion of yolk resorption. Such a mechanism may also explain the appearance of the splake pattern during this stage of development.

Since the paternal gene is expressed much later in F_1 than in F_2 splake trout, the asynchrony is gene expression probably reflects incompatibility between the maternal protein synthesizing machinery and the paternal genome. Perhaps compatibility involves recognizing more of the paternal genome than just the structural gene loci. Studies with intra-specific heterozygotes for G-6-PD such as those found in lake trout may shed more light on the complexity of the recognition mechanism(s).

ACKNOWLEDGMENTS

We thank Jean-Paul Cuerrier, J. C. Ward, and Burt Kooyman of the Canadian Wildlife Service for generously supplying the fish used in these studies.

REFERENCES

Bonsignore, A., I. Lorenzoni, R. Cancedda, and A. De Flora
    1970. Distinctive patterns of NADP binding to dimeric

and tetrameric glucose 6-phosphate dehydrogenase from human red cells. *Biochem. Biophys. Res. Comm.* 39: 142-148.

Bonsignore, A., I. Lorenzoni, R. Cancedda, A. Morelli, F. Giuliano, and A. De Flora 1969. Metabolism of human erythrocyte glucose-6-phosphate dehydrogenase. V. Exchange between free and apoenzyme-bound NADP. *Ital. J. Biochem.* 18: 439-450.

Castro-Sierra, E. and S. Ohno 1968. Allelic inhibition at the autosomally inherited gene locus for liver alcohol dehydrogenase in chicken-quail hybrids. *Biochem. Genet.* 1: 323-335.

Goldberg, E., J.-P. Cuerrier, and J. C. Ward 1969. Lactate dehydrogenase ontogeny, paternal gene activation, and tetramer assembly in embryos of brook trout, lake trout, and their hybrids. *Biochem. Genet.* 2: 335-350.

Hitzeroth, H., J. Klose, S. Ohno, and U. Wolf 1968. Asynchronous activation of parental alleles at the tissue-specific gene loci observed on hybrid trout during early development. *Biochem. Genet.* 1: 287-300.

Kamada, T. and S. H. Hori 1970. A phylogenic study of animal glucose-6-phosphate dehydrogenases. *Japan. J. Genet.* 45: 319-339.

Newburgh, R. W. and V. H. Cheldelin 1956. The intracellular distribution of pentose cycle activity in rabbit kidney and liver. *J. Biol. Chem.* 218: 89-96.

Ohno, S., H. W. Payne, M. Morrison, and E. Beutler 1966. Hexose-6-phosphate dehydrogenase found in human liver. *Science* 153: 1015-1016.

Ohno, S., C. Stenios, L. C. Christian, and C. Harris 1968. Synchronous activation of both parental alleles at the 6-PGD locus of Japanese quail embryos. *Biochem. Genet.* 2: 197-204.

Shatton, J. B., J. E. Halver, and S. Weinhouse 1971. Glucose (hexose 6-phosphate) dehydrogenase in liver of rainbow trout. *J. Biol. Chem.* 246: 4878-4885.

Shaw, C. R. 1964. The use of genetic variants in the analysis of isozyme structure. *Brookhaven Symp. Biol.* 17: 117-129.

Shaw, C. R. 1966. Glucose-6-phosphate dehydrogenase: Homologous molecules in deermouse and man. *Science* 153: 1013-1015.

Stegeman, J. J. and E. Goldberg 1971. Distibution and characterization of hexose 6-phosphate dehydrogenase in trout. *Biochem. Genet.* 5: 579-589.

Stegeman, J. J., T. Yamauchi, and E. Goldberg 1974. These Proceedings.

Whitt, G. S., P. L. Cho, and W. F. Childers 1972. Preferential inhibition of allelic isozyme synthesis in an

interspecific sunfish hybrid. *J. Exp. Zool.* 179: 271-282.
Yamauchi, T. 1972. Studies of glucose-6-phosphate dehydrogenase from brook, lake, and splake trout. Ph.D. Dissertation, Northwestern Univeristy, Evanston, Illinois.
Yamauchi, T. and E. Goldberg 1973. Glucose-6-phosphate dehydrogenase from brook, lake and splake trout: an isozymic
and immunological study. *Biochem. Genet.* 10: 121-134.
Yamauchi, T. and E. Goldberg 1974. Asynchronous expression of
glucose-6-phosphate dehydrogenase in splake trout embryos.
Devel. Biol. 39: 63-68.
Yoshida, A. and V. D. Hoagland, Jr. 1970. Active molecular
unit and NADP content of human glucose-6-phosphate dehydrogenase. *Biochem. Biophys. Res. Comm.* 40: 1167-1172.

DEVELOPMENTAL GENETICS OF TELEOST ISOZYMES

Michael J. Champion
James B. Shaklee
Gregory S. Whitt

Department of Zoology
University of Illinois
Urbana, Illinois 61801

ABSTRACT. Starch gel electrophoresis was utilized to investigate biochemical aspects of ontogeny in two distantly related teleost fishes, the lake chubsucker *(Erimyzon sucetta)* and the green sunfish *(Lepomis cyanellus)*. Isozyme patterns of embryonic stages and of various adult tissues were determined for over 15 different enzymes (representing more than 30 genetic loci).

Although exhibiting some specific differences, the ontogenetic patterns of changing isozymes in both species were generally similar and allow several general conclusions regarding the programming of gene activity during teleost development. Several of the enzymes studied were present at high levels in the earliest embryonic stages (due to maternal synthesis) but declined during subsequent development (e.g., amylase and some esterases of both species; lactate dehydrogenase B_4 and glucose-6-phosphate dehydrogenase of the lake chubsucker; adenylate kinase of the green sunfish). Many more enzymes were absent during the initial stages of ontogeny and only appeared later (e.g., glucosephosphate isomerase B_2, lactate dehydrogenase C_4, phosphoglucomutase, mannosephosphate isomerase, and hexosediphosphatase). Furthermore, apparently embryo-specific isozymes were observed in both species; the most dramatic being the highly anodal, stage-specific malate dehydrogenase present at the time of hatching in both species. Comparisons of the adult tissue-specific isozyme patterns of several enzymes with developmental changes in anatomy, physiology, and isozyme repertory suggest that biochemical differentiation is tightly coupled to cytodifferentiation and the attainment of physiological function.

INTRODUCTION

Cells of developing and adult organisms have been shown to contain, with few exceptions, a complete repertory of genet potentialities, and yet within any one cell only a small fraction of the genome is expressed at any one time. This knowled has lead to the characterization of the processes of cellular

417

differentiation which lead to the molecular heterogeneity of cells in terms of differential gene expression (Markert, 1965). Isozymes (Markert and Møller, 1959) provide powerful tools for the realization of the molecular heterogeneity of cells and thus can be used as sensitive probes of gene expression. Moreover, the examination of isozyme changes during development permits an analysis of both temporal and spatial patterns of gene activity (Markert and Ursprung, 1962, Masters and Holmes, 1972). In teleost fishes isozymes have been utilized in the examination of asynchronous allelic expression (Ohno, 1969, Yamauchi and Goldberg, 1973, Neyfakh et al., 1973, Whitt et al., 1972, 1973a), changes in levels of metabolism (Boulekbache et al., 1970), and in the correlation of specific isozyme synthesis and cellular differentiation (Whitt et al., 1973b, Miller and Whitt, 1974). The present study describes the changes in the expression of a wide variety of isozymes during the early development of two teleosts, the lake chubsucker and the green sunfish.

PROCEDURES

Ripe male and female lake chubsuckers *(Erimyzon sucetta)*, and green sunfish *(Lepomis cyanellus)* were collected in central Illinois and used in the production of normal, intraspecific crosses. Table 1 shows the time schedule and principal features of early development for both of these fishes. At numerous stages during development pooled samples of embryos were removed and later used for electrophoretic analysis. Previously published procedures were used for homogenization, centrifugation (Shaklee et al., 1974), starch gel electrophoresis (Shaklee et al., 1973), histochemical staining (Shaw and Prasad, 1970), and production and utilization of specific antisera (Shaklee et al., 1973)

RESULTS AND DISCUSSION

CHANGES IN SPECIFIC ISOZYMES DURING THE DEVELOPMENT OF THE LAKE CHUBSUCKER

Figure 1 shows the tissue distribution of LDH isozymes in the adult lake chubsucker. The use of specific antisera directed against teleost LDH-A_4 and -B_4 homotetramers permitted us to determine the subunit composition of these bands. Thus, there is an A_4 homotetramer predominant in skeletal muscle, and a series of bands containing only B subunits seen predominantly in heart muscle and in brain. This array of LDH isozymes containing different B subunits was observed in all individuals examined, and appears to be the result of the activity

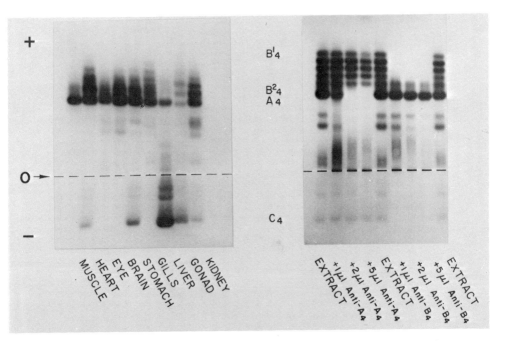

Figure 1. (left) Lactate dehydrogenase isozymes in the tissues of the adult lake chubsucker. (right) Analysis of lake chubsucker LDH isozymes through the use of specific antisera. Antisera produced against teleost LDH-A and LDH-B subunits were mixed in indicated amounts with 100μl of embryo extract before electrophoresis to precipitate isozymes containing immunochemically similar subunits.

---

of two separate LDH B loci which arose through duplication. The lake chubsucker possesses not only the usual LDH-A and-B subunits found in other vertebrates, but in addition, also has a subunit which is highly restricted it its tissue expression, specifically to the liver. Tissue-specific LDH isozymes have been reported to be found in the eye of most orders of teleosts and in the liver of some other orders. Recently it has been demonstrated that the same locus (LDH-C) encodes for both of these isozymes, and that during evolution this locus has come to be expressed in either the liver or the eye in these orders of teleosts (Shaklee et al., 1973, Whitt et al., 1973b, 1974). The ontogenetic expression of the eye-specific LDH-C gene and the coupling between this expression and the cellular and ultrastructural differentiation of the retina have been previously examined (Whitt, 1970a, Nakano and Hasegawa, 1971, Miller and Whitt, 1974). The present study examines the ontogenetic expression of the liver-specific LDH isozyme and contrasts it with the ontogeny of the retinal-specific LDH.

TABLE 1

TIME COURSE AND PRINCIPAL FEATURES OF EARLY DEVELOPMENT

Sample Number	Lake Chubsucker Hours After Fertilization[a]	Developmental Characteristics	Green Sunfish Hours After Fertilization[b]	Developmental Characteristics
0	0	unfertilized egg	0	unfertilized egg
1	5	morula	2	eight-cell stage
2	17	3/4 epiboly	5	morula
3	29	yolk plug	7	blastula
4	41	optic vesicles	8	1/3 epiboly
5	53	muscle contractions	11	yolk plug
6	64	dorsal-ventral finfold	14	embryonic shield
7	77	heart beating	15	neural groove
8	88	blood flow in heart	19	first somites
9	101	prehatch	21	12 pairs of somites
10	112	posthatch	25	head-tail buds
11	125	eyes pigmented	28	muscle movements
12	137	eyes silvery	31	heart beating
13	149	rhythmic jaw movements	38	prehatch

	a (21°C)		b (24°C)	
14	160	gall bladder	42	posthatch
15	173	pigmented air bladder	54	
16	185	active swimming	66	
17	196	air bladder inflated	78	
18	209		90	eye pigmentation
19	232		102	active swimming
20	256		114	air bladder
21	280	one week posthatch	126	
22	305		138	
23	329	yolk virtually gone	162	
24	353		186	
25	377		210	one week posthatch
26	401		234	
27	424	two weeks posthatch	258	
28	594	three weeks posthatch	282	ten days posthatch

[a] at 21°C
[b] at 24°C

LDH isozyme changes during the ontogeny of the lake chub-sucker are shown in Figure 2. In the unfertilized egg and early stages of development only LDH-B subunits are present.

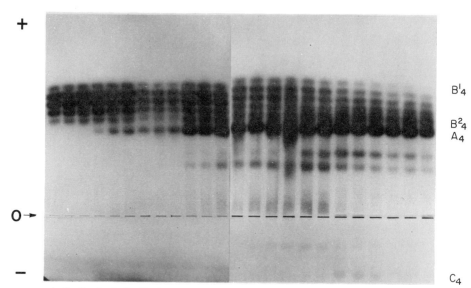

Figure 2. Lactate dehydrogenase isozyme changes during the early development of the lake chubsucker. Numbers under each slot are sample numbers. Table 1 shows the hours after fertilization and developmental features for each sample.

The normal binomial distribution of these five bands indicates that both LDH-B loci were equally and simultaneously expressed in the same cells during oogenesis. LDH-A subunits appear abruptly 40 hours after fertilization when the developing embryo has 25 pairs of somites and is exhibiting its first muscular contractions. LDH-A subunits may be essential for muscle cell differentiation and in the physiological functioning of muscle contraction.

LDH-C subunits are synthesized just prior to hatching. Initially only heteropolymers containing C subunits were detected, whereas the C_4 homotetramer was not detected until a week after hatching. The progressive appearance of C subunits first in heteropolymers and then later in the homopolymer is similar to the ontogenetic expression of the eye-specific LDH-C subunits (Miller and Whitt, 1974). The time of LDH-C gene expression coincides with the time of morphological development of the liver.

Malate dehydrogenase isozymes of teleosts have been shown to be dimeric molecules which exist in both supernatant and mitochondrial forms (Bailey et al., 1970, Whitt, 1970b, Wheat et al.,1971, 1972). The ontogenetic expression of MDH isozymes during lake chubsucker development is shown in Figure 3.

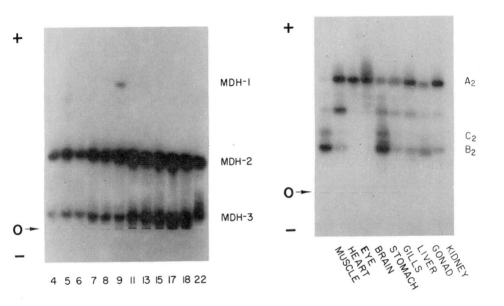

Figure 3. (left) Malate dehydrogenase isozymes during the development of the lake chubsucker. MDH-2 is the supernatant form; MDH-3 is the mitochondrial. (right) Glucosephosphate isomerase isozymes in tissues of the adult lake chubsucker.

Both MDH-2, the supernatant form, and MDH-3, the mitochondrial form, are found in nearly all tissues of the adult fish (Table 2). During development MDH-2 and MDH-3 both show a coordinate and gradual increase in activity from the unfertilized egg up to the time of hatching, and then their levels remain constant. This parallelism in pattern of expression between the supernatant and mitochondrial forms of MDH has also been observed during chick development, and suggests a coordinate regulation of these two enzymes (Greenfield and Boell, 1970). The MDH-1 band was not detected in any adult tissues and is present only for a short period just prior to hatching. The embryo-specific isozyme may be the product of a stage-specific epigenetic modification of existing MDH enzymes, or may be encoded in a third MDH locus which exhibits a very temporally restricted expression.

TABLE 2

ADULT TISSUE DISTRIBUTION OF ENZYMES AND ISOZYMES OF THE LAKE CHUBSUCKER

RELATIVE QUANTITY IN ADULT TISSUES

ENZYME	ISOZYME	muscle	heart	eye	brain	stomach	gills	liver	spleen	gonad	kidney
glucosephosphate isomerase (EC 5.3.1.9)	GPI-A	+	+++	+++	+++		+	++	+++	+	+++
	GPI-C	++				++	+	+	+	+	+
	GPI-B	+++				+++	++	++	++	++	++
lactate dehydrogenase (EC 1.1.1.27)	LDH-B	+	+++	+++	++	+++	+++	++	++	++	+++
	LDH-A	+++	+++	+++	+++	+++	+++	+++	++	++	++
	LDH-C	+	+	+	+	++	++	+++	++	++	+
phosphoglucomutase (EC 2.7.5.1)	PGM-1	++	+		+	++	+	++	+	+	++
	PGM-3	++			++			+++			
mannosephosphate isomerase (EC 5.3.1.8)	MPI	++	+	+++	+	+	++	+++	+	+	++
amylase (EC 3.2.1.1.)	AMY		++	+	+	+++	++	+++		++	+
glucose-6-phosphate dehydrogenase (EC 1.1.1.49)	G6PD-2	+	+		++		++	++		+++	+++
phosphogluconate dehydrogenase (EC 1.1.1.44)	PGD		+					+	+	+	
hexosediphosphatase (EC 3.1.3.11)	HDP	+	+	+	++		+	+++	+		+++
isocitrate dehydrogenase (EC 1.1.1.42)	IDH			+				+++			+
malate dehydrogenase (EC 1.1.1.37)	MDH-1		+++	++	++	++		+++			+++
	MDH-2	++	+++	+	++	++	++	+++	+	+	+++
	MDH-3		+	+	++	++	+			+	++
aspartate aminotransferase (EC 2.6.1.1)	AAT-2						+				+
	AAT-4		++			++		+++			++
glutamate dehydrogenase (EC 1.4.1.2)	GDH			+++	+++		++				
creatine kinase (EC 2.7.3.2)	CK-1		++	+++	+++		++				
	CK-2			++	++		+				
	CK-3			+++	++		++				
	CK-6			+	+		+				
	CK-9	+++	++								
esterase (EC 3.1.1.-)	EST-7,-6					++		++	+		++
	EST-8					+++	+++	+++	+++		++
	EST-9					+		+	+		+
alkaline phosphatase (EC 3.1.3.11)	AKP						+++	++	++	+	++
alcohol dehydrogenase (EC 1.1.1.1)	ADH					+	+++	+++			++

424

Glucosephosphate isomerase isozymes are dimeric enzymes which are encoded by three loci in the lake chubsucker (Shaklee et al., 1974) As shown in Figure 3, GPI-B_2 homodimers predominate in skeletal muscle and stomach, -A_2 homodimers are found mainly in liver, eye, and kidney, whereas the isozymes containing C subunits are present in only relatively low levels in adult tissues. The changes in the GPI isozyme pattern during development are shown in Figure 4. During the early stages

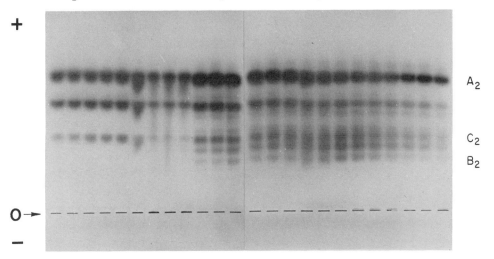

$+$

A_2

C_2

B_2

$O \rightarrow$

$-$

1 2 3 4 5 6 7 8 9 10 11 12 13 15 16 18 19 20 22 23 24 26 27 28

Figure 4. Glucosephosphate isomerase isozyme changes during the development of the lake chubsucker.

of ontogeny the GPI isozyme pattern differs from that observed in any adult tissues. Only A and C subunits are expressed (producing A_2, AC, and C_2 dimers). At the time of hatching B subunits are first detected. The appearance of B_2 homodimers concurrently with BC and AB heterodimers indicates that freshly synthesized B subunits are accessible for association with newly synthesized A and C subunits, and that in many cells of the developing embryo all three loci are simultaneously expressed.

Creatine kinase, in the tissues of the lake chubsucker, exists in a multiplicity of isozyme forms (Figure 5). Although the exact molecular and genetic basis for each of the lake chubsucker CK bands has yet to be determined, the subunit composition of several isozymes can be identified from their relative tissue distribution and electrophoretic behavior. Thus, CK-1, found in eye and brain, CK-6, found predominantly in muscle, and CK-9, distributed in stomach and kidney, appear to be

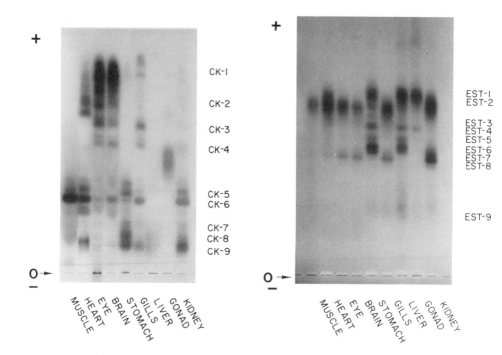

Figure 5. (left) Creatine kinase isozymes of the adult lake chubsucker, CK-1 is the brain predominant homodimer, CK-6 is the muscle predominant homodimer, and CK-9 is the stomach predominant homodimer. (right) Esterase isozymes in the tissues of the adult lake chubsucker.

homodimers composed of subunits homologous to, respectively, the brain-type, muscle-type, and stomach-type subunits seen in other teleosts (Scholl and Eppenberger, 1972). During development a number of shifts in CK isozyme expression are seen (Figure 6). Levels of CK-1 increase gradually during early stages of development and then rise sharply at the time of hatching. During this period extensive differentiation of the eye and brain is taking place. The CK-6 dimer which predominates in adult skeletal muscle initially appears 40-50 hours after fertilization, at the time the embryo is exhibiting its first muscle contractions. This muscle predominant CK isozyme has recently been shown to be intimately associated with the M-line of chick skeletal muscle and its appearance is closely coupled to myoblast fusion (Turner et al., 1974). A similar correlation between enzymatic and morphological differentiation is observed for the increase in levels of CK-9. This CK isozyme,

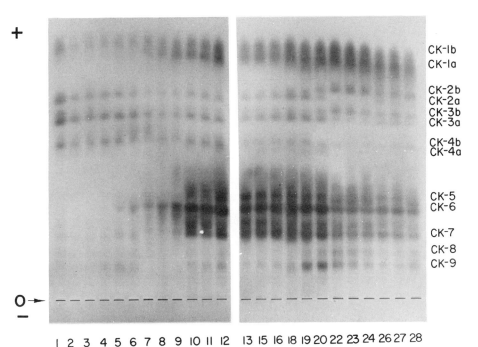

Figure 6. Creatine kinase isozyme changes during the development of the lake chubsucker.

predominant in adult stomach tissues, first appears during the cellular and functional differentiation of the digestive system.

Esterases are a complex family of hydrolytic enzymes which show a striking multiplicity of forms within many species, and which exhibit dramatically different patterns of tissue specificity between species (Masters and Holmes, 1972). The esterase isozyme pattern of the adult lake chubsucker (Figure 5) is similar to that of many other vertebrates. However, the developmental pattern for these esterases (Figure 7) is different from their ontogenesis in other vertebrates, and in particular other teleosts (Holmes and Whitt, 1970). During early development, through three weeks after hatching, only four of the nine adult esterase isozymes are detected (EST-6-7-8-9). The initially high levels of EST-8 and EST-9 activity in the early stages of development gradually decrease toward hatching. The level of EST-8 activity sharply increases 149-256 hours after fertilization. Between 256-305 hours the levels of EST-8 decrease, while levels of EST-6 and EST-7 sharply increase. However, 48 hours later a reciprocal shift is seen with the levels of EST-8 increasing while the levels of EST-6 and EST-7 decrease. These

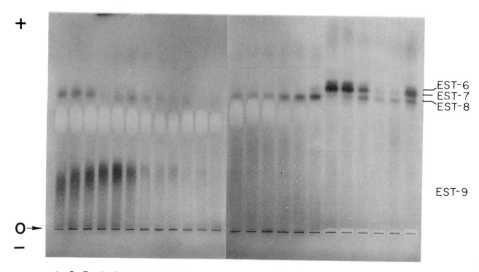

Figure 7. Esterase isozyme changes during the development of the lake chubsucker.

---

reciprocal shifts in esterase isozyme patterns during development imply changes in the rates of both enzyme synthesis and degradation which require the finely modulated control of gene expression.

OTHER ENZYME SYSTEMS OF THE LAKE CHUBSUCKER

Table 2 displays in a semi-quantitative manner the relative tissue distribution of those enzyme systems which have been previously discussed, as well as a number of other enzymes of the lake chubsucker. The ontogenetic expression of these enzymes and their relative changes in activity during development are schematically portrayed in Figure 8. A more detailed and extensive analysis of these changes has been presented elsewhere (Shaklee et al., 1974).

CHANGES IN ISOZYME SYSTEMS DURING
THE DEVELOPMENT OF THE GREEN SUNFISH

It is of interest to compare the changes in isozymes seen during lake chubsucker development to the ontogenetic changes observed for the green sunfish. In Table 3 are listed a number of enzymes of the green sunfish (primarily those also examined in the lake chubsucker), and their relative tissue distribution.

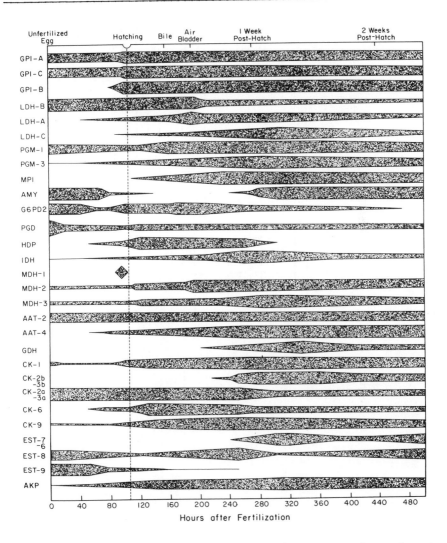

Figure 8. Lake chubsucker enzymes and isozymes and a schematic summary of their changes in activity during early development. The width of each channel represents the relative amount of activity with respect to the maximum observed for each enzyme. Activities were for the most part, determined by visual estimation of the staining intensity of bands on the gel.

TABLE 3

ADULT TISSUE DISTRIBUTION OF ENZYMES AND ISOZYMES OF THE GREEN SUNFISH

RELATIVE QUANTITY IN ADULT TISSUES

ENZYME	ISOZYME	muscle	heart	eye	brain	stomach	gills	liver	spleen	gonad	kidney
glucosephosphate isomerase (EC 5.3.1.9)	GPI-A	+	+++	+++	+++	+++	++	++	++	++	++
	GPI-B	+++	+++	+		+++	++	+			
lactate dehydrogenase (EC 1.1.1.27)	LDH-B		+++	++	+++	+++	++		++	++	++
	LDH-A	+++	++	++	+++	+++	++		++	+++	++
	LDH-C		+	+++	+	+	++				+
glyceraldehyde-3-phosphate dehydrogenase (EC 1.2.1.12)	G3PD-B		+	++	++	++	++		++	+	+
	G3PD-A	++	+++					++			
phosphoglucomutase (EC 2.7.5.1)	PGM-B	+++	++	+	++	++	++	+++	+	+	+
	PGM-A	+++	+	+	++	+++		+++	+	+	+
mannosephosphate isomerase (EC 5.3.1.8)	MPI-1,-2	+	+	+	++	+	+		+	+	+
	MPI-3		+	+	++	+	+				
glycerol-3-phosphate dehydrogenase (EC 1.1.1.8)	GPD-C							++			
	GPD-A,-B	+++						+++			
glucose-6-phosphate dehydrogenase (EC 1.1.1.49)	G6PD-1										
	G6PD-2,-3									++	
	G6PD-4					+	++	+++			++
	G6PD-5					++	++	+++			+
	G6PD-6					+	+	+++			
phosphogluconate dehydrogenase (EC 1.1.1.44)	PGD			+		+	+	+++			
hexosediphosphatase (EC 3.1.3.11)	HDP				+			+++		+	+
isocitrate dehydrogenase (EC 1.1.1.42)	IDH					+	++	+++			+
malate dehydrogenase (EC 1.1.1.37)	MDH-1				+						
	MDH-B	+++	++	++	+		+	+++		+	+
	MDH-A	+++	+++	++	+++	+++	++	+++	++	+	++

430

TABLE 3 CONT'D

ADULT TISSUE DISTRIBUTION OF ENZYMES AND ISOZYMES OF THE GREEN SUNFISH

ENZYME	ISOZYME	muscle	heart	eye	brain	stomach	gills	liver	spleen	gonad	kidney
aspartate aminotransferase (EC 2.6.1.1.)	AAT-1			+				+++			
	AAT-2,-3	++	+++	++	++	++	+	++	+	++	++
creatine kinase (EC 2.7.3.2)	CK-B		+++	+++	+++						
	CK-A	+++	++	++	+++						
	CK-C				+		+				
adenylate kinase (EC 2.7.4.3)	AK-1	++	++	++	++	+++	+++	++	+	+++	++
	AK-2	+++			++	+++		+	+	+	+
esterase (EC 3.1.1.-)	EST-1	++		+	+++	+	+	+	+	++	
	EST-3,-4	+	+	++	++	+	++	+	++	++	+
	EST-5						+		+		+
	EST-6,-7		+				++	++		+	
tetrazolium oxidase	TO		+	+	+	+	+	+++	+	+	
acid phosphatase (EC 3.1.3.2)	ACP		+	+	++	++		++	++	+++	++
alcohol dehydrogenase (EC 1.1.1.1.)	ADH							+++		+++	+++

RELATIVE QUANTITY IN ADULT TISSUES

431

A schematic summary of the ontogenetic changes in these enzymes is presented in figure 8. A more detailed presentation of these data will be published elsewhere (Champion and Whitt, 1975).

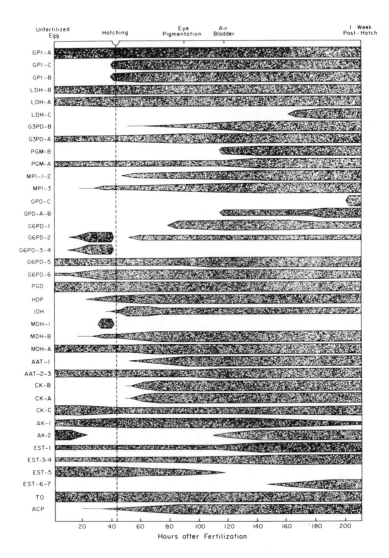

Figure 9. Green sunfish enzymes and isozymes and a schematic summary of their changes in activity during early development. This diagram was produced in a similar manner to Figure 8. Note, however, that there is a difference in time scale.

ANALYSIS OF ENZYME AND ISOZYME CHANGES
DURING LAKE CHUBSUCKER DEVELOPMENT

During the development of the lake chubsucker many different patterns of enzyme change can be observed (Figure 8). Levels of specific enzymes show gradual increases (PGM-3, LDH-C), gradual decreases (EST-9), abrupt increases (GPI-B), or abrupt decreases (AMY). Some enzymes show little or no changes in levels of activity (GPI-A,C, AAT-2), while others exhibit fluctuating levels during development (G6PD-2, EST-8). Additionally, specific isozymes are characterized by embryo-specific (PGM-1) and highly stage-specific(MDH-1)expression. In fact, nearly every conceivable pattern of enzyme change has been observed in this study.

Because each of the lake chubsucker enzymes listed in Figure 8 functions in a metabolic pathway, developmental changes in these enzyme activities permit us to infer the relative roles of these pathways during development. Two general conclusions can be drawn from metabolic changes observed during lake chubsucker ontogeny: first, there is a changing emphasis on synthesis and utilization of glycogen and glucose (see Shaklee et al., 1974); secondly, there is a tendency for enzymes that are part of the same pathway to show coordinate shifts in activity. These changes in patterns of carbohydrate metabolism are generally similar to those observed for the loach (Milman and Yurowitzky, 1973).

Several tissue-specific enzymes show a close coupling between their time of ontogenetic appearance and the morphological and physiological differentiation of the tissues in which they are found. For example both LDH-A and CK-6 which are found predominantly in adult skeletal muscle are first seen at the time the developing embryo exhibits its first muscular contractions. In addition, LDH-C and IDH which are found primarily in adult liver are first detected at the time of organogenesis of the liver. Such correlations suggest these enzymes are essential for the attainment of cellular differentiation.

During lake chubsucker development two general periods stand out as being the time intervals in which most enzyme changes take place (Figure 8). Thus, between 70 and 110 hours after fertilization there are many increases and decreases in the levels of enzyme activities and later, 240-280 hours after fertilization, another period of rapid change in enzyme levels occurs. These two periods occur at about the time the embryo is hatching from the chorion, and later when the fry first begin to derive energy from exogenous food sources.

COMPARISON OF THE DEVELOPMENTAL CHANGES FOR THE ENZYMES AND ISOZYMES OF THE LAKE CHUB-SUCKER AND GREEN SUNFISH

The lake chubsucker and green sunfish are rather distantly related, and in addition, have quite different rates of development. Nevertheless, the many different patterns of enzymatic change observed for the lake chubsucker are also seen during green sunfish ontogeny. Moreover, both species show embryo-specific isozymes (PGM-1 for the lake chubsucker and G6PD-1,2-3 for the green sunfish).

Most striking is the appearance just prior to hatching in both fish of a stage-specific isozyme of malate dehydrogenese (MDH-1). In view of the evidence for the synthesis of embryonic antigens in some tumorous tissues it would be of interest to see if embryo-specific isozymes reappear in neoplastic tissues of these or other teleosts.

The glycolytic pathways of the lake chubsucker and green sunfish show approximately the same patterns of developmental change. However, the levels of G6PD and PGD (involved in the pentose shunt) show different temporal patterns during development in these fishes. Nevertheless, the similarity of developmental progression for many enzymes of these unrelated species indicates there may be the same general pattern of metabolic change during the development of all teleosts.

Homolgous enzymes and isozymes of the lake chubsucker and green sunfish show approximately the same adult tissue distributions and tend to show roughly the same patterns of activity during development. There are, however, a number of exceptions. Although the different LDH-A isozymes of these fishes show similar adult tissue distributions, their ontogenetic changes are significantly different (Figures 8 and 9). However, these differences may be due to differences in gene activity during oogenesis rather than during early embryogenesis. The CK isozymes of the lake chubsucker and green sunfish also show very similar tissue distributions, but the CK isozymes of the green sunfish show a relatively later appearance than those of the lake chubsucker.

The liver-specific LDH isozyme of the lake chubsucker and the eye-specific isozyme of the green sunfish show strikingly different times of appearance during ontogeny. Each of these isozymes is first synthesized at the time of cellular differentiation of the tissue in which it is primarily found. Since both of these isozymes are encoded by the same locus, these results suggest that evolutionary changes in regulation have led to dramatic differences in both the adult tissue-specific expression and the ontogenetic pattern of activation of the LDH-C genes in these two teleosts.

Both the lake chubsucker and the green sunfish show two general periods during their development in which rapid changes in enzyme and isozyme levels take place. These two periods occur for both fish at around the time of hatching and later when the fry initiate swimming and feeding. Rapid changes in isozyme patterns at the time of hatching have been observed for several fishes, and are probably indicative of the extensive organodifferentiation which takes place at this time (Nakano and Whitely, 1965, Holmes and Whitt, 1970).

The extensive isozymic changes characteristic of these two distantly related species suggest that major changes in metabolism are associated with teleost embryogenesis. Further detailed analysis of many enzymes within individual pathways will permit the more precise determination of the specific roles of these pathways during development.

ACKNOWLEDGEMENTS

This research was supported by NSF grants GB 16425 and GB 43995 to G. S. W. and Cell Biology Training Grant GM 00941.

REFERENCES

Bailey, G. S., A. C. Wilson, J. E. Halver, and C. L. Johnson. 1970. Multiple forms of supernatant malate dehydrogenase in salmonid fishes. Biochemical immunological and genetic studies. *J. Biol. Chem.* 245: 5927-5940.

Boulekbache, H., A. J. Rosenberg and C. Joly. 1970. Isoenzymes de la lactico-deshydrogenase au cours des premiers stades de developement de l'oeuf de Truite (Salmo irideus, Gibb). *C. R. Acad. Sci. Paris* 271 (series D) 2414-2417.

Champion, M. J. and G. S. Whitt. 1975. Developmental genetics of green sunfish isozymes. In preparation.

Greenfield, P. C. and E. J. Boell. 1970. Malate dehydrogenase and glutamate dehydrogenase in chick liver and heart during embryonic development. *J. Exp. Zool.* 174: 115-124.

Holmes, R. S., and G. S. Whitt. 1970. Developmental genetics of the esterase isozymes of *Fundulus heteroclitus*. *Biochem. Genet.* 4: 471-480.

Markert, C. L. 1965. Mechanisms of cellular differentiation. p. 230-238. In: *Ideas in Modern Biology* (J. Moore, ed.) Natural History Press, Garden City, New York. pp. 563.

Markert, C. L. and F. Møller. 1959. Multiple forms of enzymes: Tissue, ontogenetic and species specific patterns. *Proc. Nat. Acad. Sci. U.S.A.* 45: 753-763.

Markert, C. L. and H. Ursprung. 1962. The ontogeny of isozyme patterns of lactate dehydrogenase in the mouse. *Devel. Biol.* 5: 363-381.

Masters, C. J. and R. S. Holmes. 1972. Isoenzymes and onto-geny, *Biol. Rev.* 47: 309-361

Miller, E. T. and G. S. Whitt. 1974. Lactate dehydrogenase isozyme synthesis and cellular differentiation in the tele-ost retina. *III. Isozymes: Developmental Biology*, C. L. Markert, editor, Acad. Press, New York.

Milman, L. S. and Yu. G. Yurowitzky. 1973. Regulation of gly-colysis in the early development of fish embryos. In: *Mono-graphs in Developmental Biology*, Vol. 6. S. Karger, Basel.pp. 106.

Nakano, E. and A. H. Whiteley. 1965. Differentiation of multi-ple molecular forms of four dehydrogenases in the teleost, *Oryzias latipes*, studied by disc electrophoresis. *J. Exp. Zool.* 159: 167-180.

Nakano, E. and M. Hasegawa. 1971. Differentiation of the re-tina and retinal lactate dehydrogenase isoenzymes in the teleost, *Oryzias latipes*. *Development, Growth and Differ-entiation*. 13: 351-357.

Neyfakh, A. A., M. A. Glushankova, N. S. Korobtzova and A. A. Kusakina. 1973. Expression of genes controlling FDP- aldo-lase in fish embryos. Thermostability as a genetic marker. *Devel. Biol.* 34: 309-320.

Ohno, S. 1969. The preferential activation of maternally de-rived alleles in development of interspecific hybrids. p. 137-150. In: *Heterospecific Genome Interaction* (V. Defendi, ed.) Wistar Institute Symposium Monograph No. 9. Wistar Institute Press, Philadelphia.

Scholl, A. and H. M. Eppenberger. 1972. Patterns of isoenzymes of creatine kinase in teleostean fish. *Comp. Biochem. Phys-iol.* 42B: 221-226.

Shaklee, J. B., K. L. Kepes, and G. S. Whitt. 1973. Special-ized lactate dehydrogenase isozymes: The molecular and genetic basis for the unique eye and liver LDHs of teleost fishes. *J. Exp. Zool.* 185: 217-240.

Shaklee, J. B., M. J. Champion, and G. S. Whitt. 1974. Devel-opmental genetics of teleosts: A biochemical analysis of lake chubsucker ontogeny. *Devel. Biol.* 38: 356-382.

Shaw, C. R. and R. Prasad. 1970. Starch gel electrophoresis of enzymes - A compilation of recipes. *Biochem. Genet.* 4: 297-320.

Turner, D. C., V. Maier, and H. M. Eppenberger. 1974. Creatine kinase and aldolase isoenzyme transitions in cultures of chick skeletal muscle cells. *Devel. Biol.* 37: 63-89.

Wheat, T. E., W. F. Childers, E. T. Miller, and G. S. Whitt. 1971. Genetic and *in vitro* molecular hybridization of malate dehydrogenase isozymes in interspecific bass (*Micropterus*) hybrids. *Anim. Blood Grps. Biochem. Genet.* 2: 3-14.

Wheat, T. E., G. S. Whitt and W. F. Childers. 1972. Linkage relationships between the homologous malate dehydrogenase loci of teleosts. *Genetics* 70: 337-340.

Whitt, G. S. 1970a. Developmental genetics of the lactate dehydrogenase isozymes of fish. *J. Exp. Zool.* 175: 1-35.

Whitt, G. S. 1970b. Genetic variation of supernatant and mitochondrial malate dehydrogenase isozymes in the teleost *Fundulus heteroclitus*. *Experientia* 26: 734-736.

Whitt, G. S., P. L. Cho, and W. F. Childers. 1972. Preferential inhibition of allelic isozyme synthesis in an interspecific sunfish hybrid. *J. Exp. Zool.* 179: 271-282.

Whitt, G. S., W. F. Childers, and P. L. Cho. 1973a. Allelic expression at enzyme loci in an intertribal hybrid sunfish. *J. Hered.* 64: 55-61.

Whitt, G. S., E. T. Miller and J. B. Shaklee. 1973b. Developmental and biochemical genetics of lactate dehydrogenase isozymes of fishes. p. 243-276. In: *Genetics and Mutagenesis of Fish*. Springer-Verlag, Berlin.

Whitt, G. S., J. B. Shaklee, and C. L. Markert. 1975. The evolution of the lactate dehydrogenase isozymes in fishes. *IV. Isozymes: Genetics and Evolution*. C. L. Markert, editor, Academic Press, New York.

Yamauchi, T. and E. Goldberg. 1973. Asynchronous expression of glucose-6-phosphate dehydrogenase in splaketrout embryos. *Genetics* 74: S 301. (Abstract).

BIOCHEMICAL COMPARISON OF GENETICALLY-DIFFERENT HOMOZYGOUS
CLONES (ISOGENIC, UNIPARENTAL LINES) OF THE
SELF-FERTILIZING FISH *RIVULUS MARMORATUS* POEY

EDWARD J. MASSARO, JANET C. MASSARO, and
ROBERT W. HARRINGTON, JR.
Department of Biochemistry
State University of New York at Buffalo
Buffalo, New York 14214
and
Florida Medical Entomology Laboratory
Florida Division of Health
P. O. Box 520
Vero Beach, Florida 32960

ABSTRACT: The lactate (LDH) and malate (MDH) dehydrogenases, α-naphthyl acetate (α-NA) and -butyrate (α-NB) hydrolyzing esterases, and superoxide dismutases (SOD) of three
clones of *R.marmoratus* were investigated employing starch
gel electrophoresis.

Three groups of LDH isozymes were observed: a low-
mobility group represented in all tissues, a high-mobility
group found exclusively in the eye, and an intermediate
group of variable expression.

The MDH isozymes also were classified into three groups.
Each tissue exhibited at least four MDH patterns generated
by differences in expression of certain fast-migrating
and intermediate bands. In all patterns, at least one of
the slow-group bands was expressed.

All tissues contained α-NB esterases. Variation in
isozyme expression within particular tissue types generated
numerous isozyme patterns and pattern similarities among
tissues. A heavily-staining band of intermediate mobility
was common to all tissues. All tissues also contained α-NA
esterases. As with α-NB, all tissues expressed a heavily-
staining band, the mobility of which was identical to that
of the ubiquitous heavy band hydrolyzing α-NB.

All tissues exhibited SOD activity. Liver had the most
intense activity and the largest number of isozymes (four).
The slowest-migrating isozyme was expressed by all tissues.

Employing these enzymes as gene markers, no consistent
differences among the clones of *R.marmoratus* were
discernible.

INTRODUCTION

The cyprinodontid *Rivulus marmoratus* is unique among fishes
in naturally reproducing uniparentally (Harrington, 1963; Atz,

1964), via synchronous hermaphroditism with internal self-fertilization.

Three clones of this fish were examined in the present study, two derived from genetically-different progenitors wild-caught in the same marsh, the third, from a progenitor wild-caught over 100 miles to the south. The term clone has come to be applied more broadly than as defined when it was coined (cf. Rieger, Michaelis and Green, 1968). It refers here to uniparental lines of isogenic individuals produced by selfing. Such isogenic lines (clones) could be predicted parsimoniously to be homozygous, or unparsimoniously, to be heterozygous, i.e. kept so by a cytological or genetic homozygote-destroying mechanism. The present clones were doubly verified to be homozygous (Kallman and Harrington 1964; Harrington and Kallman,1968). Intraline homozygosity and isogenicity, as well as interline heterogenicity, were demonstrated by tissue transplantation tests, alone and combined with interline hybridization. According to Billingham and Silvers (1959), the tissue transplantation test is the most sensitive indicator of homozygosity. When first used in conjunction with interclonal hybridization (Harrington and Kallman, 1968) it generated a pattern of allograft and autograft reactions with the "logical rigor of a Boolean truth table". As Lindsey and Harrington (1972) commented, "thus the evidence for utmost genetic uniformity is as cogent as possible and at best could be supplemented... by the most exhaustive survey feasible of protein electrophoretic mobilities".

We have begun such a survey (i) in search of protein polymorphism among clones, which, besides its intrinsic interest, could permit faster verification of attempted artificial hybridization, or of heterozygosity, however caused, than the highly sensitive but slow and laborious tissue transplantation test, (ii) in search of residual intra-clonal polymorphism of either genetic or epigenetic origin and (iii) to explore the little-known isozyme repertory of this unusual fish, viz. isozyme types and physicochemical relationships within and among families of specific isozymes.

In this report, we describe, the lactate (LDH) and malate (MDH) dehydrogenases, the α-naphthylacetate (α-NA) and -butyrate (α-NB) hydrolyzing esterases, and superoxide dismutases (SOD) of this fish. We have been studying also the isozymes and tissue distribution of hexokinase, phosphoglucomutase, glyceraldehyde-3-phosphate dehydrogenase, α-glycerophosphate dehydrogenase, glucose-6-phosphate dehydrogenase, creatine phosphokinase, adenylate kinase, glutamate dehydrogenase, and ethanol dehydrogenase. However, our data are incomplete but will be extended and reported in the near future.

MATERIALS AND METHODS

Animals. Rivulus marmoratus Poey is euryhaline and will tolerate fresh and sea water. The progenitor of Clone DS and that of Clone NA were captured in Vero Beach, Florida in 1961, that of Clone M, in Perrine, Florida in 1966. Most of their uniparental laboratory descendants have been reared *ab ovo* singly in jars, others in community tanks. To prolong the life of brine shrimp, their sole laboratory food, our fish have been maintained successfully without aeration in 40% sea water (sea water mixed with bottled drinking water) treated with methylene blue, which is mildly bacteriostatic. If cloudy, the water was changed completely, otherwise, only replenished to compensate for that siphoned out with eggs, feces, and uneaten food.

General Biology of Rivulus marmoratus. The eggs are emitted by their selfing parent in a range of embryonic stages, randomly. In effect, the eggs undergo intraparental incubation within the ovotestes, where they were self-fertilized, for from a few minutes to two and one-half days (Harrington, 1963), then extraparental incubation, from oviposition to hatching, for at least 10 (usually 14 or more) days (Harrington, 1967).

Although each clone of *R.marmoratus* naturally occurring,is composed of individuals as genetically uniform as is likely to be found among vertebrates, and usually comprises a generational succession of uniparental hermaphrodites, two kinds of males also can be produced from zygotes of the same genotype as their hermaphrodite parent (cf. pedigree diagram of Clone DS, Harrington, 1968, Fig. 1).

Primary males, viz. "pure males", or more accurately, true male gonochorists, can be obtained by incubating at low temperature (20° C or below) eggs that at moderate temperatures would have yielded hermaphrodites. Primary males have primary testes like the testes of the majority of fish species (Harrington, 1967; 1971).

Secondary males are derived from hermaphrodites via sex succession, the ovotestes becoming secondary testes by the involution of the ovarian area and enlargement of the testicular area (Table 1). Sex succession is a potentiality of amphisexual gonads (ovotestes) -not of unisexual gonads. It has been induced in *R.marmoratus* so as to yield secondary males at first sexual maturity (= 'false male gonochorists') by exposure to high temperature (30-31° C) during embryonic and juvenile stages (Harrington, 1971). In adult hermaphrodites of *R.marmoratus,* on the other hand, sex succession is triggered by short days after exposure to high temperature within the first six months of life (Harrington, 1971). Before

R.marmoratus: LDH

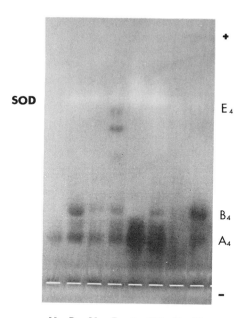

Fig. 1. TISSUE-SPECIFIC DISTRIBUTION OF THE LACTATE DEHYDRO-
GENASE ISOZYMES OF *Rivulus marmoratus*. Vertical starch gel
electrophoresis at 12V/cm for 22 hours at pH 8.6 in the ethy-
lenediamine tetraacetate (EDTA)-boric acid-tris (hydroxymethyl)
aminomethane (Tris) buffer system of Boyer et al. (1963).
Migration is toward the anode. H = heart; B = brain; SI =
stomach-intestine; E = eye; L = liver; OT = ovotestes; O = ova;
M = muscle. A_4, B_4, E_4 = subunit compositions (tentative).
Note the position of the isozymes of superoxide dismutase
(SOD).

changing to a secondary male, one hermaphrodite laid eggs over
a period of 1167 days, another changed after laying for a few
weeks and lived on as a secondary male for 918 days. Usually
hermaphrodites remain hermaphrodites throughout life. No
secondary male and only one primary male has been found in the
wild in Florida. The secondary males can be identified
histologically by either or both a vestigial oviduct and egg
residua in the secondary testes. The significance of all of
the above phenomena has been examined from a multiplicity of
standpoints (Harrington, 1971, pp. 412-428).

442

TABLE 1

INDICATIONS OF PHYSIOLOGICAL DIFFERENCES AMONG CLONES
DS, NA, AND M OF *RIVULUS MARMORATUS*

I. Deflection of primary sex determination from target phenotype (hermaphrodite) to <u>primary</u> <u>male</u> phenotype (= true
male gonochorist) by temperatures below 20°C during the
thermolabile period of sex determination and differentiation:
   A. At low temperature into Stage 31b (neural and hemal
   arches formed) but not throughout the thermolabile
   period (delimited by Harrington, 1967, 1968): M (n =
   47) 81% NA (n = 50) 52% DS (n = 43) 37% (Harrington,
   unpubl.)
   B. At low temperature throughout and beyond the thermolabile period: M (all cases so far) 100% DS (n = 30)
   67% (Harrington, 1967) NA (n = 17) 88% (Harrington,
   1967), DS (n = 107) 67% (Harrington, 1968)

II. Protogynoid sex succession (Harrington, 1971), hermaphrodite to <u>secondary</u> <u>male</u>:
   A. Juvenile: prefunctional hermaphrodite to functional
   secondary male (= false male gonochorist) induced by
   very high incubation temperature (31°C): DS (n = 46)
   13% M (n = 51) 10% NA (n = 46) 2% (Harrington,
   unpubl.)
   B. Adult: functional hermaphrodite to functional secondary male, order of frequency whether triggered by
   short days after exposure to high temperature (Clones
   DS and NA, Harrington, 1971) or spontaneous at
   moderate temperatures (Clone M, Harrington, unpubl.):
   M > DS > NA

III. Mean days, at about 25°C, to sexual maturity (1st nuptial
color in 1° or 2° male, 1st egg laid by hermaphrodite)
from dechorionation and transfer (within Stage 31b) from
the incubator (Harrington, unpubl.):

Incubated		DS	NA	M
at 31°C	Secondary males	(n=6) 110	(n=0)	(n=5) 173
	Hermaphrodites	(n=40)108	(n=37)110	(n=44)205
at 25°C	Hermaphrodites	(n=47)107	(n=52)103	(n=45)162
at 19°C	Primary males	(n=16) 84	(n=24) 88	(n=38)139
	Hermaphrodites	(n=27)119	(n=26)104	(n=9) 158

TABLE 1 (Cont.)

IV. Mortalities (Harrington, unpubl.):

	DS	NA	M
After incubation at 31°C	(n=117) 52%	(n=78) 39%	(n=79) 32%
After incubation at 25°C	(n=50) 4%	(n=52) 0%	(n=50) 4%
After incubation at 19°C	(n=50) 14%	(n=55) 7%	(n=52) 8%

Data in Table 1 indicate differences among Clones DS, NA, and M in their responses to environmental factors, not only as the latter deflect sex determination from the target (hermaphrodite) sex phenotype to the primary male phenotype, but as they induce sex succession, hermaphrodite to secondary male. Also their mortalities, especially at high temperature, are different, and the fish of Clone M reach sexual maturity far more slowly than those of Clones DS and NA.

Enzyme preparations. All fish were reared in Vero Beach, Florida. They were frozen on dry ice and shipped to Buffalo, New York, where they were stored at -86° C for varying periods of time. Thawing and tissue weighing were accomplished at room temperature. All other operations were performed at 0 - 4° C.

Skeletal muscle, ovotestes, and eyes were weighed and diluted 1:10 (w/v) with de-ionized distilled water. Liver was weighed and diluted 1:20. Because of its size, it was inconvenient to weigh the heart. Also, the time required to weigh the heart, brain, and stomach-intestine adversely affected electrophoretic resolution. Therefore, heart, brain, and stomach-intestine were not weighed routinely and their dilutions were estimated by eye (between 0.03 and 0.07 ml water added).

All tissues were hand homogenized in ground glass tissue homogenizers. Homogenization over a wide time range did not alter the tissue specific enzyme patterns. The homogenates were centrifuged at 100,000 x g for 60 minutes and the resulting supernatants (crude enzyme preparation) were analyzed by starch gel electrophoresis.

Electrophoresis. Vertical starch gel electrophoresis was employed (Buchler Instruments, Fort Lee, New Jersey). The gels were prepared with (13.7%) electrostarch in the Tris-borate-EDTA buffer system of Boyer, et al. (1963). Gel preparation has been described in detail (Massaro, 1967). **Fifty**

microliter aliquots of the enzyme preparations were placed in
the preformed sample slots of the gel. Electrophoresis was
carried out at 12 V/cm for 12-24 hours at 4° C. The gels
were sliced horizontally and stained in the dark at 37° C for
dehydrogenase activity by the formazan precipitation technique
(Markert and Møller, 1959; Massaro, 1967; Massaro and Markert,
1968). In all cases, control slabs were incubated in the
staining solution minus substrate to identify bands of "nothing
dehydrogenase" activity (Shaw and Koen, 1965). Alpha-naphthyl-
acetate and α-naphthylbutyrate hydrolyzing esterases were
detected as described previously (Holmes and Massaro, 1969).
Superoxide dismutase was detected by the method of Nishikimi
et al. (1972).

Reagents. Hydrolyzed potato starch for gel electrophoresis
was obtained from the Electrostarch Co., Madison, Wisconsin.
   All chemicals were of the highest grade available from
the Sigma Chemical Co., St. Louis, Missouri or the J. T. Baker
Chemical Co., Phillipsburgh, New Jersey.

RESULTS

   The LDH, MDH, α-NA, α-NB, and SOD isozyme patterns of the
following tissues/organs were investigated: liver, ovotestes,
skeletal muscle, heart, eye, brain, and stomach-intestine.
   Starch gel electrophoresis under our conditions separated
the LDH isozymes of *R.marmoratus* into three groups (Fig. 1):
(i) a group of low electrophoretic mobility, composed of one-
four bands (isozymes of this group were found in all tissues
studied); (ii) a group of high mobility consisting of a pair
of bands, which was found exclusively (and in all cases) in
the eye; and (iii) an intermediate group, composed of one-
three bands, found in the ovotestes, eye and brain. The
intra- and intergroup relationships of the intermediate bands
have not been resolved completely and until additional data
are obtained, classification of certain bands into this group
must be considered a matter of convenience only.
   The tissue-specific expression of the slow and intermediate
groups of LDH isozymes was variable. In the liver, two major
slow-group patterns have been discerned (Fig. 2): one consisting of two bands, the other of three. In both cases, the
slowest band was the most intense. A pattern similar to the
three-banded liver pattern was found also in the stomach-
intestine.
   Three slow-group patterns were found in skeletal muscle
(Fig. 3) consisting of two, three, or four bands. The four-
banded pattern was unique in that it was composed of a fast

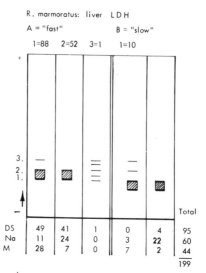

R. marmoratus: liver LDH

A = "fast"   B = "slow"

1=88   2=52   3=1   1=10

						Total
DS	49	41	1	0	4	95
Na	11	24	0	3	22	60
M	28	7	0	7	2	44
						199

Fig. 2. *R.marmoratus* LIVER LDH ISOZYME PATTERNS. Vertical starch gel electrophoresis as in Fig. 1. The numbers along the left-hand margin denote individual isozyme bands and the arrow denotes direction of migration. Note that in this particular case, bands 1 and 2 bracket a smear. DS, NA and M denote the clones investigated and the number of fish in each clone exhibiting a particular pattern. The numbers at the top margin indicate the total number of fish exhibiting each pattern. Note that in this particular illustration the patterns to the right of the vertical double line were obtained from pathological cases.

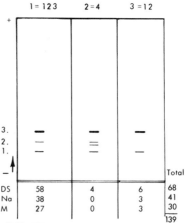

R. marmoratus: muscle LDH

1= 123   2=4   3 =12

				Total
DS	58	4	6	68
Na	38	0	3	41
M	27	0	3	30
				139

Fig. 3. *R.marmoratus* SKELETAL MUSCLE LDH ISOZYME PATTERNS. Vertical starch gel electrophoresis as in Fig. 1. Number and letter designations as in Fig. 2.

and slow band and two closely migrating intermediate bands. In all cases, the fast band of this group was the most intense and the three-banded pattern was the most prevalent. The

446

three-banded pattern, with a more intensely-staining fast band, was the only slow-group pattern found in the eye and brain and was the major pattern in the heart. Based on this data, the fast band of the slow-group has been identified tentatively as the B_4 isozyme (Fig. 1) while the slow-band has been designated the A_4 isozyme (Whitt, et al., 1973). In addition, the heart also exhibited a single-banded pattern, consisting of the slowest migrating LDH band only.

Two patterns of expression of the intermediate group of LDH isozymes were found in the eye. In all cases, a rapid migrating doublet of bands was present. Approximately 10% of the specimens of our sample also exhibited a single, slow-moving band which migrated ahead of the most anodal of the slow group bands. Brain exhibited only a single intermediate group band of low mobility which was expressed in approximately 40% of our sample. The pattern of intermediate group isozyme expression in the ovotestes was highly variable. One major and seven minor patterns were discerned. The major pattern consisted of a single relatively rapidly migrating band.

The MDH isozymes of *R.marmoratus* exhibit tissue-specific patterns (Fig. 4). Starch gel electrophoresis separated these isozymes into three groups: a fast group migrating cathodal to the superoxide dismutases and an intermediate- and a slow-migrating group. There is at least one major band in each group. Intra- and inter-group relationships, however, have not yet been established unequivocally and certain group assignments have been made purely for convenience. Since, by necessity, the tissues were frozen prior to analysis, identification, per se, of supernatant and mitochondrial forms of the enzyme and, therefore, assignment of subunit composition was not possible.

Each tissue examined exhibited at least four MDH isozyme patterns. The patterns were generated by differences in the expression (all or none) of certain bands within the groups of fast and intermediate mobility. In all patterns, at least one band of the slow-migrating group always was expressed. Also, the position of the slow-migrating group among tissues of the same type or among different tissues was variable (Fig. 4). In the liver, ovotestes, and stomach-intestine this group consisted of a major band and a smear of activity running adjacent to the anodal aspect of this band. Frequently, especially in the liver, this smear was resolved into one to four bands.

As shown in Fig. 5, ten MDH isozyme patterns were obtained from the ovotestes. Two of these were major patterns, one of which was expressed by approximately 50% of the specimens in our sample. One of the major patterns consisted

447

R. marmoratus: MDH

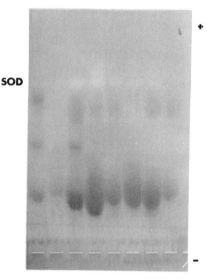

M O OT L E SI B H

Fig. 4. TISSUE-SPECIFIC DISTRIBUTION OF THE MALATE DEHYDRO-
GENASE ISOZYMES OF *R. marmoratus*. Vertical starch gel electrophoresis and letter designations as in Fig. 1.

of three more or less evenly spaced bands. The other, exhibited the same three bands plus a peculiar "U"-shaped band.
This "U"-shaped band was one of the fast-migrating intermediate bands.

Seven MDH patterns were obtained from liver (Fig. 6). Six
of these were major patterns and one pattern was shown by a
single individual only. The "U"-shaped band, which was found
in several of the ovotestes patterns, also was found in three
liver patterns (compare Figs. 5 and 6).

Skeletal muscle, heart, eye, brain and the stomach-intestine preparations each exhibited several MDH isozyme patterns
consisting, for the most part, of three or fewer bands.

All tissues of *R. marmoratus* contained esterases capable
of hydrolyzing α-NB (Fig. 7). The greatest number of such
isozymes (>seven) was found in the liver (Fig. 8). There was
considerable variation in esterase expression within a given
tissue, giving rise to numerous isozyme patterns and resulting
in certain pattern similarities among tissues. Fifteen patterns were obtained from liver and ovotestes and eight from
skeletal muscle. Except in the ovotestes, there was no

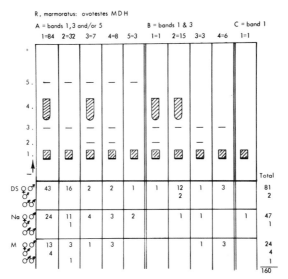

Fig. 5. *R. marmoratus* OVOTESTES MDH ISOZYME PATTERNS. Vertical starch gel electrophoresis as in Fig. 1. Number and letter designations as in Fig. 2. ♀♂ = hermaphrodite; ♂♂ = secondary male.

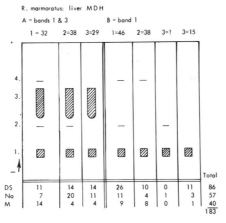

Fig. 6. *R. marmoratus* LIVER MDH ISOZYME PATTERNS. Vertical starch gel electrophoresis as in Fig. 1. Number and letter designations as in Fig. 2.

single overwhelmingly predominant pattern in any tissue. In the case of the ovotestes, more than 40% of our sample expressed the same pattern which consisted of five bands equivalent to bands one-five of the liver (Fig. 8). All patterns of all tissues contained a heavily staining band of intermediate electrophoretic mobility and this band alone was found

449

R. marmoratus

α-naphthylbutyrate

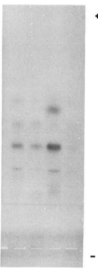

OT M L E

Fig. 7. TISSUE-SPECIFIC DISTRIBUTION OF THE α-NAPHTHYLBUTY-
RATE HYDROLYZING ISOZYMES OF *R.marmoratus.* Vertical starch
gel electrophoresis and letter designations as in Fig. 1.

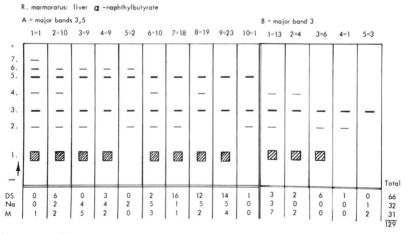

Fig. 8. *R.marmoratus* LIVER α-NAPHTHYLBUTYRATE ESTERASE ISO-
ZYME PATTERNS. Vertical starch gel electrophoresis as in Fig.
1. Number and letter designations as in Fig. 2.

in ova removed from the ovotestes.

All tissues of *R.marmoratus* also contained esterases capable of hydrolyzing α-NA. However, the number of detectable isozymes (five) was less than that hydrolyzing α-NB, as was the number of different isozyme patterns expressed by any tissue. For example, eight patterns were obtained for the liver, nine for the ovotestes (Fig. 9), and two for skeletal muscle. One of the liver patterns and two of the ovotestes patterns were distinguished only on the basis of a smear of low anodal mobility (band 1; Fig. 9). In the ovotestes three major α-NA, patterns were found (A_3, A_4, B_3; Fig. 9). These were expressed by a combined total of approximately two-thirds of our specimens. As in the case of α-NB, all tissues contained a heavily staining band with an electrophoretic mobility identical to that of the heavily staining band of intermediate mobility hydrolyzing the butyrate analogue. These bands appear to be identical.

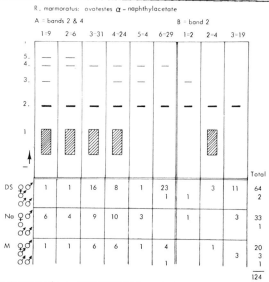

Fig. 9. *R.marmoratus* OVOTESTES α-NAPHTHYLACETATE ESTERASE ISOZYME PATTERNS. Vertical starch gel electrophoresis as in Fig. 1. Number and letter designations as in Fig. 2.

All tissues of *R.marmoratus* possessed SOD activity (Figs. 1,4). A total of four isozymes was resolved by starch gel electrophoresis and the slowest-migrating isozyme was present in all tissues examined. The most intense activity was found in the liver which also expressed the largest number of isozymes (four). In all cases, intensity of activity appeared to decrease from the slowest to the fastest migrating band.

DISCUSSION AND CONCLUSIONS

As is well known, decline in heterozygosity is a consequence of self-fertilization. In recent years, many hermaphroditic fishes have been discovered (Atz, 1964) and it has
become increasingly evident that the various types of hermaphroditism, viz.protogynous and protogynoid (with or without
diandry), protandrous, and synchronous, are inherited (see
Harrington, 1971, pp. 412-428). *Rivulus marmoratus* alone has
been found to be naturally self-fertilizing and no other hermaphroditic vertebrate seems to be represented in the wild by
either isogenic lines of descent or homozygotes (Kallman and
Harrington, 1964; Harrington and Kallman, 1968).

Using LDH, MDH, the α-NA, and α-NB esterases and SOD as
gene markers, no consistent differences among these clones of
R.marmoratus were discernible. However, each of the enzymic
activites was expressed in several isozyme patterns. In the
light of the histocompatibility data (Kallman and Harrington,
1964; Harrington and Kallman, 1968), this is of great interest.
It may be that in the particular clones examined the genes
coding for the enzymes investigated are evolutionarily conservative (i.e. their products basically are identical) in
contrast to the genes at histocompatibility loci, or real
differences existing between specific enzymes are too subtle
for discrimination by starch gel electrophoresis. In any case,
analysis of a larger number of gene products (enzymes and
other proteins) might reveal consistent differences among the
three clones complementing the existing histocompatibility
data. In our continuing effort to characterize these clones,
other enzymes (see above) are being investigated by electrophoretic and immunological methods.

REFERENCES

Atz, J. W. 1964. Intersexuality in fishes. In: *Intersexuality
in Vertebrates Including Man*. (C. N. Armstrong and A. J.
Marshall, eds.). Academic Press, New York, pp. 145-232.

Billingham, R. E. and W. K. Silvers 1959. Inbred animals and
tissue transplantation immunity. *Transplant. Bull.* 6:
399-405.

Boyer, S. H., D. C. Fainer, and E. J. Watson-Williams 1963.
Lactate dehydrogenase variant from human blood: evidence
for molecular subunits. *Science* 141: 642-643.

Harrington, R. W., Jr. 1963. Twenty-four hour rhythms of
internal self-fertilization and of oviposition by hermaphrodites of *Rivulus marmoratus*. *Physiol. Zool.* 36: 325-
341.

Harrington, R. W., Jr. 1967. Environmentally controlled induction of primary male gonochorists from eggs of the self-fertilizing hermaphroditic fish, *Rivulus marmoratus* Poey. *Biol. Bull.* 132: 174-199.

Harrington, R. W., Jr. 1968. Delimitation of the thermolabile phenocritical period of sex determination and differentiation in the ontogeny of the normally hermaphroditic fish, *Rivulus marmoratus* Poey. *Physiol. Zool.* 41: 447-460.

Harrington, R. W., Jr. 1971. How ecological and genetic factors interact to determine when self-fertilizing hermaphrodites of *Rivulus marmoratus* change into functional secondary males, with a reappraisal of the modes of intersexuality among fishes. *Copeia* No. 3: 389-432.

Harrington, R. W., Jr. and K. D. Kallman 1968. The homozygosity of clones of the self-fertilizing hermaphroditic fish *Rivulus marmoratus* Poey (Cyprinodontidae, Atheriniformes). *Am. Nat.* 102: 337-343.

Holmes, R. S. and E. J. Massaro 1969. Phylogenetic variation of rodent liver esterases. *J. Exp. Zool.* 177: 323-334.

Kallman, K. D. and R. W. Harrington, Jr. 1964. Evidence for the existence of homozygous clones in the self-fertilizing hermaphroditic teleost *Rivulus marmoratus* (Poey). *Biol. Bull.* 126: 101-114.

Lindsey, C. C. and R. W. Harrington, Jr. 1972. Extreme vertebral variation induced by temperature in a homozygous clone of the self-fertilizing cyprinodontid fish *Rivulus marmoratus*. *Can. J. Zool.* 50: 733-744.

Markert, C. L. and F. Møller 1959. Multiple forms of enzymes: tissue, ontogenetic, and species specific patterns. *Proc. Nat. Acad. Sci.* USA 45: 753-763.

Massaro, E. J. 1967. Induction of subunit reassociation among lactate dehydrogenase isozymes. *SABCO J.* (Japan) 3: 51-62.

Massaro, E. J. and C. L. Markert 1968. Isozyme patterns of salmonid fishes: evidence for multiple cistrons for lactate dehydrogenase polypeptides. *J. Exp. Zool.* 168: 223-238.

Nishikimi, M., N. A. Rao, and K. Yagi 1972. The occurrence of superoxide anion in the reaction of reduced phenazine methosulfate and molecular oxygen. *Biochem. Biophys. Res. Comm.* 46: 849-854.

Rieger, R., A. Michaelis, and M. M. Green 1968. *A Glossary of Genetics and Cytogenetics*. Springer-Verlag, Inc. New York.

Shaw, C. R. and A. L. Koen 1965. On the identity of "Nothing Dehydrogenase". *J. Histochem. Cytochem.* 13:431-433.

Whitt, G. S., E. T. Miller, and J. B. Shaklee 1973. Developmental and biochemical genetics of lactate dehydrogenase isozymes in fishes. In: *Genetics and Mutagenesis of Fish* (J. H. Schröder, ed.). Springer-Verlag, N.Y. pp. 243-276.

TISSUE DISTRIBUTION AND GENETICS OF ALCOHOL
DEHYDROGENASE ISOZYMES IN *PEROMYSCUS*

MICHAEL R. FELDER
Department of Biology
University of South Carolina
Columbia, South Carolina 29208

ABSTRACT. The genetic control of the main liver alcohol dehydrogenase (ADH) was investigated. Genetic crosses and analysis confirmed three alleles of the Adh-6 gene designated $Adh-6^F$, $Adh-6^S$ and $Adh-6^N$. These code for electrophoretically fast, slow, and not detectable (null) ADH. Heterozygotes ($Adh-6^F$/$Adh-6^S$) have a characteristic three banded pattern due to random dimerization of fast and slow polypeptides encoded by the respective alleles. In addition, $Adh-6^N$/$Adh-6^F$ and $Adh-6^N$/$Adh-6^S$ genotypes only encode slow and fast ADH respectively. This suggests that there is no polypeptide product of $Adh-6^N$ or that the product cannot dimerize with either fast or slow polypeptide to produce an active hybrid molecule. The exact mechanism is unknown but will be investigated. Older animals contain another liver ADH called Adh-5. This isozyme is not variant at all and is present in $Adh-6^N$/$Adh-6^N$ genotypes. This suggests that it is under separate genetic control. The subband of Adh-6, however, is concomitantly variant with Adh-6 fast and slow forms and is absent when Adh-6 is absent. This suggests that the subband may be epigenetically formed from the Adh-6 isozyme. Both the eye and stomach contain significant concentrations of Adh-5 but not Adh-6. The liver, kidney, small and large intestines contain Adh-6 only.

Extracts resulting in the null phenotypes on gels are quite active for ADH when measured spectrophotometrically. This may be due to contaminating enzyme systems in crude extracts which reduce pyridine nucleotides.

Starch gel electrophoresis, coupled with specific histochemical staining procedures, has been a powerful tool in the study of genetically determined variation of enzymes in plants and animals. Variant forms of enzymes provide excellent markers for genetic and developmental studies (see Scandalios, 1969 and Shaw, 1965). In a brief report, Shaw and Koen (1965) described variant isozymes of the liver alcohol dehydrogenase (ADH) in *Peromyscus* and reported that this enzyme reduces the nicotinamide adenine dinucleotide (NAD) cofactor in the absence of alcohol. This present report provides a more extensive study of the genetic control and tissue expression of the ADH isozymes.

Alcohol dehydrogenase (E.C. 1.1.1.1., alcohol:NAD oxidore-ductase) catalyzes the reversible reaction of the oxidation of alcohol to acetaldehyde with the simultaneous reduction of the NAD cofactor. The enzyme from yeast and horse liver has been extensively studied from a chemical and kinetic stand-point (Luntstorf et al, 1970; Racker, 1960; Pietruski and Theorell, 1969; Hersch, 1962; Ehrenberg and Dalziel, 1958; Wratten and Cleland, 1963; Sund and Theorell, 1963; Theorell, 1967). Alcohol dehydrogenase catalyzes the oxidation-reduction of a number of other alcohols and aldehydes (Sund and Theorell, 1963). One of the isozymes of horse liver alcohol dehydrogen-ase is of particular interest since it can utilize steroids as substrates (Pietruszko et al, 1966).

Genetic studies on alcohol dehydrogenase have been less extensive with the exception of work on *Drosophila* and maize. The genetic control of the isozymes of *Drosophila* has been studied by Courtwright et al, (1966); Grell et al, (1965); Johnson and Denniston, (1964); and Ursprung and Leone, (1965). Genetic analysis indicates that the several isozymes of ADH in crude extracts are all controlled by a single genetic locus. Grell et al, (1968) employing electrophoretic procedures re-ported on the artificial induction of ADH variants as well as the detection of mutants with very low or no ADH activity. A chemical selection method for ADH negative mutants in *Drosophila* has been used successfully by Sofer and Hatkoff, (1972).

The genetic control of ADH has also been extensively stu-died in maize (Schwartz, 1966; Schwartz and Endo, 1966; Scan-dalios, 1967). Homozygotes contain two ADH isozymes in the endosperm and these are most likely controlled by two closely linked genes (Scandalios, 1969). Efron (1970) reported on a gene which appears to lower the activity of only the slow var-iant of one isozyme. However, Felder et al., (1972) found that the purified slow isozyme had lower specific activity than the fast form.

Polymorphism and molecular heterogeneity of human ADH is known to be based upon kinetic parameters, namely altered pH optimum of an atypical isozyme (Schenker et al., 1971). The atypical enzyme was shown to be electrophoretically variant by Smith et al., (1971). Based upon a thorough developmental study and the identified polymorphism, Smith et al., (1971) hypothesized that at least three genetic loci control the ADH isozymes; one for fetal liver, another for lung, and a third for fetal and postnatal intestinal and kidney tissue ADH forms.

Electrophoretic variation cf liver ADH was reported in *Mus musculus* from Denmark by Selander et al, (1969) but no formal genetics was done. Presumed heterozygotes showed a three banded dimeric configuration. There are no known ADH variants in inbred mice lines.

This present study is intended to determine the genetic basis of ADH in an experimental mammalian system. It is hoped that this system may prove of use in eventually understanding the regulation and function of ADH in mammalian tissues.

MATERIALS AND METHODS

Animals were sacrificed by cervical dislocation and the appropriate tissue was immediately excised, washed in saline and placed on ice. Animals were generally about 40 days of age before typing for genetic studies, but occasionally animals of three months of age or more were used. Tissues were homogenized in 0.01 M phosphate buffer, pH 7.0, in a glass mortar and pestle. The homogenate was centrifuged at 27,000 x g for 1 hour and the supernatant was used for electrophoretic analysis or for quantitative assays. Less than 10 percent of the total activity could be removed from the pellet after additional suspension and centrifugation. Samples were stored in the refrigerator at 4°C overnight and then subjected to electrophoresis. Occasionally liver samples were frozen for several weeks and then homogenized with no apparent alteration in electrophoretic pattern. Very poor results were obtained if homogenates were frozen and then electrophoresed. For genetic analysis only liver tissue was used.

To determine tissue distibution of the enzymes, various tissues were excised, washed in saline and homogenized in 0.01 M phosphate buffer. The tissues studied were liver, kidney, spleen, heart, stomach, brain, eyes, large intestine, small intestine, gonads, seminal vesicle, skeletal muscle and small intestine. All were homogenized at 1:2 (w/v) except liver, brain, gonads and small intestines which were homogenized 1:1 (w/v) with buffer. The extracts were then centrifuged at 27,000 x g for 1 hour, and the supernatants were analyzed for isozyme content by electrophoresis.

Starch gel electrophoresis was carried out at 4°C using 12 percent starch gels. For tissue distribution studies a M/60 tris-citrate pH 7.0 buffer was used for preparing the gel. The electrode compartments contained M/12 tris-citrate pH 7.0 buffer (Meizel and Markert, 1967). Electrophoresis was conducted for 17 hours at 12 volts/cm. The gels were then sliced and strained.

The tris-citrate buffer system was not very good for distinguishing the ADH genetic variants although a continuous buffer system of 0.2 M tris-glycine of pH 8.3 was found to be satisfactory. Several other buffer systems were also employed,

but all were unsatisfactory for resolution of the variant isozymes. Electrophoresis was conducted for five hours at 11
volts/cm.

After the gels were sliced, enzyme was detected using the
tetrazolium method following the procedure of Scandalios (1967).
The gel slices were incubated at 37°C for approximately 1 1/2
hours for full detection of the enzyme location within the gel.

Quantitative enzyme activity was measured essentially according to the method of Bonnichsen and Brink (1955) by following spectrophotometrically at 340 nm the reduction of NAD.
The enzyme reaction mixture consisted of 2.7 ml of 0.1 M gly-
cine-NaOH buffer, pH 9.5; 0.1 ml of 17.13 M ethanol; 0.1 ml of
5 mg/ml NAD; and 0.1 ml of enzyme in a quartz cuvette with a 1
cm light path.

The *Peromyscus* stocks used in the experiments are all laboratory grown animals of two species -- *P. maniculatus* and
P. polionotus. The species are easily hybridizable when *P.
maniculatus* is the female parent, and the F_1 hybrids are fully
fertile and can be backcrossed to either species (Dawson,
1965). The reciprocal (*P. polionotus* female x *P. maniculatus*
male) cross can be made but it is less successful.

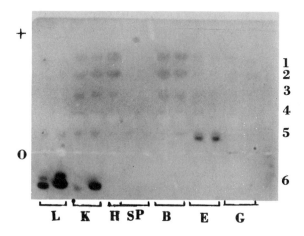

Fig. 1. Tissue distribution of ADH. Homogenates of liver (L),
kidney (K), heart (H), spleen (SP), brain (B), eye (E) and
gonads (G) are electrophoresed and stained for ADH. For each
tissue *P. maniculatus* is the sample on the left and *P. polionotus* is on the right. For heart tissue, only *P. maniculatus*
is shown. (Tris-citrate buffer system)

RESULTS

Tissue distribution of ADH: The results of the tissue distribution of ADH in four month old animals is shown in Figs. 1 and 2. The tissues studied (Fig. 1) are liver, kidney, heart, spleen, brain, eye and gonads. The kidney and brain quite clearly show anodally migrating isozymes labeled 1, 2, 3, 4 and 5 with isozyme 1 showing the most activity. There is reduced activity of these isozymes in the eye and gonads. For each tissue both *P. maniculatus* and *P. polionotus* is represented except for the heart where only a *P. maniculatus* sample is shown. There appears to be no variation between the two species in any of the anodal isozymes. In region 5 the presence of a quite active band in the eye is found that is more slowly migrating than the faint component in that region.

The most active isozyme in the tissues migrates cathodally and is designated Adh-6. This active isozyme is found in both liver and kidney tissue. Although this buffer system poorly resolves the variation, the *P. maniculatus* isozyme is faster migrating than the *P. polionotus* isozyme in both liver and kidney indicating the isozymes in the two tissues are the same.

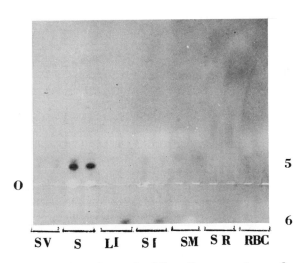

Fig. 2. Tissue distribution of ADH. Homogenates of seminal vesicle (SV), stomach (S), large intestine (LI), small intestine (SI), skeletal muscle (SM) serum (SR), and lysed red blood cells (RBC) electrophoresed and stained for ADH are shown. For each pair *P. maniculatus* is on the left and *P. polionotus* is on the right. (Tris-citrate buffer system)

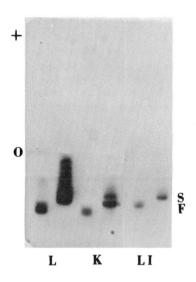

Fig. 3. ADH pattern of fast and slow phenotypes in liver (L), kidney (K), and large intestine (LI). The variation is shown in all tissues showing that the isozyme in all tissues is coded by the same gene. (Tris-glycine buffer)

The subbanding found in Adh-6 is not very consistent, but there is generally always one subband. Occasionally other subbands are seen in concentrated samples.

Fig. 2 shows the ADH pattern in seminal vesicle, stomach, large intestine, small intestines, skeletal muscle, serum and red blood cells. The stomach contains a strong band of activity which moves to the same position in the gel as the eye form of ADH. Both large and small intestines have an isozyme that moves to the same position as Adh-6 in the liver. The *P. maniculatus* (in the left slot for both large and small intestines) is a null phenotype for the liver Adh-6 and is also not present in the large and small intestines. This tends to confirm that the same isozyme is present in liver, kidney and intestinal tissue.

For further confirmation of the identity between the Adh-6

found in liver, kidney and intestinal tissue, use was made of the electrophoretic variation between *P. maniculatus* and *P. polionotus*. The result of electrophoresis of fast *(P. maniculatus)* and slow *(P. polionotus)* liver Adh-6 phenotypes for all three tissues is shown in Fig. 3. The fast and slow phenotypes are shown in all tissues confirming that the same genetic material is coding for the isozymes in the liver, kidney and intestine.

<u>Genetic control of ADH</u>: Genetic polymorphism of ADH in *Peromyscus* is of three types. In *P. maniculatus* there are two phenotypes: fast migrating Adh-6 and an absence of Adh-6. True-breeding types have been established by using siblings of typed animals for additional matings. At least 10 animals from each mating are typed. When all animals are uniformly null, or fast, heavy banded for ADH, additional offspring were used to establish true-breeding stocks. Parents are also typed after a mating is broken. Several matings out of 100 produced uniformly null offspring and several produced uniformly fast, heavy banded offspring. Additional offspring formed the basis for establishing true-breeding lines for the ADH phenotypes. Out of more than 50 animals typed *P. polionotus* only has shown the slow phenotype.

The fast, heavy banded, null, and hybrid phenotypes for Adh-6 are shown in Fig. 4. The animals used were approximately 40 days of age. The null phenotypes show no presence of enzyme activity either for the main Adh-6 isozyme or the subband. The heterozygotes (designated F/N) are clearly intermediate in activity. However, in genetic studies it was not attempted to score heterozygotes separately for fear of error in classification. Longer gels have also been used for short electrophoresis times to insure that the isozyme has not been electrophoresed off the gel.

Hybrids between *P. maniculatus* and *P. polionotus* have also been analyzed (Fig. 5). All phenotypes are illustrated here and the liver tissue is from animals approximately 40 days of age. The *P. polionotus* contains a slow migrating Adh-6 while the *P. maniculatus* is either fast or null. Hybrids between animals with fast and slow isozymes show a typical three banded pattern if only the main Adh-6 isozyme is considered, omitting the subband. This is the typical dimeric pattern assuming random association of fast and slow polypeptides into an active dimeric enzyme. A simple mixture (F+S) does not form the three banded phenotype. Also, the subband is concomitantly altered in mobility in both fast and slow phenotypes. Again, slow/null hybrids and fast/null hybrids only possess the slow or fast isozyme, respectively.

As mentioned previously, older animals contain the Adh-5

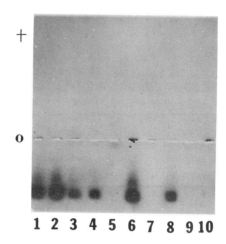

Fig. 4. Electorphoretic analysis of fast, null and the fast/
null heterozygote phenotypes. Fast samples are in slots 1, 2
and 6, null samples in 5, 7, 9, 10 and the F/N heterozygotes
in 3, 4 and 8. (Tris-glycine buffer system)

isozyme whereas its presence is not detected in younger animals.
The ADH phenotypes of older (4 months) animals is shown in
Fig. 6. Note the same three banded pattern in heterozygotes
as in younger animals. Adh-5 is not altered in electrophoretic mobility in fast versus slow animals and is present in
Adh-6 null animals. This indicates that Adh-5 is under separate genetic control.

To thoroughly elucidate the genetic control of Adh-6, a
number of mating experiments were performed. The results of
the genetic crosses are summarized in Table I. The results
were in noncompliance with a sex linked mode of inheritance
since males and females occurred in all phenotypic classes
among the offspring of each cross. Therefore, males and females are not separately indicated in the table. These results
can be explained most easily by assuming that a single genetic
locus controls the structure of Adh-6. The electrophoretic
variants of Adh-6 are controlled by two alleles acting without
dominance. Homozygotes for Adh-6^F show only the fast Adh-6

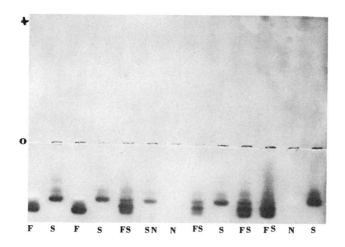

F S F S FS SN N FS S FS FS N S

Fig. 5. Electrophoretic analysis of various ADH phenotypes in 30-40 day old animals. The ADH phenotypes are given under each slot. (Tris-glycine buffer system)

band, homozygotes for the $\underline{Adh-6}^S$ allele have only the slow Adh-6 band and heterozygotes have both bands in addition to the hybrid band. The hybrid band is probably due to inter-action at the subunit level. Chi-square analysis of the data shows that the data is in good agreement with this model.

Furthermore, there is a third allele, $\underline{Adh-6}^N$, which is re-cessive to both $\underline{Adh-6}^F$ and $\underline{Adh-6}^S$ if simple presence versus absence of enzyme is scored. Without regard to mechanisms, this $\underline{Adh-6}^N$ allele codes for the absence of detectable Adh-6 activity after electrophoresis and histochemical staining. The genotypes of the parents are deductions fitting the ob-served results. The parents' phenotypes in most cases were checked after a mating was broken.

It appeared of value to determine quantitatively the total ADH activity in $\underline{Adh-6}^F/\underline{Adh-6}^F$, $\underline{Adh-6}^N/\underline{Adh-6}^N$, and $\underline{Adh-6}^S/\underline{Adh-6}^S$ genotypes. Male animals of 40 days of age were used and the assay was as described in Materials and Methods. Surprisingly, the null animals demonstrated quite high ADH activity in the spectrophotometric assay (Table II). This could be due to the ability of other dehydrogenases to utilize ethanol as the sub-strate. In addition, a microsomal fraction (not ADH) has been

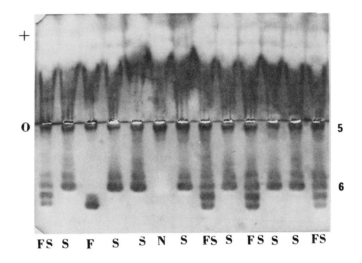

FS S F S S N S FS S FS S S FS

Fig. 6. Electrophoretic analysis of 4 month old animals of varying ADH phenotype. The phenotypes are under each slot. Notice that the mobility of Adh-5 (near the origin) is not affected by alterations in mobility of Adh-6. Also observe that the null animal for Adh-6 contains isozyme Adh-5. (Tris-glycine buffer system)

implicated in the metabolism of ethanol using NAD in animals (Lieber and DeCarli, 1968). However, there is still the possibility that there is active enzyme produced by the $\underline{Adh-6}^N$ allele, but that this enzyme is destroyed upon electrophoresis. This would also have to assume that an F/N or S/N hybrid enzyme would not have activity since only the fast and slow forms are found in $\underline{Adh-6}^F/\underline{Adh-6}^N$ and $\underline{Adh-6}^S/\underline{Adh-6}^N$ genotypes, respectively. This assumes that the product of the $\underline{Adh-6}^N$ allele is electrophoretically different from both the slow and fast enzyme. It must be different from at least one, of course.

The quantitative data is confusing since the null phenotypes so clearly lack enzyme after starch gel electrophoresis and staining. The quantitative assay, however, was performed at 24°C while the gel was incubated at 37°C. The possibility, therefore, exists that in vitro the enzyme is rapidly inactivated at 37°C. Null and fast phenotypes were electrophoresed and stained at 37°C and 24°C (Fig. 7A). It is clearly seen that the null form does not stain at either temperature. Being a zinc containing enzyme (Theorell et al, 1955), ADH null forms could be loosening zinc from the active site upon

TABLE I

SUMMARY OF GENETIC DATA FOR Adh-6

CROSS		OFFSPRING PHENOTYPES				χ^2	P VALUE
FEMALE	MALE	F	N	S	FS		
FF*	FF	68	0	0	0	-	-
SS	SS	0	0	37	0	-	-
NN	FN	60	57	0	0	0.076	0.9>p>0.7
FN	NN	29	25	0	0	0.396	0.7>p>0.5
FN	FN	33	11	0	0	-	-
FF	SS	0	0	0	33	-	-
FS	NN	33		30		0.142	0.9>p>0.7
NN	FS	13		12		0.04	0.9>p>0.7
NN	SN		4	5		0.110	0.9>p>0.7
SN	NN		34	36		0.057	0.9>p>0.7
FS	FS	6		7	11	0.253	0.9>p>0.7
SN	FN	13	7	7	7	3.172	0.3>p>0.2
FN	SS			26	15	2.94	0.1>p>0.05

*Shortened notation for the postulated parental genotypes.

electrophoresis. Zinc (10^{-3}M in $ZnCl_2$ form) was incorporated into the gel and into the staining reaction in a similar experiment. The null forms were still not detectable on the gel. The null enzyme might be a very unstable form destroyed by the electrophoresis procedure itself. However, the fast and null isozymes appear to have approximately equal sensitivities to urea (Fig. 8).

The enzyme as reported by Shaw and Koen (1964) does possess considerable "nothing dehydrogenase" activity (Fig. 7B). However, upon spectrophotometric analysis after dialysis the enzyme has lost considerable activity when ethanol is omitted from the reaction mixture (Table III). Dialysis of both null

TABLE II

ADH ACTIVITY OF VARIOUS ADH GENOTYPES

Genotype	ΔA_{340}/min/g tissue
Adh-6^F/Adh-6^F	18.2 ∓ 1.8*
Adh-6^N/Adh-6^N	18.4 ∓ 3.0
Adh-6^S/Adh-6^S	18.6 ∓ 4.2

*∓ the total variance

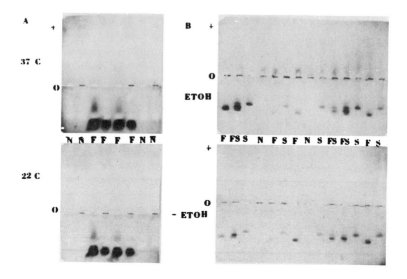

Fig. 7A. Electrophoretic analysis and staining at 37°C and 22°C of fast and null ADH phenotypes. Staining at 22°C did not detect the presence of any enzyme in the $Adh-6^N$ phenotype. (Tris-glycine buffer system)

Fig. 7B. Histochemical staining in the absence (top) and presence of ethanol (bottom). The ADH activity is detected in the absence of added substrate. (Tris-glycine buffer system)

and fast isozymes results in some loss of activity, but both isozyme types still possess considerable activity when measured spectrophotometrically. Undialyzed extracts must contain substrates utilized by ADH in the histochemical stain and this perhaps accounts for the "nothing dehydrogenase" activity present in crude, undialyzed extracts.

DISCUSSION

The results presented indicate that ADH in *Peromyscus* is controlled by more than a single genetic locus. Genetic variants of the Adh-6 isozyme do not affect the mobility or activity on gels of the Adh-5 isozyme. This would imply that Adh-6 and Adh-5 do not share common subunits. Based upon the genetic variants, it was found that the protein product of the Adh-6 gene is found in liver, kidney, and intestinal tissue.

Thorough genetic analysis of the Adh-6 isozyme has detected

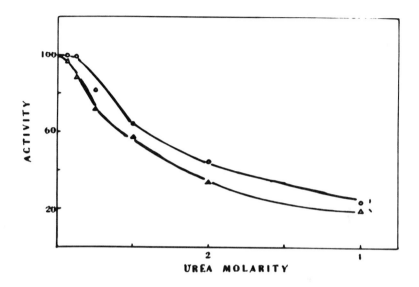

Fig. 8. Urea sensitivity of the fast and null ADH activities. A constant amount of enzyme activity is incubated for one hour at various urea concentrations. Spectrophotometric assays then determine relative activity remaining.

three alleles of the Adh-6 gene. The alleles are Adh-6^F, Adh-6^S and Adh-6^N coding for fast migrating, slow migrating, and absence of the isozyme, respectively. Furthermore, these results indicate that Adh-6 exists as a dimer, and the hybrid enzyme formed in the Adh-6 zone of Adh-6^F/Adh-6^S heterozygotes is most likely the result of random association of two differ-

TABLE III

EFFECTS OF DIALYSIS ON ADH ACTIVITY

		Change A_{340}/min/5µl enzyme
Non dialyzed	(+ ethanol)	0.036
	(- ethanol)	0.033
Dialyzed	(+ ethanol)	0.021
	(- ethanol)	0.002

ent monomeric subunits. Thus two homodimers (FF and SS) and a heterodimer (FS) are formed by random association of subunits which differ in net charge. Presumed heterozygotes found in a wild population of *Mus musculus* also possess the three banded ADH pattern (Selander et al, 1969). In a genetic polymorphism found in the kidney and intestine of human fetuses, presumed heterozygotes also show a three banded configuration for ADH (Smith et al, 1971).

It can be inferred from these findings that the subband(s) of the Adh-6 isozyme are controlled by the same Adh-6 gene since the subband is also altered in mobility in the variant phenotypes. The subband is also missing in the null phenotypes. This would make it appear that the subband is altered epigenetically or is a conformational isozyme. Inbred strains of *Drosophila melanogaster* have at least five electrophoretically distinguishable forms (Johnson and Denniston, 1964; Jacobson, 1968; Ursprung and Carlin, 1968; Grell et al, 1968). It is known in the *Drosophila* system that dialysis against high NAD concentration will cause interconversion of isozymes (Ursprung and Carlin, 1968; Jacobson, 1968; Knapp and Jacobson, 1972). Horse liver ADH may also be altered in electrophoretic mobility by treatment with NAD (McKinley-McKee and Moss, 1965). In the present study, simple preincubation with various concentrations of NAD did not cause any shift in migration of the main or subband of Adh-6. Prolonged dialysis has not been attempted because of rather large losses in activity encountered.

Grell et al (1968) recovered five chemically induced ADH negative mutations in *Drosophila*. Two of the ADH null alleles produced hybrid enzymes of altered electrophoretic mobility in the genetic hybrid. This suggests that the null alleles produced a product with little catalytic activity, but these polypeptides can associate with a fully active subunit to form an active hybrid molecule. In the *Peromyscus* system $Adh-6^S$/$Adh-6^N$ and $Adh-6^F$/$Adh-6^N$ heterozygotes produce only slow and fast isozymes, respectively. This could mean that the $Adh-6^N$ allele does not produce a polypeptide product or the polypeptide is so altered that it cannot dimerize. Alternatively, it may dimerize, but the hybrid could be inactive and thus undetectable. The spectrophotometric analysis showing high enzyme activity in null extracts before electrophoretic analysis may be erroneous due to many other contaminating enzyme systems that catalyze the oxidation and reduction of pyridine nucleotides. It may be possible to overcome this by using the assay system of Raskin and Sokoloff, (1968) because it does not depend on the content of NADH. This would be less sensitive to the effects of contaminating pyridine nucleotide oxidoreductase systems. Alternatively, the null form could be destroyed by electrophoresis. This is not corroborated by urea sensitivity experiments. The fast and null activity is about equally destroyed by the same urea concentration.

The several anodally migrating bands that are rather faintly staining in brain and kidney extracts (numbered 1-5) may be lactate dehydrogenase instead of ADH. Koen and Shaw, (1965) showed that lactate dehydrogenase in *Peromyscus* does stain when ethanol is substituted for lactate in the staining reaction. In addition, the heavily staining band in the five region for eye and stomach tissue does stain much more intensely and appears to be a true ADH isozyme. It would be of interest to compare substrate specificities of Adh-5 and Adh-6 since ADH is reported to be able to oxidize the alcohol form of vitamin A which is important in the visual process (Sund and Theorell, 1963; Koen and Shaw, 1966). Unfortunately, variants of Adh-5 have not been detected in this system so linkage relationships cannot be determined.

Attempts are being made to purify the fast form of Adh-6 so that monospecific antibodies to it can be produced. Hopefully, this can be used to determine the nature of the null Adh-6 variant.

ACKNOWLEDGEMENTS

This research was supported by a grant from the Committee on Research and Productive Scholarship, University of South Carolina.

REFERENCES

Bonnichsen, R. K. and N. G. Brink 1955. Liver alcohol dehydrogenase. *Methods in Enzymology* I: 495-503.

Courtwright, J. B., R. B. Imberski, and H. Ursprung 1966. The genetic control of alcohol dehydrogenase and octonal dehydrogenase isozymes in *Drosophila*. *Genetics* 54: 1251-1260.

Dawson, W. D. 1965. Fertility and size inheritance in a *Peromyscus* species cross. *Evol.* 19: 44-55.

Efron, Y. 1970. Alcohol dehydrogenase in maize: Genetic control of enzyme activity. *Science* 170: 751-753.

Ehrenberg, A. and A. Dalziel 1958. The molecular weight of horse liver alcohol dehydrogenase. *Acta Chem. Scand.* 12: 465-469.

Felder, M. R., J. G. Scandalios, and E. H. Liu 1972. Purification and partial characterization of two genetically defined alcohol dehydrogenase isozymes in maize. *Biochim. Biophys. Acta* 318: 149-159.

Grell, E. H., K. B. Jacobson, and J. B. Murphy 1965. Alcohol dehydrogenase of *Droshophila melanogaster:* Isozymes and genetic variants. *Science* 149: 80-82.

Grell, E. H., K. B. Jacobson, and J. B. Murphy 1968. Alterations of genetic material for analysis of alcohol dehydrogenase isozymes of *Drosophila melanogaster.* *Ann. N.Y. Acad. Sc.* 151: 441-455.

Hersch, R. T. 1962. Effect of sodium dodecyl sulfate on yeast alcohol dehydrogenase. *Biochim. Biophys. Acta* 58: 353-354.

Johnson, F. M. and C. Denniston 1964. Genetic variation of alcohol dehydrogenase in *Drosophila melanogaster.* *Nature* 204: 906-907.

Koen, A. L. and C. R. Shaw 1965. Multiple substrate specificities of some dehydrogenase molecules. *Biochem. Biophys. Res. Commun.* 15: 92-99.

Koen, A. L. and C. R. Shaw 1966. Retinol and alcohol dehydrogenase in retina and liver. *Biochim. Biophys. Acta* 128: 48-54.

Lieber, C. and L. DeCarli 1968. Ethanol oxidation by hepatic microsomes: adaptive increase after ethanol feeding. *Science* 162: 917-918.

Luntstorf, U. M., P. M. Schurch, and J. P. von Wartburg 1970. Heterogeneity of horse liver alcohol dehydrogenase: Purification and characterization of multiple molecular forms. *Eur. J. Biochem.* 17: 497-508.

McKinley-McKee, J. S. and D. W. Moss 1965. Heterogeneity of liver alcohol dehydrogenase on starch-gel electrophoresis. *Biochem. J.* 96: 583-587.

Meizel, S. and C. L. Markert 1967. Malate dehydrogenase isozymes of the marine snail, *Ilyanassa obsoleta.* *Arch. Biochem. Biophys.* 122: 753-765.

Pietruszko, R., A. F. Clark, J. M. Granes, and H. J. Ringold 1966. The steroid activity and multiplicity of crystalline horse liver alcohol dehydrogenase. *Biochem. Biophys. Res. Commun.* 23: 526-534.

Pietruszko, R. and H. Theorell 1969. Subunit composition of horse liver alcohol dehydrogenase. *Arch. Biochem. Biophys.* 131: 288-298.

Racker, E. 1950. Crystalline alcohol dehydrogenase from Baker's yeast. *J. Biol. Chem.* 184: 313-319.

Raskin, N. H. and Sokoloff, 1968. Brain alcohol dehydrogenase. *Science* 162: 131-132.

Scandalios, J. G. 1967. Genetic control of alcohol dehydrogenase isozymes in maize. *Biochem. Genet.* 1: 1-9.

Scandalios, J. G. 1969. Alcohol dehydrogenase in maize: genetic basis for isozymes. *Science* 166: 623-624.

Scandalios, J. G. 1969. Genetic control of multiple molecular forms of enzymes in plants: a review. *Biochem. Genet.* 3: 37-59.

Schenker, T., L. J. Teeple, and J. P. von Wartburg 1971. Heterogeneity and polymorphism of human liver alcohol dehydrogenase. *Eur. J. Biochem.* 24: 271-279.

Schwartz, D. 1966. The genetic control of alcohol dehydrogenase in maize: gene duplication and repression. *Proc. Natl. Acad. Sci. USA* 56: 1431-1436.

Schwartz, D. and T. Endo 1966. Alcohol dehydrogenase polymorphism in maize--simple and compound loci. *Genetics* 53: 709-715.

Selander, R. K., W. G. Hunt, and S. Y. Yang 1969. Protein polymorphism and genic heterozygosity in two European subspecies of the house mouse. *Evol.* 23: 379-390.

Shaw, C. R. 1965. Electrophoretic variation in enzymes. *Science* 149: 936-942.

Shaw, C. R. and A. L. Koen 1965. On the identity of "nothing dehydrogenase." *J. Histochem. and Cytochem.* 13: 431-433.

Smith, M., D. A. Hopkinson, and H. Harris 1971. Developmental changes and polymorphism in human liver alcohol dehydrogenase. *Ann. Hum. Genet. Lond.* 34: 251-271.

Sofer, W. A. and M. A. Hatkoff 1972. Chemical selection of alcohol dehydrogenase negative mutants in *Drosophila*. *Genetics* 72: 545-549.

Sund, H. and H. Theorell 1963. Alcohol dehydrogenases. In *The Enzymes* (ed. P. D. Boyer, H. Lardy, and K. Myrback) 7: 25-83.

Theorell, H., S. Taniguchi, A. Akeson, and L. Skursky 1966. Crytallization of separate steroid-active liver alcohol dehydrogenase. *Biochem. Biophys. Res. Comm.* 24: 603-610.

Theorell, H., A. P. Nygaard, and R. K. Bonnichsen 1955. Studies on liver alcohol dehydrogenase. III. The influence of pH and some anions on the reaction velocity constants. *Acta Chem. Scand.* 9: 1148-1165.

Ursprung, H. and L. Carlin 1968. *Drosophila* alcohol dehydrogenase: In vitro changes of isozyme patterns. *Ann. N.Y. Acad. Sc.* 151: 456-475.

Ursprung, H. and J. Leone 1965. Alcohol dehydrogenase: A polymorphism in *Drosophila melanogaster*. *J. Exptl. Zool.* 160: 147-154.

Wratten, C. and W. W. Cleland 1963. Product inhibition studies on yeast and liver alcohol dehydrogenase. *Biochemistry* 2: 935-941.

INVESTIGATION OF HUMAN PLASMA DNASE(S)
BY DNA-POLYACRYLAMIDE GEL ELECTROPHORESIS

JAMES E. STRONG and ROGER R. HEWITT
Department of Biology,
The University of Texas System Cancer Center
M. D. Anderson Hospital and Tumor Institute
Houston, Texas 77025

ABSTRACT. At least five electrophoretically distinct forms of endodeoxyribonuclease activity have been observed in human blood plasma by a fluorometric method utilizing polyacrylamide gels containing covalent circular DNA. Each form required divalent cation for activity in the neutral pH range. Optimal activity was observed with 5mM Mg^{++} plus 1mM Ca^{++}, or 5 mM Mn^{++} as cofactors. Patterns of activity were not affected by varying the conditions of electrophoresis or the amount of plasma analyzed. Re-electrophoresis of individual gel fractions containing plasma proteins yielded DNase activity which had the mobility expected of proteins contained in the fraction analyzed. None of the fractions yielded all of the forms observed in unfractionated plasma. The number and proportion of activities were not influenced by prolonged storage or by several treatments, including dialysis and incubation of the plasma before analysis. Some of the activities exhibited sensitivity to 65°C.

Our investigation suggests that the pattern of DNase activity observed in DNA-polyacrylamide gels is not an artifact of sample preparation or electrophoresis conditions. Rather, the pattern appears to be representative of multiple molecular forms of plasma DNase in vivo.

INTRODUCTION

Previous studies on DNase activities of human blood plasma and serum and on variations associated with the donor's state of health have yielded apparently contradictory results. Wroblewski and Bodansky (1950) were first to detect an acid DNase in human serum which was decreased from normal levels in a group of 50 patients having a variety of cancers. Kurnick (1953) also detected DNase activity in normal sera, but found no significant differences among sera from normal persons and from patients with infectious, malignant, or miscellaneous diseases. Kowlessor and McEvoy (1956) determined serum DNase I activities in normal subjects, patients with various diseases, and patients with pancreatic disease. Significant elevation of enzyme activity was observed only in patients with acute hemorrhagic pancreatitis. Gavosto, Buffa, and Moraini (1959) examined serum DNases I and II in a group of patients with liver disease, carcinomas, and miscellaneous conditions. No correlation was found between DNase I and DNase II. Significantly lower levels of DNase I were observed in patients with malignant diseases, whereas higher levels were occasionally observed in patients with advanced liver insufficiency.

These conflicting results may be attributed to differences in the sensitivities of the assay methods employed. Conflicting findings can also be expected from differences in blood collection and fractionation procedures. Potential problems are illustrated by the detailed studies of Herriott, Connolly, and Gupta (1961), Connolly, Herriott, and Gupta (1962) and Gupta and Herriott (1963). In these studies, DNase was assayed by inactivation of transforming *Hemophilus influenzae* DNA. They found that whole blood contains an acid and a neutral DNase, as well as at least one nuclease inhibitor. Platelets are rich in DNase activity. Thus, serum usually contains more activity (4 to 6-fold) than plasma due to platelet destruction during clotting. White cells are rich in nuclease inhibitor, and therefore, DNase levels may be decreased by white cell destruction. In view of the number of interactions among blood components that can influence DNase activity, a clear need exists for identifying and distinguishing the various activities in blood components.

We have investigated the DNase activity in human plasma utilizing an electrophoretic method which has been shown to be sensitive to as little as 5 picograms of pancreatic DNase I (Grdina, Lohman, and Hewitt, 1973). The rationale for utilizing an electrophoretic method for this study is based on the assumption that plasma DNase activity is derived from many cell types in the blood and tissues. If the DNases from various cell types are structurally or functionally distinct, changes in their contribution to the plasma might be identified by changes in the level of particular electrophoretic forms of DNase. In this paper we wish to report our observation of multiple electrophoretic forms of DNase activity in human plasma. We have adopted many of the recommendations of Ressler (1973) in our investigation of the nature of these electrophoretically distinct DNase activities and present evidence that these forms are present in human plasma and do not arise as artifacts of sample preparation or the conditions of electrophoresis.

MATERIALS AND METHODS

PM2 DNA purification. Covalent circular DNA is included in polyacrylamide gels to serve as a substrate for the detection of endoDNase activity as described by Grdina et al. (1973). The marine bacteriophage PM2 (Espejo and Canelo, 1968a) is a convenient source of DNA. Its genome is a covalently closed double-stranded DNA molecule with a molecular weight of about 6×10^6 daltons (Espejo, Canelo, and Sinsheimer, 1969). PM2 phage was produced by lytic infection of *Pseudomonas* BAL-31 (Espejo and Canelo, 1968b) by a procedure adapted from that of Salditt et al. (1972). After complete lysis of infected cells, polyethylene glycol (Carbowax 6000, Union Carbide) and sodium dextran sulfate 500 (Pharmacia) were added to final concentrations of 4.3% (w/v) and 0.235% (w/v), respectively. Separation of two phases was complete after 16 hr at 4°C. After recovery of the bottom phase, which contains the virus and cell debris, virus was partially purified by a modified method of Salditt et al. (1972). Our modifications include the use of AMSE buffer (1 M NaCl; 0.02 M Tris; 0.005 M EDTA; pH 8.1) for dilution of the dextran sulfate fractions, extraction of dextran sulfate from crude virus pellets, and for suspension of virus prior to ultracentrifugation (Spinco A30 rotor, 28,000 rpm, 2 hr., 4°C). Virus pellets were overlaid with NTE buffer (0.2 M NaCl; 0.02 M Tris, 0.02 M EDTA; pH 8.1) for 14-16 hr before DNA extraction. After dilution with 3 ml NTE per liter of initial lysate, the solution was lysed at 60°C by the addition of Sarkosyl NL97 (0.5%) and incubated for 10 min. The clear solution was chilled to 4^{d}C and extracted repeatedly with chloroform-isoamyl alcohol (24:1). Lower phases were extracted with NTE buffer. Aqueous phases were combined and the DNA precipitated by the addition of 2 volumes of ethanol. DNA was recovered by centrifugation and dissolved in TNE buffer. Covalently closed PM2 DNA was purified by equilibrium centrifugation in gradients of CsCl (0.803 gm/ml) containing ethidium bromide (0.39 mg/ml) in a Spinco 50 Ti rotor (42,000 rpm, 60 hr, 15°C).

Preparation of blood plasma: Whole blood was obtained by venipuncture into chilled heparinized vacutainer tubes (Becton-Dickinson). After centrifugation (3,000 xg, 5 min, 4°C) plasma was removed and centrifuged again (12,000 xg, 10 min, 4°C). Aliquots were analyzed immediately or stored frozen (-20°C).

DNA-Polyacrylamide gel electrophoresis: Electrophoresis was performed on 8 cm polyacrylamide gels (7%) containing covalent circular PM2 DNA (20 µg/ml) under conditions similar to those of Davis (1964). Plasma was diluted 1:10 in 20% sucrose containing 0.02% bromphenol blue and layered directly onto the stacking gel. A constant current of 1 ma/tube (∿ 100 V) was applied for 15 min and then 3 ma/tube (200-300 V) for 4 to 5 hr. Electrophoresis temperature was maintained at 4°C by cir-

culating coolant through the jacketed glass lower chamber
(Buchler Instr.).

In some cases a Tris-borate continuous electrophoresis system was used in which both acrylamide gels and the electrophoresis buffer contained 0.37 M Tris-borate (pH 9.5). Stacking
gels were not used with this system.

DNase assay: Analyses were performed essentially as described
by Grdina et al. (1973). The steps involved in this analysis
include: (1) electrophoresis of plasma proteins of PM2-poly-
acrylamide gels, (2) extrusion of the gels into test tubes containing appropriate buffers, (3) incubation to allow DNase
activities to react with the PM2 DNA by producing single or
double strand breaks, (4) staining the gels for about 16 hr in
a solution containing 0.01 M Tris (pH 8.1), 0.02 M NaCl, 0.005
M EDTA, and 5 μg/ml ethidium bromide, and (5) scanning the gels
for fluorescence. Identification of regions of DNase activity
within gels relies on the increased binding of ethidium bromide
by broken circular or linear DNA molecules compared to the more
limited binding of dye by covalent circular DNA. Thus, regions
of increased fluorescence result wherever DNase activity introduced breaks into PM2 DNA. This assay is illustrated in Fig.
1 and described further in Results.

RESULTS

PM2-Polyacrylamide gel electrophoresis: This method was developed by combining the DNA-polyacrylamide gel electrophoresis
method of Boyd and Mitchell (1965) and the fluorescence assay
for endoDNase developed by Paoletti, LePecq, and Lehman (1971).
This latter assay was based on the differential binding of
ethidium bromide (EB) by covalent circular and nicked circular
DNA, as illustrated in Fig. 1. Covalent circular DNA can bind
one EB molecule per 8 base pairs, whereas nicked circles or
linear DNA can bind one EB molecule per 5 base pairs. Binding
of EB to DNA increases the quantum efficiency of fluorescence
by a factor of 20 to 25. Thus, at saturating concentrations of
bound EB the relative fluorescence can serve as an indicator of
DNase activity. The lower illustration in Fig. 1 includes a
scan of fluorescence of a PM2-gel in which 50 picograms of
pancreatic DNase I was electrophoresed and then incubated for
105 min prior to staining with EB. The PM2 DNA in the region
containing the DNase was nicked, thus yielding increased
fluorescence in that region.

DNases of human plasma: Six regions of nuclease activity were
observed when a gel containing the equivalent of 2.5 λ of plasma

476

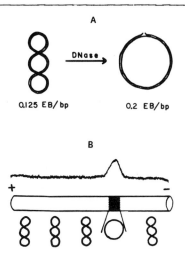

Fig. 1. Fluorometric analysis of endoDNase activity on covalent circular PM2 DNA. A: The reaction of DNase on PM2 DNA which produces nicked circular DNA is illustrated. The approximate numbers of ethidium bromide molecules bound per base pair are indicated for each form of DNA. B: The fluorometric assay of DNase in a PM2-polyacrylamide gel is illustrated. The fluorescence scan was obtained after a gel containing 50 picograms of bovine pancreatic DNase I was incubated for 105 min in 0.1M Tris(pH 7) and 5mM Mg^{++}. The form of the DNA within various gel regions is indicated.

was incubated for 30 min in a buffer containing 5 mM Mg^{++} and 1 mM Ca^{++} (Fig. 2C). The relative mobilities of the regions indicated by the arrows are 0.15, 0.32, 0.4, 0.51, 0.59, and 0.68. We found the combination of Mg^{++} and Ca^{++} to be nearly optimal for the expression of the plasma activities. Substitution of 5 mM Mn^{++} or 5 mM Mn^{++} and 1 mM Ca^{++} yielded nearly identical results to those presented in Fig. 2C. Mg^{++} (5 mM) was a less effective cofactor (Fig. 2B). No activity was observed in the presence of EDTA (Fig. 2A). Ca^{++} was a poor cofactor (data not shown). DNase activity was observed in most of the same regions when the amount of plasma analyzed was reduced 2 1/2-fold (Fig. 2D), indicating that the proportions of the different electrophoretic forms of DNase were not influenced by the amount of plasma protein analyzed.

Figure 3 demonstrates the effect of varying the incubation time on the pattern of DNase activity. No activity was observed without incubation at 37°C. Only one region was observed after 10 min, whereas the activities in 5 or 6 regions were

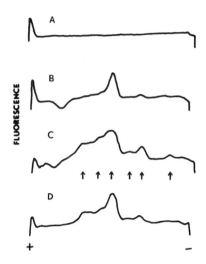

Fig. 2. Effect of cofactor and amount of protein on plasma DNase activity. Plasma (25λ in A, B, and C; 10λ in D) was electrophoresed in the Tris-glycine discontinuous system and then incubated for 30 min at 37°C in 0.1M Tris (pH 7) buffer containing, A: 5mM EDTA; B: 5mM MgCl$_2$; C and D: 5mM MgCl$_2$ and 1mM CaCl$_2$.

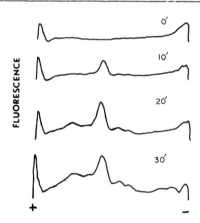

Fig. 3. Plasma (10λ) was electrophoresed in the Tris-glycine system and then incubated for the indicated times at 37°C in 0.1M Tris (pH 7), 5mM MgCl$_2$, and 1mM CaCl$_2$.

observed after 30 min. A similar analysis has been performed
using only 5 mM Mg^{++} as cofactor and increased amounts of plasma
or longer incubation times (data not shown). THe results were
similar to those of Fig. 3, suggesting that each of the electrophoretic forms of DNase responds similarly to cofactor. No
differential responses to cofactor have been observed.

The effect of gel and electrophoresis buffer composition
was examined by comparing the Tris-glycine HCl discontinuous
system (Fig. 3) with a Tris-borate continuous system (Fig. 4).

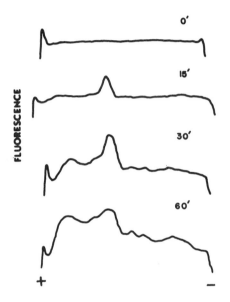

Fig. 4. Plasma (10λ) was electrophoresed in the 0.37M Tris-
borate (pH 9.5) continuous system and then incubated as
described in Fig. 3.

The same number of regions of activity and the same proportions
of activity were observed after electrophoresis under both
conditions.

The possibility that the multiple forms were an artifact
of electrophoresis was additionally investigated by fractionation and re-electrophoresis of the plasma proteins. A plasma
sample was electrophoresed and the gel was sliced in 3 mm
sections. Each section was then placed on top of a new gel,
electrophoresed and then examined for DNase activity. The results of analysis of 15 fractions are presented in Fig. 5.
Little DNase activity was observed in the first six fractions.

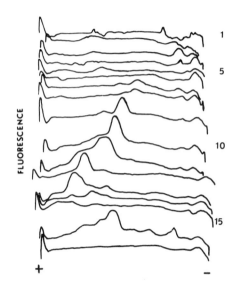

Fig. 5. Fractionation and re-electrophoresis of plasma proteins. After the electrophoresis of 25λ of plasma, the gel
was cut into 3mm slices. Each slice was placed on a new gel,
electrophoresed, and then incubated for 60 min as described in
Fig. 3. Scans 1 through 15 are presented sequentially from
the top (cathodal) to the bottom (anodal) slices from the
fractionated gel. Scan 16 (10λ of plasma, 30 min incubation)
was a reference gel which was electrophoresed with the fractionated gel. Scan 17 (no plasma, 60 min incubation) was a control
gel, which was electrophoresed with the slices.

Most activity was observed in fractions 9 through 13. The major
activity in each of these fractions had a relative mobility
similar to that expected for its location from the original gel.
In no case were all electrophoretic forms produced by the
electrophoresis of any fraction.

The results of a study on the heat sensitivity of plasma
DNases after electrophoresis is presented in **Fig. 6**. The major
activity near Rm=0.5 was more sensitive to heating at 65°C
than the faster electrophoresing form.

Summary of additional results (data not shown): The DNase
activity patterns were not affected by the following treatments
of plasma prior to electrophoresis: storage at 4°C for 10 days
or at -20°C for at least six months, dialysis against 5 mM Tris
(pH 8.3) with or without 10 mM mercaptoethanol, dialysis
against 0.37 m Tris (pH 8.3), incubation at 37°C for 1 hr with

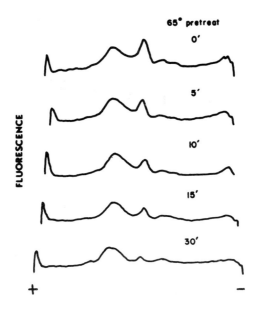

Fig. 6.   Plasma (10λ) was electrophoresed and then gels were
incubated at 65°C for the indicated times in 0.1M Tris (pH 7)
without cofactor.   DNase activity was then examined by incubation for 30 min as described in Fig. 3.

or without 10 mM mercaptoethanol, and addition of Triton X-100
(1.0%).

DISCUSSION

We have demonstrated that human plasma contains 5 or 6
electrophoretic forms of DNase activity.   These multiple forms
do not appear to be artifacts of the electrophoretic method
because: (1) the DNase patterns were not influenced by the
concentration or amount of plasma protein analyzed or by the
composition of the gel and electrophoresis buffer, and (2)
re-electrophoresis of individual gel fractions containing
plasma proteins yielded no more than two DNase forms which had
mobilities similar to that expected of proteins contained in
the fraction analyzed.   None of the fractions yielded all of
the forms observed in unfractionated plasma.

It appears that the electrophoretic patterns are representative of multiple forms of plasma DNase in vivo, because their
number and proportional activities were not influenced by
prolonged storage or by several treatments of the plasma before analysis, including dialysis or incubation.   In addition,

similar DNase patterns have been observed in the plasma and/or sera from several donors (data not shown). The nature of our analysis in PM2-polyacrylamide gels insures that all the activities we detect have similar specificity. They must be endonucleases, since covalent circular DNA provides no site for the initiation of exonucleolytic activity. All the activities we have observed require divalent cation as cofactor and they are active in the neutral pH range. Thus, each of the forms has enzymatic properties in common with bovine pancreatic DNase I (Laskowski, 1971) and the DNase previously purified from human serum by Doctor (1963).

ACKNOWLEDGEMENTS

The interest and support of Dr. David Anderson is gratefully acknowledged. This research has been supported in part by NIH grant GM 19513-C1 and CA 5831 to D. A. and grant G-315 from the Robert A. Welch Foundation to R. H. J. S. is a Rosalie B. Hite Predoctoral Fellow. Cathy Ketterman and Marie Hokanson have provided excellent technical assistance in the preparation of PM2 DNA.

REFERENCES

Boyd, J. B. and H. K. Mitchell 1965. Identification of deoxyribonucleases in polyacrylamide gel following their separation by disk electrophoresis. *Anal. Biochem.* 13: 28-42.

Connolly, J. H., R. M. Herriott, and S. Gupta 1962. Deoxyribonuclease in human blood and platelets. *Brit. J. Exptl. Path.* 43: 392-401.

Davis, B. J. 1964. Disc electrophoresis II. Method and Application to human serum proteins. *Ann. N. Y. Acad. Sci.* 121: 404-427.

Doctor, V. M. 1963. Studies on the purification and properties of human plasma deoxyribonuclease I. *Arch. Biochem. Biophys.* 103: 286-290.

Espejo, R. T. and E. S. Canelo 1968a. Properties of bacteriophage PM2: A lipid containing bacterial virus. *Virol.* 34: 738-747.

Espejo, R. T. and E. S. Canelo 1968b. Properties and characterization of the host bacterium of bacteriophage PM2. *J. Bact.* 95: 1887-1891.

Espejo, R. T., E. S. Canelo, and R. L. Sinsheimer 1969. DNA bacteriophage PM2: A closed circular double-stranded molecule. *Proc. Nat. Acad. Sci. U. S. A.* 63: 1164-1168.

Gavosto, F., F. Buffa, and G. Maraini 1959. Serum deoxyribonuclease I and II in pathologic conditions other than pancreatic diseases. *Clin. Chim. Acta* 4: 192-196.

Grdina, D. J., P. H. M. Lohman, and R. R. Hewitt 1973. A fluorometric method for the detection of endodeoxyribonuclease on DNA-polyacrylamide gels. *Anal. Biochem.* 51: 255-264.

Gupta, S., and R. M. Herriott 1963. Nucleases and their inhibitors in the cellular components of human blood. *Arch. Biochem. Biophys.* 101: 88-95.

Herriott, R. M., G. H. Connolly, and S. Gupta 1961. Blood nucleases and infectious viral nucleic acids. *Nature* 189: 817-820.

Kowlessor, O. D., and R. K. McEvoy 1956. Deoxyribonuclease I activity in pancreatic disease. *J. Clin. Invest.* 35: 1325-1330.

Kurnick, N. B. 1953. Deoxyribonuclease activity of sera of man and some other species. *Arch. Biochem. Biophys.* 43: 97-107.

Laskowski, M. 1971. Deoxyribonuclease I. In: *The Enzymes.* 4: 289-311.

Paoletti, C., J. B. LePecq, and I. R. Lehman 1971. The use of ethidium bromide-circular DNA complexes for the fluorometric analysis of breakage and joining of DNA. *J. Mol. Biol.* 55: 75-100.

Ressler, N. 1973. A systematic procedure for the determination of the heterogeneity and nature of multiple electrophoretic bands. *Anal. Biochem.* 51: 589-610.

Salditt, M., S. N. Braunstein, R. D. Camerini-Otero, and R. M. Franklin 1972. Structure and synthesis of a lipid-containing bacteriophage X. Improved techniques for the purification of bacteriophage PM2. *Virol.* 48: 259-262.

Wroblewski, F., and O. Bodansky 1950. Presence of deoxyribonuclease activity in human serum. *Proc. Roy. Soc. Exp. Biol. Med.* 74: 443-445.

ISOELECTRIC FOCUSING ON POLYACRYLAMIDE GEL
AND STARCH GEL ELECTROPHORESIS OF SOME
GADIFORM FISH LACTATE DEHYDROGENASE ISOZYMES

PAUL H. ODENSE and TED C. LEUNG
Halifax Laboratory
Fisheries and Marine Service
Research and Development Directorate
Environment Canada

ABSTRACT. The lactate dehydrogenase (LDH) isozymes of the tissues of fourteen species of gadiform fishes were examined by starch gel electrophoresis and by isoelectric focusing on polyacrylamide gel by a modification of the method of Awdeh, Williamson, and Askanas (1968).

Gadiform fish have three LDH loci producing A, B, and C subunits respectively. The B_4 isozyme activity predominates in heart tissue, C_4 in liver, but A_4 was generally more active in gas gland than muscle tissue. The electrophoretic mobility of C_4 was always more cathodal than the A_4 or B_4 isozyme of the species.

During isoelectric focusing isozymes are concentrated rather than diffused. Weaker isozyme activity is thereby detected. Thus B_4 activity was found in all tissues, A_4 in nearly all tissues, and C_4 activity was detected in more tissues by this technique than by gel electrophoresis.

Extra bands were resolved in retinal and liver tissue extracts. Their identity must await subunit composition analysis.

The presence of the A_2B_2 isozyme was found in cod and haddock heart extracts. A cusk A subunit mutant and a haddock B subunit mutant were found.

Approximate isoelectric point values for A_4, B_4, and C_4 isozymes were determined. The C_4 value was always higher than the corresponding A_4 or B_4 isoelectric point. The isoelectric point values are preferable for comparison purposes to relative electrophoretic mobilities. Mutant subunits produced relatively large shifts in isoelectric points.

INTRODUCTION

Fishes of the order Gadiformes have been classified by various workers, among them Svetovidov (1948). The species referred to or examined in the present study of gadiform lactate dehydrogenase (E.C. 1.1.1.27) are listed according to his classification in Table I. The common names which are given correspond to those approved by the American Fisheries Society (1970).

Like most teleosts, the gadiform species have LDH isozyme patterns which seem to be the tetrameric products of three LDH loci, A, B, and C. (Shaklee et al., 1973). The gadiform A and B subunits conform in character to the mammalian A and B subunits and to the corresponding subunits of other fish (Markert and Faulhaber, 1965). The C subunit appears to have arisen from a duplication of the LDH B gene (Horowitz and Whitt, 1972). Previous authors (Odense et al., 1969; Utter and Hodgins, 1969; Lush, 1970; Odense et al., 1971; Sensa-baugh and Kaplan, 1972) have reported unique retinal or liver LDH subunit activity among gadiform species and have employed various nomenclatures to describe subunit activities which may now be attributed to and designated as the C subunit. The gadiform C_4 isozyme is characterized by its cathodal elec-trophoretic mobility. In all cases in the above studies and among gadiform species examined by Shaklee et al., 1973, the A_4 and B_4 isozymes showed greater anodal electrophoretic mo-bility than the C_4 isozyme.

Mutant forms of all these subunits have been reported. These are listed in Table II. Examination of the mutant form reported in Pacific hake by Utter and Hodgins (1969) suggests that this is a case of a C subunit mutation. Thus there are mutant forms of B subunits in six species, A mutants in three species, and C mutants in two species. Not all reports include the number of specimens nor the number of loci studied. It is probable that more mutants will be found in a systematic survey of all loci of these species. These mutant alleles are useful in the determination of the subunit composition of isozyme bands. In addition, protein polymorphisms have been useful in characterizing and identifying different populations of a species, such as demonstrated by Sick (1965, 1965a) in his use of hemoglobin polymorphisms to compare cod populations.

In the majority of isozyme studies and especially in LDH studies, the separation of the isozymes has been based on Smithies (1955, 1959) starch gel electrophoresis techniques or on some modification thereof. Recently Awdeh et al. (1968) described a method of isoelectric focusing in polyacrylamide gel, and we have adapted this procedure for the study of LDH isozymes. In the present paper isoelectric focusing and starch gel electrophoresis techniques have been used to study gadiform LDH isozymes.

TABLE   I

CLASSIFICATION OF SOME OF THE FISHES OF THE ORDER *GADIFORMES*[1]

ORDER GADIFORMES

Family *Moridae*
Genus *Antimora* Günther
A. *microlepis Beau*.........................Longfin cod[2]

Family *Gadidae*
Subfamily *Lotinae*
Genus *Brosme (Cuvier)* Oken
B. *brosme* (Müll.).........................Cusk
Genus *Enchelyopus* Bloch and Schneider
E. *cimbrius* (L.)..........................Fourbeard rockling
Genus *Phycis* Rose
Ph. *chesteri* (Goode & Bean)..............Longfin hake
Genus *Urophycis* Gill
U. *tenuis* (Mitch.)........................White hake
Genus *Lota (Curvier)* Oken
L. *Lota lota (L.)*.........................Burbot

Subfamily *Merlucciinae*
Genus *Merluccius* Rafinesque
M. *productus* (Ayres).....................Pacific hake
M. *bilinearis* (Mitch.)...................Silver hake

Subfamily *Gadinae*
Genus *Pollachius* (Nilsson) Bonaparte
P. *virens* (L.)...........................Pollock
Genus *Melonogrammus* Gill
M. *aeglefinus* (L.).......................Haddock
Genus *Gadus* Linné
G. *morhua morhua* L. ....Atlantic cod
G. *morhua macrocephalus Til.* ....Pacific cod
Genus *Microgadus* Gill
M. *tomcod* (Walb.)........................Atlantic tomcod
M. *proximus* (Girard).....................Pacific tomcod
Genus *Boreogadus* Günther
B. *saida* (Lep.)..........................Arctic cod
Genus *Theragra* Lucas
Th. *chalcogramma chalcogramma* (Pall.)....Walleye pollock

[1] Classification of Svetovidov 1948
[2] Species not examined in present study.

TABLE II
REPORTED MUTANT FORMS OF GADIFORM LDH SUBUNITS

Species	Mutant Subunits		
Cusk	A,A' (1),	B,B' (2)	
Pacific hake			C,C' (3)
Silver hake		B,B', B" (4)	
Pollock	A,A' (5) (2)	B,B' (2)	
Haddock	A,A'A" (2) (6)	B,B' (1)	
Atlantic cod		B,B', B', B'" (7)	
Pacific cod			C,C' (6)
Tomcod		B,B' (2)	

1 - Present study
2 - Odense et al. 1971
3 - Utter and Hodgins 1969
4 - Markert and Faulhaber 1965
5 - Lush 1970
6 - Sensabaugh and Kaplan 1972
7 - Odense et al. 1969

METHODS AND MATERIALS

Specimens. The fish species which were studied are indicated in Table I. Frozen, samples were kindly supplied by Dr. H. Tsuyuki, Vancouver Laboratory, British Columbia (Pacific hake, Pacific cod, and walleye pollock), Dr. J. Clayton, Freshwater Institute, Winnipeg, Manitoba (turbot), and E. J. Sandeman, St. John's Biological Station, Newfoundland, (arctic cod and fourbeard rockling). The remaining species were caught alive near Halifax and were held in marine aquaria until used.

Preparation of tissue extracts. Tissue samples, dissected from freshly killed or thawed specimens, were blended in a Sorvall microhomogenizer for 1 min together with 1-5 volumes of cold 0.25M sucrose, 0.001M EDTA solution at pH 7.0. The resultant homogenate was centrifuged at 30,000 x g for 20 min at 2°C and the supernatant was used directly for electrophoresis.

Electrophoresis. A Buchler (Fort Lee, N.J.) vertical starch gel apparatus was used. A gel of 12% starch (Connaught Laboratories, Toronto) was prepared using a Tris-EDTA-borate buffer adjusted to pH 8.7 (Odense et al., 1966a). Only the haddock muscle isozymes were not separated by this buffer system and a Tris-citrate buffer (Syner and Goodman 1966) was used to resolve these bands. The electrophoresis runs were made in the coldroom at 2°C and lasted 18-20 hours at a voltage gradient of 10-15 volts/cm. The gels were sliced and stained as described previously (Odense et al., 1966a).

Isoelectric focusing on polyacrylamide gels. The method used for isoelectric focusing on polyacrylamide gels was a slight modification of the procedure described by Awdeh et al. (1968). To prepare the gel slab, 75 ml of acrylamide solution (5.5%) containing catalyst and 0.7% ampholyte carrier (Ampholine, LKB, Sweden) was poured into a shallow lucite tray (inside dimensions 28 cm x 12 cm x 0.8 cm deep) carefully levelled on a leveling table. The tray was covered with a glass plate which was lowered slowly from one end, allowing the slight excess of solution to spill over the edge of the tray. After 1 hour of photopolymerization under a fluorescent lamp, the gel was cooled 2 hours in the refrigerator. The gel was then inverted and the lucite tray lifted off, leaving the slab on the glass plate. Sample slots were cut in the gel by using a scalpel or in some cases slot formers formed part of the lucite gel mold and produced a row of sample slots near one end of the gel. Approximately 20µl of sample was applied to each slot.

For the isoelectric focusing the glass plate was rested horizontally on two of the electrode vessels used for the starch gel electrophoresis. The sample slots were placed near the anode end. One electrode vessel (anode) was filled with 1.4% phosphoric acid solution, the other with 2% ethanolamine solution (cathode). Fiberglass cloth wicks were used to make contact between the electrode solutions and the ends of the gel. During the run the gel, with the exception of the slots, was covered with a plastic film (Saran wrap, Dow Chemical, Sarnia). Runs lasted 19-20 hours at a current of 18 milliamps and 450 volts.

At the end of the run the pH gradient was determined by removing gel samples along one side of the gel with a small cork borer. After soaking in 1 ml of water the pH of each gel sample was determined.

To stain the gel the glass plate was placed on a sheet of Saran wrap resting on a piece of aluminum foil. The foil was folded up around the edges of the plate, thus forming a disposable staining tray. The same staining solution used for LDH in the starch gels was employed. The gel was incubated in the dark at room temperature until all the bands appeared. To record the results the staining solution was washed off and the gel slab, still adhering to the glass plate, was placed on a lightbox and photographed.

RESULTS AND DISCUSSION

The results of starch gel electrophoresis of the LDH isozymes of three individual cusk specimens is shown in Fig. 1. The homozygote individual of the AABB genotype is seen in Fig. la. The A_4 band in this species migrates faster towards the anode than the B_4 band. The C_4 isozyme is active in liver but scarcely visible in other tissues. Its characteristic position near the origin is seen. The activity of the B subunit is present in all tissues except liver and kidney. However the A_4 activity is greatest in the gas gland, and not in skeletal muscle tissue. This was usually found to be the case among the gadiforms studied. Fish in this order have a closed swim bladder and the gas gland is an active tissue, playing an important role in maintaining neutral buoyancy.

In Fig. lb the isozymes of an AABB' heterozygote individual are shown. The B'_4 band is nearly as cathodal as the C_4. Less B or B' subunit activity is seen in the other tissues as the activity is diluted by distribution over the five B and B' tetramers.

The final figure in this group, Fig. lc, shows the isozymes of an individual heterozygous at the A locus. The small amount of A or A' subunit activity does not show up in any tissue except skeletal muscle and gas gland where the five A and A' tetramer combinations are seen.

The isoelectric focusing LDH zymograms of the heart, muscle, liver, retina, gonad, and gas gland of these individuals are shown in Fig. 2 for comparison. The heart band B_4 isozyme activity is seen in all these tissues and has the lowest isoelectric point (pI), approximately pH 5.4. It may be identified because it is the dominant band in the heart tissue and five bands are present when the locus is heterozygous, Fig. 2b. The B'_4 band, also identified by its tissue activity and its presence in the heterozygous individual, has a pI of about 6.0. The A_4 band has an intermediate pI of 5.8, while the A'_4 isozyme pI is 7.0, Fig. 2c. In these runs the A_4 band may be seen in all tissues although barely present in the gonad tissue extract. The dominant liver isozyme C_4 has a pI of 6.2, intermediate between A_4 and A'_4. Generally the more cathodal isozymes have a higher pI but as shown in this case the pI does not always determine the mobility on the starch gel.

As noted by Sensabaugh and Kaplan (1972), the liver bands have a tendency to streak in starch gel electrophoresis. In all the gadiform species run on polyacrylamide gels the isoelectric focusing procedure produced a series of liver isozyme bands. Whether these are partially degraded tetramers

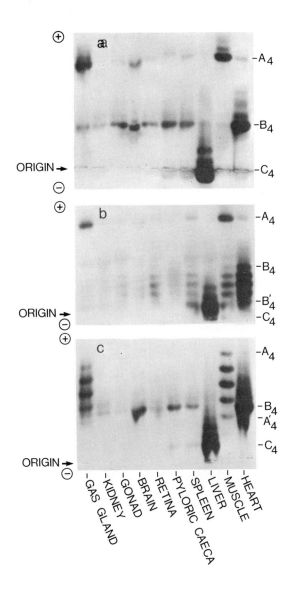

Fig. 1: Tissue LDH isozymes of cusk after starch gel electro-phoresis in a Tris-EDTA-borate buffer, pH 8.7. Fig. 1a is a homozygous AABB genotype, 1b is a heterozygous AABB' genotype and 1c is a heterozygous AA'BB genotype.

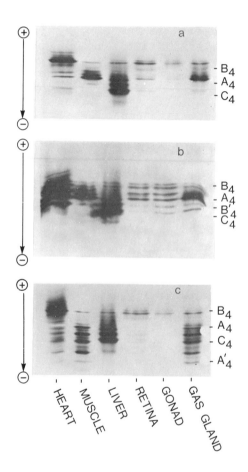

Fig. 2: Tissue LDH isozymes of cusk after isoelectric focusing on polyacrylamide gel using an ampholyte with a range of
pH 3-10. These are the same fish examined by starch gel electrophoresis (fig. 1). Their genotypes are: fig. 2a, AABB,
fig. 2b, AABB', and fig. 2c, AA'BB.

possibly produced by the action of lysosomal enzymes in the liver or whether they arise from some other artifactual cause is not known. However the isoelectric focusing technique does resolve the liver isozymes into discrete bands and makes them amenable to other means of subunit analysis, such as immunological techniques using specific subunit antisera, temperature inhibition studies, or differential stains (Whitt and Booth, 1970; Whitt, 1970).

The presence or absence of various subunit activities in different tissue extracts depends upon the same factors in isoelectric focusing as in starch gel electrophoresis. These factors include the freshness and homogeneity of the tissue, the duration of staining, the stain buffer composition, etc. Nevertheless, there is an advantage to the isoelectric focusing technique in that the proteins are literally focused at their isoelectric point and rather than diffusing during the run they become concentrated. Thus weak isozyme activity is more likely to be detected in this system. This is illustrated by an examination of the tissue runs in Fig. 2. The C_4 isozyme activity is present in heart, retina, and gas gland, and is sometimes seen in gonad and skeletal muscle. This activity was not detected in the starch gel runs.

The isoelectric focusing runs of the heart, skeletal muscle, liver, retina, gonad, and gas gland tissues of white hake and of longfin hake are shown in Fig. 3. These fish have C_4 bands which migrate towards the cathode in starch gel electrophoresis and are seen here to possess high isoelectric points (white hake 8.2, longfin hake 8.3). The B_4 and A_4 bands are seen in each tissue as well as an intermediate band which may be the A_2B_2 isozyme. Additional bands are present in the retinal tissue and are especially seen in the white hake. In many of the gadiforms the isoelectric focusing runs revealed extra bands between the A_4 and B_4 isozymes which did not correspond to an A_2B_2 isozyme band. Previous studies (Odense et al, 1966; 1971; Lush, 1970; Shaklee et al., 1973) have shown the presence of B_4, C_4 and the intermediate tetramers, B_3C, B_2C_2, and BC_3 in eye and brain tissues in starch gel runs. However, with the notable exception of the silver hake tissue isozymes (Shaklee et al., 1973), these intermediate bands do not seem to occur in other tissues even when a considerable amount of B_4 and C_4 isozyme activity is present in the tissue. Whether or not the extra band or bands seen in retinal extracts in most of the gadiform species isoelectric focusing runs are heterotetramers of B and C subunits is uncertain and must await subunit composition analysis.

In haddock muscle three mutant A subunits have been described, A, A' and A" (Odense et al., 1971; Sensabaugh and

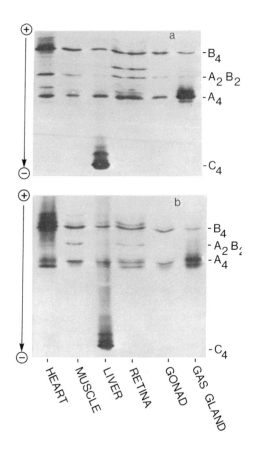

Fig. 3: Tissue LDH isozymes of a) white hake and B) long-fin hake, after isoelectric focusing on polyacrylamide gel using a pH 3-10 ampholyte.

Kaplan, 1972). A starch gel separation of the six possible
phenotypes is shown in Fig. 4a. An isoelectric focusing run
of the AA, AA", and AA' genotypes, using an ampholyte with a
pH range of 3-10 is shown in Fig. 4b. This separation was
improved by using ampholytes with a pH range of 5-10, and
the resolution achieved is shown in Fig. 4c. Although several sub-bands or satellite bands are present, it is easy to
pick out the five tetramer isozymes in each heterozygote
phenotype pattern. The approximate pI's of the A_4, A'_4, and
$A"_4$ isozymes are 7.1, 6.7, and 6.5 respectively.

During the survey of haddock muscle phenotypes a B subunit mutant was found. In Fig. 5a the isoelectric focusing
run of the heart extracts of a haddock AABB and the heterozygous AABB' genotype are shown, together with some haddock
muscle extract runs. The B_4 isozyme band is seen in the
muscle extracts as well as in the heart extracts. An inter-
mediary band is visible between the A_4 and B_4 isozymes of the
heart extract. This can be identified as an A_2B_2 isozyme
because in the heterozygous heart extract the three isozymes
A_2B_2, A_2BB', and $A_2B'_2$ are present, as well as the five B and
B' tetramers. This was the first haddock B mutant found after
typing over 700 haddock heart and muscle tissues. A similar
situation in cod is shown by the isoelectric focusing run of
three cod heart extracts, a BB, a BB', and a B'B' genotype,
seen in Fig. 5b. In tomcod and cod starch gel electrophoresis
runs a band is sometimes seen between the A_4 and B_4 isozymes.
The isoelectric focusing run clearly indicates this is the
A_2B_2 band as all three bands, A_2B_2, A_2BB', and $A_2B'_2$ are
present in the heterozygous BB' genotype, as in the case of
the haddock.

The isoelectric focusing and the starch gel electrophoresis runs described above were typical of the patterns obtained with the other gadiform species examined in this study.
The relative mobilities of the A_4, B_4, and C_4 isozymes in the
starch gel runs are listed in Table III. In such a comparison it is necessary to specify the buffer and pH used. In
the case of isoelectric focusing, the isoelectric points of
the isozymes are fixed values. However the pI values obtained in the present study must be considered as approximate
values only, since the method for determining the pH gradient
is still somewhat crude. Despite this, by comparison runs and
by repeated runs, a table of approximate pI values of A_4,
B_4, C_4, and mutant tetramer isozymes for several species of
gadiforms has been established (Table IV). In all cases the
B_4 isozyme of the common B allele has a lower pI than the
corresponding A_4 isozyme of the common A allele, and in all
but the haddock the pI of the A_4 isozyme is less than the

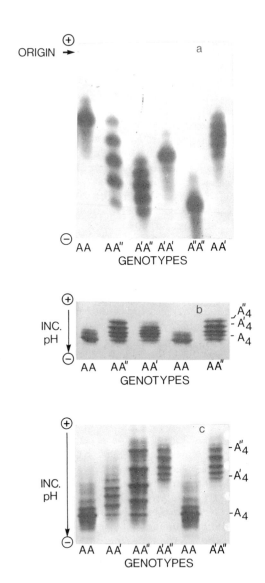

Fig. 4: The LDH isozyme patterns of some haddock muscle
extracts after starch gel electrophoresis or isoelectric focusing on polyacrylamide gel. Fig. 4a shows the starch gel
electrophoresis patterns of muscle extracts of the six genotypes arising from the three A subunit mutants, A, A' and A".
A Tris-citrate buffer was used in the gel. Fig. 4b shows

Fig. 4 Continued.   the isoelectric focusing separation of
the muscle isozymes of three genotypes AA, AA', and AA".
The ampholyte used had a range of pH 3-10. Fig. 4c is a
similar separation of the isozymes of different muscle genotypes. This time the isoelectric focusing was effected with
an ampholyte of pH 5-10. Use of an ampholyte with a narrow
pH range produces a better separation of isozymes with close
isoelectric points.

TABLE III

RELATIVE ELECTROPHORETIC MOBILITIES

OF COMMON A_4, B_4, and C_4

ISOZYMES OF GADIFORM SPECIES[1]

Species	Relative Isozyme Bobility				
Cusk	A_4	>	B_4	>	C_4
Fourbeard rockling	B_4	>	A_4	>	C_4
Longfin hake	B_4	=	A_4	>	C_4
White hake	B_4	>	A_4	>	C_4
Pacific hake	B_4	>	A_4	>	C_4
Silver hake	B_4	>	A_4	>	C_4
Pollock	B_4	>	A_4	>	C_4
Haddock	B_4	>	A_4	>	C_4
Atlantic cod	A_4	=	B_4	>	C_4
Pacific cod	A_4	=	B_4	>	C_4
Atlantic tomcod	B_4	>	A_4	>	C_4
Arctic cod	A_4	>	B_4	>	C_4
Walleye pollock	A_4	=	B_4	>	C_4

[1] Buffer = Tris, EDTA, borate, pH 8.7 (Odense et al., 1966a)

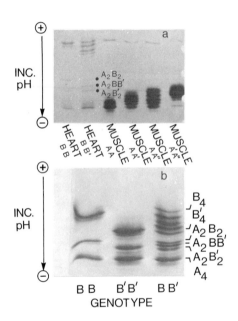

Fig. 5: The LDH isozyme patterns of haddock and cod heart
and of haddock muscle after separation by isoelectric focusing with an ampholyte of pH 3-10. Fig. 5a shows a haddock
heart BB and BB' genotypes. The A_2B_2 isozyme is seen in the
homozygote and the A_2B_2, A_2BB', and $A_2B'_2$ isozymes are seen in
the heterozygous pattern. Fig. 5b shows a similar situation
in cod. The five B and B' tetramer combinations forming five
isozymes are seen in the heterozygote BB' genotype. The B_4
and B'_4 isozymes have different isoelectric points. The A_2B_2'
isozyme and the $A_2B'_2$ isozymes appear respectively in the

Fig. 5 continued.   BB and B'B' homozygous genotype patterns, while the A_2B_2, A_2BB' and $A_2B'_2$ isozymes are all present in the heterozygous BB' genotype pattern of cod heart.

TABLE IV
APPROXIMATE ISOELECTRIC POINTS OF GADIFORM LDH ISOZYMES

| Isozyme | Isoelectric Point | | | | | |
Species	A_4	A'_4	A''_4	B_4	B'_4	C_4
Cusk	5.8	7.0		5.4	6.0	6.2
Fourbeard rockling	6.9			6.2		8.0
Longfin hake	6.4			5.6		8.3
White hake	6.5			5.5		8.2
Pacific hake	6.9			5.5		7.0
Silver hake	6.4			5.3		7.9
Pollock	6.3			6.2		6.4
Haddock	7.1	6.7	6.5	5.3		6.5
Atlantic cod	6.3			5.3	5.4	6.6
Pacific cod	6.9			5.9		7.0
Arctic cod	6.7			5.8		7.9
Walleye pollock	6.0			5.4		7.4

corresponding pI of the C_4 isozyme of the same species.   The pI of the C_4 isozymes of members of the subfamily Lotinae are, with the exception of the cusk, higher than the values found for the other species.   No other subfamily characteristics are apparent.

In conclusion, the technique of isoelectric focusing on polyacrylamide gel is seen as a useful complement to the study of LDH isozymes by the technique of starch gel electrophoresis.   In some respects the resolving power is greater since the proteins are concentrated at their isoelectric point and by a choice of ampholyte ranges good separations of proteins with nearly identical pI values may be obtained. The values of pI are absolute values and more useful for comparison purposes than relative mobilities. In the present study the A_4, B_4, and C_4 isozyme bands were identified by their relative activities in specific tissues or by making use of heterozygous individuals to identify the subunit activity. However, the identity of some bands remains unknown and the application of the techniques of subunit composition analysis mentioned previously should be tried.

REFERENCES

American Fisheries Society 1970. A list of common and scientific names of fishes from the United States and Canada. Special Publication No. 6, third edition.

Awdeh, Z. L., A. R. Williamson, and B. A. Askanas 1968. Isoelectric focusing in polyacrylamide gel and its application to immunoglobulins. *Nature* 219 (5149): 66-67.

Horowitz, J. J. and G. S. Whitt 1972. Evolution of a nervous system specific lactate dehydrogenase isozyme in fish. *J. Exp. Zool.* 180: 13-31.

Lush, I. E. 1970. Lactate dehydrogenase isoenzymes and their genetic variation in coalfish *(Gadus virens)* and cod *(Gadus morrhua)*. *Comp. Biochem. Physiol.* 32: 23-32.

Markert, C. L. and I. Faulhaber 1965. Lactate dehydrogenase isozyme patterns of fish. *J. Exp. Zool.* 159: 319-332.

Odense, P. H., T. C. Leung, T. M. Allen, and E. Parker 1969. Multiple forms of lactate dehydrogenase in the cod *Gadus morhua L.* *Biochem. Genet.* 3: 317-334.

Odense, P. H., T. M. Allen, and T. C. Leung 1966a. Multiple forms of lactate dehydrogenase and aspartate aminotransferase in herring *(Clupea harengus harengus L)*. *Can. J. Biochem.* 44: 1319-1326.

Odense, P. H., T. C. Leung, and Y. M. MacDougall 1971. Polymorphism of lactate dehydrogenase (LDH) in some gadoid species. Conseil International Pour L'exploration de la Mer. Extrait du *Rapports et Proces Verbaux*, Vol. 161. Special Meeting on the Biochemical and Serological Identification of Fish Stocks, Dublin, 1969.

Sensabaugh, G. F. Jr. and N. O. Kaplan 1972. A lactate dehydrogenase specific to the liver of gadiod fish. *J. Biol. Chem.* 247: 585-593.

Shaklee, B., K. L. Kepes, and G. S. Whitt 1973. Specialized lactate dehydrogenase isozymes: the molecular and genetic basis for the unique eye and liver LDHs of teleost fishes. *J. Exp. Zool.* 185: 217-240.

Sick, K. 1965. Haemoglobin polymorphism of the cod in the Baltic Sea and the Danish Belt Sea. *Hereditas* 54: 19-48.

Sick, K. 1965a. Haemoglobin polymorphism of cod in the North Sea and the North Atlantic ocean. *Hereditas* 54: 49-69.

Smithies, O. 1955. Zone electrophoresis in starch gels: Group variation in the serum proteins of normal human adults. *Biochem. J.* 61: 629-641.

Smithies, O. 1959. An improved procedure for starch-gel electrophoresis: Further variations in the serum proteins of normal individuals. *Biochem. J.* 71: 585 - 587.

Svetovidov, A. N. 1948. Gadiformes. *Fauna of the USSR*, Vol. IX, No. 4.

Syner, F. N. and M. Goodman 1966. Polymorphism of lactate dehydrogenase in gelada baboons. *Science* 151: 206-208.

Utter, F. M. and H. O. Hodgins 1969. Lactate dehydrogenase isozymes of Pacific hake (*Merluccius productus*). *J. Exp. Zool.* 172: 59-68.

Whitt, G. S. and G. M. Booth 1970. Localization of lactate dehydrogenase activity in the cells of the fish (*Xiphophorus helleri*) eye. *J. Exp. Zool.* 174: 215-224.

Whitt, G. S. 1970a. Developmental genetics of the lactate dehydrogenase isozymes of fish. *J. Exp. Zool.* 175: 1-35.

GENETIC AND STRUCTURAL BASIS
FOR ANIMAL HEMOGLOBIN HETEROGENEITY

HYRAM KITCHEN
Division of Comparative Medical Research
Center for Laboratory Animal Resources and
Department of Biochemistry
Michigan State University
East Lansing, Michigan 48824

ABSTRACT. The genetic and environmental basis of animal heterogeneity will be compared to and contrasted with the heterogeneity seen with the human hemoglobins and their abnormal forms. Animal hemoglobin heterogeneity due to multiple, polymorphic and abnormal hemoglobins will be illustrated. The ontogeny of animal hemoglobin will be compared to the well established developmental sequence of the emergence of embryonic, fetal and adult hemoglobin in man. It is quite clear that in certain animal species the genetic basis of hemoglobin heterogeneity is limited to as few as three pairs of alleles for the various hemoglobin polypeptide chain types. In these examples only a single hemoglobin type may be prominent at any one stage of life. On the other hand, the great and remarkable degree of hemoglobin heterogeneity may be due to numerous duplications of genetic material.

There is a remarkable degree of hemoglobin polymorphism in white-tailed deer. The numerous structurally different hemoglobins which are found in one species will be illustrated and compared to other animal species. Variation in the proportions of hemoglobin components within a species and variation in hemoglobin due to environmental changes will be discussed.

INTRODUCTION

Numerous abnormal hemoglobins have been reported in man. However, the number of abnormal hemoglobins in a frequency great enough to be considered polymorphic is very few. For example: S, C, D, and E Punjab. In addition, the great majority of the abnormal hemoglobins of man are not associated with pathological conditions. There has not been strong evidence for genetic selection of most of the human variant polypeptide chains.

Of considerable interest has been the finding that the proportions of the abnormal and the normal hemoglobin component within a single heterozygote individual are often unequal.

TABLE I
HEMOGLOBIN COMPARISON AMONG SPECIES

Human	Other Animals
Numerous variant hemoglobins Few polymorphic forms	Numerous polymorphic forms
Some associated with pathological consequences	No documented hemoglobin-opathies
Variable expression of single point mutation variants which are products of single pairs of alleles, i.e. sickle cell trait	Products of single pair of alleles although different greatly in primary structure are produced in nearly equal proportions
Duplication of α and γ genes	Duplication of α genes

This variation in the quantitative control of hemoglobins in the heterozygote can be considered the variable expressions of normal and abnormal genes. In many cases, particularly as explained by the human alpha chain point mutations, distribution of the variant hemoglobins in the heterozygote were consistent with the interpretation that the alpha chains are products of at least two pairs of non allelic genes. However, most beta chain variants are consistent with the supposition that they are controlled by a single pair of allelic genes. The variant polypeptide chains are hardly ever produced in equal proportion to the normal or wild type component. All these conclusions can be contrasted to the genetic and stuctural basis for heterogeneity in other animals (Table I). Two broad classifications of animal hemoglobin heterogeneity can be described (Kitchen, 1969; Huisman et al, 1971; Gratzer et al, 1960):

    I. Gene related heterogeneity; and,
    II. Non-genetic heterogeneity
which can be further identified as given in Table II.

As given in I. A. (Table II) heterogeneity of human hemoglobins can be related to differential gene activity related to development. In the gene related category, the most obvious heterogeneity related to gene activity is the occurrence of structurally different hemoglobin at various times during development (see Fig. 1). As well, heterogeneity seen in the normal adult individual is related to the proportion of non-alpha-polypeptide chains produced such as in hemoglobin A, A_2 and fetal seen in the normal adult human (I.B.3., Table II).[2]

Heterogeneity due to mutation in the structural genes can

TABLE II
HEMOGLOBIN HETEROGENEITY

I. Gene-Related

   A. Differential gene activity related to development. (The ontogeny of a hemoglobin with a sequential emergence and arrest of respective polypeptide chain types.)

   B. Structural genes

      1. Abnormal forms due to recent mutations

      2. Polymorphic forms, products of alleles having different amino acid sequence in a frequency greater than would be expected by re-mutation alone.

      3. Multiple forms - usually as a result of gene duplication

   C. Variable expressions of genes or differences in quantitative production of polypeptide chain types.

II. Non-Genetic

   A. Environmentally related in vitro

      1. Isolation procedures

         a. Enzymatic degradation

         b. Polymerization of hemoglobin

      2. Methemoglobin

   B. Chemical modification in vivo

      1. Hexose moiety attached to a polypeptide chain

      2. Acetylation of one of the polypeptide chain types

be categorized as in Table III.

The purpose of this discussion will be to illustrate the genetic and structural basis for some examples of the animal hemoglobin heterogeneity. First, by comparison and generalization of the ontogeny of hemoglobins; second, a consideration of polymorphic hemoglobins within a species; third, a discussion of heterogeneity due to multiple hemoglobins within a species; and finally the variable expression of hemoglobin genes.

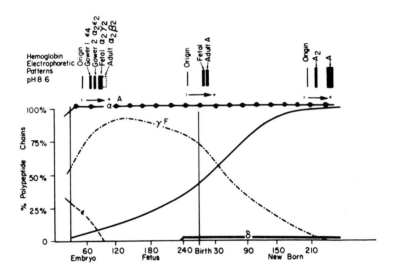

Transition Time Comparison

Fig. 1. Proportions of various hemoglobin polypeptide chains through development for humans. The expected hemoglobin patterns for several time periods are demonstrated.

GENE RELATED HETEROGENEITY

I. Comparison of the Ontogeny of Animal Hemoglobin

Numerous hemoglobin components can be correlated with cell type and site of synthesis during development. However, not all these components have had complete structural comparisons to verify their genetic relationships or define their genetic control. Hemoglobin components seen in early erythropoiesis and associated with nucleated red blood cells or embryonic erythrocytes derived from the yolk sac have been identified as embryonic hemoglobins. All mammals thus far studied regardless of species have had one or more discernible embryonic hemoglobin components (Kitchen et al, 1974; Huenn et al, 1964; Kleihauer et al, 1968). At a specific time of development, dependent upon the species, a change of the site of erythropoiesis from the yolk sac to the liver occurs. This is usually associated with the production of non-nucleated biconcave disc shaped cells characteristic of normal adult

TABLE III
HETEROGENEITY DUE TO MUTATION
IN THE STRUCTURAL GENES

A. Single amino acid substitutions resulting in changes in
the amino acid sequence of one or more polypeptide chain
type.

B. Deletion mutations resulting in the loss of one or more
amino acids in a given polypeptide chain type.

C. Chain elongation or amino acid additions to one of the
polymorphic peptide chain types.

D. Hybridization of genes for different chains. For
example, the Lepore Type.

E. Variation in the quantitative proportions of the
individual hemoglobin components within the erythrocytes.

erythrocytes.

In many species, hemoglobin components associated with the
non-nucleated red blood cells of liver origin as distinct
from an adult component seen later on in life are referred to
as fetal hemoglobin. Not all animals have a distinguishable
fetal hemoglobin component. Components seen during adult life
and associated with the typical biconcave disc and derived
from myeloid tissue in the normal adult organism are considered
adult hemoglobins.

Let me contrast the ontogeny of hemoglobins as seen in man
with the ontogeny as seen in the dog. Then let me quickly
compare the occurrence of hemoglobin components in other species. In Fig. 2 the minimum number of stuctural genes for
hemoglobin in man and dog are compared. However, the number
of structural genes and their expression as related to development vary greatly. The occurrence of embryonic hemoglobin
in various species is given in Figure 3. The presence or absence of fetal hemoglobin is also given for comparison.

II. Polymorphic Forms

Many examples of hemoglobin polymorphism can be given for
numerous animal species. The structural basis for the polymorphisms can be a single amino acid substitution like those
that account for most of the human hemoglobin variants (see
Table III. A.). Or, as in the hemoglobin polymorphism in
other animal species structural differences may be due to multiple amino acid substitutions, many of which would seem to
represent amino acid changes that are due to more than a

GENETIC CONTROL OF HEMOGLOBIN

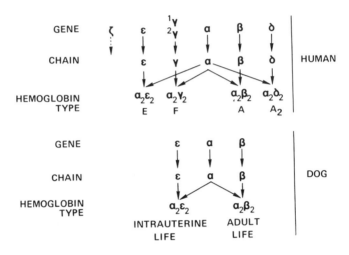

Fig. 2. Speculation of genetic control of hemoglobin in human and dog. Note the differences in the number of postulated genes between these two species.

single base change (Dayhoff, 1972; Kitchen et al, 1968; Balani et al, 1968; Boyer et al, 1966; Wrightstone et al, 1970; Kitchen et al, 1972; Taylor et al, 1972; Adams et al, 1968; personal communication with W. Jape Taylor). That is, there is evidence for multiple mutations at a single genetic point. Often these multiple amino acid substitutions are seen in products of the same pair of alleles. These, in fact, are the predominant mutations which account for ruminant hemoglobin heterogeneity.

The remarkable degree of hemoglobin polymorphism seen in the white-tailed deer is apparent in Fig. 4 (Kitchen et al, 1972). Structural comparisons of the polymorphic deer hemoglobins have demonstrated multiple amino acid differences in the beta chain and alpha chains. From one to three alpha chains and either one or two of seven identifiable beta chains may be present in a single animal (Kitchen et al, 1968; Kitchen et al, 1972; Taylor et al, 1972) (see Table IV). Despite the fact that many of the beta chains of the white tailed deer hemoglobins differ by several amino acids, pedigree studies of up to five generations indicate that a single pair of allelic genes control beta chain synthesis. In contrast, the

ONTOGENY of HEMOGLOBIN: A SPECIES COMPARISON

	Mouse	Human	Sheep	Horse
EMBRYONIC	ε_I $x_2 y_2$ ε_{II} $\alpha_2 y_2$ ε_{III} $\alpha_2 z_2$	ε_I ε_4 ε_{II} $\alpha_2 \varepsilon_2$ ζ $\alpha_2 \zeta_2$	ε ε_4 $\alpha_2 \varepsilon_2$	ε $\alpha_2^S \varepsilon_2$ $\alpha_2^F \varepsilon_2$
FETAL	?	$\alpha_2 y_2^G$ $\alpha_2 y_2^A$	$\alpha_2 y_2$	none
ADULT	$\alpha_2 \beta_2$ 100 % at Birth	$\alpha_2 \beta_2$ $\alpha_2 \delta_2$ 20-40 % at Birth	$\alpha_2 \beta_2$ 0-10 % at Birth	$\alpha_2^S \beta_2$ $\alpha_2^F \beta_2$ 100 % at Birth

Fig. 3. Hemoglobin subunit comparison of the structure of embryonic, fetal and adult hemoglobin in various species. The amount of adult hemoglobin seen in each respective species at birth is also estimated.

α chain heterogeneity in white-tailed deer is due to gene duplication (non-allelic α genes) as well as primary structural differences in the products produced (Taylor et al, 1972).

In goats the hemoglobin polymorphism is due to multiple structural differences in beta chains which are products of allelic genes and in alpha chains which are products of non-allelic genes very similar to the heterogeneity of deer hemoglobins (Adams et al, 1968). However, in sheep there are a limited number of hemoglobins (A and B), the beta chains of which are products of a single pair of alleles (Boyer et al, 1966). Five polymorphic hemoglobins have been found in cattle (Schroeder et al, 1967; Carr, 1964; Efremov et al, 1965). Both of these species share a common genetic basis for the structural differences in their respective hemoglobins. Both are the result of multiple amino acid substitutions. One interpretation is that mutations have occurred resulting in products that have been maintained in the population long enough for remutation to occur again and again. One questions what mechanism is operative in ruminants which allows selection and maintenance of the hemoglobin gene but rigorously resists selection of hemoglobin variants in human beings and the hemoglobins of other animals such as the cat, dog, or wolf (Dresler et al, 1974). Therefore, the description of a unique

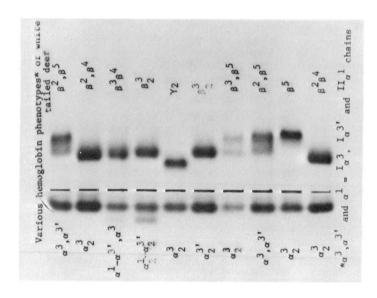

Various hemoglobin phenotypes* of white tailed deer

β^2, β^5

β^2, β^4

$\beta^3 \beta^4$

β_2^3

γ_2

β_2^3

β^3, β^5

β^2, β^5

β^5

$\beta^2 \beta^4$

$\alpha^3, \alpha^{3'}$
α_2^3

$\alpha^1 - \alpha^{3}, \alpha^3$

$\alpha_2^1 - \alpha_2^{3'}$

α_2^3

$\alpha_2^{3'}$

α_2^3

$\alpha^3, \alpha^{3'}$

α_2^3

α_2^3

$*\alpha^3, \alpha^{3'}$ and $\alpha^1 = I_\alpha^3, I_\alpha^{3'}$ and II_α^1 chains

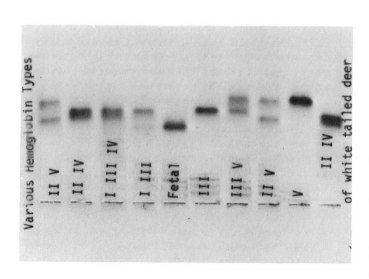

Various Hemoglobin Types

II V

II IV

I III IV

I III

Fetal

III

III V

II V

V

II IV

of white tailed deer

Fig. 4. Electrophoretic Separation of Deer Polymorphic Hemoglobin.
A. Left: Vertical starch gel electrophoresis of deer hemoglobin types at pH 8.6 demonstrating some of the polymorphic patterns.

Fig. 4 legend continued
B. Right: 6 M urea starch gel electrophoresis pH 8.1, demonstrating the electrophoretic separation of the various alpha-
and beta-like polypeptide subunits that account for some of
the deer polymorphic hemoglobins. Variation occurs both in
the alpha and beta chains; not all the structurally different
deer hemoglobins or their corresponding alpha or beta chains
are distinguishable by electrophoretic behavior. (Courtesy
of W. J. Taylor, University of Florida)

---

genetic event which would account for the heterogeneity of
ruminant hemoglobin has been under exploration by many. Perhaps similar mutation mechanisms required for immunoglobins
could be included for speculation. The simple explanation of
multiple point mutations has been questioned because of the
evolutionary time dimension.

There are a few examples of animal hemoglobin structural
variants due to mutational events B and no examples of C or
D such as classified in Table II. Hemoglobin C in sheep and
goats may be controlled by a gene which is non-allelic with
the normal beta gene of goats and sheep. Recent evidence
supplied by structural studies of deer Hemoglobin V indicates
that this beta chain allelic mutation is 143 amino acids long.
Selection of polypeptide chains with apparent multiple amino
acid substitutions in animal hemoglobins could be the result
of crossing over events within homologous, non-allelic genes
(Event C, Table II). However, evidence is not apparent now
for this type of event.

III. Multiple Forms

The term multiple hemoglobins will be reserved for hemoglobin heterogeneity in species whose members always have
more than a single hemoglobin other than fetal hemglobin. The
respective components must have identical physical and chemical
characteristics in every member of the species.

This leads to the subject of gene duplication. Because of
the tetrameric form normal for all the mammalian hemoglobins,
subunit variations can easily account for multiple forms.
Hemoglobin is composed of two pairs of non identical polypeptide chains; variations of the individual components such as
the non-alpha in humans are easily visualized. In adult human hemoglobin, the non-allelic beta and delta chains account
for hemoglobin A and A_2 when they are combined with a structurally identical alpha chain. Such an example of multiple
hemoglobins accounts for common human hemoglobin heterogeneity. In this example the beta and delta differences in

TABLE IV
ALPHA AND BETA POLYPEPTIDE CHAINS
IN WHITE-TAILED DEER

	$\alpha^3\alpha^3$	$\alpha^3\alpha^{3'}$	$\alpha^{3'}\alpha^{3'}$	α^1 $\alpha^{3'}\alpha^{3'}$	α^1 $\alpha^{3'}\alpha^3$	$\alpha^1\alpha^1$ $\alpha^{3'}\alpha^{3'}$
$\beta^2\beta^2$	1	2			3	
$\beta^2\beta^3$	4	2			2	
$\beta^2\beta^{4d}$						
$\beta^2\beta^{4e}$	1					
$\beta^2\beta^5$	3	1				
$\beta^3\beta^3$	4	1	1	2	3	2
$\beta^3\beta^{4d}$	2				1	
$\beta^3\beta^{4e}$	3					
$\beta^3\beta^5$	4	1				
$\beta^{4d}\beta^{4d}$						
$\beta^{4d}\beta^{4e}$						
$\beta^{4d}\beta^5$						
$\beta^{4e}\beta^{4e}$						
$\beta^{4e}\beta^5$						
$\beta^5\beta^5$	1					

Either 1 or 2 α chain gene loci are present.
The β chain variants are alleles, but have multiple differences.
An individual animal may have from 1 to 6 different hemoglobins.
Hemoglobin genotypes in the current research of White-tailed
  deer, March, 1974.

The various alpha and beta polypeptide chains, allelic and
non-allelic are given for Florida white-tailed deer. The
various tetrameres combination are given as found in the
current research herd of white-tailed deer, University of
Florida, March 1974 (Taylor).

structure present evidence for early divergence of the beta
gene, and the quantitative distribution between these cellular
products is 40 beta to 1 delta. However, in other animals,
examples of multiple hemoglobins which result from combining

tetramere components occur in such quantitative proportions and structural homology that the chains are considered homologous as the result of recent mutations. Therefore, in many mammals examples of beta and alpha chain duplication are frequent. Beta-δ gene duplication has been reported only in some primates. However, duplication of a beta chain gene which has evolved to analogous delta chains is not recognized in all primates (Boyer et al, 1969; Boyer, 1971).

The analogous human major and minor adult hemoglobin A and A_2 are present in New World monkeys and apes (Kunkel et al, 1957). Extensive studies by Boyer et al (1969 and 1971) have given structural evidence for δ gene duplication preceding the divergence of the Apes and New World Primates. A relative abundance of δ chain polymorphism which are attributed to non-adaptive factors are presented in non-human primates (Boyer, 1971).

Heterogeneity due to multiple hemoglobins in animals may be due to the existence of more than one structural gene for a polypeptide chain type as in some examples in man. Hollan et al (1972) have shown that the alpha chain is coded by at least two pairs of non-allelic genes. Schroeder and Huisman et al (1968) have given extensive evidence for their conclusion that multiple non-allelic gamma chains best explain the heterogeneity and variable expression of the human fetal hemoglobin. Many examples of animal heterogeneity can be explained by multiple non-allelic genes. No matter what nomenclature is used, the logic is clear. Gene duplication resulting in the production of identical proteins by more than a single gene loci occurs. This evolutionary event may be followed by sequential mutation which accounts for two or more loci which will produce polypeptide chains differing by one or several amino acids, while maintaining sufficient homology for the polypeptide chains to be easily assigned as either alpha, beta or gamma etc. Clearly, these duplicate chains are homologous in function and participate in tetramere formation with the respective polypeptide chain types. Multiple non-allelic genes provide the logical genetic explanation of multiple hemoglobins in most animals. By comparing the number of amino acid substitutions between the products of gene duplication one could speculate whether these are recent or old mutational events.

The occurrence of hemoglobin heterogeneity in the rabbit which was found to be due to structural differences in the alpha chain was a stimulus to study a similar heterogeneity due to intra species differences in the structure of alpha chains of mice, goats, and horses (Rifkin et al, 1966; Adams et al, 1969; Clegg et al, 1970). Early speculation for the

genetic basis included ambiguity of translation, genetic polymorphism due to alleles for a single pair of structural genes
for a given polypeptide chain type, and the postulation of
numerous structural gene loci for structurally different but
very homologous polypeptide chains.

However, the very detailed structural and genetic studies
of the alpha chains from goat hemoglobins (Adams et al, 1968)
and the precise amino acid sequence determinants of the human
gamma chain (Schroeder et al, 1968), have both contributed to
the conclusion of multiple non-allelic genes. Structural and
genetic studies have established confirmatory findings for
chain duplication in the rabbit and mouse δ chains. Hollan
et al (1972) have reported that alpha chain variations in
humans could best be explained by at least two non-allelic
pairs of genes. From structural comparisons and limited genetic studies of gorilla hemoglobins Boyer et al (1973) invokes
the possibility that there are three genetic loci for α chains
in this species. Clearly genetic duplication is a very acceptable evolutionary mechanism, and the presence of more non-allelic genes for a given protein instills biological flexibility
for evolutionary events. As well as explaining heterogeneity
of animal hemoglobin, gene duplication offers the additional
explanation of differences in quantitative distribution of
hemoglobin chains.

Multiple non-allelic gamma chains best explain the heterogeneity and variable expressions of the human fetal hemoglobins
and have certainly stimulated the inclusion of similar explanations for variations in hemoglobin expression in other species.

Hemoglobin heterogeneity in horses was first described by
Cabannes and Serain in 1955. Most horses have two distinct
hemoglobins on starch gel electrophoresis. The fast component accounts for approximately 60% of the total hemoglobin
(Bangham et al, 1958). A few horses have a higher proportion
of the fast component 80 F/20S (Braend, 1967; Kitchen et al,
1966). Braend also noted that in a population of Norwegian
horses a third phenotype with only the fast component was
present. Therefore, three hemoglobin phenotypes can be identified in the horse: 60F/40S, 80F/20S, and 100F/0 (Fig. 5).
The genetic explanation given by Clegg (1974) has been based
upon multiple alleles due to α gene duplication. The horse
hemoglobin heterogeneity is an example of gene duplication
with subsequent mutation. Because α gene duplication is not
present in all individuals hemoglobins occur in a polymorphic
distribution within the species.

HORSE HEMOGLOBIN PHENOTYPES

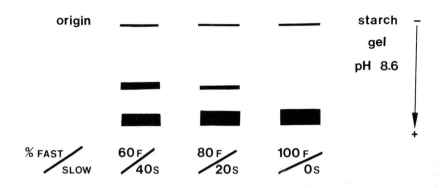

Fig. 5. An illustration of the various phenotypes of hemoglobin polymorphism of the horse.

IV. Variable Expressions of Genes

An intriguing finding has been the variable expression of
hemoglobin components within individuals of a species (an
example is the horse). As previously presented, the distribution of the mutant hemoglobin components in the heterozygote
is not always equal in concentration within erythrocytes to
the wild type component.

In an analysis given by Boyer (1972), the distribution of
the portion of the variant hemoglobins in human heterozygotes
was consistent with the interpretation that alpha chains are
produced by two loci whereas most beta chain variants are controlled by single loci. Upon comparison of the available
data for animal hemoglobins similar conclusions can be drawn
for the alpha chain; that is, alpha chain duplication. However, in contrast to the human heterozygotes cited by Boyer
which were due to single amino acid substitutions, most of
the polymorphism of animal hemoglobin is structurally due to
multiple amino acid interchanges.

Sheep present an excellent example of beta chain polymorphism, and a primate, the stump-tail macaque, an example of
alpha chain polymorphism in animals. In sheep hemoglobins a
heterozygote for hemoglobin A and B will have a structurally
common alpha chain, and the hemoglobins within an erythrocyte
will be present in nearly equal concentrations. This certainly

implicates another level of selection for functional proteins. In the majority of beta chain polymorphism in animals the polypeptide chains are produced in a nearly 1:1 ratio, regardless of whether the structural differences are due to single amino acid or multiple differences. Such examples are cattle, goats, deer. This situation is true of most beta chain hemoglobin heterogeneity in animals with the exception of the beta chains of cats and galagoes (Taketa et al, 1968; Nute et al, 1969). The ratio varies from a trace to 50% in the cat, and the galago chains occur in a 3:1 ratio. In these cases, based on limited studies, it is fair to conclude that this heterogeneity is most likely due to multiple hemoglobin components, and therefore due to beta chain duplication in these two species.

In almost all cases of animal hemoglobin heterogeneity due to structural differences in the alpha chain, genetic studies have concluded that there are more than a single pair of alleles governing alpha chain production. Non-allelic alpha chain loci is the most plausible explanation for many of these animal hemoglobin heterogeneities. Many times the alpha chains will differ by only a single amino acid interchange such as the case of stump-tail macaque and horses. Or there may be multiple amino acid differences between alpha chains suggesting longer periods of evolutionary time between duplication. In the case of the stump-tail macaque, a primate having two different alpha chains and a structurally common beta chain existing in all individuals studied so far, the two hemoglobins are produced in unequal concentration (Oliver et al, 1968; Kouba et al, 1970; Kouba et al, 1971). The three phenotypes may be represented by the relative amounts of the two hemoglobins in these animals (Fig. 6). Some animals have a distribution of 60-40 fast to slow hemoglobin type. Others exhibit a 30-70 fast to slow, nearly the reverse. A few others have a nearly 50-50 distribution. All evidence indicates that there is duplication in the alpha chains. The difference between the two alpha chains is a single amino-acid substitution, a glycine substitution by aspartic acid, alpha 15, at an identical sequence position to a structural mutant, J_{oxford} in humans.

However, in contrast to the analogous single amino acid substitutions in man, there has been gene duplication in the non-human primate whereas the single amino acid substitution in man is the result of a single base mutation occurring at a single pair of alleles. Just as the unequal expression of the fetal hemoglobin in humans may be explained in part by unequal number of structural genes for the γ chain, a possible explanation for the difference in proportions of

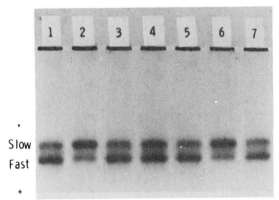

HEMOGLOBIN PHENOTYPES IN STUMP TAILED MACAQUE

Fig. 6. Vertical Starch Gel Electrophoresis of Stump-Tail
Macaque Hemoglobins at pH 8.6.
Selected samples from seven adult animals of both sexes, 3-5
years old. Fast hemoglobin, 60-65%; slow hemoglobin 35-40%
in positions 1, 3, 4, 5, and 7. Fast hemoglobin, 20-25%; slow
hemoglobin 75-80% in positions 2 and 6.

---

hemoglobin types within an individual of other animal species
may be due to the number of non-allelic genes for a polymorphic
polypeptide chain (Kitchen et al, 1970).

Another level of biosynthetic control or variable expression
of genes has been the unusual finding that in some species
that become anemic or are made anemic by acute hemolytic processes such as phenylhydrazine, rapid recovery by increased
erythropoiesis is accompanied by the emergence of a new
structurally distinct hemoglobin. Extensive studies of this
hemoglobin, Hemoglobin C, in sheep and goats have been reported. Recent reports of a similar heterogeneity of hemoglobins
after acute anemia have been reported for the cat. Additional
hemoglobin components have been seen during subacute anemia
and during the development of an organism. With the sophisticated techniques and methods now available for identifying
polymorphisms it is possible to identify multiple hemoglobin
components as structurally distinct and not simply due to another level of genetic control, to artifacts, or to environmental changes.

NON-GENETIC HETEROGENEITY

Non-genetic heterogeneity can be broken down into environmentally related and chemically modified products (Table II).
Examples of environmentally related heterogeneity are structural modifications that occur during isolation of the hemoglobin components such as enzymatic degradation reported both
in humans (Marti et al, 1967) and in animals (Hammerberg et
al, 1974) due to a peptidase of tissue or plasma origin. A
carboxypeptidase has been described which is specific for
basic amino acids and therefore results in the cleavage of
the arginine from only the alpha chain during isolation procedures. This reaction can lead to misinterpretation. In the
isolation of rodent hemoglobin (Ahl et al, 1973), the polymerization of the hemoglobin components can lead to erroneous
interpretations of starch gels and acrylamide gels. Such
artifacts which can lead one to presume heterogeneity when it
does not exist can be eliminated by the techniques described
for mouse, turtles and from hemoglobins (Riggs, 1965). Although
polymerization such as disulfide interaction in rodent hemoglobins is common in other animals, chain aggregation
α_4, γ_4, δ_4, and β_4 found in the human has not been reported for
animals with the exception of ε_4 in the embryo of some species
(Hammerberg et al, 1974; Hunt et al, 1972; Kleihauer et al,
1968). Another possibility of environmentally related heterogeneity can be the formation of oxidized products such as
methemoglobin. The presence of methemoglobin can account for
an apparent hemoglobin heterogeneity due to the electrophoretically distinguishable components depending upon the valence
state of iron. Chemically modified hemoglobins in vivo such
as the occurrence of hexose moiety associated with hemoglobin
from diabetics (Rahbar et al, 1969), acytelation of cat Beta
chains (Taketa et al, 1972) and human gamma chains (Schroeder
et al, 1962) also can account for apparent heterogeneity.

SUMMARY

Although multiple point mutations of a polypeptide chain
which are products of alleles may account for hemoglobin heterogeneity in many animal species, numerous examples of multiple hemoglobin within species may be due to gene duplication. The quantitative distribution of hemoglobin type within
an individual may be explained by variation in the number of
multiple non allelic genes for a polypeptide chain type in a
variety of species. No examples of hemoglobinopathies have
yet been identified in animals, other than man. Nor can
functional hemoglobin abnormalties, or examples of chain

elongation, be given at the present time in other species.

ACKNOWLEDGMENTS

This work was supported in part by United States Public Health Services Grant HD 07894-01.

I am grateful to Yvonne Kitchen and Coral Johnson for the final preparation of this paper.

REFERENCES

Adams, H. R., E. M. Boyd, J. B. Wilson, A. Miller, and T. H. J. Huisman 1968. The structure of goat hemoglobins. III. Hemoglobin D, a β chain variant with one apparent amino acid substitution (21 Asp→His). *Arch. Biochem. Biophys.* 127: 398-405.

Adams, H. R., R. N. Wrightstone, A. Miller, and T. H. J. Huisman 1969. Quantitation of hemoglobin chains in adult and fetal goats; Gene duplication and the production of polypeptide chains. *Arch. Biochem. Biophys.* 132: 223-236.

Ahl, A., I. J. Brett, and H. Kitchen 1973. Systematic implication of some hemoglobin studies in sigmodon. *Theriology,* (in press).

Balani, A. S., P. K. Ranjekar, and J. Barnabas 1968. Structural basis for genetic heterogeneity in haemoglobins of adult and new-born ruminants. *Comp. Biochem. Physiol.* 24: 809-815.

Bangham, A. D. and H. Lehmann 1958. Multiple haemoglobin in the horse. *Nature* 181: 267-268.

Boyer, S. H., E. F. Crosby, T. F. Thurmon, A. M. Noyes, G. F. Fuller, S. E. Leslie, M. K. Shepard, and C. M. Herndon 1969. Hemoglobins A and A$_2$ in New World primates: Comparative variation and its evolutionary implications. *Science* 166: 1428-1431.

Boyer, S. H., P. Hathaway, F. Pascasio, J. Bordley, C. Orton, and M. A. Naughton 1966. Differences in the amino acid sequences of tryptic peptides from three sheep hemoglobin β chains. *J. Biol. Chem.* 242: 2211-2232.

Boyer. S. H. 1971. Primate hemoglobins: Some sequences and some proposals concerning the character of evolution and mutation. *Biochem. Genet.* 5: 405-448.

Boyer, S. H. 1972. From *Atlas of Protein Sequence and Structure 1972*. National Biomedical Research Foundation, Washington, D.C. 5: 77.

Boyer, S. H., A. N. Noyes, M. L. Boyer, and K. Marr 1973.

Hemoglobin α chains in apes: Primary structures and the
presumptive nature of back mutation in a normally
silent gene. *J. Biol. Chem.* 248: 992-1003.

Braend, M. 1967. Genetic variation of horse haemoglobin.
Hereditas 58: 385-392.

Cabannes, R. and C. Serain 1955. Etude electrophoretique des
hemoglobines des mammiferes domestiques d'Algerie.
C.r. Scanc. Biol. 149: 1193.

Carr, W. R. 1964. The hemoglobins of indigenous breeds of
cattle in Central Africa. *Rhod. J. Agric. Res.* 2: 93-94.

Clegg, J. B. 1974. Horse hemoglobin polymorphism. Presented
at Conference on Hemoglobins: Comparative Molecular
Biology Models for the Study of Disease. To be published
in *Ann. N.Y. Acad. Sc.*, 1974.

Clegg, J. B., M. Hosseinion, and J. Clegg 1970. Horse haemoglobin polymorphism: Evidence for two linked nonallelic
α-chain genes. *Proc. Roy. Soc. B.* (Lond.) 176: 235.

Dayhoff, M. O. 1972. *Atlas of Protein Sequence and Structure.*
1972. 5. National Biomedical Research Foundation,
Washington, D.C.

Dresler, S. L., D. Runkel, P. Stenzel, B. Brimhall, and R. T.
Jones 1974. Multiplicity of the hemoglobin α chain in
dogs and variations among related species. Presented
at Conference on Hemoglobins: Comparative Molecular
Biology Models for the Study of Disease. To be published
in *Ann. N.Y. Acad. Sc.*, 1974.

Efremov, G. and M. Braend 1965. A new hemoglobin in cattle.
Act. Vet. Scand. 6: 109-111.

Gratzer, W. B. and A. C. Allison 1960. Multiple haemoglobins.
Biological Reviews of the Cambridge Philosophical
Society 35: 459-506.

Hammerberg, B., I. Brett, and H. Kitchen 1974. Ontogeny of
hemoglobin in sheep. Presented at the Conference on
Hemoglobins: Comparative Molecular Biology Models for
the Study of Disease. To be published in *Ann. N.Y.
Acad. Sc.*, 1974.

Hollan, S. R., R. T. Jones, and R. D. Koler 1972. Duplication of haemoglobin genes. *Biochimie* 54: 639.

Huenn, E. R., N. Dance, G. H. Beaven, F. Hecht, and A. G.
Motulsky 1964. Human embryonic hemoglobins. *Symp.
Quant. Biol.* 29: 327.

Huisman, T. H. J. and W. A. Schroeder 1971. *New Aspects of
the Structure, Function, and Synthesis of Hemoglobins.*
CRC Press, Cleveland, Ohio.

Hunt, L. T., M. R. Sochard, and M. O. Dayhoff 1972. Mutations
in human genes: Abnormal hemoglobins and myoglobins.
In *Atlas of Protein Sequence and Structure 1972.*

Edited by M. O. Dayhoff. 5: 75.

Kitchen, H. 1969. Heterogeneity of animal hemoglobin. *Advances in Veterinary Science*, C. A. Brandly and C. E. Cornelius, editors. 13. Academic Press, New York.

Kitchen, H., E. O. Kouba, and C. W. Easley 1970. Animal models for the study of hemoglobinopathies. *Animal Models for Biomedical Research* III, pp. 52-69. National Academy of Sciences. Washington, D.C.

Kitchen, H. and I. Brett 1974. Embryonic and fetal hemoglobin in animals. Presented at the Conference on Hemoglobins: Comparative Molecular Biology Models for the Study of Disease. To be published in *Ann. N.Y. Acad. Sc.*, 1974.

Kitchen, H., C. W. Easley, F. W. Putnam, and W. J. Taylor 1968. Structural comparison of polymorphic hemoglobins of deer with those of sheep and other species. *J. Biol. Chem.* 243: 6: 1204-1211.

Kitchen, H. and W. J. Taylor 1972. The sickling phenomenon of deer erythrocytes. In *Hemoglobin and Red Cell Structure and Function*. Edited by G. J. Brewer, Plenum Pub. Co., New York, New York.

Kitchen, H., W. F. Jackson, and W. J. Taylor 1966. Hemoglobin and hemodynamics in the horse during physical training. Proc. 11th Amer. Assoc. Equine Practitioners, p. 97, Miami, Florida.

Kleihauer, E. and G. Stoffler 1968. Embryonic hemoglobins of different animal species - quantitative and qualitative data about production and properties of hemoglobins during early developmental stages of pig, cattle, and sheep. *Mol. J. Genet.* 101: 59-69.

Kouba, E. O. and H. Kitchen 1970. A comparative structural study on hemoglobins in *Macaca speciosa*. *Arch. Biochem. Biophys.* 140: 415-424.

Kouba, E. O. and H. Kitchen 1971. Hemoglobins in *Macaca speciosa:* Quantitative differences. *Arch. Biochem. Biophys.* 145: 1: 283-289.

Kunkel, H. G., R. Ceppellini, U. Muller-Eberhard, and J. Wolf 1957. Observations on the minor basic hemoglobin component in the blood of normal individuals and patients with thalassemia. *J. Clin. Invest.* 36: 1615-1625.

Marti, H. R., D. Beale, and H. Lehmann 1967. Haemoglobin Koelliker: A new acquired haemoglobin appearing after severe haemolysis: α_2 minus 141 argβ_2. *Acta Haemat.* 37: 174-180.

Nute, P. E., V. Buettner-Janusch, and J. Buettner-Janusch 1969. Genetic and biochemical studies of transferrins and hemoglobins of galago. *Folia. Primat.* 10: 276-287.

Oliver, E. and H. Kitchen 1968. Hemoglobins of adult

Macaca speciosa: An amino acid interchange
(α-15 (gly→asp)). *Biochem. Biophys. Res. Commun.*
31: 749-754.

Rahbar, S., O. Blumenfeld, and H. M. Ranney 1969. Studies
of an unusual hemoglobin in patients with *Diabeties
mellitus. Biochem. Biophys. Res. Commun.* 36: 838-843.

Rifkin, D. B., D. I. Hirsh, M. R. Rifkin, and W. Konigsberg
1966. A possible ambiguity in the coding of mouse
hemoglobin. Cold Spring Harbor. *Symp. Quant. Biol.*
31: 715-718.

Riggs, A. 1965. Hemoglobin polymerization in mice. *Science*
147: 621-623.

Schroeder, W. A., T. H. J. Huisman, J. R. Shelton, J. B.
Shelton, E. F. Kleihauer, A. M. Dozy, and B. Robberson
1968. Evidence for multiple structural genes for the
γ chain of human fetal hemoglobin. *Proc. Natl. Acad.
Sci. USA* 60: 2: 537-544.

Schroeder, W. A., J. T. Cua, G. Matsuda, and W. D. Fenninger
1962. Hemoglobin F, an acetyl-containing hemoglobin.
Biochim. Biophys. Acta 63: 532.

Schroeder, W. A., J. R. Shelton, J. B. Shelton, B. Robberson,
and D. R. Babin 1967. A comparison of amino acid
sequences in the β chains of adult bovine hemoglobins
A and B. *Arch. Biochem. Biophys.* 120: 124-135.

Taketa, F., M. R. Smits, and J. L. Lessard 1968. Hemoglobin
heterogeneity in the cat. *Biochem. Biophys. Res. Commun.*
30: 30: 219.

Taketa, F., M. H. Attermeier, and A. G. Mauk 1972. Acetylated
hemoglobins in feline blood. *J. Biol. Chem.* 247: 1: 33-35.

Taylor, W. J., C. W. Easley, and H. Kitchen 1972. Structural
evidence for heterogeneity of two hemoglobin α chain
gene loci in white-tailed deer. *J. Biol. Chem.* 247: 22:
7320-7324.

Personal communication with W. Jape Taylor.

Wrightstone, R. M., J. B. Wilson, A. Miller, and T. H. J.
Huisman 1970. The structure of goat hemoglobins. IV.
A Third β Chain Variant (β^{D}) with Three Apparent Amino
Acid Substitutions. *Arch. Biochem. Biophys.* 138:451.

MECHANISMS OF INTRACELLULAR ENZYME LOCALIZATION

RICHARD T. SWANK, SHIRO TOMINO, AND KENNETH PAIGEN
Department of Molecular Biology
Roswell Park Memorial Institute
666 Elm Street
Buffalo, New York 14203

ABSTRACT: Mutation at the Eg gene in the mouse destroys virtually all binding of β-glucuronidase to microsomal membranes. Normally, 30-50% of the enzyme is specifically located in microsomes in hepatic cells. Genetic studies have mapped the Eg gene to a site on chromosome 8, where it is genetically unlinked to the enzyme structural gene, Gus, on chromosome 5. Physical-chemical studies on electrophoretically separable forms of the enzyme revealed that the microsomal components are X, a tetramer of 260,000 molecular weight, which is especially prominent after induction of the enzyme in kidney by androgens and M_1, M_2, M_3, and M_4 which differ from X in that they contain 1,2, 3, and 4 copies of an enzymatically inactive protein of about 55,000 molecular weight. Lysosomal component L like X is a tetramer of 260,000 daltons though it has a more negative charge than the microsomal components. The Eg^o mutant, while containing detectable levels of the microsomal tetramer X, lacks microsomal polymers M_1,M_2,M_3, and M_4. The presence of an additional protein in the microsomal glucuronidase of wild type mice was confirmed in immunoprecipitates obtained with the F (ab)$_2$ fragment of specific anti-mouse glucuronidase. The new protein, named egasyn, is noncovalently associated with microsomal glucuronidase, is not present in immunoprecipitates of lysosomal enzyme, and is not detectable in immunoprecipitates of glucuronidase from mutant Eg^o mice. Specific membrane attachment of glucuronidase thus is effected through complex formation with a noncovalently associated protein, egasyn, whose attachment requires the Eg gene. The use of hydrophobic peptide sequences may be a general mechanism for specifically associating enzymes and membranes.

INTRODUCTION

The analysis of inbred mouse lines has identified genetic variants of β-glucuronidase altered in enzyme structure (Paigen, 1961a; Lalley and Shows, 1974), developmental pattern (Paigen, 1961b), hormonal inducibility (Dofuku, et al., 1971a,b; Swank, et al., 1973), and subcellular localization (Paigen, 1961a; Ganschow and Paigen, 1967; Swank and Paigen, 1973). The study of these variants has begun to provide an understanding of some of the factors controlling the final realization of one enzyme activity in a mammalian organism, and has permitted us to determine how the responsible genetic factors are organized within

523

the mammalian chromosome (reviewed in Paigen, et al., 1974).

Murine β-glucuronidase has proved to be a useful model enzyme for these purposes. Although it is an acid hydrolase the enzyme is not found solely in lysosomes; a large microsomal complement is likewise present. For example, in the livers of mice of various inbred strains from 30-50% of the enzyme is present in the microsomal fraction, the remainder being lysosomal (Ganschow and Paigen, 1968). Kidney, especially after induction of the enzyme by androgens, contains a similarly high complement of microsomal enzyme. The glucuronidase found in both microsomes and lysosomes of several organs contains the same polypeptide chain. The lysosomal and microsomal activities show identical rates of heat denaturation, substrate affinity, and kinetic parameters (Paigen, 1961a; Van Lancker and Lentz, 1971; Kato, et al., 1972). More importantly, two types of mutation in the amino acid sequence of the enzyme, one affecting heat lability (Paigen, 1961a) and the other affecting electrophoretic mobility (Lalley and Shows, 1974), simultaneously alter the properties of glucuronidase at both subcellular sites.

When purified from mouse liver the enzyme is a glycoprotein of molecular weight 280,000. In sodium dodecyl sulfate it dissociates into apparently identical subunits of 70-75,000 molecular weight, suggesting that the parent molecule is a tetramer (Ganschow, 1973; Tomino, unpublished). In these respects its properties closely resemble those reported for purified rat glucuronidase (Stahl and Touster, 1971).

One inbred line of mice, the YBR strain, unlike all others that have been examined, lacks microsomal glucuronidase while retaining a normal complement of enzyme in lysosomes as the result of a single gene mutation (Ganschow and Paigen, 1967). This gene, Eg, (for ergastoplasmic glucuronidase) has recently been mapped to a site on chromosome 8 quite distinct from the glucuronidase structural gene, Gus, on chromosome 5 (Karl and Chapman, 1974). We have been especially interested in the analysis of this mutation because so little is presently known about the mechanisms by which enzymes are assigned to specific subcellular sites and because the availability of enzyme with a dual intracellular location provides a favorable experimental situation. Our results indicate that an important factor in the binding of glucuronidase to microsomal membranes is its non-covalent association with another hydrophobic protein. This protein, egasyn[1], has now been purified, and is devoid of catalytic activity. The attachment of egasyn to β-glucuronidase is under the control of the Eg gene.

[1] Egasyn (from Eg, endoplasmic glucuronidase and syn, the Greek root for holding together). We suggest that should proteins

MATERIALS AND RESULTS
PHENOTYPE OF THE Eg^O MUTATION

Mice carrying the $\underline{Eg}^O$ mutation lack microsomal but not lysosomal glucuronidase (Ganschow and Paigen, 1967). This is illustrated in Fig. 1, depicting a sucrose density gradient after
centrifugation of particles from wild type $\underline{Eg}^a$ and mutant $\underline{Eg}^O$
mouse liver. The peak to the left corresponds to lysosomes and
the one to the right to microsomes. The peaks were identified
by their centrifugal characteristics and their content of appropriate marker enzymes; however, only the glucuronidase activity
is shown. A quantitative study by Karl and Chapman (1974) has
confirmed that the $\underline{Eg}^O$ mutation affects only microsomal glucuronidase leaving $\underline{Eg}^O$ animals with normal levels of lysosomal
glucuronidase.

The $\underline{Eg}^O$ mutation does not affect other microsomal proteins.
Other microsomal marker enzymes are not diminished in activity
and acrylamide gel electrophoresis of microsomal proteins from
normal and affected mice showed no differences (Ganschow and
Paigen, 1967). At least some aspects of the molecular mechanism
building glucuronidase into membranes are therefore unique to
this enzyme. Thus, glucuronidase must possess recognition
features not shared by other proteins of the endoplasmic reticulum.

Genetically, the $\underline{Eg}^O$ mutation segregates as a simple autosomal Mendelian trait (Ganschow and Paigen, 1967). Furthermore,
the mutation is recessive. F_1 mice have the same quantity of
microsomal glucuronidase as wild type mice, indicating the function of the $\underline{Eg}$ gene product is not rate limiting in enzyme attachment. Early experiments had shown that the site of the mutation
is not linked to the $\underline{Gus}$ locus on chromosome 5 of the mouse
which determines the amino acid sequence of glucuronidase. More
recently, Karl and Chapman (1974) have determined that the $\underline{Eg}$
locus maps within one centimorgan of the esterase-1 gene on
chromosome 8. The $\underline{Eg}$ site, therefore, is completely distinct
from the chromosome 5 locus coding for the structure of the
glucuronidase enzyme subunit.

MULTIPLE FORMS OF GLUCURONIDASE

When extracted from liver and several other organs with the
mild detergent Triton X-100 glucuronidase exists in a number of
forms that are separable by polyacrylamide gel electrophoresis
at pH 8.1. Gels run in this way and specifically stained for
glucuronidase activity with the substrate napthol-AS-BI-glucur-
onide are shown in Fig.2. Over half of the total liver enzyme
activity is contained in band L closest to the anode. Migrat-
with homologous function serve in the attachment of other enzymes to membrane that the suffix syn be used in naming them
to indicate their functional relationship.

ing cathodally to L are 4 bands M_1, M_2, M_3 and M_4 with bands M_2 and M_3 in highest concentration. Uninduced kidney and lung contain the same enzyme forms though the proportion of the various M bands is altered. However, spleen, brain, and heart contain only the L form of the enzyme; no M forms are seen.

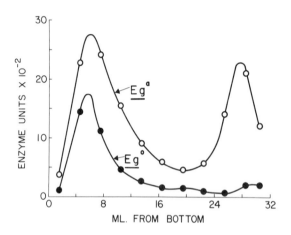

Fig. 1. Distribution of glucuronidase after density gradient centrifigation of cytoplasmic particles from $\underline{Eg}^a$ and $\underline{Eg}^o$ liver.

The cytoplasmic particle fraction of a DBA/2 Ha ($\underline{Eg}^a$ /$\underline{Eg}^a$) liver and of a YBR (Eg^o/Eg^o) liver were layered over a non-linear sucrose concentration gradient and sedimented at 24,000 x g for 30 min. Fractions (3 ml each) were collected and assayed for glucuronidase.

Occasionally, a sixth electrophoretically separable form of glucuronidase is visible in liver extracts as a minor component migrating between forms L and M_1. We refer to this band as component X. X is especially prominent in extracts of kidney after the rate of synthesis of glucuronidase has been stimulated in kidney proximal tubule cells by administration of androgen.

The subcellular location of each of the electrophoretically separable glucuronidase forms has been determined. Microsomal and lysosomal fractions were prepared from induced kidneys by an osmotic shock technique which solubilizes enzyme from osmotically fragile lysosomes, but not that from microsomes (Ganschow and Paigen, 1968). Microsomal enzyme can then be

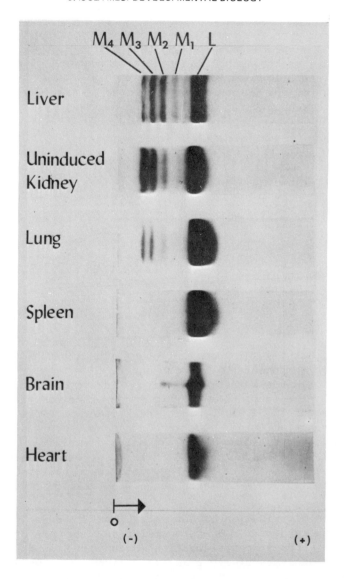

Fig. 2. Organ specificity of glucuronidase electrophoretic forms. Samples were of a Triton X-100 treated organ homogenate of DBA/2 Ha (Eg^a/Eg^a) female mice (Swank and Paigen, 1973).Gels were stained with naphthol-AS-BI-β-D-Glucuronide. Total units of glucuronidase from each organ were: liver, 0.107; kidney, 0.034; lung, 0.060; spleen 0.042; brain, 0.008; heart, 0.003. All gels were stained 4 hr. except the gel of liver homogenate, which was stained for 1 hr.

extracted from the residue with detergent.   Fig. 3 is a tracing
of the analytical gels showing that forms M_1, M_2, M_3 and M_4
and X are exclusively localized in microsomes. Approximately
95% of the L form is contained in the lysosomal fraction. The
other 5% of the L form is firmly bound to the microsome fraction
where it accounts for 20-25% of the enzyme present in microsomes

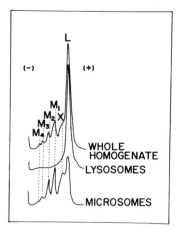

Fig. 3. Subcellular localization of glucuronidase electrophoretic components by osmotic shock.
      Homogenates were prepared from kidneys of female A/Ha
(Eg^a /Eg^a) mice to whom dihydrotestosterone had been administered for 10 days. Osmotic shock fractions were subjected
to polyacrylamide gel electrophoresis and then scanned at
550 nm after staining for glucuronidase activity with naph-
thol-AS-BI-β-D-glucuronide. Units of glucuronidase electrophoresed for each fraction were whole homogenate, 0.120;
lysosomes, 0.083 and microsomes 0.083. All gels were stained
for 1 hr.

This distribution of enzyme was confirmed in induced kidney when lysosomal and microsomal fractions were prepared by
sucrose gradient centrifugation (Swank and Paigen, 1973).
Similar analyses of liver and lung likewise showed that the
M forms in these organs are exclusively microsomal in location
and that 90-95% of form L is lysosomal.

STRUCTURAL DIFFERENCES AMONG GLUCURONIDASE ISOZYMES

In order to analyze the detailed charge and size relationships of each of the lysosomal and microsomal glucuronidase
components we measured their electrophoretic mobilities in
poly-acrylamide gels of varying concentration following the
method of Ferguson (1964) as modified by Hedrick and

Smith (1968). When mobilities determined in this way are plotted as a function of acrylamide concentration the slopes of the resulting lines are proportional to molecular weights and the intercepts to molecular charge. When this procedure was carried out for the glucuronidase activity forms all of the microsomal components, X and M_1 to M_4, gave lines intersecting at a single point, but with different slopes (Fig. 4).

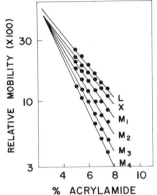

Fig. 4. Ferguson plots of glucuronidase components. Log of enzyme electrophoretic mobility relative to that of the dye bromphenol blue in gels of defined acrylamide concentration is plotted against acrylamide concentration. The experimental values are from a representative experiment using androgen-induced kidney of A/Ha females.

Thus, they comprise a family of isozymes that differ in molecular weight but have a constant surface charge. The lysosomal form L, on the other hand, gave a line that did not intersect at this common point, but was parallel to that of the microsomal form X. Thus forms X and L are equal in size, but differ in charge.

The acrylamide gel system was calibrated for molecular weight estimation by measuring the slopes of similar plots for 10 standard proteins ranging in molecular weight from 45,000 to 960,000 daltons (Fig. 5). The molecular weights of the glucuronidase forms(solid circles)have been calculated by a regression line through the standards. The molecular weights of forms L(whether extracted from lysosomes or microsomes)and X were both estimated to be 260,000 and those of the microsomal M components to increase stepwise from 310,000 to 470,000 in increments of 50,000-55,000 daltons.

These differences in molecular weight were confirmed when triton X-100 treated extracts of organs of wild type mice were chromatographed on Sephadex G-200 (Swank and Paigen, 1973). Although the columns were not capable of completely resolving the individual M forms, two clear peaks of activity were separated. The higher molecular weight peak contained, by acrylamide gel analysis, predominantly forms M_1 to M_4 while the low molecular weight peak contained only forms X and L.

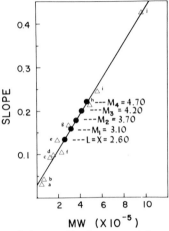

Fig. 5. Molecular weights of glucuronidase forms on poly-acrylamide gels.

Plots of log relative mobility (χ 100) against percentage acrylamide were constructed as described in the legend to Fig. 4 for 10 standard proteins and the six glucuronidase components. The slopes of the resulting curves were calculated and are plotted here against molecular weight of the standards. The plotted slope of each glucuronidase component is the mean value from 5 experiments. The standardization curve was obtained by the method of least squares. The standard proteins used and their molecular weights were (a) ovalbumin, 45,000; (b) bovine serum albumin monomer, 65,000; (c) bovine serum albumin dimer, 130,000; (d) alcohol dehydrogenase, 150,000; (e) bovine serum albumin trimer, 195,000; (f) catalase, 240,000; (g) leucine amino peptidase, 300,000; (h) apoferritin monomer, 480,000; (i) β-galactosidase, 520,000; (j) apoferritin dimer, 960,000.

Taken together, these findings suggest that the microsomal M forms carry additional material increasing their molecular weights over that of forms X and L. Forms L and X have molec-ular weights of 260,000, very close to the molecular weight of the isolated enzyme tetramer. The microsomal forms M_1 to M_4 have a surface charge like form X, from which they are pre-sumably derived, but carry from one to four additional pro-tein chains. The molecular weight of these additional chains, 50-55,000, appears to be significantly lower than the molecular weight of the glucuronidase subunit, which is 65,000-70,000.

The presence of additional protein chains in the M forms increasing their molecular weight was confirmed by dissociating this polypeptide from the M forms and recovering the basic

microsomal tetramer, X (Fig. 6). This occurred without any
loss of activity when enzyme extracts were exposed to 6 M-urea
at 0°. Dissociation appeared to be a sequential process. After
the first hour, M_1 as well as X increased in activity as though
both were derived from M_2, M_3, and M_4. Subsequently, M_1 was
converted to X. Lysosomal form L was not affected by this
treatment. There was no change in total activity during urea
treatment, indicating that the additional protein chains are
not catalytically active and their presence does not alter
the activity of the enzyme tetramer. Other agents effecting
the depolymerization are heating to 56° for 15 min or at 37° for
4-6 hours. The 37° conversion is accelerated by trypsin, acid
pH and chelating agents. However, microsomal polymers are
stable if stored at -20°.

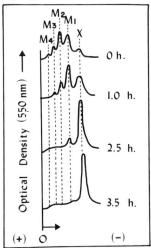

Fig. 6. Time course of dissociation of microsomal polymers in
6M-urea.

Partially purified microsomal polymers were isolated by
preparative polyacrylamide gel electrophoresis of liver extracts of DBA/Ha (Eg^a/Eg^a) males. They were suspended in 6 M
urea at 0o C for the indicated times and then dialyzed against
1 liter of pH 8.1 buffer for 2 hours in 1 cm-flat-width tubing. The samples were then electrophoresed, stained for glucuronidase activity and scanned at 550 nm. The control sample
at time 0 hr. contains some form X due to partial depolymerization of the polymers during preparative steps.

MOLECULAR BLOCK IN THE Eg^O MUTANT

Polyacrylamide gel electrophoresis was used to determine
the molecular nature of the defect in Eg^O mice (Fig. 7).
Whereas induced kidneys of wild type mice contain all six

electrophoretically separable glucuronidase components, the kidneys of Eg^O mice contain only forms L and X with L comprising 85-90% of the total activity. We have been unable to detect microsomal components M_1-M_4 in any tissue of this strain even when very large amounts of tissue were electrophoresed. Livers of Eg^O mice contain very small amounts of microsomal form X (about 2-4% of total activity). All other tissues of Eg^O, including uninduced kidney, contain only the lysosomal form L.

The X form of glucuronidase seen in the Eg^O mutant is indistinguishable from the X form seen in induced kidneys of normal mice. Analysis of the subcellular location of form X in Eg^O mice by either the osmotic shock or sucrose gradient techniques gave results similar to wild type. Band X was found exclusively in microsomes and over 90% of band L was found in lysosomes. No difference was found in the molecular weights of the X forms present in induced kidney of Eg^O and Eg^a mice when they were compared by Sephadex chromatography or acrylamide gel electrophoresis (Swank and Paigen, 1973).

Thus Eg^O mice differ from their normal counterparts in lacking demonstrable amounts of forms M_1, M_2, M_3 and M_4 in any tissue and in possessing detectable amounts (2-4% of total) of form X in liver. The Eg^O mutation appears to be a block in the intracellular synthesis of M_1 from X.

Combining our data on the physical properties and structures of the various intracellular forms of glucuronidase and the function of the Eg^O gene suggests a model for the intracellular maturation of glucuronidase enzyme (Fig. 8). The glucuronidase structural gene on chromosome 5 codes for the 65-70,000 molecular weight subunit of the enzyme, here designated G_O. The microsomal polymers M_1-M_4 are formed sequentially from X in a process requiring the presence of the Eg gene product which is missing in Eg^O.

IDENTIFICATION AND ISOLATION OF THE GLUCURONIDASE ANCHOR PROTEIN, EGASYN

The presence of an additional protein chain of predicted molecular weight has been demonstrated in the M forms of glucuronidase. It was not possible to do this by chemical isolation of the M forms, since they are too unstable and spontaneously degrade to form X before chemical isolation is possible. We therefore purified the glucuronidase-anchor protein complex by antibody precipitation using an antiserum prepared against mouse glucuronidase. If the anchor protein remains associated with the glucuronidase tetramer during immunoprecipitation, it should be present after SDS-gel electrophoresis of the immunoprecipitate. However, since the

estimated molecular weight of the peptide is 50-55,000, it
is quite likely that it will be masked on the gel by the presence of IgG heavy chains which have a similar molecular
weight. To overcome this difficulty we used the F (ab)$_2$
fragment of IgG instead of intact IgG (Fig. 9). If the IgG
molecule is cleaved just below the interchain disulfide bonds,
the resulting F (ab)$_2$ fragment is still held together by 2
disulfide bridges. Possessing 2 antigen binding sites it is
still capable of precipitin formation. However, it will dis-
sociate in SDS-mercaptoethanol into chains of 22-25,000 molecular weight and leave the 50-60,000 region of the gel open.

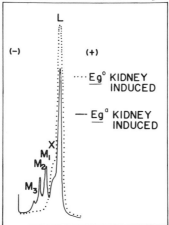

Fig. 7. Electrophoretically separable components of Eg^a and
Eg^o induced kidney.
    Females of strains A/Ha (Eg^a/Eg^a) and YBR (Eg^o/Eg^o) were
killed 8 days after induction with dihydrotestosterone. Whole
kidney homogenates were prepared, extracted with Triton X-100,
electrophoresed on polyacrylamide and stained for glucuronidase. The Eg^a gel (————) contained 0.190 units glucuronidase
and was stained 30 min; the Eg^o gel (·····) contained 0.136
units glucuronidase and was stained 45 min.

    For this purpose specific anti-mouse glucuronidase IgG was
digested with pepsin and the F (ab)$_2$ fragment was purified by
CM-cellulose chromatography. SDS gel electrophoresis showed
that no interfering heavy chain was present in the purified
F (ab)$_2$ fragments (Fig. 9). This F (ab)$_2$ fragment was used
to purify glucuronidase from various sources. All steps of
the immunoprecipitation were conducted in an ice bath to avoid
the temperature sensitive dissociation of the anchor protein
from glucuronidase. The immunoprecipitates were then analyzed
by SDS-gel electrophoresis.
    The upper part of Fig. 10 shows the pH 8.1 glucuronidase

isozyme patterns of various enzyme sources after the gels had been stained with naphthol-AS-BI glucuronide. The lower portion of the figure shows the Coomassie blue stained SDS gels of immunoprecipitates of the glucuronidase present in the same enzyme sources.

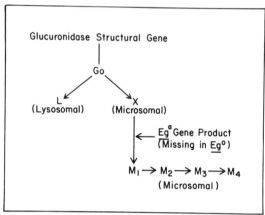

Fig. 8. Proposed model for the synthesis of the various intra-cellular forms of glucuronidase.

The total riton extract of a DBA ($\underline{Eg}^a$) mouse liver shows the previously described M and L forms of glucuronidase. Visible in the immunoprecipitate are the 75,000 MW subunit of glucuronidase (above), 23,000 molecular weight chains of the $F(ab)_2$ fragment (below) and an additional polypeptide that migrated slightly faster than the 75,000 subunit at about 64,000 MW. We designate this protein egasyn. If the liver extract was heated to convert all M forms to X and then immunoprecipitated, egasyn was no longer present in the immunoprecipitate. Likewise the immunoprecipitate from strain YBR, which carries the $\underline{Eg}^o$ mutation, lacked egasyn.

From these observations we conclude that there is, in fact, a polypeptide, which is non-covalently associated with glucuronidase. Heating at 56° removes egasyn and the $\underline{Eg}^o$ mutation in YBR causes loss of egasyn.

To investigate whether egasyn is bound to microsomal or lysosomal glucuronidase, microsomes and lysosomes were separated by differential centrifugation and glucuronidase was immunoprecipitated from each fraction. The upper gels show the enzymically active lysosomal L form and microsomal M forms. The SDS gels of the glucuronidase immunoprecipitates show that egasyn is associated only with microsomal glucuronidase.

These experiments were repeated with another wild type mouse, C57BL/6J (Fig. 11) with essentially the same results. There was an additional polypeptide in enzyme precipitated

from the total extract. Prior treatment of the extract with
deoxycholate which, like heat treatment, converts M forms to
X, abolished the polypeptide from the immunoprecipitate. In
C57BL/6J, as in DBA/2J mice, also, the additional polypeptide
was associated with microsomal but not lysosomal glucuronidase.

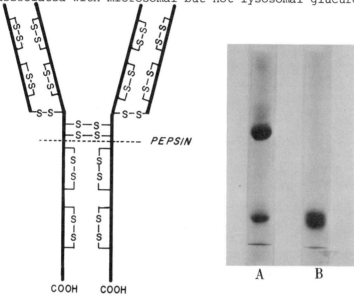

A   B

Fig. 9.   Schematic illustration of digestion of IgG by pepsin.
Whole globulin fraction was purified from goat anti-mouse
β-glucuronidase anti-serum by ammonium sulfate precipitation
at 33% saturation and incubated 24 hr. with crystalline pepsin
in 0.1 M acetate buffer pH 4.0 (globin/pepsin = 100/1). After
dialysis against 0.01M acetate buffer pH 5.5, the mixture was
applied to a column of CM-Cellulose (CM-52) preequilibrated
with the same buffer. The F(ab)$_2$ fragment was eluted from the
column with 0.1M Na-acetate buffer pH 5.5 and concentrated by
ultra filtration. SDS-gel electrophoresis was run following
the method of Laemmli (1970). Proteins were stained with
Coomassie blue (Burgess, 1969) overnight and destained by
diffusion.

    A)   Purified anti-mouse β-glucuronidase IgG
    B)   F(ab)$_2$ fragment of anti-mouse β-glucuronidase IgG.

The SDS gels of the immunoprecipitates indicate another
fact of possible importance for understanding the mechanism of
subcellular localization of glucuronidase in the mouse. The
mobilities of the microsomal and lysosomal enzyme subunits are
slightly different. Although it is not readily apparent in the
photograph, careful examination of the gels themselves indicated

that the subunits migrated as a doublet. Comparison of the

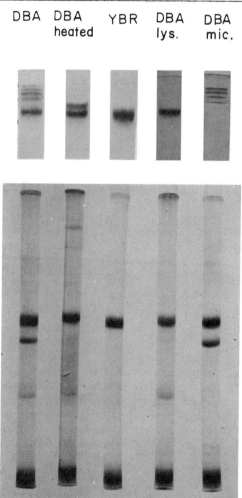

Fig. 10. Gel electrophoresis of immunoprecipitates.
    Microsomal and lysosomal glucuronidase were prepared by
differential centrifugation after osmotic shock treatment as
in Fig. 3. All preparations were adjusted to contain 0.02 M
tris-HCL, 0.15 M NaCl, 1% Triton X-100 pH 7.4 and mixed with
equivalent amount of anti β-glucuronidase F(ab)2 fragment.
Mixtures were incubated in an ice bath overnight. Immunoprecipitates were collected by centrifugation and washed 3
times with cold tris-saline pH 7.4 To each pellet was added
100 μl of sample buffer (Laemmli, 1970) containing 2% SDS and
5% mercaptoethanol and heated in a boiling water bath for 2 min.

Fig. 10 Legend continued
 SDS gel electrophoresis was carried out by the method of
Laemmli (1970).
Upper gels: Electrophoretic separation (See Fig. 2) of β-
glucuronidase before immunoprecipitation and stained for activity with naphthol AS-BI glucuronide. Total β-glucuronidase
was extracted from high speed pellets with Triton X-100
(See Fig. 2). One portion of the DBA/Ha extract were heated
at 56° for one hour and centrifuged before electrophoresis.

Lower gels: SDS gel electrophoresis of immunoprecipitates
stained for protein with Coomassie blue. Liver extracts from
5 animals were used for each gel:

glucuronidase subunit from microsomes and from lysosomes
showed that the upper band of the doublet is the microsomal
subunit and the lower band is the lysosomal subunit. The
difference of molecular weight of these subunits is very
small -at most 2-3,000 daltons. It likely reflects the post-
translational modification responsible for the difference in
electrophoretic mobility of the native microsomal and lysosomal glucuronidase tetramers in the pH 8.1 system. This difference suggests the microsomal X subunit contains some additional region covalently bonded to the subunit molecule. This
extra region on the microsomal subunit may participate in binding egasyn.

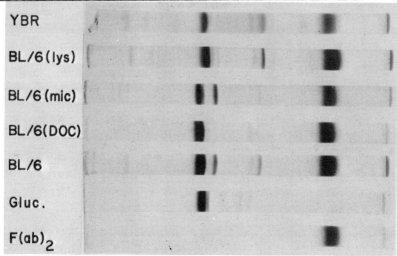

Fig. 11. SDS gel electrophoresis of glucuronidase immunoprecipitates. Conditions were same as described under Fig. 10 except that the livers of C57BL/6 male mice were used for enzyme
preparation. Protions of extract were mixed with Na-deoxycholate at a final concentration 0.5% and incubated for 5 min at

Fig. 11, continued. 0°C. The mixture was adjusted to pH 4.8 and centrifuged. The supernatant was titrated to pH 7.4 and immunoprecipitated. YBR-Total YBR extract; B46(lys)-Lysomal fraction from BL/6; B46(mic)-Microsomal fraction from BL/6; BL/6(DOC)-Total BL/6 extract treated with Na-deoxycholate; BL/6-Total BL/6 extract; Glu-Purified β-glucuronidase from BL/6; F(ab)₂-Purified **F**(ab)₂ fraction from anti-β-glucuronidase serum.

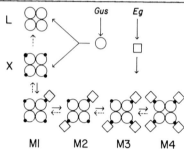

Fig. 12. Proposed interaction of the products of the glucuronidase structural gene and localization gene in bringing about specific membrane attachment of the enzyme.

The solid lines represent the direction of in vivo synthesis of the glucuronidase subcellular forms while the broken lines represent the direction of their in vitro interconversion.

DISCUSSION

Our present ideas on the maturation of intracellular glucuronidase are illustrated in Fig. 12. The structural gene Gus codes for the 75,000 molecular weight peptide sequence of the glucuronidase monomer subunit. Post-translational processing converts some of G_O to the lysosomal tetramer form L. The sugar side chains are presumably added during these steps. Post-translational processing also converts some of G_O to form X, the microsomal tetramer. In normal mice carrying the Eg^a allele the anchor protein egasyn then associates with the X-tetramer to give the higher molecular weight microsomal M forms of glucuronidase which contain from 1 to 4 units of egasyn. Mice homozygous for the Eg^O allele are unable to form the M complexes. As a result they lack microsomal glucuronidase, presumably because complexing with egasyn is required to bind or stabilize glucuronidase to the membranes. The defect in Eg^O mice must be either in the production of egasyn or in its binding to the microsomal tetramer X.

The depolymerization of the M forms to X can be accomplished non-enzymatically. However, the in vitro conversion of X to L probably is catalyzed by a cellular hydrolase, since it can be halted by prior heating of organ extracts (S. Tomino, unpublished).

In summary, then, although the microsomal and lysosomal tetramers are encoded by the same structural gene on chromosome 5 and are synthesized in parallel from the same 75,000 MW subunit, post-translational mechanisms differentiate the glucuronidase components in the two subcellular compartments. These mechanisms include covalent modifications causing small differences in size and charge of the tetramers. Furthermore, it is evident that the polypeptide product of at least one other gene, the Eg locus on chromosome 8, is necessary for the proper attachment and/or stabilization of glucuronidase in microsomes. The membrane binding is achieved through complex formation with a catalytically inactive protein egasyn.

The participation of hydrophobic anchor polypeptide sequences may be a general mechanism for the attachment of enzymes to specific subcellular membranes. Mitochondrial ATPases of both beef and yeast have been shown to be bound to the inner mitochondrial membrane by covalently attached protein of 18,000 molecular weight, the so-called oligomycin-sensitivity-conferring-protein (Kagawa and Racker, 1966; MacLennan and Tzagaloff, 1968; Tzagoloff, 1970). More recent experiments indicate that noncovalently bound polypeptides of low MW, encoded by the mitochondrial genome may serve in turn to anchor the OSCP protein to the inner mitochondrial membrane (Sierra and Tzagoloff, 1973). A low molecular weight hydrophobic peptide of 4-5,000 daltons covalently bound to microsomal cytochrone b5 also serves to anchor this enzyme to microsomal membranes (Strittmatter, et al., 1972).

Such anchor peptides may serve a dual role, not only providing a means of attaching enzymes to specific subcellular membranes, but in some cases also providing a mechanism to regulate the concentration of enzyme at separate intracellular sites. Thus the availability of egasyn may be the critical factor determining the presence or absence of microsomal glucuronidase among different types of specialized cells.

REFERENCES

Burgess, R. D. 1969. Separation and characterization of the subunits of ribonucleic acid polymerase. *J. Biol. Chem.* 244: 6168-6176.

Dofuku, R., U. Tettenborn, and S. Ohno. 1971a. Testosterone regulon in the mouse kidney. *Nature New Biology* 232: 5-7.

Dofuku, R., U. Tettenborn, and S. Ohno. 1971b. Further characterization of o^S mutation of mouse β-glucuronidase locus. *Nature New Biology* 234: 259-261.

Ferguson, K. A. 1964. Starch-gel electrophoresis - - application to the classification of pituitary proteins and polypeptides. *Metabolism (Clin. Exp.)* 13: 985-1002.

Ganschow, R. and K. Paigen. 1967. Separate genes determining the structure and intracellular location of hepatic glucuronidase. *Proc. Nat. Acad. Sci.* 58: 938-945.

Ganschow, R. and K. Paigen. 1968. Glucuronidase phenotypes of inbred mouse strains. *Genetics* 59: 335-349.

Ganschow, R. E. 1973. The genetic control of acid hydrolases. In: *Metabolic Conjugation and Metabolic Hydrolysis* (Fishman, W. M. ed.) Vol. 2, Academic Press, pp. 189-207.

Hedrick, J. L. and A. J. Smith. 1968. Size and charge isomer separation and estimation of molecular weights of proteins by disc gel electrophoresis. *Arch. Biochem. Biophys.* 126: 155-164.

Kagawa, Y. and E. Racker. 1966. Reconstitution of oligomycin-sensitive adenosine triphosphatase. *J. Biol. Chem.* 241: 2467-2474.

Karl, T. R. and V. M. Chapman. 1974. Linkage and expression of the Eg locus controlling inclusion of β-glucuronidase into microsomes. *Biochem. Genet.* 11:367-372.

Kato, K., I. Hirohata, W. H. Fishman,and H. Tsukamoto. 1972. Intracellular transport of mouse kidney β-glucuronidase induced by gonadotropin. *Biochem. J.* 127: 425-435.

Laemmli, U. K. 1970. Cleavage of structural proteins during the assembly of the head of bacteriophage T_4. *Nature.* 227: 680-685.

Lalley, P. and T. Shows. 1974. Lysosomal and microsomal glucuronidase: genetic variant alters electrophoretic mobility of both hydrolases. *Science* 185:442-444.

MacGregor, C. H. and C. A. Schnaitman. 1973. Reconstitution of nitrate reductase activity and formation of membrane particles from cytoplasmic extracts of chlorate-resistant mutants of *Escherichia coli*. *J. Bacteriol.* 114: 1164-1176.

MacLennan, D. H. and A. Tzagaloff. 1968. Purification and characterization of the oligomycin sensitivity conferring protein. *Biochemistry* 7: 1603-1610.

Paigen, K. 1961a. The effect of mutation on the intracellular location of β-glucuronidase. *Expt. Cell Res.* 25: 286-301.

Paigen, K. 1961b. The genetic control of enzyme activity during differentiation. *Proc. Nat. Acad. Sci.* 47: 1641-1649.

Paigen, K., R. Ganschow, R. T. Swank,and S. Tomino. 1975. The molecular genetics of mammalian glucuronidase. *J. Cell. Physiol.*, (in press).

Sierra, M. F. and A. Tzagaloff. 1973. Assembly of the mitochondrial membrane system: purification of a mitochondrial product of the ATPase. *Proc. Nat. Acad. Sci.* 70: 3155-3159.

Stahl, P. D. and O. Touster. 1971. β-glucuronidase of rat liver lysosomes. *J. Biol. Chem.* 246: 5398-5406.

Strittmatter, R., M. J. Rogers, and L. Spatz. 1972. The binding of cytochrome b5 to liver microsomes. *J. Biol. Chem.* 247: 7188-7194.

Swank, R. T. and K. Paigen. 1973. Biochemical and genetic evidence for a macromolecular β-glucuronidase complex in microsomal membranes. *J. Mol. Biol.* 77: 371-389.

Swank, R. T., K. Paigen, and R. E. Ganschow. 1973. Genetic control of glucuronidase induction in mice. *J. Mol. Biol.* 81: 225-243.

Tzagaloff, A. 1971. Properties of a dispersed preparation of the rutamycin-sensitive adenosine triphosphatase of mitochondria. *J. Biol. Chem.* 246: 7328-7336.

Van Lancker, J. L. and P. L. Lentz. 1970. Study on the site of biosynthesis of β-glucuronidase and its appearance in lysosome in normal and hypoxic rat. *J. Histochem. Cytochem.* 18: 529-540.

REGULATION OF ISOZYMES
IN INTERSPECIES SEA URCHIN HYBRID EMBRYOS

HIRONOBU OZAKI[1]
Kitasato University
School of Medicine, JAPAN

ABSTRACT. In order to analyze the genome-phenotypic product relationship in ontogeny, isozymes of esterases were examined in hybrids of three species of sea urchins: *Strongylocentrotus purpuratus*, *Strongylocentrotus droebachiensis*, and *Dendraster excentricus*. The crosses between the two strongylocentrotids form hybrids which develop pluteus larvae, while the hybrids between the strongylocentrotids and *Dendraster* terminate in blocked development at the gastrula stage.

The enzyme activity increases through development. The complex isozyme pattern obtained by disc electrophoresis of saline extracts is identical and stage specific for the two strongylocentrotids. For *Dendraster*, however, the pattern changes little up to the pluteus stage. Substrate specificity studies have shown that the enzymes are carboxylesterase (E.C.3.1.1.1.) and arylesterase (E.C.3.1.1.2.). Only in advanced embryos does acetylcholinesterase (E.C. 3.1.1.7.) appear.

In the compatible hybrid *S.purpuratus* ♀ x *S. droebachiensis* ♂, a paternal influence was demonstrated by the precocity of the elevated activity and the appearance of isozymic components. In the blocked hybrids between both species of *Strongylocentrotus* and *Dendraster*, no evidence was obtained as to the presence of paternal enzymes or any other enzyme which is unique to the hybrids. Since these hybrids contain both parental genomes, the lack of paternal esterases in this study and of other molecular markers indicates a lesion in the transcription-translation process of the paternal genome.

INTRODUCTION

The recognition of isozymes (Markert and Møller, 1959) has enabled one to explore developmental processes in finer detail and to resolve these processes into better defined individual components. The change in the total activity of many enzymes in the course of development was found to be a composite of

---

[1]On leave from the Department of Zoology, Michigan State University, East Lansing, Michigan.

several isozymes each of which undergoes a change. These
isozymes permit one to ask questions not merely about quantitative changes in total enzyme activity during development, but
also about qualitative changes. How many isozymes exist? How
do the isozymes change in space and in time during development;
and what controls these changes? As a consequence, many
characteristic ontogenetic changes in the isozyme patterns
have been found to reflect the states of differentiation. Isozymes serve as good gene markers for analyzing cellular differentiation. The analysis of isozymes, therefore, is one important aspect of the studies of the genome-phenotypic product
relationship which is the central problem in current developmental biology.

A useful approach to study this relationship is the analysis
of interspecies sea urchin hybrids. In the hybrids, the nucleus
of one species is introduced into the egg of another species
by cross fertilization. This adds new interactions to the
normal nucleo-cytoplasmic interactions of the host cell. The
new interactions are not only between the introduced genome
and the cytoplasm of the host, but also between the two genomes
of different origins. The foreign genome may also cause alterations in the normal nucleo-cytoplasmic interactions of the
host. In some crosses, fairly normal development ensues; in
others, developmental arrest takes place. It is not clear how
such failure is brought about. A search for possible causes
would provide an insight into the normal function of genes that
are important in development. The present study constitutes
an effort toward this goal and is concerned with the analysis
of esterase isozymes in the development of three sea urchin
species and their reciprocal hybrids. The isozyme patterns
of one group of sea urchins are stage-specific, while the
pattern from a sea urchin belonging to another group apparently
lacks such stage specificity. This difference in ontogenetic
expression of isozymes undoubtedly reflects the evolutionary
divergence between the species. The observations of the
present study and of others on blocked development in hybrids
suggest a lesion in the transcription-translation process of
the paternal genome.

EXPERIMENTAL ANIMALS

Three sea urchins, *Strongylocentrotus purpuratus*,
Strongylocentrotus droebachiensis, and *Dendraster excentricus*,
obtained from the coast of the Pacific Northwest were used in
the experiment. *Strongylocentrotus* and *Dendraster* belong to
different Superorders (Moore, 1966). For the sake of convenience, the following abbreviations will be used in the

illustrations: P for *S.purpuratus*; D for *S.droebachiensis*; and De for *Dendraster excentricus*. The nature of the species cross is identified by listing the species of the female parent first and then the species of the male parent as in PDe, indicating that an *S.purpuratus* egg is fertilized by *D.excentricus* sperm.

After a rapid succession of divisions of the fertilized egg, the embryo attains the blastula stage, then an invagination of the larval gut takes place to give rise to the gastrula stage. The development of some interspecies hybrids become arrested at this stage. The last larval stage is the pluteus stage. These stages will be identified in the figures by the notations of Whiteley and Baltzer (1958).

All three species show clear species specific characteristics, such as pigmentation, and rate of development. Table 1 compares the rates of development of the three species at a controlled temperature of 10°C.

Hybrids between the two *Strongylocentrotus* species have compatible genomes. They develop pluteus larvae. The hybrids between either of the two *Strongylocentrotus* species and *Dendraster* are incompatible. Their development is blocked at the gastrula stage. Judging from cytological observations on the chromosomal behavior (Moore, 1957) and by molecular hybridization of DNA-DNA (Whiteley and Whiteley,1972) and by other studies (Flickinger, 1957; Griffiths, 1965; Ozaki, 1965; Ozaki and Whiteley, 1970; and Brookbank and Cummins, 1972) all of these hybrids are true hybrids since no elimination of the paternal genome occurs.

Interspecific fertilizations, with the exception of the *S.purpuratus* ♀ x *D.excentricus* ♂ hybrid, were brought about by inseminating the eggs with a heavy suspension of sperm. For *S.purpuratus* ♀ x *D.excentricus* ♂ , the eggs were treated with 0.05% trypsin in sea water at 10°C for 5 min prior to fertilization. In this instance, the eggs treated with trypsin were also fertilized with homologous sperm as a control.

ESTERASES IN THE DEVELOPMENT OF HOMOLOGOUS CROSSES

Saline extracts cleared by centrifugation were subjected to the standard polyacrylamide gel disc electrophoresis of Ornstein and Davis (1961). The esterase activity in the supernate represents some 60-70% of the total activity in the homogenate. The enzyme activity was demonstrated on the gel by an azo dye method with l-naphthol esters as substrates.

Isozyme patterns of *Dendraster* embryos at various stages of development are shown in figure 1. Dark bands of dye deposition, indicated by lines to the right of each zymogram, result from esterase activity when l-naphthyl acetate is

TABLE 1

TIME TABLE FOR NORMAL DEVELOPMENT OF *S.PURPURATUS*,
S.DRÖBACHIENSIS and *D.EXCENTRICUS* AT 10°C

Age in Hours	PP	DD	DeDe
0	UF	UF	UF
6	2-cell		
12	16-cell		
18			
24	M	M	M
			h
30			
36	h	h	BlMy-1
42	BlMy-1	Ga-1	Ga Jl/4
48	BlMy-2		
54	Ga-Jl/4		
60			Pr
66		Pr	
72	Ga J-4/5		Pl-I
78			
84			
90			Pl-II
96	Pr	Pl-I	
102			
108			
114	Pl-I		
120			

UF = Unfertilized eggs
M = Morulae
Bl = Blastulae   My-1 mesenchyme cells start migrating and
                  complete migration in My-2
Ga = Gastrulae   Jl/4 shows invagination 1/4 of the complete
                  blastocoele diameter.
Pr = Prism

Pl = Plutei   I is young pluteus with short arm hump,
                  II in De with long arms.

employed as a substrate. There are four stained bands after
electrophoresis of the extract of the unfertilized egg. These
isozymes persist until the pluteus stage when one additional
band is detected.

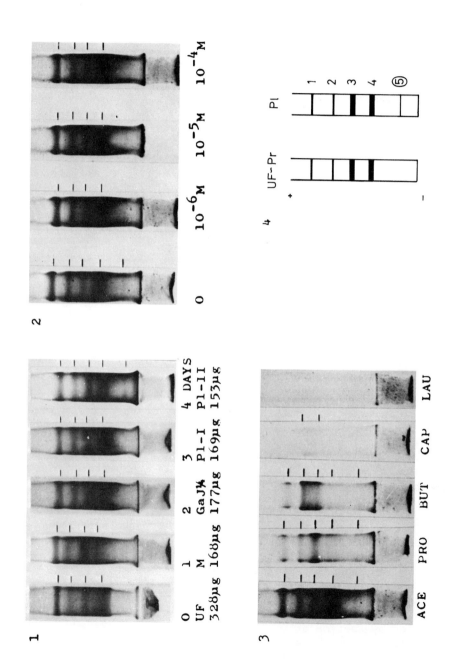

Figs. 1-4 legends. **Fig. 1.** Ontogenetic changes in esterases during development of DeDe. Jelly free eggs and embryos were homogenized with an equal volume of cold 0.5M KCl. The supernatant obtained after centrifugation of the homogenate at 6,250xg for 1 hr was mixed with an equal volume of 1 M sucrose solution and was subjected to acrylamide gel disc electrophoresis (Ornstein and Davis, 1961). Electrophoresis was conducted at 2 mA/tube for 50 min at 10°C. The gels were rinsed with 0.1 M phosphate buffer pH 6.8 for 20 min, then incubated with a reaction mixture that consisted of 2 ml 1% acetone solution of 1-naphthyl acetate and 50 mg Fast Garnet GBC(Dajac) in 100 ml of the above buffer. Stained gels were stored in 7% acetic acid, and photographed with transmitted light using a Kodak filter no. 15(G) and High Contrast Copy Film. Enzyme stains are marked by bars. The direction of protein migration is from the bottom to the top. Ages in days are shown. Developmental stages are indicated by the notations of Whiteley and Baltzer (1958). Amounts of protein used in the electrophoresis are given and were determined by the method of Lowry et al. (1951) with bovine serum albumin as standard. **Fig. 2.** Inhibition of esterases of DeDe plutei by eserine. Gels were rinsed and stained in the presence of physostigmine salicylate at the concentrations indicated. Other conditions as described for Fig. 1. **Fig. 3.** Substrate specificity of esterases of DeDe plutei. In addition to acetate (ACE), the following 1-naphthol esters were used as substrates: propionate (PRO), butyrate (BUT), caprylate (CAP), and laurate (LAU). Other conditions as described for Fig. 1. **Fig. 4.** A diagram summarizing ontogeny of esterase isozymes in DeDe.

In order to determine whether any of these bands were due to cholinesterase activity, the specific inhibitor eserine was added in increasing concentrations to the pluteus enzyme extracts (figure 2). The esterase band that appears last during development is sensitive and is inhibited at a concentration as low as 10^{-6}M. The other esterases are not inhibited.

Substrate specificities were also examined by use of 1-naphthol esters of carboxylic acids with varying carbon-chain lengths (Holmes and Masters, 1967). It is clear from figure 3 that the enzymes act primarily upon relatively short carbon-chain esters, and react most actively with the acetate moiety.

The isozyme patterns exhibited during the ontogeny of *Dendraster* are summarized in figure 4. The last band to appear during development, band 5, is identified as acetylcholinesterase (E.C.3.1.1.7.) because of its: (1) specific inhibition by eserine at very low concentrations, (2) and its high affinity for the acetyl substrate. Furthermore Augustinsson and Gustafson (1949) identified the sea urchin cholinesterase as the same enzyme as the one in the nervous system and red blood cells of vertebrates. Band 4 is identified tentatively as arylesterase (E.C.3.1.1.2.), since it does not utilize butyrate as a substrate as well as the other esterases which are identified as carboxylesterases (E.C.3.1.1.1.). From these analyses, it can be concluded that *Dendraster* contains an acetylcholinesterase which appears late in development. All the other esterases are present from the very beginning of embryonic development. This pattern sharply contrasts with that of the *Strongylocentrotus* species.

The results obtained from similar analyses of *Strongylocentrotus purpuratus* are presented in figures 5 - 7. In the early stages of development of this species, there was difficulty in resolving the esterases due to the presence of large amounts of yolk protein which caused a high background stain, but basically two isozymes can be discerned (figure 5). Their presence becomes clearer by the blastula stage. From this stage on, an increasingly complex pattern evolves. At the final pluteus stage, there are six stained bands. The 4th band is sensitive to 10^{-6}M eserine as shown in figure 6. Studies of substrate specificity are shown in figure 7.

The second species of *Strongylocentrotus*, *S. droebachiensis*, was similarly examined. Ontogenetic change, eserine sensitivity, and substrate specificities are shown in figures 8 - 10. These results are identical with the observations made on *S. purpuratus* at the same stages of development. The results on the two *Strongylocentrotus* species are summarized in figure 11. The stage-specific developmental progression of the esterases are identical for *S. purpuratus* and *S. droebachiensis*. The combination of eserine inhibition and substrate specificity studies permits the following tentative identifications: band 1 is an arylesterase, 2 and 3 are carboxylesterases, 4 is an acetylcholinesterase, and 5 and 6 are carboxylesterases. The patterns are similar to the group banding patterns of esterases of *Arbacia punctulata* during ontogeny (O'Melia, 1972) with some differences in the assignment of acetylcholinesterase activity.

In all three species examined in the present study, the total esterase activity increases 3 to 4 fold during development, primarily after gastrulation (Ozaki, 1965).

The ontogenetic changes of the three species may be summed up as follows:

Acetylcholinesterase. Acetylcholinesterase is first detected electrophoretically at an advanced stage. This observation is in agreement with the increase in the activity of this enzyme later in development (Augustinsson and Gustafson, 1949).

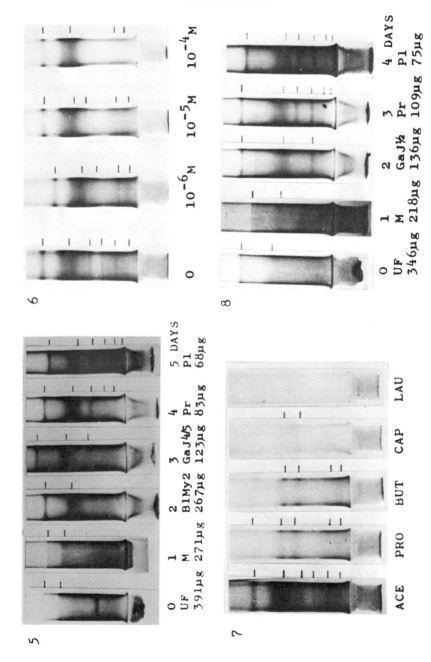

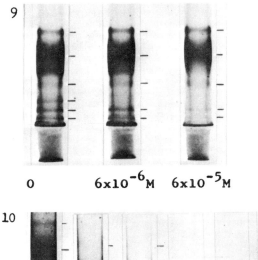

9

0　　　　$6 \times 10^{-6}M$　$6 \times 10^{-5}M$

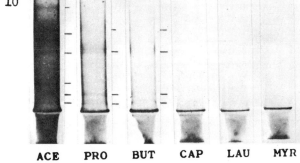

10

ACE　　PRO　　BUT　　CAP　　LAU　　MYR

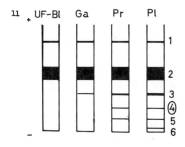

Figs. 5-11 legends. <u>Fig. 5.</u> Ontogenetic changes in esterases
during development of PP. Conditions as described for Fig. 1.
One dark line in the lower portion of the UF gel is an uniden-
tified yellow substance and is not an enzyme stain. <u>Fig. 6.</u>
Inhibition of esterases of PP plutei by eserine. Conditions
as described for Fig. 2. <u>Fig. 7.</u> Substrate specificity of
esterases of PP plutei. Conditions as described for Fig. 3.
<u>Fig. 8.</u> Ontogenetic changes in esterases during development
of DD. Conditions as described for Fig. 1.

551

Fig. 9. Inhibition of esterases of DD plutei by eserine. Conditions as described for Fig. 2. Fig. 10. Substrate specificity of esterases of DD plutei. Myristate (MYR) is included. Other conditions as described for Fig. 3. Fig. 11. A diagram summarizing ontogeny of esterase isozymes in PP and DD.

However, a little activity is detected in the unfertilized egg and embryos prior to gastrulation (Ozaki, unpublished result). The "ghosts" prepared from unfertilized and fertilized eggs of sea urchins were found to have the acetylcholinesterase activity (Barber and Foy, 1973). The low levels of enzyme activity found in early stages may be due to a membrane bound enzyme, and therefore was not observed in the current study.

Other esterases. For *Dendraster*, the increase in activity does not accompany qualitative increases in the number of isozyme bands; instead the same bands present in the unfertilized egg are found throughout development to the pluteus stage. In the case of the *Strongylocentrotus* species, the increase in the activity accompanies the appearance of new isozyme bands. The pattern is stage specific, but not species specific.

Although at this early phase in ontogeny, when physiological requirements of these embryos are seemingly close for both *Strongylocentrotus* and *Dendraster*, the above remarkable difference is surprising. It must be that the difference is an indication of extensive evolutionary divergence between the two sea urchins. Such divergence has also been shown in another study. DNA-DNA molecular hybridization between *Strongylocentrotus purpuratus* and *Dendraster excentricus* shows that 10% or less of the redundant portion of the DNA is common to both these species, whereas the two *Strongylocentrotus* species share roughly 70% (Whiteley et al., 1970).

ESTERASES IN THE DEVELOPMENT OF HYBRID CROSSES

Compatible hybrids. Compatible hybrids are formed between *S.purpuratus* and *S.droebachiensis*. The isozymic patterns of both species are not distinguishable at the same stage of development. However, the two species have significantly different rates of development, *S.droebachiensis* developing more rapidly. For the hybrid, *S.purpuratus* ♀ x *S.droebachiensis* ♂, this faster rate of development of the paternal species is reflected in the precocious appearance of the isozyme patterns (presumably paternal) (figures 12 and 13), and the elevated level of the enzyme activity of the hybrid over that of the maternal control (Ozaki, 1965).

Since the isozyme patterns are not distinguishable between

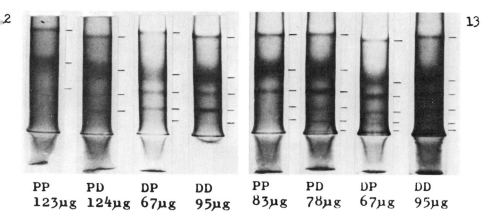

PP	PD	DP	DD	PP	PD	DP	DD
123µg	124µg	67µg	95µg	83µg	78µg	67µg	95µg

Fig. 12. Zymograms of 72 hr hybrids PD and DP and the controls. Fig. 13. Zymograms of 96-hr hybrids PD and DP and the controls.

the two species, one cannot be certain whether or not the precociously appearing component was made on the messenger RNA transcribed from the paternal genome. However, in the same cross, echinochromes of the paternal type have been clearly demonstrated (Griffiths, 1965). Thus, it is possible that paternal genes for the esterases are also active in this hybrid.

Blocked hybrids. Hybrids between *Strongylocentrotus* and *Dendraster* form embryos in which the development is blocked. The isozyme pattern of the two sea urchins is different. When the extracts from each species are mixed and then subjected to electrophoresis, some of the esterases are clearly identifiable with a specific parental species (figure 14). In all possible crosses between *Strongylocentrotus* and *Dendraster* no evidence was obtained as to the presence of paternal enzymes, nor were enzymes unique to the hybrid embryos detected (figures 15-18).

OTHER STUDIES

For the blocked hybrids between *Stroygylocentrotus* and *Dendraster*, other molecular markers have been examined. The results given below confirm that these paternal markers fail to appear in the hybrids.[2]

Malate dehydrogenase. Malate dehydrogenase activity increases throughout development. The activity is due to the presence of two isozymes in the cell. They differ in cellular localizations: one in the cytoplasm and the other in the mitochondria. They are further distinguished by differential substrate

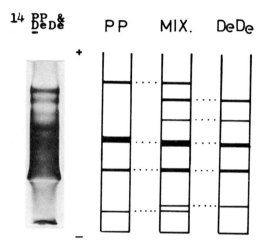

Fig. 14. A zymogram of a mixture of PP and DeDe, and diagrams to show species origins of individual isozymes.

---

inhibition, by electrophoresis in acrylamide gel, and by ion exchange chromatography. *Strongylocentrotus* and *Dendraster* possess different malate dehydrogenase isozyme patterns. Hybrids formed by crossing the two lack the paternal isozymes (Ozaki and Whiteley, 1970).

Hatching enzymes. When sea urchin embryos hatch from the fertilization membrane, they do so by releasing hatching enzymes which hydrolyze the membrane. The enzyme is synthesized on a messenger RNA transcribed after fertilization and during cleavage (Yasumasu, 1963). In the hybrid *D. excentricus* ♀ x *S. purpuratus* ♂, the enzyme that is released by the hybrid is strictly maternal. It can attack only the membrane of the maternal species, but not that of the paternal species (Whiteley and Whiteley, personal communication).

Pigmentation. Pigmentation in all three species is different, yet no paternal influence is detected in the blocked hybrids between *Strongylocentrotus* and *Dendraster* (Ozaki, 1965; Whiteley and Whiteley, 1972). This may suggest that the enzymes needed to elaborate the paternal pigmentation are absent.

Antigens. In the hybrid *S. purpuratus* ♀ x *D. excentricus* ♂

---

[2] A recent report (Easton, D.P. et al., 1974, *Biochem. Biophys. Res. Commun.* 57: 513-519) shows that paternal F1 histone, unlike enzymes, appears in the blocked hybrids DeP, DeD, and DDe.

no antigens specific to the paternal species have been detected (Badman and Brookbank, 1970).

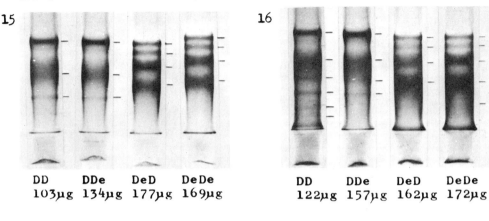

15   DD   DDe   DeD   DeDe
         103μg   134μg   177μg   169μg

16   DD   DDe   DeD   DeDe
         122μg   157μg   162μg   172μg

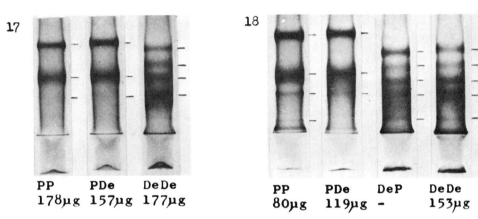

17   PP   PDe   DeDe
         178μg   157μg   177μg

18   PP   PDe   DeP   DeDe
         80μg   119μg   -   153μg

Fig. 15. Zymograms of 68-hr hybrids DDe and DeD and the controls. Fig. 16. Zymograms of 92 hr hybrids DDe and DeD and the controls. Fig. 17. Zymograms of 66-hr hybrids PDe and the controls. Fig. 18. Zymograms of 91-hr hybrids PDe and DeP and the controls.

CONCLUSION

Considering that a wide array of molecular markers of paternal species was not manifested in the blocked hybrids, and in view of the presence of favorable evidence for the replication of both parental genomes, the chance of the above failures due to chromosomal delection is unlikely. Evidently, messenger precursor heterogeneous nuclear RNA molecules are transcribed from both parental genomes (Whiteley and Whiteley, 1972). Thus the reasonable interpretation of the data is that in the blocked hybrids a lesion in the post-transcription process exists. It may be in the failure of transport of messenger RNA molecules to the cytoplasmic site of protein synthesis as suggested by Ficq and Brachet (1963) from their autoradiographic studies on another blocked sea urchin hybrid *Paracentrotus lividus* ♀ x *Arbacia lixula* ♂. Alternatively, it may be in the failure of the host translational mechanism such as t-RNA and its synthetase (Ilan et al., 1970) or in an inititation factor (Ilan and Ilan, 1971) which is necessary for the synthesis of stage specific protein. These and others are possible causes of the observed failure of the appearance of paternal phenotypic products in the hybrids. Further studies are needed for the identification of the exact nature of the lesion.

It should be borne in mind also that the mere failure of paternal gene activity would not account for the arrest of development in these hybrids. Obviously the introduced foreign genome does disturb the normal nucleo-cytoplasmic interactions in the host cell. The level at which this perturbation occurs requires further study.

In order to understand the genome-phenotypic product relationship during development, it ultimately becomes necessary to select a suitable protein marker, identify its messenger RNA, and then trace the message from the site of transcription on the DNA to the synthesis of the protein in the cytoplasm. Recent progress in the isolation of specific messenger RNA molecules and the synthesis of their complementary DNA molecules indicates that this is a feasible approach (Gilmour and Paul, 1973; Ruderman et al., 1974). Isozymes, particularly allelic isozymes have been shown to be excellent gene markers. This present study indicates that acetylcholinesterase in sea urchins may serve as a useful gene marker (Ozaki, 1974). The increase in its activity occurs more rapidly than most other enzymes do during early sea urchin development, and the activity is localized primarily in a specific region of the embryo. Our current research is focusing on these issues.

ACKNOWLEDGEMENTS

I am most grateful to Dr. Arthur H. Whiteley for his invaluable guidance throughout the investigation. This research was conducted at the University of Washington, and Friday Harbor Laboratories, and was partly supported by a National Science Foundation grant to A. H. W. The manuscript was prepared while the author was on sabbatical leave. I express my appreciation to my host, Dr. Masaya Kawakami, for his hospitality.

REFERENCES

Augustinsson, K-B, and T. Gustafson 1949. Cholinesterase in developing sea-urchin eggs. *J. Cell. Comp. Physiol.* 34: 311-321.

Badman, W. S. and J. W. Brookbank 1970. Serological studies of two hybrid sea urchins. *Dev. Biol.* 21: 243-256.

Barber, M. L. and J. E. Foy 1973. An enzymatic comparison of sea urchin egg ghosts prepared before and after fertilization. *J. Exp. Zool.* 184: 157-166.

Brookbank, J. W. and J. E. Cummins 1972. Microspectrophotometry of nuclear DNA during the early development of a sea urchin, a sand dollar, and their interordinal hybrids. *Dev. Biol.* 29: 234-240.

Ficq, A. and J. Brachet 1963. Metabolisme des acides nucleiques et des proteines chez les embryons normaux et les hybrids letaux entre echinodermes. *Exp. Cell Res.* 32: 99-108.

Flickinger, R. A. 1957. Evidence from sea urchin-sand dollar hybrid embryos for a nuclear control of alkaline phosphatase activity. *Biol. Bull.* 112: 21-27.

Gilmour, R. S. and J. Paul 1973. Tissue-specific transcription of the globin gene in isolated chromatin. *Proc. Nat. Acad. Sci. USA* 70: 3440-3442.

Griffiths, M. 1965. A study of the synthesis of naphthaquinone pigments by the larvae of two species of sea urchins and their reciprocal hybrids. *Dev. Biol.* 11: 433-447.

Holmes, R. S. and C. J. Masters 1967. The developmental multiplicity and isoenzyme status of cavian esterases. *Biochim. Biophys. Acta* 132: 379-399.

Ilan, J., J. Ilan, and N. Patel 1970. Mechanism of gene expression in *Tenebrio molitor*. Juvenile hormone determination of translation control through transfer ribonucleic acid and enzyme. *J. Biol. Chem.* 245: 1275-1281.

Ilan, J. and J. Ilan 1971. Stage-specific initiation factors for protein synthesis during insect development. *Dev. Biol.* 25: 280-292.

Lowry, O. H., N. J. Rosebrough, A. L. Farr, and R. J. Randall 1951. Protein measurement with the Folin phenol reagent. *J. Biol. Chem.* 193: 265-275.

Markert, C. L. and F. Møller 1959. Multiple forms of enzymes: Tissue, ontogenetic, and species specific patterns. *Proc. Nat. Acad. Sci. USA* 45: 753-763.

Moore, A. R. 1957. Biparental inheritance in the interordinal cross of sea urchin and sand dollar. *J. Exp. Zool.* 135: 75-83.

Moore, R. C. (ed.) 1966. *Treatise on Invertebrate Paleontology.* Part U, Echinodermata, 3: two volumes. Univ. of Kansas Press, Lawrence, Kansas.

O'Melia, A. F. 1972. Changes in esterase and cholinesterase isozymes in normally developing, animalized and radialized embryos of *Arbacia punctulata. Exp. Cell. Res.* 73 :469-474.

Ornstein, L. and B. J. Davis 1961. *Disc Electrophoresis.* Distillation Products Industries.

Ozaki, H. 1965. Differentiation of esterases in the development of echinoderms and their hybrids. University of Washington Ph.D. Thesis.

Ozaki, H. and A. H. Whiteley 1970. L-malate dehydrogenase in the development of the sea urchin *Strongylocentrotus purpuratus. Dev. Biol.* 21: 196-215.

Ozaki, H. 1974. Localization and multiple forms of acetylcholinesterase in sea urchin eggs. *Dev. Growth Differ.* (in press).

Ruderman, J., C. Baglioni, and P. R. Gross 1974. Histone mRNA and histone synthesis during embryogenesis. *Nature.* 247: 36-38.

Whiteley, A. H. and F. Baltzer 1958. Development, respiratory rate and content of deoxyribonucleic acid in the hybrid *Paracentrotus* ♀ x *Arbacia* ♂. *Pubbl. Staz. Zool. Napoli.* 30: 402-457.

Whiteley, H. R., B. J. McCarthy, and A. H. Whiteley 1970. Conservatism of base sequences in RNA for early development of echinoderms. *Dev. Biol.* 21: 216-242.

Whiteley, A. H. and H. R. Whiteley 1972. The replication and expression of maternal and paternal genomes in a blocked Echinoid hybrid. *Dev. Biol.* 29: 183-198.

Yasumasu, I. 1963. Inhibition of the hatching enzyme formation during embryonic development of the sea urchin by chloramphenicol, 8-azaguanine and 5-bromouracil. Sci. Papers Coll. Edu., Univ. Tokyo 13: 241-246.

SEX-LINKED ISOZYMES AND SEX CHROMOSOME EVOLUTION AND INACTIVATION IN KANGAROOS

D. W. COOPER, P. G. JOHNSTON, CAROLYN E. MURTAGH,
G. B. SHARMAN, J. L. VANDEBERG, AND W. E. POOLE
School of Biological Sciences,
MacQuarie University, North Ryde, N.S.W., 2113
and
CSIRO Division of Wildlife Research
Lyneham, A.C.T., 2602, AUSTRALIA

ABSTRACT: The structural genes for the enzymes glucose-6-phosphate dehydrogenase (G6PD) and phosphoglycerate kinase (PGKA) are sex-linked in marsupials. This accords with Ohno's thesis of the conservative nature of the X-linkage group in mammals, since those two genes are also sex-linked in Man. The data on the inheritance of isozymic forms of these enzymes and the phenotypes of females heterozygous for them suggest that the cells and tissues of kangaroos may be divided into at least two kinds. One kind, represented by the blood, has dosage compensation achieved by paternal X-inactivation. The other kind, represented by primary uncloned cultures of fibroblasts, may have both chromosomes active though whether within the same cell is not known. The G6PD patterns of heterozygotes seemingly have interaction products, a result which is compatible either with activity of both alleles within the same cell or transfer of gene product between cells within a mixture of cells with contrasting types active. PGKA patterns of lysates of primary uncloned cultures of fibroblasts and also muscle homogenates, both derived from heterozygotes, show expression of both isozymes with the maternally derived one always predominant.

INTRODUCTION

In eutherian mammals such as man and mouse sex chromosome dosage compensation is achieved by random X inactivation or Lyonization, as originally proposed by Lyon (1961). According to her hypothesis, in any one somatic cell of an adult female only one X is active. Each cell possesses either an active maternal or an active paternal X, but no cell has both active. Inactivation of one or the other is thought to take place early in development, possibly at the late blastula stage, so that the female becomes a mixture of approximately equal numbers of the two types of cells. For eutherian mammals the Lyon hypothesis has been verified for individual loci whenever it has been possible to test it critically. X-linked genes in eutherians which escape inactivation may possibly exist (e.g. the Xg locus in man; Lyon, 1972), but their existence has yet to be conclusively demonstrated. Lyon has written several comprehensive summaries of the data and ideas concerning her hypothesis, the most recent being Lyon (1972).

Marsupials are the nearest living mammalian group to eutherian mammals (Tyndale - Biscoe, 1973). Like eutherians they are XX in the female, XY in the male, with the Y being male determining (Sharman et al., 1970). The structural genes for the enzymes glucose-6-phosphate dehydrogenase (G6PD) and phosphoglycerate kinase (PGKA) are sex linked in man and in kangaroos. These facts suggest that the sex chromosomes of eutherians and marsupials are at least partially homologous, in agreement with Ohno's thesis that the X has been a linkage group conserved as a single entity during the evolution of mammals (Ohno, 1967). It might then have been expected that both groups would have had X chromosome dosage compensation and that it would have been brought about in the same way. But kangaroos and therefore perhaps other marsupials seem to be different. In female kangaroos one kind of cell, of which the best investigated are the cells of the blood, possess dosage compensation which is achieved by paternal X-inactivation. There is only one active X and it is always that derived from the mother. The ^{3}H thymidine labelling patterns of the X chromosomes of hybrid kangaroos (Sharman, 1971) are in accord with this hypothesis. Mass lysates of the other kind, notably primary uncloned cultures of fibroblasts, show activity of both X chromosomes, as evidenced by G6PD and PGKA isozymes. The purpose of this paper is to summarize the allelic isozyme data for these conclusions.

MATERIALS AND METHODS

Animals. The kangaroos and wallabies used in this study are listed in Table 1, together with their G6PD and PGKA types. Since the distinction between kangaroo and wallaby is a popular one based solely on size and has no basis in phylogeny, for convenience in this paper, we shall use the term "kangaroo" to cover all species listed in Table 1.

Gel Electrophoresis. G6PD was usually electrophoresed on Cellogel a cellulose acetate supporting medium, using a modification of the buffer of Rattazzi et al. (1967), in which the amount of citric acid had been reduced from 3.28 g/l to 1.60 g/l to bring the pH to 8.0. Runs were from 1 - 2 hours at room temperature using hemolysates of red cells. PGK was electrophoresed by the method of Beutler (1969), as modified by

TABLE 1

The kangaroo species and subspecies used in this study with their G6PD and PGKA types.

SPECIES	ISOZYMES OF: G6PD	PGKA	APPROXIMATE FREQUENCY OF VARIANT
Macropus robustus robustus Gould, 1841 (Wallaroo)	F	N	–
Macropus robustus erubescens Gould, 1841 (Euro)	S	N	–
Macropus rufogriseus banksianus Quoy and Gaimard, 1825 (Red-necked wallaby)	F,S	N	S varies from 0 to 50% in different populations
Macropus rufogriseus rufogriseus Desmarest, 1817 (Bennett's wallaby)	F	N	–
Megaleia rufa = Macropus rufus Desmarest, 1822 (Red Kangaroo)	S	N	–
Macropus giganteus Shaw, 1790 (Eastern Grey Kangaroo)	S	N,VE	VE 10%
Macropus fuliginosus Desmarest, 1817 (Western Grey Kangaroo)	S	N,VW	VW 3%
Macropus parryi Bennett, 1835 (Prettyface or Whiptail Wallaby)	S	N,VP	VP 55%

S = Slow   F = Fast   N = Normal

VE = Eastern variant   VW = Western variant

VP = Prettyface variant

The common name of each species is given in parentheses.

Cooper et al. (1971). For PGK from 1-3 x 10^6 fibroblasts were used to type cell cultures and for G6PD 1 - 5 x 10^5 were used. Cell cultures were initiated from excised pieces of ear, skin, or fascia from living animals or from heart or body wall of freshly killed animals and examined electrophoretically in the first one, two, and three passages.

RESULTS

KANGAROOS. G6PD IN BLOOD.

One of the two isozymic forms of G6PD may be found in the erythrocytes of kangaroos of the genus *Macropus* listed in Table 1. They are designated F and S and are illustrated in Figure 1. It is important to note that all kangaroos so far examined have one and only one of the two types in their ery-throcytes. The alternative type has not been found even in trace amount. Mixing experiments indicate that on cellogel this means that at least 90-95% of erythrocyte G6PD must be of one type, i.e. the minimum detectable minor or trace amount of the alternative type which would be detectable is about 5-10% (Johnston, unpublished data). The evidence that the differ-ence between these two variants is inherited in a sex linked manner with inactivation of the paternally derived X comes mainly from interspecies and interracial crosses involving wallaroos (F), euros (S) and red kangaroos (S). Richardson, Czuppon, and Sharman (1971) examined three $M.r.robustus$ (♀) X $M.rufus$ (♂) hybrids, two male and one female. All had only the F type of their female parent. They also examined three interracial hybrids between $M.r.erubescens$ (♀) and $M.r.robustus$ (♂). Again there were two males and one female, all with the S type of the female parent. Studies involving both these races and their hybrids have now been greatly extended (Shar-man, 1973; Sharman and Johnston, 1973). The cross between the races has been made reciprocally and hybrids have been back-crossed to both parents. All the results are consistent with the hypothesis of paternal X-inactivation. Furthermore the pat-ernal gene which is not expressed in the blood of a female may be transmitted by her to the next generation in a form which is expressed. Inactivation either does not occur in the germ line or, if it does, reactivation must occur at some point.

An alternative explanation of these results is that the failure of the paternal gene to be expressed is a character-istic of hybridity rather than of kangaroos. This explanation is made implausible by the existence of a true polymorphism for G6PD within populations of the red-necked wallaby $M.r.$ $banksianus$. All animals are either F or S in their erythro-cytes. Breeding data are consistent with X linkage with

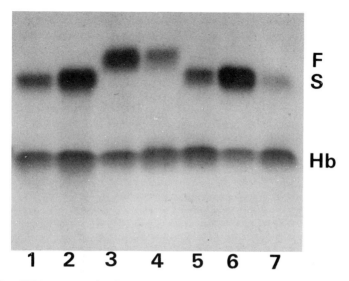

Fig. 1. G6PD types in kangaroos. F = fast, S = slow, Hb = hemoglobin. 1 *M.parryi*; 2 *M.r.erubescens*; 3 *M.r.robustus*; 4 *M.r.banksianus*; 5 *M.r.banksianus*; 6 *M.giganteus*; 7 *M.rufus*. Note the lower activity of *M.rufus*. Origin (not shown) is toward the bottom, and the anode towards the top, in this and the other three figures.

paternal X-inactivation, although it is not possible rigorously to exclude the hypothesis of autosomal dominance of one of the forms (Johnston, VandeBerg, and Sharman, in preparation).

There is considerable variation in erythrocyte G6PD activity between races and species in the genus *Macropus*. Richardson, Czuppon, and Sharman (1971) present data which suggest that in the hybrids they examined this too is under control of the X chromosome and is consistent with paternal X-inactivation.

PGKA IN BLOOD

Mammals have two isozymes of PGK, PGKA, and PGKB, which are specified by two separate loci. That for PGKA is sex linked and that for PGKB is autosomal (Cooper et al., 1971; VandeBerg, Cooper and Close, 1973). PGKB is the major isozyme in the testis in both marsupials and eutherians and is probably the only PGK of sperm (VandeBerg, Cooper and Close, 1973). In all eutherian mammals so far examined, except dogs and foxes, it is absent from other tissues. In some marsupials, it is found in erythrocytes and other tissues but has very much less activity on gels than does PGKA (Figure 2).

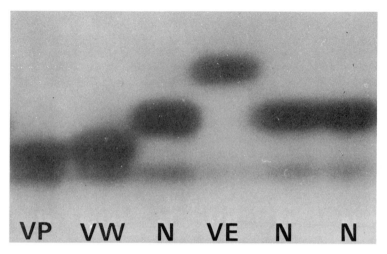

Fig. 2. PGKA types in kangaroos. VP = prettyface variant
found in *M.parryi;* VW = western variant found in *M.fuliginosus;*
VE = eastern variant found in *M.giganteus;* N = normal type
found in most animals of the *Macropus* genus. VP and VW cannot
be distinguished in mobility. The fainter staining band just
behind VP and VW and common to all samples is PGKB.

We are concerned here with the sex-linked form PGKA. One
form of this, designated N, occurs in all the species listed
in Table 1. In three of these species individuals may occur
which do not possess N, but one of three variant forms (Figure
2). As with G6PD, no kangaroo with more than one allelic
isozyme form of PGKA in its erythrocytes has been found,
even in trace amounts. Mixing experiments indicate that at
least 98-99% of all erythrocyte PGKA must be of the one
form (VandeBerg unpublished). The evidence that the PGKA is
sex-linked is from both family and population data. In
M.giganteus, reciprocal crosses between N and VE yielded
significantly different ratios, with the offspring having
their mother's type in 13 out of 16 cases (Cooper et al., 1971).
In *M.parryi,* it has been found possible to distinguish heterozygotes from homozygotes because the second (presumably
paternal) X is expressed to some degree in muscle tissue.
Seeming heterozygotes were found only among females and the
population ratios in both sexes were consistent with sex-
linkage (VandeBerg, Cooper and Close, 1973). The behavior of
the PGK paternal allele in the germ line parallels that of
G6PD; despite its lack of expression in the blood, it may be
contributed to the next generation in active form (Cooper et
al., 1971; VandeBerg, Cooper, and Sharman, 1973).

G6PD IN TISSUE HOMOGENATES AND CULTURED FIBROBLASTS

Sharman and Johnston (1973) and Johnston (unpublished) have examined tissue homogenates for the G6PD types of four heterozygotes at the pouch young stage for this enzyme. They examined ovaries, uteri, kidney, liver, heart, and skeletal muscle. They found only the maternally contributed allele expressed. The tissues had the same pattern as the blood, indicating paternal X-inactivation. We report here that lysates of cultured fibroblasts have also been examined. For these neither F nor S has been found in heterozygotes but a type to which we have given the provisional designation of "Intermediate" (Figure 3). This type is peculiar to the cultured cells of females known to be heterozygous. It was not found in the cultured cells of one male produced from a cross between the two alternative types, nor in the cultured cells of five females which were probably homozygous for their G6PD type (one *M.r.rufogriseus*, three *M.giganteus*, and one *M.rufus*). Our interpretation of the fibroblast results is that heteropolymers are being formed within these cells. Human G6PD can be either an inactive monomer, an active dimer, or an active tetramer depending upon pH, ionic strength, and concentrations of $NADP^+$ and NADPH (Bonsignore et al., 1971; Cancedda, Ogunmala, and Luzzatto, 1973; Yoshida, 1973). Under the conditions of pH, ionic strength, etc. used for electrophoresis in doves G6PD is a tetramer of subunits coded for by the one locus (Cooper and Irwin, 1968) while it is a dimer in man (Gartler et al., 1973) and in some galliform birds (Bhatnagar, 1969). In trout, it is a tetramer composed of two subunits coded for by different loci (Yamauchi and Goldberg, 1973). We do not know if kangaroo G6PD is a dimer or a tetramer. In either case, the difference in mobility between the two isozymes is insufficient for us to resolve them and their putative interaction product or products. All three (dimer) or all five (tetramer) bands would appear to coalesce into the "Intermediate" pattern, which is significantly broader than either F or S, as would be expected on the basis of this interpretation.

The formation of heteropolymers can be explained by two hypotheses. The uncloned primary cultures could consist of a mixture of cells, some with only the allele F active and some with only the allele for S active. The transfer of an mRNA or monomers between such cells would result in the formation of heteropolymers. The other possibility is that both alleles are active within the one cell. We regard this second possibility as the more likely but only cloned populations of cells will settle the matter. The minimal conclusion from the G6PD patterns of fibroblasts is that the paternal gene is

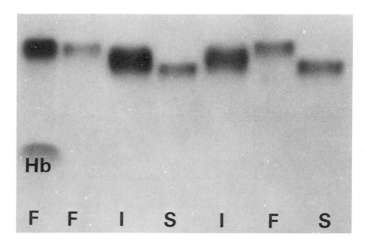

Fig. 3. G6PD types in cultured fibroblasts. F = fast, S = slow, I = Intermediate, Hb = haemoglobin. Note that the Intermediate type is broader than either the F or S type and presumably represents interaction products as well as F and S homopolymers. The sample on the extreme left is an F blood standard.

expressed, in contrast to the blood where it is not.

PGKA IN TISSUES AND CULTURED FIBROBLASTS

The results for the expression of PGKA in heterozygous females are as follows. In one pouch young *M.giganteus* Cooper et al. (1971) found that two tissues, heart and the reproductive tract composed of uterus and ovaries, gave evidence of expression of both alleles. In a further investigation in the species *M.parryi* the tissues of seven known heterozygotes were examined (VandeBerg, Cooper, and Sharman, 1973). These animals were known to be heterozygotes because they had pouch young with the alternative PGKA type to their own. We assume this to be the product of the allele contributed by the female's paternal parent. All seven heterozygotes had some expression of the presumed paternal type in skeletal and cardiac muscle, and in one smooth muscle, the bladder. The degree of expression was very much less than that of the presumed maternal allele. The same sort of result has been obtained for cultured fibroblasts in *M.giganteus* (Figure 4). Two known *M.giganteus* heterozygotes from reciprocal matings have been cultured and examined. The paternal allele is expressed in both but to a lesser degree than the maternal. The comparison between these two cultures clearly shows that the reduced

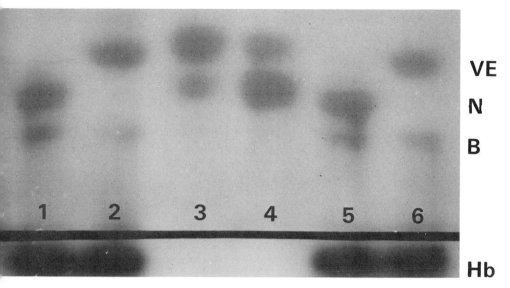

Fig. 4. PGKA types in cultured fibroblasts and blood. 1 and 5 are N blood, 2 and 6 are VE blood, 3 and 4 are PGKA^N PGKA^{VE} heterozygotes with PGKA^{VE} being maternally contributed in 3 and PGKA^N being maternally contributed in 4. Note how the paternally contributed allele has less activity. B = PGKB and Hb = haemoglobin. Note the virtual absence of PGKB from fibroblasts.

expression is a parental source effect and has nothing to do with the properties of the isozyme itself. The fibroblasts of a male from an N (♀) x VE (♂) mating had only an N isozyme type. A female *M.giganteus* for whom no parentage or progeny information was available likewise had only N, the type found in her blood.

For both PGKA and G6PD we wish to emphasize that our failure to find a second type in tissue homogenates or cell lysates does not preclude its being there in small amounts. Some tissue homogenates have low activity, e.g. lung in *M.parryi*, where we suspect activity of the second PGK type, but cannot be certain.

DISCUSSION

Paternal X-inactivation is a finding so much without precedent within mammals that it is perhaps not surprising that at least two groups of authors have cautiously suggested explanations in more conventional terms, such as cell selection (e.g. Rattazzi and Cohen, 1971; Mukherjee and Milet, 1972). It is true that a number of examples of either cell selection or

preferential inactivation of one X chromosome have been found
in eutherians. The normal form of the sex-linked enzyme hypoxanthine guanine phosphoriboxyl transferase is selected for
(Dancis et al., 1968; Nyhan et al., 1970), as is the horse X
as opposed to the donkey X in mules and hinnies (Giannelli and
Hamerton, 1971; Hamerton et al., 1971; Rattazzi and Cohen,
1971). Ohno and his co-workers have recently shown the existence of a gene O^{hv} in the mouse, which can cause preferential
activation of the X which carries it (Ohno et al., 1973; Ohno,
Geller and Kan, 1974; Drews et al.,1974). The O^{hv} gene is an
extreme form of the "controlling element" described by Cattanach (see Cattanach, 1970, for references). We would emphasize
that the selection or preferential inactivation demonstrated
in eutherians is a result of the influence of the marker being
used to follow the process and is not influenced by whether the
marker came from the father or mother. In kangaroos, whether
selection or preferential inactivation is responsible, the
process occurs in spite of the marker being used and solely
upon the basis of whether the X chromosome came from the father
or mother. Our results for both blood samples where inactivation is complete, or in other cells or tissues where it is not,
are clearly explicable only in terms of a parental source
effect.

For tissues other than blood, the two enzymes give different results. The paternal allele for PGKA is expressed to a
slight degree in the skeletal, cardiac, and bladder muscle of
M.parryi and the heart of *M.giganteus*, while no evidence of
such partial activity could be found for G6PD in any tissues
including muscle of *Macropus robustus* interracial hybrids. It
is not clear what this difference between the two enzymes is
due to. It could be that the stain for PGKA is more sensitive
than that for G6PD, and that we are not detecting small levels
of paternal G6PD activity. It could also reflect differences
in behavior between the two gene loci, or a difference between
species of kangaroo. The possibility that it is a developmental difference cannot be ruled out. VandeBerg, Cooper, and
Sharman (1973) found no evidence of expression of the paternal
PGKA allele in *M.parryi* pouch young, although they could not
exclude the possibility that this was because there were lower
levels of activity in pouch young as opposed to adults. Double
heterozygotes would be of great value, but so far, no species
in which two sex linked polymorphisms exist has been found. It
might be thought that it would have been very informative to
have looked for sex chromatin in the *M.parryi* tissues used in
our enzyme analysis. Unfortunately, it appears that Australian
marsupials, unlike many eutherians, do not have easily detectable sex chromatin (Robinson, personal communication).

The results for reproductive organs call for special comment. Both PGKA genes were expressed in the uterus and ovary of one 35 day old *M. giganteus* individual examined for these tissues. This may mean that the smooth muscle of uterus has both alleles expressed at this stage of development. The degree of egg cell contribution to the pattern is unknown, but it was probably higher at this stage than at any other (Alcorn, personal communication). We would expect that if egg cells alone were examined, we would find both alleles active for both PGKA and G6PD, as found by Gartler, Liskay, and Grant (1973) for G6PD in human fetal oocytes and Epstein (1969,1972) for mouse oocyte G6PD and HGPRT.

The results for cultured fibroblasts for the two enzymes are in partial agreement. The G6PD results suggest very strongly that heteropolymers are being formed, and raise the possibility that both genes may be active within the one cell. This is in clear contradistinction to the human G6PD locus in cultured human fibroblasts (Davidson, Nitowsky, and Childs, 1963; DeMars and Nance, 1964; Steele and Migeon, 1973), and for HGPRT (Migeon et al., 1968; Salzman, DeMars, and Benke, 1968) and PGKA (Deys et al., 1973) in the same type cell. It is not possible to say categorically from our results whether both genes are fully active. There is certainly a high degree of activity of the paternal allele and we cannot exclude the possibility that it is as active as the maternal one.

PGKA in fibroblasts represents expression of the paternal allele, though not to the same degree as the maternal. Since PGKA is a monomer (Scopes, 1971; Blake, Evans, and Scopes, 1972) interaction products do not occur. Therefore, we cannot tell from the electrophoretic pattern whether both genes are expressed within one cell or whether we are dealing with a mosaic. The mosaic could be composed of a majority of cells with a maternal X active and a minority with the paternal X active, or it could be composed of a majority with the maternal X active and a minority with both active. If the cell cultures are not mosaic we must conclude that there is a fully active maternal X and a partially active paternal X. We are currently attempting to carry out cloning experiments to decide between these possibilities.

If cells without dosage compensation or a partially active X do exist, it will be necessary to alter the various models of X-inactivation which have been put forward connecting the eutherian and marsupial systems of X chromosome dosage compensation (Cooper, 1971; Lyon, 1971; Brown and Chandra, 1973). None of them specifically accommodates either of these two phenomena. The closest apparent parallel to our results is the heterochromatization of the entire paternal set of chromosomes in male mealybugs (Brown and Nur,, 1964; Brown, 1966, 1969). Not only is this a parental source effect, but in addition, Nur (1967) has shown that in these organisms reversal of heterochromatization of the paternal set may take place in some tissues so that it becomes euchromatic and presumably active.

ACKNOWLEDGEMENTS

This work was supported by grants from the Australian Research Grants Committee to two of us (G. B. S. and D. W. C.). The presentation of this paper was made possible by a travel grant from the Australian American Educational Foundation to D. W. C.

REFERENCES

Beutler, E. 1969. Electrophoresis of phosphoglycerate kinase. *Biochem. Genet.* 3: 189-195.

Bhatnager, M. K. 1969. Autosomal determination of erythrocyte glucose-6-phosphate dehydrogenase in domestic chickens and ring-necked pheasants. *Biochem. Genet.* 3: 85-90.

Blake, C. C. F., P. R. Evans and R. K. Scopes 1972. Structure of horse muscle phosphoglycerate kinase at 6A resolution. *Nature New Biol.* 235: 195-198.

Bonsignore, A., A. Cancedda, A. Nicolini, G. Damiani, and A. De Flora 1971. Metabolism of human erythrocyte glucose-6-phosphate dehydrogenase. *Arch. Biochem. Biophys.* 147: 493-501.

Brown, S. W. 1966. Heterochromatin. *Science* 151: 417-425.

Brown, S. W. 1969. Developmental control of heterochromatization in coccids. *Genetics* 61: Suppl 1: 191-198.

Brown, S. W. and H. S. Chandra 1973. Inactivation system of the mammalian X chromosome. *Proc. Nat. Acad. Sci.* 70: 195-199.

Brown, S. W. and U. Nur 1964. Heterochromatic chromosomes in the coccids. *Science* 145: 130-136.

Cancedda, R., G. Ogunmala and L. Luzzatto 1973. Genetic variants of human erythrocyte glucose-6-phosphate dehydrogenase. Discrete conformational states stabilized by NADP$^+$ and NADPH. *Eur. J. Biochem.* 34: 199-204.

Cattanach, B. M. 1970. Controlling elements in the mouse X-chromosome III. Influence upon both parts of an X divided by rearrangement. *Genet. Res.* 16: 293-301.

Cooper, D. W. 1971. A directed genetic change model for X-inactivation in eutherian mammals. *Nature* 230: 292-294.

Cooper, D. W. and M. R. Irwin, 1968. Glucose-6-phosphate dehydrogenase: Evidence for tetrameric structure and possible inherited deficiency of enzyme activity in a dove. *Proc. Nat. Acad. Sci. U. S.* 61: 979-981.

Cooper, D. W., J. L. VandeBerg, G. B. Sharman and W. E. Poole 1971. Phosphoglycerate kinase polymorphism in kangaroos provides further evidence for paternal X inactivation. *Nature New Biol.* 230: 155-157.

Dancis, J., P. H. Berman, V. Jensen and M. E. Balis 1968. Absence of mosaicism in the lymphocyte in X-linked congenital hyperuricosuria. *Life Sci.* 7 (ii): 587-591.

Davidson, R. G., H. M. Nitowsky and B. Childs 1963. Demonstration of two populations of cells in the human female heterozygous for glucose-6-phosphate dehydrogenase variants. *Proc. Nat. Acad. Sci. U. S. A.* 50: 481-485.

De Mars, R. and W. E. Nance 1964. Electrophoretic variants of glucose-6-phosphate dehydrogenase and the single-active-X in cultivated human cells. *Wistar Institute Symposium Monograph* 1: 35-46.

Deys, B. F., K. H.Grzeschik, A. Grzeschik, E. R. Jaffe, and M. Siniscalco 1972. Human phosphoglycerate kinase and inactivation of the X chromosome. *Science* 175: 1002-1003.

Drews, U., S. R. Blecher, D. A. Owen, and S. Ohno 1974. Genetically directed preferential X-inactivation seen in mice. *Cell* 1: 3-8.

Epstein, C. J. 1969. Mammalian oocytes: X chromosome activity. *Science* 163: 1078-1079.

Epstein, C. J. 1972. Expression of the mammalian X chromosome before and after fertilization. *Science* 175: 1467-1468.

Gartler, S. M., R. M. Liskay and N. Gant 1973. Two functional X chromosomes in human fetal oocytes. *Exp. Cell Res.* 82: 464-466.

Giannelli, F. and J. L. Hamerton 1971. Non-random late replication of X chromosomes in mules and hinnies. *Nature* 232: 315-319.

Hamerton, J. L., B. J. Richardson, P. A. Gee, W. R. Aden and R. V. Short 1971. Non-random X chromosome expression in female mules and hinnies. *Nature* 232 312-315.

Johnston, P. G., J. L. VandeBerg, and G. B. Sharman 1974. Inheritance of glucose-6-phosphate dehydrogenase in the red-necked wallaby, *Macropus rufogriseus*. In preparation.

Lyon, M. F. 1961. Gene action in the X-chromosome of the mouse. *(Mus musculus L.) Nature* 190: 372-373.

Lyon, M. F. 1971. Possible mechanisms of X-chromosome inactivation. *Nature New Biology.* 232: 229-232.

Lyon, M. F. 1972. X-chromosome inactivation and developmental patterns in mammals. *Biol. Rev.* 47: 1-35.

Migeon, B. R., V. M. der Kaloustian, W. L. Nyhan, W. J. Young, and B. Childs 1968. X-linked hypoxanthine-guanine phosphoribosyl transferase deficiency : heterozygote has two populations. *Science* 160: 425-427.

Mukherjee, B. B. and R. G. Milet 1972. Nonrandom X-chromosome inactivation - an artifact of cell selection. *Proc. Nat. Acad. Sci. U..S. A.* 69: 37-39.

Nyhan, W. L., B. Bakay, J. D. Connor, J. F. Marks and D. K. Keele 1970. Hemizygous expression of glucose-6-phosphate dehydrogenase in erythrocytes of heterozygotes for the Lesch-Nyhan syndrome. *Proc. Nat. Acad. Sci. U. S. A.* 65: 214-218.

Nur, U. 1967. Reversal of heterochromatization and the activity of the paternal chromosome set in the male mealy bug. *Genetics* 56: 375-389.

Ohno, S. 1967. *Sex Chromosomes and Sex Linked Genes.* Springer-Verlag Berlin. Heidelberg and New York.

Ohno, S., L. Christian, B. J. Attardi, and J. Kan 1973. Modification of expression of the Testicular feminization (Tfm) gene of the mouse by a "controlling element" gene. *Nature New Biol.* 245: 92-93.

Ohno, S., L. N. Geller, and J. Kan 1974. The analysis of Lyon's hypothesis through preferential X-inactivation. *Cell* 1: 175-181.

Rattazzi, M. C. and M. M. Cohen 1971. Further proof of genetic inactivation of the X chromosome in the female mule. *Nature* 237: 393-396.

Richardson, B. J., A. B. Czuppon and G. B. Sharman 1971. Inheritance of glucose-6-phosphate dehydrogenase variation in kangaroos. *Nature New Biol.* 230: 154-155.

Salzman, J., R. De Mars and P. Benke 1968. Single-allele expression at an X-linked hyperuricemia locus in heterozygous human cells. *Proc. Nat. Acad. Sci.* 60: 542-552.

Scopes, R. K. 1971. An improved procedure for the isolation of 3-phosphoglycerate kinase from yeast. *Biochem. J.* 122: 89-92.

Sharman, G. B. 1971. Late DNA replication in the paternally derived X chromosome of female kangaroos. *Nature* 230: 231-232.

Sharman, G. B. 1973. "The Chromosomes of Non Eutherian Mammals". In *Cytotaxonomy and Vertebrate Evolution*. A. B. Chiarelli and E. Capanna (Editors). Academic Press, London and New York.

Sharman, G. B. and P. G. Johnston 1973. X chromosome and X-linked gene inactivation in marsupials. Thirteenth International Congress of Genetics. Berkeley. 1973. (abstract). *Genetics* 74: s250.

Sharman, G. B., E. S. Robinson, S. M. Walton, and P. J. Berger 1970. Sex chromosomes and reproductive anatomy of some intersexual marsupials. *J. Reprod. Fert.* 21: 57-68.

Steele, M. W. and B. R. Migeon 1973. Sex differences in activity of glucose-6-phosphate dehydrogenase from cultured fetal lung cells despite X-inactivation. *Biochem. Genet.* 9: 163-168.

Tyndale - Biscoe, H. 1973. *Life of Marsupials.* Edward Arnold, London.

VandeBerg, J. L., D. W. Cooper and P. J. Close 1973. Mammalian testis phosphoglycerate kinase. *Nature New Biol.* 243: 48-50.

VandeBerg, J. L., D. W. Cooper and G. B. Sharman 1973. Phosphoglycerate kinase A polymorphism in the wallaby *Macropus parryi* : activity of both X chromosomes in muscle. *Nature New Biol.* 243: 47-48.

Yamauchi, T. and E. Goldberg 1973. Glucose-6-phosphate dehydrogenase from brook, lake and splake trout : an isozymic and immunological study. *Biochem. Genet.* 10: 121-134.

Yoshida, A. 1973. Hemolytic anemia and G6PD deficiency. *Science* 179: 532-537.

PHYLOGENY, ONTOGENY, AND PROPERTIES OF THE
HEXOKINASES FROM VERTEBRATES

TITO URETA

Departamento de Bioquimica, Facultad de Medicina Sede Norte,
Universidad de Chile, Santiago-4, Chile

ABSTRACT. Four hexokinases (A, B, C, and D) are present
in rat tissues. Isozymes A, B, and C have low Km glucose
values ($\sim 10^{-5}$M), molecular weights of 100,000 daltons,
and broad sugar specificities and tissue distributions.
Isozyme D has a high Km for glucose ($\sim 10^{-2}$M), a
molecular weight of about 55,000 daltons, a narrow substrate specificity, and is restricted to the liver.
The isozymes seem to be products of independent genes.

Isozyme D levels in rat liver depend on the availability of dietary carbohydrate and the interplay of
insulin, glucagon, and catecholamines. The low-Km
isozymes in liver are not subjected to this type of
regulation, but hexokinase B levels from several tissues
are under the control of insulin.

The high-Km isozyme is present in the liver of most
mammals, turtles, and amphibians, but is absent in
birds and higher reptiles. Isozyme C has been found
only in some mammals and amphibians. Some properties of
the low-Km hexokinases from birds and lizards are very
different from those of other vertebrates.

A sequential pattern of development of the hexokinases
in rat liver has been observed. Each isozyme reaches
peak values at different ages. Isozyme A levels show a
maximum at birth, isozyme B during the first week, isozyme
C at the second week, and isozyme D from the third week
onwards.

THE HEXOKINASES

The importance of the hexokinase reaction (ATP: -hexose
6-phosphotransferase, EC 2.7.1.1) for the regulation of
glucose utilization in mammals has been demonstrated by a
variety of studies, but the special features of the reaction
in liver were not recognized until Weinhouse and his colleagues established the fact that the liver enzyme has an
unusually high Km for glucose (DePietro and Weinhouse 1960;
DiPietro et al., 1962). Subsequently, Walker (1962, 1963)
and Viñuela et al., (1963) could distinguish at least two
different glucose phosphorylating enzymes, and shortly

thereafter, the isolation of four such isozymes was achieved (Ureta et al., 1963; Gonzalez et al., 1964).

The glucose phosphotransferases from rat liver can be resolved by DEAE-cellulose chromatography (Fig. 1). Electrophoresis followed by activity staining in starch gel (Katzen et al., 1965), cellulose acetate (Sato et al., 1969), and polyacrylamide (Bernstein and Kipnis 1973) have also been used. The rat isozymes were designated either as A, B, C, and D according to their order of elution from DEAE-cellulose columns, or as I, II, III, and IV in the sequence of increasing electrophoretic mobility. Several properties of the isozymes are listed in Table I.

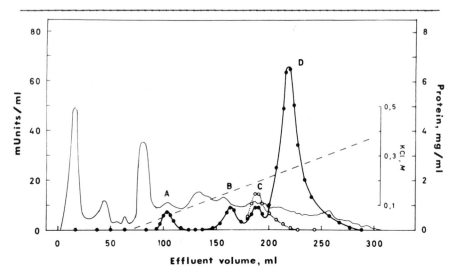

Fig. 1. Glucose phosphorylating isozymes from rat liver. A fifty per cent homogenate was prepared in Tris buffer (0.01 M Tris-0.001 M EDTA, pH 7). After centrifugation at 105,000 x g for one hour, the supernatant liquid was treated batchwise with CM-cellulose in Tris buffer. The non-absorbed fraction was then chromatographed on a DEAE-cellulose column (1.2 x 21 cm) and eluted with a linear gradient (---) of NaCl in Tris buffer. Aliquots of each fraction were assayed by a spectrophotometric method at 100 mM (●——●) and 0.5 mM (○-----○) glucose. The weak line indicates protein content.

SUBSTRATE SPECIFICITIES AND KINETIC PROPERTIES

TABLE 1
Some properties of the glucose phosphorylating
isozymes from rat liver[1]

Parameter	Isozymes			
	A	B	C	D
Substrate specificity (Relative V_{max})				
Glucose	1.0	1.0	1.0	1.0
Mannose			1.0	0.8
2-Deoxyglucose			1.0	0.4
Fructose	1.1	1.2	1.3	0.2
Michaelis constants (mM)				
Glucose	0.044	0.130	0.020	18
Fructose	3.1	3.0	1.2	12
ATP	0.42	0.70	1.29	0.49
Chromatographic mobility (mM KCl)	74	163	218	253
Relative proportion(%)	4	4	7	85
Molecular weight (daltons x10^{-3})	100	100	100	55

[1]   Data from González et al., (1964, 1967), Grossbard and Schimke (1966) and Pilkis et al., 1968a).

A description of the detailed kinetic studies by several authors, aiming to unravel the mechanism of the hexokinase reaction, is outside the scope of this paper, but may be found in reviews by Crane (1962), Walker (1966), Purich et al., (1973), and Colowick (1973).

Isozymes A, B, and C show broad specificities being able to phosphorylate glucose, fructose, mannose, and 2-deoxy-glucose at approximately the same rate (Table 1) (Gonzalez et al., 1964, 1967; Grossbard and Schimke 1966). Isozyme D utilizes glucose as the best sugar acceptor, but mannose, 2-deoxyglucose, and fructose are also substrates albeit at a lower rate (González et al., 1964, 1967; Salas et al., 1965; Parry and Walker 1966, 1967). Partly because of this fact, isozyme D has been called "glucokinase" but, on nomenclatural grounds, the name is not proper and should be reserved for phosphotransferases strictly specific for glucose such as those described in several species (cf. Colowick 1973).

Isozymes A, B, and C have low Km values for glucose while that of isozyme D is much higher (Table 1 and Fig. 2). Thus, at normal levels of blood glucose (depicted in the shaded area in Fig. 2) hexokinases A and B are fully saturated, whereas isozyme D is working at a submaximal velocity and this fact may be of importance for the regulation of carbohydrate utilization. Moreover, Cárdenas et al., (1974) have reported that isozyme D from rat and other species shows a sigmoidal saturation curve for glucose with a Hill coefficient of about 1.6 and a $K_{0.5}$ of about 7.5 mM glucose. Isozyme C is strongly inhibited by high levels of glucose, the inhibition being most marked precisely at physiological levels of glucose (Fig. 2).

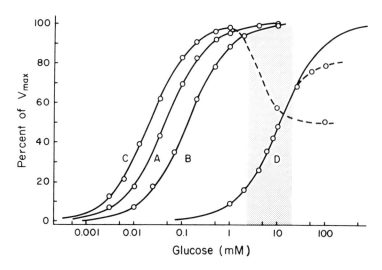

Fig. 2. Relative activities of the glucose phosphotransferases from rat liver as function of glucose concentration. The shaded area corresponds to an estimation of the range of glucose concentrations in the blood reaching the hepatocyte.

TISSUE AND SUBCELLULAR DISTRIBUTION

When assessed by chromatography, the hepatic levels of the isozymes A, B, C, and D are about 4, 4, 7, and 85 per cent, respectively, of the total glucose phosphorylating activity of normal, adult, well fed rats (González et al., 1964). However, because of its high Km for glucose, the actual in vivo contribution of isozyme D may be significantly less than the above quoted figure.

Isozyme D has been thought to be restricted to hepatocytes but several reports indicate its presence in other cells as well, e.g., human brain (Bachelard 1967a), rat kidney (Pilkis and Hansen 1968), mouse pancreatic islets (Matchinsky and Ellerman 1968; Ashcroft and Randle 1970), rabbit reticulocytes (Gerber et al., 1970), and rat intestinal epithelium (Weiser et al., 1971; Anderson and Tyrrell 1973).

Sols et al., (1964) suggested that the liver parenchymal cells have only the high-Km isozyme whereas bile duct and other liver cells contain exclusively low-Km isozymes. The increase of the latter in the liver of rats fed 3'-methyldimethylaminoazobenzene, which produced cholangiocarcinomas, supported that suggestion (Weinhouse et al., 1963). Also, the separation of hepatic parenchymal and non-parenchymal cells from rat or mice by mechanical disruption or by the use of collagenase and hyaluronidase resulted in a marked enrichment of the high-Km isozyme in the parenchymal fractions (Sapag-Hagar et al., 1969; Crisp and Pogson 1972). However, other authors claim that parenchymal cells isolated by the collagenase-hyaluronidase procedure or by microdissection do contain low-KM hexokinases in addition to isozyme D (morrison 1967; Werner et al., 1972; Bonney et al., 1973). Furthermore, the isozymic electrophoretic patterns from the parenchymal fraction were described as almost identical to those from total liver homogenates (Werner et al., 1972). The problem will remain unsettled until the unambiguous isolation of the different cell types of the liver can be accomplished without the drawbacks of present techniques. Alternatively, the application of immunofluorescent antibodies to histological sections of liver should solve the problem. The controversy is not purely academic since its elucidation may simplify the interpretation of data about glucose utilization obtained with liver slices, perfusion systems, or even the whole animal.

Extra-hepatic tissues present up to three low-Km phosphotransferases in different and characteristic proportions (see Katzen 1967). Isozyme A predominates in brain (Grossbard and Schimke 1966), kidney (Grossbard and Schimke 1966; Pilkis and Hansen 1968), and testes (Fig. 3). Isozyme B is the most prominent form in adipose tissue, heart muscle, and skeletal muscle (Grossbard and Schimke 1966; Katzen et al., 1970).

Rat testes are said to contain a low-Km hexokinase of

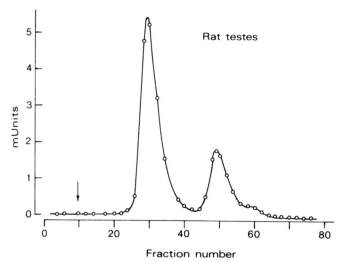

Fig. 3. Glucose phosphotransferases of rat testes. One
hundred microliters of a high-speed supernatant liquid
from rat testes were chromatographed on a 2-ml DEAE-cellulose
column and eluted with a linear gradient from 0 to 0.5 M
KCl (Total gradient volume, 34 ml). One hundred fractions of
0.3 ml were collected and assayed at 100 mM glucose by a
sensitive radioassay. From T. Ureta, J. Davagnino and M.
Boric (unpublished results).

---

very low electrophoretic mobility in addition to isozymes
A and B (Katzen 1967; Pilkis et al., 1968). This special
isozyme called "sperm-type" hexokinase has not been characterized. Chromatography of supernatant liquids of rat testes
(Fig. 3) shows only two isozymes with mobilities and kinetic
properties identical to those of isozymes A and B. However,
the possibility that at pH 7 the sperm-type hexokinase and
isozyme A elute together has not been excluded.

Most tissues show a variable proportion of glucose
phosphorylating activity loosely associated with mitochondria
(see Purich et al., 1972, and Colowick 1973). The particulate
material seems to be composed of the same set of isozymes
present in cytoplasm, except for isozyme C which does not
show up in particles (Bachelard 1967b; Katzen et al., 1970;
Thompson and Bachelard 1970; Mayer and Hubscher 1971; Saito
and Sato 1971).

High-speed centrifugation of adult rat liver extracts
results in the quantitative recovery of the phosphorylating

activity in the supernatant liquids (Siekevitz and Potter
1955; Niemeyer et al., 1965) implying that the four liver
isozymes are cytoplasmic enzymes. However, Berthillier
et al., (1970) have reported that a high proportion of the
activity from rat liver is associated with microsomes.
Actually, two high-Km enzymes of different molecular weights
(50,000 and 120,000 daltons) were purified from the microsomes.
The substrate specificities and Km values of these enzymes
are quite different from those of the "classical" soluble
isozyme D. Our attempts to isolate the mcrosomal enzymes
have been so far unsuccessful.

MOLECULAR WEIGHTS AND SUBUNIT STRUCTURE

The molecular weights of isozymes A, B, and C from
various sources are about 96,000 daltons as measured by
several procedures (Grossbard and Schimke 1966; Easterby 1971;
Chou and Wilson 1972; Neumann et al., 1974). Homogeneous
isozyme A from pig heart has the same molecular weight by
analytical centrifugation and after SDS-polyacrylamide
electrophoresis of the heat-treated enzyme (Easterby 1971).
SDS-acrylamide electrophoresis of homogeneous isozyme A from
rat brain in the presence of 7 M urea or 5.5 M guanidine-HCl
resulted also in a single band of 96,000 to 98,000 daltons.
In addition, a single N-terminal amino acid, glycine, was
found (about 0.8 moles per 98,000 g) (Chou and Wilson 1972).
These results indicate that isozyme A from mammals is a
single polypeptide chain.

The molecular weight of isozyme D is still uncertain.
Parry and Walker (1967) reported values of 48,000 daltons
employing Sephadex G-100. On the other hand, Pilkis et al.,
(1968) found 48,000 daltons by Sephadex G-100 only at low
ionic strength, but 68,000 daltons either by Sephadex G-100
filtration at high ionic strength or at any ionic strength
by BioGel P-100 chromatography or sucrose gradient centrifugation.

Measurements of the molecular weight of purified
isozyme D preparations by Sephadex G-75 or G-100 in our laboratory gave values of about 55,000 daltons (Fig. 4).
Occasionally a heavier fraction of about 112,000 daltons
was also observed. No effect of ionic strength on the
elution position of the enzyme was observed. Sucrose
gradient centrifugation gave approximately the same result.

IMMUNOLOGY OF THE GLUCOSE PHOSPHOTRANSFERASES

Pilkis et al., (1968) and Clark-Turris et al., (1974)

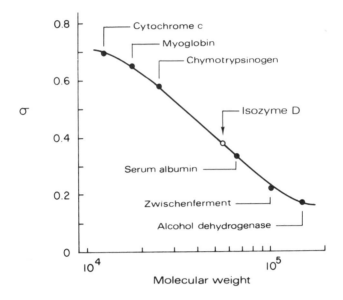

Fig. 4. Molecular weight determination of partially purified isozyme D. A Sephadex G-100 column (2.4 x 54 cm) was used. The distribution coefficient (σ) of the several standard proteins are plotted against their molecular weights. From T. Ureta (unpublished results).

have prepared antibodies against rat isozyme D. These antibodies did not inhibit the activity of the isolated rat isozymes A, B, or C (Clark-Turri et al., 1974) or of isozyme D from the bullfrog *Rana catesbeiana* (Pilkis et al., 1968) or from the Chilean leptodactylid frog *Calyptocephalella caudiverbera* (Clark-Turri et al., 1974). The antibodies cross-react with isozyme D from other mammalian sources (Pilkis et al., 1968; Clark-Turri et al., 1974).

Neumann et al., (1974) and Neumann and Pfeiderer (1974) have recently reported than an antibody against human isozyme A did not inhibit the activity of human isozyme C. The antibody against human isozyme C also failed to cross-react with isozyme A from the same source. Furthermore, an antibody against isozyme B from rat muscle was not able to inhibit the activity of isozyme A from rat adipose tissue (Creighton et al., 1972).

The implications of these results for the understanding of the relationships between the glucose phosphotransferases are far reaching since the most reasonable interpretation of the data so far available suggests that each isozyme is the product of a separate gene.

ADAPTIVE BEHAVIOR OF THE RAT GLUCOSE PHOSPHOTRANSFERASES

Rat isozyme D (the so-called glucokinase) has been long recognized as an adaptive enzyme. It is the only liver phosphotransferase with variable levels in the rat after modification of the amount or nature of the food (Fig. 5). The enzyme activity decreases exponentially with a half-life of about 32 hours during fasting or after feeding carbohydrate-free diets (Chamorro and Schilkrut 1969). The administration of glucose to rats with low levels of isozyme D brings about a rapid increase of the enzyme activity (Niemeyer et al., 1963, 1965) and the effect can be blocked by inhibitors of protein synthesis (Niemeyer et al., 1962, 1965; Sharma et al., 1963; Salas et al., 1963). Since the increased levels of activity are accompanied by an increase in enzyme quantity as shown by immunological titration (Clark-Turri et al.,1974), the effect is considered to be an exampe of exzyme induction.

The adaptability of rat liver isozyme D to diet is not a general phenomenon in mammals. For instance, isozyme D levels of the rodents guinea pig and *Octodon degus* were not modified by 96 hours of fasting. Also, isozyme D from BALB or C57BL mice did not show the same dependency upon the presence of carbohydrate in the diet (Ureta et al., 1971a).

Isozyme D levels are also controlled by the endocrine system, especially by insulin which is necessary for the glucose-mediated induction (Salas et al., 1963; Sharma et al., 1963; Niemeyer et al., 1966, 1967). Glucagon (Niemeyer et al., 1966) and epinephrine are powerful inhibitors of isozyme D induction and probably act through 3', 5'-cyclic AMP (Ureta et al., 1970). Glucocorticoids and a pituitary hormone also modulate the rate of isozyme D induction by glucose (Niemeyer 1971; Niemeyer and Ureta 1972).

The dietary and endocrine manipulations mentioned above seem not to affect the levels of the hepatic low-Km isozymes. However, high carbohydrate diet, fasting, and diabetes markedly affect the levels of hexokinase B of several tissues, e.g., adipose tissue (Moore et al., 1964; Katzen 1966; Hansen et al., 1970; Borrebaek 1970), muscle (Katzen et al., 1970), intestinal mucosa (Weiser et al., 1971), and mammary gland (Walters and McLean 1967).

PHYLOGENETIC STUDIES

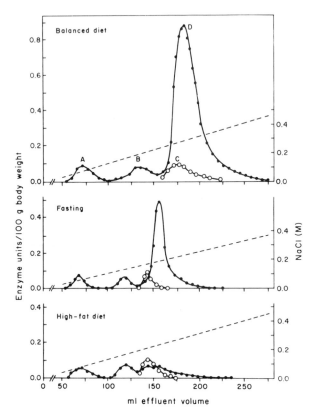

Fig. 5. Liver glucose phosphotransferases from rats under different dietary conditions. The extracts were prepared as described in the legend to Fig. 1. The enzyme activities are expressed as units/100 g body weight. Data from Gonzáles et al., (1964).

NOMENCLATURAL COMMENTS

Two features have been found useful for identifying some glucose phosphotransferases: the high Km for glucose of isozyme D and the inhibition of isozyme C by excess glucose. Other characteristics, such as order of elution, chromatographic mobilities, other Km values, or ability to phosphorylate fructose, may be confusing. For instance, the inhibited isozyme C may elute as the second peak in some rodents or even as the first in some anurans (*vide infra*). The ratio

$V_{fructose}$: $V_{glucose}$ that distinguishes the high-Km from the low-Km isozymes in rat liver (0.2 vs 1.2 respectively) may be about 0.8 in some amphibian high-Km phosphotransferases. In mammals, the Km glucose of isozyme A (0.05 mM) is lower than that of isozyme B (0.13 mM) but no such difference is observed in non-mammalian species (Ureta et al., 1973, 1974).

Pending further work, the denomination "isozyme D" will be used for the high-Km enzyme even in species with only three isozymes (e.g., amphibians). The isozyme inhibited by excess of glucose will be referred to as "isozyme C" regardless of its elution position. The names A and B will be given respectively to first and second eluting isozymes, provided they are not either high-Km or inhibited isozymes. The provisional nature of this nomenclature is acknowledged and comments will be welcomed.

THE GLUCOSE PHOSPHOTRANSFERASES OF MAMMALIAN LIVER

Chromatographic profiles from the liver of several selected mammalian species are shown in Fig. 6. The presence of isozyme A is constant in all species studied which include representatives of six orders of mammals. The other isozymes may be lacking. For instance, isozyme B is absent in the monkey, the goat, and the pig. Isozyme C is absent in several rodents (*Abrothrix longipilus, Akodon olivaceus, A. xanthorhinus, Spalacopus cyanus, and Ctenomys maulinus*), the dog, and a marsupial (*Marmosa elegans*). Isozyme D was not found in the goat, a fact which agrees with the report by Ballard (1965) for ruminants and ruminant-like mammals and tentatively ascribed to the fact that those animals do not absorb glucose from the intesting. The cat also lacks the high-Km isozyme.

An interesting feature of the mammals studied is that the isozyme inhibited by excess glucose appears as the second peak in the mouse, the hamster, and the guinea pig (Ureta et al., 1971b, see also Fig. 6).

THE GLUCOSE PHOSPHOTRANSFERASES OF AVIAN LIVER

The isozymes from avian liver (Fig. 7) are very different from those of mammalian liver. Generally two isozymes are found, but in some species, a third isozyme is clearly present either as a separate peak (Australian parakeet, *Melopsittacus undulatus*) or as a shoulder in the trailing edge of the second isozyme (Passeriformes). The mobilities of these two isozymes (117 and 175 mM KCl, respectively) are very different from those of isozymes A and B from mammalian liver (70 and 152 mM KCl). Neither the high-Km isozyme nor isozyme C were found in Aves (Ureta et al., 1972, 1973).

The avian isozymes also differ from the mammalian ones in their kinetic properties. The Michaelis constants for glucose of the two isozymes isolated from several avian species have been found to be 0.113±0.026 and 0.083±0.016 mM, respectively (Ureta et al., 1973). These values may be compared to those for the mammalian isozymes A and B which are 0.048 ±0.006 and 0.130-0.240 mM glucose, respectively (Ureta et al., 1971b). Furthermore, the ratio $V_{fructose}:V_{glucose}$ is about 2 in the case of the two avian isozymes. The molecular weights of the two chicken isozymes are about 96,000 daltons (Ureta et al., 1972).

THE GLUCOSE PHOSPHOTRANSFERASES OF REPTILIAN LIVER

Fig. 8 shows representative chromatographic profiles from lower (Chelonia) and higher (Squamata) reptiles. In turtles (*Geochelone, Hydromedusa*) four isozymes with chromatographic mobilities very similar to those of rat liver were found. The last eluting isozyme is of the high-Km type and has restricted substrate specificity. Isozyme C was not detected (Ureta et al., 1974).

On the other hand, the patterns of higher reptiles are quite different. In most cases three isozymes were found (*Liolaemus gravenhorstii, L. tennuis, L. monticola, Anolis auratus, Iguana iguana, Callopistes maculatus, and Tropidurus peruvianus*) but in the snakes *Dromicus chamissonis* and *Tachimenis peruviana*, as well as in *Amphisbaena alba* only two isozymes were discernible. In all the higher reptiles so far studied, the chromatographic mobilities of the isozymes were similar to those of the avian isozymes. Also, neither the high-Km isozyme nor isozyme C could be found (Ureta et al., 1974)

THE GLUCOSE PHOSPHOTRANSFERASES OF AMPHIBIAN LIVER

About twenty species of anurans and two urodeles have been studied in our laboratory (manuscript in preparation). Some examples are show in Fig. 9. In the case of Bufonids (*B. spinulosus, B. arenarum, B. marinus, B. rubropunctatus, and B. Paracnemis*) and in some leptodactylid frogs (*P. thaul, O. americanus, and T. halli*) the first eluting isozyme is almost identical to isozyme A from rat liver with respect to chromatographic mobilities (about 50 mM KCl) and kinetic

MAMMALIA

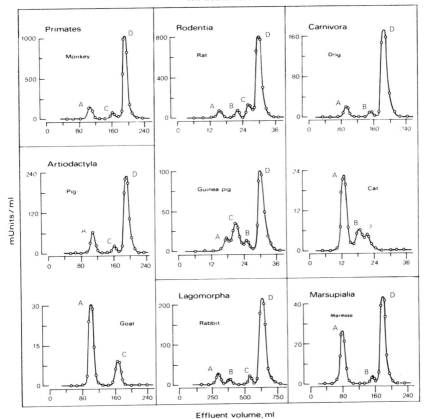

Effluent volume, ml

Fig. 6. Liver glucose phosphotransferases of some mammals.
The isozymes were separated by chromatography on DEAE-cellulose
columns using a linear gradient from 0 to 0.5 M KCl at pH 7.
The detailed procedure of separation as well as the methods
for enzyme assay were those of Ureta et al., (1973, 1974).
The ordinate values are not comparable since the amount of
liver used varied in each case. The species used were:
monkey, *Saimiri scuirea*; albino rat, *Rattus norvegicus*;
guinea pig, *Cavia porcellus*; pig; goat; dog; cat; rabbit;
mouse opposum, *Marmosa elegans*. From T. Ureta, J.C. Slebe
and J. Radojkovic (unpublished experiments).

features. In other species (*T. torosa, A. mexicanum, X.
laevis, R. pipiens, C. caudiverbera*) the first peak of activity
elutes at a higher KCl concentration about 120 mM KCl) and, in

587

AVES

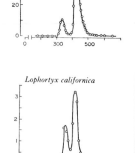
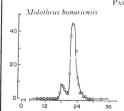
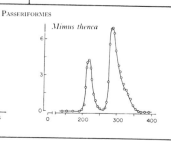

mUnits / ml

Effluent volume, ml

Fig. 7. Liver glucose phosphotransferases of some birds.
For details see legend to Fig. 6. From Ureta et al., (1973).

some species, this isozyme is inhibited by excess glucose,
i.e., is very similar to isozyme C from rat liver. All the
amphibians studied present a high-Km isozyme, although in
some (*O. americanus*, *T. halli*) it is barely detectable. The
molecular weight of isozyme D isolated from *B. spinulosus*
and *A. mexicanum* was found to be similar to that of the
mammalian high-Km isozyme, e.g., about 55,000 daltons. Not-
withstanding, its ability to phosphorylate fructose is higher
than that of the rat isozyme and its activity is not inhibited
by antibodies against rat isozyme D (Clark-Turri et al.,
1974).

REPTILIA

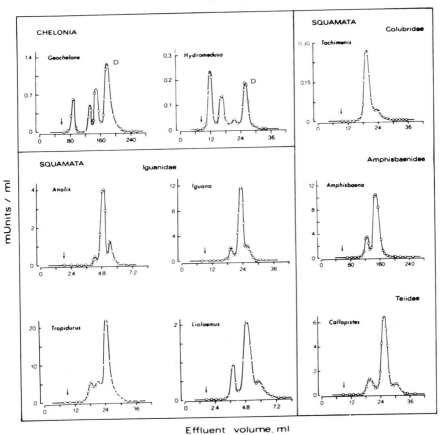

Effluent volume, ml

Fig. 8. Liver glucose phosphotransferases of some reptiles. For details see legend to Fig. 6. The species shown are: garden turtle, *Geochelone chilensis;* side-necked turtle, *Hydromedusa tectifera;* Colombian anole, *Anolis auratus;* Colombian iguana, *Iguana iguana;* beach lizard, *Tropidurus peruvianus;* garden lizard, *Liolaemus gravenhorstii;* short-tailed snake, *Tachimenis peruviana;* amphisbaenid lizard, *Amphisbaena alba;* Chilean "iguana", *Callopistes maculatus.* From Ureta et al., (1974).

AMPHIBIA

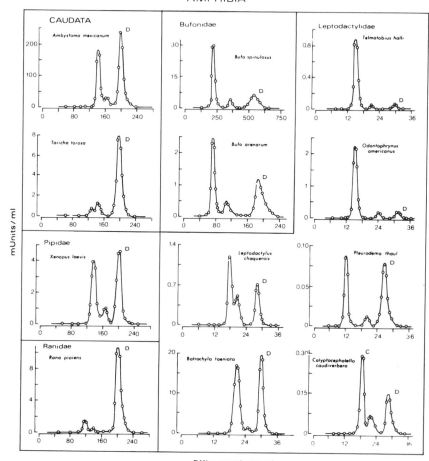

Fig. 9. Liver glucose phosphotransferases of some amphibians. For details see legend to Fig. 6. From T. Ureta, J.C. Slebe, J. Radojković, and C. Lozano (manuscript in preparation).

THE GLUCOSE PHOSPHOTRANSFERASES OF FISH LIVER

Only a few experiments have been performed in the case of fishes, including (not shown) *Cyprinus carpio*, *Halaerurus chilensis*, *Salmo gairderii*, *Salmo trutta*, and *Galaxias maculatus*. The isozymic pattern of *S. gairdnerii* is almost indistinguishable from that of a turtle (*Gaeochelone*) but the profile of *Salmo trutta* is very similar to that of a lizard.

ONTOGENETIC STUDIES

The high-Km isozyme is absent in the liver of fetal rats. Actually, it was this very fact that led Walker (1962, 1963) to discover the multiplicity of the phospho-transferases in liver. Isozyme D appears only after the 15th day of extrauterine life, i.e., at the time of weaning (Walker, 1963, 1965; Ballard and Oliver, 1964a, 1964b; Walker and Holland, 1965; Walker and Eaton 1967). Repeated injections of glucose and hydrocortisone brings about the premature appearance of the enzyme at the 9th day (Jamdar and Greengard 1970).

The levels of the low-Km hexokinases are very high before birth (Walker and Holland 1965; Walker and Eaton 1967; Jamdar and Greengard 1970) but studies on the contribution of each isozyme at different ages have been only of semiquantitative nature (Katzen and Schimke 1965; Farron 1972). Our chromatographic procedure was used to separate the isozymes of individual rats of different ages and the profiles obtained were quantitated by integrating the areas under each peak and correcting the values to gram of liver tissue (Fig. 10). The focus was on the low-Km isozymes inasmuch as a fair amount of information is already available about the development of the high-Km isozyme as discussed in the preceding paragraph.

The sum of the activities of isozymes A, B, and C, as determined by chromatography, closely agree with the values found by assay of the total low-Km activity in the high-speed supernatant fluids. The data show that the four isozymes appear in a sequential manner reaching peak values at different ages (Fig. 10).

The levels of isozyme A rise from a low value of 0.16 units/ g liver at day -5 (earliest age studied) to a maximum value of about 0.35 around birth. The levels fall during days +1 to +5 to about 0.2 units, and during the second week of life they further decrease to about 0.08 units, i.e., the normal adult levels.

Isozyme B levels are very low at day -5 and rise progressively to a high value of about 0.18 units/g liver during the first week of extrauterine life. The levels decrease then to reach the low adult values approximately at day +10.

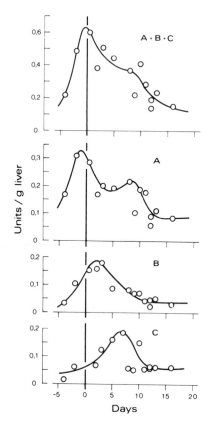

Fig. 10. Levels of the low-Km isozymes from rat liver as function of age. The isozymes were separated in small columns of DEAE-cellulose. The areas under each peak were integrated and expressed as units/g of liver. Each point represents a single animal. From T. Ureta, J. Babul, and R. Bravo (manuscript in preparation).

Isozyme C, on the other hand, which is all but absent before and around birth, shows a peak of activity during the second week of life and declines rather abruptly after +10 days.

Isozyme D is present in the liver of newborn rats although in very small amounts (not shown). As already mentioned, its levels rise abruptly from day +16 onwards to reach the adult values at about 30 days of age. A detailed account of these studies will be published elsewhere.

A few preliminary studies have been performed on the development of the glucose phosphotransferases during the metamorphosis of *Calyptocephalella caudiverbera*, a Chilean leptodactylid frog (not shown). In this case, the high-Km isozyme is already present at the tadpole stages, its levels being actually higher than in adults (unpublished).

CONCLUDING REMARKS

The data summarized in this paper indicate that glucose phosphorylation has evolved in vertebrates to a very high degree. At least four such enzymes catalyze that reaction and each differs from the others in several ways enabling the organism to adjust the use of sugar in the most efficient manner by utilizing the differences among the isozymes.

Each glucose phosphotransferase seems to be the product of a separate gene. Firstly, the complete lack of immunological cross reaction suggests wide structural differences between the isozymes. Secondly, at least one of the low-Km isozymes seems to be composed of a single polypeptide chain, thus precluding its participation in the formation of "hybrid" molecules even though the three low-K_m isozymes have the same molecular weight. Thirdly, the different molecular weight, strikingly different kinetic features, and the very special tissue distribution of isozyme D make its inclusion as a member of a "family" of related enzymes very unlikely. Lastly, as shown by the phylogenetic studies, each isozyme may be absent in a given species.

It is tempting to speculate about the events responsible for the very different patterns of phosphotransferases observed in the different classes of vertebrates, and the diagram of Fig. 11, which is "more an expression of feeling than a painting" is an effort in that direction. It simply states that the genetic information for the synthesis of isozyme D was lost, or became permanently repressed, in the evolutionary line leading to modern higher reptiles and birds, probably as the result of a mutational even occurring during the late Permian period. It is not yet possible to apply a similar reasoning to the low-Km hexokinases.

ACKNOWLEDGEMENTS

Several colleagues have participated in the experiments reported in this paper, particularly Jasna Radojković, Juan Carlos Slebe, Samuel B. Reichberg, Carlos Lozano, Rodrigo Bravo, and Dr. Jorge Babul. Several friends contributed the animals used and their participation is acknowledged in the

THE EVOLUTION OF HEXOKINASES

"Mehr Ausdruck der Empfindung als Mahlerei"

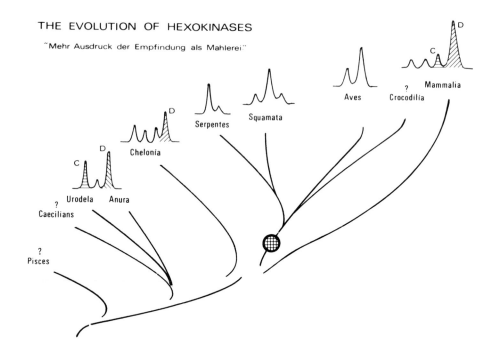

Fig. 11. A simplified scheme of the evolution of the liver glucose phsophorylating isozymes. The idealized patterns at top are intended to illustrate examples of chromatographic profiles of <u>some</u> species, but within each taxa the relative proportion of the isozymes may vary considerably. Example shown: Mammalia, *Rattus norvegicus*; Aves, *Gallus domesticus*; Squamata, *Liolaemus gravenhorstii*; Serpentes, *Tachimenis peruviana*; Chelonia, *Geochelone chilensis*; Anura, *Calyptocephalella caudiverbera*; Urodela, *Ambystoma mexicanum*. The inclined shades represent high-Km isozymes and the horizontal ones depict inhibition by excess of glucose. The different inclination of the shades under the high-Km peaks illustrate the lack of immunologic cross reaction (see text). The circle on the line leading to Aves and higher reptiles marks the postulated mutational event resulting in the loss or permanent repression of glucokinase. No temporal scale is intended.

---

original publications. This line of research began (and still continues) under the inspiring guidance of Dr. Hermann Niemeyer to whom my sincerest thanks are due. Grants from the Faculty of Medicine and from the Comisión de Investigación of the University of Chile made this work possible.

REFERENCES

Anderson, J.W. and J.B. Tyrrell 1973. Hexokinase activity of rat intestinal mucosa: demonstration of four isozymes and of changes in subcellular distribution with fasting and refeeding. *Gastroenterology* 65: 69-76.

Ashcroft, S.J.H. and P.J. Randle 1970. Enzymes of glucose metabolism in normal mouse pancreatic islets. *Biochem. J.* 119: 5-15.

Bachelard, H.S. 1967a. Glucose-phosphorylating enzyme with high Km in human brain. *Nature* 215: 959-960.

Bachelard, H.S. 1967b. The subcellular distribution and properties of hexokinases in the guinea pig cerebral cortex. *Biochem. J.* 104: 286-292.

Ballard, F.J. 1965. Glucose utilization in mammalian liver. *Comp. Biochem. Physiol.* 14: 437-443.

Ballard, F.J. and I.T. Oliver 1964a. Ketohexokinase, isozymes of glucokinase and glycogen synthesis from hexoses in neonatal rat liver. *Biochem. J.* 90: 261-268.

Ballard, F.J., and I.T. Oliver 1964b. The effect of concentration on glucose phosphorylation and incorporation into glycogen in the livers of foetal and adult rats and sheep. *Biochem. J.* 92: 131-136.

Berstein, R.S. and D.M. Kipnis 1973. Regulation of rat hexokinase isozymes. I- Assay and effect of age, fasting and refeeding. *Diabetes* 22: 913-922.

Berthillier, G., L. Colobert, M. Richard, and R. Got 1970. Glucokinases due foie de rat. Purification et propriétés des formes particulées. *Biochim. Biophys. Acta* 206: 1-16.

Bonney, R.J., H.A. Hopkins, P.R. Walker, and V.R. Potter 1973. Glycolytic isozymes and glycogen metabolism in regenerating liver from rats on controlled feeding schedules. *Biochem. J.* 136: 115-124.

Borrebaek, B. 1970. Mitochondrial-bound hexokinase of rat epididymal adipose tissue and its possible relation to the action of insulin. *Biochem. Med.* 3: 485-497.

Cárdenas, M.L., E. Rabajille, and H. Niemeyer 1974. Sigmoidicidad de la función de saturación de la glucoquinasa con glucosa. *XVI Annual Meeting of the Sociedad de biología de Chile. Abstracts* R-12.

Chamorro, G., and R. Schilkrut 1969. Cinética de inducción y degradación de glucoquinasa en higade de rata. Algunos factores que influyen en el proceso. *Tesis, Facultad de Medicina, Universidad de Chile, Santiago, Chile.*

Chou, A.C., and J.E. Wilson 1972. Purification and properties of rat brain hexokinase. *Arch. Biochem. Biophys.* 151: 48-55.

Clark-Turri, L., J. Penaranda, E. Rabajille, and H. Niemeyer 1974. Immunochemical titration of liver glucokinase from normal, fasted, and diabetic rats. *FEBS Letters* 41: 342-344.

Colowick, S.P. 1973. The hexokinases. In: P.D. Boyer (Editor) *The Enzymes.* Vol. IX, Part B. Third Edition, Academic Press, New York and London. pp. 1-48.

Crane, R.K. 1962. Hexokinases and pentokinases. In: P.D. Boyer, H. Lardy, and K. Myrback (Editors) *The Enzymes.* Vol. VI, Academic Press, New York. pp. 47-66.

Creighton, S.R., A.M. McClure, B.J. Watrous, and R.J. Hansen 1972. Hexose-ATP phosphotransferases: comparative aspects- III. Interrelationships of animal hexokinases. *Comp. Biochem. Physiol.* 42: 509-516.

Crisp, D.M. and C.I. Pogson 1972. Glycolytic and gluconeo-genic enzyme activities in parenchymal and non-parenchymal cells from mouse liver. *Biochem. J.* 126: 1009-1023.

DiPietro, D.L., and S. Weinhouse 1960. Hepatic glucokinase in the fed, fasted and alloxan-diabetic rats. *J. Biol. Chem.* 235: 2542-2545.

DiPietro, D.L., C. Sharma, and S. Weinhouse 1962. Studies on glucose phosphorylation in rat liver. *Biochemistry* 1: 455-462.

Easterby, J.S. 1971. The polypeptide chain molecular weight of a mammalian hexokinase. *FEBS Letters* 18: 23-26.

Farron, F. 1972. The isozymes of hexokinase in normal and neoplastic tissues of the rat. *Enzyme* 13: 223-237.

Gerber, G.K.G., M. Schultze, and S.M. Rapoport 1970. Occurrence and function of a high-Km hexokinase in immature red blood cells. *Eur. J. Biochem.* 17: 445-449.

González, C., T. Ureta, J. Babul, E. Rabajille, and H. Niemeyer 1967. Characterization of isozymes of adenosine tri-phosphate: D-hexos 6-phosphotransferase from rat liver. *Biochemistry* 6: 460-468.

González, C., T. Ureta, R. Sánchez, and H. Niemeyer 1964. Multiple molecular forms of ATP: hexose 6-phosphotrans-ferase from rat liver. *Biochem. Biophys. Res. Commun.* 16: 347-352.

Grossbard, L., and R.T. Schimke 1966. Multiple hexokinases of rat tissues. Purification and comparison of soluble forms. *J. Biol. Chem.* 241: 3546-3560.

Hansen, R.J., S.J. Pilkis, and M.E. Krahl 1970. Effect of insulin on the synthesis in vitro of hexokinase in rat epididymal adipose tissue. *Endocrinology* 86: 57-65.

Jamdar, S.C. and O. Greengard 1970. Premature formation of

glucokinase in developing rat liver. *J. Biol. Chem.*
245: 2779-2783.

Katzen, H.M. 1966. The effect of diabetes and insulin in
vivo and in vitro on a low-Km form of hexokinase from
various rat tissues. *Biochem. Biophys. Res. Commun.*
24: 531-536.

Katzen, H.M. 1967 The multiple forms of mammalian hexokinase
and their significance to the action of insulin. *Adv.
Enzyme Regulat.* 5: 335-356.

Katzen, H.M. and R.T. Schimke 1965. Multiple forms of
hexokinase in the rat: tissue distribution, age dependency, and properties. *Proc. Natl. Acad. Sci. U.S.A.* 54:
1218-1225.

Katzen, H.M., D.D. Soderman, and H.M. Nitowsky 1965. Kinetic
and electrophoretic evidence for multiple forms of
glucose-ATP phosphotransferase activity from human cell
cultures and rat liver. *Biochem. Biophys. Res. Commun.*
19: 377-382.

Katzen, H.M., D.D. Soderman, and C.E. Wiley 1970. Multiple
forms of hexokinase. Activities associated with subcellular particulate and soluble fractions of normal and
streptozotocin diabetic rat tissues. *J. Biol. Chem.*
245: 4081-4096.

Matchinsky, F.M. and J.E. Ellerman 1968. Metabolism of
glucose in the islets of Langerhans. *J. Biol. Chem.*
243: 2730-2736.

Mayer, R.J. and G. Hubscher 1971. Mitochondrial hexokinase
from small-intestinal mucosa and brain. *Biochem. J.*
124: 491-500.

Moore, R.O., A.M. Chandler, and N. Tettenhorst 1964. Glucose-
ATP transferases in adipose tissue of fasted and refed
rats. *Biochem. Biophys. Res. Commun.* 17: 527-531.

Morrison, G.R. 1967. Hexokinase and glucokinase activities
in bile duct epithelial cells and hepatic cells from
normal rat and human livers. *Arch. Biochem. Biophys.*
122: 569-573.

Neumann, S., F. Falkenberg, and G. Pfleiderer 1974. Purification and immunological characterization of the human
hexokinase isozymes I and II (ATP-D-hexose 6-phospho-
transferase EC 2.7.1.1). *Biochim. Biophys.Acta* 334:
328-342.

Neumann, S. and G. Pfleiderer 1974. Immunological specificity
of the isozymes I and III of human hexokinase (ATP-D-
hexose 6-phosphotransferase EC 2.7.1.1). Estimation of
isozyme pattern by quantitative immunotechniques.
Biochim. Biophys. Acta 334: 343-353.

Niemeyer, H. 1971. Dietary and hormonal effects on liver

glucokinase. In: *Metabolic Adaptation and Nutrition,* Pan American Health Organization, Scientific Publication No. 222, Washington, U.S.A. pp. 36-44.

Niemeyer, H., L. Clark-Turri, N. Pérez, and E. Rabajille 1965. Studies on factors affecting the induction of ATP: D-hexose 6-phosphotransferase in rat liver. *Arch. Biochem. Biophys.* 109: 634-645.

Niemeyer, H., L. Clark-Turri, and E. Rabajille 1963. Induction of glucokinase by glucose in rat liver. *Nature* 198: 1096-1097.

Niemeyer, H., N. Pérez, and R. Codoceo 1967. Liver glucokinase induction in acute and chronic insulin insufficiency in rats. *J. Biol. Chem.* 242: 860-864.

Niemeyer, H., N. Pérez, E. Garcés, and F.E. Vergara 1962. Enzyme synthesis in mammalian liver as a consequence of refeeding after fasting. *Biochim. Biophys. Acta* 62: 411-413.

Niemeyer, H., N. Pérez, and E. Rabajille 1966. Interrelation of actions of glucose, insulin and glucagon on induction of adenosine triphosphate: D-hexose phosphotransferase in rat liver. *J. Biol. Chem.* 241: 4055-4059.

Niemeyer, H. and T. Ureta 1972. Enzyme adaptation in mammals. In: K. Gaede, B.L. Horecker, and W.J. Whelan (editors) *Molecular Basis of Biological Activity.* PAABS Symposium Volume I, 221-273. Academic Press, New York and London.

Parry, M.J. and D.G. Walker 1966. Purification and properties of adenosine 5'-triphosphate-D-glucose-6-phosphotransferase from rat liver. *Biochem. J.* 99: 266-274.

Parry, M.J. and D.G. Walker 1967. Further properties and possible mechanism of action of adenosine 5'-triphosphate-D-glucose-6-phosphotransferase from rat liver. *Biochem. J.* 105: 473-482.

Pilkis, S.J. and R.J. Hansen 1968. Resolution of two high-K_m ATP: D-hexose 6-phosphotransferase bands by starch-gel electrophoresis. *Biochim. Biophys. Acta* 159: 189-191.

Pilkis, S.J., R.J. Hansen, and M.E. Krahl 1968a. Apparent molecular weights of some ATP: D-hexose 6-phosphotransferases: specific effects of Sephadex G-100. *Biochim. Biophys. Acta* 154: 250-252.

Pilkis, S.J., R.J. Hansen, and M.E. Krahl 1968b. Hexose-ATP phosphotransferases: comparative aspects. *Comp. Biochem. Physiol.* 25: 903-912.

Purich, D.L., H.J. Fromm, and F.B. Rudolph 1973. The hexokinases: kinetic, physical, and regulatory properties. *Adv. Enzymol.* 35: 249-326.

Saito, M. and S. Sato 1971. Studies on particle-bound hexokinase in rat ascites hepatoma cells. *Biochim. Biophys. Acta* 227: 344-353.

Salas, M., E. Viñuela, and A. Sols 1963. Insulin-dependent synthesis of liver glucokinase in the rat. *J. Biol. Chem.* 238: 3535-3538.

Salas, J., E. Vinuela, M. Salas, and A. Sols 1965. Glucokinase of rabbit liver. Purification and properties. *J. Biol. Chem.* 240: 1014-1018.

Sapag-Hagar, M., R. Marco, and A. Sols 1969. Distribution of hexokinase and glucokinase between parenchymal and non-parenchymal cells of rat liver. *FEBS Letters* 3: 68-71.

Sato, S., T. Matsushima, and T. Sugimura 1969. Hexokinase isozyme patterns of experimental hepatomas of rats. *Cancer Res.* 29: 1437-1446.

Sharma, C., R. Manjeshwar, and S. Weinhouse 1963. Effects of diet and insulin on glucose-adenosine triphosphate phosphotransferases of rat liver. *J. Biol. Chem.* 238: 3840-3845.

Siekevitz, P. and V.R. Potter 1955. Biochemical structure of mitochondria. II. Radioactive labeling of intramitochondrial nucleotides during oxidative phosphorylation. *J. Biol. Chem.* 215: 237-255.

Sols, A., M. Salas, and E. Viñuela 1964. Induced biosynthesis of liver glucokinase. *Adv. Enzyme Regulat.* 2: 177-188.

Thompson, F.M. and H.S. Bachelard 1970. Cerebral cortex hexokinase. Comparison of properties of solubilized mitochondrial and cytoplasmic activities. *Biochem. J.* 118: 25-34.

Ureta, T., C. González, S. Lillo, and H. Niemeyer 1971a. Comparative studies on glucose phosphorylating isozymes of vertebrates- I. The influence of fasting and the nature of the diet on liver glucokinase and hexokinases of rodents. *Comp. Biochem. Physiol.* 40: 71-80.

Ureta, T., C. González, and H. Niemeyer 1971b. Comparative studies on glucose phosphorylating isozymes of vertebrates- II. Chromatographic patterns of glucokinase and hexokinases in the liver of rodents. *Comp. Biochem. Physiol.* 40: 81-91.

Ureta, T., J. Radojković, and H. Niemeyer 1970. Inhibition by catecholamines of the induction of rat liver glucokinase. *J. Biol. Chem.* 245: 4819-4824.

Ureta, T., J. Radojković, J.C. Slebe, and S.B. Reichberg 1972. Comparative studies on glucose phosphorylating isozymes of vertebrates- III. Isolation and properties of two hexokinases from chick liver. *Int. J. Biochem.*

3: 103-110.

Ureta, T., S.B. Reichberg, J. Radojković, and J.C. Slebe 1973. Comparative studies on glucose phosphorylating isozymes of vertebrates- IV. Chromatographic profiles of hexokinases from the liver of several avian species. *Comp. Biochem. Physiol.* 45: 445-461.

Ureta, T., R. Sánchez, C. González, and H. Niemeyer 1963. Isozimas de hexoquinasa de hígado de tata y su variación por efecto de la dieta. First Congress on Nutrition, Bromatology and Toxicology. Santiago, Chile, October 28. *Proceedings*, p. 20.

Ureta, T., J.C. Slebe, J. Radojković, and C. Lozano 1974. Comparative studies on glucose phosphorylating isozymes of vertebrates- V. Glucose phosphotransferases in the liver of Reptiles. *Comp. Biochem. Physiol.* In the press.

Viñuela, E., M. Salas, and A. Sols 1963. Glucokinase and hexokinase in liver in relation to glycogen synthesis. *J. Biol. Chem.* 238:1175-1177.

Walker, D.G. 1962. The development of hepatic hexokinases after birth. *Biochem. J.* 84: 118P

Walker, D.G. 1963. On the presence of two soluble glucose-phosphorylating enzymes in adult liver and the development of one of these after birth. *Biochim. Biophys. Acta* 77: 209-226.

Walker, D.G. 1965. Development of hepatic enzymes for the phosphorylation of glucose and fructose. *Adv. Enzyme Regulat.* 3: 163-184.

Walker, D.G. 1966. The nature and function of hexokinases in animal tissues. *Essays Biochem.* 2: 3-67.

Walker, D.G. and S.W. Eaton 1967. Regulation of development of hepatic glucokinase in the neonatal rat by the diet. *Biochem. J.* 105: 771-777.

Walker, D.G. and G. Holland 1965. The development of hepatic glucokinase in the neonatal rat. *Biochem. J.* 97: 845-854.

Walters, E. and P. McLean 1967. Multiple forms of glucose-adenosine triphosphate phosphotransferase in rat mammary gland. *Biochem. J.* 104: 778-783.

Weinhouse, S., V. Cristofalo, C. Sharma, and H.P. Morris 1963. Some properties of glucokinase in normal and neoplastic liver. *Adv. Enzyme Regulat.* 1: 363-371.

Weiser, M.M., H. Quill, and K.J. Isselbacher 1971. Isolation and properties of intestinal hexokinases, fructokinase, and N-acetylglucosamine kinase. *J. Biol. Chem.* 246: 2331-2337.

Werner, H.V., J.C. Bartley, and M.N. Berry 1972. Glucose-adenosine 5'-triphosphate 6-phosphotransferases of

isolated rat liver parenchymal cells. *Biochem. J.*
130: 1153-1155.

EFFECTS OF GENE DOSAGE ON PEROXIDASE ISOZYMES
IN *DATURA STRAMONIUM* TRISOMICS

H. H. SMITH and M. E. CONKLIN
Department of Biology
Brookhaven National Laboratory
Upton, New York 11973 and
Department of Biology
San Diego State University
San Diego, California 92182

ABSTRACT. Uniform leaf samples of eight of the trisomic
(2n + 1) types of *Datura stramonium*, as well as the diploid (2n) and tetraploid (4n), were analyzed for peroxidase isozymes. Six isoperoxidases appeared in each type;
however, the staining intensity of specific bands was
altered from the 2n according to the genotype. At least
one trisomic gave a significant increase in intensity of
a particular band; others were associated with significant decreases in intensity of specific bands. The tetraploid zymogram did not differ from that of the diploid.
On the assumption that isoperoxidases represent direct
gene products, speculations about the results in terms
of gene regulatory systems are discussed.

INTRODUCTION

The first trisomic type discovered was the "Globe" mutant
of *Datura stramonium* L. Its aberrant morphological characteristics and breeding behavior were described in 1916 (*Datura*
investigations, 1916); and in 1920 the true cytological nature
of this, and of the eleven other primary trisomics, was reported (Blakeslee et al., 1920; Khush, 1973). The distinct
changes in the appearance of the trisomic plants were attributed to the specific extra genetic material in each of the
so-called "unbalanced" types (Blakeslee, 1922, 1930). This
study of the effects of gene dosage on peroxidase isozymes in
Datura stramonium trisomics was begun in 1970 largely as an
effort to aid in the understanding of genetic "unbalance" at
the molecular level.

MATERIALS AND METHODS

All lines of *Datura stramonium* used in these experiments
were derived from the standard diploid isogenic line 1A of
the late Dr. A. F. Blakeslee. The trisomic types, the diploid and the tetraploid stocks were obtained from, and iden-

603

tified by, Mr. Amos Avery. Supplementary stocks of certain
of the types, derived from the same source, were furnished
by the National Seed Storage Laboratory, Fort Collins, Colorado.

The seedlings were started in a growth chamber under
controlled conditions at 24°C and 2200 ± 300 ft-c of light
(cool white fluorescent with 25% incandescent) for an 18 hr
day. They were later transplanted into pots and grown to
maturity in a greenhouse maintained at a constant temperature
of approximately 24°C and 18 hr of light. The trisomic types
were selected as early as possible on the basis of leaf shape,
and later confirmed according to growth habit, morphology and
capsule characteristics (Avery et al., 1959). Verification
of the chromosome number, 2n + 1 = 25, was made for one or
more plants, typical of each trisomic type, from aceto-orcein
squash preparations of pollen mother cells.

Although seed stocks were furnished for eleven of the
twelve possible primary trisomic types, we were able to identify
with certainty (Fig. 1) only eight for use in these studies.
The trisomics are designated by a descriptive name and by the
chromosome nomenclature, adopted by the Blakeslee group,
whereby each arm of the twelve chromosomes of the haploid
complement is numbered (Blakeslee and Cleland, 1930; Satina
et al., 1941). For example, the trisomic with the largest
single extra chromosome was designated 2n + 1.2 and named
Rolled.

The material selected for electrophoresis was the fourth
leaf down from the shoot apex, which was harvested as the
plant reached maturity. Leaf samples so selected have previously been found to be suitable for comparative zymogram studies
(Conklin and Smith, 1971).

Electrophoretic separation of a crude extract of equivalent weights of leaf blade samples was carried out on a horizontal starch gel at 4°C and pH 8.3. The procedure for separation and staining was that described by Conklin and Smith
(1971). The appearance of the isoperoxidase migration sites
depends on the production of a colored form of ortho-dianisi-
dine (3-3' dimethoxybenzidine) from the oxidation of the colorless form in the presence of hydrogen peroxide and peroxidase.
The bands that stain represent multiple molecular forms that
are related on the basis of this simple, not highly specific,
enzyme assay. From four to 19, and on an average 13, runs
were made of each trisomic type. For purposes of comparison,
samples of D. stramonium 2n and 4n were usually included on
each gel.

Fig. 1. Leaves typical of *Datura stramonium* diploid (2n), tetraploid (4n), and seven of the trisomic (2n + 1) types.

After clearing in 50% glycerol for at least four days, the gels were scored for position and staining intensity of each band (Fig. 2). The peroxidase zymogram of the fourth apical leaf of *D. stramonium* had previously been characterized as having six distinct repeatable bands (Conklin and Smith, 1971). Five are anodal and were designated Nos. 3, 4, 5, 8, and 9. The sixth band is cathodal and was designated No. 19. Other poorly stained regions appeared in some gels between the origin and bands 9 or 19 (Fig. 2), but these were not consistent and were therefore not included in this study.

The intensity of each band was scored visually and was shown to be related to the level of peroxidase enzyme activity in the following way. Densitometer tracings (made with a Joyce and Loebl double beam recording microdensitometer) of photographic negatives of typical zymograms (Fig. 3), were calibrated on the basis of absolute density readings (made with a Kodak ASA Visual Diffuse Density Photographic Step Tablet No. 3 and corrected for film density background). These densities could be approximated by visual ratings of band intensities made by using a quantitative scoring of: 4-strong, 3-moderate, 2-light, 1-faint, and 0-absent. There is a high degree of correlation between the two methods, correlation coefficient $= -0.97$, so that absolute density is reasonably predictable by visual rating.

It was then shown (Fig. 4) that the intensity (optical density) of the o-dianisidine color reaction is directly proportional to the concentration of peroxidase, or level of enzyme activity. Therefore, a comparison between the intensity of a particular isozyme band in the diploid vs. trisomic or tetraploid was considered to measure the relative amount of enzyme at that migration site. The visual rating scale used tends to underestimate, in terms of fold differences, the absolute density scale, which in turn, is directly proportional to enzyme activity. Our inferences, then, on the statistical significance of differences in intensity between corresponding isozyme bands, are conservative. Enzymatic efficiency per molecule and degradation rate would be expected to be the same in corresponding isozyme bands, and therefore not contributory to the differences observed.

RESULTS

A mean intensity rating for each band in each genotype was computed as the summed products of each rating class times its frequency divided by the total number of observations. The results for each of the six peroxidase bands, expressed as a mean value, are shown in Table 1 for the diploid (2n), tetraploid (4n) and the eight trisomic types.

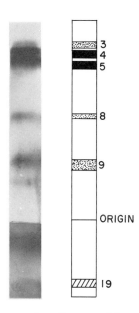

Fig. 2: Left: Photograph of peroxidase zymogram of diploid (2n) *Datura stramonium*, leaf 4. Right: Diagrammatic represent-ation of the number, position and relative intensity of the 2n leaf 4 peroxidase isozyme bands.

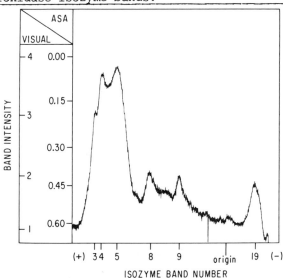

Fig. 3: Microdensitometer tracing of a typical peroxidase zymo-

gram of diploid *Datura stramonium* leaf extract, made from a photographic negative of a stained gel. Isozyme band numbers are shown on the horizontal axis. On the vertical axis the left scale approximates the visual ratings used; the scale to the right shows absolute densitites measured by corrected readings from a calibrated ASA diffuse density photographic step tablet.

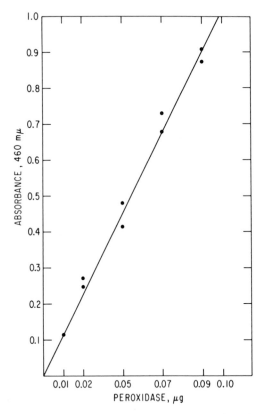

Fig. 4: Plot showing the direct correlation between absorbance and amount of peroxidase. The concentration of substrate and stain was similar to that used in the zymograms. The peroxidase (horseradish, Worthington, 0.23 mg in 500 ml H_2O) was added and color development followed for five minutes. The assays were carried out using a Cary 15 recording spectrophotometer in 3 ml cuvettes at 25°C. All readings were taken against a blank and duplicate samples were run at each of five concentrations.

Table 1
Peroxidase isozyme band intensity in diploid, tetraploid and trisomic types of *Datura stramonium*.

Type Name	Chrom. arms	No. observed	Band 3 $\overline{m}$*	Band 3 P**	Band 4	Band 5	Band 8 $\overline{m}$	Band 8 P	Band 9 $\overline{m}$	Band 9 P	Band 19 $\overline{m}$	Band 19 P
Diploid	2n	54	2.7		4.0	4.0	2.3		2.2		1.4	
Rolled	2n + 1.2	4	3.0	.548	4.0	4.0	2.2	.917	2.8	.224	0.8	.149
Buckling	2n + 5.6	15	**1.4**	**.000**	4.0	4.0	2.1	.362	**1.7**	**.050**	1.5	.876
Echinus	2n + 9.10	13	**1.7**	**.001**	4.0	4.0	**1.5**	**.006**	**0.5**	**.000**	0.9	.073
Cocklebur	2n + 11.12	19	**0.1**	**.000**	4.0	4.0	2.3	.932	2.3	.583	**0.7**	**.003**
Microcarpic	2n + 13.14	12	**1.2**	**.000**	4.0	4.0	**1.5**	**.005**	**1.3**	**.044**	1.2	.366
Reduced	2n + 15.16	9	2.6	.665	4.0	4.0	1.8	.098	2.6	.251	1.1	.331
Globe	2n + 21.22	14	2.9	.591	4.0	4.0	2.6	.183	2.4	.364	**3.4**	**.000**
Ilex	2n + 23.24	18	2.8	.671	4.0	4.0	**1.6**	**.002**	2.8	.703	**2.1**	**.006**
Tetraploid	4n	13	2.7	.969	4.0	4.0	2.8	.079	2.5	.318	1.6	.495

*Mean value of band expression.
**Probability that mean band expression is the same as in the diploid (2n).

Bands 4 and 5 always showed the maximum amount of staining (rating of 4.0) for each genotype. Therefore, these two bands could not be used in quantitative comparisons among genotypes. The intensity rating of the remaining four bands in the diploid, based on 54 separate runs was: Band 3, 2.7; Band 8, 2.3: Band 9, 2.2; and Band 19, 1.4 (Table 1). Whether or not each of these bands in the trisomics and tetraploid differed significantly in intensity from the identical band in the diploid was then considered. The hypothesis that the mean band intensity was equal to that of the diploid for each band was tested by the use of a nonindependent Student's t test. Probabilities of .05 or smaller are taken as evidence that the mean of a particular trisomic band differs significantly from the diploid mean. These cases are shown in bold type in Table 1. Of the thirteen significant differences, eleven were less intense and two were more intense. From this observation alone, we conclude that peroxidase band intensity is not directly correlated with, or a reflection of, some general growth phenomenon since all the trisomics are less vigorous in growth habit than the diploid.

A comparison between genotypes revealed that the tetraploid (4n) zymogram did not differ significantly from the diploid in spite of the fact that the gene dosage was double for the entire genome. Furthermore, the zymogram of two of the trisomic types, Rolled and Reduced (Table 1), did not differ in any band intensity from the diploid. Since these three genotypes (4n, 2n + 1.2, 2n + 15.16) have markedly different phenotypes and differ conspicuously in appearance from the diploid (Fig. 1), it is clear that not every morphological change is correlated with a change in peroxidase isozyme activity.

The remaining six trisomic types did show significant differences from the diploid in the intensity of certain isozyme bands. In the following trisomics the bands noted were less intense: Buckling, Bands 3 and 9; Echinus, Bands 3, 8, and 9; Cocklebur, Bands 3 and 19; and Microcarpic, Bands 3, 8, and 9. In Globe, Band 19 was more intense. In Ilex, Band 8 was less intense and Band 19 was more intense. Each of these 2n + 1 types has a characteristic average peroxidase zymogram pattern. Consequently, the peroxidase differences observed are due not to altered byproducts resulting from changes in environmental conditions, but rather to a difference in genetic constitution.

Other ways of examining the results are summarized in Table 2. A rough estimate for comparing total peroxidase activities was made by adding the mean value for bands 3, 8, 9, and 19 as given in Table 1. This sum, for each derived genotype, was then expressed as a per cent of the 2n level, defined as 100%. Where the sum was within $\pm$ 1 unit of the 2n level, it was rounded off to 100%. One of the trisomics, Globe (2n + 21.22), gave a higher level, 133%, compared to the diploid. This type of result has been observed in other systems and has been attributed to the increased dose, 3/2, of a structural gene in the extra chromosome coding for the particular protein. In four of the trisomics (Buckling, Echinus, Cocklebur and Microcarpic) there was a reduction in the summed band intensity. These averaged 64%, (77 + 54 + 63 + 60)/4, and were tentatively considered to be about 2/3, or 67% of the diploid level. Three other trisomics (Rolled, Reduced and Ilex) and the tetraploid did not differ from the diploid.

The inadequacy of using overall intensity as an enzyme assay in this material for genetic studies is demonstrated in the results on Ilex (Table 2). In this genotype the summed band intensity did not differ significantly from the diploid, but we can see that this was because band No. 8 was less intense and band No. 19 was more intense, thus demonstrating the importance of electrophoretic separation of isozymes for significant analysis.

A band by band comparison somewhat different from the method in Table 1, is summarized to the right in Table 2. Each band mean was compared by a Student's t test to 67, 100, and 150% of the diploid mean. The probabilities computed, by utilizing the pooled error variance, permit us to decide which of the three levels of diploid intensity best approximates the observed mean in each type. An X in the table indicates the most probable percent of the 2n level.

Table 2

Peroxidase isozyme band intensity (summed and individual) in zymograms of tetraploid and trisomic types compared to the diploid, defined as 100% level.

Type name	Chrom. arms	Summed* Band intensity Σ	% of 2n	67% Band No. 3	67% 8	67% 9	67% 19	100% Band No. 3	100% 8	100% 9	100% 19	150% Band No. 3	150% 8	150% 9	150% 19
Diploid	2n	8.6	100					×	×	×	×				
Rolled	2n+ 1.2	8.8	100				×	×	×	×					
Buckling	2n+ 5.6	6.7	77	×		×			×		×				
Echinus	2n+ 9.10	4.6	54	×	×	×	×								
Cocklebur	2n+11.12	5.4	63	×			×		×	×					
Microcarpic	2n+13.14	5.2	60	×	×	×	×								
Reduced	2n+15.16	8.1	100		×		×	×		×					
Globe	2n+21.22	11.4	133					×	×	×					×
Ilex	2n+23.24	9.3	100	×				×		×					×
Tetraploid	4n	9.6	100					×	×	×	×				

*Summation of mean intensity of bands nos. 3, 8, 9 and 19.

The results show five different kinds of effects: 1) The tetraploid did not differ from the diploid. 2) In two trisomic types (Echinus and Microcarpic) all bands were nearer to 67% than 100% of the 2n level. 3) In four trisomics (Rolled, Buckling, Cocklebur, and Reduced) some bands were unchanged (100%) while others were less intense (67%). 4) In one trisomic (Globe) three bands were unchanged and one was more intense (150%). 5) In one trisomic (Ilex, as noted above) two bands were unchanged, one was decreased and one was increased in intensity. The results as presented in Table 1 vs. Table 2 are in general agreement except that, in the latter, additional significant reductions from the diploid level are indicated for Band 8 in Reduced and for Band 19 in Rolled, Echinus, Microcarpic and Reduced.

It is not our intention to emphasize the 67% and 150% amounts as such, particularly in view of the rough quantitive estimates used; nevertheless, the departures from 100% are so substantial that they are more likely to be of this magnitude, rather than chance deviations from the 2n level.

DISCUSSION

Although there is at present no conclusive proof that the isozyme bands are direct gene products, it is evident that each genotype has a recognizable isozyme pattern. Mendelian inheritance of peroxidase isozymes has been demonstrated in a number of plants, including species of the related solanaceous genus, *Nicotiana* (Hoess et al., 1974).

With the assumption that zymogram bands and their intensity are a direct product of gene activity, certain speculations follow. We tentatively consider that our results are consistent with a genetic interpretation that a structural gene for peroxidase is located on chromosome 21.22, and possibly also on chromosome 23.24. In the trisomics that show substantial reduction in staining of particular isozyme bands, we tentatively consider this as a possible indication of an extra dose of a "regulatory gene(s)" producing repressor (s) in greater amount than in the diploid.

The data with selected leaves of *Datura* trisomics are in close agreement with those of Carlson (1972), who used tissue of cultures derived from hypocotyl callus of the same trisomics, and assayed the level of total activity per mg of protein of fifteen enzymes. For each specific enzyme the addition of one particular chromosome gave an increased activity of about 150% (average 144%) of the diploid (100%) level. This result was considered to demonstrate that structural genes for any one of nine enzymes showing increased activity could be located on chromosomes by their dosage effect. It should be noted, however, that the evidence for assigning human genes to particular chromosomes on the basis of elevated enzyme activities in trisomic individuals (e.g. trisomy of chromosome 21, Down's syndrome) has been questioned (Hsia et al., 1971; Layzer and Epstein, 1972), particularly from negative results of subsequent isozyme studies (Monteleone et al., 1967; Benson and deJong, 1968).

Carlson's results also showed a reduction in enzyme activity in the tissue cultures for most trisomics. In the absence of statistical tests, if levels of less than 90% are considered to be below the diploid amount, then from 2 to 5 of the trisomics gave reduced activity for each of the nine enzymes tested. This, also, is consistent with our results on peroxidases in differentiated leaf tissue.

Trisomic types in barley, as in *Datura*, have been used to detect gene dosage effects associated with extra chromosomes. Nielson and Frydenberg (1971) found that in plants trisomic for chromosome 3 there was an increase in esterase activity associated with an increased dose of a particular allele

located on that chromosome. Earlier, McDaniel and Ramage
(1970) had reported an electrophoretic analysis of seed proteins in the seven primary trisomic types of barley. Each of
the trisomic zymograms could be distinguished from the diploid
and from each other by both qualitative and quantitative
differences in bands. The dosage effects included: 1) Extra
chromosome 5 resulted in the production of 140-150% of the
diploid quantity of the protein in one particular band. 2)
Suppression was manifested by the loss of a specific protein
band in particular trisomics. The zymograms illustrated
appear to show, though this was not noted, a reduction in
intensity of certain protein bands in certain trisomics
(notably trisomics 2 and 7). This is the same phenomenon
that we commonly observe in our *Datura* zymograms, and which
we speculate may be due to a "regulator gene(s)" dose effect
and consequent increased production of a repressor substance(s).

Correlations between a greater number of structural gene
copies (gene dosage) and an increase in enzyme level have now
been established for various enzymes in other eukaryotes.
These include *Drosophila* (Grell, 1962; Stewart and Merriam,
1974), yeast (Nelson and Douglas, 1963), wheat (Brewer et al.,
1969) and mouse (Epstein, 1969, 1972).

Although these dosage-activity relationships have been
found to hold when the number of gene copies is altered on a
constant residual genetic background, in tetraploids or higher polyploids, where multiplicity of the whole genome accompanies the increasing dosage of a particular enzyme-governing
gene, conflicting results have been reported. In tetraploid
Chinese hamster cell lines enzyme activities were approximately double that in the corresponding diploid line (Puck and
Kao, 1968; Westerveld et al., 1972) and a similar result
had been reported by Priest and Priest (1969) for the rate
of synthesis of collagen in diploid vs. tetraploid rat cells.
On the other hand, Ohno (1970) proposed that with increased
ploidy the regulator genes act in greater than linear fashion
so that, "it is likely that all the regulated structural genes
are subjected to superrepression in a newly arisen tetraploid".
Effects attributable to increased repression or diminished
dosage effect at higher ploidy or gene dosage levels have
been reported in hypotetraploid mouse myeloma cell lines
(Cohn, 1967), *Saccharomyces* tetraploids (Nelson and Douglas,
1963), tetraploid amphibian species (Becak, 1969), and higher
ploidy *Triticale* (Mitra and Bhatia, 1971).

In the *Datura* tetraploid, an isogenic and chromosomally
"balanced" type, we find no change from the diploid in peroxidase activity. A genetic interpretation consistent with the
results would be that the enhanced activity due to an increase

613

in dosage of the structural gene(s) (Table 2, Band 19, 150%)
is balanced by an increased production of repressor(s) of gene
activity, due, we speculate, to an increased dosage of "regulatory genes". The latter appear to be repeated in a number
of different chromosomes (Table 2, 67%) an observation that
could conceivably relate to the proposed regulatory function
of repetitive DNA in the genome of eukaryotes (Britten and
Davidson, 1969, 1971; Britten, 1972). A striking feature
of the system is that the various chromosomally situated
factors, which individually have different dosage effects on
the amount of enzyme activity, when double *in toto* (tetraploid),
are so integrated as to give the same level of activity as in
the diploid.

There is a paucity of evidence in erkaryotic systems
for regulatory genes that control the production of enzymes
by structural genes. Shows and Ruddle (1968) presented
evidence for a regulatory gene in the mouse that controls the
presence and absence of B subunits of lactate dehydrogenase
in erythrocytes. In the Japanese quail one of the alleles
at a gene locus coding for liver alcohol dehydrogenase is
associated with a capability of suppressing the wild type
chicken allele in quail-chicken hybrids (Castro-Sierra and
Ohno, 1968). Nishikawa and Nobuhara (1971) concluded from
studies on alpha-amylase isozymes of wheat that, "Presence
or absence of a given isozyme band is not necessarily due to
change of a particular structural gene, but sometimes depends
on the regulator system." Regulatory loci controlling isozymes of esterases (Macdonald, 1969) and root peroxidases
(Macdonald and Smith, 1972) have been suggested for maize and
wheat, respectively.

Model systems for gene regulation in eukaryote differentiation (Britten and Davidson, 1969, 1971; Schwartz, 1971)
have tended to stress positive control (gene activation) for
specific sites that are generally held in a state of nonspecific repression due to association of DNA with histones. Our
findings, along with similar results cited, lead to the speculation that in eukaryote cells with an increase in number of
structural genes there is a concomitant increase in amount of
transcription which is reflected in a higher level of enzyme
activity shown by a greater intensity of a specific isozyme
band. In addition, we find indications of negative controls
(gene repression) in that an increase in gene dosage, through
trisomy for several chromosomes added singly, is accompanied
by decreased intensity in one or more specific isozyme bands.
In the diploid genotype these controls are "balanced" to
produce the species phenotype. In genetically "unbalanced"
trisomic types, we speculate that extra amounts of structural

or regulatory gene products are produced, the isozyme patterns thereby become altered, and the aberrant phenotypes result.

Although this interpretation in terms of gene action is consistent with the data, and there is a clear association of isozyme band pattern with genotype, nevertheless other explanations are possible. One of these is that the addition of the extra chromosome imposes a "stress" condition which is reflected in the altered zymogram. Genetic tests are required to gain further insight into the problem.

ACKNOWLEDGEMENTS

We thank Ms. Alexandra Jahn for extensive technical assistance in the preparation and scoring of zymograms;   Mr. Keith Thompson for the statistical analyses;   Mrs. Jeanne Wysocki for determining the relation between color density and peroxidase activity levels;   Mr. Kenneth Grist for making the densitometer tracings;   and Drs. Peter Carlson, T.-Y. Cheng, Iris Mastrangelo, H. James Price and John G. Scandalios for helpful comments and discussions.

Research was carried out at Brookhaven National Laboratory under the auspices of the U.S. Atomic Energy Commission.

ADDENDUM

Since the increase in cell volume from 2n to 4n *D. stramonium* is at least two-fold (Sinnott, et al., 1934) the cell number per unit fresh weight would be correspondingly fewer and the isoperoxidase activities per cell correspondingly greater (De Maggio and Lambrukos, 1974). On a cell basis, the computed activity for each isoperoxidase was found to be proportional to gene dosage, a consistency which implies the same gene-for-gene dose effect for both the structural and the postulated regulatory genes in diploids and tetraploids.

REFERENCES

Avery, A. G., S. Satina, and J. Rietsema 1959. *Blakeslee; The Genus Datura,* Ronald Press, New York.

Beçak, W. 1969. Genic action and polymorphism in polyploid species of amphibians. *Genetics* S61: 183-190.

Benson, P. F. and M. deJong 1968. Leucocyte-hexokinase isoenzymes in Down's syndrome. *Lancet* 2: 197-198.

Blakeslee, A. F. 1922. Variations in *Datura* due to changes in chromosome number. *Amer. Naturalist* 56: 16-31.

Blakeslee, A. F. 1930. Extra chromosomes, a source of variations in the Jimson weed. *Smithsonian Inst. Ann. Rpt.*
Publ. 3077: 431-450.

Blakeslee, A. F., J. Belling, and M. E. Farnham 1920. Chromosomal duplication and Mendelian phenomena in *Datura*
mutants. *Science* 52: 388-390.

Blakeslee, A. F. and R. E. Cleland 1930. Circle formation in
Datura and *Oenothera*. *Proc. Nat. Acad. Sci. U.S.A.*
16: 177-183.

Brewer, G. J., C. F. Sing, and E. R. Sears 1969. Studies of
isozyme patterns in nullisomic-tetrasomic combinations
of hexaploid wheat. *Proc. Nat. Acad. Sci. U.S.A.* 64:
1224-1229.

Britten, R. J. 1972. DNA sequence interspersion and a speculation about evolution. *Brookhaven Symp. Biol.* 23: 80-94.

Britten, R. J. and E. H. Davidson 1969. Gene regulation for
higher cells: a theory. *Science* 165: 349-357.

Britten, R. J. and E. H. Davidson 1971. Repetitive and non-
repetitive DNA sequences and a speculation on the origins
of evolutionary novelty. *Quart. Rev. Biol.* 46: 111-138.

Carlson, P. S. 1972. Locating genetic loci with aneuploids.
Molec. Gen. Genetics 114: 273-280.

Castro-Sierra, E. and S. Ohno 1968. Allelic inhibition of
the autosomally inherited gene locus for liver alcohol
dehydrogenase in chicken-quail hybrids. *Biochem. Genetics*
1: 323-335.

Cohn, M. 1967. The natural history of myeloma. *Cold Spring
Harbor Symp. Quant. Biol.* 32: 211-221.

Conklin, M. E. and H. H. Smith 1971. Peroxidase isozymes: a
measure of molecular variation in ten herbaceous species
of *Datura*. *Amer. J. Botany* 58: 688-696.

Datura Investigations 1916. *Carnegie Inst. Wash. Yearbook*
15: 131.

De Maggio, A. E. and J. Lambrukos 1974. Polyploidy and gene
dosage effects on peroxidase activity in ferns.
Biochem. Genetics (in press).

Epstein, C. J. 1969. Mammalian oocytes: X chromosome activity.
Science 163: 1078-1079.

Epstein, C. J. 1972. Expression of the mammalian X chromosome before and after fertilization. *Science* 175:
1467-1468.

Grell, E. H. 1962. The dose effect of $ma\text{-}1^+$ and ry^+ on xanthine dehydrogenase activity in *Drosophila melanogaster*.
Z. Vererbungsl. 93: 371-377.

Hoess, R. H., H. H. Smith, and C. P. Stowell 1974. A genetic
analysis of peroxidase isozymes in two species of *Nicotiana*. *Biochem. Genetics* 11: 319-323.

Hsia, D. Y.-Y., P. Justice, G. F. Smith, and R. M. Dowben 1971. Down's syndrome. A critical review of the biochemical and immunological data. *Amer. J. Diseases Children* 121: 153-161.

Khush, G. S. 1973. *Cytogenetics of Aneuploids*, Academic Press, New York.

Layzer, R. B. and C. J. Epstein 1972. Phosphofructokinase and chromosome 21. *Amer. J. Human Genetics* 24: 533-543.

Macdonald, T. 1969. Esterase isozymes: new loci. *Maize Genet. Coop. Newsletter* 43: 31-34.

Macdonald, T. and H. H. Smith 1972. Variation associated with an *Aegilops umbellulata* chromosome segment incorporated in wheat. II. Peroxidase and leucine aminopeptidase isozymes. *Genetics* 72: 77-86.

McDaniel, R. G. and R. T. Ramage 1970. Genetics of a primary trisomic series in barley: identification by protein electrophoresis. *Canad. J. Genet. Cytol.* 12: 490-495.

Mitra, R. and C. R. Bhatia 1971. Isoenzymes and polyploidy I. Qualitative and quantitative isoenzyme studies in the Triticinae. *Genet. Res.* 18: 57-69.

Monteleone, P. L., H. L. Nadler, C.-S. Pi, and D. Y.-Y. Hsia 1967. Isoenzymes in Down's syndrome. *Lancet* 2: 367-368.

Nelson, N. M. and H. C. Douglas 1963. Gene dosage and galactose utilization by Saccharomyces tetraploids. *Genetics* 48: 1585-1591.

Nielsen, G. and O. Frydenberg 1971. Chromosome localization of the esterase loci *Est-1* and *Est-2* in barley by means of trisomics. *Hereditas* 67: 152-154.

Nishikawa, K. and M. Nobuhara 1971. Genetic studies of alpha-amylase isozymes in wheat. I. Location of genes and variations in tetra- and hexaploid wheat. *Japan. J. Genetics* 46: 345-353.

Ohno, S. 1970. *Evolution by Gene Duplication*, Springer-Verlag, Berlin.

Priest, R. E. and J. H. Priest 1969. Diploid and tetraploid clonal cells in culture: gene ploidy and synthesis of collagen. *Biochem. Genetics* 3: 371-382.

Puck, T. T. and F.-T. Kao 1968. Genetics of somatic mammalian cells. VI. Use of an antimetabolite in analysis of gene multiplicity. *Proc. Nat. Acad. Sci. USA* 60: 561-568.

Satina, S., A. D. Bergner, and A. F. Blakeslee 1941. Morphological differentiation in chromosomes of *Datura stramonium*. *Amer. J. Botany* 28: 383-390.

Schwartz, D. 1971. Genetic control of alcohol dehydrogenase--a competition model for regulation of gene action. *Genetics* 67: 411-425.

Shows, T. B. and F. H. Ruddle 1968. Function of the lactate dehydrogenase B gene in mouse erythrocytes: evidence for control by a regulatory gene. *Proc. Nat. Acad. Sci. USA* 61: 574-581.

Sinnott, E. W., H. Houghtaling, and A. F. Blakeslee 1934. The comparative anatomy of extra-chromosomal types in *Datura stramonium*. Carnegie Inst. Wash., Publ. 451.

Stewart, B. R. and J. R. Merriam 1974. Segmental aneuploidy and enzyme activity as a method for cytogenetic localization in *Drosophila melanogaster*. *Genetics* 76: 301-309.

Westerveld, A., R. P. L. S. Visser, M. A. Freeke, and D. Bootsma 1972. Evidence for linkage of 3-phosphoglycerate kinase, hypoxanthine-guanine-phosphoribosyl transferase, and glucose-6-phosphate dehydrogenase loci in Chinese hamster cells studied by using a relationship between gene multiplicity and enzyme activity. *Biochem. Genetics.* 7: 33-40.

GENETICS, EXPRESSION, AND CHARACTERIZATION OF ISOZYMES IN SOMATIC CELL HYBRIDS

THOMAS B. SHOWS
Roswell Park Memorial Institute
New York State Department of Health
Buffalo, New York 14203

ABSTRACT. Thirty-eight enzymes were examined by gel electrophoresis in proliferating man-rodent somatic cell hybrids. Since chromosomes and enzymes from both species are present and functional in these cell hybrids and since differences occur in the electrophoretic mobility of enzymes between species, cell hybrids are particularly suited for a study of the genetics, expression, and structural characteristics of mammaliam enzymes. Chromosome loss observed in interspecific hybrids offers the possibility of dissecting complex systems and isolating the individual components. Gene assignment is based on concordant loss of an enzyme and a specific chromosome. Following this reasoning, a human gene map of genes coding for enzymes has been compiled and is presented. Seventeen of the 24 human chromosomes have enzyme markers assigned to them. Assignment of the gene coding for human esterase D on chromosome 13 or 16 is presented and utilizing an (Xq-;9p+) translocation, glucose-6-phosphate dehydrogenase, hypoxanthine-guanine phosphoribosyl transferase, and phosphoglycerate kinase have been assigned to the long arm of the human X chromosome. The presence of two different mammalian genomes in the same nucleus can influence the expression of a gene as demonstrated by the activation of mouse esterase-2 and the extinction of mouse acid phosphatase in cell hybrids. Aspects of enzyme structure are presented such as the subunit structure of 29 enzymes, the phenotypic change of human adenosine deaminase in cell hybrids, a structural relationship of human hexosaminidase A and B, and the complementation of human enzyme deficiencies.

INTRODUCTION

Aspects of the genetics, expression, and structural characteristics of enzymes can be uniquely studied in man-mouse somatic cell hybrids. This is possible by combining procedures for identifying enzymes and fusing somatic cells, usually from different species. After cell fusion with inactive Sendai virus, it is possible to isolate clones with a single nucleus composed of chromosomes of both parental cells (Ephrussi and Weiss, 1965). In man-mouse cell hybrids there is a progressive loss of human but not mouse chromosomes (Weiss and Green, 1967) which makes it possible to dissect the human genome and its complex systems such as the multiple forms of an enzyme. Significantly, this loss of human chromosomes makes it feasible to map human genes coding for enzymes by analyzing the joint segregation of enzyme markers and chromosomes in man-mouse cell hybrids with different numbers and combinations of human chromosomes (cf. Ruddle, 1972).

The interaction of two genomes which are under different regulatory controls and which function in a single hybrid nucleus has stimulated the study of gene expression in cell hybrids (Davidson, 1974). Enzymes have been observed to be activated or extinguished in cell hybrids as a result of genetic information influencing the genes coding for enzymes rather than mutations to the genes themselves.

Since electrophoretic differences often occur between homologous enzymes of different species, enzymes and their subunits from each species are easily recognized in interspecific cell hybrids. This permits the study of an enzyme's subunit structure by the formation of hybrid enzymes, the study of secondary structural modifications, and the complementation and characterization of enzyme deficiencies.

These features of somatic cell hybrids are of extraordinary importance for characterizing an enzyme and they will be utilized to describe the genetics, expression, and structural characteristics of selected enzymes of the 38 enzymes we have analyzed in cell hybrids. A human gene map of enzyme markers will be presented together with the probable gene assignment of esterase D (Es-D); the regional mapping of sex linked genes; the activation of an esterase; the extinction of an acid phosphatase; the phenotypic change of adenosine deaminase; the hexosaminidase A and B association; the subunit structure of enzymes; and the complementation of enzyme deficiencies.

MATERIALS AND METHODS

The procedures we employ for somatic cell fusion with inactivated Sendai virus, isolation of man-mouse somatic cell hybrids in HAT selection medium, and growth of cells have been reported (Shows, 1972). The human parental cell lines include the diploid fibroblasts WI-38, TS-408, and SH-421 and cell lines JoVa, AnLy, and CaVa each with a balanced translocation. These parental cells have been described (Lalley et al., 1974). The mouse parental cell lines include A9, RAG, LM/TK⁻, and LTP and have been described (Littlefield, 1966; Shows, 1972; Lalley et al., 1974). These cell lines do not

survive in the HAT selection system (Littlefield, 1964). Somatic cell hybrid combinations of mouse-mouse, human-human,
mouse-rat, and human-Chinese hamster have been described
(Shows et al., 1972; Glaser et al., 1973; Shows and Lalley,
1974; Sun et al., 1974). Cell hybrids are grown to confluency
and harvested at $0.70 - 1.0 \times 10^8$ cells/ml for enzyme analysis
as described (Shows, 1972).

The enzymes that were analyzed in cell hybrids are listed
in Table 3 and the starch-gel electrophoretic conditions and
specific histochemical staining are reported by Shows and
Bias (1974). With reference to this discussion, adenine phosphoribosyl transferase was determined by the procedure of Mowbray et al. (1972); esterase D after Hopkinson et al. (1973);
glucose-6-phosphate dehydrogenase, phosphoglycerate kinase,
and hypoxanthine-guanine phosphoribosyl transferase after
Shows (1974), and Shows and Brown (1974); esterase-2 after
Shows (1972); pyruvate kinase after Shows (1974); hexosaminidase and mannosephosphate isomerase after Lalley et al. (1974);
and adenosine deaminase after Edwards et al. (1971) in the
electrophoretic system reported by Shows and Ruddle (1968).

RESULTS AND DISCUSSION

GENETICS OF ENZYMES IN CELL HYBRIDS

When enzyme markers and specific chromosomes are present
or absent together in man-mouse cell hybrids, genes coding for
enzymes can be assigned to specific chromosomes. This strate-
gy has led to the construction of a somatic cell genetic map
of the human genome. A map featuring the chromosomal assignment of genes coding for 44 enzymes is presented in Table 1.
Enzyme markers have now been assigned to 1 of the 24 human
chromosomes (cf. New Haven Conference, 1973). A probable
gene assignment is described here for esterase D (Es-D).
Regional mapping, utilizing chromosome rearrangements, makes
it possible to assign genes to regions of chromosomes and thi
this is described for three sex linked genes.

The Es-D Gene Assignment. Adenine phosphoribosyl transferase
(APRT) was assigned to chromosome 16 in man-mouse hybrids by
Tischfield et al. (1974) using mouse L-cells (A9) deficient
in APRT. In these hybrids only human APRT was expressed when
chromosome 16 was present (Figure 1). When a mouse parental
line (RAG) which expresses APRT is hybridized with human cells,
both parental enzymes and a single heteropolymer are expressed
in hybrid cell homogenates (Fig.1). This indicates that APRT

TABLE 1
THE HUMAN GENE MAP OF ENZYME MARKERS
Determined by Somatic Cell Hybridization

Chromosome	Enzyme Marker
1	AK-2, FH, Guk, Pep-C, PGD, PGM-1, PPH, UGPP
2	ACP-1, Gt, IDH-1, MDH-1
3	No assignment
4 or 5	Ade$^+$B, Esterase activator
5	HexB
6	ME-1, PGM-3, SOD-2
7	MDH-2
8	No assignment
9	No assignment
10	GOT-1, HK-1*
11	LDH-A, Es-A$_4$, ACP-2
12	LDH-B, CS, Pep-B, Gly$^+$A, TPI
13	Es-D?
14	NP
15	MPI, PK-3, HexA
16	APRT, Es-D?
17	TK
18	Pep-A
19	PHI
20	ADA
21	SOD-1
22	No assignment
Y	No assignment
X	PRT, G6PD, PGK, α-Gal, TATr

Table 1. The enzymes, their symbols, and procedures for analysis have been described by Shows and Bias (1974). Details of somatic cell hybridization and the assignment of enzyme markers to specific chromosomes are summarized in the New Haven Conference (1973): First International Workshop on Human Gene Mapping
*Hexokinase in human fibroblasts has been identified as HK-1 observed in mammalian tissues and tissue culture cells (Katzen et al., 1965; Schimke and Grossbard, 1968).

is a dimer.

When APRT was present or absent in cell hybrids another enzyme, esterase D (Figure 2), was found to segregate concordantly (Table 2). The concordant loss of Es-D and APRT in cell hybrids suggested gene linkage. Preliminary chromosome

Adenine Phosphoribosyl Transferase

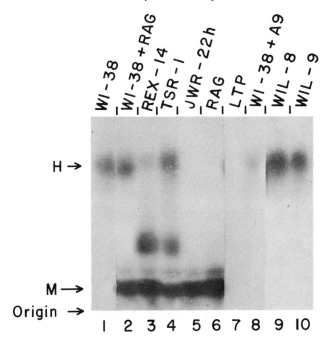

Figure 1. Starch-gel electrophoresis of adenine phosphoribosyl transferase in cell homogenates of human, mouse, and somatic hybrid cells. 1. Human diploid WI-38 fibroblasts; 2. Homogenate mixture of WI-38 and mouse RAG cells; 3. Cell hybrid of human lymphocytes (CaVa) and RAG cells; 4. Cell hybrid of human fibroblast (TS-408) and RAG; 5. Cell hybrid of human lymphocytes (JoVa) and RAG; 6. Mouse RAG(APRT+); 7. Mouse L-cell LTP(APRT−); 8. Homogenate mixture of WI-38 and mouse A9 (APRT−) L-cell; 9 and 10. Hybrids of WI-38 and LTP. Human APRT is expressed in cell hybrids in 3, 4, 9, and 10. However, an APRT hybrid enzyme is only expressed only in 3 and 4.

studies have indicated that APRT and Es-D segregated concordantly with chromosome 16 but chromosome 13 could not be excluded (Shows and Brown, in press). Exceptions in the segregation of APRT and Es-D occur (10%) which could indicate chromosome breakage or, in fact, non-linkage. If genes for APRT and Es-D are not linked then the Es-D gene must be located on 13.

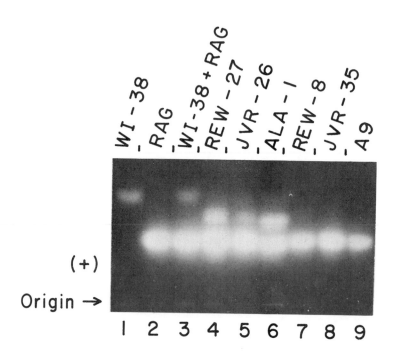

Figure 2. Starch-gel electrophoresis of esterase D cell homogenates of human, mouse, and hybrid cells. 1. Human WI-38;
2. Mouse RAG; 3. Homogenate mixture of WI-38 and RAG cells;
4 and 7. WI-38 human x mouse RAG cell hybrids; 5 and 8.
JoVa human fibroblasts x RAG hybrids; 6. AnLy human fibroblasts x mouse A9 hybrid; 9. Mouse A9 cells. An Es-D hybrid
enzyme was present in hybrids in 4, 5, and 6. Human Es-D was
only faintly expressed in these hybrids.

A single hybrid esterase D (Fig. 2) was observed in positive
hybrids suggesting a dimeric structure which confirms the observation of Hopkinson et al. (1974) in human heterozygotes
polymorphic for Es-D variants.

Regional mapping of the X chromosome. Although assignment of
a gene to a specific chromosome location is not yet possible,
assignment to a region of a chromosome is by employing

TABLE 2

Segregation of Adenine Phosphoribosyl Transferase
and Esterase D in Somatic Cell Hybrids

Hybrids Sets	APRT$^+$/Es-D$^+$	APRT$^+$/Es-D$^-$	APRT$^-$/Es-D$^+$	APRT$^-$/Es-D$^-$
WIL(WI-38 x LTP)	4	1	0	8
REW(WI-38 x RAG)	8	1	2	5
JVR(JoVa x RAG)	7	0	1	3
TSR(TS-408 x RAG)	6	0	1	0
TSA(TS-408 x A9)	0	0	0	2
REX(CaVa x RAG)	28	1	2	2
	53	3	6	20 /82

Table 2. Independent hybrid clones from each hybrid set were
scored for the presence (+) or absence (-) of APRT and Es-D.
The hybrids were derived from four different human lines.

naturally occurring or induced chromosome rearrangements.

We have employed an inherited X/9 balanced translocation
expressed in fibroblasts to regionally map the X-linked enzymes hypoxanthine-guanine phosphoribosyl transferase (HPRT),
glucose-6-phosphate dehydrogenase (G6PD), and phosphoglycerate
kinase (PGK) (Shows, 1974; Shows and Brown, 1974). Almost all
the long arm (q) of the X chromosome has been translocated to
the short arm (p) of chromosome 9. Those X-linked genes coding
for the enzymes that segregate concordantly with the (Xq-;9p+)
translocation must be encoded on the long arm (q) of the X
chromosome. The enzyme phenotypes which are present in cell
hybrids which possess the remaining X fragment must be coded
by genes located on the Xq-. Analyzing man-mouse cell hybrids
derived from human fibroblasts possessing the X/9 translocation,
it was shown that HPRT, G6PD, PGK, and the X/9 translocation
segregated concordantly indicating that genes coding for HPRT,
G6PD, and PGK are located on the long arm of the X chromosome
(Shows and Brown, 1974). This confirmed a report by Ricuitti
and Ruddle (1973) employing an X/14 translocation.

Several X-autosome translocations involving different
lengths of the X chromosome have been employed in hybrid
studies for determining the gene order of X-linked enzymes (cf.
Gerald and Brown, 1974) and the results predict that the order
of genes on the long arm of the X chromosome from the centro-
mere to the distal end is PGK, HPRT, and G6PD.

EXPRESSION OF ENZYMES IN CELL HYBRIDS

The presence of two different mammaliam genomes in the
same nucleus of proliferating somatic cell hybrids has proved
important for understanding the expression of a gene when influenced by different genetic information (Davidson, 1974;
Shows and Lalley, 1974). Homologous enzymes that are expressed
in both parental tissue culture cells before cell hybridization
tend to be expressed in the cell hybrids. However, the interaction of information from two genomes, one of which expresses
a differentiated function, has resulted in the activation of
certain enzymes and proteins and the extinction of others in
cell hybrids.

Esterase-2 activation. Esterase-2 (Es-2) is expressed in
certain mouse tissues and the gene coding for the phenotype
has been assigned to linkage group 18 in the mouse (Ruddle et
al., 1969). In man-mouse cell hybrids Klebe et al. (1970)
described the apparent existence of a human regulator gene
capable of extinguishing mouse Es-2 activity. In this work
the mouse parent was Es-2 positive. In mouse L-cell (Es-2
negative) x human hybrids, we observed the activation of
murine Es-2. Cloned mouse LTP cells do not express Es-2 even
when examined at a concentration of 3×10^8 cells/ml which is
three times greater than observed in Fig. 3 (channel 6). In
three of 13 primary hybrid clones (LTP x WI-38), an esterase
migrating to the same position as Es-2 was expressed (Fig. 3).
The esterase does not correspond to a human esterase in mobility. If the activation of this mouse esterase is influenced
by human genetic information, then mouse Es-2 should disappear
as human chromosomes are lost in man-mouse hybrids. Fifteen
subclones from two Es-2 positive primary clones were tested.
Eight subclones were positive while 7 subclones were in fact
negative for mouse Es-2, demonstrating the genetic nature of
this extinction. It is therefore concluded that the activation of mouse Es-2 is influenced by the human genome. It
should be noted that when cells were grown in fetal calf serum,
Es-2 disappeared after several passages. The loss of Es-2 was
not observed when gamma globulin free newborn calf serum
(Grand Island Biological Co.) was employed for growth. The
experiments described here utilized GG-free serum in the
growth medium. Kao and Puck (1972) have described a human
activator of a Chinese hamster esterase and the gene was
tentatively assigned to chromosome 4. It is not known if the
esterase activator we describe is related.

Acid phosphatase extinction. Lysosomal acid phosphatase activity in human and mouse cells was separated into multiple zones
by starch gel electrophoresis. The cathodal enzyme of the two

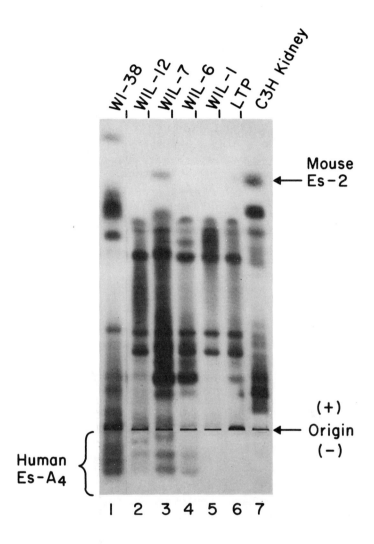

Fig. 3. Esterase zymogram of human, mouse, and hybrid cell extracts employing α-naphthyl acetate as substrate. 1. WI-38 human fibroblast; 2-5. WI-38 x mouse LTP cell hybrid homogenates; 6. Mouse LTP; 7. Mouse (C3H/J kidney) strain from which LTP originated. Es-2 was activated in WIL-7 in channel 3. Human Es-A$_4$ segregates independently of the activation of esterase-2.

major zones of activity in the mouse was apparently extinguished in proliferating man-mouse somatic cell hybrids (Shows and Lalley, 1974). If the absence of this mouse acid phosphatase was caused by a specific human gene product, then the presence or absence of the mouse acid phosphatase should segregate in hybrid clones since human, but not mouse, chromosomes are lost in man-mouse hybrids. This was, in fact, observed suggesting that the absence of the mouse lysosomal acid phosphatase (homologous to human ACP_2) was influenced by the human genome. The structural gene coding for human acid phosphatase$_2$ was shown to be unlinked to the presumed human component which extinguished the mouse acid phosphatase. The mechanism of extinction is postulated to be a modification in the processing of the mouse lysosomal enzyme. A dimeric structure was suggested for acid phosphatase$_2$ of man, mouse, and rat since a single hybrid enzyme was expressed in man-mouse and mouse-rat somatic cell hybrids.

STRUCTURAL CHARACTERIZATION OF ENZYMES IN CELL HYBRIDS

Subunit structure. The subunit structure of an enzyme can be determined in cell hybrids by the formation of heteropolymers if the rodent and human enzymes have different electrophoretic mobilities. No heteropolymer in a cell hybrid would indicate a monomeric structure, while a single heteropolymer and the parental forms predict a dimeric structure. Two heteropolymers would predict a trimer, and three heteropolymers a tetramer.

Table 3 lists 29 enzymes that we have tested in man-mouse, man-Chinese hamster, mouse-mouse, human-human, mouse-rat, and Muntjac deer-mouse somatic cell hybrids. Their subunit structure was predicted by the above criteria. It should be noted that heteropolymers have been observed for all enzymes tested which are known to have a polymeric structure. Species differences have not been a barrier in heteropolymer formation although in the case of APRT the heteropolymer does not migrate intermediate to the two parental types (Fig. 1).

Adenosine deaminase phenotypic change in cell hybrids. The phenotype of human adenosine deaminase (ADA) in cell hybrids suggests that the electrophoretic mobility of an enzyme can be changed in the environment of a hybrid cell (Fig. 4). Human red blood cells possess only ADA 1 while tissues express red cell ADA in addition to other molecular forms (Fig. 4, channel 2). There is biochemical and genetic evidence to sug-

628

TABLE 3
SUBUNIT STRUCTURE OF ENZYMES AS DETERMINED
IN SOMATIC CELL HYBRIDS

Monomer	Dimer
Acid phosphatase-1 E.C. 3.1.3.2	Acid phosphatase-2 E. C. 3.1.3.2
Adenylate kinase-1 E. C. 2.7.4.3	Adenine phosphoribosyl transferase E.C. 2.4.2.7
Adenylate kinase-2 E.C. 2.7.4.3	Enolase E.C. 4.2.1.11 Esterase D E.C. 3.1.1.-
Mannosephosphate isomerase E.C. 5.3.1.8	α-galactosidase E.C. 3.2.1.22 Glucose-6-phosphate dehydrogenase E.C. 1.1.1.49
Peptidase-B E.C. 3.4.3.-	Glutamic oxaloacetic transaminase E.C. 2.6.1.1 (cytoplasmic)
Peptidase-C E.C. 3.4.3.-	Galactose-1-phosphate uridyltransferase E.C. 2.7.7.10
Phosphoglucomutase-1 E.C. 2.7.5.1	Hypoxanthine-guanine phosphoribosyl E. C. 2.4.2.8 transferase
Phosphoglycerate kinase E.C. 2.7.2.3	Isocitrate dehydrogenase E.C. 1.1.1.42 (cytoplasmic)
	Malate dehydrogenase E.C. 1.1.1.37 (cytoplasmic)
	Peptidase-A E.C. 3.4.3.-
	6-Phosphogluconate dehydrogenase E.C. 1.1.1.43
	Phosphohexose isomerase E.C. 5.3.1.9
	Superoxide dismutase-1 E.C. 1.15.1.1

Trimer	Tetramer
Nuceoside phosphorylase E.C. 2.4.2.1	Lactate dehydrogenase E.C. 1.1.1.27 Malic enzyme E.C. 1.1.1.40 (cytoplasmic) Peptidase-S E.C. 3.4.3.- Pyruvate kinase-3 E.C. 2.7.1.40 Superoxide dismutase-2 E.C. 1.15.1.1

Table 3. Electrophoretic procedures for testing these enzymes have been referenced in Materials and Methods and by Shows and Bias (1974).

gest that perhaps all forms of ADA are coded by a single gene. For example, absence of all forms of ADA has been associated

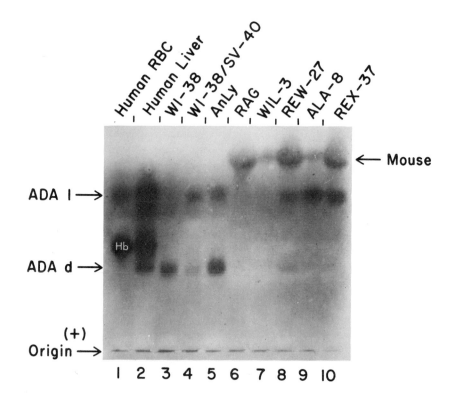

Fig. 4.   Adenosine deaminase starch-gel electrophoretic patterns of human, mouse, and hybrid cell homogenates. 1. Human
red blood cells containing hemoglobin (Hb) and ADA 1; 2.
Human liver expressing ADA 1 and ADA d; 3. Human WI-38 fibroblast line expressing ADA d; 4. SV40 transformed WI-38 expressing ADA 1 and ADA d faintly; 5. Human AnLy fibroblast
line which expresses ADA 1 and d; 6. Mouse heteroploid parental line, RAG; 7. WI-38 x L-cell (LTP) cell hybrid negative for human ADA; 8. WI-38 x RAG cell hybrid that is ADA
1 positive and weakly positive for ADA d; 9. ADA 1 positive
AnLy x L-cell (A9) cell hybrid; 10. ADA 1 positive human
lymphocyte x RAG cell hybrid.

with combined immunodeficiency disease (Hirshhorn et al.,1973),
and the multiple forms of ADA can be interconverted (Nishihara
et al., 1973).

Human fibroblasts present three phenotypes of ADA (Fig. 4);
a cathodal form (ADA d); both ADA d and ADA 1; and only ADA 1

which may relate to culture conditions (Edwards et al., 1971).
When WI-38 (Fig. 4, channel 3) expressing only ADA d is hybridized with RAG the majority of hybrids express ADA l, and in
some, both ADA l and ADA d are present but only a very few
express the ADA d phenotype of the human parental cell. This
is the same result of LTP x WI-38 cell hybrids. Curiously,
when WI-38 expressing only ADA d is transformed with SV40
virus, both forms of ADA are present. When AnLy expressing
ADA l and d are hybridized with L-cells, only ADA l is expressed. However, in RAG x AnLy hybrids some express only ADA
d. In human lymphocytes (which express both ADA's) and RAG
hybrids, the ADA l is observed predominantly.

Apparently, in the presence of the cell hybrid environment
the large 280,000 molecular weight ADA d component is reversibly converted into the smaller ADA l 35,000 molecular weight
component. It is possible that this conversion is coded for
by a human gene segregating in cell hybrids. This phenotypic
interconversion supports the findings that both isozymes are
coded by the same gene.

Dissection of complex enzyme systems. If an enzyme is composed of subunits coded by different genes, secondarily
modified by independent genes, or controlled by a regulator
gene, the loss of human chromosomes in man-rodent hybrids
makes it feasible to dissect complex enzyme systems by removing genes which individually contribute to the phenotype
of a structural gene. Hexosaminidase (Hex) is a lysosomal
enzyme composed of two major molecular forms, HexA and HexB.
Hexosaminidase A is deficient in Tay Sachs disease and HexA
and HexB are deficient in Sandhoff-Jatzkewitz disease. Biochemical, immunological, and genetic data indicate that the
two hexosaminidase forms are structurally related, that
several genes may be involved in the phenotypic expression of
each form, and that the two diseases are inherited as autosomal recessive disorders (cf. Lalley et al., 1974).

We (Lalley et al., 1974) have evidence to suggest that in
man-mouse somatic cell hybrids HexA expression is dependent
on HexB expression. In 60 primary clones, 34 expressed HexA
and HexB; 12 were negative for both markers; and 14 clones
were negative for HexA but positive for HexB. Hybrids were
never observed that were positive for HexA but negative for
HexB. This distribution would indicate that HexA expression
is dependent on the presence of the HexB gene product. This
finding is consistent with an explanation for the pattern of
hexosaminidase deficiency in Tay Sachs disease and Sandhoff-
Jatzkewitz disease and would indicate that a common subunit
is shared by HexA and HexB.

HexA expression was shown to be linked to MPI in cell hybrids (Lalley et al., 1974; Gilbert et al., 1974). Therefore if HexA is dependent on HexB, then hybrids should be found that are HexA negative and HexB negative but which are MPI positive, indicating that the gene for HexA expression is present but not expressed. Two such hybrids have been found among the 12 hybrids negative for HexA and HexB.

This pattern of HexA and HexB segregation could of course change as more hybrids are examined. Other reports employing gel electrophoresis to score hexosaminidase in man-mouse and man-Chinese hamster hybrids have obtained segregation patterns which indicate the independence of HexA and HexB (van Someren et al., 1973; Gilbert et al., 1974). We have evidence that the expression of hexosaminidase can be altered in hybrids making it difficult to score by gel electrophoresis (Rattazzi et al., 1973). Our results are compiled from starch, agar, and cellogel electrophoresis and from immunodiffusion and immunoelectrophoresis with anti-human hexosaminidase antibodies and specific anti-human HexA antibodies.

Enzyme complementation. Since hexosaminidase A (HexA) is deficient in Tay Sachs disease (TSD) and both HexA and HexB are deficient in Sandhoff-Jatzkewitz disease (SJD), fibroblasts from patients with the diseases were fused to determine if the deficiencies could complement each other. The appearance of HexA in cell hybrids would indicate that the mutation in TSD is different than in SJD. HexA was observed after fusion of deficient TSD and SJD fibroblasts with inactive Sendai virus (Rattazzi et al., submitted). Although a selection system was not available to select TSD-SJD cell hybrids from the deficient parental fibroblasts, a combination of cellogel electrophoresis in three buffer systems, immunoelectrophoresis employing a specific anti-human HexA antibody, and heat sensitivity of HexA demonstrated that HexA was present in cell homogenates containing proliferating cell hybrids and deficient parental cells. Control experiments demonstrate that HexA was absent in (a) TSD-TSD and SJD-SJD cell fusions; (b) homogenates of parental cells that had been grown together and (c) just after fusion before cell division. Heterokaryons of TSD and SJD cells occurred after fusion judged by autoradiography of ^{3}H thymidine labeled TSD nuclei and unlabeled SJD nuclei in the same cytoplasm. These findings indicate that the two diseases are the result of different mutations which affect the same enzyme activity.

Enzyme deficiencies. Cell hybrids are also important for structurally characterizing enzyme deficiencies as for example

the adenine phosphoribosyl transferase (APRT) (Fig. 1) and
dipeptidase-2 (Dip-2) (Shows et al., 1972) deficiencies observed in mouse L-cells (A9 and LM/TK⁻). In man-mouse hybrids
between APRT positive human and APRT deficient mouse cells,
only human APRT is observed (Fig. 1). In APRT positive human
and APRT positive mouse hybrids, both parental enzymes and a
single heteropolymer are expressed (Fig. 1) indicating that
APRT is a dimer. The absence of a mouse or a hybrid APRT in
L-cell x human hybrids suggests either the absence of a mouse
APRT polypeptide in L-cells, or the synthesis of a CRM positive
but defective polypeptide which may or may not form an inactive
hybrid enzyme. We have previously shown that the formation of
such a hybrid enzyme between normal human and deficient mouse
peptidase polymers is possible in hybrids between peptidase
deficient mouse and peptidase positive human cells (Shows et
al., 1972). The peptidase heteropolymer possessed activity
and demonstrated that the mouse cells deficient for Dip-2
activity continue to produce polypeptides which are capable
of forming a man-mouse hybrid peptidase.

CONCLUSIONS

The advantage of somatic cell hybrids is that two genomes
can be combined in a single cell making it possible to analyze
and dissect mammalian cells in ways similar to classical techniques for genetically analyzing microorganisms such as E. coli,
bacteriophage, and fungus.

Characterization of a mammalian enzyme at structural, regulatory, and genetic levels is an important feature of intra-
and inter-specific somatic cell hybrids. Genes coding for
enzymes can now be readily assigned to specific chromosomes
and regions of chromosomes. Studies on the genetic control
of enzyme expression in mammals is possible when cells expressing differentiated functions are hybridized. Important
features of enzyme structure such as number of subunits, formation of hybrid enzymes, secondary modifications of enzymes,
number of genes involved in an enzyme's structure, and complementation of enzyme deficiencies are uniquely studied in
somatic cell hybrids.

ACKNOWLEDGEMENTS

The excellent technical assistance of L. Haley, A. Goggin,
R. Eddy, and S. Baumgartel is acknowledged. This work was
supported by NIH grants HD 05196 and GM 20454 and NSF grant
GB-39273.

REFERENCES

Davidson, R. L. 1974. Control of expression of differentiated functions in somatic cell hybrids. In: *Somatic Cell Hybridization*, R. L. Davidson and F. de la Cruz eds., Raven Press, New York, pp. 131-146.

Edwards, Y. H., D. A. Hopkinson, and H. Harris 1971. Adenosine deaminase isozymes in human tissues. *Ann. Hum. Genet.* 35: 207-219.

Ephrussi, B. and M. C. Weiss 1965. Interspecific hybridization of somatic cells. *Proc. Natl. Acad. Sci.* 53: 1040.

Gerald, P. S. and J. A. Brown 1974. Report of the committee on the genetic constitution of the X chromosome. *Cytogenet. Cell Genet.* 13: 29-34.

Gilbert, F., R. Kucherlapati, M. J. Murnane, G. J. Darlington, R. Creagan, and F. H. Ruddle 1974. Assignment of a locus involved in the expression of hexosaminidase A to chromosome 7 in man. *Cytogenet. Cell Genet.* 13: 96-99.

Glaser, R., B. Decker, R. Farrugia, T. Shows, and F. Rapp 1973. Growth characteristics of Burkitt somatic cell hybrids in vitro. *Cancer Research* 33: 2026-2029.

Hirschhorn, R., V. Levytska, B. Pollars, and H. J. Meuwissen 1973. Evidence for control of several different tissue specific isozymes of adenosine deaminase by a single genetic locus. *Nature New Biology* 246: 200-202.

Hopkinson, D. A., M. A. Mestriner, J. Cortner, and H. Harris 1973. Esterase D: a new human polymorphism. *Ann. Hum. Genet.*, Lond. 37: 119-137.

Kao, F. -T. and T. T. Puck 1972. Genetics of somatic mammalian cells: demonstration of a human esterase activator gene linked to the Ade B gene. *Proc. Natl. Acad. Sci.* 69; 3273.

Katzen, H. M., D. D. Soderman, and H. M. Nitowsky 1965. Kinetic and electrophoretic evidence for multiple forms of glucose-ATP phosphotransferase activity from human cell cultures and rat liver. *Biochem. Biophys. Res. Commun.* 19: 377-382.

Klebe, R. J., T. -R. Chen, and F. H. Ruddle 1970. Mapping of a human genetic regulator element by somatic cell genetic analysis. *Proc. Natl. Acad. Sci.* U.S.A. 66: 1220-1227.

Lalley, P. A., M. C. Rattazzi, and T. B. Shows 1974. Human β-D-N acetylhexosaminidase A and B; expression and linkage relationships in somatic cell hybrids. *Proc. Natl. Acad. Sci.* U.S.A. 71: 1569-1573.

Littlefield, J. S. 1974. Selection of hybrids from matings of fibroblasts in vitro and their presumed recombinants. *Science* 145: 709-710.

Littlefield, J. W. 1966. The use of drug-resistant markers to study the hybridization of mouse fibroblasts. *Exp. Cell Res.* 41: 190-196.

Mowbray, S., B. Watson, and H. Harris 1972. A search for electrophoretic variants of human adenine phosphoribosyl transferase. *Ann. Hum. Genet.*, Lond. 36: 153-162.

New Haven Conference (1973): First International Workshop on Human Gene Mapping. *Cytogenet. and Cell Genet.* 13: 1-216 (1974).

Nishihara, H., S. Ishikawa, K. Shinkai, and H. Akedo 1973. Multiple forms of human adenosine deaminase II. Isolation and properties of a conversion factor from human lung. *Biochim. Biophys. Acta* 302: 429-442.

Rattazzi, M. C., P. A. Lalley, P. J. Carmody, and T. B. Shows 1973. Tay-Sachs disease: characterization of β-D-N acetylhexosaminidase in man-mouse somatic cell hybrids. *Amer. J. Hum. Genet.* 25: 63a.

Rattazzi, M. C., J. A. Brown, R. G. Davidson, and T. B. Shows Complementation of hexosaminidase A deficiency in somatic cell hybrids of Tay Sachs and Sandhoff disease fibroblasts. (Submitted for publication; also presented at this Conference.).

Ricuitti, F. C., and F. H. Ruddle 1973. Assignment of three gene loci (PGK, HGPRT, G6PD) to the long arm of the human X chromosome by somatic cell genetics. *Genetics* 74: 661-678.

Ruddle, F. H. 1972. Linkage analysis using somatic cell hybrids. *Advances in Human Genetics* 3: 173-226.

Ruddle, F. H., T. B. Shows, and T. H. Roderick 1969. Esterase genetics in *Mus musculus:* expression, linkage, and polymorphism of locus Es-2. *Genetics* 62: 393-399.

Schimke, R. T. and L. Grossbard 1968. Studies on isozymes of hexokinase in animal tissues. *Ann. New York Acad. Sci.* 151: 332-350.

Shows, T. B. 1972. Genetics of human-mouse somatic cell hybrids: linkage of human genes for lactate dehydrogenase-A and esterase-A$_4$. *Proc. Natl. Acad. Sci.* U.S.A. 69: 348-352.

Shows, T. B. 1974. Somatic cell genetics of enzyme markers associated with three human linkage groups. In: *Somatic Cell Hybridization*, R. L. Davidson and F. de la Cruz eds., Raven Press, New York, pp. 15-23.

Shows, T. B. and F. H. Ruddle 1968. Function of the lactate dehydrogenase B gene in mouse erythrocytes: evidence for control by a regulatory gene. *Proc. Natl. Acad. Sci.* 61: 574-581.

Shows, T. B., J. May, and L. Haley 1972. Human-mouse cell hybrids: a suggestion of structural mutation for dipeptidase-2 deficiency in mouse cells. *Science* 178: 58-60.

Shows, T. B. and W. B. Bias 1974. Gene markers for mapping the human genome. *Cytogenet. Cell Genet.* 13: 35-48.

Shows, T. B. and J. A. Brown 1974. An (Xq-;9p+) translocation suggests the assignment of G6PD, HPRT, and PGK to the long arm of the X chromosome in somatic cell hybrids. *Cytogenet. Cell Genet.* 13: 146-149.

Shows, T. B. and P. A. Lalley 1974. Control of lysosomal acid phosphatase expression in man-mouse cell hybrids. *Biochemical Genetics* 11: 121-139.

Shows, T. B. and J. A. Brown 1974. Segregation of esterase-D and adenine phosphoribosyl transferase in somatic cell hybrids: gene linkage or chromosome association. (in press press).

Someren, H. van and H. B. van Henegouwen 1973. Independent loss of human hexosaminidases A and B in man-Chinese hamster somatic cell hybrids. *Humangenetik* 18: 171-174.

Sun, N. C., C. C. Chang, and E. H. Y. Chu 1974. Chromosome assignment of the human gene for galactose-1-phosphate uridyltransferase. *Proc. Natl. Acad. Sci.* 71: 404-407.

Tischfield, J. A. and F. H. Ruddle 1974. Assignment of the gene for adenine phosphoribosyl transferase to human chromosome 16 by mouse-human somatic cell hybridization. *Proc. Natl. Acad. Sci.* U.S.A. 71: 45-49.

Weiss, M. C. and H. Green 1967. Human-mouse hybrid cell lines containing partial complements of human chromosomes and functioning human genes. *Proc. Natl. Acad. Sci.* U.S.A. 58: 1104.

GLUTAMATE OXALOACETATE TRANSAMINASE ISOZYMES OF *TRITICUM:*
EVIDENCE FOR MULTIPLE SYSTEMS OF TRIPLICATE STRUCTURAL
GENES IN HEXAPLOID WHEAT

GARY E. HART
Genetics Section, Plant Sciences Department
Texas A&M University
College Station, Texas 77843

ABSTRACT. The glutamate oxaloacetate transaminase (GOT)
zymogram phenotypes of 54 wheat strains, each possessing a
distinctive chromosomal constitution, were determined.
Three of the several genetically independent GOT systems
expressed were analyzed. The results obtained support the
hypothesis that the active GOT isozymes of each system are
dimers composed of the six possible combinations of subunits
coded by triplicate structural genes. Sets of triplicate
GOT structural genes were linked to the chromosome arms
3Aα, 3BL, and 3Dα, to 6Aα, 6BS, and 6Dα, and to 6Aβ, 6BL,
and 6Dβ. Genes involved in the production of GOT isozymes
were also located in the chromosomes of homoeologous group
7. Tissue specificity in the expression of the GOT systems
was observed. The results of this study suggest that genetic regulation of the expression of the GOT isozymes
occurs at the level of gene transcription and that the rate
of transcription is the same for each copy present of each
of the three possible members of the homoeologous sets of
GOT structural genes.

INTRODUCTION

The possession of duplicate gene loci on homoeologous
chromosomes confers upon polyploid organisms tolerance to a
wide variety of aneuploid conditions. This has allowed the
construction of a large number of aneuploid strains of diverse
chromosomal constitution in several polyploid species of
plants. A powerful method for the study of the basic genetics
of polyploids consists of the analysis of these strains with
the zymogram technique. To date this method has been used
extensively only in the study of hexaploid wheat (Barber et
al., 1968; Shepherd, 1968; Hart, 1970; Nishikawa and Nobuhara,
1971; and several reports in Sears and Sears, 1973).

This paper reports the results of a study of the genetic
control and subunit structure of the glutamate oxaloacetate
transaminase (GOT; E.C.2.6.1.1.) isozymes of *Triticum*, utilizing principally as experimental materials a compensating
nullisomic - tetrasomic series and several ditelosomic

637

strains of hexaploid wheat (*Triticum aestivum* L. cultivar Chinese Spring) developed by Sears (1966a, 1966b). This study is part of an ongoing investigation of genic, enzymic, and chromosomal evolution in *Triticum* (Hart, 1973). GOT catalyzes the reversible conversion of aspartate and α-ketoglutarate to oxaloacetate and glutamate.

MATERIALS AND METHODS

The 21 different chromosomes of hexaploid wheat have been classified into seven homoeologous (related) groups of three (Sears, 1954), based primarily on the finding that the deleterious effects of nullisomy for each chromosome of each genome can be reduced or eliminated by making tetrasomic a specific (homoeologous) chromosome belonging to either of the two other genomes (Sears, 1966a). There are 42 possible compensating nullisomic - tetrasomic combinations, six for each of the seven homoeologous groups. All, with two exceptions (nulli-2A tetra-2B and 2A-2D, which are sterile or nearly so) were analyzed in this study. Ditelosomic 2AS and monotelo-2AS ditelo-2AL plants (the former are deficient for the long arm of chromosome 2A and the latter possess only one dose of the short arm of 2A) were analyzed to determine the possible effect of a reduction in the dosage of chromosome 2A. In addition, each of the six possible ditelosomic strains of chromosome group 3, five of the six possible ditelosomic strains of chromosome group 6 (ditelo-6Aβ is not available) and the variety Chinese Spring were studied.

Except where otherwise noted, the genetic analyses described in this paper utilized extracts of the blade of the first foliage leaf of seven-day-old etiolated seedlings grown in moist toweling at $23^{\circ}C$. Each leaf blade was macerated in sand in a mortar with pestle in 0.3 ml of a pH 7.5 buffer described by Carlson (1972), made to 0.1 M sucrose. The slurry was centrifuged (30,000 x g for 15 min) and the entire quantity of supernatant obtained electrophoresed in one gel. Extraction and centrifugation were carried out at $4^{\circ}C$. A minimum of three plants of each aneuploid type was analyzed. Extracts of individual three-day-old scutella and endosperms were prepared in the same manner.

Disc electrophoresis was performed in 8% acrylamide gels in a Buchler apparatus (Buchler Instruments, Inc.), using procedures adapted from those described by Davis (1964). Riboflavin was used in "solution E" at a concentration of 0.5 mg/100 ml. A pH 8.57 Tris-glycine buffer (5.0 g Tris and 28.8 g glycine/liter) was used in the reservoirs. The differences in relative electrophoretic mobility among the isozymes of

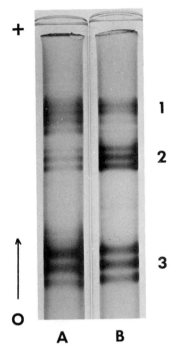

Figure 1. Photograph of the GOT zymogram phenotypes of the scutellum (A) and of the blade of the first foliage leaf (B) of seven-day-old etiolated seedlings of Chinese Spring. The major zones of GOT activity are identified by the numbers to the right. Migration was toward the anode from the origin, as indicated by the arrow.

---

zone 1 and, to a lesser degree, of zone 2 (see RESULTS) are very small. To aid in the resolution of these isozymes and to make it possible to readily discern differences in relative staining intensities between bands within these zones, electro-phoresis was performed in gel tubes 17 cm in length, using a 14 cm small-pore gel and a 1 cm spacer gel. A current of 1 ma/tube was used for the first three hours of electrophoresis and of 2 ma/tube thereafter until the Bromophenol Blue mi-grated out of the gels. As a further aid to the resolution of the isozymes of zones 1 and 2, electrophoresis was then continued for an additional two hours at 3 ma/tube. Both the anodal and cathodal buffer reservoirs were maintained at 2°C with a refrigerant during electrophoresis.

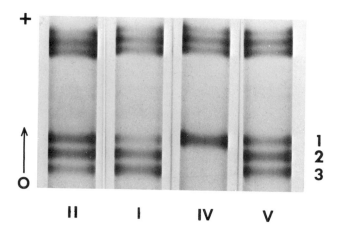

Figure 2. Photograph of the four zone 3 GOT zymogram phenotypes observed (phenotype I of zone 2 is visible in the upper
portion of the photograph in each gel). II. Phenotype of each
strain examined which was disomic for each of the chromosomes
of group 3, of the nulli-3B tetra-3D and nulli-3D tetra-3B
strains, and of the ditelo-3Aα, -3BL, and -3Dα strains. I: Phenotype of the nulli-3B tetra-3A and nulli-3D tetra-3A strains.
IV: Phenotype of the nulli-3A tetra-3B and nulli-3A tetra-3D
strains and of the ditelo-3Aβ strain. V: Phenotype of the
ditelo-3BS and -3Dβ strains. The bands of GOT activity are
identified by the numbers to the right. Migration was toward
the anode from the origin, as indicated by the arrow.

The gels were stained for GOT activity using the procedure of Yang (1971).

RESULTS

TISSUE EXPRESSION OF GOT ISOZYMES

Certain of the GOT isozymes contained in young seedlings
of *T. aestivum* are tissue specific in their expression (Hart,
1974). The isozymes contained in the first foliage leaf,
root, coleoptile, scutellum, and endosperm of three and seven-
day-old seedlings of Chinese Spring are resolved by the electrophoretic procedures described above into three major zones,
each zone composed of three or more isozymes (Fig. 1). However, only in the zone of slowest electrophoretic mobility

(zone 3) are the zymogram phenotypes produced by each of these tissues indistinguishable. In the zone of intermediate electrophoretic mobility (zone 2), the zymogram phenotype produced by each of the tissues is composed of three bands. However, the bands produced by the first foliage leaf, coleoptile, and root stain intensely while those produced by the scutellum and particularly by the endosperm stain only lightly. In the zone of greatest electrophoretic mobility (zone 1), the zymogram phenotype of the first foliage leaf, root, and coleoptile is composed of three intense anodal and two light cathodal bands. The zymogram phenotypes produced by the scutellum and endosperm differ from those of the other tissues in that the bands these tissues produce in the cathodal portion of zone 1 stain intensely.

CHROMOSOMAL CONTROL OF VARIATION IN ZONE 3

Four distinct zone 3 GOT zymogram phenotypes, differing with respect to the presence or absence and the relative staining intensities of their bands, were observed among the stains examined (Fig. 2). Chinese Spring produces a phenotype on which three bands are observable (II, Figs. 2 and 4). The two more anodal bands (bands 1 and 2) stain much more intensely than the cathodal band (band 3). Each aneuploid type ex-

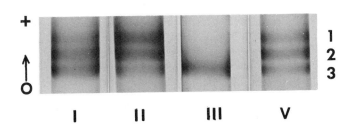

Figure 3. Photograph for the four zone 2 GOT zymogram phenotypes observed. I: Phenotype of each strain examined that was disomic for each of the chromosomes of group 6, of the nulli-6A tetra-6B and nulli-6B tetra-6A strains, and of the ditelo-6BL and -6Dβ strains. II: Phenotype of the nulli-6A tetra-6D and nulli-6B tetra-6D strains. III: Phenotype of the nulli-6D tetra-6A and nulli-6D tetra-6B strains and of the ditelo-6Dα strain. V: Phenotype of the ditelo-6Aα and -6BS strains. The bands of GOT activity are identified by the numbers to the right. Migration was toward the anode from the origin, as indicated by the arrow.

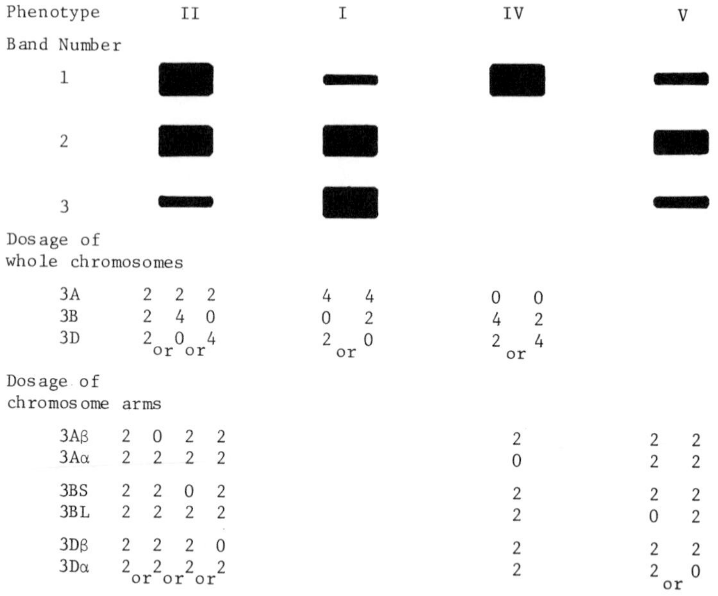

Figure 4 diagram:

Phenotype	II	I	IV	V
Band Number				
1	▆	▬	▆	▬
2	▆	▆		▆
3	▬	▆		▬

Dosage of whole chromosomes

	II	I	IV
3A	2 2 2	4 4	0 0
3B	2 4 0	0 2	4 2
3D	2 0 4	2 0	2 4
	or or	or	or

Dosage of chromosome arms

	II				IV	V	
3Aβ	2	0	2	2	2	2	2
3Aα	2	2	2	2	0	2	2
3BS	2	2	0	2	2	2	2
3BL	2	2	2	2	2	0	2
3Dβ	2	2	2	0	2	2	2
3Dα	2	2	2	2	2	2	0
	or	or	or				or

Figure 4. Diagram showing the relationships between dosages of the group 3 chromosomes and chromosome arms and the zone 3 GOT zymogram phenotypes produced.

---

amined which was disomic for each of the three group 3 chromosomes produced a phenotype indistinguishable from that of Chinese Spring.

Differences between the zymogram phenotypes occurred when the chromosomes of group 3 were varied. When chromosome 3A is absent, bands 2 and 3 are missing (IV, Figs. 2 and 4). This indicates that 3A contributes to the production of both bands 2 and 3. That 3A alone of the three homoeologues contributes to band 3 is indicated by the finding that tetrasomy for 3A, combined with nullisomy for either 3B or 3D, produces a phenotype in which the intensity of band 3, relative to that of bands 1 and 2, is greatly increased over that which occurs when there is disomy for each of the homoeologues of group 3 (I, Figs. 2 and 4).

Band 1 of the phenotype produced when there is nullisomy for either 3B or 3D, and tetrasomy for 3A, is much less intense, relative to the other bands, than in Chinese Spring (I, Figs. 2 and 4). This indicates that 3B and 3D are both involved in the production of band 1, and that they make an approximately equal contribution. The finding that nulli-3B tetra-3D and nulli-3D tetra-3B plants each produce a phenotype indistinguishable from that of Chinese Spring is consistent with this conclusion.

It was noted above that 3A contributes to the production of band 2, based on the absence of this band when 3A is absent. 3A cannot be the only chromosome of group 3 involved in the production of band 2, however, for the intensity of band 2 has been observed to vary independently of that of band 3 (increasing the dosage of 3A from disomy to tetrasomy, while either 3B or 3D is decreased from disomy to nullisomy, results in a large increase in the intensity of band 3 relative to that of band 2) and it was shown above that 3A is the only chromosome of group 3 involved in the production of band 3. It may thus be concluded that chromosomes 3B and 3D also contribute to the production of band 2. Further, they must contribute equally, since nullisomy or tetrasomy for 3B produces an effect indistinguishable from that of nullisomy or tetrasomy for 3D.

The relationships just described are confirmed by the results of analyses of the phenotypes produced by the six group 3 ditelosomic strains. Furthermore, these analyses establish that among the arms of the group 3 chromosomes, it is specifically the α arm of chromosome 3A which contributes to the production of bands 2 and 3, and the long arm of 3B and the α arm of 3D which contribute to the production of bands 1 and 2. In the absence of 3Aβ or 3BS or 3Dβ the zymogram phenotype produced is indistinguishable from that of Chinese Spring (II, Figs. 2 and 4). However, when 3Aα is absent, as when chromosome 3A is absent, bands 2 and 3 are missing (IV, Figs. 2 and 4). And in the absence of either 3BL or 3Dα band 1 in particular and band 2 to a lesser degree are less intense relative to band 3 than in Chinese Spring, so that band 2 stains somewhat more intensely than the approximately equally intense bands 1 and 3 (V, Figs. 2 and 4). This is consistent with the conclusion that 3BL and 3Dα contribute equally to the production of band 1 and also contribute, along with 3Aα, to the production of band 2.

CHROMOSOMAL CONTROL OF VARIATION IN ZONES 1 AND 2

Analysis, in a manner analogous to that described above for zone 3, of the relationships between the chromosomal constitution and the zone 2 zymogram phenotypes of each of the strains examined establishes that a gene (or genes) involved in the production of band 1 is located in 6Dβ, of band 2 in

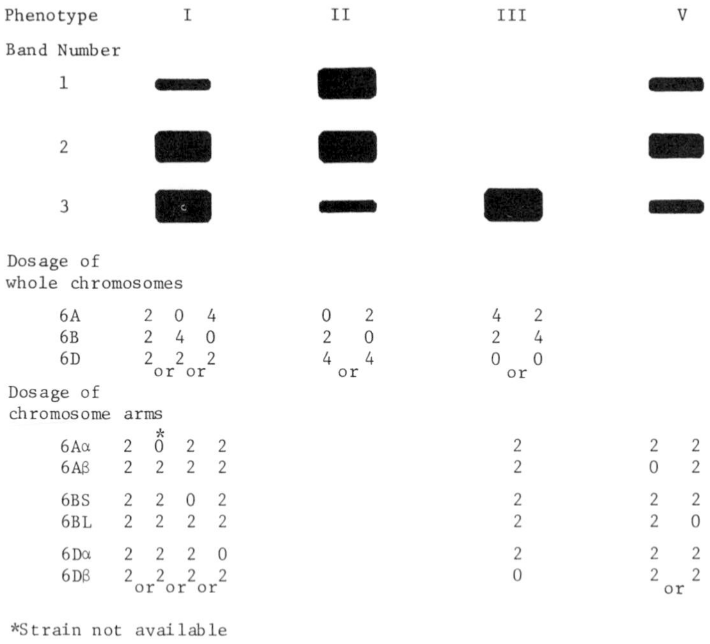

Phenotype   I   II   III   V

Band Number

	I	II	III	V
1	▬	■		▬
2	■	■		■
3	■	▬	■	▬

Dosage of
whole chromosomes

	I			II		III		
6A	2	0	4	0	2	4	2	
6B	2	4	0	2	0	2	4	
6D	2	2	2	4	4	0	0	
		or or			or		or	

Dosage of
chromosome arms

	I				III	V	
6Aα	2	0*	2	2	2	2	2
6Aβ	2	2	2	2	2	0	2
6BS	2	2	0	2	2	2	2
6BL	2	2	2	2	2	2	0
6Dα	2	2	2	0	2	2	2
6Dβ	2	2	2	2	0	2	2
		or or or					or

*Strain not available

Figure 5. Diagram showing the relationships between dosages of the group 6 chromosomes and chromosome arms and the zone 2 GOT zymogram phenotypes produced.

6Aβ, 6BL and 6Dβ, and of band 3 in 6Aβ and 6BL (Figs. 3 and 5).

A similar analysis of zone 1 establishes that a gene (or genes) involved in the production of band 1 is located in 6Dα, of band 2 in 6Aα, 6BS, and 6Dα, and of band 3 in 6Aα and 6BS (Fig. 6). The genetic analysis of the zone 1 leaf blade isozymes was confined to this system. An as yet incomplete genetic analysis of three-day-old scutella and endosperms has established (1) that genes involved in the production of other zone 1 isozymes are located in the chromosomes of homoeologous group 7 and (2) that there are in this zone several isozymes of independent genetic origin which have coincident electrophoretic mobility (Hart, unpublished results).

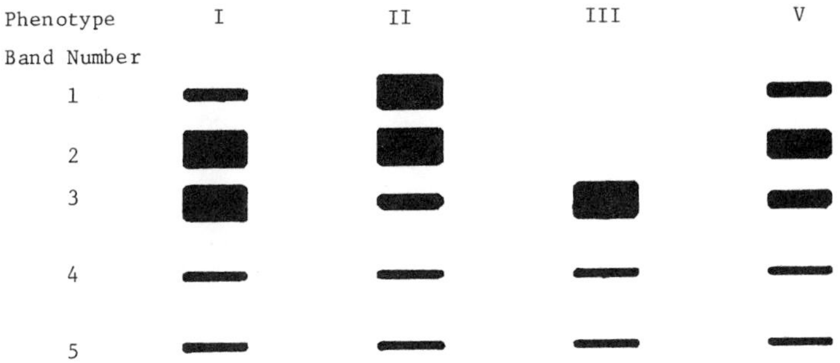

	I	II	III	V
Dosage of whole chromosomes				
6A	2 0 4	0 2	4 2	
6B	2 4 0	2 0	2 4	
6D	2 or 2 or 2	4 or 4	0 or 0	
Dosage of chromosome arms				
6Aα	2 2 2 2		2	0* 2
6Aβ	2 0 2 2		2	2 2
6BS	2 2 2 2		2	2 0
6BL	2 2 0 2		2	2 2
6Dα	2 2 2 2		0	2 2
6Dβ	2 or 2 or 2 or 0		2	2 or 2

*Strain not available

Figure 6. Diagram showing the relationships between dosages of the group 6 chromosomes and chromosome arms and the zone 1 GOT zymogram phenotypes produced.

DISCUSSION

GENETIC CONTROL OF THE ISOZYMES OF ZONE 3

The results clearly demonstrate that the chromosome arms 3Aα, 3BL, and 3Dα each possess a gene (or genes) involved in the production of the zone 3 GOT isozymes. Dimeric structures for the active GOT isozymes of several species have been reported (Davidson et al., 1970; Chapman and Ruddle, 1972; MacDonald and Brewbaker, 1972; Gottlieb, 1973). The simplest hypothesis which is in full agreement with the observed relationships between the chromosomal constitution and the zone 3 zymogram phenotypes of each of the strains examined in this study is one which assumes, (1) that 3Aα, 3BL, and 3Dα each

Figure 7

ISOZYMES	CHINESE SPRING	NULLI-3B TETRA-3D	NULLI-3D TETRA-3B	NULLI-3B TETRA-3A	NULLI-3D TETRA-3A	NULLI-3A TETRA-3B or 3D
GOT-3a	4/9 $\beta^3\beta3$, $\delta^3\delta3$, $\beta^3\delta3$	4/9 $\delta^3\delta3$	4/9 $\beta^3\beta3$	1/9 $\delta^3\delta3$	1/9 $\beta^3\beta3$	$\beta^3\beta3$, $\delta^3\delta3$, $\beta^3\delta3$
GOT-3b	4/9 $\alpha^3\beta3$, $\alpha^3\delta3$	4/9 $\alpha^3\delta3$	4/9 $\alpha^3\beta3$	4/9 $\alpha^3\delta3$	4/9 $\alpha^3\beta3$	
GOT-3c	1/9 $\alpha^3\alpha3$	1/9 $\alpha^3\alpha3$	1/9 $\alpha^3\alpha3$	4/9 $\alpha^3\alpha3$	4/9 $\alpha^3\alpha3$	
DOSAGE OF						
Got-A3	2	2	2	4	4	0
Got-B3	2	0	4	0	2	2
Got-D3	2	4	0	2	0	2 or 4

Schematic model for the subunit composition of the zone 3 GOT isozymes produced by Chinese Spring and by each of the group 3 nulli-tetra types. Dimers on the same line in the figure have coincident electrophoretic mobility. The expected quantitative distribution of the isozymes is indicated by the ratios preceeding the dimers. The expected quantitative distribution of the isozymes is indicated by the ratios preceeding the dimers.

646

possess a GOT structural gene, (2) that the active GOT enzymes are dimers produced by the random association of subunits, and (3) that each GOT structural gene produces an approximately equal quantity of its respective GOT subunit.

A schematic model for the subunit composition of the zone 3 GOT isozymes of Chinese Spring and of plants aneuploid for group 3 chromosomes and chromosome arms based on this hypothesis is summarized in Figs. 7 and 8. The GOT structural genes located in 3Aα, 3BL, and 3Dα are designated, respectively, as Got-A3, Got-B3 and Got-D3, and the subunits for which they code as α^3, β^3, and δ^3, respectively.

The hypothesis predicts that in strains disomic for the chromosomes of group 3 (e.g., Chinese Spring), random association of the α^3, β^3, and δ^3 subunits results in the production of six possible dimers. The three forms of GOT composed of these dimers are designated GOT-3a ($\beta^3\beta^3$, $\delta^3\delta^3$, and $\beta^3\delta^3$ dimers), GOT-3b ($\alpha^3\beta^3$ and $\alpha^3\delta^3$), and GOT-3c ($\alpha^3\alpha^3$). These isozymes produce bands 1, 2, and 3, respectively.

The association of the $\alpha^3\alpha^3$ dimer with band 3 is consistent with the finding that 3Aα is involved in the production of band 3. Likewise, the designation of the $\beta^3\beta^3$, $\delta^3\delta^3$, and $\beta^3\delta^3$ dimers as the enzymes responsible for the production of band 1 is in agreement with the finding that 3BL and 3Dα are involved in the production of band 1. Finally, the association of the $\alpha^3\beta^3$ and $\alpha^3\delta^3$ dimers with band 2 is based on the observation that 3Aα, 3BL, and 3Dα each contribute to the production of this band.

This study has shown that nullisomy for either 3B or 3D does not produce any detectable change in terms of the presence or absence of zymogram bands. The model is consistent with this finding. The $\delta^3\delta^3$ and $\alpha^3\delta^3$ isozymes cause the production of bands 1 and 2, respectively, when the β^3 subunit is absent as a result of nullisomy for 3B, while nullisomy for 3D results in the production of bands 1 and 2 by the $\beta^3\beta^3$ and $\alpha^3\beta^3$ dimers, respectively. With nullisomy for 3A or with 3Aα absent, only band 1 is produced, however. This is in agreement with the linkage of Got-A3 to 3Aα and with the α^3 subunit being a necessary component of the isozymes which produce both bands 2 and 3.

Strong support for the hypothesis, and particularly for the assumptions of random association of subunits and of production of an approximately equal amount of subunits by each of the GOT structural genes, comes from observations on the relative staining intensities of the zymogram bands. Based on the hypothesis, the expected distribution of the six possible dimeric enzyme molecules will be based on $(p + q + r)^2$, where p, q, and r represent, respectively, the frequen-

Figure 8

ISOZYMES	DITELO-3Aα or 3BL or 3Dα	DITELO-3Aβ	DITELO-3BS	DITELO-3Dβ
GOT-3a	$4/9\ \beta^3\beta^3,\ \delta^3\delta^3,\ \beta^3\delta^3$	$\beta^3\beta^3,\ \delta^3\delta^3,\ \beta^3\delta^3$	$1/4\ \delta^3\delta^3$	$1/4\ \beta^3\beta^3$
GOT-3b	$4/9\ \alpha^3\beta^3,\ \alpha^3\delta^3$		$2/4\ \alpha^3\delta^3$	$2/4\ \alpha^3\beta^3$
GOT-3c	$1/9\ \alpha^3\alpha^3$		$1/4\ \alpha^3\alpha^3$	$1/4\ \alpha^3\alpha^3$
DOSAGE OF				
Got-A3	2	0	2	2
Got-B3	2	2	0	2
Got-D3	2	2	2	0

Figure 8. Schematic model for the subunit composition of the zone 3 GOT isozymes produced by each of the group 3 ditelosomic strains. See legend for Fig. 7 for further explanation.

cies of the α^3, β^3, and δ^3 subunits. When there is disomy for each of the homoeologues of group 3, $p = q = r = 1/3$, and the expected trinomial proportions are $1/9$ $\alpha^3\alpha^3 : 1/9$ $\beta^3\beta^3 : 1/9$ $\delta^3\delta^3 : 2/9$ $\alpha^3\beta^3 : 2/9$ $\alpha^3\delta^3 : 2/9$ $\beta^3\delta^3$. Combining the proportions for those dimers with coincident electrophoretic mobility, the expected distribution of the isozymes that are assumed to be responsible for the production of bands 1, 2, and 3 is 4:4:1, respectively. This proportion is in good agreement with the observed staining intensities of the zymogram bands of Chinese Spring and other strains disomic for the group 3 homoeologues (II, Figs. 2 and 4).

For those strains which are nullisomic for either 3B or 3D or which lack either 3BS or 3Dβ, the expected distribution of the three possible dimers will be based on $(p + q)^2$, where p represents the frequency of the α^3 subunit and q the frequency of either the β^3 or δ^3 subunit, depending upon which gene is absent. $p = 1/3$ and $q = 2/3$ when 3A is disomic and either 3B or 3D is tetrasomic, and $p = 2/3$ and $q = 1/3$ when 3A is tetrasomic and either 3B or 3D is disomic. In the former case, the expected distribution of the isozymes assumed to produce bands 1, 2 and 3 is 4:4:1, respectively (as for Chinese Spring), but in the latter it is 1:4:4, respectively. For each of the four nulli-tetra combinations in which there is nullisomy for either 3B or 3D, the observed staining intensities of the zymogram bands are in good agreement with the expected distribution of the isozymes that are assumed to be responsible for the production of the bands (I and II, Figs. 2 and 4). When either 3BL or 3Dα is absent, $p = q = 1/2$, and the expected distribution of the isozymes assumed to be responsible for the production of bands 1, 2, and 3 is 1:2:1. The observed staining intensities of the bands on zymograms produced by ditelo-3BS and ditelo-3Dβ plants are in good agreement with this distribution (V, Figs. 2 and 4). When 3Aβ or 3BS or 3Dβ is absent, given linkage of the GOT genes to 3Aα, 3BL, and 3Dα, production of a phenotype indistinguishable from that of Chinese Spring is expected and observed (II, Figs. 2 and 4).

The zone 3 zymogram phenotypes of nulli-3B tetra-3D and nulli-3D tetra-3B plants are identical, as are those of nulli-3D tetra-3A and nulli-3B tetra-3A plants and of ditelo-3BS and ditelo-3Dβ plants, in agreement with the quantitative distribution predicted by the model. However, the model does predict differences between the members of these pairs in terms of the specific subunit composition of the isozymes that produce bands 1 and 2. But, since the β^3 and δ^3 subunits are differentiated solely on the basis of their genetic site of origin, it is clear that the model does not predict

Figure 9

ISOZYMES	CHINESE SPRING	NULLI-6A TETRA-6B	NULLI-6B TETRA-6A	NULLI-6A TETRA-6D	NULLI-6B TETRA-6D	NULLI-6D TETRA-6A or 6B
GOT-2a	$1/9\ \delta^2\delta^2$	$1/9\ \delta^2\delta^2$	$1/9\ \delta^2\delta^2$	$4/9\ \delta^2\delta^2$	$4/9\ \delta^2\delta^2$	$\alpha^2\alpha^2,\ \beta^2\beta^2,\ \alpha^2\beta^2$
GOT-2b	$4/9\ \alpha^2\delta^2,\ \beta^2\delta^2$	$4/9\ \beta^2\delta^2$	$4/9\ \alpha^2\delta^2$	$4/9\ \beta^2\delta^2$	$4/9\ \alpha^2\delta^2$	
GOT-2c	$4/9\ \alpha^2\alpha^2,\ \beta^2\beta^2,\ \alpha^2\beta^2$	$4/9\ \beta^2\beta^2$	$4/9\ \alpha^2\alpha^2$	$1/9\ \beta^2\beta^2$	$1/9\ \alpha^2\alpha^2$	

DOSAGE OF

	CHINESE SPRING	NULLI-6A TETRA-6B	NULLI-6B TETRA-6A	NULLI-6A TETRA-6D	NULLI-6B TETRA-6D	NULLI-6D TETRA-6A or 6B
Got-A2	2	0	4	0	2	2
Got-B2	2	4	0	2	0	4
Got-D2	2	2	2	4	4	0 or 0

Figure 9. Schematic model for the subunit composition of the zone 2 GOT isozymes produced by Chinese Spring and by each of the group 6 nulli-tetra types. See legend for Fig. 7 for further explanation.

a structural difference between these subunits. No evidence has been obtained to date to suggest such a difference and it is thus quite possible that Got-B3 and Got-D3 are identical so that the isozymes within each of the pairs of strains mentioned above are structurally and functionally identical.

GENETIC CONTROL OF THE ISOZYMES OF ZONES 1 AND 2

A hypothesis identical to that described above for the genetic control of the GOT isozymes of zone 3, except for differences in the chromosome arms and therefore in the genes involved, is in full agreement with the observed relationships between the chromosomal constitution and the zone 2 zymogram phenotypes of each of the strains examined. The GOT structural genes located in 6Aβ, 6BL, and 6Dβ are designated, respectively, as Got-A2, Got-B2, and Got-D2, and the subunits for which they code as α^2, β^2, and δ^2. The three forms of GOT produced by the six types of dimers which result from random associations of these subunits are designated GOT-2a, GOT-2b, and GOT-2c. These isozymes produce bands 1, 2, and 3, respectively (Figs. 9 and 10).

Similarly for zone 1, the GOT structural genes located in 6Aα, 6BS, and 6Dα are designated, respectively, as Got-A1, Got-B1, and Got-D1, and the subunits for which they code as α^1, β^1, and δ^1. The three forms of GOT produced by the six types of dimers which result from random associations of these subunits are designated GOT-1a, GOT-1b, and GOT-1c. These isozymes are located at the sites of bands 1, 2, and 3, respectively (Figs. 11 and 12).

An analysis of additional isozymes which resolve in zone 1 is as yet incomplete, although the chromosomes of group 7 have been shown to possess genes involved in the production of these isozymes. Under the assumption that these additional active GOT isozymes are dimers, the pattern of variation observed suggests that the additional isozymes are the products of a minimum of two independent genetic systems (Hart, unpublished results). The available evidence thus suggests that the several isozymes of zone 1 are the products of three independent genetic systems, each of which may consist of triplicate homoeologous structural genes.

GENE AND CHROMOSOME HOMOEOLOGY

It is highly probable that the structural genes for the three groups of GOT isozymes analyzed are located in the chromosomes of homoeologous groups 3 and 6. The strongest evidence for this comes from the demonstration that with nullisomy or

651

Figure 10

ISOZYMES	DITELO-6Aβ or 6BL or 6Dβ	DITELO-6Aα	DITELO-6BS	DITELO-6Dα
GOT-2a	$1/9\ \delta^2\delta^2$	$1/4\ \delta^2\delta^2$	$1/4\ \delta^2\delta^2$	$1/4\ \delta^2\delta^2$
GOT-2b	$4/9\ \alpha^2\delta^2,\ \beta^2\delta^2$	$2/4\ \beta^2\delta^2$	$2/4\ \alpha^2\delta^2$	$2/4\ \beta^2\beta^2,\ \alpha^2\beta^2$
GOT-2c	$4/9\ \alpha^2\alpha^2,\ \beta^2\beta^2,\ \alpha^2\beta^2$	$1/4\ \beta^2\beta^2$	$1/4\ \alpha^2\alpha^2$	$1/4\ \alpha^2\alpha^2,\ \beta^2\beta^2,\ \alpha^2\beta^2$
DOSAGE OF				
Got-A2	2	0	2	2
Got-B2	2	2	0	2
Got-D2	2	2	2	0

Figure 10. Schematic model for the subunit composition of the zone 2 GOT isozymes produced by each of the group 6 ditelosomic strains. See legend for Fig. 7 for further explanation.

Figure 11

ISOZYMES	CHINESE SPRING	NULLI-6A TETRA-6B	NULLI-6B TETRA-6A	NULLI-6A TETRA-6D	NULLI-6B TETRA-6D	NULLI-6D TETRA-6A or 6B
GOT-1a	$1/9\ \delta^1\delta^1$	$1/9\ \delta^1\delta^1$	$1/9\ \delta^1\delta^1$	$4/9\ \delta^1\delta^1$	$4/9\ \delta^1\delta^1$	
GOT-1b	$4/9\ \alpha^1\delta^1,\ \beta^1\delta^1$	$4/9\ \beta^1\delta^1$	$4/9\ \alpha^1\delta^1$	$4/9\ \beta^1\delta^1$	$4/9\ \alpha^1\delta^1$	
GOT-1c	$4/9\ \alpha^1\alpha^1,\ \beta^1\beta^1,\ \alpha^1\beta^1$	$4/9\ \beta^1\beta^1$	$4/9\ \alpha^1\alpha^1$	$1/9\ \beta^1\beta^1$	$1/9\ \alpha^1\alpha^1$	$\alpha^1\alpha^1,\ \beta^1\beta^1,\ \alpha^1\beta^1$
DOSAGE OF						
Got-A1	2	0	4	0	2	2
Got-B1	2	4	0	2	0	4
Got-D1	2	2	2	4	4	0 or 0

Figure 11. Schematic model for the subunit composition of the zone 1 GOT isozymes produced by Chinese Spring and by each of the group 6 nulli-tetra types. See legend for Fig. 7 for further explanation.

tetrasomy for any chromosome (for 2AS, monosomy but not nullisomy has been tested) outside of the 3 group (zone 3) or 6
group (zones 1 and 2) the zymogram phenotype produced is indistinguishable from that of Chinese Spring, while with nullisomy for 3Aα or 3BL or 3Dα or for an arm of a group 6 chromosome changes in phenotype are produced consistent with those
expected to result from the absence from the complement of a
GOT structural gene.

Evidence has been presented that the active enzymes of
each of the three groups of GOT isozymes are dimers formed by
the random association of the products of three genes, that
the genes are located on homoeologous chromosomes, and that
the subunits of each group do not interact with the subunits
of either of the other two groups. It is therefore reasonable
to conclude that these three groups of isozymes are the products of three independent sets of triplicate homoeologous
GOT structural genes.

The linkages reported here provide direct evidence of
homoeology between 3Aα, 3BL, and 3Dα, between 6Aα, 6BS, and
6Dα, and between 6Aβ, 6BL, and 6Dβ. These relationships are
consistent with those suggested by other studies (for a summary, see the report by McIntosh in Sears and Sears, 1973).

The evolutionary implications of the results of this
study will be discussed elsewhere.

REGULATION OF GOT GENE ACTIVITY

For each of the three GOT systems analyzed herein, evidence for a linear relationship between the quantity of each
of the three possible types of subunits contained in the
active enzymes and the dosage of the chromosomes and chromosome arms which carry the structural genes for the subunits
has been obtained. This suggests that the rate of transcription is the same for each copy present of each of the three
possible members of the homoeologous sets of GOT structural
genes. It further implies that genetic regulation of the
expression of the GOT isozymes occurs at the level of gene
transcription. That is, it implies that a given rate of
transcription of a structural gene is what is genetically
determined in a given tissue at a given point in development
so that, with a reasonably constant environment, variation
in the level of enzyme activity will be linearly proportional
to variation in the number of structural genes present.

Evidence for a linear relationship between the quantity
of subunits contained in active enzymes and the dosage of
chromosomes and chromosome arms carrying structural genes for
the subunits has also been obtained for the alcohol dehydro-

Figure 12

ISOZYMES	DITELO-6Aα or 6BS or 6Dα	DITELO-6Aβ	DITELO-6BL	DITELO-6Dβ
GOT-1a	$1/9\ \delta^1\delta^1$	$1/4\ \delta^1\delta^1$	$1/4\ \delta^1\delta^1$	
GOT-1b	$4/9\ \alpha^1\delta^1,\ \beta^1\delta^1$	$2/4\ \beta^1\delta^1$	$2/4\ \alpha^1\delta^1$	
GOT-1c	$4/9\ \alpha^1\alpha^1,\ \beta^1\beta^1,\ \alpha^1\beta^1$	$1/4\ \beta^1\beta^1$	$1/4\ \alpha^1\alpha^1$	$\alpha^1\alpha^1,\ \beta^1\beta^1,\ \alpha^1\beta^1$
DOSAGE OF				
Got-A1	2	0	2	2
Got-B1	2	2	0	2
Got-D1	2	2	2	0

Figure 12. Schematic model for the subunit composition of the zone 1 GOT isozymes produced by each of the group 6 ditelosomic strains. See legend for Fig. 7 for further explanation.

genase (ADH), aminopeptidase (AMP), and acid phosphatase (ACPH) isozymes of hexaploid wheat (Hart, 1970, 1973, and in preparation). These several studies provide no evidence that the quantity of subunit produced or the quantity of active GOT, ADH, AMP, and ACPH isozymes present in tissues is genetically regulated other than at the level of gene transcription.

ACKNOWLEDGEMENTS

I am very grateful to Dr. E. R. Sears for generously supplying me with seed stocks of the strains used in this study and to Pat Langston for excellent technical assistance. This paper is technical article number 11253 of the Texas Agricultural Experiment Station.

REFERENCES

Barber, H. N., C. J. Driscoll, P. M. Long, and R. S. Vickery 1968. Protein genetics of wheat and homoeologous relationships of chromosomes. *Nature* 218: 450-452.

Carlson, P. S. 1972. Locating genetic loci with aneuploids. *Mol. Gen. Genet.* 114: 272-280.

Chapman, V. M. and F. H. Ruddle 1972. Glutamate oxaloacetate transaminase (GOT) genetics in the mouse: Polymorphism of GOT-1. *Genetics* 70: 299-305.

Davidson, R. G., J. A. Cortner, M. C. Ratazzi, F. H. Ruddle, and H. A. Lubs 1970. Genetic polymorphisms of human mitochondrial glutamate oxaloacetate transaminase. *Science* 169: 391-392.

Davis, B. J. 1964. Disc electrophoresis. II. Method and application to human serum proteins. *Ann. N.Y. Acad. Sci.* 121: 404-427.

Gottlieb, L. D. 1973. Genetic control of glutamate oxaloacetate transaminase isozymes in the diploid plant *Stephanomeria exigua* and its allotetraploid derivative. *Biochem. Genet.* 9: 97-107.

Hart, G. E. 1970. Evidence for triplicate genes for alcohol dehydrogenase in hexaploid wheat. *Proc. Nat. Acad. Sci.* 66: 1136-1141.

Hart, G. E. 1973. Homoeologous gene evolution in hexaploid wheat. *Proc. 4th Int. Wheat Genet. Symp.* (Mo. Agr. Exp. Sta., Columbia): 805-810.

Hart, G. E. 1974. Glutamate oxaloacetate transaminase: Tissue specificity and genetic control in hexaploid wheat. *Isozyme Bulletin* 7: 17.

MacDonald, T. and J. L. Brewbaker 1972. Isoenzyme polymorphism in flowering plants VIII. Genetic control and dimeric nature of transaminase hybrid maize isoenzymes. *J. Hered.* 63: 11-14.

Nishikawa, K. and M. Nobuhara 1971. Genetic studies of α-amylase isozymes in wheat. I. Location of genes and variation in tetra- and hexaploid wheat. *Jap. J. Genet.* 46: 345-353.

Sears, E. R. 1954. The aneuploids of common wheat. *Mo. Agr. Exp. Sta. Res. Bull.* 572: 58 pp.

Sears, E. R. 1966a. Nullisomic-tetrasomic combinations in hexaploid wheat. In *Chromosome Manipulations and Plant Genetics* (Eds. R.Riley and K. R. Lewis; Oliver and Boyd, London): 29-45.

Sears, E. R. 1966b. Chromosome mapping with the aid of telocentrics. *Proc. 2nd Int. Wheat Genet. Symp. Hereditas Suppl.* 2: 370-381.

Sears, E. R. and L. M. S. Sears (Ed.) 1973. *Proc. 4th Int. Wheat Genet. Symp.* (Mo. Agr. Exp. Sta., Columbia).

Shepherd, K. W. 1968. Chromosomal control of endosperm proteins in wheat and rye. *Proc. 3rd Int. Wheat Genet. Symp.* (Aust. Acad. Sci., Canberra): 86-96.

Yang, S. Y. 1971. *Appendix. Stud. Genet.* (Univ. Texas Publ.) 6(7103): 85-90.

POLYMORPHISMS OF THE MAJOR PEROXIDASES OF MAIZE

JAMES L. BREWBAKER and YOICHI HASEGAWA
Department of Horticulture
University of Hawaii
3190 Maile Way
Honolulu, Hawaii 96822

ABSTRACT. Thirteen major peroxidases of maize are described.
Genetic loci have been identified that govern isozymic polymorphisms of nine of these, with 26 alleles presently known.
Levels of activity of the 13 peroxidases are reported for
21 major tissues. None of the peroxidases was active in all
tissues. Six enzyme loci appeared to be derepressed during
ontogeny at or near stages of intensive cell elongation and
lignification. The more basic peroxidases were highly
associated with cell-wall fractions of mature tissues. They
utilized eugenol efficiently as hydrogen donor, and are suggested to be competent lignifying enzymes. Substrate
studies revealed that four of the enzymes had high activity
on IAA. Two of these were characteristic of leafy tissues
and one of roots and tissue cultures; each increased greatly
in activity in diseased or aging tissues. Other peroxidatic
functions in maize are intimated by the specific association
of three enzymes with photosynthetically active tissues and
by the presence of one pollen-specific enzyme and one root-
specific enzyme.

INTRODUCTION

Peroxidases are highly polymorphic and ubiquitous in plant
tissues, and their intimated functions are many and diverse.
The peroxidases of maize were first studied electrophoretically
by McCune (1961), who emphasized their potential role as
indoleacetic acid oxidases (Lee, 1972; Raa, 1973). A second
major role of peroxidases is probably that of lignification,
a role suggested by biochemical studies and by the histochemical observations of plant cell walls and highly lignified
tissues (Stafford and Bravinder-Bree, 1972). The rapid increase
of peroxidase activity incited by diseases, irradiation, and
ethylene (Farkas and Stahmann, 1966; Endo, 1967), the association of certain peroxidases with photosynthetic tissues (Macnicol, 1966), and the ability of peroxidases to utilize a
wide variety of phenolic and indole compounds as hydrogen
donor substrates imply other important roles in plant development and differentiation for these versatile enzymes.
Genetic polymorphisms have been observed for most peroxi-

dases that have been studied intensively. Attempts to discern
the precise roles of peroxidases must contend fully with these
genetic and epigenetic polymorphisms. Much of the classic
plant peroxidase research has been with a genetic unknown, the
horseradish (Shannon 1968). We have therefore set out method-
ically to locate genetic polymorphisms for the peroxidases of
maize and to establish homozygous marker stocks for the Maize
Genetics Cooperative and for our own physiological and developmental investigations (Hamill and Brewbaker 1969, Macdonald
and Brewbaker 1972, Brewbaker and Hasegawa 1974, Hasegawa 1974).

Thirteen distinct peroxidases have been identified in maize,
most of them having null or co-dominant variants. Genetically-
marked stocks have been used in the present study to assess
the tissue, ontogenetic, intracellular, and physiological
polymorphisms of the maize peroxidases, and to establish relationships between these polymorphisms and suggested functions
of the enzymes.

MATERIALS AND METHODS

Peroxidase polymorphisms have been studied in an extremely
diverse collection of maize seedstocks, and in three strains of
teosinte, *Zea mexicana*. Weekly plantings of field and sweet
corns are made throughout the year in Hawaii for our breeding
research, providing a regular source of tissues at all stages
in plant development.

Electrophoretic procedures were generally similar to those
described by Brewbaker et al. (1968), with the exception that
gels were routinely made of acrylamide. Vertical acrylamide
preparations were used to verify results from the horizontal
gels. The standard horizontal gels were 7% polyacrylamide and
were electrophoresed using lithium-borate buffers at pH 8.1 for
several hours at 8V/cm and at 4°C. Both pH and gel concentrations were varied to facilitate band resolution; in particular,
pH was lowered to aid resolution of cathodal and rapidly-
migrating anodal bands and increased to aid resolution of slow-
migrating anodal bands. Benzidine dihydrochloride and o-dianisidine were used interchangeably as hydrogen donors for
staining gels, the latter for densitometric readings.

Most gels were made with crude macerates of tissue in
saline or calcium buffer solutions. Little improvement in
background readings for densitometry was provided by proteins
precipitated with ammonium sulfate or subjected to gel filtration. Extraction of wall-bound peroxidases from most
tissues was accomplished satisfactorily in 30 minutes with
0.2 M calcium chloride (cf. Haard 1973, who used 0.8 M).
Ascorbate, polyvinylpyrolidone, and urea treatments were not

used regularly, although resolution of certain peroxidases is improved by such techniques.

GENETIC POLYMORPHISMS OF MAIZE PEROXIDASES

Thirteen major peroxidases have been identified in *Zea spp.*, three cathodal and ten anodal when electrophoresed at pH 8.1 (Figure 1). Two regions involve overlapping isozymes. Enzymes 6 and 12 overlap in one region but are entirely exclusive in their tissue specificities, a fact confirmed by the use of Px_6-null homozygotes. Anodal enzymes 2 and 9 also have overlapping isozymes, but nulls for Px_9 permitted the definitive localization of peroxidase 2 activity to the pollen grains.

Genetic polymorphisms have been discerned for nine of the maize peroxidases, and the pertinent genetic studies have been summarized by Brewbaker and Hasegawa (1974). Nine loci governing variations of these regions have been designated with the symbols Px_1, Px_2....Px_9 (Table 1). Allelic variants of these loci include co-dominant positional isozymes at six loci and null variants at five. In no instances have hybrid bands been observed. Multiple bands are observed for the Px_3 alleles, however, especially in senescing and diseased tissues. Each Px_3 allele is represented by four bands, as discussed later under ontogenetic polymorphisms. Two closely-paired isozymes are regularly observed in three regions, including the 10 region, the 6 region (allele Px_6^1), and the allele Px_3^6 of enzyme 3.

TABLE 1

ALLELIC ISOZYME VARIANTS OF THE NINE PEROXIDASE LOCI IN MAIZE

Locus	Alleles
Px_1	1, 2, 3, null
Px_2	1, 2
Px_3	1, 2, 3, 4, 5, 6
Px_4	1, 2, 3,
Px_5	1, null
Px_6	1, null
Px_7	1, 2, null
Px_8	1, 2
Px_9	1, null

Genetic polymorphisms have not been shown convincingly for four maize peroxidases, enzymes 10, 11, 12, and 13. Enzyme 10 is represented in our seedstocks by two isozymes, whose control by alleles of a single locus is inferred but not yet proven. Enzyme 10 was referred to by Chenchin and Yamamoto

Fig. 1. Approximate positions on 7% acrylamide gels at pH 8.1 of the major allelic isozymes of 13 maize peroxidases.

(1973) as A-5 and its exceptional heat lability reported. We have referred previously to this enzyme as peroxidase D (Brewbaker and Hasegawa 1974). No genetic variants have been identified for root enzymes 11 and 12, previously called peroxidases B and A, respectively. Enzyme 11 represents over half the peroxidase activity of roots, while weakly-staining enzyme 12 is represented by two closely-paired, rapidly migrating anodal isozymes. Enzyme 13 has been discerned clearly only in developing endosperm and in endosperm tissue cultures, and is poorly studied.

Much of our isozymic information on tissue polymorphisms has been from vigorous plants derived from hybrids, races, or varieties in which slow-fast heterozygotes at several loci prevail. Knowledge of the genetics has been basic to unequivocal statements about such polymorphisms. Critical quantitative studies and enzyme purifications have been made using appropriate genotypes, usually allelic homozygotes. Many of these were derived from our tropical seedstocks crosses, which are available to interested scientists (Brewbaker 1972).

TISSUE POLYMORPHISMS OF MAIZE PEROXIDASES

None of the maize peroxidases proved to be detectable in all tissues studied (Table 2). Enzymes 1 and 5 were very widely represented in maize tissues, but most peroxidases were rather tissue-specific, notably enzyme 2 of pollen and enzymes 11 and 12 of roots and tissue cultures. The observations summarized in Table 2 represent 21 different tissues or organs. The data are presented on a scale from 0 to 3, ranging from no activity to intense activity on 7% acrylamide gels at pH 8.1 stained with benzidine or o-dianisidine. An internal standard for each tissue, therefore, was its major isozyme, designated 3. Where possible, all observations reflect the performance of allelic homozygotes for all loci. Readings for the leaf, mesocotyl, and root were derived linearly from densitometric peaks, while the others were derived empirically from gels. The data cannot be applied directly to comparisons of different samples; mature roots, leaves and silks were among the highest in activity.

TABLE 2

TISSUE POLYMORPHISMS OF 13 MAJOR MAIZE PEROXIDASES

RELATIVE ACTIVITY OF PEROXIDASE[1]

TISSUE	1	2	3	4	5	6	7	8	9	10	11	12	13
Leaf	2	0	3	1	1	1	3	0	0	0	0	0	0
Husk	1	0	3	1	1	2	3	0	1	0	0	0	0
Glume (Tassel)	1	0	2	1	0	1	3	0	0	0	0	0	0
Coleoptile	2	0	3	2	2	3	2	1	1	0	0	0	0
Mesocotyl	2	0	3	2	1	3	2	2	1	0	0	0	0
Cortex	1	0	3	2	1	2	1	1	0	0	0	0	0
Stele	1	0	0	0	1	0	0	3	1	0	0	0	0
Pith	1	0	0	2	1	2	0	3	0	0	0	0	0
Stem Apex	0	0	0	0	2	0	0	3	0	2	0	0	0
Tassel Initial	0	0	0	0	1	0	0	3	0	3	0	0	0
Ear Initial	0	0	0	0	2	0	0	3	0	1	0	0	0
Pericarp (20 D)	1	0	2	2	2	0	1	1	3	2	0	0	0
Embryo "	0	0	0	1	0	0	0	2	1	3	0	0	0
Endosperm "	0	0	0	0	3	0	0	1	1	2	0	0	1
Root	2	0	1	0	2	0	2	0	0	0	3	1	0
Brace Root	1	0	2	0	1	1	3	0	0	0	3	0	0
Callus, Stem Apex	2	0	0	0	2	0	0	0	0	1	3	1	0
Callus, Endosperm	1	0	0	0	0	0	0	0	0	1	3	1	2
Silk	1	0	0	3	0	0	3	0	0	0	0	0	0
Anther (2nd Mit)	0	0	0	0	3	0	0	0	0	0	0	2	0
Pollen	0	3	0	0	1	0	0	0	0	0	0	3	0

[1]Scale: 0 = no activity; 1 = low, 2 = medium, 3 = high activity

Tissues and genotypes were selected at specific stages for Table 2: leaves and pith were chosen subtending the ear at anthesis; coleoptiles, roots, and mesocotyls of week-old seedlings; husk, silk, anther, and glumes at anthesis; cob and seed tissues at sweet corn stage; anther at second mitosis of the pollen; ear and tassel initials at 1 cm and 2 cm length, respectively; stem apex of 4 week old seedlings. Homozygotes were used for loci having null alleles, as the heterozygote nulls do not show full enzyme activity, closely approximating 50% of the homozygote activity for enzymes 1, 6, and 9.

The peroxidase isozyme patterns can be used to distinguish at least seven basic groups of maize tissues (Table 2). The first of these comprises leafy tissues with dominating enzymes

1, 3, 4, 6, and 7, almost all of them absent in juvenile tissues. Among these, the enzymes 3, 6, and 7 appear almost exclusively in leafy tissues, the notable exceptions being the activity of peroxidase 3 in pericarp and of peroxidase 7 in silks. Macnicol (1966) also observed the association of specific pea peroxidases C_1 and C_2 with photosynthetic capability.

A second group of tissues includes the lignified stele of the mesocotyl (the first internode of maize stalks) and the internodal pith. The stele is distinguished by its absence of 3, 4, 6, and 7 activity and presence of the 8 and 9 enzymes. The pith also lacked enzymes 3 and 7 and showed high activity of 8. Its enzymes 1, 4, and 5 decreased markedly during lignification of the stem (cf. Fig. 6, Hamill and Brewbaker 1969), an observation to be related later to wall-binding. Further studies of excised conducting and lignified tissues may give more unity to this class of tissues. The peroxidatic distinction in maize of stele and cortical tissues is pronounced (Table 2); in sorghum also, the stele lacks a soluble peroxidase present in the cortex (Stafford and Bravinder-Bree 1972).

A third class of tissues of identical peroxidase zymograms includes the meristematic initials of stem, tassel, and ear, with enzymes 5, 8, and 10 active. Differentiation from these initials of leafy tissues like glumes, husks, and leaves involves synthesis of at least five enzymes--1, 3, 4, 6, and 7. Enzymes 1 and 3 are synthesized in ear and tassel within a week of the primordial stages chosen here.

The major tissues of seeds are treated here as a fourth but obviously non-coherent class. They do share activity of enzymes 8, 9, and 10. However, the pericarp is a tissue with unusually high peroxidase activity, involving eight of the 13 peroxidases. The endosperm and endosperm tissue cultures appear unique in showing activity of enzyme 13, a weakly-staining band in the region of the Px_3 alleles.

Roots are distinguished by the activity of enzymes 11 and 12 that are inactive in all other tissues. Enzyme 11 is the dominant root enzyme and is also active in brace roots. The brace roots have chlorophyll in their cortical cells and are distinguishable from true roots by the activity of enzymes 3 and 7, characteristic of leafy tissues. Dvorak and Cernohorska (1967) also reported a root-specific peroxidase in *Cucurbita spp.*

The peroxidases of tissue culture calluses, derived either from stem apex or juvenile endosperm, bore some resemblance to those of roots. Enzymes 11 and 12 were active in the cultures, together with enzymes 1 and 10. Peroxidases 5 of endosperm and 8 of stem apex were not seen in the pertinent cultures. Both cultures are auxin requiring, the stem apex cultures rooting more or less freely; neither callus differentiates stem or leafy tissues. Some doubt exists as to the activity of enzyme 4 in these calluses and in roots.

The silks (elongated styles) have a distinct set of peroxidases, dominated by enzymes 4 and 7 that are active in leafy tissues. Upon aging of silks, enzymes 1 and 3 are activated, the latter during senescence. The silk is one of the most rapidly growing tissues in maize and is rich in phenolics (Martin and Brewbaker 1971) that are suitable hydrogen-donors for peroxidases 4 and 7 but of unknown significance during pollen germination and growth through the style.

The pollen grains are distinguished by the activity of tissue-specific peroxidase 2, a broad-substrate enzyme active in no other maize tissue that we have studied. Peroxidase 2 is not observed in whole anthers prepared at the time of second mitosis in the pollen, prior to pollen shed. Enzymes 5 and 10 characterize both anthers and pollen, as they do also the tassel initial.

ONTOGENETIC POLYMORPHISMS OF MAIZE PEROXIDASES

Extensive empirical observations have been made of peroxidases during maize development, and essentially all tissues show a general pattern of increasing number and intensity of isozymes (Hamill and Brewbaker 1969). Critical comparisons of peroxidase isozyme activities during ontogeny, however, must be done on an isonitrogenous basis from samples treated to extract both the wall-bound and soluble proteins. Current studies of this type in maize include the developing seedling tissues (Hasegawa 1974) and will be illustrated here with data from the developing mesocotyl (Figure 2).

Apical meristems show little or no activity of enzymes 1, 3, 4, 6, and 7 that occur in the mature mesocotyl (Table 2). During mesocotyl development these enzymes progressively increase in activity with most of the increase concentrated between 5 and 7 days after the initiation of germination (Figure 2). A similar increase occurs between 3 and 6 days in coleoptile and root, most notably for enzymes 1, 3, and 7. These are stages of most active cell elongation and maturation in these tissues.

Juvenile tissues rarely show activity of enzyme 1 which increases and subsequently decreases during development, as evident for mesocotyl in Figure 2. This apparent drop in total activity is believed to be largely the result of the progressive increase in the wall-bound fraction of enzymes like 1, 4, and

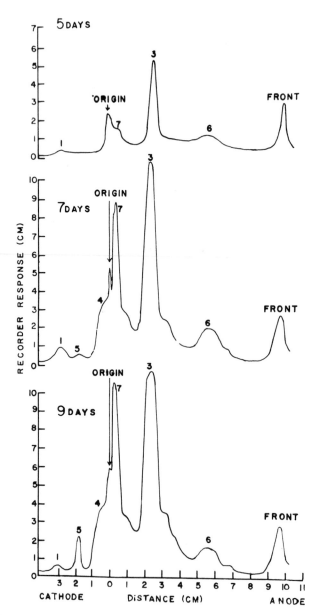

Fig. 2. Densitometric readings of activity on o-dianisidine of peroxidases in developing mesocotyl of maize.

5, to be discussed later. Juvenile tissues provide especially clean preparations for enzyme isolation and purification. Among these are the coleoptile, silk, root tip, and meristematic initials of stem, ear, and tassel.

Ontogenetic data from zymograms can be biased considerably if care is not taken to correct for protein concentrations during development. However, an apparently valid generalization is that maize peroxidases are highly temperature-stable and long-lived, remaining at high concentrations throughout the maturation of many tissues. Somewhat similar observations have been made in other plants, including tomatoes (Gordon and Alldridge 1971), beans (Racusen and Foote 1966), and cottonwood (Gordon 1971).

Isozymes arising under epigenetic control characterize peroxidase 3 of leafy tissues. During maturation, peroxidase 3 develops three regularly spaced, rapidly-migrating isozymes in most tissues. These can be seen as shoulders distal to the 3 peak in 9-day old mesocotyl (Figure 2); they are also shown in Figure 2 of Peirce and Brewbaker (1973) and for maize esterase 4 in Brewbaker et al. (1968). These epigenetic isozymes increase to a concentration equal to that of the original or parental isozyme in aged or diseased maize leaves. All 6 alleles of the Px_3 locus develop isozymes, producing 8 bands in heterozygotes. Similar enzymes characterize other grasses and are notable in the highly polyploid sugarcane where a major leaf enzyme similar to maize 3 is represented by a host of closely spaced bands.

Diseases and injuries to plant tissues are almost inevitably associated with increases in peroxidase activity (Farkas and Stahmann 1966; Yu and Hampton 1964). Maize diseases that lead to increase in leafy tissue peroxidase activities, especially enzymes 3 and 7, are the dwarfing Maize Mosaic Virus I, the *Helminthosporium turcicum* and *H. maidis* blights, and the *Puccinia sorghi* rust. There is no evidence in maize that these diseases derepress specific genes without affecting others, nor does this appear true upon treatments with auxins; rather, all peroxidases of the infected tissues increase proportionately.

INTRACELLULAR LOCALIZATION OF MAIZE PEROXIDASES

Peroxidases are often found in significant association with cell wall and particulate fractions of plant tissues (Stafford and Bravinder-Bree 1972; Raa 1973). Divalent cations are among the best agents for release of the peroxidases from the wall pellets after centrifugation (Lipetz and Garro 1965). Several maize tissues have been assayed for the proportions of wall-bound and cytoplasmic or soluble peroxidase (Hasegawa, 1974), commonly extracting with 0.2 M $CaCl_2$ for 30 minutes. Variations in the wall-bound fraction occur between tissues and during ontogeny, with a typical pattern of increasing wall-binding with increasing age of tissue.

The more acidic enzymes were recovered largely as soluble proteins in tissues studied(Table3). All cathodal enzymes (1, 4,and5) were highly associated with wall pellets and were often difficult to visualize in the soluble fraction. Enzymes 3, 6, 7, and 8 were found in limited association with the wall pellet.

TABLE 3

Relative activities of 12 major peroxidases on several hydrogen donors, and relationship of wall-bound and soluble fractions (enzymes arranged from most basic to most acidic).

Substrate	Relative Activity Scores of Peroxidases:*											
	1	5	4	7	8	11	3	10	6	2	9	12
benzidine	2	3	3	4	4	5	4	5	1	4	1	1
o-dianisidine	2	3	3	4	4	5	4	5	1	4	1	1
guaiacol	2	1	2	4	1	1	2	0	0	3	0	0
pyrogallol	2	3	0	3	0	0	2	0	0	4	0	0
catechol	1	1	1	2	2	2	3	0	0	3	0	0
caffeic acid	2	3	1	1	1	1	1	0	0	3	0	0
eugenol	4	4	3	5	3	3	3	0	0	5	0	0
indoleacetic acid	1	1	2	4	4	4	5	1	1	3	0	1
Relationship of wall (W) and soluble (S) fractions	W>S	W>S	W>S	S=W	S>W	S	S>W	S	S>W	S	S	S

*Scale ranges from 0 = no activity to 5 = most intense activity on each substrate.

Week-old coleoptilar peroxidases were about equally distributed in wall and soluble fractions (Figure 3). Cathodal enzymes 1, 5, and 4 were highly associated with the wall fraction and the non-migrating 7 was about equally distributed between the two fractions. At an earlier stage, the proportion of soluble protein was higher. Similar evidence for increasing association of peroxidases with the wall fraction has been obtained in banana by Haard (1973). In critical electrophoretic studies of peroxidases, caution must be exercised to release these wall-bound fractions.

The maize data suggest a general correlation between the basic charge of a peroxidase and its degree of association, largely through covalent bonding, with the wall pellets.

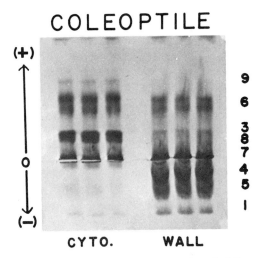

Fig. 3. Zymogram of the wall-bound and soluble or cytoplasmic fractions of peroxidases from the maize coleoptile.

Peroxidases of very young sorghum seedlings generally conformed to this pattern (Stafford and Bravinder-Bree 1972). The two major wall-bound peroxidases of cucurbits included one cathodal and one near-neutral at pH 8, while two other cathodal peroxidases were found largely in soluble fractions (Denna 1974). Observations on wall binding in banana fruit peroxidases (Haard 1973) lend no support to the charge hypothesis.

SUBSTRATE SPECIFICITIES OF MAIZE PEROXIDASES

Peroxidases generally have a broad substrate specificity. Benzidine and o-dianisidine are the most practical substrates for gel electrophoresis, despite critics (van Loon 1971); they give identical results for the maize enzymes (Table 3). Data in Table 3 were derived empirically from acrylamide gels stained in 0.01 M substrate in phosphate buffer at pH 6; both substrate concentration and pH can affect these values. Guaiacol and pyrogallol were not suitable substrates for visualization of all maize peroxidases on acrylamide and staining was very slow on each (Table 3). In the absence of hydrogen peroxide, none of the peroxidases could utilize substrates such as catechol or caffeic acid, although enzymes 1, 2, 3, and 6 showed some laccase activity on p-phenylenediamine in the absence of H_2O_2.

Eugenol is a substrate for flourescent, lignin-like materials, providing an indication of lignification capability by peroxidases (Liu 1971). Most of the maize peroxidases utilize eugenol well as a substrate, including peroxidase 2 from a non-lignified tissue, the pollen (Table 3). The cathodal enzymes 1, 4, and 5 were especially active on eugenol; these are enzymes commonly associated with the cell wall fraction. A highly-lignified tissue like the stele (Table 2) thus has one major peroxidase (8) and two minor ones (1 and 5) that utilize eugenol and are at least partly wall-associated; it is noteworthy that the latter two are often absent in null allele homozygotes of the two loci. Preliminary studies of the brown midrib-3 mutant, which has a 30% reduction in leaf and stalk lignins, revealed a substantial decrease in the wall-binding of lignifying enzymes 1, 4, and 5 (Hasegawa 1974).

Indoleacetic acid was utilized most efficiently as a sub-strate by enzymes 3, 7, 8, 6, and 11, all slow-moving anodal proteins (Table 3). The remaining peroxidases can utilize IAA, but less efficiently. Colorimetric verification of the gel data has been obtained for enzyme 3, a most competent IAA oxidase. Enzymes 3 and 7 are almost always found together in maize, especially in leafy tissues. Their activities increase greatly upon maturation (Figure 2), are elevated by many leaf diseases, and are very high in mutants such as slashed leaf and Knotted leaf. Both enzymes are also very active on eugenol, and assignment of a specific auxin-control function would be premature. Among the other IAA oxidases, enzyme 8 is strongest in meristematic tissues, stele, and pith; enzyme 11 is restric-ted to roots and tissue cultures, increasing greatly in aged tissues. Auxin control in the roots could also be attributed to the single enzyme 11. Two rootless mutants reducing secondary rooting in maize have been studied for 11 activity and both were entirely normal; one of these, rootless-1, is auxin-reversible.

Peroxidase 2 was distinguished both by a restricted tissue specificity, to the pollen grains, and a broad substrate specificity (Table 3). It showed some laccase activity but no polyphenoloxidase activity in the absence of H_2O_2 and utilized many substrates efficiently in its presence. Activity of this peroxidase disappeared rapidly from pollen that were lyophilized or air-dried. Since maize and other grass pollen are extremely short-lived, peroxidase 2 may hold a clue to this fragility.

DISCUSSION

The 13 major maize peroxidases have entirely distinct tissue specificities in 21 surveyed tissues. No tissue was free of peroxidase activity. Several mature tissues (anther,

pith, embryo, endosperm, silk, and pollen) had 4 or fewer
active enzymes, but none of the tissues displayed more than
8 of the 13. Two similar patterns of gene action were those
of loci Px_3 and Px_7 in leafy tissues and of the enzymes 11 and
12 of roots (Table 2); other enzymes had entirely unique
specificities.

Tissue-specific gene repression is thus intimated for most
maize peroxidases in most tissues (63% of the 273 associations
in Table 2). Only 2 peroxidase loci, Px_1 and Px_5, appeared to
be generally active in all maize tissues. The ontogenetic
evidence suggests genetic derepression of several enzymes
during maturation at stages corresponding to the onset of
lignification and to the later stages of cell elongation and
auxin control. The derepression of three loci Px_3, Px_7, and
Px_8, during development of leafy tissues might be related to
auxin stimulus, since all are active IAA oxidases. Loci Px_3
and Px_7 are also activated by diseases and injuries to leaves,
perhaps due to ethylene stimulus (Farkas and Stahmann 1966).
The 11 enzyme (gene locus not known) that dominates peroxidase
activity in maize roots and tissue cultures is the only
effective IAA oxidase of these tissues. It seems improbable
that such an enzyme is restricted in function to IAA, but
auxins may play a major role in derepression of the controlling
locus in roots and in tissue cultures. In the tissue cultures,
2,4D may promote synthesis of peroxidase 11, an enzyme not
normally found in endosperm or stem apex, source of the calluses
studied. Auxins are associated both with increase and repression of peroxidases in tobacco callus and other plant tissues,
and the concentration of the hormones is critical (Lee 1971,
Lavee and Galston 1968).

The control of lignification in maize must be resident
especially in enzymes like peroxidase 11 of roots and 8 of the
stele and pith. The presence of high activity of 8 in juvenile
tissues like the stem apex (Table 2) is enigmatic but it is
perhaps poorly associated with the walls in these tissues.
The ability of most maize peroxidases to utilize eugenol as a
substrate suggests that lignification often involves several
enzymes, perhaps in direct relation to their availability at
the sites of lignification.

Genetic nulls exist for at least 5 of the 13 maize peroxidases (Tabel 1), and the nulls have been associated with no
adverse morphological or physiological effects. In appropriate
null genotypes, for example, silks can be obtained with only
the enzyme 4, steles with only enzyme 8, anthers with only
enzyme 10, and roots having only enzymes 11 and 12. Further
study of such genotypes may aid in resolving the principal
functions of these peroxidases.

REFERENCES

Brewbaker, James L. 1972. Genetic marker stocks in tropical flint background. *Maize Genetics Cooperative Newsletter* 46: 33-37.

Brewbaker, James L., Mahesh D. Upadhya, Yrjo Makinen, and Timothy Macdonald 1968. Isoenzyme polymorphism in flowering plants. III. .Gel electrophoretic methods and applications. *Physiol. Plant.* 21: 930-940.

Brewbaker, James L. and Y. Hasegawa 1974. Nine maize peroxidase loci and their tissue specificities. *Maize Genetics Cooperative Newsletter* 58: (in press).

Denna, Donald W. 1974. The isoperoxidases of *Cucurbita pepo* L. *II. Isozymes: Physiology and Function.* C. L. Markert, editor. Academic Press, New York.

Dvorak, M. and J. Cernohorska 1967. Peroxidases of different parts of the pumpkin plant (*Cucurbita pepo* L.). *Biol. Plant.* 9: 308-316.

Endo, T. 1967. Comparison of the effects of gamma-rays and maleic hydrazide on enzyme systems of maize seed. *Radiation Bot.* 7: 35-40.

Farkas, G. L. and Mark A. Stahmann 1966. On the nature of changes in peroxidase isoenzymes in bean leaves infected by southern bean mosaic virus. *Phytopathology* 56: 669-677.

Gordon, Albert R. and N. A. Alldridge 1971. Cytochemical localization of peroxidase A in developing stem tissues of extreme dwarf tomato. *Can. J. Bot.* 49: 1487-1496.

Gordon, John C. 1971. Changes in total nitrogen, soluble protein, and peroxidases in the expanding leaf zone of eastern cottonwood. *Plant Physiol.* 47: 595-599.

Haard, N. F. 1973. Upsurge of particulate peroxidase in ripening banana fruit. *Phytochemistry* 12: 555-560.

Hamill, D. E. and James L. Brewbaker 1969. Isoenzyme polymorphism in flowering plants. IV. The peroxidase isoenzymes of maize (*Zea mays*). *Physiol. Plant.* 22: 945-958.

Hasegawa, Y. 1974. Polymorphism of peroxidase isozymes and their biochemical properties in maize (*Zea mays* L.). Ph.D. thesis, University of Hawaii, Department of Horticulture, Honolulu, Hawaii.

Lavee, S. and A. W. Galston 1968. Structural, physiological, and biochemical gradients in tobacco pith tissue. *Plant Physiol.* 43: 1760-1768.

Lee, T. T. 1971. Promotion of indoleacetic acid oxidase isoenzymes in tobacco callus cultures by indoleacetic acid. *Plant Physiol.* 48: 56-59.

Lee, T. T. 1972. Interaction of cytokinin, auxin, and gibberellin on peroxidase isoenzymes in tobacco tissues cultured in vitro. *Can. J. Bot.* 50: 2471-2477.

Lipetz, Jacques and Anthony J. Garro 1965. Ionic effects on lignification and peroxidase in tissue cultures. *J. Cell Biol.* 25: 109-116.

Liu, Edwin H. 1971. The use of eugenol and tyrosine as zymogram stains specific for the peroxidase-catalyzed condensation of phenolic compounds. *Isozyme Bulletin* 4: 41.

Macdonald, Timothy and James L. Brewbaker 1972. Isoenzyme polymorphism in flowering plants. VIII. Genetic control and dimeric nature of transaminase hybrid maize isoenzymes. *J. Heredity* 63: 11-14.

Macnicol, P. K. 1966. Peroxidases of the alaska pea (*Pisum sativum* L.). Enzymatic properties and distribution within the plant. *Arch. Biochem. Biophys.* 117: 347-356.

Martin, Franklin W. and James L. Brewbaker 1971. The nature of the stigmatic exudate and its role in pollen germination. In: *Pollen: Development and Physiology.* J. Heslop-Harrison ed., Butterworths, London.

McCune, D. C. 1961. Multiple peroxidases in corn. *Ann. N. Y. Acad. Sci.* 94: 723-730.

Peirce, L. C. and J. L. Brewbaker 1973. Applications of isozyme analysis in horticultural science. *Hort. Science* 8: 17-22.

Raa, J. 1973. Cytochemical localization of peroxidase in plant cells. *Physiol. Plant.* 28: 132-133.

Racusen, D. and M. Foote 1966. Peroxidase isozymes in bean leaves by preparative disc electrophoresis. *Can. J. Bot.* 44: 1633-1638.

Shannon, L. M. 1968. Plant isoenzymes. *Ann. Rev. Plant Physiol.* 19: 187-210.

Stafford, H. A. and S. Bravinder-Bree 1972. Peroxidase isozymes of first internodes of sorghum. *Plant Physiol.* 49: 950-956.

Van Loon, L. C. 1971. Tobacco polyphenoloxidases: A specific staining method indicating non-identity with peroxidases. *Phytochemistry* 10: 503-507.

Yu, Leona M. and Raymond E. Hampton 1964. Biochemical changes in tobacco infected with *Collectotrichum destructivum*. II. Peroxidases. *Phytochemistry* 3: 499-501.

DIFFERENTIAL EXPRESSION OF PARENTAL
ALLELES FOR AMYLASE IN MAIZE

SU-EN CHAO AND JOHN G. SCANDALIOS
Division of Biochemical Genetics School of Medicine
State University of New York at Buffalo
Buffalo, New York 14207
and
Genetics Laboratory, Department of Biology
University of South Carolina
Columbia, South Carolina 29208

ABSTRACT. Three major classes of starch degrading enzymes
have been identified in *Zea mays* (L.): α-amylase, β-amylase, and phosphorylase. The α- and β-amylases have been
shown to be controlled by independent loci; each locus consists of two codominant alleles. Developmental studies
have shown that the alleles at the *Amy-1* gene locus, coding
for the α-amylase, are differentially expressed temporally
and spatially. A model is proposed where we invoke a modulator gene as the most likely mechanism to account for the
differential expression of the *Amy-1* locus.

INTRODUCTION

We have identified two major amylases in maize; α-amylase
and β-amylase, each controlled by an independent structural
locus with a pair of co-dominant alleles (Chao and Scandalios,
1969; 1971; Scandalios et al., 1974). In the course of developmental studies, we found both amylases to be abundant in
the seed endosperm, but their expression differed considerably
during development. If the expression of the maize amylases
is dependent on gibberellic acid, as in the case of barley
α-amylase in excised aleurone layers (Varner and Chandra,
1964), our findings may provoke interest in examining how the
hormone differentially activates these two genes. Furthermore, we report here the detection of differential allelic
expression of the α-amylase gene in various developmental
stages of maize; a phenomenon which may be common, but overlooked, in other systems in a variety of organisms.

MATERIALS AND METHODS

The following fourteen maize inbred lines were obtained
from Dr. E.C. Rossman, Crop and Soil Science Department,
Michigan State University: M14, W64A, 58-3-5, 58-3-6, 58-3-9,
38-11, Oh51A, In2, MS206, MS215, CMD5, A509. Golden Cross
Bantam, and Hawaiian Sugar. These have been inbred for at

at least 10 generations. Methods of growth, tissue extract, amylase assay, and polyacrylamide gel electrophoresis have been reported by us earlier (Chao and Scandalios, 1972).

RESULTS

DEVELOPMENTAL PROFILES OF AMYLASES IN MAIZE

According to the electrophoretic pattern and the developmental time of their appearance, starch-degrading enzymes in maize can be grouped into four zones (Figure 1). The enzyme near the origin (Zone 3) has been identified as phosphorylase (Chao and Scandalios, 1969). Among the other three zones, the most anodal zone (Zone 1) consists of three α-amylases, zone 2 consists of four β-amylases and zone 4 contains a single amylase; whether the latter is an α- or β-amylase is not yet clear. Zone 4 is only detected in the leaves, roots and residual seed proper of the young seedlings; the other three zones are present in essentially all tissues from gametophytic pollen extract to sporophytic immature kernels and germinating seeds. In alignment with the zymogram are the quantitative measurements of amylase activity before seed maturation and following seed germination (bottom Figure 1).

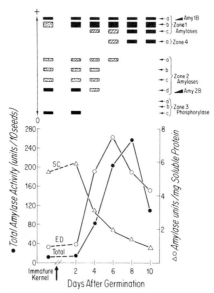

Fig. 1. Developmental profiles of maize starch degrading enzymes; qualitatively shown following polyacrylamide gel electrophoresis, and quantitatively assayed for amylase activity by the starch-iodine method.

When amylase units per mg of soluble protein are plotted
against time in days of germination, the level in endosperm
extract climbs steeply to a peak at the 6th day, while that
in scutellar extract declines sharply and levels off at the
6th day. From the general developmental profile of amylases,
we can conclude tentatively, without the quantitation of each
one of the amylase isozymes, that α-amylase is responsible for
the major increase of amylolytic activity during seed germination, and it is mostly localized in the endosperm. That is,
during seed germination α-amylase is the predominant species
of soluble proteins in the endosperm but not in the scutellum.
The appearance of the β-amylases is limited to the first 6
days of seed germination.

GENETIC VARIANTS OF AMYLASE ISOZYMES

The complexity of the developmental profile of the amylases
suggests that there may be more than two structural genes controlling the amylases and that the isozymes may be resulting
from protein-protein interactions. To avoid such complica-
tions, we screened maize inbred lines for genetic variants by
using the liquid endosperm of immature kernels which are
known to have consistent isozyme patterns throughout the 30
days of the seed maturation period. Among the 14 unrelated
inbred lines investigated, electrophoretic variants were
found in the most anodic α-amylase (Zone 1 a) and in the least
anodic β-amylase (Zone 2 d). Henceforth, the α-amylase
variants are designated as Amy-1A and Amy-1B for the fast and
slow electrophoretic isozymes, respectively, and likewise for
the β-amylase variants as Amy-2A and Amy-2B. For genetic
analysis, we used three out of the fourteen inbred lines which
have the advantage of having high yield following artificial
pollination and are thus most desirable for these studies.
The three inbred lines are: W64A (Amy-1A, Amy-2B), 58-3-6
(Amy-1A, Amy-2A), and 38-11 (Amy-1B, Amy-2B).

INHERITANCE OF AMY-2 β-AMYLASE

The genetics of Amy-2 was elucidated by scoring the Amy-2
phenotypes of liquid endosperm in immature kernels resulting
from backcrosses and F_2 progeny of W64A and 58-3-6. The mode
of segregation among 657 backcross progeny and 508 F_2 progeny
fits single-gene Mendelian inheritance (Chao and Scandalios,
1969). A typical gel with 9 F_2 segregants is shown in
Figure 2. Thus, Amy-2 as demonstrated is controlled by two
co-dominant alleles. Since Amy-2 is abundant in the triploid
endosperm, it might be expected to show a gene-dosage effect
if Amy-2 is primarily synthesized in this tissue. However, a

gene-dosage effect was not detected in the liquid endosperm of F_1 of either $Amy\text{-}2^A/Amy\text{-}2^A/Amy\text{-}2^B$ or $Amy\text{-}2^B/Amy\text{-}2^B/Amy\text{-}2^A$. This finding suggests that the endosperm tissue may not be the site for the synthesis of Amy-2 β-amylase. An alternative explanation is that the $Amy\text{-}2$ gene-dosage effect was not detected by the gel assay method used. This is quite unlikely for two reasons: (1) concentrated F_1 endosperm extract failed to affect the relatively equivalent band intensities of the two allelic β-amylases; (2) the same gel assay method clearly demonstrated the gene-dosage effect of Amy-1 α-amylases; results of which will be presented in the following sections.

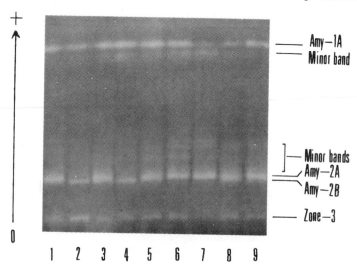

Fig. 2. Electrophoretic analysis of Amy-2 in individual kernels from a F_2 cross between W64A ($Amy\text{-}2^B/Amy\text{-}2^B$) and 58-3-6 ($Amy\text{-}2^A/Amy\text{-}2^A$). 0 = point of sample insertion. Samples 1, 6, and 7 are A^2A^2; 2 and 4 are B^2B^2; 3, 5, 8, and 9 are A^2B^2.

INHERITANCE OF AMY-1 α-AMYLASE

Like Amy-2, Amy-1 was shown to be controlled by two co-dominant alleles at a single-gene locus (Chao and Scandalios, 1971). However, the mode of its inheritance was obscure in our early attempts to assay the Amy-1 pattern in immature kernels from F_1, backcrosses and F_2 progeny of W64A and 38-11. In particular, in F_1 only Amy-1A phenotype (A^1) was observed when the maternal parent was A^1. Conversely, Amy-1B was observed in the reciprocal cross (Table 1). The puzzle of the apparent maternal inheritance of Amy-1 was not resolved until

TABLE 1

THE PATTERN OF AMY-1 AMYLASE IN GERMINATING F_1 MAIZE OFFSPRING

Crosses and Tissues	Immature* Kernel	Days of Germination					
		2	4	6	8	10	12**
W64A x 38-11							
Endosperm	A^1	A^1	A^1	A^1	A^1	A^1	A^1
Scutellum		A^1	A^1	A^1	A^1	A^1	A^1
Shoot	A^1	A^1B^1	A^1B^1	A^1B^1	A^1B^1	A^1B^1	A^1B^1
Root				A^1B^1	A^1B^1	A^1B^1	A^1B^1
38-11 x W64A							
Endosperm	B^1	A^1B^1	A^1B^1	A^1B^1	A^1B^1	A^1B^1	A^1B^1
Scutellum			A^1	A^1	A^1	A^1B^1	A^1B^1
Shoot	A^1	A^1	A^1	A^1B^1	A^1B^1	A^1B^1	A^1B^1
Root				A^1B^1	A^1B^1	A^1B^1	A^1B^1
58-3-6 x 38-11							
Endosperm	A^1	A^1	A^1	A^1	A^1		
Scutellum			A^1	A^1	A^1		
Shoot	A^1		A^1B^1	A^1B^1	A^1B^1		
Root				A^1B^1	A^1B^1		
38-11 x 58-3-6							
Endosperm	B^1		A^1B^1	A^1B^1	A^1B^1		
Scutellum			A^1B^1	A^1B^1	A^1B^1		
Shoot	A^1		A^1B^1	A^1B^1	A^1B^1		
Root			A^1B^1	A^1B^1	A^1B^1		

*pooled supernatant of endosperm and embryo extract from twenty 20-day old immature kernels.
**Extracts were made from pooled tissues of five seedlings. W64A and 58-3-6 = *Amy-1A/Amy-1A*; 38-11 = *Amy-1B/Amy-1B*.

the tissues of germinating seedlings were assayed; it was
then suggested that the expression of the two alleles controlling Amy-1 might be temporally and spatially dependent. The
results on tissue distribution of the two allelic isozymes
Amy-1A and Amy-1B are shown in Table 1. Leaf tissue of two-
week old seedlings were preferred for the purpose of elucida-
ting genetic segregation because Amy-1 was at that time the
only amylase present in that tissue. A typical gel assay of
F_2 segregation of Amy-1A and Amy-1B is shown in Figure 3. A
total of 336 progeny including 87 from the backcross B^1 x
(A^1 x B^1) and 249 from the selfing F_1 (B^1 x A^1) were scored
for Amy-1 phenotypes in their leaf tissues. The results given
in Table 2 seem to fit monogenic inheritance. However, this
supposition became questionable when we began to look into the
Amy-1 phenotypes in endosperm of the same germinating seedlings
from which the leaf assays were made.

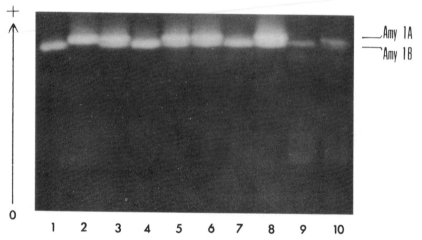

Fig. 3. Electrophoretic analysis of Amy-1 in leaf extracts
of 14-day-old F_2 seedlings of 38-11 ($Amy-1^B/Amy-1^B$) and W64A
($Amy-1^A/Amy-1^A$). 0 = point of sample insertion. Sample numbers 1, 4, 7, and 9 are B^1B^1; 2 is A^1A^1; 3, 5, 6, 8, and 10
heterozygotes (A^1B^1 and B^1A^1 are not distinguishable).

DIFFERENTIAL ALLELIC EXPRESSION OF AMY-1

In contrast to the leaf extract, the F_2, endosperm were
shown to segregate into 4 phenotypes; two parental and two
heterozygous. The latter two can be distinguished from each
other on the basis of the relative isozyme band intensity.
There are either a strong Amy-1A coupled with a weak Amy-1B
or the two bands have equal intensity as shown in Figure 4.

TABLE 2

MODE OF INHERITANCE OF AMY-1 AMYLASE VARIANTS IN
LEAF TISSUE OF 6-22 DAY-OLD MAIZE SEEDLINGS

| Genetic cross | | Days after | Amy-1 patterns in offspring | | | | χ^2 | P |
Female	Male	Germination	A1	A1B1	B1	Total		
B1	(A1 x B1)	6	0	46	41	87	0.28	> 0.50
(B1 x A1)	(B1 x A1)	6	11	21	8	40	0.55	> 0.70
		8-10	36	50	27	113	0.89	> 0.20
		14-22	26	53	17	96	2.73	> 0.20
		F_2 Sum	73	124	52	249	3.57	> 0.10

Note: The χ^2 values were calculated on the basis of monogenic inheritance, A^1B^1 = Amy-1A and Amy-1B are of equal staining intensity.

681

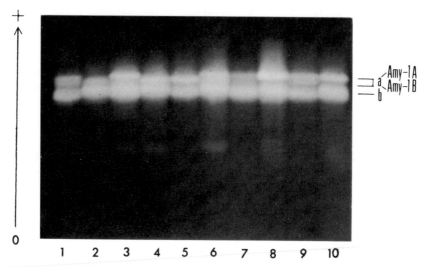

Fig. 4. Electrophoretic analysis of Amy-1 in individual endosperm of 10-day-old F_2 seedlings of 38-11 (Amy-1^B/Amy-1^B) and W64A (Amy-1^A/Amy-1^A). 0 = point of sample insertion. Samples numbers 2, 4, and 5 are B^1B^1; 8, and 10 are A^1A^1; 3 is A^1B^1; and 1, 6, 7, and 9 are B^1A^1. Electrophoretic variants of the broad band traveling behind the Amy-1 are detected in the endosperm extracts of these F_2 seedlings suggesting that it is genetically independent from Amy-1. The genetics of this particular band of amylase activity, however, was not studied in depth.

When 238 selfing progeny of (A^1 x B^1) and 201 selfing progeny of (B^1 x A^1) were sorted out for their Amy-1 phenotypes in the endosperm of 8-10 day old seedlings, the Chi-square values we calculated for the 1 : 1 : 1 : 1 ratio significantly exceeding the statistically acceptable level (Chao and Scandalios, 1971). The discrepancy of the scores between leaf extract and endosperm extract obtained from the seedlings was further strikingly demonstrated in the backcross A^1 x (B^1 x A^1) as shown in Table 3 where A^1 dominates the scores obtained from endosperm while A^1 and A^1B^1 are equally represented in the leaf. The possibility of erroneous classification of A^1B^1 for A^1 in endosperm was ruled out by the fact that both endosperm and leaf extracts from 72 seedlings were assayed in pairs and yet the frequent observation of A^1B^1 from leaf in no way helped the scoring of A^1B^1 from endosperm. This was observed in the crosses wherever B^1 was contributed only by the participating paternal heterozygote, and there was no evidence

TABLE 3

GENETIC SEGREGATION OF *AMY-1* IN BACKCROSS PROGENY

Genetic cross		Amy-1 patterns in offspring					χ^2	P
Female	Male	A^1	$\underline{A^1B^1}$	$A^1\underline{B^1}$	B^1	Total		
$(A^1 \times B^1)$	B^1	0	34	0	32	66	0.06	> 0.80
$(B^1 \times A^1)$	A^1	49	0	56	0	105	0.46	> 0.30
$(B^1 \times A^1)$	B^1	0	43	0	44	87	0.02	> 0.80
A^1	$(B^1 \times A^1)$							
Endosperm		237	17	0	0	254	> 500	< 0.01
Leaf		39	0	33	0	72	0.50	> 0.30

Note: Endosperm from 8-10 day old seedlings were used in the first three backcrosses. In the 4th backcross, endosperm and leaf tissue from 72 out of 273 twelve-day old seedlings were assayed in pairs. The χ^2 values were calculated on the basis of monogenic inheritance $\underline{A^1B^1}$ = the staining intensity of Amy-1A is stronger than the Amy-1B. $A^1\underline{B^1}$ = Amy-1A and Amy-1B are of equal staining intensity.

of any selective survival of the kernels from the randomly
picked ears. The Amy-1 phenotype in endosperm of three other
backcrosses shown in Table 3 all yielded normal monogenic
segregation ratios.

The finding of the latent expression of $Amy-1^B$ in germinating heterozygotes (Table 1) and the deficiency of $\underline{A}^1B^1$ in the
endosperm of A^1 x (A^1 x B^1) backcross (Table 3) led us to the
belief that the aforementioned aberrant F_2 endosperm segregation ratios could be explained by offspring showing an A^1
phenotype with maternal $Amy-1^A$ and paternal $Amy-1^B$. Besides,
the latent expression of $Amy-1^B$ genotype is tissue specific
and age dependent. If this explanation is valid we can look
upon the selfing progeny of (B^1 x A^1) as the sum progeny of
two crosses namely: B^1 x (B^1 x A^1) and A^1 x (B^1 x A^1). To
verify this hypothesis we studied the segregation of Amy-1
allelic amylases in the endosperm of F_2 seedlings on sequential
days after germination until the endosperm was totally disin-
tegrated at the expense of seedling growth. The putative
segregation ratio in the cross B^1 x (B^1 x A^1) would be normal
and that in the cross of A^1 x (B^1 x A^1) would be distorted but
the distorted ratio would be less severe when $Amy-1^B$ was finally expressed as development of the seedlings progressed. The
results of this experiment (Table 4) are consistent with our
hypothesis in the following two aspects: (1) the segregation
ratio of A^1, $\underline{A}^1B^1$, A^1B^1, and B^1 is generally in agreement with
the supposition that the F_2 progeny in question are the sum of
B^1 x (B^1 x A^1) and A^1 x (B^1 x A^1) with the latter segregating
with an apparent deviation from 1 : 1 ratio; (2) the expression
of $Amy-1^B$ in the triploid endosperm of $Amy-1^A/Amy-1^A/Amy-1^B$
genotype was detected only in older seedlings. This late
manifestation of $\underline{A}^1B^1$ in endosperm is apparent in comparing the
8-10 day old with the 4-day old seedlings. It is sustained by
the fact that the Chi-square value calculated for each age
group decreases with the age of the seedlings assayed. In
fact, the segregation ratio fits better with 3 : 1 : 2 : 2
than with 1 : 1 : 1 : 1, even in the older seedlings. This
leads us to the speculation for the existence of another gene
modulating the allelic expression of $Amy-1$ which will be discussed later. With respect to the 3 : 1 : 2 : 2 segregation,
the Chi-square contributions from A^1 and $\underline{A}^1B^1$ phenotypes
dropped from 93.8% at the 4th day to less than 50% at the 10th
day of growth. In fact, the endosperm segregation ratio
starting at the 4th day with a misfit ends up with a good fit
to the ratio 3 : 1 : 2 : 2. This ratio can be accounted by
the $Amy-1^A/Amy-1^A/Amy-1^B$ endosperm having the A^1 phenotype
with certain exceptions which will be explained in our discussion. It might also be noted that by the 10th day of growth
the F_2 segregation would also fit the normally expected

TABLE 4

GENETIC SEGREGATION OF AMY-1 IN ENDOSPERM TISSUES OF F_2 SEEDLINGS ASSAYED AT
VARIOUS DAYS AFTER SEED IMBIBITION AND GERMINATION

| Days of Germination | Amy-1 phenotype in F_2 segregants | | | | | χ^2 and (P) values for ratios | | | % χ^2 attributed to A^1 and A^1B^1 | |
	A^1	$\underline{A}^1B^1$	A^1B^1	B^1	Total	1:1:1:1	3:1:2:2	1:1:1:1	3:1:2:2
4	54	0	17	21	92	66.6 (<.01)	14.4 (<.01)	96.9	93.8
6	27	2	10	7	46	30.7 (<.01)	11.6 (<.01)	93.5	67.7
8	47	17	24	30	118	16.7 (<.01)	1.4 (>.7)	94.0	26.9
10	28	13	18	25	84	6.5 (>.05)	2.2 (>.5)	81.6	46.2

Note: $\underline{A}^1B^1$ = the staining intensity of Amy-1A is stronger than that of Amy-1B. A^1B^1 = Amy-1A and Amy-1B are of equal staining intensity.

1 : 1 : 1 : 1 ratio for a single gene with two co-dominant alleles, although the fit was statistically a borderline case.

DISCUSSION

THE IMPLICATION OF THE GENE-DOSAGE EFFECTS ON TISSUE ORIGINS OF AMY-1 α-AMYLASE AND AMY-2 β-AMYLASE

The asymmetric distribution of the amounts of two allelic gene products in the triploid heterozygous maize endosperm was observed with Amy-1 but not Amy-2. The simplest interpretation is that the α-amylase, Amy-1, from the endosperm is endogenously synthesized but the β-amylase is not. Nevertheless, this observation should not be taken to mean that Amy-1 and Amy-2 are each synthesized in exclusive tissues: for instance Amy-1 from endosperm and Amy-2 from scutellum. Tissue specific allelic expression of *Amy-1* is evident in our results presented in Table 1. Comparing our observation on the amylase gene-dosage effect in maize with those in barley (Carlson, 1973; Allison, 1973) where gene-dosage effects were observed with certain α- and β-amylase isozymes from aleurone layers and endosperm, one must reach the conclusion that the ontogeny of these enzymes in maize and barley is different. This is not surprising since the analogy between the two plants exists only at the anatomical level.

THE DIFFERENTIAL EXPRESSION OF AMY-1^A AND AMY-1^B

When the entire span of maize seed development is examined from maturation to germination, there is no preferential expression of either one of the two alleles, although their expressions are not in synchrony (Table 1). The repression of Amy-1^B, however, is most pronounced in the triploid endosperm of Amy-1A/Amy-1A/Amy-1B genotype regardless whether it is in F_1, backcross, or F_2 progeny.

While the phenomenon of differential allelic expression is just being unveiled, we can try to relate our observations to similar findings reported by others and speculate on some conceivable underlying mechanisms. Recently, Davies (1973) had observed the preferential expression of a maternal allele for the embryo globulins in heterozygotes between certain strains of *Pisum sativum*. Among animal species, similar cases have been reported in Drosophila (Courtright, 1967), frog (Wright and Moyer, 1966), trout (Hitzeroth, et al., 1968; Goldberg, et al., 1969), chicken-quail hybrids (Castro-Sierra and Ohno, 1968), and carp (Klose and Wolf, 1970). In all cases the observations were limited to interspecific hybrids whose viability was low and it was always the onset of the paternal isozyme allele being delayed. Our observation is specifically for a single gene locus, *Amy-1*, and is irrespective of its parental origin. It would be interesting if the same investigation can be extended into somatic cell hybrids in plants and animals. The inherent handicap in recognizing such differential allelic expression in somatic cell hybrids might be that the hybrid cells are fixed in their timing of differentiation. Hence, the two allelic isozymes gene expression might either have been both activated prior to cell hybridization, or one of the two alleles might remain silent throughout the limited development of the cell hybrid and does not get the chance to be expressed. Our proposal is perhaps a mere reemphasis in part of Nelson and Burr's comment (1973) on comparing gene action in parasexual hybrid cells with sexual hybrid cells in plants, particularly since interspecific hybrids of tobacco have been produced by Carlson et al. (1972). Furthermore, we would like to see plant somatic cell hybrids made between differentiated and undifferentiated cells in a fashion analogous to the hamster melanoma x mouse fibroblast hybrids studied by Davidson (1973) so that one can perhaps manipulate the mechanisms modulating differential allelic expression of several isozyme loci.

A model is needed to accommodate and explain the various aspects of differential allelic expression of Amy-1 which we have observed in genetic crosses as well as in triploid and diploid tissues. Let us assume that there is a factor for promoting the synthesis of Amy-1 α-amylase, and it exists in two allelic forms encoded by a gene not linked to *Amy-1*. The allele, p^+, codes for the promotion factor for Amy-1B and is present in the genome of 38-11 and distinguished by its affinity to the gene product of $Amy-1^B$ versus that of $Amy-1^A$ for which the promotion factor is coded by the p^- allele. Our data would require us to impose two restrictions on the mode of this modulator gene: (1) recombination between this gene and *Amy-1* occurs during microsporogenesis but not megasporogenesis. In other words, a heterzygote (A,-/B,+) resulting from W64A (A,-) and 38-11 (B,+) produced four types of pollen; two parental and two recombinants with equal frequencies, but would yield only two parental types of megaspores; (2) the promotion factor affects the two allelic products of $Amy-1^A$ a and $Amy-1^B$ by the balance of its gene dosage. The allele, $p-$, favors the expression of $Amy-1^A$ whereas the, $p+$, favors that of $Amy-1^B$. In the heterozygous triploid endosperm where the allelic dosage is imbalanced; the genotype A-/A-/B+ would give rise to the A^1 phenotype, but the genotype B+/B+/A- would have the phenotype A^1B^1. The recombinant B+/B+/A+ would be phenotypically A^1B^1 because a basal level of Amy-1A is still being synthesized in the absence of Amy-1A promotion factor.

TABLE 5

THE PATERNAL GENETIC SEGREGATION OF A HYPOTHETICAL GENE FOR MODULATING THE DIFFERENTIAL ALLELIC EXPRESSION OF *AMY-1* AND ITS EFFECT ON THE PHENOTYPES IN THE OFFSPRING

Genetic cross Female	Male	Genotype	Endosperm A^1	$\underline{A^1B^1}$	A^1B^1	B^1	Leaf A^1	A^1B^1	B^1
$(A,-)$	$(B,+)$	$(A,-/B,+)$	1					1	
$(B,+)$	$(A,-)$	$(B,+/A,-)$			1			1	
$(A,-)$	$(B,+ \times A,-)$	$(A,-/A,-)$	1				1		
		$(A,-/A,+)$	1				1		
		$(A,-/B,-)$		1*				1	
		$(A,-/B,+)$	1					1	
		Phenotypic Ratio	3	: 1			2	: 2	
$(B,+)$	$(B,+ \times A,-)$	$(B,+/A,-)$			1			1	
		$(B,+/A,+)$			1			1	
		$(B,+/B,-)$				1			1
		$(B,+/B,+)$				1			1
		Phenotypic Ratio			2	: 2		2	: 2
$(B,+ \times A,-)$	$(B,+ \times A,-)$	Phenotypic ratio is equivalent to the sum of the two backcrosses given above	3 :	1 :	2 :	2	2 :	4 :	2

*Phenotypically, the endosperm is A^1 in young seedlings but becomes A^1B^1 in older seedlings.

The phenotype of the other recombinant A-/A-/B- is developmentally dependent; it is A^1 in the endosperm of 4-day old
seedlings but becomes $\underline{A}^1B^1$ in 8-10 day old seedlings. The
rationale for this change would be that the promotion factor
p- is rather labile in the decaying endosperm as compared to
the p+ promotion factor. The expression of $Amy-1^B$ is delayed
in development in the heterozygous diploid tissue with the
balanced allelic dosages A-/B+. With these assumptions, we
can satisfy all of our observations including the difference
seen between the two crosses, (B^1 x A^1) x A^1 and A^1 x (B^1 x A^1)
shown in Table 3. Accordingly, the offspring genotypes from
the factor cross would have been (B, +/A, -) and (A, -/A, -)
whereas (A, -/B, +) and (A, -/B, -) would be expected from the
latter cross. For clarity, we present our genetic model in
Table 5; note that the outcome of the F_2 results in a
3 : 1 : 2 : 2 segregation ratio.

We could enumerate a number of alternative mechanisms by
which the promotion factor modulates the allelic expression at
the $Amy-1$ gene locus. The simplest mechanism requiring the
least speculation is that the modulation takes place at the
post-transcriptional level where the translation of the mRNAs
specified for Amy-1A and Amy-1B are promoted each by its own
promotion factor, i.e. p- for Amy-1A and p+ for Amy-1B. The
two allelic factors differ in their affinity for the two
allelic mRNAs and their efficiency in promoting the transla-
tions even in the diploid tissue of A-/B+ where Amy-1A mRNA is
translated with higher efficiency than Amy-1B mRNA.

REFERENCES

Allison, M.J. 1973. Genetic studies on the β-amylase iso-
    zymes of barley malt. *Genetica* 44: 1-15.
Carlson, P.S., H.H. Smith, and R.D. Dearing. 1972. Para-
    sexual interspecific plant hybridization. *Proc. Natl.
    Acad. Sci. U.S.* 69: 2292-2294.
Carlson, P.S. 1973. Somatic cell genetics of higher plants.
    In *Genetic Mechanisms of Development, 31st Symp. Soc.
    Dev. Biol.* (Ruddle, F.H., ed.). Academic Press.
    pp. 329-353.
Castro-Sierra, E., and S. Ohno. 1968. Allelic inhibition at
    the autosomally inherited gene locus for liver alcohol
    dehydrogenase in chicken-quail hybrids. *Biochem. Genet.*
    1: 323-325.
Chao, S.E., and J.G. Scandalios. 1969. Identification and
    genetic control of starch-degrading enzymes in maize
    endosperm. *Biochem. Genet.* 3: 537-547.
Chao, S.E., and J.G. Scandalios. 1971. Alpha-amylase of

maize: Differential allelic expression at the *Amy-1* gene locus, and some physiochemical properties of the isozymes. *Genetics* 69: 47-61.

Chao, S.E., and J.G. Scandalios. 1972. Developmentally dependent expression of tissue specific amylases in maize. *Mol. Gen. Genet.* 115: 1-9.

Courtright, J.B. 1967. Polygenic control of aldehyde oxidase in Drosophila. *Genetics* 57: 25-39.

Davidson, R.L. 1973. Control of the differentiated state in somatic cell hybrids. In *Genetic Mechanisms of Development, 31st Symp. Soc. Dev. Biol.* (Ruddle, F.H., ed.). Academic Press. pp. 295-328.

Davies, D.R. 1973. Differential activation of maternal and paternal loci in seed development. *Nature New Biol.* 245: 30-32.

Goldberg, E., J.P. Cuerrier, and J.C. Ward. 1969. Lactate dehydrogenase ontogeny, parental gene activation and tetramer assembly in embryos of Brook trout, Lake trout, and their hybrids. *Biochem. Genet.* 2: 335-350.

Hitzeroth, H., J. Klose, S. Ohno, and U. Wolf. 1968. Asynchronous activation of parental alleles at the tissue-specific gene loci observed on hybrid trout during early development. *Biochem. Genet.* 1: 287-300.

Klose, J., and U. Wolf. 1970. Transitional hemizygosity of the maternally derived alleles at the 6GPD locus during early development of the cyprinid fish *Rutilys ratilus*. *Biochem. Genet.* 4: 87-92.

Nelson, O.E., and B. Burr. 1973. Biochemical genetics of higher plants. *Ann. Rev. Plant Physiol.* 24: 493-518.

Scandalios, J.G., S.E. Chao, and C. Melville. 1974. Characterization of Amy-1: The major form of amylase in maize, in preparation.

Varner, J.E., and G.R. Chandra. 1964. Hormonal control of enzyme synthesis in barley endosperm. *Proc. Natl. Acad. Sci. U.S.* 52: 100-106.

Wright, D.A., and F.H. Moyer. 1966. Parental influences on lactate dehydrogenase in the early development of hybrid frogs in the genus Rana. *J. Exptl. Zool.* 163: 215-230.

REGULATION OF THE INTERCONVERSION
OF LIVER GLYCOGEN SYNTHETASE a AND b

Harold L. Segal
Biology Department, State University of New York
Buffalo, New York 14214

ABSTRACT.   A number of enzyme systems are now known to
exist in interconvertible forms differing markedly in
their physiological activity. The unique regulatory capability which this device provides allows enzymes to be
switched from a form whose properties are adapted to intracellular homeostasis to one in which these restraints are
overriden in response to external signals.

Glycogen synthetase b has been isolated from rat liver
and found to be composed of subunits of molecular weight
85,000. The enzyme as isolated appears to be a trimer.
The subunits contain 12 alkali-labile phosphate groups
per subunit. The enzyme undergoes reversible cold inactivation, which may explain the observations frequently
reported of inactive forms of the synthetase in tissues.

The interconversions of liver glycogen synthetase a
and b are mediated by a kinase and a phosphatase, both
of which are sites of control of the glycogen synthetase
system. The kinase is the cyclic adenosine 3',5'-phos-
phate (cyclic AMP) -dependent protein kinase and thus
responds to hormones such as glucagon and epinephrine
which elevate cyclic AMP levels. The phosphatase, which
appears to be the dominant control site of the system,
is subject to regulation by a number of factors, including
glucocorticoids and glucose. The phosphatase has not
yet been purified and its properties defined, so that
many aspects of its regulation remain undetermined.

INTERCONVERTIBLE FORMS OF ENZYMES

Glycogen synthetase is one of several key metabolic enzymes now known to exist in covalently distinct, interconvertible
forms. The prototype of this group is the phosphorylase pair,
which has been known since the pioneering work of the Coris
and their collaborators 25 to 30 years ago to consist of two
forms, a and b, differing by the presence or absence of es-
terified phosphate groups and in their kinetic properties.
Other systems of this type that have been identified are
triglyceride lipase, pyruvate dehydrogenase, glutamine synthetase, and phosphorylase kinase, the last of which is another
component of the phosphorylase system. All these interconversions involve a phosphorylation-dephosphorylation cycle except

691

glutamine synthetase, where the adduct is an adenylyl group. In some cases the substituted form is the more active one and in other cases the reverse. Some of the phosphorylations are catalyzed by the non-specific, cyclic AMP[1]-dependent protein kinase and some by apparently specific, cyclic AMP-independent kinases (for review, see Segal 1973). More or less persuasive reports have also appeared of covalent interconvertibility of other enzymes and proteins, some of which are discussed in the above reference and by Holzer and Duntze (1971), and more recently of acetyl CoA carboxylase (Carlson and Kim, 1973), adenylate cyclase (Constantopaulos and Najjar, 1973), and 3-hydroxy, 3-methylglutaryl CoA reductase (Beg et al., 1973) via phosphorylation-dephosphorylation cycles.

The regulatory significance of this process lies in the fact that the two forms differ markedly in their activity under physiological conditions, so that interconversion represents an on-off switch of metabolic flow. This is illustrated for glycogen synthetase in Table I, where at physiological concentrations of IDP-glucose (UDPG), glucose 6-phosphate (G-6-P), and P_i, the activity of the a form is seen to be about 15 times that of the b form. If the differential sensitivity of these forms to ATP inhibition (Gold, 1970a) is also taken into account, the ratio is even greater.

TABLE I.

Activity of liver glycogen synthetase b and a under physiological conditions

Ligand concentrations (mM)[a]	Activity (cpm)	
	b form	a form
P_i, 3.0; G-6-P, 0.05; UDPG, 0.25	357	4,950
P_i, 3.0; G-6-P, 0.25; UDPG, 0.50	480	8,190

[a]For other experimental details, see Mersmann and Segal (1967).

Interconversion of active and inactive forms and modulation by ligand binding are in some aspects complementary and mutually reinforcing. They have in common rapidity of response and the potentiality for direct, multisite integrative effects. As has been pointed out, however, (Segal, 1973), in multicellular organisms the interconversion device provides certain unique properties. It allows metabolic reactions to be switched on and off as a result of situations external to the responsive cell, which are most often relayed by hormonal mediators. The effect is to switch enzymes from a form whose properties are adapted to serve the need for intracellular homeostasis to one which overrides these constraints to serve a more urgent need of the whole organism.

GLYCOGEN SYNTHETASE

With the discovery of separate pathways for glycogen break-down to glucose 1-phosphate (G-1-P) via the phosphorylase system and glycogen synthesis from G-1-P via the glycogen synthetase system, it became evident that control of glycogen breakdown and synthesis at this point in the pathways was feasible, and perhaps even indispensable to avoid energy short-circuiting or futile cycling.

Early in the investigation of the glycogen synthetase system it was found that glucose 6-phosphate (G-6-P) stimulated the reaction in tissue extracts (Leloir et al., 1959). Larner and his coworkers subsequently established that the degree of G-6-P dependent and G-6-P independent activity was a reflection of the relative amounts of two forms of the enzyme (Villar-Palasi and Larner, 1961), which they refer to as the D, or G-6-P dependent form, and the I, or G-6-P independent form, respectively (Roselle-Perez et al., 1962). While it is well established that tissue extracts exhibit G-6-P dependent and independent glycogen synthetase activities as assayed in vitro, it is not precisely correct to refer to G-6-P dependent and independent eyzyme forms. Both forms are stimulated by G-6-P but with different quantitative and, in the case of muscle, qualitative characteristics (Rosell-Perez et al., 1962; Rosell-Perez and Larner, 1964; Mersmann and Segal, 1967). The increased activity with G-6-P present is a result of the stimulation of both forms, one considerably more than the other under assay conditions. For this reason we have preferred the more non-committal terminology, a and b, to refer to the physiologically active and inactive forms, respectively (Mersmann and Segal, 1967).

With the establishment that the two forms were enzymatically interconvertible (Larner, 1966; Larner and Villar-Palasi, 1971), it became apparent that regulation of the amount of the active form, and thus of metabolic flow through the pathway, may be modulated via effects on either of the interconverting enzymes. Some of the mechanisms by which this potentiality is realized with liver glycogen synthetase are discussed in the following sections.

THE LIVER SYSTEM

The liver system consists of four components thus far identified, viz, the a and b forms of the synthetase, the kinase that phosphorylates a to form b, and the phosphatase that catalyzes the reverse reaction.

THE SYNTHETASE

Apparently homogeneous, polysaccharide-free preparations

of synthetase b have been obtained from trout liver (Lin et al., 1972) and rat liver (Lin and Segal, 1973). Steiner et al. (1965) have purified a material from rat liver that consists largely of the a form and contains polysaccharide, and Sevall and Kim (1970) have reported the purification of a synthetase that also contained substantial amounts of glycogen from tadpole liver.

The purification from rat liver involved centrifugal separation of the glycogen pellet fraction, to which most of the synthetase is attached, followed by phosphorolysis of the glycogen by the endogenous phosphorylase, which released the synthetase in soluble form. The procedure is summarized in Table II.

Polyacrylamide gel electrophoresis showed a single band corresponding to the synthetase activity (Fig. 1). Electrophoresis in the presence of SDS for determination of subunit size gave a molecular weight of 85,000 (Fig. 2).

The pattern in a sedimentation equilibrium experiment to determine the molecular weight of the oligomer was heterdisperse with a value extrapolated to the meniscus of 260,000 and a value extrapolated to the base of 500,000. Since the time required to complete this experiment produces an aggregated form of the enzyme (Fig. 3), we can tentatively conclude that the enzyme as isolated has a molecular weight of about 260,000 and aggregates to a form double the original size (Lin and Segal, 1973). This molecular weight value corresponds to that obtained by us for the trout liver enzyme (Lin et al., 1972) and by Brown and Larner for muscle synthetase b (1971). On the other hand, Soderling et al. (1970) have reported a value of 400,000 for muscle synthetase a, and a value of 370,000 has been reported for both forms of kidney synthetase (Issa and Mendicino, 1973). There is general agreement, however, on the value of 85,000 to 90,000 daltons for the subunit size, indicating 3 subunits if the molecular weight is in the 260,000 range and 4 subunits if the molecular weight is in the 370,000 to 400,000 range.

Determination of the phosphate content of the purified rat liver synthetase gave the values shown in Table III. Crystalline rabbit muscle phosphorylase and bovine serum albumin were included in the determinations as checks on the procedure. Smith et al. (1971) have reported a value of about 7 for the alkali-labile phosphate content of the muscle enzyme (corrected to a subunit size of 85,000).

Incubation of the purified enzyme at 0° led to a progressive decline in activity to an equilibrium value (Fig. 4).

TABLE II

Purification of Glycogen Synthetase b from Rat Liver

Fraction	Volume	Total Units	Specific Activity	a Form	Recovery
	ml		units/mg protein	%	%
Crude extract	645	232	0.016	8	100
Glycogen pellet	61	169	1.57	0	73
Soluble extract	70	95	2.00	0	41
$Ca_3(PO_4)_2$ gel eluate	1.5	47	35.40	0	20

from Lin and Segal (1973)

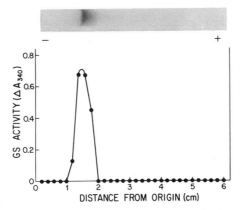

Fig. 1. Disc gel electrophoresis of purified glycogen synthetase b. Electrophoresis was at 2 ma per gel for 2 hrs. Gels
were run in duplicate and contained 8.9 µg of freshly prepared
enzyme. One gel was stained for protein, and the other was
cut into 2-mm segments that were assayed for glycogen synthetase activity as previously described (Lin et al., 1972),
except incubation was at 37° for 90 min. Activity is expressed
as total absorbance change per segment. The stained gel is
shown at the top.

from Lin and Segal (1973)

Rewarming at 25° restored the original activity. The presence
of glycogen in the enzyme solution prevented this cold inactivation. The existence of an additional form of glycogen synthetase has been reported in chloroma tumors; this form was
active only in the presence of extraordinarily high concentrations of glucose-6-P (Assaf and Yunis, 1971). The cold-
inactivated form described here did not exhibit this property.

In fact, there were no kinetic differences from the fully
active preparation, except, of course, for the diminished
specific activity. Inactive forms of synthetase have also
been reported to exist in muscle and other tissues, i.e., forms
not active even in the presence of glucose-6-P, but which
yield active species under certain conditions of incubation
(Steiner et al., 1965; Rosell-Perez, 1972; Lin et al., 1972).
Rosell-Perez (1972) has proposed that the inactive species is
a hyperphosphorylated form of synthetase and that the conversion reflects the action of a phosphatase. However, the
possibility of a conformational change as the basis of the
genesis of active forms, as is evidently the case here, does
not seem to be ruled out.

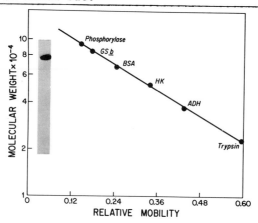

Fig. 2. Subunit molecular weight of glycogen synthetase b.
The procedure was that of Weber and Osborn (1969). Monomer
molecular weights of the reference proteins are: bovine serum
albumin (BSA), 68,000 (Weber and Osborn, 1969); trypsin,
23,000 (Weber and Osborn, 1969); phosphorylase, 94,000 (Ulmann
et al., 1968); yeast hexokinase (HK), 52,000 (Derechin et al.,
1972); yeast alcohol dehydrogenase (ADH), 37,000 (Tanford et
al., 1967). The electrophoretogram of the synthetase (40 µg)
is shown in the inset.

from Lin and Segal (1973)

THE KINASE BRANCH

The phosphokinase that catalyzes the transfer of phosphate
from ATP to glycogen synthetase b has been identified as the
cyclic AMP-dependent protein kinase that also phosphorylates
phosphorylase kinase, as well as certain other proteins (Soderling et al., 1970). In consequence the synthetase system is

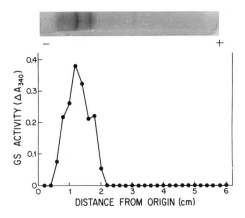

Fig. 3. Disc gel electrophoresis of aged glycogen synthetase b. The procedure was as in Fig. 1 except the enzyme (17.8 μg) was aged at 4° for 3 days.

from Lin and Segal (1973)

TABLE III.

Phosphate content of glycogen synthetase b of rat liver

Protein	Phosphate equivalents/subunit[a]	
	Alkali-labile	Total
Glycogen synthetase b	12.4 ± 0.6 (8,9)[b]	17.3 ± 1.1 (4,4)
Phosphorylase a	0.74 ± 0.1 (2,5)	1.02 ± 0.1 (1,4)
Bovine serum albumin	0.10	

[a]Subunit sizes: Glycogen synthetase b, 85,000 (Lin and Segal 1973); phosphorylase a, 94,000 (Ulmann et al., 1968); bovine serum albumin, 68,000 (Weber and Osborn, 1969).

[b]Numbers in parentheses are the number of preparations assayed and the total number of analyses, respectively.

from Lin and Segal (1973)

responsive to those hormonal and other factors that affect cyclic AMP levels, notably glucagon and epinephrine. It is well established that insulin also affects the a and b inter-conversion in muscle (in the opposite direction from glucagon and epinephrine) (Villar-Palasi and Larner, 1960; Shen et al., 1970), and this appears to be the case in liver as well (Miller

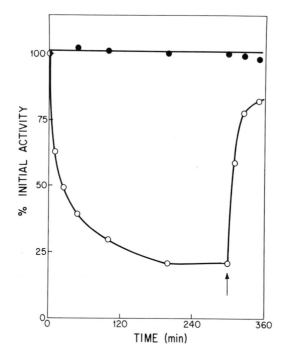

Fig. 4. Reversible cold inactivation of glycogen synthetase b. The purified enzyme (0.66 mg per ml) was removed from liquid nitrogen and placed at 0° with no additions (O) or with 5 mg per ml of glycogen present. At the point marked by the arrow, the solutions were transferred to a 25° bath.

and Larner, 1973). However, it is as yet unclear whether its affect is on the kinase branch, the phosphatase branch, or both. Insulin appears not to alter the cyclic AMP level of tissues under conditions where it produces an increase in synthetase a (Craig et al., 1969; Larner, 1972; Miller and Larner, 1973). However, it has been reported that exposure to insulin may decrease tissue protein kinase activity nevertheless, by a mechanism not involving cyclic AMP (Shen et al., 1970; Miller and Larner, 1973).

THE PHOSPHATASE BRANCH

There is a close correlation in individual livers between the rate of glycogen synthesis and the level of glycogen synthetase a (De Wulf and Hers, 1968a). The latter is normally

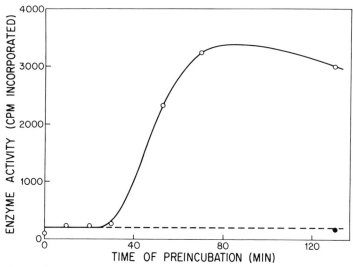

Fig. 5. Time-dependent increase in glycogen synthetase activity at 20°. At zero time a portion of the enzyme was transferred to a 20° bath and aliquots were taken for assay periodically thereafter. The filled symbols are activities of the enzyme preparation maintained at 0°. Assays were in the absence of G-6-P.

from Gold and Segal (1967)

---

quite low in the liver, as is the activity of the phosphatase that produces it from the b form (Fig. 5) (Gold and Segal, 1967). Furthermore, a number of factors that stimulate hepatic glycogen synthesis have now been shown to be effectors of the phosphatase branch of the glycogen synthetase system. Thus it can be concluded that promotion of the phosphatase reaction, with the consequent elevation of synthetase a levels and glycogen synthesis, underlies in large measure the glycogenic effect of these factors.

Glucocorticoids - The long-known effect of glucocorticoids in promoting liver glycogen deposition can be attributed, in part at least, to a direct or indirect influence of these hormones on the phosphatase branch of the glycogen synthetase system. In adrenalectomized rats deprived of food, glycogen levels, synthetase a levels, and phosphatase activity are virtually nil. Several hours after glucocorticoid administration there is a restoration of phosphatase activity, synthetase a levels, and rates of glycogen synthesis in a sequence consistent with a causal relationship in that order (Mersmann and Segal, 1969). This effect, which appears to be mediated by insulin (Nichols

and Goldberg, 1972), is blocked by cycloheximide or actinomycin D (Gruhner and Segal, 1970), and thus may be concluded
to depend upon an induction of the phosphatase or some other
component that leads indirectly to an elevation of phosphatase
activity.

In differently designed experiments, De Wulf et al. (1970)
have also found an effect of glucocorticoids on the phosphatase
They observed that pretreatment with the hormone led to a phosphatase activity in liver extracts with a greatly reduced lag
period prior to the expression of full activity (cf. Fig. 5).
Phosphorylase a is an inhibitor of synthetase phosphatase
(Stalmans et al., 1971), and glucocorticoids increase the activity of phosphorylase phosphatase (Stalmans et al., 1970),
which removes phosphorylase a. Therefore, these workers propose that the lag period represents the time required for removal of phosphorylase a and that the glucocorticoid response
they observed depends on the increased rate of phosphorylase
a removal. It seems unlikely that this explanation underlies
the glucocorticoid effect in the experiments of Mersmann and
Segal (1969), where synthetase phosphatase activity in adrenalectomized animals deprived of food was totally lacking, even
after lengthy periods of incubation.

Glucose - Two types of effect of glucose on the phosphatase
system have been observed. In the starved, adrenalectomized
animals referred to above, intubation of glucose restored synthetase phosphatase activity similar to but independent of
the glucocorticoid effect (Fig. 6). In this case the response
was also blocked by cycloheximide, but not by actinomycin
(Gruhner and Segal, 1970). A possible explanation for the
glucocorticoid and glucose effects is that in the starved,
adrenalectomized liver a factor required for phosphatase stabilization (glycogen?) is lacking, and that this factor is restored by glucose administration, or by glucocorticoids via a
process depending upon enzyme induction.

In contrast to the glucose effect illustrated in Fig. 6,
which requires a minimum of about 1½ hours to become manifest,
a very rapid response of a different sort to glucose has also
been observed. In normal animals (De Wulf and Hers, 1967) or
perfused livers therefrom (Buschiazzo et al., 1970; Glinsmann
et al., 1970), within a few minutes after glucose administration there was a rise in the levels of active synthetase (a
form), together with a fall in phosphorylase a levels. Given
the inhibition of synthetase phosphatase by phosphorylase a
discussed above and the stimulatory effect of glucose on the
phosphorylase phosphatase reaction which removes phosphorylase
a (Holmes and Mansour, 1968; Stalmans et al., 1970), this
acute glucose effect on synthetase activation appears to be

700

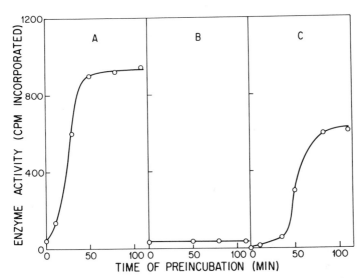

Fig. 6. Effect of glucose administration on activation of glycogen synthetase in vitro. The abscissa is incubation time at 20° of the 8500 x g supernatant fraction of liver homogenate. Assays were in the absence of G-6-P. A. Liver from normal, fed animal. B. Liver from adrenalectomized, 48hr-starved animal. C. Liver from adrenalectomized, 48hr-starved animal given 2.5 g/kg glucose by stomach tube 4hr before sacrifice.

from Gruhner and Segal (1970)

explainable by its effect on the latter reaction. In livers from starved, adrenalectomized or diabetic animals, where phospatase activity has disappeared (Mersmann and Segal, 1969; Gold, 1970b), the acute elevation of synthetase a in response to glucose does not occur (Buschiazzo et al., 1973).

Insulin - As stated above, phosphatase activity is greatly reduced in the diabetic state and restored several hours after insulin administration (Gold, 1970b; Bishop, 1970; Nichols and Goldberg, 1972).

Some evidence exists which suggests that the phosphatase exists in interconvertible forms differing in their sensitivity to Mg^{2+} and that insulin promotes conversion to the more active form, (Bishop, 1970).

Glucagon - Glucagon, in addition to its well established effect on the kinase via its elevation of cyclic AMP levels (see above), has also been reported to lead to a reduction of phosphatase activity (Bishop, 1970).

Glycogen - Although glycogen promotes the phosphatase reaction in vitro (Masaracchia and Segal, unpublished), it is

inhibitory in higher concentrations (De Wulf and Hers, 1968b).
There is a correlation between the sensitivity of the liver
and muscle (Larner, 1966) phosphatase reactions to glycogen
and the level of glycogen which these tissues can accumulate,
suggesting that glycogen inhibition of the phosphatase may
reflect a feed-back regulation by glycogen of its own synthesis.

DISCUSSION

It appears from current knowledge of the hepatic glycogen
synthetase system that its regulation is dominated by the
phosphatase branch which converts the synthetase to the active
form. The aspects that support this conclusion are the predominance of the inactive form of synthetase in normal, un-
stimulated liver, and the concomitance of stimulation of glycogen synthesis and activation of the phosphatase reaction by
a number of factors. While regulation of the protein kinase
reaction, via hormone-mediated elevations of cyclic AMP, is
also a well established concept, the synthetase unlike phosphorylase, is already predominantly in the phosphorylated
(inactive) state in such livers, so that little effect on this
system would be expected from further stimulation of the kinase.
The basis of this differential susceptibility of the synthetase
and the phosphorylase system to phosphorylation under the same
conditions in the liver is an interesting, but a separate
question.

Progress in identifying the nature of the liver phosphatase system and the molecular mechanisms underlying its regulation has been slow as a result of the difficulties that have
been encountered in isolating it and defining its components.
At least two major questions wait to be addressed. 1) Whether
the phosphatase is interconvertible between active and inactive
forms, and, if separate forms exist, the chemical and kinetic
difference between them and the identity of the interconverting
enzymes. 2) Whether the lack of phosphatase activity in
starved, adrenalectomized animals and diabetic animals is a
result of the absence of the enzyme or its existence in an inactive state; if the former, whether there is a failure of
phosphatase synthesis or an increased rate of degradation, if
the latter, what factors are lacking for its activity.

It may be hoped that with the isolation of the substrate
of the phosphatase reaction, synthetase b, and the elucidation
of some of its properties, one of the obstacles to further
progress in this area has been overcome.

REFERENCES

Assaf, S. A. and A. A. Yunis. 1971. Identification of a third form of glycogen synthetase in rat chloroma tumors. *FEBS Letters*. 19: 22-26

Beg, Z. H., D. W. Allman, and D. M. Gibson. 1973. Modulation of 3-hydroxy-3-methylglutaryl coenzyme A reductase activity with cAMP and with protein fractions of rat liver cytosol. *Biochem. Biophys. Res. Commun.* 54: 1362-1369

Bishop, J. S. 1970. Inability of insulin to activate liver glycogen transferase D phosphatase in the diabetic pancreatectomized dog. *Biochim. Biophys. Acta*. 208: 208-218.

Brown, N. E. and J. Larner, 1971. Molecular characteristics of the totally dependent and independent forms of glycogen synthase of rabbit skeletal muscle. I. Preparation and characteristics of the totally glucose 6-phosphate dependent form. *Biochim. Biophys. Acta*. 242: 69-80

Buschiazzo, H., J. H. Exton, and C. R. Park. 1970. Effects of glucose on glycogen synthetase, phosphorylase, and glycogen deposition in the perfused rat liver. *Proc. Nat. Acad. Sci. U. S. A.* 65: 383-387

Carlson, C. A. and K. -H. Kim. 1973. Regulation of hepatic acetyl coenzyme A carboxylase by phosphorylation and dephosphorylation. *J. Biol. Chem*. 248: 378-380

Constantopoulos, A. and V. A. Najjar. 1973. The activation of adenylate cyclase. II. The postulated presence of (A) adenylate cyclase in a phospho (inhibited) form (B) a dephospho (activated) form with a cyclic adenylate stimulated membrane protein kinase. *Biochem. Biophys. Res. Commun*. 53: 794-799

Craig, J. W., T. W. Rall, and J. Larner. 1969. The influence of insulin and epinephrine on adenosine 3', 5'-phosphate and glycogen transferase in muscle. *Biochim. Biophys. Acta*. 177: 213-219

Derechin, M., Y. M. Rustum, and E. A. Barnard. 1972. Dissociation of yeast hexokinase under the influence of substrates. *Biochemistry* 11: 1793

De Wulf, H. and H. G. Hers. 1967. The stimulation of glycogen synthesis and of glycogen synthetase in the liver by the administration of glucose. *Eur. J. Biochem*. 2: 50-56

De Wulf, H. and H. G. Hers. 1968a. The role of glucose, glucagon, and glucocorticoids in the regulation of liver glycogen synthesis. *Eur. J. Biochem*. 6: 558-564

De Wulf, H. and H. G. Hers. 1968b. The interconversion of liver glycogen synthetase a and b in vitro. *Eur. J. Biochem*. 6: 552-557

De Wulf, H., W. Stalmans, and H. G. Hers. 1970. The effect of glucose and of a treatment by glucocorticoids on the activation in vitro of liver glycogen synthetase. *Eur. J. Biochem.* 15: 1-8

Glinsmann, W.,G. Pauk, and E. Hern. 1970. Control of rat liver glycogen synthetase and phosphorylase activities by glucose. *Biochem. Biophys. Res. Commun.* 39: 774-782

Gold, A. H. 1970a. On the possibility of metabolite control of liver glycogen synthetase activity. *Biochemistry* 9: 946-952

Gold, A. H. 1970b. The effect of diabetes and insulin on liver glycogen synthetase activation. *J. Biol. Chem.* 245: 903-906

Gold, A. H. and H. L. Segal. 1967. Time-dependent increase in rat liver glycogen synthetase activity in vitro. *Arch. Biochem. Biophys.* 120: 359-364

Gruhner, K. and H. L. Segal. 1970. Effects of glucose and inhibitors of protein synthesis on the liver glycogen synthetase-activating system. *Biochim. Biophys. Acta.* 222 : 508-514

Holmes, P. A. and T. E. Mansour. 1968. Glucose as a regulator of glycogen phosphorylase in rat diaphragm. II. Effect of glucose and related compounds on phosphorylase phosphatase. *Biochim. Biophys. Acta.* 156: 275-284

Holzer, H. and W. Duntze. 1971. Metabolic regulation by chemical modification of enzymes. *Ann. Rev. Biochem.* 40: 345-374

Issa, H. A. and J. Mendicino. 1973. Role of enzyme-enzyme interactions in the regulation of glycolysis and gluconeogenesis. Properties of glycogen synthetase isolated from swine kidney. *J. Biol. Chem.* 248: 685-696

Larner, J. 1966. Hormonal and nonhormonal control of glycogen metabolism. *Trans. N. Y. Acad. Sci. Ser. 2.* 29: 192-209

Larner, J. 1972. Insulin and glycogen synthase. *Diabetes.* 21 (Suppl. 2): 428-438

Larner, J. and C. Villar-Palasi. 1971. Glycogen synthase and its control. *Curr. Topics Cell Regul.* 3: 195-233

Leloir, L. F., J. M. Olavarria, S. H. Goldemberg, and H. Carminatti. 1959. Biosynthesis of glycogen from uridine diphosphate glucose. *Arch. Biochem. Biophys.* 81: 508-520

Lin, D. C. and H. L. Segal. 1973. Homogeneous glycogen synthetase b from rat liver. *J. Biol. Chem.* 20: 7007-7011

Lin, D. C., H. L. Segal, and E. J. Massaro. 1972. Purification and properties of glycogen synthetase from trout liver. *Biochemistry.* 11: 4466-4471

Mersmann, H. J. and H. L. Segal. 1967. An on-off mechanism for liver glycogen synthetase activity. *Proc. Nat. Acad. Sci. U. S. A.* 58: 1688-1695

Mersmann, H. J. and H. L. Segal. 1969. Glucocorticoid control of the liver glycogen synthetase-activating system. *J. Biol. Chem.* 244: 1701-1704

Miller, T. B., Jr. and J. Larner. 1973. Mechanism of control of hepatic glycogenesis by insulin. *J. Biol. Chem.* 248: 3483-3488

Miller, T. B., Jr., R. Hazen, and J. Larner. 1973. An absolute requirement for insulin in the control of hepatic glycogenesis by glucose. *Biochem. Biophys. Res. Commun.* 53: 466-474

Nichols, W. K. and N. D. Goldberg. 1972. The relationship between insulin and apparent glucocorticoid-promoted activation of hepatic glycogen synthetase. *Biochim. Biophys. Acta.* 279: 245-259

Rosell-Perez, M. 1972. Inactive forms of UDPG: α-1, 4-glucan α-4-glucosyltransferase. *Ital. J. Biochem.* 21: 34-69

Rosell-Perez, M. and J. Larner. 1964. Studies on UDPG-α-glucan transglucosylase. IV. Purification and characterization of two forms from rabbit skeletal muscle. *Biochemistry.* 3: 75-81

Rosell-Perez, M., C. Villar-Palasi, and J. Larner. 1962. Studies on UDPG-glycogen transglucosylase. I. Preparation and differentiation of two activities of UDPG-glycogen transglucosylase from rat skeletal muscle. *Biochemistry.* 1: 763-768

Segal, H. L. 1973. Enzymatic interconversion of active and inactive forms of enzymes. *Science.* 180: 25-32

Sevall, J. S. and K. -H. Kim. 1970. Purification and properties of hepatic glycogen synthetase of *Rana catesbeiana*. *Biochim. Biophys. Acta.* 206: 359-368

Shen, L. C., C. Villar-Palasi, and J. Larner. 1970. Hormonal alteration of protein kinase sensitivity to 3' ,5'-cyclic AMP. *Physiol. Chem. Physics.* 2: 536-544

Smith, C. H., N. E. Brown, and J. Larner. 1971. Molecular characteristics of the totally dependent and independent forms of glycogen synthase of rabbit skeletal muscle. II. Some chemical characteristics of the enzyme protein and of its change on interconversion. *Biochem. Biophys. Acta.* 242: 81-88

Soderling, T. R., J. P. Hickenbottom, E. M. Reimann, F. L. Hunkeler, D. A. Walsh, and E. G. Krebs. 1970. Inactivation of glycogen synthetase and activation of phosphorylase kinase by muscle adenosine 3' ,5' - monophosphate-dependent protein kinases. *J. Biol. Chem.* 245: 6317-6328

Stalmans, W., H. De Wulf, and H. G. Hers. 1971. The control of liver glycogen synthetase phosphatase by phosphorylase. *Eur. J. Biochem.* 18: 582-587

Stalmans, W., H. De Wulf, B. Lederer, and H. G. Hers. 1970. The effect of glucose and of a treatment by glucocorticoids on the inactivation in vitro of liver glycogen phosphorylase. *Eur. J. Biochem.* 15: 9-12

Steiner, D. F., L. Younger, and J. King. 1965. Purification and properties of uridine diphosphate glucose-glycogen glucosyltransferase from rat liver. *Biochemistry.* 4: 740-751

Tanford, C., K. Kawahara, and S. Lapanje. 1967. Proteins as random coils. I. Intrinsic viscosities and sedimentation coefficients in concentrated guanidine hydrochloride. *J. Amer. Chem. Soc.* 89: 729

Ullman, A., M. E. Goldberg, D. Perrin, and J. Monod 1968. On the determination of molecular weight of protein and protein subunits in the presence of 6 M guanidine hydrochloride. *Biochemistry* 7: 261-265.

Villar-Palasi, C. and J. Larner. 1960. Insulin-mediated effect on the activity of UDPG-glycogen transglucosylase of muscle. *Biochim. Biophys. Acta.* 39: 171-173

Villar-Palasi, C. and J. Larner. 1961. Insulin treatment and increased UDPG-glycogen transglucosylase activity in muscle. *Arch. Biochem. Biophys.* 94: 436-442.

Weber, K. and M. Osborn 1969. The reliability of molecular weight determinations by dodecyl sulfate-polyacrylamide gel electrophoresis *J. Biol. Chem.* 244: 4406-4412.

STUDIES ON THE CARBOXYLESTERASES THAT CATABOLIZE
THE JUVENILE HORMONE OF INSECTS

DONALD H. WHITMORE JR.[1], ELAINE WHITMORE,
LAWRENCE I. GILBERT, AND P. I. ITTYCHERIAH
Department of Biological Sciences
Northwestern University
Evanston, Illinois 60201

ABSTRACT: Within six hours after an injection of juvenile
hormone (JH) into saturniid pupae (a stage devoid of JH),
several carboxylesterases (EC 3.1.1.1) which are sensitive
to diisopropylfluorophosphate appear in the hemolymph. Although several other JH mimics were tested, only one was able
to "induce" these enzymes. Their induction was prevented
by the administration of puromycin, cycloheximide, or actinomycin D. These esterases are capable of rapidly degrading
juvenile hormone in vitro and are thus referred to as JH-
esterases.

Fat body tissue cultured in vitro released esterases into
the culture medium when stimulated with JH. These released
esterases reacted with antiserum prepared against the induced
JH-esterases of the hemolymph. However, no JH-esterases
appeared in the culture medium when treated with puromycin.
All attempts to specifically label these JH-esterases have
failed. A model is presented in which it is suggested that
this JH-dependent phenomenon may be the result of (1) *de novo*
synthesis of JH-esterases by the fat body; (2) release of
previously existing particle-bound JH-esterases from the fat
body; or (3) modification of an inactive proenzyme in the
fat body.

INTRODUCTION

The juvenile hormone (JH) of insects has been described as
a modulator that determines the quality of the molt, while
molting itself is initiated by the molting hormone, ecdysone.
JH concentration is high at a larval-larval molt, low at the
larval-pupal molt, and must be absent if the molt from pupa to
adult is to be initiated normally (Gilbert, 1964). Insect
development is therefore controlled by both the titer of JH and
the precise timing of its secretion by the corpora allata (Gilbert and King, 1973). The titer of JH throughout these developmental stages is regulated by biosynthetic activity of the

---

[1] Present address: Biology Department, The University of Texas,
Arlington, Texas 76019.

corpora allata and degradative mechanisms.

Mechanisms of JH degradation were first described by Slade and Zibitt (1971 and 1972) and White (1972). Degradation primarily involves (1) hydrolysis of the methyl ester by carboxylesterases; (2) hydration of the epoxide function by epoxide hydrase; (3) sulfate conjugation of the acid diol. This report is concerned with the insect carboxylesterases which are responsible for hydrolyzing the methyl ester group.

MATERIALS

Diapausing *Hyalophora gloveri* and *Hyalophora cecropia* pupae were obtained commercially and stored at 6°C for 2-6 months. The unlabeled juvenile hormone was a mixture of the eight possible isomers of the cecropia C_{18} juvenile hormone (courtesy of Ayerst Corporation), and was diluted in peanut oil to give a final concentration of 10 µg/µl. Radioactive juvenile hormone was either [2-^{14}C] (25.3 Ci/mmole, courtesy of Zoecon Corporation) or (7-ethyl-1, 2-^{3}H(N)(14.1 Ci/mmole; New England Nuclear Corporation). [^{3}H] leucine (58 Ci/mmole and [^{14}C] amino acids (57 mCi/mAtom Carbon) were obtained from Amersham/Searle.

METHODS

Diapausing pupae were injected with JH or peanut oil (controls), incubated for various periods of time and bled through an incision in the pupal wing. Hemolymph could be used immediately or stored for long periods of time at -20°C.

Electrophoretic analysis of hemolymph JH-esterases has previously been described (Whitmore et al., 1972 and 1974). JH-esterases were removed from the gels by carefully slicing bands and grinding the gel in cold distilled water using a Sorvall Omnimixer. After centrifugation of the gel slurry at 27,000xg for 15 minutes at 4°C, the supernatant was removed. It was used as the enzyme source for studying the breakdown of JH in vitro (Whitmore et al., 1972) and for the preparation of antibodies (Whitmore et al., 1974).

In short-term organ culture studies, *H. gloveri* pupae were surface sterilized. Fat bodies were dissected into sterile dishes containing Mark's M-20 medium (Grand Island Biological Co.). Portions of the fat body were then suspended in a hanging drop of 100 µl of medium. JH (10 µg/µl) and 9,10-epoxyhexadecanoic acid methyl ester (10 µg/µl) were sonicated into culture medium. One µl quantities were used to treat the cultured fat bodies. Puromycin was used in 1 µl quantities (0.05 mg). The cultures were incubated at 27°C for 18 hours, after which the medium was collected and analyzed.

RESULTS

Injection of JH into *H. gloveri* or *H. cecropia* (pupae which are devoid of juvenile hormone) results in the appearance of several fast migrating hemolymph esterases as revealed by gel electrophoresis (Figure 1:Whitmore et al, 1972). These esterases can be visualized using α-naphthyl proprionate, α-naphthyl acetate or β-naphthyl.acetate. Diisopropylfluorophosphate inhibited their enzymatic activity, but they appeared to be insensitive to both eserine and p-chloromercuribenzoate.

By analyzing hemolymph from pupae at various time intervals following injection of JH, we determined that the appearance of the JH-esterases was both time and dose dependent. The JH-esterases were first detected in the hemolymph 6-8 hours after JH injection and their enzymatic activity continued to increase for at least the next ten hours, resulting in hemolymph with a total esterase specific activity of 2-3 times the control hemolymph. Figure 2 reveals that a 20 fold increase in dose (from 2 µg to 40 µg) yields an approximate doubling of enzyme activity, although in the semi-log plot there is an apparent straight line dose response up to 80 µg of JH. The 160 µg dose causes some inhibition of response analogous to the phenomenon of substrate inhibition in enzyme kinetics.

Since JH is known to stimulate the prothoracic glands to initiate development by increasing the titer of molting hormone, the possibility existed that this enzyme induction was only an indirect result of JH. However, when β-ecdysone (5 µg/g) was injected into the pupae, there was only a slight stimulation of JH-esterase activity over the basal level within 24 hours. In addition, JH is capable of stimulating the brain neurosecretory cells (K. Davey, personal communication), but injection of JH into brainless *H. gloveri* pupae yielded the same results as injection into animals with brains intact. These results strongly suggest that 'JH is acting directly, not secondarily, through another hormone.

To further examine the specificity of this phenomenon, we injected pupae with 9, 10-epoxyhexadecanoic acid methyl ester, which is chemically similar to JH but without hormonal activity (Williams and Law, 1965); or with juvabione, which has JH activity in Hemiptera bugs but not in *H. gloveri* or *H. cecropia;* or with the hydrochlorination product (Law, Yuan, and Williams, 1966), which has potent JH activity in *H. gloveri*, but is chemically very different from cecropia juvenile hormone. The results demonstrated that 9, 10-epoxyhexadecanoic acid methyl ester stimulated the appearance of JH-esterase only slightly, even at the maximum dose of 50 µg/g of fresh weight. Ten µg/g of the hydrochlorination product stimulated the appearance of

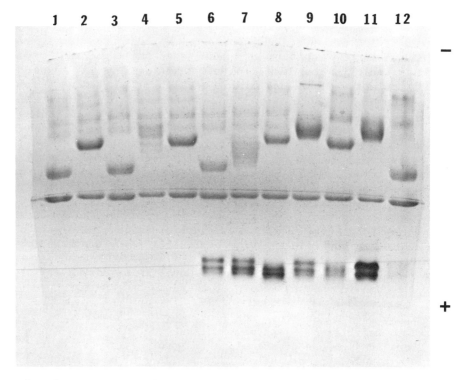

Fig. 1. Zymogram of hemolymph esterases. 5 μl samples of
H. gloveri hemolymph were subjected to electrophoresis on a
vertical acrylamide slab gel. The gel was a 3 1/2, 5, 7%
discontinuous acrylamide gradient gel (Ortec). The buffer
system was 0.75M Tris-sulfate gel buffer and 0.65M Tris-
borate tank buffer pH 8.4. The gels were stained for esterase
activity in a reaction mixture containing 100 mg fast blue RR,
2 ml α-naphthyl acetate (1% in acetone) in 0.1M sodium phosphate buffer, pH 6.8. Wells 1-5 contained hemolymph from
peanut oil controls and wells 6-12 contained hemolymph from
JH-injected animals (Whitmore et al., 1974).

the JH-esterases to some extent, but 50 μg/g was comparable to
JH in inducing the appearance of the enzymes. Juvabione on the
other hand was completely ineffective. It has recently been
suggested that JH mimics may not really possess inherent JH
activity, but may exert their effects by inhibiting the degradation of endogenous JH (Slade and Wilkinson, 1973).

When JH-esterases were eluted from preparative gels and
incubated with [2-14C]-JH for 40 minutes at 30°C and the incubation mixture analyzed, it was readily seen that the JH was
rapidly and completely degraded (Figure 3). The radiochroma-

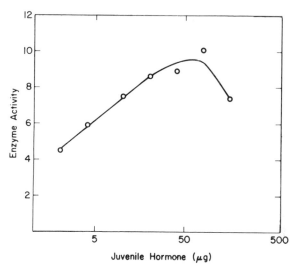

Fig. 2. Juvenile hormone esterase in response to varying concentrations of juvenile hormone. Pupae·were injected with
juvenile hormone and hemolymph withdrawn 18 hours later. Hemolymph samples (10 µl) were subjected to electrophoresis and the
gels stained for esterase activity. The gels were scanned at
500 nm and the area under the JH-esterase peaks was integrated
and expressed as arbitrary units of enzyme activity (Whitmore
et al., 1974).

togram (Figure 3b) indicates the presence of two major degradation products, the slower migrating one possibly being a 10,
11-dihydroxy acid (Slade and Zibitt, 1971; Siddall, Anderson,
Henrick, 1971). If so, one or more of the induced enzymes may
have epoxide hydrase activity. The fast-migrating fat body
carboxylesterases are also capable of breaking down JH, but
are not nearly as active.

The time and dose dependency of the JH-esterase appearance
in response to JH was strongly suggestive of an enzyme induction phenomenon. In order to examine this possibility, we
relied on the use of cycloheximide, puromycin, actinomycin D,
and radioactive amino acids. When pupae were injected with
puromycin (0.2 mg/g) or cycloheximide (0.2 mg/g) simultaneously
with JH or at any time up to six hours after the hormone injection, JH-esterases failed to appear in the hemolymph. Actinomycin D (6 µg/g) prevented the appearance of JH-esterases
if injected two hours before the hormone. Our attempts to
specifically label the JH-esterases in vivo by injecting 25
µCi [^{3}H] leucine or 2.5 µCi [^{14}C] amino acids 6 hours prior to,

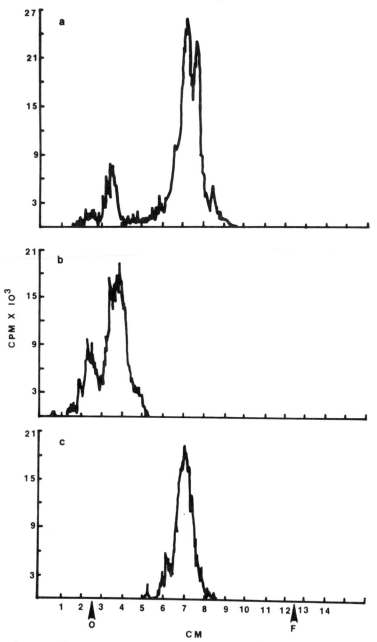

Fig. 3. Radiochromatoscans of thin-layer plates illustrating the breakdown of [^{14}C] juvenile hormone by JH-esterases.

Fig. 3. continued. (a) From incubation mixture of fat body carboxylesterases and [2-^{14}C] juvenile hormone (b) From incubation mixture of JH-esterases (c) [2-^{14}C] juvenile hormone. O=point of application. F=front. Protein concentration of fat body enzyme was 1.3 mg/ml, while that of the hemolymph was 2.2 mg/ml.

---

simultaneously with, 6 hours after, or 12 hours after the hormone injection failed. The JH-esterases had approximately the same number of counts as a portion of the control gel which exhibited no JH-esterase activity.

For in vitro labeling studies it was necessary to locate the source of the JH-esterases. Antibodies to the JH-esterases were prepared. Examination of such tissues as pupal midgut, fat body, and testes revealed esterases with electrophoretic mobilities similar to those of the JH-esterases. However, when these extracts were pre-incubated with JH-esterase antiserum and then analyzed by electrophoresis, the antiserum had no noticeable effects on the esterase patterns. At this point we turned to short-term organ culture.

A series of hanging drop cultures containing *H. gloveri* and *H. cecropia* fat body or midgut were studied for the ability to release JH-esterases into the incubation medium. All experiments with the midgut revealed the nonspecific leaking of esterases into the incubation medium, none of which appeared to be identical with any JH-esterases. However, fat body incubated with JH (10 µg), at 27°C for 18 hours released what appeared to be JH-esterases into the medium (Figure 4a). A very slight reaction was obtained when the fat body was incubated in the presence of 10 µg of 9, 10-epoxyhexadecanoic acid methyl ester. It is of interest that the fat body incubated simultaneously with JH and 0.05 mg puromycin failed to release JH-esterases into the medium.

Further evidence that these released enzymes are JH-esterases comes from studies with the antibodies against these enzymes. Following fat body incubation, antiserum was added to the medium. Figure 4b reveals that the induced JH-esterases in the culture medium reacted with the antiserum and thus prevented their appearance upon subsequent electrophoresis.

Attempts to label these esterases in culture failed. Either [^{3}H] leucine was added at the time of JH treatment or the fat body was pre-labeled with [^{3}H] leucine 18 hours prior to JH treatment. The resulting incubation media showed no differences in labeling pattern after analysis.

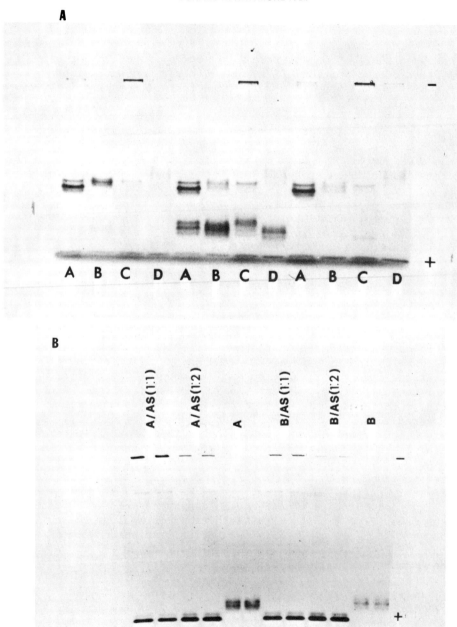

Fig. 4. (A) Hanging drop culture. Pupal fat bodies from four
animals (A - D) were divided into thirds, placed in hanging
drop culture and treated as follows: Left (A-D) 9, 10 - epoxy-
hexadecanoic acid methyl ester; Center (A-D) juvenile hormone
treated;

Fig. 4. continued.   Right (A-D) treated with juvenile
hormone and puromycin. After incubation for 18 hours at 27°C,
50 µl samples of the media were analyzed for esterase activity
by electrophoresis: (B) Effect of antiserum on JH-esterases
released in vitro. JH antiserum was added to medium taken from
hanging drop cultures treated with JH. This mixture was then
analyzed for esterase activity by electrophoresis. The portions
of media A and B mixed 1:1 or 1:2 with antiserum exhibited no
JH-esterase activity (Wells 1-4). Wells 5 and 6 showed untreated medium with JH-esterases present (Whitmore et al., 1974).

DISCUSSION

A basic question in insect physiology which now has great
practical significance is how insects regulate their hormone
titers. With the suggested use of JH mimics as pesticides, it
has become imperative that we understand the metabolism of such
substances not only in insects but in other animals as well.
The data described in this report are concerned with the action
of hemolymph carboxylesterases which apparently play a vital
role in the degradation of endogenous JH. Our results describe
an effective and novel means by which some insects may regulate
their JH titer. The hormone titer induces a group of carboxylesterases whose function is mediating the breakdown of the
unbound hormone. Recently, Terriere and Yu (1973) have
described a similar induction phenomenon with the microsomal
epoxidase system of flies when exposed to JH.

In vitro experiments strongly indicate that in saturniid
pupae, the fat body is a source of JH-esterases. It is, however, paradoxical that extracts of the fat body do not exhibit
JH-esterase activity, but fat body tissue challenged with JH
releases the JH-esterases into the medium (or hemolymph). We
have presented three possible explanations for this phenomenon
which may serve as guides for future research (Figure 5).

I. JH injected into pupae is carried to the fat body as a
component of a hemolymph lipoprotein complex (Whitmore and
Gilbert, 1972; Trautmann, 1972; Kramer et al., 1974). Within
a six hour period, the hormone stimulates fat body mRNA and
protein synthesis, including *de novo* synthesis of the JH-
esterases. This hypothesis is supported by the inhibitor
studies and our observations that JH-esterase activity is time
and dose dependent. Two lines of evidence oppose this hypothesis: (1) all attempts to specifically label JH-esterases,
both in vitro and in vivo, have been unsuccessful; (2) homogenates of JH-treated fat bodies do not yield esterases immunologically similar to the JH-esterases.

II. The second hypothesis is based upon the observations

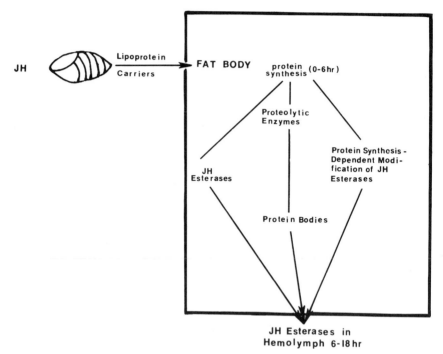

Fig. 5. Summary diagram of the three alternatives for the JH "induction" of hemolymph JH-esterases (Whitmore et al., 1974).

that the insect fat body sequesters hemolymph proteins in the form of granules (protein bodies) during larval life and that this process may be under hormonal control. In several insects, including *Hyalophora,* protein body formation has been correlated with decreasing JH titer suggesting that the converse may also be true; i.e., a rising JH titer may cause protein body breakdown (for a comprehensive review of this literature, see Price, 1973).

It is suggested that larval JH-esterases, which reach their peak activity at about the time that protein body formation begins, are sequestered in the pupal fat body granules. This occurs during larval-pupal development in an environment of declining JH titer and would explain the absence of JH-esterases in normal pupal hemolymph. Near the end of adult development, the JH titer increases profoundly (Gilbert and Schneiderman, 1961), protein bodies are virtually absent from the fat body (Bhakthan and Gilbert, 1972) and JH esterases reappear in the hemolymph. We postulate that JH induces an enzyme (proteolytic?) which breaks down the protein bodies, allowing the

716

release of the previously confined JH-esterases into the hemo-lymph where they mediate the hydrolysis of unbound JH. Although we have not been able to demonstrate any JH-esterase activity associated with isolated *H. gloveri* fat body granules, prelim-inary results revealed a 10-25% decrease in their number after exposure of the fat body tissue to JH.

III. JH may function in an in direct manner by stimulating fat body mRNA and protein synthesis in order to initiate the activiation or modification of pre-existing protein. Numerous examples of proenzyme modification by attachment or removal of small molecules have been described (see Segal, 1971). It has been shown that carboxylesterases can exist in an active state as dimers or monomers (Benöhr and Krisch, 1967; Ecobichon, 1969; Ljundquist and Augustinsson, 1971). Substrate activation of liver carboxylesterases is due in part to the presence of two types of active sites (Hofstee, 1972) while a serum carboxyl-esterase exists as a dimeric lipoprotein that undergoes a marked change in conformation upon delipidation (Kingsbury and Masters, 1972). There are many such examples in esterase enzymology, and the possibility exists that such changes may represent control points for hormone action.

We are currently examining the relationship between the JH-esterases and the lipoprotein complex which carries the hormone to the fat body. Yet a great deal more information is needed about the physical-chemical characteristics of the JH-esterases, their ability to be epigenetically modified, and JH effects on fat body metabolism in order to elucidate the mechanism of JH-esterase induction.

REFERENCES

Benöhr, H. C. and K. Krisch 1967. Carboxylesterase aus rinder-lebermikrosdem. 2. Dissoziation und assoziation molekular-wicht reaktion mit E600. *HoppeSeyler's Z. Physiol. Chem.* 348: 1115-1119.

Bhakthan, N. M. G. and L. I. Gilbert 1972. Studies on the cyto-physiology of the fat body of the American Silkmoth. *Z. Zellforsch.* 124: 433-444.

Ecobichon, D. J., 1969. Bovine hepatic carboxylesterases: chromatographic fractionation, gel filtration, and mole-cular weight estimation. *Can. J. Biochem.* 47: 799-805.

Gilbert, L. I. 1964. Physiology of growth and development: Endocrine aspects. In: *Physiology of Insecta,* ed.: Rockstein, M. (Acedemic Press, N. Y.) 1: 149-225.

Gilbert, L. I. and D. S. King 1973. Physiology of growth and development: Endocrine aspects. In: *Physiology of Insecta,* ed.: Rockstein, M. (Academic Press, N. Y.)

Vol. 1: 249-370 2nd. ed.

Gilbert, L. I. and H. A. Schneiderman 1961. The content of juvenile hormone and lipid in Lepidoptera: sexual differences and development changes. *Gen. Comp. Endocrinol.* 1: 453-472.

Hofstee, B. H. J., 1972. On the substrate activation of liver esterase. *Biochim. Biophys. Acta* 258: 446-454.

Kingsbury, N. H. and C. J. Masters 1972. On the immunological interrelationships of the vertebrate esterases. *Biochim. Biophys. Acta* 289: 331-346.

Kramer, K. J., L. L. Sanburg, F. J. Kezdy and J. H. Law 1974. The juvenile hormone binding protein in the hemolymph of *Manduca sexta* Johannson. *Proc. Nat. Acad. Sci. U. S. A.* 71: 493-497.

Law, J. H., C. Yuan and C. M. Williams 1966. Synthesis of a material with high juvenile hormone activity. *Proc. Nat. Acad. Sci., U. S. A.* 55: 576-578.

Ljungquist, A. and K.-B. Augustinsson 1971. Purification and properties of two carboxylesterases from rat-liver microsomes. *Eur. J. Biochem.* 23: 303-313.

Price, G. M., 1973. Protein and nucleic acid metabolism in insect fat body. *Biol. Rev.* 48: 333-375.

Segal, H. L., 1973. Enzymatic interconversion of active and inactive forms of enzymes. *Science* 180: 25-32.

Siddall, J. B., R. J. Anderson, and C. A. Henrick 1971. Studies on insect hormones. *Proc. 23rd Int. Congr. Pure and Appl. Chem.* 3: 17-25.

Slade, M. and C. F. Wilkinson 1973. Juvenile-hormone analogs: A possible case of mistaken identity? *Science* 181: 672-674.

Slade, M. and C. H. Zibitt 1971. Metabolism of exogenous cecropia juvenile hormone in lepidopterans. *Proc. 2nd Int. IUPAC Congr. Pest Chem.* 3: 45-58.

Slade, M. and C. H. Zibitt 1972. Metabolism of cecropia juvenile hormone in insects and in mammals. In: *Juvenile Hormones,* ed.: Menn, J. and M. Beroza (Academic Press, N. Y.) pp. 155-176.

Trautmann, K. H. 1972. In vitro studium der Träger proteine von ³H-markierten juvenilhormonwirksamen verbindungen in der hämolymphe von *Tenebrio molitor* L. Larren. *Z. Naturforsch.* 27B: 263-273.

White, A. F. 1972. Metabolism of the juvenile hormone analogue methyl farnesoate 10, 11-epoxide in two insect species. *Life Sci.* 11: 201-210.

Whitmore, D., E. Whitmore, L. I. Gilbert 1972. Juvenile hormone induction of esterases: A mechanism for the regulation of juvenile hormone titer. *Proc. Nat. Acad. Sci. U. S. A.*

69: 1592-1595.

Whitmore, L., L. I. Gilbert, P. I. Ittycheriah 1974. The origin of hemolymph carboxylesterases "induced" by the insect juvenile hormone. *Mol. and Cell. Endocrin.* 1: 37-54.

Whitmore, E. and L. I. Gilbert 1972. Hemolymph lipoprotein transport of juvenile hormone. *J. Insect Physiol.* 18: 1153-1167.

Yu, S. J. and L. C. Terriere 1971. Hormonal modifications of microsomal oxidase activity in the house fly. *Life Sci.* 10: 1173-1185.

LOCALIZATION AND SUBSTRATE-INHIBITOR SPECIFICITY
OF INSECT ESTERASE ISOZYMES

G. M. BOOTH[1], C. L. STRATTON[2], AND J. R. LARSEN[3]
[1]Department of Zoology, 575 WIDB
Brigham Young University, Provo, Utah 84601,
[2]Medical School, University of Nevada
Reno, Nevada 89503
[3]Department of Entomology, University of Illinois
Urbana, Illinois 61801

ABSTRACT. Acetylcholinesterase (AChE) was histochemically localized in the central nervous system (CNS) and musculature of the house cricket, *Acheta domestica*. Both light and electron microscopy (EM) techniques were used. In vivo light microscopy inhibition of AChE in crickets occurred in selective areas of the CNS, particularly in the interganglionic connectives and peripheral areas of the thoracic ganglion. Inhibition also seemed to be dependent on the type of inhibitor used. Electrophoresis data showed that cricket femur-muscle homogenates had one enzyme band when acetylthiocholine (ATCh) was used as the substrate, while there were 5 enzyme bands when α-naphthylacetate was employed as the substrate. EM histochemical procedures showed AChE localized in axonal membranes, inside of axons, glial membranes, sarcoplasmic reticulum, T-system, and neuromuscular junctions. All of the enzyme activity in these structures was inhibited with eserine at 10^{-4}M. Cricket muscle homogenates hydrolyzed the substrate 2-naphthylthioacetate 4.8 times greater than either ATCh or acetyl-β-methylthiocholine (AMTCh). The substrates ATCh and AMTCh were hydrolyzed at identical rates. This is the first time the presence of AChE in insect muscles has been demonstrated at the ultrastructural level.

INTRODUCTION

Acetylcholinesterase (AChE) is an enzyme that is responsible for hydrolyzing acetylcholine (ACh) in vertebrate and invertebrate tissues. This enzyme is of particular significance in insect toxicology since AChE is the target site of organophosphorous and carbamate insecticides (Booth and Metcalf, 1970a). These compounds are often used against insects in laboratory and field studies to determine the total amount of AChE that is inhibited after treatment. However, such data may not be very meaningful since multiple forms of AChE and other esterases with differing inhibitor sensitivities have

721

recently been reported in insects (Eldefrawi et al., 1970; Booth et al., 1973; Tripathi et al., 1973; Tripathi and O'Brien, 1973). Evidence of the presence of an AChE-type enzyme in cricket muscle has also been reported by Booth and Lee (1971) and Lee, Metcalf, and Booth (1973). The presence of this enzyme in insect muscle suggests that precautions should be taken in homogenizing different body segments of insects when one is looking for isozymes of the central nervous system (CNS).

The objectives of the present study are: (1) To demonstrate the histochemical localization of AChE and other esterases in the cricket CNS and musculature using light and electron microscopic techniques. (2) To determine if more than one form of AChE exists in cricket muscle. (3) To determine hydrolysis rates of cricket muscle AChE on selective and non-selective substrates.

METHODS AND MATERIALS

Gel electrophoresis. Acrylamide disc and slab gel techniques were utilized according to published procedures. (Booth and Metcalf, 1970b; Booth et al., 1973). Esterase and AChE substrates included α-naphthylacetate and acetylthiocholine (ATCh).

Light microscopy. Histochemical methods for demonstrating AChE activity were identical to those previously published (Booth and Metcalf, 1970a). The substrate ATCh was used in all experiments.

Electron microscopy. Karnovsky's (1964) method was used in this part of the study. The ganglionic or muscle tissue was fixed in 2.5% glutaraldehyde and the tissues were incubated with ATCh for 15-90 minutes depending on the thickness of the tissue.

Substrate hydrolysis. Cricket legs were ground in 0.1 M phosphate buffer, pH 7.8 with a concentration of 102 mg/ml. Assay of substrate hydrolysis was done in a 1 ml cuvette with 1 ml of 0.1 M phosphate buffer, 0.05 ml of 5,5'-dithio-bis-2-nitrobenzoate (Ellman et al., 1961) and 0.01 ml of 0.075 M ATCh, acetyl-β-methylthiocholine (AMTCh) or 2-naphthylthioacetate (2-NTA). A total of 0.1 ml of the enzyme solution was pipetted into the assay vessel and hydrolysis was followed with time using a Perkin-Elmer 360 recording spectrophotometer.

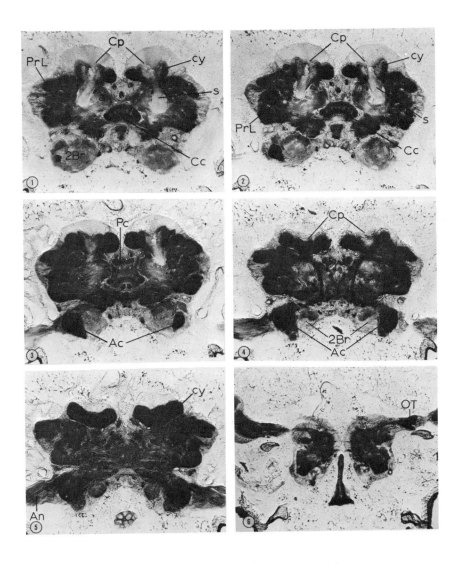

Fig. 1-6. Normal distribution of AChE in frozen sections of the brain of the house cricket, *Acheta domestica* using ATCh as the substrate. (12 μ M, 40X). Ac, antennal center; An,

Fig. 1-6 (legend cont.) antennal nerve; 2Br, deutocerebrum; Cc, corpus centrale; Cp, corpora pedunculata; (a, alpha lobe; b, beta lobe; cy, calyx; s, stalk); OT, optical tract; Pc, pons cerebralis; PrL, protocerebral lobe; Bs, brain stem; Fg, frontal ganglion; FGC, frontal ganglion connective; Me, lamina ganglionaris; Mi, medulla interna; SG, suboesophageal ganglion; (Lee et al., 1973).

RESULTS AND DISCUSSION

Figures 1-10 show control frozen sections of the brain and nerve cord of *Acheta domestica* (House cricket) that have been stained with ATCh. Figures 11 and 12 show AChE activity in peripheral nerves in frozen muscle sections of the house cricket and figures 13 and 14 show the AChE activity in the peripheral nerves of the field cricket, *Gryllus assimilus*. Figures 15 and 16 are frozen sections of mouse intercostal muscle stained for AChE as a comparison with the insect muscle.

The data in Fig 1-12 are representative of the many studies demonstrating that AChE is found in high concentrations in insect nervous tissue but is not necessarily uniformly distributed in the CNS. (Lee et al., 1973; Booth and Metcalf, 1970a, 1970b). These data suggest that some structures of the insect CNS may contain differing titers of AChE and perhaps different isozymes.

House crickets were treated with 0,0-dimethyl S-phenyl phosphorothioate and 0,0-dimethyl S-o-methylphenyl phosphorothioate on the tip of the abdomen and when the animals were in the hyperactive state, i.e. just prior to knockdown, they were sacrificed and sectioned. The CNS AChE distribution is shown in Fig. 17-20. Note that very specific areas of the CNS are inhibited. The interganglionic connections are completely inhibited by both compounds and peripheral areas of the thoracic ganglion were inhibited. Inhibition of brain AChE varied according to which compound was used. These data confirm other published results showing that localized effects of inhibitors on insects appear to be more important than total measurements of inhibition (Booth and Metcalf, 1970; Ramade, 1965). Recent data published on the fate of paraoxon 0,0-diethyl-p-nitrophenyl phosphate in the brain of the American Cockroach (*Periplaneta americana*) has shown that complete inhibition of AChE in the corpora pedunculata was always a consistent feature of poisoning regardless of where the toxicant was topically placed (Booth and Metcalf, 1972).

The interesting work by Tripathi and O'Brien (1973) has shown the presence of 4 soluble AChE isozymes in housefly

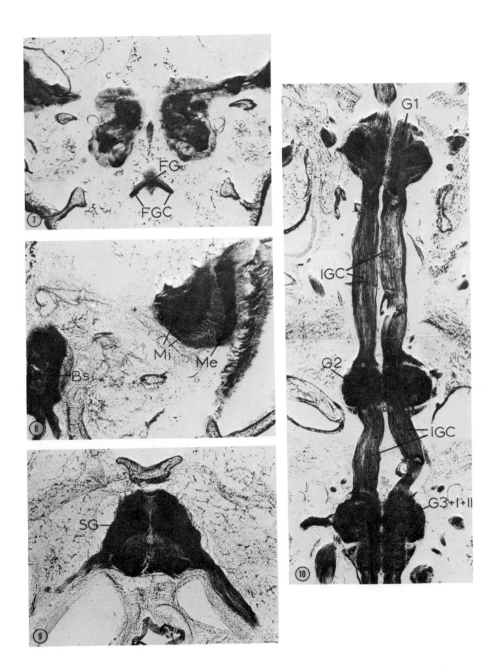

Figs. 7-10. For legend for Figs. 7, 8, and 9, see legend for

Fig. 7-10 (legend cont.) Figs. 1-6. Fig. 10: Normal distribution of AChE in a frozen section of the house cricket thoracic ganglion (12 μM, 20X). G1, prothoracic ganglion; G2,
mesothoracic ganglion; G3 + I + II, compound ganglion of meta
thoracic, 1st and 2nd abdominal segments. (Lee et al., 1973).

---

heads and 3 in the housefly thorax. They found varying sensitivities of each isozyme to selected organophosphorous compounds. Thoracic isozymes were found to be more sensitive to
inhibition than head isozymes, suggesting again the importance
of localized inhibition in the total poisoning process.

The data shown in Figs. 11-14 are particularly interesting
since AChE was thought not to be present in peripheral nerves
of insect muscles (Usherwood, 1973). Fig. 21A shows a dia-
gramatic sketch of 5 isozymes separated from cricket femurs
on acrylamide gels using α-naphthylacetate. When ATCh was
used as the substrate, only 1 enzyme could be detected (Fig.
21B). The single band using ATCh was almost completely inhibited by eserine at 10^{-4}M. A series of experiments were
conducted to ultrastructurally localize AChE in the ganglia
and muscles of insects. Very little information is available
on the electron microscopy (EM) of AChE in insects. In fact
only one insect (honeybee, *Apis mellifera*) has been investigated for localizing AChE using ATCh and EM techniques (Booth
and Lee, 1971).

Figs. 22 and 23 show control EM sections of a cricket ganglion with typical structures that have been reported in other
insects (Smith and Treherne, 1965). Figs. 24-26 show EM micrographs of the cricket ganglion which has been stained for
AChE.

The arrow in Fig. 24 shows the intense AChE activity in
the axonal membranes (15000X magnification). The "point" in
the lower portion of the micrograph shows activity in the
glial membranes distributed among the axons. However, not
all of the axonal membranes nor glial structures stained
equally in the section. Fig. 25 shows intense activity at a
probable synapse (S) between 2 axons (24000X). It is interesting that AChE activity was not limited to membranes. Fig. 26
(16000X) shows an axon that is "peppered" with AChE in the
central portion of the ganglion, while the axonal membranes
(arrow) are relatively devoid of activity. Treatment with
10^{-4}M eserine inhibited all activity in these structures.

Figs. 27-32 show micrographs of peripheral nerves and
muscle taken from the femur of the house cricket. Fig. 27
(20000X) shows AChE in the T-system and sarcoplasmic reticulum (points) of the muscle. Fig. 28 (12500X) shows highly
localized activity in the sarcoplasmic reticulum (points)
along side of a tracheole (T). Fig. 29 (20000X) shows

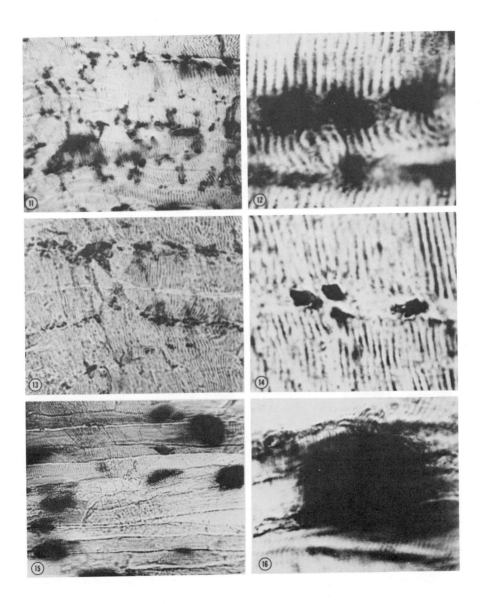

Figs. 11 and 12. House cricket thoracic muscle; 250X + 900X respectively. The dark spots show the presence of an AChE-type enzyme in the peripheral nerves. (Lee et al., 1973). Figs. 13 and 14. Field cricket, *Gryllis assimilis* thoracic

Figs. 11-16. (legend cont.) muscle sections, 250X and 900X respectively. (Lee et al., 1973). Dark spots show AChE activity. Figs. 15 and 16. Mouse intercostal muscle, 250X and 900X respectively. Dark spots show ACHE activity (Lee et al., 1973).

---

activity in the sarcolemma (point) and glial membranes (arrow). Fig. 30 (12500X) shows the cricket femur after 30 minutes treatment with eserine. Axons (A), axonal membranes, glial membranes, sarcolemma, and T-system are devoid of activity. Fig. 31 (50000X) shows a typical neuromuscular junction (NMJ) between an axon (A) and muscle (Mu) that has AChE localized in the synaptic cleft (single arrow). This micrograph fits all of the criteria for a typical NMJ, i.e. (1) the glial sheath is absent or incomplete at the synaptic terminal. This is true for the terminal shown between the point and arrow in Fig. 31. This allows the membranes of muscle and nerve to lie close together separated by a narrow gap or cleft. The 2 points shown at the top of Fig. 31 show the glial and axonal membranes together. (2) The membranes of the muscle and nerve are separated by a narrow cleft of 100-200 angstroms. This is the approximate distance shown in Fig. 31. (3) The presence of synaptic vesicles (SV) in the axon proper. These may be either solid or clear depending on the type of SV. The SV are not always demonstrated in treatments for showing AChE because of incubation. The membranes of the SV shown in Fig. 31 are not clearly differentiated in every SV shown. Fig. 32 (25000X) shows a NMJ after treatment with 10^{-4} M eserine. All activity was absent. The SV (point) were destroyed in the incubation process.

It is interesting to compare the AChE found in mammals with that found in crickets using ATCh as the substrate. Reported sites of activity in mammalian systems are: axonal membranes, axonal vesicles of all unmyelinated nerve fibers, NMJ, sarcolemma, sarcoplasmic reticulum, nuclear envelope, T-system, A and M bands in muscle, and mitochondria (Tennyson et al., 1968, 1973). In the cricket CNS and femur, AChE has now been demonstrated for the first time in axonal membranes, glial membranes, NMJ, sarcolemma, sarcoplasmic reticulum, and T-system.

Table 1 shows the hydrolysis rates of 3 thiol substrates by cricket muscle extracts. It can be seen that 2-NTA is hydrolyzed 4.8X faster than either ATCh or AMTCh. Both ATCh and AMTCh were hydrolyzed at identical rates.

The presence of only one electrophoretic band using ATCh and its sensitivity to eserine suggests one enzyme in cricket muscles. The question of the exact function of the enzyme,

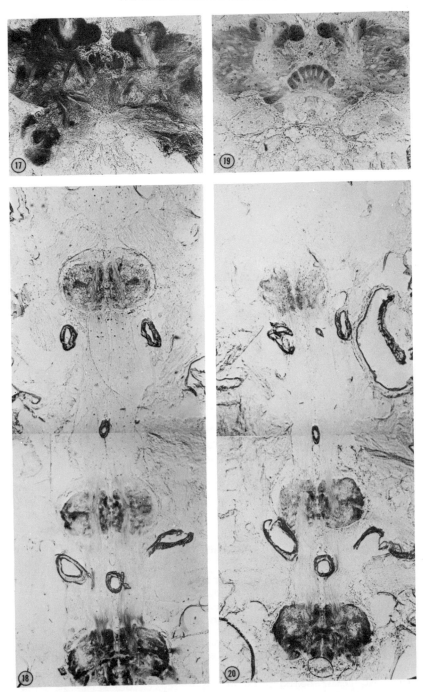

Figs. 17 and 18. In vivo Inhibition of house cricket AChE in

Figs. 17-20 (legend cont.) the head (Fig. 17) and thoracic ganglia (Fig. 18) by 0,0-dimethyl S-phenyl phosphorathioate. Inhibition is shown by the lack of staining for AChE. (Lee et al., 1973). Fig. 19 and 20. Inhibition of house cricket AChE in the head (Fig. 19) and thoracic ganglia (Fig. 19) by 0,0-dimethyl S-o-methylphenyl phosphorothioate. Chemicals were applied topically on the 7th abdominal tergum at 10 µg/ adult female. The animals were hyperactive. (Lee et al., 1973).

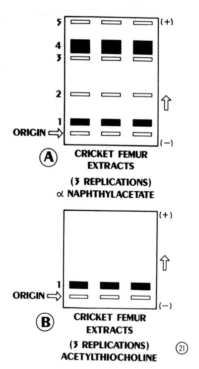

Fig. 21. (A) Electrophoretic zymogram of cricket femur homogenate using α-naphthylacetate as substrate (3 replications); 5 esterase isozymes are shown. (B) Electrophoretic zymogram of cricket femur homogenate using acetylthiocholine as substrate (3 replications); 1 enzyme is shown.

FIGURES 22 and 23

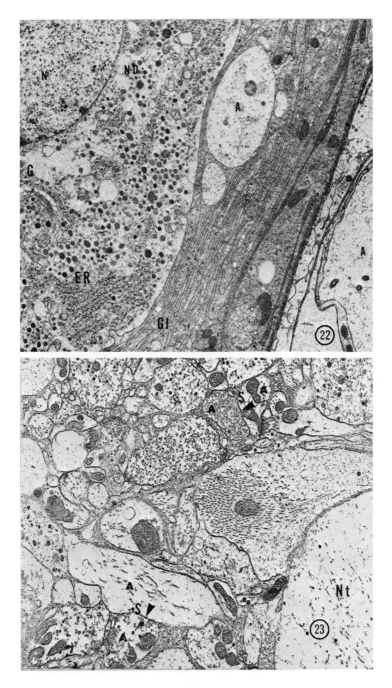

FIGURES 24 to 26

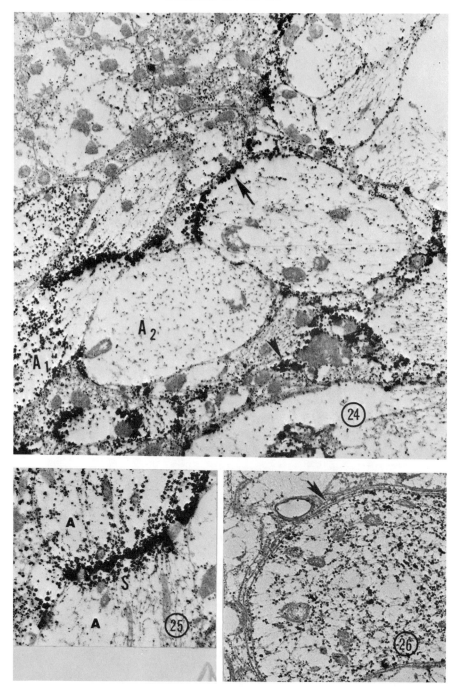

FIGURES 27 to 29

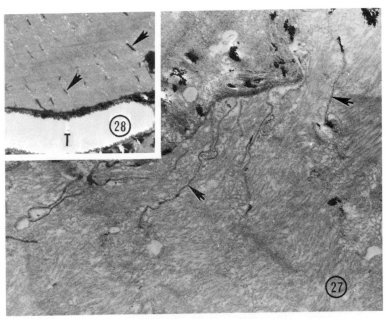

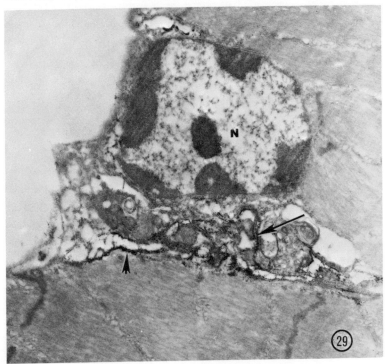

FIGURES 30 to 32

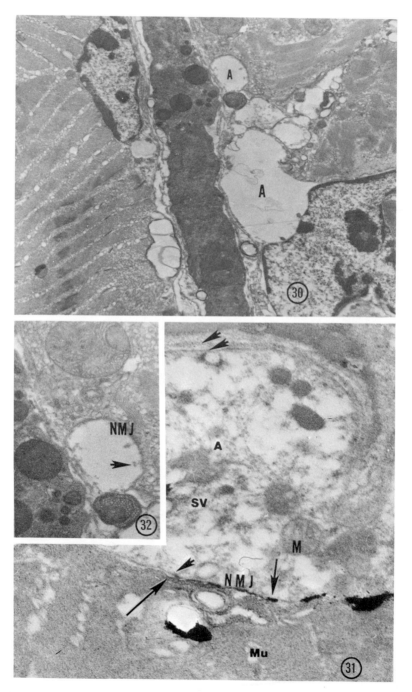

LEGENDS FOR FIGURES 22 TO 32

Fig. 22. Cricket ganglion control, no pre- or post- stain. 8000X. A, axon; N, nucleus; ND, neurosecretory droplets; G, golgi; ER, endoplasmic reticulum; Gl, glial membranes.

Fig. 23. Cricket ganglion control, no pre- or post- stain. 15000X. A, axon; S, synapse with accompanying synaptic vesicles; Nt, neurotubules.

Fig. 24. Cricket ganglion incubated with ATCh; no pre- or post- stain. 12500X; Arrow shows probable synapse with high activity; A_1, axon with light activity; point shows AChE in glial membranes. A_2, axon with little or no activity.

Fig. 25. Cricket ganglionic synapse (S) with AChE; 24000X; A, axon incubated with ATCh.

Fig. 26. Cricket ganglionic axon incubated with ATCh. AChE activity inside axon, while arrow shows axonic membrane relatively devoid of activity. 16000X.

Fig. 27. Cricket femur incubated with ATCh. 20000X; AChE in T-system and sarcoplasmic reticulum (arrows) no pre- or post-stain.

Fig. 28. Cricket femur incubated with ATCh; Post stained with lead citrate and uranyl acetate. 12500X; AChE (points) in sarcoplasmic reticulum. T, tracheole.

Fig. 29. Cricket femur incubated with ATCh; 20000X; no pre- or post- stain. Activity along sarcolemma (point) and glial membranes (arrow). N, nucleus.

Fig. 30. Cricket femur incubated with ATCh; 12500X; eserine treated; post-stained with lead citrate and uranyl acetate; A, axons; activity in all structures inhibited by eserine.

Fig. 31. Cricket femur incubated with ATCh; 50000X; no pre- or post- stain; Glial membrane and sarcolemma (2 points); neurolemma and sarcolemma (point and arrow); NMJ, neuromuscular junction with enzyme activity (single arrow); A, axon; SV, synaptic vesicle; M, mitochondria; Mu, muscle.

Fig. 32. Cricket femur incubated with ATCh after eserine treatment; 25000X; post-stained with lead citrate and uranyl acetate; NMJ, neuromuscular junction shows no activity; synaptic vesicles destroyed in incubation process (point).

TABLE I

COMPARISON OF THE HYDROLYSIS RATES OF 3 THIOL SUBSTRATES
BY CRICKET MUSCLE ESTERASES

Substrate	μ-moles substrate hydrolyzed/min/g	RR**
Acetylthiocholine	0.5017 ± 0*	1
Acetyl-β-methythiocholine	0.5017 ± 0*	1
2-naphthylthioacetate	2.4250 ± 0.30*	4.8

* ± standard deviation
** Relative rate of hydrolysis with ATCh = 1.

however, remains to be solved. If AChE is involved in neuro-musclar transmission, then one may wonder what this enzyme does in the structures found in this report. If ACh or choline acetylase can be demonstrated in cricket muscle, then one would have to reconsider the possibility that the neuromuscu-lar transmitter in house crickets may be cholinergic.

To date we have found this AChE-type enzyme in the muscle tissue of house crickets, field crickets, and green lace wings (Lee et al., 1973). In addition there is sufficient light and EM evidence for the presence of AChE in the muscles of *Periplaneta americana* (data to be published elsewhere). More insects ought to be examined in the future for the presence of AChE in the peripheral nervous system. Certainly the present data suggest muscle tissue needs to be examined as carefully as nervous tissue for AChE isozymes. In order to clarify these various issues we are investigating the presence or absence of acetylcholine or choline acetylase in cricket and other insect muscles. In addition we are studying the effects of selected inhibitors on hydrolysis rates of thiol substrates by insect muscle homogenates.

REFERENCES

Booth, G. M., J. Connor, R. A. Metcalf, and J. R. Larsen 1973. A comparative study of the effects of selective inhibi-tors on esterase isozymes from the mosquito *Anopheles punctipennis*. *Comp. Biochem. Physiol.* 44B: 1185-1195.
Booth, G. M. and R. L. Metcalf 1972. The histochemical fate of paraoxon in the cockroach (*Periplaneta americana*) and honey bee (*Apis mellifera*) brain. *Israel J. Entomol.* 7: 143-156.

Booth, G. M. and A-H Lee 1971. Distribution of cholinesterases in insects. *Bull. Wld. Hlth. Org.* 44: 91-98.

Booth, G. M. and R. L. Metcalf 1970a. Histochemical evidence
for localized inhibition of cholinesterase in the house
fly. *Ann. Entomol. Soc. Amer.* 63(1): 197-204.

Booth, G. M. and R. L. Metcalf 1970b. Phenylthioacetate: A
useful substrate for the histochemical and colorimetric
detection of cholinesterase. *Science* 170: 455-457.

Eldefrawi, M. E., R. K. Tripathi, and R. D. O'Brien 1970.
Acetylcholinesterase isozymes from the housefly brain.
Biochem. Biophys. Acta 212: 308-314.

Ellman, G. L., K. D. Courtney, V. Andres, Jr., and R. M. Featherstone 1961. A new and rapid colorimetric determination
of acetylcholinesterase activity. *Biochem. Pharmacol.*
7: 88-95.

Karnovsky, M. J. 1964. The localization of cholinesterase
activity in rat cardiac muscle by electron microscopy
J. Cell. Biol. 23: 217-232.

Lee, A-H, R. L. Metcalf, and G. M. Booth 1972. House cricket
acetylcholinesterase: Histochemical localization and
in situ inhibition by 0,0-dimethyl S-aryl phosphorothioates.
Ann. Entomol. Soc. Amer. 66(2): 333-343.

Ramade, F. 1965. L'action anticholinestérasique de quelques
insecticides organophorés sur le système nerveux central
de Musca domestica. *Ann. Soc. Ent. Fr. (N.S.)* 1: 549-
566.

Smith, D. S. and J. E. Treherne 1965. The electron microscopic localization of cholinesterase activity in the central
nervous system of an insect, *Periplaneta americana L.*
J. Cell. Biol. 26: 445-459.

Tennyson, V. M., M. Brzin, and L. T. Kremzer 1973. Acetylcholinesterase activity in the myotube and muscle satellite
cell of the fetal rabbit: An electron microscopic and
biochemical study. *J. Histochem. Cytochem.* 21(7): 634-
652.

Tennyson, V. M., M. Hagopian, and D. Spiro 1968. Ultracyto-
chemical studies of cholinesterase in rabbit embryonic
cardiac muscle. *J. Cell. Biol.* 39: 134A.

Tripathi, R. K. and R. D. O'Brien 1973. Effect of organophos-
phates in vivo upon acetylcholinesterase isozymes from
housefly head and thorax. *Pestic. Biochem. Physiol.*
2(4): 418-424.

Tripathi, R. K., Y. C. Chiu, and R. D. O'Brien 1973. Reactivity in vitro toward substrate and inhibitors of acetylcholinesterase isozymes from electric eel electroplax
and housefly brain. *Pestic. Biochem. Physiol.* 3(1): 55-
60.

Usherwood, P. N. R. 1973. Action of iontophoretically applied gamma-aminobutyric acid on locust muscle fibers. *Comp. Biochem. Physiol.* 44A: 663-664.

ACETYLCHOLINESTERASE ISOZYMES AND MUSCLE DEVELOPMENT:
NEWLY SYNTHESIZED ENZYMES AND CELLULAR SITE OF ACTION OF
DYSTROPHY OF THE CHICKEN

BARRY W. WILSON, THOMAS A. LINKHART, CHARLES R. WALKER
AND G. WENDEL YEE
Department of Avian Sciences
University of California, Davis, California   95616

ABSTRACT. Previous studies have shown that chick embryo
muscles contain 3 isozymes of AChE, two of which become
undetectable after hatching. These isozymes are associated with activity outside the neuromuscular junction and
with release of AChE from the muscle in embryo, denervated,
and dystrophic muscle. The studies led to the hypothesis
that neural activity represses embryo AChE forms in normal
but not in genetically dystrophic chicken muscles. Limb
bud transplantation of wings between $3\frac{1}{2}$ day chick embryos
tested whether the dystrophic nerve or muscle was defective.
The results conclusively show that the gene defect of AChE
regulation is expressed in the limb and not in the nerve
of the dystrophic chicken. The regulation of AChE in the
muscle fibers themselves is under study with cell cultures.
In the experiments presented here, cells were briefly treated with an irreversible inhibitor of AChE, DFP (diisopropylfluorophosphate), to permit study of the mobilization of
newly synthesized enzyme. Newly produced AChE rapidly returned in the cells and behaved as if it moved and as-
sembled from low molecular weight forms.

INTRODUCTION

Multiple molecular forms of many proteins such as hemoglobin (Ingram, 1972), creatine phosphokinase (Dawson et al.,
1968), and lactate dehydrogenase (Markert and Møller, 1959)
change during development. We have been studying one such
isozyme system, acetylcholinesterase (AChE E.C. 3.1.1.7) in
chick embryo muscle, trying to understand its regulation and
developmental program and its relationship to a single gene
abnormality, inherited muscular dystrophy of the chicken.
The results of previous studies are summarized in Figure
1. AChE is found in mononucleated myoblasts before they fuse
to form myotubes (Engel, 1961; Tennyson et al., 1971; Wilson
et al., 1973a; Fluck and Strohman, 1973). In addition, AChE
is released in quantity from the myotubes into the medium and
appears in plasma from chick embryo (Wilson et al., 1973b).

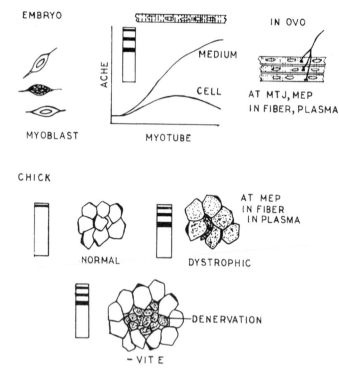

Fig. 1.   Development and regulation of AChE in fast twitch
muscle of the chicken.

AChE activity is due to at least 3 isozymes with molecular
weights of 420,000, 293,000, and 219,000 (Wilson et al., 1969).
Their pattern on acrylamide gels is shown in Figure 2.

In ovo , motor end plate AChE appears after two weeks of
incubation.   At this time AChE is found at myoneural and
myotendon junctions and at the surface and within the embryo
muscle fibers (Mumenthaler and Engel, 1961; Filogamo and
Gabella, 1967; Wilson et al., 1970).   After hatching, AChE
decreases in activity in twitch fibers, extrajunctional activity falls, and the two lowest molecular weight isozymes become virtually undetectable in muscle homogenates (Wilson
et al., 1970).

Fast twitch muscles from birds with inherited muscular
dystrophy maintain relatively high AChE levels, extrajunctional fiber and plasma AChE activity, and the 3 AChE isozymes
found in embryo muscle (Wilson et al., 1970; 1973b).   The
results led to the conclusion that muscles afflicted with
inherited muscular dystrophy were defective in a system that
regulated the development of muscle AChE.   The fact that

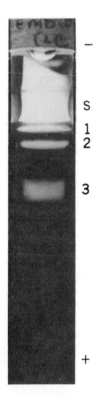

Fig. 2. Isozymes of AChE in chick embryo muscle. A homogenate of pectoral muscle from a 14 day old embryo was exposed to
electrophoresis on a 10% acrylamide gel, 0.047 M TRIS-glycine
buffer, pH 8.3 and stained with acetylthiocholine, $CuSO_4$,
$MgCl_2$ and Maleate-glycine buffer, pH 6.0 according to Maynard
(1966). Opaque Bands 1, 2, and 3 are enzyme activity. Spacer
gel (S) light color is not enzymatic. Substrate hydrolysis
and selective inhibitors show Bands 2 and 3 contain only AChE,
while Band 1 is a mixture of both AChE and non-specific ChE.
Migration is from cathode (top) to anode (bottom). (For specific methods see Wilson et al., 1969; 1973a).

---

denervation, but not tenotomy, of normal muscle brought about
an increase in AChE and a return of extrajunctional activity
and embryonic AChE isozymes led to the conclusion that neural
activity in some way repressed embryonic AChE properties
(Wilson et al., 1970). Vitamin E deficiency yielded similar
results in atrophying muscle fibers (Wilson and Viola, 1972).
In toto, the results suggested that muscles with inherited
muscular dystrophy were defective in a neurally mediated

repression and localization of AChE activity and regulation
of its isozymes. However, the data did not indicate whether
the genetic defect was in the nerves or in the muscles of dystrophic chickens. Was the nerve unable to communicate to the
muscle or was the muscle unable to respond to the nerve?

LIMB BUD TRANSPLANTS

We attacked the problem using embryonic limb bud transplantation (Eastlick, 1943). In our experiments, wing limb
buds were surgically removed from embryos before nerves had
innervated the muscles and transferred to the body of another
host. Transplanted limb buds grew, differentiated, became
innervated, and ultimately produced functional wings. Right
wing buds from normal and dystrophic embryos were reciprocally
transplanted at 3½ days of incubation, the embryos were incubated to hatching, and the chicks were examined when 5-12 weeks
old. Ten to twenty percent of the embryos treated survived
past hatching. Specifically, we asked the question whether the
AChE pattern of limb muscles would be dependent upon the genotype or the limb of the host. Levels of AChE and AChE isozyme
patterns as revealed by acrylamide gel electrophoresis were all
determined in each experiment.

The embryos that hatched provided an unequivocal answer
to whether the defect in AChE regulation was expressed in the
limb and its muscles or in the host and its nerves. Normal
limbs remained normal when grown on a dystrophic host and
dystrophic limbs remained dystrophic when transplanted to
a normal host. Whatever the host-donor combination, the
genotype of the embryo limb and not that of the host and
its nerves determined the pattern of AChE activity of the
developed wing.

The AChE levels of the various combinations are shown in
Table 1. Normal limbs on normal hosts had AChE levels similar
to those of the host limbs, dystrophic limbs on normal hosts
had more than 10 times the AChE activity of the normal limbs,
and normal limbs on dystrophic hosts had AChE levels less than
that of the dystrophic limbs of the hosts.

Figure 3 is a composite of photomicrographs of frozen sections of various limb transplants stained for AChE activity.
It shows that normal wing muscles, whether from transplants
or hosts have AChE activity only at their motor end plates
and that dystrophic wings, whether from transplants or hosts,
had AChE in many muscle fibers. All transplants showed AChE
staining of motor end plates and muscle spindle fibers.

Figure 4 is a diagram of the electrophoresis results. Normal limbs grafted on normal or dystrophic hosts exhibited one
AChE form. Dystrophic limbs grafted on normal hosts had the

TABLE I

AChE activity of biceps muscles from limb bud transplants of
normal and dystrophic genotype. (Activity expressed as mean
OD/min/gm wet weight.) AChE determined by the method of
Ellman et al. (1961) with 10^{-4} iso-OMPA to inhibit nonspecific
cholinesterases.

Normal donor → Normal host	AChE
Normal host limb	1.03
Normal donor limb	1.50
(5 - 10 weeks, 4 birds)	

Dystrophic donor → Normal host	AChE
Normal host limb	1.16
Dystrophic donor limb	16.6
(5-12 weeks, 6 birds)	

Normal donor → Dystrophic host	AChE
Dystrophic host limb	13.1
Normal donor limb	2.0
(5-6 weeks, 5 birds)	

3 AChE isozymes expected of abnormal muscle.

Thus, whether with regard to AChE levels, localization, or
isozymes, the limb bud transplants behaved as if inherited
muscular dystrophy of the chicken was a property of the limb,
and by inference, the muscles, and did not depend upon the
genome of the host or its nerves. Theoretically, it is possible that tissues other than muscle in the limb may be responsible for the dystrophic pattern of AChE. However, it is more
plausible to proceed with the idea that the genetic defect is
expressed in the muscle fibers so that dystrophic muscles are
unable to respond to instructions from their nerves to repress
levels, extrajunctional localization, and low molecular weight
isozymes of AChE.

The nature of the defect in dystrophic muscle that makes
it unable to repress these embryo AChE properties is not known.
One of the difficulties in studying it is that the mechanisms
regulating AChE in normal muscle fibers are themselves unknown.

AChE IN CELL CULTURES

We have turned to muscle cell cultures to study the ways
in which AChE is mobilized in embryo muscle, first examining
the properties that are intrinsic to embryo muscle fibers in

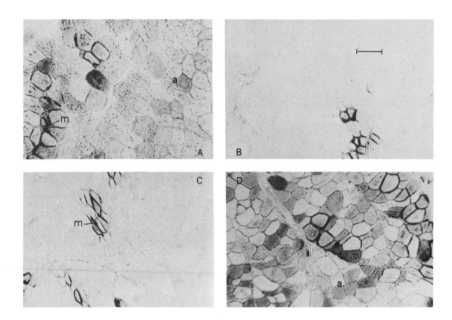

Fig. 3. AChE activity of limb bud transplants. Reciprocal transplants of wing buds between normal and dystrophic embryos were performed at 3½ days of incubation. The hatched chicks were sacrificed at 5-12 weeks and the biceps muscles were frozen in iso-pentane-liquid N_2, stained for AChE using acetylthiocholine, selective inhibitors, and the method of Karnovsky and Roots (1964) and examined under the light microscope. The photomicrographs show: a dystrophic wing on a normal host (A), and its contralateral normal control wing (B); a normal wing on a dystrophic host (C) and its contralateral dystrophic control wing (D). Note motor end plate AChE (m) and extrajunctional AChE activity (a). Bar shown is 100 μ.

the absence of nerves (Wilson et al., 1973a). Recently we have been examining the fate of newly synthesized AChE, measuring its synthesis, degradation, localization, and release into the medium (Wilson and Walker, 1974). The experiments involve inhibiting previously synthesized enzyme with DFP (diisopropylfluorophosphate), an irreversible inhibitor of AChE activity. The results show that recovery of AChE in DFP-treated muscle fibers requires protein synthesis and suggest that newly synthesized enzyme undergoes a complex process of movement, degradation, release, and shift in isozyme forms.

LIMB BUD TRANSPLANTS BETWEEN NORMAL AND

DYSTROPHIC CHICK EMBRYOS

Acetylcholinesterase Isozymes in Biceps of Hatched Chicks

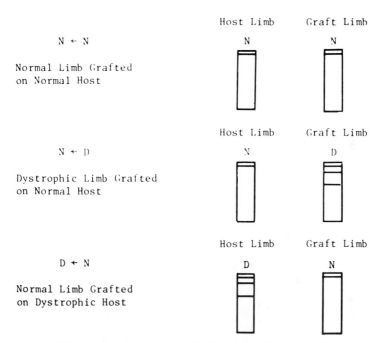

Fig. 4.   AChE isozymes of limb bud transplants.   5-12 week old chicks.   Methods as described in previous figures.

Table 2 shows an example of the return of AChE activity in DFP treated cells. AChE activity returned rapidly, reaching more than 70% of its initial level in 2 hours. No activity returned when the cells were incubated in the continuous presence of 10^{-5}M cycloheximide.

The isozyme pattern of DFP-treated cells suggested that an assembly process for AChE may have occurred. Figure 5 shows the results of one experiment with 14-day old cultures. Untreated control cultures exhibited high levels of Bands 1 and 2 and a trace of activity of Band 3 (Gel A). No activity appeared immediately after DFP treatment. Two (not shown) and 4 hours (Gel B) after DFP all three AChE forms were present; Band 1 was lower in activity than the untreated controls

TABLE II

Recovery of AChE in chick embryo muscle after DFP treatment
(14 day cultures, briefly treated with 10^{-4}M DFP and returned
to normal medium or incubated in 10^{-5}M cycloheximide. AChE
activity expressed as OD/min/dish. Medium activity is corrected for AChE initially in the medium).

	AChE					
	Untreated		DFP Treated		Cycloheximide	
Hours	Cell	Medium	Cell	Medium	Cell	Medium
0	0.437	0	0.021	0	0.021	0
2	0.459	0	0.203	0	0.041	0
4	0.310	0.017	0.262	0.060	0.016	0
8	0.372	0.063	0.188	0.076	-	-
24	0.280	0.369	0.129	0.492	0.021	0

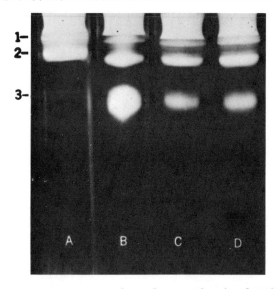

Fig. 5. Isozyme pattern of newly synthesized AChE from pectoral muscle cultures. Single cells dissociated from 11-day old
chick embryo pectoral muscles were grown for 14 days in vitro,
treated for 10 minutes with 10^{-4}M DFP and returned to their
normal growth medium. Untreated control cultures (A) and cultures 4 hours (B), 8 hours (C) and 24 hours (D) after DFP treatment were homogenized, subjected to electrophoresis on 10%
acrylamide gels, .047M TRIS-glycine buffer, pH 8.3 and stained
for AChE activity. Note low activity in Band 3 and A and activity between Bands 1 and 2 in B,C, and D (Wilson and Walker,
1974).

and Band 3 was much more intense. After 8 (Gel C) and 24
hours (Gel D) Band 3 was still present but at lower intensity
than earlier and Band 1 seemed to have increased in activity.
It was as if newly synthesized AChE was relatively rich in a
low molecular weight form, and that with time, the low molecular weight form increased in activity. One possible interpretation is that the low molecular weight AChE aggregates to
form higher molecular weight isozymes.

DISCUSSION

AChE IN EMBRYO MUSCLE

The results of these and other investigations reviewed
in the introduction support the idea that embryo muscle fibers
synthesize low molecular weight forms of AChE that assemble
together, bind to cell constituents, and eventually are either
released from the cells or degraded. Later in development
their synthesis is repressed throughout much of the length of
normal fibers by neural activity. It is possible that these
AChE forms continue to be synthesized near myoneural junctions
and are released into the space between the nerve and muscle
to form the functional AChE of the motor end plate.

Multiple molecular forms of AChE have been described in
the muscle of several animals (Barron et al., 1968; Davis and
Arganoff, 1968; Tripathi et al., 1973). The nature and number
of polypeptides that make up the active proteins are unclear.
For example, Leuzinger et al. (1969) isolated an AChE form of
250,000 molecular weight from electric eel electroplax composed
of 2 pairs of polypeptides. However, Dudai et al. (1973)
found several AChE forms from the same tissue with molecular
weights up to and above 1×10^{6}. Interestingly, the high
molecular weight forms seemed composed of aggregates of molecules shaped as if they had heads and tails.

The worth of hypotheses are in the experiments they generate. Studies are underway to examine the ultrastructural
localization of newly synthesized AChE forms in chick muscle
cultures, their binding to cell fractions, particularly membranes, and the number of polypeptides that compose them. In
addition, we are also studying the effect of chronic electri-
cal stimulation on the AChE level of muscle cultures to find
out whether contraction of the muscle fibers represses their
AChE activity. Cohen and Fishbach (1973) reported that ACh
receptor decreased in stimulated muscle cultures and our preliminary results (Walker and Wilson, in preparation) suggest
that stimulation reversibly represses AChE activity.

AChE AND LIMB BUD TRANSPLANTS

There has been much interest in finding out whether nerves or muscles are the primary sites of the genetic lesions of inherited neuromuscular abnormalities (McComas et al., 1971; Hofmann et al., 1973; Peterson, 1974). Three lines of evidence suggested that the neuromuscular junction was abnormal in dystrophic chicken muscles: 1) the results reviewed earlier that showed there was a defect in a neurally regulated system that suppressed embryo AChE patterns, 2) the finding of Jedrez-ejczyk et al. (1973) of abnormally low AChE at dystrophic neuromuscular junctions, and 3) the report of Albuquerque and Warnick (1971) that suggested that a neural defect might be responsible for the abnormalities.

The limb-bud transplantation experiments demonstrate that the defect in AChE regulation of dystrophic muscle resides in the limb tissue and not in the nerves or in the rest of the body of the dystrophic chicken. The lesion is most likely within the muscle fibers themselves, although other limb cells cannot be logically excluded by the experiments. Other muscle properties known to be altered in dystrophic muscle were also examined in the transplantation experiments. Determinations of cytochemical succinic dehydrogenase, lactate dehydrogenase activity, electromyographic potentials, muscle fiber size, and ultrastructure all supported the conclusion that inherited muscular dystrophy of the chicken is caused by a muscle defect (Linkhart, in preparation).

Previous reports have suggested that muscular dystrophy of the chicken is myogenic. In a "preliminary" report, Cosmos and Butler (1972) transplanted minced muscle fragments between normal and dystrophic chicks. They found that regenerated normal muscle retained its fiber size, oil-red-O staining and succinic dehydrogenase activity regardless of the genotype of the host. However, dystrophic muscle did not regenerate as well as normal muscle, and the muscles that were transplanted had already been innervated by nerves of their own genotype during development. Peacock and Nelson (1973) found that spinal nerves from normal and dystrophic chick embryos formed functional synapses equally well with cultured dystrophic muscle cells and suggested this supported the existence of a muscle lesion. However, the study did not include normal muscle cultures nor is there any evidence that normal and dystrophic embryo muscles or nerves differ in vivo and in vitro.

Cross innervation and transplant studies on adult mice (for example; Hironaka and Miyata, 1973), cell culture studies (Gallup and Dubowitz, 1973), and one report using allophenic mouse embryos (Peterson, 1974) have led to suggestions that dystrophy of the mouse is due to a neural lesion. However some workers have suggested the opposite based on similar experiments (Law and Atwood, 1972; Douglas and Cosmos, 1974). The site of the lesions in human muscle abnormalities is not known. Evidence for impaired neural function has led to the proposal that Duchenne dystrophy is due to a neural lesion (McComas et al., 1971). However, the fact that muscles and the nerves that innervate them develop and mature as partners raises the possibility that abnormalities in one will be reflected by changes in the other.

The nature of the defect causing inherited muscular dystrophy of the chicken (as well as those causing other dystrophies) is not known. It is possible, but unproven, that the error in AChE regulation in dystrophic chick muscle is the primary site of action of the dystrophic gene.

Regardless, the experiments presented here demonstrate the complicated nature of the pattern of development of AChE in embryo muscle and the intimate involvement of AChE in a muscle abnormality.

ACKNOWLEDGEMENTS

The authors are grateful to Dr. Ursula K. Abbott for her contributions to the limb bud transplantation studies. Research reviewed and reported here supported in part by NIH grants NS-10957, ES-00202 and, with regard to dystrophy, the Muscular Dystrophy Associations of America.

REFERENCES

Albuquerque, Edson X. and J.E. Warnick 1971. Electrophysiological observations in normal and dystrophic chicken muscles. *Science* 172:1260-1263.

Barron, Kevin D., A.T. Ordinario, J. Bernsohn, A.R. Hess,and M.T. Hedrick 1968. Cholinesterases and nonspecific esterases of developing and adult (normal and atrophic) rat gastrocnemius. I. Chemical assay and electrophoresis. *J. Histochem. Cytochem.* 16:346-361.

Cohen, Stephen A. and G.D. Fischbach 1973. Regulation of muscle acetylcholine sensitivity by muscle activity in cell culture. *Science* 181:76-78.

Cosmos, Ethel and J. Butler 1972. Differentiation of muscle transplanted between normal and dystrophic chickens. In: *Research in Muscle Development and the Muscle Spindle.* B.Q. Barker, R.J. Przybylski, J.P. Vander Meulen,and M. Victor, eds. *Excerpta Medica,* Amsterdam. pp. 149-162.

Davis, Gary A. and B.W. Arganoff 1968. Metabolic behavior of isozymes of acetylcholinesterase. *Nature* 220:277-278.

Dawson, David M., H.M. Eppenberger,and M.E. Eppenberger 1968. Multiple molecular forms of creatine kinases. *Ann. N.Y. Acad. Sci.* 151:616-626.

Douglas, W. Bruce and E. Cosmos 1974. Histochemical responses of murine dystrophic muscles cross-innervated by sciatic nerves of normal mice. In: *Exploratory Concepts in Muscle (II)*. A.T. Milhorat, ed. In press.

Dudai, Yadin, M. Herzberg, and I. Silman 1973. Molecular structures of acetylcholinesterase from electric organ tissue of the electric eel. *Proc. Nat. Acad. Sci. U.S.A.* 70:2473-2476.

Eastlick, Herbert L. 1943. Studies on transplanted embryonic limbs of the chick. *J. Exp. Zool.* 93:27-49.

Ellman, George L., K.D. Courtney, V. Andres,and R.M. Featherstone 1961. A new and rapid colorimetric determination of acetylcholinesterase activity. *Biochem. Pharmacol.* 1: 88-95.

Engel, W. King 1961. Cytological localization of cholinesterase in cultured skeletal muscle cells. *J. Histochem. Cytochem.* 9:66-72.

Filogamo, Guido and G. Gabella 1967. The development of neuromuscular correlations in vertebrates. *Arch. Biol.* (Liege) 78:9-60.

Fluck, Richard A. and R.C. Strohman 1973. Acetylcholinesterase activity in developing skeletal muscles in vitro. *Develop. Biol.* 33:417-428.

Gallup, Belinda and V. Dubowitz 1973. Failure of dystrophic neurones to support functional regeneration of normal or dystrophic muscle in culture. *Nature* 243:287-289.

Hironaka, Tetsuji and Y. Miyati 1973. Muscle transplantation in aetiological elucidation of murine muscular dystrophy. *Nature New Biology* 244:221-223.

Hofmann, William W., K.L. Birnberger, A. Harlacher,and M. Schreiber 1973. Normal and dystrophic muscle in vitro: an argument against denervation. *Exp. Neurol.* 39:249-260.

Ingram, Vernon M. 1970. Embryonic red blood cell formation. *Nature* 235:338-339.

Jedrezejczyk, J., J. Wieckowski, T. Rymaszewska,and E.A. Barnard 1973. Dystrophic chicken muscle: altered synaptic acetylcholinesterase. *Science* 180:406-408.

Karnovsky, Morris J. and L. Roots 1964. A direct coloring thiocholine method for cholinesterases. *J. Histochem. Cytochem.* 12:219-221.

Law, Peter K. and H.L. Atwood 1972. Nonequivalence of surgical and natural denervation in dystrophic mouse muscles. *Exp. Neurol.* 34:200-209.

Leuzinger, Walo, M. Goldberg, and E. Cauvin 1969. Molecular properties of acetylcholinesterase. *J. Mol. Biol.* 40: 217-225.

Markert, Clement L. and F. Møller 1959. Multiple forms of enzymes: tissue, ontogenetic and species specific patterns. *Proc. Nat. Acad. Sci. U.S.A.* 45:753-763.

Maynard, Edith A. 1966. Electrophoretic studies of cholinesterase in brain and muscle of developing chicken. *J. Exp. Zool.* 161:319-336.

McComas, A.J., R.E.P. Sica, and M.J. Campbell 1971. Sick motoneurones, a unifying concept of muscle disease. *Lancet* 1:321-325.

Mumenthaler, M. and W.K. Engel 1961. Cytological localization of cholinesterase in developing chick embryo skeletal muscle. *Acta. Anat.* 47:274-299.

Peacock, John H. and P.G. Nelson 1973. Synaptogenesis in cell cultures of neurones and myotubes from chickens with muscular dystrophy. *J. Neurol. Neurosurg. and Psychiatry* 36:389-398.

Peterson, Allen C. 1974. Chimera mouse study shows absence of disease in genetically dystrophic muscle. *Nature* 248: 561-564.

Tennyson, Virginia M., M. Brzin, and P. Slotwiner 1971. The appearance of acetylcholinesterase in the myotome of the embryonic rabbit. *J. Cell Biol.* 51:703-721.

Tripathi, Ram K., Y.C. Chiu, and R.D. O'Brien 1973. Reactivity in vitro toward substrate and inhibitors of acetylcholinesterase isozymes from electric eel electroplax and housefly brain. *Pesticide Biochem. and Physiol.* 3:55-60.

Wilson, Barry W., M.A. Mettler, and R.V. Asmundson 1969. Acetylcholinesterase and non-specific esterase in developing avian tissues: distribution and molecular weights of esterases in normal and dystrophic embryos and chicks. *J. Exp. Zool.* 172:49-58.

Wilson, Barry W., M.A. Kaplan, W.C. Merhoff, and S.S. Mori 1970. Innervation and the regulation of acetylcholinesterase activity during the development of normal and dystrophic chick muscle. *J. Exp. Zool.* 174:39-54.

Wilson, Barry W., T.A. Linkhart, C.R. Walker, and P.S. Nieberg 1973b. Tissue cholinesterase in plasma of chick embryos and dystrophic chickens. *J. Neurol. Sci.* 18:333-350.

Wilson, Barry W., P.S. Nieberg, C.R. Walker, T.A. Linkhart, and D.M. Fry 1973a. Production and release of acetylcholinesterase by cultured chick embryo muscle. *Develop. Biol.* 33:285-299.

Wilson, Barry W. and G.A. Viola 1972. Multiple forms of acetylcholinesterase in nutritional and inherited muscular dystrophy of the chicken. *J. Neurol. Sci.* 16:183-192.

Wilson, Barry W. and C.R. Walker 1974. Regulation of newly synthesized acetylcholinesterase in muscle cultures treated with diisopropylfluorophosphate. *Proc. Nat. Acad. Sci. U.S.A.* 71:3194-3198.

SPECIFIC ISOZYME PROFILES OF ALKALINE PHOSPHATASE IN PREDNISOLONE-TREATED HUMAN CELL POPULATIONS

ROBERT M. SINGER and WILLIAM H. FISHMAN
Tufts Cancer Research Center
Tufts University School of Medicine
Boston, Massachusetts 02111

ABSTRACT. Using polyacrylamide disc gel electrophoresis, we examined the isozyme profiles of alkaline phosphatase in four human cell lines and determined the effect of hormone treatment. The isozymes of alkaline phosphatase were identified on the basis of their inhibition properties, heat stability, and cross reactivity to specific antisera. HeLa TCRC-1, a cell line monophenotypic for the Regan isozyme, and HeLa TCRC-2, monophenotypic for non-Regan alkaline phosphatase, each had a single specific band of enzyme activity. Fl amnion and HEp-2 cell lines each possessed their own characteristic isozyme pattern containing two bands of activity. We observed a marked alteration in the isozyme profiles after 72 hours of growth in prednisolone containing media. The Fl amnion, HEp-2, and HeLa TCRC-1 all showed an elevation of enzyme activity at the Regan migration position. The HEp-2 cell line exhibited a band at that position, which was enhanced by hormone treatment. The Fl amnion did not show a band at the Regan position; however, hormone treatment caused its induction. The HeLa TCRC-1 band of enzyme activity was significantly enhanced by hormone treatment. The enhanced band in all cell lines was shown to have all the properties of the Regan isozyme. Since HeLa TCRC-2 did not respond to hormone treatment, it seems that hormone enhancement is restricted to those cell lines which are producing or have the potential to produce the Regan isozyme. In addition to enhancing the Regan isozyme, this hormone was shown to diminish an intestinal component which was present in the Fl amnion and HEp-2 cell lines. In the lines studied, it appears that prednisolone regulates alkaline phosphatase activity by altering the activities of different isozymes.

INTRODUCTION

We have become most interested in evaluating the control of gene expression of a carcinoplacental isozyme, the Regan isozyme, which is regarded as placental alkaline phosphatase (Fishman et al., 1968b, 1968c, 1971). The fact that HeLa cells produce this isozyme (Fishman et al., 1968b; Elson and Cox,

1969; Beckman et al., 1970) has stimulated us to examine a number of factors which others have reported in connection with the expression of alkaline phosphatase in cells growing in culture. The first one we have chosen to study is the prednisolone-enhancement of HeLa cell alkaline phosphatase and the first emphasis has been placed on the characterization of the isozymic form(s) produced in this way.

In order to simplify the experimental approach, we have made use of two sub-lines of HeLa cells which were cloned to a monophenotypic state with regard to HeLa alkaline phosphatase (Singer and Fishman, 1974). Thus, TCRC-1 produces only the Regan isozyme and TCRC-2 only the non-Regan isozyme. Thus, in this laboratory the HeLa cells were selected for phenotype rather than karyotype.

The alkaline phosphatase in the Regan cell line was found to be inducible by adding prednisolone to the growth media, the non-Regan line was not. From this information it is possible to suggest that all isozymes of alkaline phosphatase may not respond similarly to hormone treatment. The question which we have chosen to ask is whether prednisolone specifically affects certain isozymes of alkaline phosphatase, and in particular is hormone induction restricted to the Regan isozyme?

Using the HeLa TCRC-1 and TCRC-2 as well as FL amnion and HEp-2 cell lines for purposes of comparison we studied the effect of prednisolone on the gene expression of isozymes of alkaline phosphatase in cultured cells. FL amnion and HEp-2 cell lines were included because we found them to possess heat stable, L-phenylalanine sensitive alkaline phosphatase.

Briefly, the three cell lines which underwent modification by prednisolone, all showed isozyme appearing or being enhanced in the Regan isozyme position on acrylamide gel electrophoresis. Unexpected findings relate to the apparent presence of intestinal isozyme components in HEp-2 and FL-amnion cells, which were suppressed as a result of prednisolone in the medium.

The isozyme identity following polyacrylamide gel electrophoresis was based on the results of a series of tests including L-phenylalanine inhibition (Fishman and Ghosh, 1967), heat-stability (Fishman and Ghosh, 1967), neuraminidase treatment (Fishman et al., 1968a), and exposure to specific antisera to placental and intestinal isozymes (Inglis et al., 1971; Fishman, L. et al., 1972).

Accordingly, this paper reports the results of experiments which succeeded in characterizing the gel isozymic forms of alkaline phosphatase of several relevant cell lines and relates these findings to current concepts.

MATERIALS AND METHODS

Culture methods: Cells are grown in monolayer in Eagle's minimum essential medium with Earle's salts (Gibco, Grand Island, N.Y., Cat. #143EG) as previously described (Singer and Fishman, 1974).

Cells: FL Amnion and HEp-2 cells were purchased from the American type Culture Collection, Rockville, Maryland. HeLa TCRC-2 was cloned by Dr. Robert Rustigian, Veterans Administration Hospital, Brockton, Mass. (Rustigian et al., 1974). HeLa TCRC-1 was cloned in this laboratory by the present authors (Singer and Fishman, 1974).

Neuraminidase digestion: Neuraminidase digestion was accomplished using a method similar to that described by Fishman et al (1968a). To 0.1 ml of enzyme sample, the following were added and agitated in sequence: 0.1 ml 1M Acetate Buffer pH 5.5, 0.2 ml 0.1M $CaCl_2$ and 0.05 ml neuraminidase (Vibricholerae, from Calbiochem catalog #480717 grade B, 500 u/ml). This solution was stoppered and incubated for 3 hours at 37°C.

Electrophoresis: Polyacrylamide gel electrophoresis was performed as described by L. Fishman (1974). The method requires that the gels be prepared with Triton X-100 (0.5%) detergent. The cells were sonicated in Hanks' balanced salt solution and diluted 1:1 (with a 20% sucrose, 1.0% Triton X-100 solution) prior to application. 50λ of diluted sample is applied to each gel. Best results were obtained using a sample which was found to contain 30 K.A.'s of alkaline phosphatase activity. The electrophoresis (in an 8 tube Shandon Chamber) was run for 5 minutes at 1mA per tube, followed by 25 minutes at 4mA per tube. After this first 30 minutes, the Vokam power supply (Shandon Southern, Sewickley, Pa.) usually read 250 volts; at this time, the machine was adjusted to constant voltage at 250, and run for another 60 minutes. Alkaline phosphatase activity is visualized through its phosphohydrolase activity using α-naphthyl phosphate substrate with simultaneous coupling of the product with 4 amino-diphenylamine diazonium salt, as previously described (Fishman, L., 1974).

Electrophoretic inhibition studies: L-phenylalanine inhibition by first pre-incubating gels in 20 mM phenylalanine in 1M propanedial buffer pH 9.68, 0.233 mM $ZnSO_4 \cdot 7H_2O$ and

3.0 mM $MgCl_2 \cdot 6H_2O$, for 30 minutes prior to staining for phosphohydrolase activity. Next, the usual staining procedure was carried out by using buffers which contained 20 mM L-phenylalanine. The control mixture was identical except that the non-inhibitor, D-phenylalanine, was substituted for L-phenylalanine.

Biochemical inhibition studies: Inhibition studies were performed using 18 mM disodium phenylphosphate in 0.05 M carbonate-bicarbonate buffer pH 9.8. D-and L-phenylalanine were dissolved in buffer for a working stock solution of 5.0 mM as described by Ghosh and Fishman (1968). Heat inactivated samples (5 min. at 65°C were assayed using 72 mM disodium phenylphosphate in carbonate-bicarbonate buffer pH 10.7 (Fishman and Ghosh, 1967).

Immunologic studies: Rabbit antisera against partially purified human placental alkaline phosphatase was used.

The antisera was diluted 1:1000 in 20% sucrose, 1% Triton X-100 solution, and when mixed 1:1 with the sample the final antisera dilution was 1:2000.

Rabbit antisera to partially purified human intestinal alkaline phosphatase (Fishman, L. et al., 1972) was also employed in these studies. This antisera was diluted 1:25 in the 20% sucrose 1% Triton X-100 diluting solution. The final antisera dilution when mixed 1:1 with the sample was 1:50. The antisera were kindly made available to us by L. Fishman.

Prednisolone treatment: Prednisolone (from Calbiochem, Los Angeles, Calif., 90054, Cat. #5296) was dissolved in Hanks' balanced salt solution (50 mg/500 ml) and was stirred overnight at room temperature. 1 ml of this solution was brought up to 100 ml. with growth media. Final concentration of compound was 1 ug/ml. Cells were plated out at 5×10^5 cells per T-25 Falcon tissue culture flask, and allowed to grow for 24 hours. At this time, the media was replaced with prednisolone containing media and incubated for 72 hours. Cells were harvested by trypsinization.

Densitometry: Gels were scanned at 455 mu using a Beckman DU monochromator with a Gilford 2220 adaptor and a Gilford Model 2410-S linear transport (from Gilford Instrument Laboratories, Oberlin, Ohio).

Heat stability studies: Aliquots were heated in stoppered test tubes for 5 minutes at 65°C. They were then added to 20% sucrose, 1% Triton X-100 detergent solution 1:1, as were

the unheated sample which defines the sample specimen for
polyacrylamide gel electrophoresis as done in this paper.

Radial immunodiffusion: Single radial immunodiffusion by the
unpublished method of Doellgast et al., was performed in 1.5%
agarose gels containing 12 ug/ml of lyophilized rabbit antiserum to highly-purified human placental alkaline phosphatase
(Doellgast and Fishman, 1974). The agarose contained 0.05 M
veronal - HCl buffer, pH 8.6, 0.25% Triton X-100 detergent,
and 0.04% sodium azide as a preservative. Known activities
of control and prednisolone-induced alkaline phosphatase from
HeLa TCRC-1, and of placental membrane preparations (Doellgast
and Fishman, these Proceedings) were applied to the wells and
allowed to diffuse for four days. The alkaline phosphatase
activity in the precipitin ring was developed using 2 mM naphthol AS-MX phosphate in the 1 M propanediol HCl buffer described previously and photographed under ultraviolet light.
The results from this experiment are given in Figure 8.

RESULTS

This is divided into two sections, one on biochemical
studies, and the other on immunologic studies.

BIOCHEMICAL STUDIES

The biochemical properties of the cell lines are presented in Table 1. The specific activity ranges from 0.83 for
HeLa TCRC-1 to 3.47 for the HEp-2 cell line. The most significant increase in enzyme activity after growth in medium
containing prednisolone was observed in HeLa TCRC-1, although
the FL-amnion cell line showed a slight but reproducible elevation in activity after treatment. The enzyme activity of
HeLa TCRC-2 was not changed during hormone treatment. All
cell lines except the non-Regan line (TCRC-2) were found to
possess a significant level of heat stable L-phenylalanine
sensitive alkaline phosphatase.

The isozyme profiles of the untreated cell lines studied
with disc gel electrophoresis are presented as controls in
Figure 1. It is seen that the monophenotypic HeLa cell lines
each have a single isozyme form. The HeLa TCRC-2 band is
anodally somewhat faster migrating and more diffuse than that
of HeLa TCRC-1. The Fl amnion and the HEp-2 cell lines each
show two bands of activity. The HEp-2 has a slow band of
activity which corresponds to that found in the HeLa TCRC-1,
and an additional faster moving band. The FL amnion contains
two bands, which both migrate more rapidly than the Regan

757

TABLE I

PROPERTIES OF CELL LINES

CELL LINE	SPECIFIC ACTIVITY[a] (μMoles/min/mg)	% INHIBITION BY L-PHENYLALANINE	L-HOMOARGININE	% RESIDUAL ENZYME ACTIVITY AFTER 5 min. at 65°C
HeLa TCRC-1	.83	76.7	13.7	91.5
+ Prednisolone	1.53	76.5	9.8	94.2
HeLa TCRC-2	1.4	0	77.5	0
+ Prednisolone	1.7			2.6
HEp-2	3.47	70.5	8.7	80.7
+ Prednisolone	3.61	68.7	4.4	80.8
FL Amnion	1.59	77.6	18.4	51.6
+ Prednisolone	2.04	79.1	21.1	57.4

a) Values after 96 hours in culture, hormone was introduced after 24 hours.

758

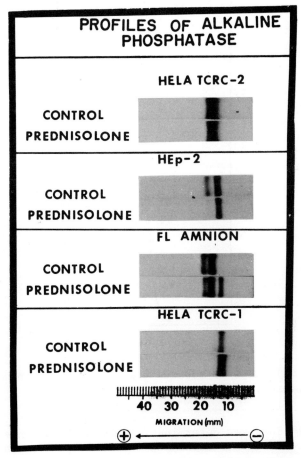

Figure 1.   Isozyme profiles and alteration by prednisolone treatment.

cell line.   These cell lines are clearly quite different in their isozyme profiles of alkaline phosphatase.   The control gels presented in this figure will reappear throughout this report for comparison to the results obtained under different experimental conditions.

The effect of hormone treatment on the isozyme profiles

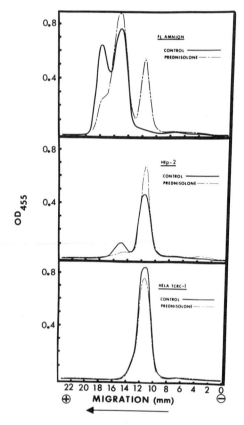

Figure 2. Gel scans of hormone treated cultures, compared to control cultures.

of the cell lines was determined. Cells were grown for 72 hours in prednisolone containing media, cells were harvested, and the electrophoretic patterns are presented in Figure 1. HeLa TCRC-2, monophenotypic for non-Regan isozyme, and HeLa TCRC-1, monophenotypic for Regan isozyme, are not altered with respect to isozyme profiles. The Regan producing cell line, however, reflects a marked elevation in enzyme activity while the non-Regan cell line is unchanged. The hormone treated HEp-2 is found to contain only the slower band which corresponds to that found in HeLa TCRC-1. Upon prednisolone treatment, the faster band becomes faint, while the slower band remains strong. The addition of prednisolone to the media of

760

the FL amnion cell line resulted in a dual effect on the isozyme profile of this cell line: a discrete strong band appearing at the Regan position and the fastest moving band
being concurrently reduced in intensity.

Gels were scanned in order to obtain a more accurate representation of the alteration in isozyme profiles resulting
from hormone treatment (Figure[2]). All samples for these gels
were diluted to approximately 30 K.A. units prior to electrophoresis which explains why the control and experimental profiles for the HeLa TCRC-1 appears to be equal in activity.

Two cell lines exhibited a change in profile. The isozyme
pattern of FL amnion is markedly altered by prednisolone which
caused a clear shift of activity from the fast band to a newly
formed slow band. In the HEp-2 cell line, hormone treatment
had no effect on total activity, but the enzyme profile reveals
that enzyme activity is decreased in the faster band and is
correspondingly increased in the slow band.

On the other hand, the isozyme profile of HeLa TCRC-1 remained unchanged, although the total activity was increased
by 2-fold. It is apparent from this figure that enhancement
of alkaline phosphatase activity when it occurs is found to
involve a single electrophoretic band position and no single
band position seems to be consistently repressed. The band
of activity in FL amnion, which is significantly diminished
by hormone treatment, migrates faster than the band which is
eliminated in the HEp-2 cell line after exposure to hormone.

Alteration in isozyme profiles may be due to alteration
of terminal sialic acid residues associated with the enzyme.
As neuraminidase is known to cleave terminal sialic acid
residues (Moss et al., 1966) HEp-2 and FL amnion cells were
treated with neuraminidase for 3 hours as described in the
Materials and Methods section. As seen in Figure 3, although
neuraminidase treatment reduced the migration of all the bands
in both these cell lines, the basic electrophoretic pattern
is maintained in both the control and induced cultures. Also,
similar results were observed when neuraminidase digestion
was continued for 24 hours. It is clear from this experiment
that although all isozymes in these cell lines contain significant terminal sialic acid residues, this property is not
responsible for modification of the isozyme pattern observed
in the induced cultures.

The differential heat sensitivity of isozymes of alkaline
phosphatase is well established. Bone and liver type alkaline
phosphatase were shown by Moss and King (1962) to be most
heat labile. The placental enzyme form has been shown to be
remarkably heat stable (Neale et al., 1965), In order to
characterize the heat stability of the isozymes visualized on

761

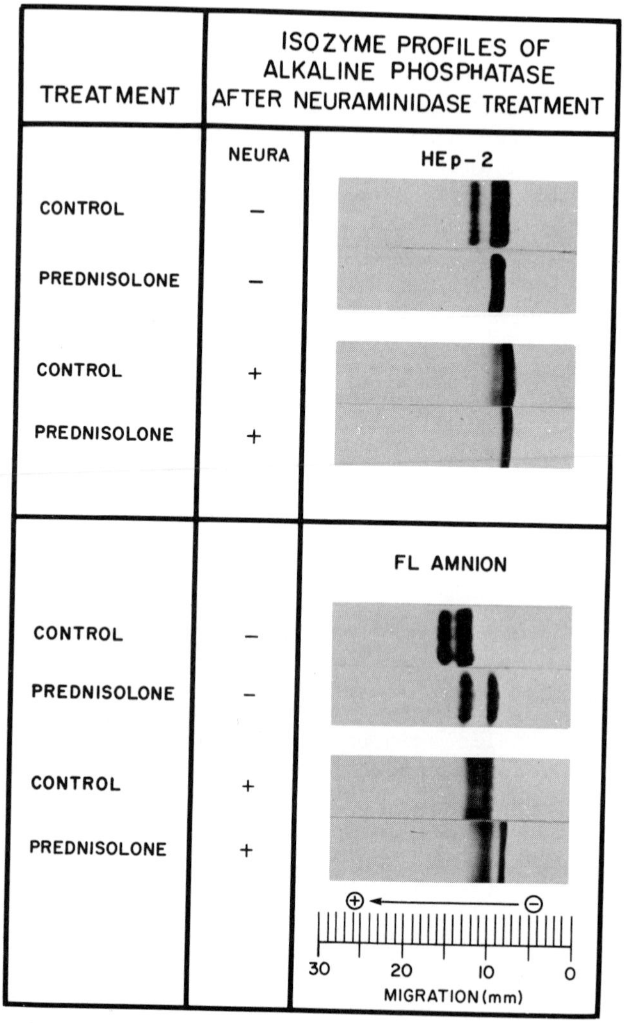

Figure 3. Effect of neuraminidase on isozyme profiles.

electrophoresis the samples were heated for 5 min at 65°C
prior to applying to the gel.

The gels presented in Fig. 4 demonstrate the heat stability characteristics of the alkaline phosphatase isozymes in
both the control and hormone treated cell lines. As one

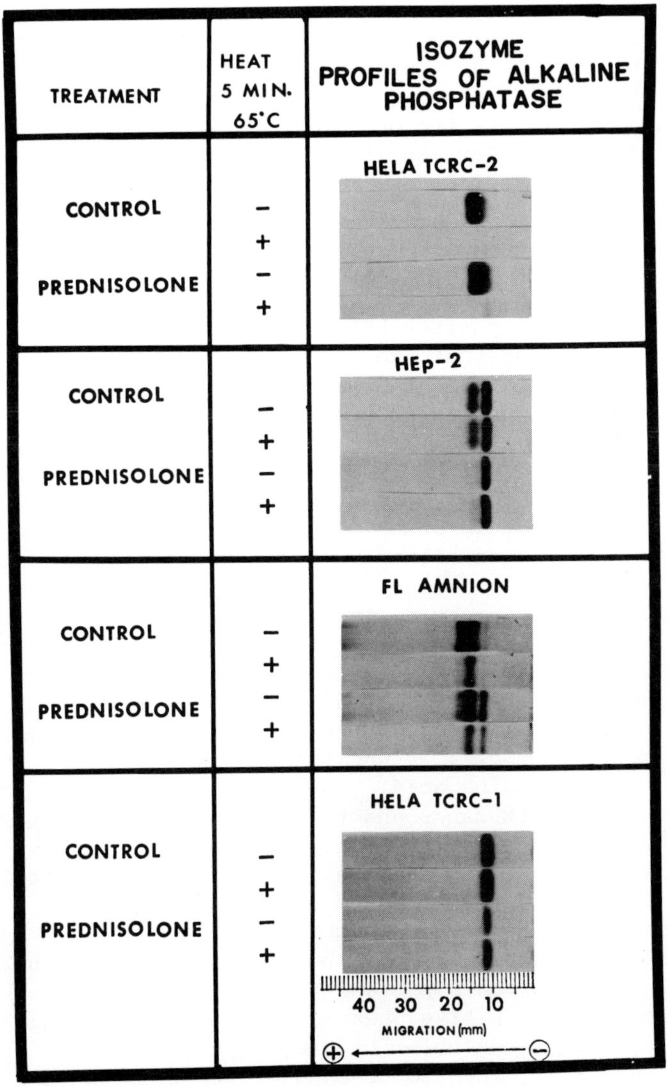

Figure 4.　Heat stability of isozyme bands.

would expect from the biochemical data (Table I), the band of
activity in the HeLa TCRC-2 cell line is heat labile in both
the control and hormone treated cultures while the isozyme

band of the Regan HeLa line maintains its heat stability characteristics and migration distance after hormone treatment. It
is apparent, however, that the HEp-2 cell line has two isozymes
which differ in their heat stabilities. The fastest band in
the control is somewhat inhibited by heat, while the slower
band is unchanged. After hormone treatment, the slow band at
the Regan position is the only band present and it is heat
stable. Also, a heterogeneity of isozymes is apparent in the
FL amnion cell line. It seems that the control has a heat-
labile fast band and a partially heat-stable slow band. After
hormone treatment, a triple banded pattern is present due to
the appearance of a slower migrating heat stable component at
the Regan position. The fastest band appears to retain its
heat-labile property after hormone treatment.

It is striking that the bands which are enhanced by
hormone treatment are heat stable while those bands which were
partially or completely reduced in intensity were heat labile
to a variable extent.

L-phenylalanine inhibition has been directed towards isozymes of intestinal and placental alkaline phosphatase (Fishman et al., 1963). In order to determine whether bands of
activity in completed electrophoresis could be inhibited by L-
phenylalanine, the gels, prior to staining, were cut in half
longitudinally and treated simultaneously with staining solutions which contained D or L phenylalanine respectively, as
described in the Materials and Methods section.

The results (Figure 5) indicate that HeLa TCRC-1, hormone
treated and control, both show significant L-phenylalanine
inhibition. HeLa TCRC-2 is unaltered in staining intensity after
exposure to this amino acid. The HEp-2 control underwent loss
of activity in both bands and the hormone treated HEp-2 also
showed significant L-phenylalanine inhibition. Likewise, the
FL amnion cell line exhibited a loss of activity in all bands
in both the control and treated cultures.

From this and previous experiments, a tentative identification of isozymes involved in hormone regulation can be made.
The fast bands in the HEp-2 and FL amnion cell lines which are
L-phenylalanine-sensitive, heat-labile, and disappear after
hormone treatment resemble the properties of intestinal alkaline phosphatase. The slower band of HEp-2 and a similar one
appearing in the hormone treated FL amnion cells has the heat-
stability and L-phenylalanine inhibition of the placental isozyme.

It seems in these cell lines that prednisolone may regulate alkaline phosphatase activity by suppressing intestinal
type alkaline phosphatase and enhancing Regan type alkaline
phosphatase. Additional immunological evidence is needed.

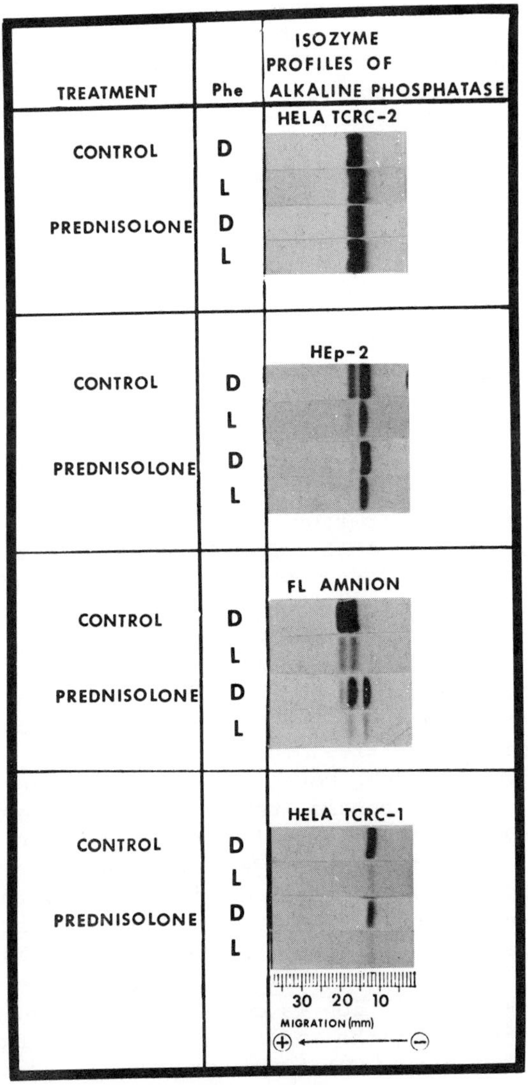

Figure 5.   L-phenylalanine sensitivity of isozyme profile.

IMMUNOLOGIC STUDIES

Since the previous information indicated that the isozymes in these cell lines had properties of placental or intestinal type alkaline phosphatase, specific antisera to these forms were used to further characterize them.

It has been previously shown (Inglis et al., 1971) that addition of antisera against human placental alkaline phosphatase to the specimen prior to electrophoresis on cellulose acetate resulted in a decrease in migration and in activity of placental alkaline phosphatase. Accordingly, prior to application on acrylamide gel electrophoresis, the samples were mixed with antisera to placental alkaline phosphatase and the results are shown in Figure 6. It can be seen that the single band of activity in both control and hormone treated HeLa TCRC-1 is prevented from entering the gel, the activity in the antiserum treated sample remaining closer to or at origin. On the other hand, the HeLa TCRC-2 profile is unchanged by the antiserum treatment as expected, since it is essentially non-Regan. The specificity of this procedure is thus illustrated. No attempt was made to quantitate possible diminution of isozyme activity in the antigen-antibody complex.

A differential effect of the antisera to placental alkaline phosphatase is apparent with the HEp-2 cell line. In the control, the band at the Regan position is almost completely removed by the antiserum while the fast moving band is not affected. In the hormone treated HEp-2, the single band present at the Regan position without antiserum is prevented from entering the gel with antiserum. Thus, in this cell line, the enzyme band which is enhanced by hormone treatment has the antigenic determinants of the Regan isozyme. The band which disappears on hormone treatment is heat labile, L-phenylalanine-sensitive, but does not cross react with antisera to placental alkaline phosphatase, thus typing it as non-placental.

The isozyme bands of the Fl amnion cell line also show a differential reactivity with antisera to human placental alkaline phosphatase. In the control culture, the slower band cross reacts with the antisera while the faster band which is diminished after hormone treatment is hardly affected. In the hormone treated Fl amnion cells, the newly formed band at the Regan position is completely removed by the antisera, indicating that the induced band has all the properties of the Regan isozyme. The two fastest bands of the hormone treated cultures respond in a manner similar to those of the control, the fastest band being unchanged while the middle band was partially retarded by antisera treatment. Thus, the bands of enzyme activity which disappear after hormone treatment in the Fl amnion and the HEp-2 cell lines have similar characteristics. They both were heat stable, L-phenylalanine sensitive and did not cross react with antisera against placental alkaline phosphatase. Thus, while the enhanced bands

766

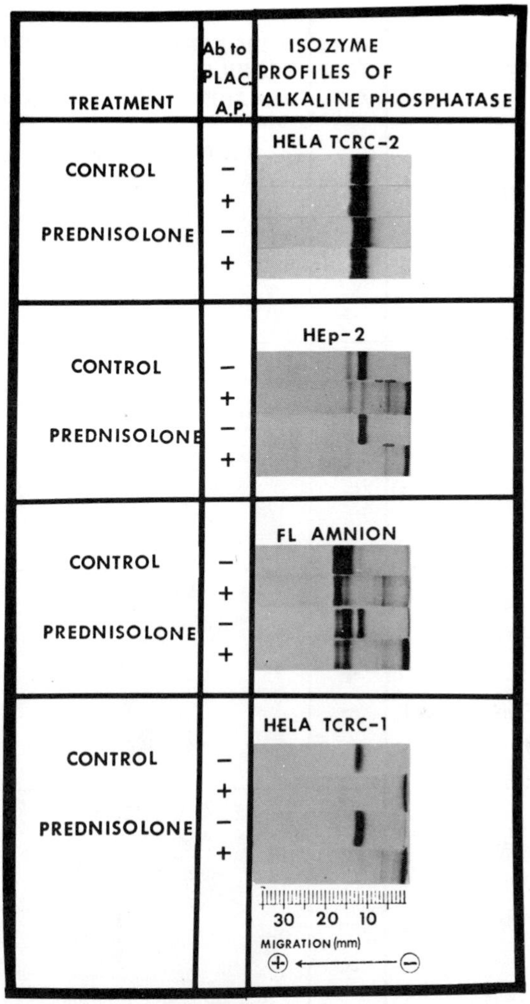

Figure 6. Isozyme profile following pre-treatment with anti-body to placental alkaline phosphatase.

of activity in all cell lines where it occurred appeared to be Regan, the bands which were weakened by hormone treatment appeared to be non-placental. Their L-phenylalanine inhibition and partial heat-sensitivity do suggest intestinal type.

In order to ascertain the enzyme form which is reduced in intensity after hormone treatment an experiment similar to

767

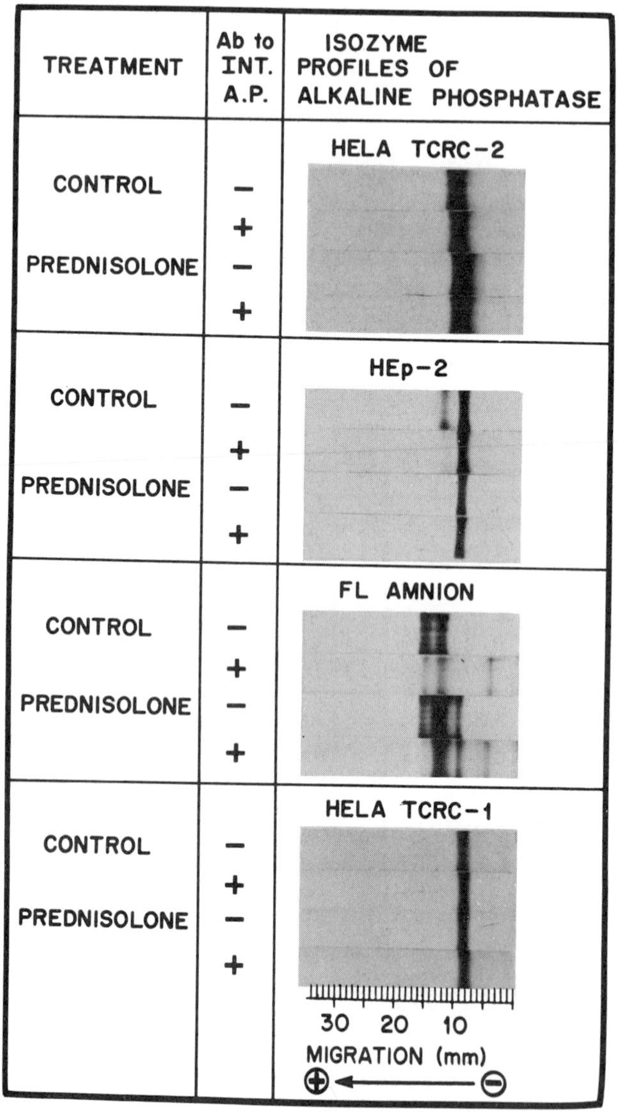

Figure 7. Isozyme profile following pre-treatment with antibody to intestinal alkaline phosphatase.

the previous one was performed using an antisera prepared against human intestinal alkaline phosphatase.

The data in Figure 7 demonstrates that the isozyme bands
of HeLa TCRC-1 and HeLa TCRC-2 are unaffected by the antisera
to intestinal alkaline phosphatase. Neither the control nor
the hormone treated cultures appear to have bands of activity
which cross-react with intestinal antisera. These cell lines
do not possess the intestinal from of alkaline phosphatase.

The HEp-2 cell line appears to have a fast migrating alkaline phosphatase isozyme which cross reacts with intestinal
antisera whereas the slower moving band which has been shown
to be similar to the Regan isozyme does not react. Thus, the
fast moving band in the HEp-2 cell line, which disappears
after hormone treatment, is intestinal type alkaline phosphatase. The single band of enzyme activity at the Regan position
after hormone treatment contains only the Regan isozyme.

From the F1 amnion control culture, it is apparent that
the fast moving as well as the slower moving bands cross
react with antisera to intestinal alkaline phosphatase. It
seems that in this cell line, the fastest band which is diminished by hormone treatment most strongly cross-reacts with
this antisera and appears to be intestinal alkaline phosphatase. The slower band in the control cross-reacted with antisera to placental as well as intestinal alkaline phosphatase,
indicating that this may be a hybrid enzyme form or a migration position shared by two isozymes. The newly appearing
Regan band after hormone treatment does not cross-react with
the antisera to intestinal alkaline phosphatase, whereas the
fastest band in the hormone treated FL amnion was found to
maintain its cross-reactivity with the antisera to intestinal
alkaline phosphatase.

Thus, in these cell lines it appears that prednisolone
enhances the Regan isozyme, while causing at the same time
the disappearance of intestinal type bands of alkaline phosphatase activity.

We observed that in HeLa TCRC-1 the control and induced
enzymes were similar in many respects, i.e. electrophoretic
pattern, L-phenylalanine sensitivity, heat stability, and
cross-reactivity with antisera to placental alkaline phosphatase.

An additional comparison was performed using an immunological method of quantitation based on antigenic determinants
(Mancini et al., 1965). Using radial immunodiffusion, modified for placental alkaline phosphatase (Doellgast, et al.,
unpublished data), we quantitated the levels of enzyme protein relative to enzyme activity for the induced and the control HeLa TCRC-1 and compared these results to those of placental alkaline phosphatase. All three enzyme sources were
diluted to equal activity and applied in graded dilutions to

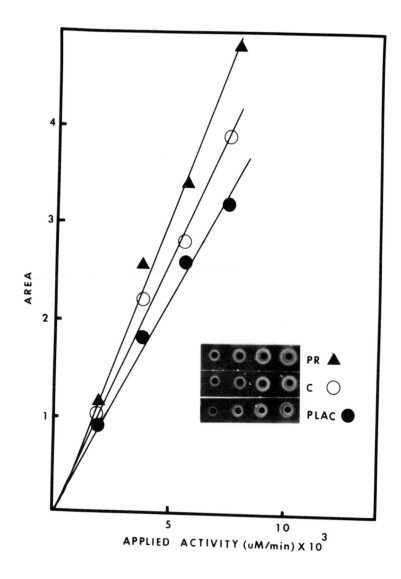

Figure 8. Single radial immunodiffusion of enzyme from control and hormone treated HeLa TCRC-1 compared to placental alkaline phosphatase. Equal activities from all three enzyme sources are applied in graded dilutions. Well number 1 (on the left), contains 25% of the activity in well number 4 (on the far right). Well number 2 contains 50% of the activity of well numer 4, and well number 3 contains 75% of the activity of well number 4.

wells in agarose which contained antiserum to placental alkaline phosphatase (See Fig. 8). The area encircled by the precipitin rings visualized as the fluorogenic products of hydrolysis of naphthol AS-MX phosphate was determined and the area
of the well was subtracted from each value. In Fig.8 the
applied enzyme activity is plotted against the area.

For HeLa TCRC-1, it can be seen that the control and
hormone enhanced alkaline phosphatases are similar in the
amount of enzyme activity per unit of protein present, and
both are comparable to placental alkaline phosphatase by the
radial immunodiffusion technique within the limits of experimental error.

DISCUSSION

It is apparent from this report that prednisolone regulates
levels of alkaline phosphatase by the enhancement or suppression of specific isozymic forms. In the cell lines studied,
hormone enhancement appears to be specific for the Regan isozyme. Cell lines which possess the Regan isozyme were shown
to have elevated activity at this band position (HeLa TCRC-1
and HEp-2). In the FL amnion cell line, however, the hormone
caused the appearance of a Regan band, which was not present
in the control. It seems that in some cell lines hormonal
enhancement occurs while in another cell line (FL amnion)
hormone induction takes place. An additional effect of hormone
treatment of the FL amnion and the HEp-2 cell lines was the
considerable decrease in enzyme activity of certain other
bands (See Fig. 1). The bands affected in this manner were
shown to be intestinal type alkaline phosphatase. Hormonal
regulation in these cell lines clearly appears to be a specific process of altering the relative activity of various isozymic forms produced by the cell population.

The mechanism(s) which may be involved in this process
are not yet clear. The production of the observed isozyme
patterns may be the result of the gene expression of an individual cell or may reflect the contribution of different cell
types, each producing a separate isozyme.

If the isozyme pattern is due to diferent cell populations
each producing its own characteristic enzyme form, then the
levels at which the hormone may act are numerous. Hormone
treatment could have a differential effect on the proliferation rate of a Regan isozyme producing population versus an
intestinal isozyme porducing one. Prednisolone has been previously reported to have an effect on the proliferation rate
in cultured human cells (Miedema et al., 1967). However, a
detailed study on the cell types affected has not been

carried out. Another possibility is that this hormone may have a direct effect on the activation of genes which code for each isozyme. We are currently pursuing this possibility.

Hormone mediated alterations in isozyme profiles of alkaline phosphatase have been observed by several investigators. Beckman and Regan (1964) using RA human amnion cells found an alteration in the zymogram pattern on starch gel electrophoresis in cells grown in the presence of cortisone. Cortisone interfered with the synthesis of a fast-moving enzyme while the slower band appeared to be elevated. Later studies by Griffin and Bottomley (1969) on HeLa cells revealed that hydrocortisone caused the appearance of a new isozyme band of alkaline phosphatase midway between the two pre-existing main bands. Further study of this phenomenon brought the authors to the conclusion that the appearance of this band was not simply a function of hydrocortisone treatment but a function of the state of confluency of the cell culture. At 80% confluency, both control and hormone treated cultures were identical. The control cultures in our experiments showed no significant alteration in isozyme profiles throughout the growth cycle.

Spencer (1970) using HeLa S$_3$ cells found an altered isozyme pattern after growth of cells in prednisolone. This alteration of isozyme profile, as well as the effect of the hormone in elevating levels of alkaline phosphatase activity, appeared dependent on the type of serum present in the growth media. The effect of prednisolone on alkaline phosphatase was markedly increased when calf serum rather than fetal calf serum was used in the media. A more recent study by F. Herz (1973) using KB cells further demonstrated the differential effect of prednisolone on different isozymic forms of alkaline phosphatase. The authors grew these cells in prednisolone containing or hyperosmolar media and observed a decrease in total alkaline phosphatase activity. The proportion of heat-stable enzyme, however, was elevated while the heat-labile form was greatly reduced. Their report demonstrated the differential action of prednisolone on the heat-labile and heat-stable alkaline phosphatase activity in KB cells. Is this heat stable moiety Regan isozyme?

Finally, two interesting observations deserve attention: First, the lack of an increase in specific activity in cells growing in the presence of hormone may obscure alterations in phenotype expression unless isozyme characterization is carried out. Second, we must now consider the intestinal isozyme as an ectopic carcinoamnion phenotype in the HEp-2 cancer cell line.

ACKNOWLEDGEMENTS

This research was supported by Grants-in-aid (CA-12924; CA-13332) of the National Cancer Institute, National Institutes of Health, Public Health Service, Bethesda, Maryland. One of us (W. H. Fishman) is the recipient of a Career Research Award (K6-CA-18543) of the N.I.H.
Grateful appreciation is expressed for the valuable help of Loris J. White and James E. Perry.

REFERENCES

Beckman, B., L. Beckman, and E. Lundgren 1970. Isoenzyme variations in human cells grown in vitro. IV. Identity between alkaline phosphatase from HeLa cells and placenta. *Human Heredity* 20: 1-5.

Beckman, L. and J. D. Regan 1964. Isozyme studies of some human cell lines. *Acta Path. et Microbiol. Scand.* 62: 567-574.

Doellgast, G. J. and W. H. Fishman 1974. Purification of human placental alkaline phosphatase; Salt effects in affinity chromatography. *Biochem. J.*, in press.

Elson, N. A. and R. P. Cos 1969. Production of fetal-like alkaline phosphatase by HeLa cells. *Biochem. Genet.* 3: 549-561.

Fishman, L. 1974. Acrylamide disc gel electrophoresis of alkaline phosphatase of human tissues, serum and ascites fluid using triton X-100 in the sample and gel matrix. *Biochem. Med.* 9: 309-315.

Fishman, L., N. R. Inglis, and W. H. Fishman 1972. Preparation and characterization of human intestinal alkaline phosphatase antigens. *Clin. Chim. Acta* 38: 75-83.

Fishman, W. H. and N. K. Ghosh 1967. Isoenzymes of human alkaline phosphatase. *Adv. in Clin. Chem.* 10: 256-370.

Fishman, W. H., S. Green, and N. R. Inglis 1963. L-phenylalanine; an organ specific stereospecific inhibitor of human intestinal alkaline phosphatase. *Nature* 198: 685-686.

Fishman, W. H., N. R. Inglis, and N. K. Ghosh 1968a. Distinction between intestinal and placental isoenzymes of alkaline phosphatase. *Clin. Chim. Acta* 19: 71-79.

Fishman, W. H., N. R. Inglis, and S. Green 1971. Regan isoenzyme: A carcinoplacental antigen. *Cancer Res.* 31: 1054-1057.

Fishman, W. H., N. R. Inglis, S. Green, C. L. Anstiss, N. K. Ghosh, A. E. Reif, R. Fustigian, M. J. Krant, and L. L. Stolbach 1968b. Immunology and biochemistry of Regan isoenzyme of alkaline phosphatase in human cancer. *Nature* 219: 697-699.

Fishman, W. H., N. R. Inglis, L. L. Stolbach, and M. J. Krant 1968c. A serum alkaline phosphatase isoenzyme of human neoplastic cell origin. *Cancer Res.* 28: 150-154.

Ghosh, N. K. and W. H. Fishman 1968. Purification and properties of molecular-weight variants of human placental alkaline phosphatase. *Biochem. J.* 108: 779-792.

Griffin, M. J. and R. H. Bottomly 1969. Regulation of alkaline phosphatase in HeLa clones of differing modal chromosome number. *Annals of the N.Y. Acad. of Sci.* 166: 417-432.

Herz, F. 1973. Alkaline phosphatase in KB cells: influence of hyperosomolarity and prednisolone on enzyme activity and thermostability. *Arch. of Biochem. and Biophys.* 158: 225-235.

Inglis, N. R., D. T. Guzek, S. Kirley, S. Green, and W. H. Fishman 1971. Rapid electrophoretic microzone membrane technique for Regan isoenzyme (placental type alkaline phosphatase) using fluorogenic substrate. *Clin. Chim. Acta* 33; 287-292.

Mancini, G., A. P. Carbonara, and J. F. Heremans 1965. Immunochemical quantitation of antigens by single radial immunodiffusion. *Immunochemistry* 2: 235-254.

Miedema, E. and P. F. Kruse 1967. Effect of Prednisolone and contact phenomena on the alkaline phosphatase activity of HEp-2 cells. *Biochem. and Biophys. Res. Commun.* 26: 704-711.

Moss, D. W., R. H. Eaton, J. K. Smith, and L. G. Whitby 1966. Alteration in the electrophoretic mobility of alkaline phosphatases after treatment with neuraminidase. *Biochem. J.* 98: 32c-33c.

Moss, D. W. and E. J. King 1962. Properties of alkaline phosphatase fraction separated by starch-gel electrophoresis. *Biochem. J.* 84: 192-195.

Neale, F. C., J. S. Clubb, D. Hotchkis, and S. Posen 1965. Heat stability of human placental alkaline phosphatase. *J. Clin. Path.* 18: 359-363.

Rustigan, R., J. P. W. Kelley, D. A. Ellis, L. A. Clark, N. Inglis, and W. H. Fishman 1974. Regan type of alkaline phosphatase in a human heteroploid cell line. *Cancer Res.* (in press)

Singer, R. M. and W. H. Gishman 1974. Characterization of two HeLa sublines: TCRC-1 produces Regan isoenzyme and TCRC-2 non-Regan isoenzyme. *J. Cell Biol.* 60 #3: 777-780.

Spencer, T. 1970. Some factors controlling alkaline phosphatase isoenzymes in HeLa cells. *Biochem. J.* 116: 927-928.

ESTERASE ISOZYMES AS MARKERS IN NORMAL
AND DISEASE PROCESSES

ROBERT SCHIFF
Hyland Division Travenol Laboratories
Costa Mesa, California   92626

ABSTRACT. This paper reviews the work conducted in the
author's laboratory on the role of the non-specific esterases as markers to trace and identify genetic loci, to
assay the therapeutic effects of surgery (creation of
artificial joints), to study the effects of basic disease
processes (glaucoma), to monitor transplanted cells in
radiation chimeras (repopulation of lung cells by bone
marrow cells), and as tools for observing the molecular
interaction of genome upon genome in somatic cell hybrids.
Starch and polyacrylamide gel electrophoresis were employed to separate the isozymes from mouse and rabbit
tissues which included red cells, platelets, synovial
fluid, ocular tissue, bone marrow cells, alveolar macrophages, liver, kidney, spleen, and serum. Activity
measurements were made by spectrophotometry and densitometry.
Various inhibitors, activators, and substrates were used
to differentiate and classify the enzymes separable by
starch gel electrophoresis. The studies show that due to
their diversity and ability for simultaneous analysis the
use of nonspecific esterases in several disciplines has
rapidly expanded.

INTRODUCTION

Carboxylic acid esters of naphthols, phenols, and alcohols
are hydrolyzed by esterases. In addition to their hydrolytic
activity, some esterases are capable of catalyzing synthetic
substrates. Alpha and beta naphthol substrates have been
used to distinguish between the simple fatty acid esterases
and true lipases (Hofstee, 1952).

In order to distinguish lipases, cholinesterases, and
sterolesterases from those enzymes which preferentially act
on the simple aliphatic esters and glycerides, Richter and
Croft (1942) coined the term "aliesterase" to refer to the
latter enzymes. This group of enzymes was found in the serum
and red cells of the horse, human, cat, dog, and rabbit. There
was very little activity with fats, acetylcholine, or choles-
terol acetate as substrates.

When conventional histochemical staining procedures for
esterases were combined with the method of starch gel electro-

phoresis (Hunter and Markert, 1957), the study of electrophoretic enzyme variation was made possible. The non-specific
esterase isozymes were found to be excellent biochemical markers because their large numbers facilitated simultaneous analysis. Their activity was observed in many species, primarily
by spectrophotometric, electrophoretic, and histochemical
techniques. These biochemical markers have been used extensively to study the genetic and developmental control mechanisms of mammalian enzymes (Markert and Humter, 1959; Harris
et al., 1962; Popp and Popp, 1962; Wright, 1963; Grunder et
al., 1965; Tashian, 1965; Tucker et al.,1967; Giblett, 1969;
Schiff and Stormont, 1970). For further details, see: Shaw
(1965), Beckman (1966), and Lush (1967).

In addition, these isozymes have been used to demonstrate
the embryonic origin and development of lung cells (Schiff et
al., 1970; Brunstetter et al., 1971; Schiff et al., 1972), the
phylogenetic relationships among vertebrates (Holmes et al.,
1968), and have been used to explore the normal and diseased
mammalian central nervous system (Barron et al., 1963; Schiff
and Jacobson, 1970).

In the present report, we shall survey the work conducted
in the author's laboratory over a six year period. The purpose is to demonstrate how non-specific esterase markers have
contributed to our understanding of genetic mechanisms, have
been used to assay the therapeutic effects of surgery, have
been used to study the effects of basic disease processes, have
been used to monitor the effects of transplantation, and have
demonstrated the reaction of genomes on diverse genetic backgrounds. This paper will cover such experimental animals as
the rabbit and mouse and the unique cellular enigmas of somatic
cell hybrids. We will also examine such diverse tissues as
red cells, platelets, synovial fluid, aqueous and vitreous
humor, bone marrow cells, and alveolar macrophages to mention
just a few.

METHODS AND MATERIALS

The methods for the demonstration of the non-specific
esterases were primarily electrophoretic and spectrophotometric
in nature. Starch gel electrophoresis was performed after
the methods of Schiff (1970), Schiff et al., (1970); Schiff
and Stormont, (1970); and Schiff and Sonnenschein, (1972).
Polyacrylamide gel electrophoresis and spectrophotometry was
performed after Schiff et al., (1970) and Schiff and Sonnenschein, (1972). The acrylamide studies were based on the
methodolgy developed by Robert Allen (Ortec, Inc.).

STARCH GEL ELECTROPHORESIS

Horizontal starch gel electrophoresis was performed by two different procedures. In the first, the gel and electrolyte buffers were 0.03 M and 0.3 M borate, respectively (both pH 8.5). Each gel contained 14% starch (Connaught Medical Re - search Laboratory) and was prepared after Kristjansson (1960). Sample materials were absorbed on 6 X 10 mm wicks (Beckman Instruments, Inc., Paper No. 319329) and placed in a cut made 5 cm from the cathodal end of the gel, perpendicular to the long axis. Electrophoresis was begun by applying 6 V/cm along the gel for 30 minutes, then removing the wick and continuing the voltage for 4½ hours. The milliamperage during this period ranged from 15 to 19 ma. The tissues studied with this procedure consisted of somatic cell hybrid extracts, and mouse and rat serum.

Routine staining for acetyl- and butyrylesterases was accomplished by the method of Petras (1963). The staining solution consisted of 50 mg Fast Blue RR salt, 90 ml distilled water, 8 ml 0.1 M phosphate buffer (pH 6.8), and 2 ml 1% acetone solution of α-naphthyl acetate or α-napthyl butyrate. The gel was sliced horizontally and each slice immersed in the respective staining solution. The slices were incubated in the staining solution at 37°C for 1½ hours until the esterase zones were of sufficient intensity for examination. The zones developed in acetate were usually the first to appear. For preservation, the gels were rinsed twice in distilled water and fixed in 50% methanol overnight.

The second method of starch gel electrophoresis employed an electrolyte buffer (pH 8.2), consisting of 2.4 g lithium hydroxide, 23.6 g boric acid, and 2000 ml distilled water, and a second buffer (pH 6.6) containing 17.32 g tris (Sigma 7-9), 9.68 g citric acid (monohydrated), and 2000 ml distilled water. These were the same as those described by Grunder et al (1965). However, in the present studies, the buffer (pH 7.3) for each gel was prepared by adding 40 ml of the electrolyte buffer and 70 ml of the second stock buffer to 140 ml of distilled water. Each gel contained 34-40 g of starch depending upon the starch lot used. Horizontal electrophoresis was begun by applying 150 v for 30 minutes at which time the wicks were removed and an ice pack was placed on the upper surface of the gel. Then 350 V were applied until the borate boundary had migrated 9 cm from the origin. The total time of electrophoresis was approximately 3½ hours.

After electrophoresis, each gel slice was immersed in 100 ml of preheated (37°C) distilled water containing 3×10^{-3} M disodium ethylenediamine tetraacetate (EDTA) for 30 minutes.

This concentration of EDTA activates most of the esterase
zones with which we were concerned without inhibiting the
remainder. Two milliliters of a 1% solution of substrate in
acetone and 100 mg of Fast Blue BB salt were then added to
each tray. The gels were left in this staining solution at
37°C for 1½ hours. Routinely the top slice of the gel was
immersed in a solution containing α-naphthyl acetate and the
bottom slice in α-naphthyl butyrate. Occasionally α-naphthyl
propionate and β-naphthyl acetate were also used as substrates.
For preservation, the gels were also rinsed in distilled water
and fixed in 50% methanol overnight.

The tissues subject to this type of starch gel electrophoresis were rabbit red cells, platelets, serum, synovial
fluid, synovial membrane, aqueous and vitreous fluids, iris,
lens, and retina.

ACRYLAMIDE GEL ELECTROPHORESIS

A vertical flat-bed gel system was cast at a continuous
pH utilizing a discontinuous tris-SO_4 (or tris citrate) and
tris-borate buffer according to previously described procedures (Allen et al., 1969,Ortec, Inc.). The gels were cast in
a vertical position utilizing either a two or three layer gel
pore size gradient. The gels and electrolyte solutions were
prepared from combinations of the following stock solutions
according to the instructions of Ortec, Inc. Stock solution 1:
36.3 g tris (Sigma 7-9), 62.0 ml 1 N H_2SO_4, 0.48 ml N, N, N',
N'-tetramethylethylene diamine (TEMED) (Eastman Organic Chemicals), and H_2O to a final volume of 200 ml. Stock solution
2: 20 ml stock solution 1, 0.19 ml TEMED and H_2O to a final
volume of 100 ml. Stock solution 3: 6.25 ml stock solution
1, 20.0 g sucrose, 7.4 ml H_2O, 0.1 ml 0.1% aqueous bromophenyl
blue, the final volume 25 ml. Stock solution 4 (unrecrystalized
acrylamide purchased from American Cyanamide): 4.8 g twice
recrystalized monomer, 0.12 g methylene bisacrylamide and H_2O
to a final volume of 15 ml (prepared fresh daily). Stock
solution 5: 61.3 g tris (Sigma 7-9), 395 ml 0.35 M boric acid,
0.78 g methiolate and H_2O to make up 7.8 liters. Stock solution 6: 0.105 g ammonium persulfate/100 ml H_2O. Stock solution 7: 0.21 g ammonium persulfate/100 ml H_2O. When tris-
citrate buffers were used stock solution 1 consisted of 36 g
tris, 28 ml 1 M citric acid (monohydrate), 0.98 ml TEMED and
H_2O to 200 ml.

The gels were cast starting from the anodal end of the
cell. The two-layer gradient system was prepared in the
following manner. A 7.5% gel consisted of 5 ml stock solution
1, 4.7 ml stock solution 4, 0.3 ml H_2O and 10 ml stock solution
6; the 4.5% gel layer was comprised of 0.8 ml stock solution

1, 0.45 ml stock solution 4, 0.35 ml H_2O and 1.6 ml stock solution 7; and the 8% well and cap gel contained 0.9 ml stock solution 2, 0.9 ml stock solution 4 and 1.8 ml stock solution 7. The cap gel cast on top of the sample in sucrose provided a 1.5 cm column of gel containing only the leading ion and counter ion between the sample and trailing ion at an ionic strength one-fifth that of the separating gel.

The three-layer system consisted of an 8% gel prepared by adding 4.5 ml stock solution 1, 4.5 ml stock solution 4, and 9 ml stock solution 6; the 6% gel was prepared with 1.2 ml stock solution 1, 0.9 ml stock solution 4, 0.3 ml H_2O, and 2.4 ml stock solution 6; a 4.5% gel was prepared with 0.8 ml stock solution 1, 0.45 ml stock solution 4, 0.35 ml H_2O, and 1.6 stock solution 7. The well and cap gel was prepared in the manner as for the two-layer system. All gel layers were made to have optically flat surfaces through polymerization with a 1 ml H_2O layer on the top surface of the gel. The cell in which these gels were cast had interior cell dimensions of 3.5 mm deep by 100 mm wide and 80 mm in length.

It should be noted that when the 0.075 M 8% well gel was poured a Teflon well former consisting of 12 finger-like projections was inserted with the bottom edges of the well former resting on the top of the preceding layer. After polymerization of this 8% layer the Teflon well former was removed and the wells were rinsed with distilled water. The samples were then placed in the wells by means of a capillary tube or Hamilton syringe. The 8% cap gel was then layered carefully on top of each sample.

The amount of sample used depended upon the material tested. With serum samples 4-10 µl serum were first mixed with 10-26 µl 50-80% sucrose solution (stock solution 3) and then applied to the well. For alveolar macrophage and cell clone extracts, 10-15 µl were mixed with the sucrose solutions to yield a final volume of 30 µl. Just before electrophoresis, the entire cell with the polymerized gel and samples was cooled for 30 minutes at 4^OC.

The electrolyte buffer (stock solution 5) was precooled to 4^OC. The tank and cell assembly was similar to that previously reported (Allen and Moore, 1966) with the exception of the wider cell noted above. Power was supplied by the Ortec model 4100 pulsed constant power supply. The protocol on pulse rates and voltage used in both the two-layer and three-layer gradient gel systems is given in Schiff et al (1970) and Schiff and Sonnenschein (1972).

The approximate time of the run was from 47-55 minutes or in the case of the two-gradient pore size system when the tracking dye had migrated 5.5 cm from the sample wells or

4.5 cm for the three-layer system. After completion of the run, the gel slab was equilibrated for 5 minutes in 100 ml 0.04 M tris-HCl buffer (pH 6.6) at 37°C. It was then stained in a solution containing 100 mg Fast Blue RR salt, 100 ml 0.04 M tris buffer and 2 ml α-naphthyl butyrate prepared from a 1% acetone solution for 15 minutes at 37°C. After staining, the gel was rinsed with distilled water and fixed in a solution contianing 90 ml 7.5% glacial acetic acid and 10 ml absolute ethanol.

SPECTROPHOTOMETRIC ANALYSIS

Mouse serum esterase activity was determined by the method of Gomori (1953) as modified by Popp and Popp (1962). To 1 ml 1:20,000 dilution of serum in distilled water were added 5 ml freshly prepared substrate solution containing 1 ml 1% α-naphthyl butyrate in acetone, 20 ml 0.2 M phosphate buffer (pH 6.8) and 80 ml distilled water. Cell clone activity was determined with 1 ml 1:100 dilution of extract in distilled water. The mixture was gently agitated for 10 seconds and incubated in a 37°C water bath for 30 minutes. One milliliter of freshly prepared dye coupler solution consisting of 100 mg Fast Red salt ITR (C.I. No. 37150, Matheson, Coleman and Bell), 15 ml 5% sodium lauryl sulfate and 7 ml distilled water which had been filtered twice was added to the reaction mixture after 30-minute incubation. The color development due to the binding of released α-naphthol with the dye coupler was carried out for 10 minutes, at which time the optical density at 540 nm was recorded.

Serum protein determinations were performed after the method of Campbell et al (1963). To 1.5 ml 1:250 or 1:150 dilution of serum in distilled water were added 1.5 ml biuret reagent. Cell clone extracts were diluted 1:20 in saline. The mixture was allowed to stand at room temperature for 30 minutes and then read at 520 nm. Esterase and protein determinations were both recorded on a Beckman DB spectrophotometer. The specific esterase activity is expressed in milligrams of α-naphthol released per milligram of protein during 30-minute incubation.

DENSITOMETRY

Relative isozyme activity of the samples was determined with a Joyce Loebl Chromoscan densitometer. An optical reflectance system with a complementary filter of 610 nm wave length was used to scan the zymogram through a 1 mm X 2 mm aperture. A 10D wedge and a B cam with a specimen record

780

ratio of 3:1 provided a suitable range of expansion. Areas under the curves were automatically integrated. The tracings were then scaled down for publication to represent three-fourths of the zymogram length.

RESULTS

GENETICS

Starch gel zymograms of rabbit red cell and platelet lysates gave remarkable results when developed with α-naphthyl acetate, propionate, or butyrate as substrate. Figure 1 is a diagrammatic representation of the zones of esterase activity

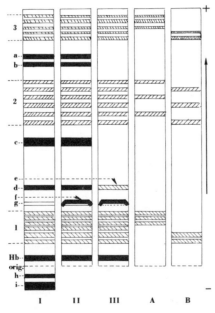

Fig. 1. A diagram of the zones of esterase activity obtained with α-naphthyl acetate (I), propionate (II), and butyrate (III) as substrates. Those zones that were invariably present with any one substrate are designated by lowercase letters and are shown in solid bars. The zones which exhibited phenotypic variation are cross-hatched and are shown by systems designated 1, 2, and 3. Columns I, II, and III contain the heterozygote phenotype AB in each of the three systems, whereas columns A and B show, respectively, the A and B homozygotes in each of the three systems. The symbol Hb indicates hemoglobin. The letter e indicates a zone of esterase activity

Fig. 1 legend, continued.
which is associated with phenotypes A and AB of system 1 and
is derived from the red cell stroma. (Modified from Schiff
and Stormont, 1970).

with these substrates. Those zones that were invariably
present with any one substrate were designated by lowercase
letters and are shown as solid bars. Phenotypic variation
is crosshatched and shown by systems 1, 2, and 3. The heterozygote phenotype AB in each of the three systems is found in
columns I, II, and III. Columns A and B show respectively,
the A and B homozygotes for each of the three systems. The
isozymic diversity observed presented an interesting challenge
for the dissection and classification of these zones into
discrete genetic systems. Grunder et al (1965) had previously
demonstrated the existence of system 1 which was controlled by
autosomal codominant alleles. We were able to show that systems 2 and 3 both were products of codominant autosomal alleles (Figure 2). Breeding data supported this conclusion
(Table 1˙

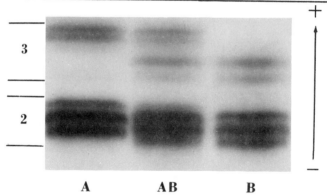

Fig. 2. A portion of a gel developed with α-naphthyl butyrate
as substrate showing the appearance of the esterase zones in
phenotypes A, AB, and B for systems 2 and 3. (Reprinted with
permission from Schiff and Stormont, 1970).

Although we had been studying the products of one species,
a unique opportunity presented itself for the study of interspecies hybridization. Dr. Carlos Sonnenschein and myself
(Schiff and Sonnenschein, 1972) were interested in observing
the effects of placing genes on foreign backgrounds and studying their products (i.e., nonspecific esterases). Electrophoretic and spectrophotometric techniques were used to study
the esterase isozymes of rat x mouse somatic cell hybrids.

TABLE I

INHERITANCE OF ESTERASE PHENOTYPES IN EACH OF
THREE GENETIC SYSTEMS[a]

MATINGS	NO. OF LITTERS	NO. OF PROGENY OF PHENOTYPES		
		A	AB	B
System 1				
A X A	2	13	0	0
A X B	7	0	37	0
A X AB	6	11	14	0
B X B	13	0	0	67
B X AB	18	0	37	46
AB X AB	9	8	18	16
System 2				
A X A	8	32	0	0
A X B	5	0	28	0
A X AB	14	26	45	0[b]
B X B	2	0	0	14
B X AB	13	0	38	31
AB X AB	12	11	23	15
System 3				
A X A	19	107	0	0
A X B	5	0	28	0
A X AB	15	28	31	0
B X B	9	0	0	36
B X AB	1	0	4	3
AB X AB	6	11	11	8

[a] In compiling the data shown in this table, care was taken
to exclude any litters where there was no doubt on the part
of the authors that one or more of the offspring were
illegitimate. This policy was necessitated largely because of the widespread practice of fostering.

[b] Chi-square $\simeq$ 5.0; $p < 0.05 > 0.02$.
Reproduced with permission from Schiff and Stormont, 1970.

It should be pointed out that somatic cell hybridization was
chosen for study since inter-and intraspecific combinations
which have undergone zymogram analysis have proven useful for
the demonstration of parental genome input.

Electrophoretic separation of esterases from a mouse cell
clone (Clone 1 D), a clone derived from rat pituitary cells
($GH_1 2C_1$) and two somatic cell hybrid clones (α-RST and I-RST)
established from mixed cultures of Clone 1 D and $GH_1 2C_1$ showed
marked variation in acetyl- and butyrylesterase migration rates.
These differences were seen when starch gel and polyacrylamide
gel electrophoresis were employed (Figures 3 and 4). An

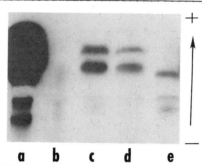

Fig. 3. Starch gel esterase zymogram of cell clone extracts
and serum developed with α-naphthyl butyrate as substrate.
Origin is at the bottom of picture. Mouse serum (a), Clone
1 D (b), 1-RST(C), α-RST(d), and $GH_1 2C_1$(e). Note the presence
in the somatic cell hybrids (c,d) of the fastest anodally
migrating zones which have no parental counterparts.
(Reprinted with permission from Schiff and Sonnenschein, 1972).

anodally migrating zone in the hybrid clones had no electrophoretic counterpart in the parental lines. Differences in
intensity of the zones were recorded. Spectrophotometric
analysis of specific butyrylesterase activity showed significantly higher activity in the hybrids than either parental
line (Table 2). This could not be accounted for by simple
additive or average genome effects. Interpretations of the
control mechanisms involved in these phenomena consisted of
gene derepression, increased substrate affinity of the hybrid
enzymes, and/or increased quantity of a preexisting enzyme.

HISTOCHEMISTRY

In order to better understand and classify the nonspecific
esterases, we have employed various inhibitors, activators,
and substrates to differentiate the enzymes separable by
starch gel electrophoresis (Schiff, 1970).
Substances such as eserine sulfate, disodium ethylenedi-
aminetetraacetate (EDTA), p-chloromercuibenzoic acid, p-hydroxy-
mercuribenzoate (PHMB), diisopropylfluorophosphate (DFP),

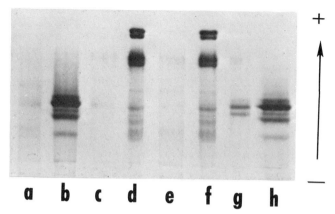

Fig. 4. Polyacrylamide gel esterase zymogram of concentrated
cell clone extracts developed with α-naphthyl butyrate as
substrate. Origin is at the bottom of the picture. Clone 1D
(a,c,e,g), GH_12C_1 (b,h), α-RST (d), and I-RST (f). Note the
most intensely stained zones of the hybrid cell samples (d,f).
These represent the sets of anodally migrating doublets.
(Reprinted with permission from Schiff and Sonnenschein, 1972).

TABLE II

PROTEIN AND ESTERASE CONTENT OF SOMATIC
CELL HYBRIDS AND PARENTAL LINES

Cell Line	Specific Esterase Activity[a]	Protein/Cell[a] ($\times 10^{-7}$ mg)	α-Naphthol Liberated/Cell[a] ($\times 10^{-8}$ mg)
Cl_1D	0.015 ± 0.002	3.28 ± 0.53	0.53 ± 0.12
GH_12C_1	0.026 ± 0.003	5.04 ± 1.19	1.12 ± 0.17
α-RST	0.039 ± 0.004	5.98 ± 0.52	2.21 ± 0.25
1-RST	0.110 ± 0.015	8.04 ± 0.58	8.61 ± 0.97

[a] ± vales indicate S.E.M.
Reproduced with permission from Schiff and Sonnenschein, 1972.

manganous chloride, magnesium chloride, acetylcholine chloride,
acetazolamide, and the follwoing substrates have been examined:

α-naphthyl acetate (C2)   α-naphthyl caproate (C6)
β-naphthyl acetate (β-C2)   α-naphthyl caprylate (C8)
α-naphthyl propionate (C3)   α-naphthyl nonanoate (C9)
α-naphthyl butyrate (C4)   α-naphthyl caprate (C10)
α-naphthyl valerate (C5)   β-carbonaphthoxycholine iodide
                                     (β-CCI).

As an example Figure 5 shows the effects of EDTA concentration on rabbit red cell and platelet esterases. The
numbers 1, 2, and 3 represent the polymorphic isozyme systems.
The normal intensity of staining is at a level of 3 whereas
complete inhibition is "O" on a relative unit scale. The

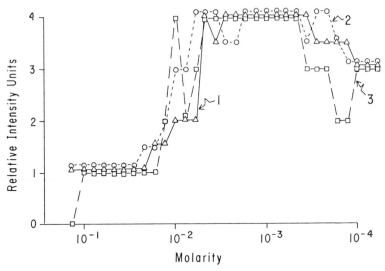

Fig. 5. The effects of EDTA concentration on rabbit red cell
and platelet esterases separated by starch gel electrophoresis.
Numbers 1, 2, and 3 represent the polymorphic isozyme systems.
(Reprinted with permission from Schiff, 1970).

abscissa contains the molarity of the test reagent while the
relative unit scale is found on the ordinate. Whereas, all
three systems are activated at 10^{-3}M, the organophosphorus
compound, DFP, inhibits all three systems at 10^{-4}M (Figure 6).
   Employing this type of approach, the esterases can be
placed into major categories by their substrate affinity and
subdivided into groups by their responses to activator and
inhibitor substances (Table 3).

ESTERASES IN DISEASE

In 1971-1972, Dr. Lewis Estabrooks (an oral surgeon) and
I investigated the use of the nonspecific esterases as markers
following joint surgery. When the temporomandibular joint
(TMJ) becomes ankylosed (fused) to the temporal or zygomatic
bone as a result of trauma, infection, or congenital disease,

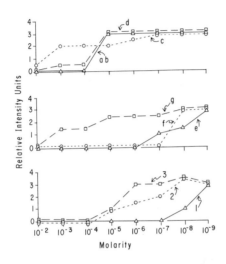

Fig. 6. The effects of DFP concentration on rabbit red cell and platelet esterases separated by starch gel electrophoresis. Letters and numbers represent the esterase zones of Figure 1. (Reprinted with permission from Schiff, 1970).

correction involves the creation of a surgical fracture at some point distal to the fixed joint and insertion of a material to prevent reunion. Dr. Estabrooks created artificial joints in rabbits by surgically inserting silastic in the TMJ. We therefore had the opportunity to examine the butyrylesterases derived from the synovial fluid of the knee, normal TMJ, artifically created joints, and the synovial membrane. Starch gel electrophoresis was the method of enzyme separation. All fluids examined showed two major regions of esterase activity that migrated at rates comparable to systems 1 and 2 of the red cell (Figure 1). However, inhibitor tests differentiated the body fluid esterases from those found in the red cell. The use of densitometer tracings of the zymograms (Figure 7) showed an excess of region II activity in the

TABLE III

CLASSIFICATION OF RABBIT RED CELL AND PLATELET ESTERASES

Esterase Type	Naphthol Esters Hydrolyzed	Diagnostic Features
Acetylesterases		
Group 1 (zones a and b)	C2 > C3 β-C2	Activated by 10^{-3} M EDTA
Group 2 (zones c and g)	C2 ⩾ β-C2 > C3	Not activated by 10^{-3} M EDTA
Group 3 (zone d)	C2 > β-C2 > C3	Inhibited by 10^{-4} M PCMBA
Group 4 (system 1)	C2 = C3 ⩾ C4 > β-C2 >	Inhibited by 10^{-7} M DFP
	C5 > C6 > C8	Activated by 10^{-3} M EDTA
		Not affected by PCMBA
Propionylesterases		
Group 1 (system 2)	C4 ⩾ C3 > C5 ⩾ C2 > β-	Activated at 10^{-3} M EDTA
	C2 ⩾ C6 > C8	Partially inhibited by eserine at 10^{-2}–10^{-6} M
		Inhibited by 1 M $MnCl_2$
		Not affected by PCMBA
		Not affected by 10^{-2} M acetylcholine

TABLE III, Classification of Rabbit Red Cell and Platelet Esterases, continued.

Esterase Type	Naphthol Esters Hydrolyzed	Diagnostic Features
Group 2 (system 3)	C3 > C2 > C4 > β-C2 > C5	Inhibited by 10^{-1} M $MnCl_2$ Inhibited by 10^{-4} M DFP Activated by 10^{-3} M EDTA
Group 3 (zone e)	C3 = C4 > C5	Inhibited by 10^{-6} M DFP Not affected by 10^{-3} M EDTA Not affected by 1 M $MgCl_2$
Butyrylesterases		
Group 1 (zone f)	C4 > C3 > C5 > C6 > C2	Activated at 10^{-4}-10^{-7} M PCMBA Activated at 10^{-3}-10^{-4} M eserine Activated at 10^{-3} M EDTA Inhibited by 10^{-4} M DFP Inhibited by 1 M $MnCl_2$ Activated at pH 3
Carbonic anhydrases		
Group 1 (zones h and i)	β-C2 > C2	Inhibited by 10^{-5} M acetazolamide

Reproduced with permission from Schiff, 1970.

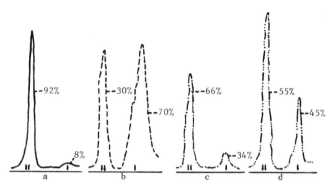

Fig. 7.   Densitometer recordings that show the esterase activity of knee (a), homogenized synovial membrane (b), TMJ (c)
and false joint (d) synovial fluids. Tracings are three-fourths
of the length of the zymogram.   The numbers next to the curves
of regions I and II represent the percentage area of that region
to the whole.   (Reprinted with permission from Estabrooks and
Schiff, 1973).

artificial joint fluids which appeared to be contributed by
the synovial membrane. Because increased activity of the
membrane is common during a repair process, analysis of the
false joint fluid reinforced light microscopic findings that
not only does synovial membrane differentiate *de novo*, but
the membrane produces or maintains a fluid similar in esterase
activity to the normal TMJ.

In conjunction with the study conducted previously by Dr.
Estabrooks, we also investigated the nonspecific esterases and
cholinesterases of various portions of the rabbit eye. Although the cholinesterases have been studied extensively by
histochemical means in various regions of the eye, there is
still uncertainty about their localization.

We examined the aqueous and vitreous humor, iris, lens ,
retina, serum, and red cells from normal rabbits and animals
with inherited glaucoma (Estabrooks and Schiff, in press).
This glaucoma which is called buphthalmia in rabbits is controlled by homozygous recessive alleles at an autosomal locus.
In the study, New Zealand White (NZW), AX bu bu (strain with
buphthalmia) and AX rabbit (normal strain)tissues were compared
by starch and polyacrylamide gel electrophoresis for the non-
specific esterases and cholinesterases. Figure 8 is a densitometer tracing of the vitreous humor non-specific esterases in
starch gel samples from these rabbits. Note that the AX and
NZW results are essentially the same, whereas the region 2
activity is significantly higher in the bupthalmic strain.
This may be the direct result of increased intraocular pressure,

VITREOUS FLUID

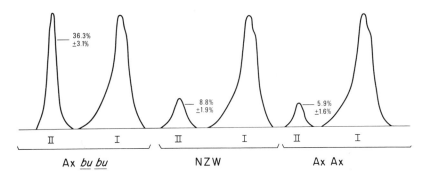

Fig. 8. Densitometer recordings of the region II vitreous fluid esterase activity of AX bu bu, NZW and AX rabbits. The numbers next to the curves represent the relative activity of that region in comparison to the whole. The standard error of the mean is also given. (Reprinted with permission from Estabrooks and Schiff, in press).

which causes a compression between the aqueous and vitreous chambers. Prior to this study no enzyme differences were demonstrated to exist between AX and AX bu bu strains. Cholinesterase activity was markedly increased in buphthalmic retinas. This may be due to retinal cell destruction caused by compression.

TRANSPLANTATION MARKERS

In an ongoing study of cellular esterases we were interested in the pulmonary alveolar macrophage (PAM) which is part of the host defense mechanism directed against inhaled particulate and infectious materials. We undertook the study of PAM esterases for several reasons. There remains a conflict of opinion about the PAM life cycle, origin, and ultimate fate (Pinkett et al., 1966; Virolinen, 1968; Bowden et al., 1969). In addition, the PAM's relationship with bone marrow progenitor cells has been particularly unclear.

The use of nonchromosomal marker techniques such as the ES-2 esterase (Petras, 1963) found in one of two sublines of the RFM strain of mice (Hard et al., 1967) would serve as a useful tool for the determination of PAM origin if the marker esterase could be demonstrated in the PAMs of the RF/AL (+) subline. In addition, the RF positive and RF negative sublines are homozygous with respect to the presence and absence of the Es-2 esterase respectively, and appear to be congenic at this locus. Thus, these animals seemed suitable for our studies of alveolar macrophage esterase activity and for transplantation experiments which utilized the esterase markers. We hypothesized that application of high resolution electrophoresis may be useful for understanding the PAM origin (Schiff et al., 1970; Brunstetter et al., 1971).

Figures 9 and 10 demonstrate the Es-2 butyrylesterases found in the serum and PAMs of RF positive mice but absent in

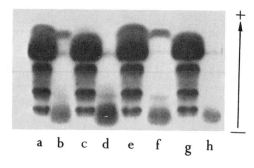

Fig. 9. Starch gel zymograms of mouse serum (a,c,e,g) and macrophage esterases (b,d,f,h) developed with α-naphthyl butyrate. Samples a, b, e, and f are from RF/Al (+) mice; whereas, c, d, g, and h represent RF/Al (-) animals. Note the fastest anodally migrating esterase found only in (+) animals. (Reprinted with permission from Schiff et al., 1970).

negative animals. We subsequently transplanted bone marrow from positive animals to irradiated RF negative mice, thereby creating radiation chimeras. Figure 11 shows the presence of the marker esterase in various tissues and PAMs of the recipient. Therefore, 8 weeks post transplant, the lung PAMs of the recipient were repopulated from donor bone marrow cells. The marker could also be found in the spleen and bone marrow of the recipient (Schiff et al., 1972).

DISCUSSION

Although this report has concentrated on the esterases of the mouse, rabbit, and somatic cell hybrids, genetically controlled electrophoretic variants have been reported in maize (Schwartz, 1960), Tetrahymena (Allen, 1960), and fruit flies

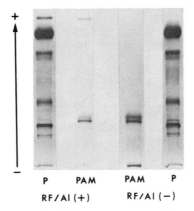

Fig. 10. Typical acrylamide esterase pattern of serum and
PAMs. RF/Al (+) serum and PAM zymograms are on the left.
RF/Al (-) PAM and plasma zymograms are on the right. The
marker Es-2 esterase is apparent as the leading anodal band
and is present only in cells from RF/Al (+) animals. (Reprinted with permission from Brunstetter et al., 1971).

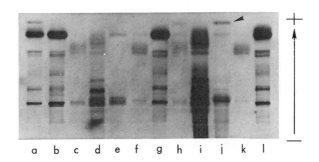

Fig. 11. A butyrylesterase zymogram on polyacrylamide gel of
various tissues from unirradiated and irradiated mice. Samples a and b are serum controls from an untreated RF/Al (+)
and (-) animal, respectively. Samples c - g are derived from
spleen, liver, PAM, bone marrow and serum of an irradiated
RF/Al (-) mouse transplanted with (-) bone marrow. Samples
h - l represent the same tissues as c - g although they were
taken from an irradiated (-) mouse given (+) bone marrow.
Note the marker esterase in the PAM sample (j, arrow). (Reprinted with permission from Schiff et al., 1972).

(Wright, 1963) among other species.
    An interesting application of esterase enzymology has
recently been demonstrated by Matteo et al (in press) with

the Marine Periwinkle, *Littorina littorea*. The object of her study was to identify and characterize genetically determined biochemical markers. Dr. Matteo investigated the nonspecific esterases in this snail with respect to substrate and inhibitor affinity. This was the first step toward correlating these activities with the products of distinct genetic loci. In her paper she discusses the possible action of *L. littorea* as an indicator of the genetic effects of environmental factors on the existence of polymorphic loci which show inter-population variation.

To the biochemist, an investigation which utilizes an enzyme whose function is not known or understood is difficult to pursue although not impossible. To the geneticist a nonspecific enzyme can be a very useful tool for the study of phylogenetic and linkage relationships, and to the histochemist the tissue distribution of these enzymes may lead to a better understanding of metabolic processes in the cell. Thus the simple demonstration of nonspecific esterases has resulted in the field of esterase isozymology expanding rapidly within several disciplines.

What is apparent from this study is the diverse usage of the esterase isozymes. Although we have touched upon such broad areas as genetics, histochemistry, and disease the common thread is the nonspecific esterases. These enzymes first demonstrated histochemically by Gomori have subsequently been used in the present review as markers: Markers to trace and identify genetic loci, markers to determine post-surgical processes, markers to follow transplanted bone marrow cells, markers in glaucoma, and finally as tools for studying molecular interaction of genome upon genome.

ACKNOWLEDGEMENT

The author would sincerely like to thank Mrs. Lynnette Lowry for her assistance in the typing of the manuscript.

REFERENCES

Allen, R.C. and D.J. Moore 1966. A vertical flat-bed discontinuous electrophoresis system in polyacrylamide gel. *Anal. Biochem.* 16:457-465.

Allen, R.C., D.J. Moore, and R.H. Dilworth 1969. A new rapid electrophoresis procedure employing pulsed power in gradient gels at a constant pH: The effect of various discontinuous buffer systems on esterase zymograms. *J. Histochem. Cytochem.* 17:189.

Allen, S.L. 1960. Inherited variations in the esterases of *Tetrahymena*. *Genetics* 45:1051-1070.

Barron, K.D. and J.J. Bernsohn 1968. Esterases of developing human brain. *Neurochem*. 15:273-284.

Beckman, L. 1966. *Isozyme Variations in Man*. S. Karger, N.Y.

Bowden, D.H., I.Y.R. Adamson, W. Grantham, and J.P. Wyatt 1969. The origin of the lung macrophage: Evidence derived from radiation injury. *Arch. Path*. 88:540-546.

Brunstetter, M.A., J.A. Hardie, R. Schiff, J.P. Lewis, and C.E. Cross 1971. The origin of pulmonary alveolar macrophages: Studies of stem cells using the Es-2 marker of mice. *Arch. Int. Med*. 127:1064-1068.

Campbell, D.H., J.S. Garvey, N. Cremer, and D.H. Sussdorf 1963. *Methods in Immunology*. W.A. Benjamin, New York, p. 51.

Estabrooks, L. and R. Schiff 1972. Esterase isozymes from rabbit synovial fluid: Normal and artificial joints. *J. Histochem. Cytochem*. 20:211-219.

Estabrooks, L. and R. Schiff (In press). Comparison of esterase in normal and buphthalmic rabbit ocular tissues and fluids. *J. Histochem. Cytochem*.

Giblett, E.R. 1969. *Genetic Markers in Human Blood*. Blackwell Scientific Publications, Oxford.

Gomori, G. 1953. Human esterases. *J. Lab. Clin. Med*. 42:445-453.

Grunder, A.A., G. Sartore, and C. Stormont 1965. Genetic variation in red cell esterases of rabbits. *Genetics* 52:1345-1353.

Hard, R.C. Jr., B. Kullgren, D.J. Moore, and R.C. Allen 1967. Esterase isozyme changes in radiation chimeras. *Fed. Proc*. 26:755.

Harris, H.D., A. Hopkinson, and E.B. Robson 1962. Two dimensional electrophoresis of pseudocholinesterase components in normal human serum. *Nature* 196:1296-1298.

Hofstee, B.H.J. 1952. Specificity of esterases I. Identification of two pancreatic aliesterases. *J. Biol. Chem*. 199:357-364.

Holmes, R.S., C.J. Masters, and E.C. Webb 1968. Comparative study of vertebrate esterase multiplicity. *Comp. Biochem. Physiol*. 26:837-852.

Hunter, R.L. and C.L. Markert 1957. Histochemical demonstration of enzymes separated by zone electrophoresis in starch gels. *Science* 125:1294-1295.

Kristjansson, F.K. 1960. Genetic control of two blood serum proteins in swine. *Can. J. Genet. Cytol*. 2:295-300.

Lush, I.E. 1967. *The Biochemical Genetics of Vertebrates Except Man*. Wiley, New York.

Markert, C.L. and R.L. Hunter 1959. The distribution of esterases in mouse tissues. *J. Histochem and Cytochem.* 7:42-49.

Matteo, M., R. Schiff, and L. Garfield (in press). The nonspecific esterases of the marine snail, *Littorina littorea.* Histochemical characterization. *Comp. Biochem. Physiol.*

Ortec Instructional Manual. Ortec Model 4100. Pulsed Constant Power Supply. Ortec, Oak Ridge, Tenn.

Ortec Model 4200 Electrophoresis System. Ortec, Oak Ridge, Tenn.

Petras, M.L. 1963. Genetic control of a serum esterase component in *Mus musculus. Proc. Nat. Acad. Sci. U.S.A.* 50:112-116.

Pinkett, M., C.R. Cowdrey, and P.C. Nowell 1966. Mixed hematopoietic and pulmonary origin of "alveolar macrophages" as demonstrated by chromosome markers. *Amer. J. Path.* 48:859-867.

Popp, R.A. and D.M. Popp 1962. Inheritance of serum esterases having different electrophoretic patterns among inbred strains of mice. *J. Heredity* 53:111-114.

Richter, D. and P.G. Croft 1942. Blood esterases. *Biochem. J.* 36:746-757.

Schiff, R. 1970. The biochemical genetics of rabbit erythrocyte esterases: Histochemical classification. *J. Histochem. Cytochem.* 18:709-721.

Schiff, R., M.A. Brunstetter, R.L. Hunter, and C.E. Cross 1970. Electrophoretic separation of esterases of pulmonary alveolar cells. *J. Histochem. Cytochem.* 18:167-177.

Schiff, R., J.P. Lewis, and C.E. Cross 1972. Esterase markers in mouse radiation chimeras. *J. Histochem. Cytochem.* 20:472-473.

Schiff, R. and S. Jacobson 1970. Genetic control of esterase isozymes from rabbit brain. *Genetics* 64:s56-s57.

Schiff, R. and C. Sonnenschein 1972. Somatic cell hybridization: Cellular interaction upon esterase isozymes. *Exp. Cell Res.* 70:269-278.

Schiff, R. and C. Stormont 1970. The biochemical genetics of rabbit erythrocyte esterases: Two new esterase loci. *Biochem. Genet.* 4:11-23.

Schwartz, D. 1960. Genetic studies on mutant enzymes in maize. Synthesis of hybrid enzymes by heterozygotes. *Proc. Nat. Acad. Sci. U.S.A.* 46:1210-1215.

Shaw, C.R. 1965. Electrophoretic variation in enzymes. *Science* 149:936-943.

Tashian, R.E. 1965. Genetic variation and evolution of the carboxylic esterases and carbonic anhydrases of primate erthrocytes. *Am. J. Human Genet.* 17:257–272.

Tucker, E.M., Y. Suzuki, and C. Stormont 1967. Three new phenotypic systems in the blood of sheep. *Vox Sanguinis* 13:246–262.

Virolainen, M. 1968. Hematopoietic origin of macrophages as studied by chromosome markers in mice. *J. Exp. Med.* 127:943–952.

Wright, T.R.F. 1963. The genetic control of an esterase in *Drosophila melanogaster*. *Genetics* 48:787–801.

THE CHOLESTATIC DOUBLET OF ALKALINE PHOSPHATASE: ITS ORIGIN AND CLINICAL SIGNIFICANCE

MARC E. DE BROE and ROGER J. WIEME
Department of Clinical Chemistry
University Hospital Gent
De Pintelaan 135
9000 Gent, Belgium

ABSTRACT. Agar gel electrophoresis is found to be a simple and satisfactory procedure for separating liver, placental, bone, and intestinal alkaline phosphatase (A.P.) isozymes. In serum of patients with cholestasis, an isozyme doublet appears consisting of the liver fraction (L) and a rapidly migrating fraction (R). This isozyme doublet can be observed even in serum with normal total A.P. activity.

The R fraction remains at the origin when other gelified media are used (starch and polyacrylamide gel). The high molecular weight of the R fraction is confirmed by ultra-centrifugation. The R fraction is never found in normal bile, but can be seen in fresh bile of patients with the doublet in their serum. This argues against a bile origin for the R fraction. The R and L fractions behave in an identical way when submitted to physico-chemical tests such as heat inactivation, L-phenylalanine, and levamisole inhibition. Upon n-butanol treatment of serum with the isozyme doublet, the R fraction is converted into an L fraction.

Bile duct ligation of rats results in the appearance in the serum of a doublet of A.P. very similar to that seen in man.

It is our conclusion that both isozymes (L and R) originate in the liver and not in the bile. The R fraction corresponds to L type of A.P. which is in a macromolecular state, associated with fragments of liver cell membranes.

We propose a modified regurgitation theory for the origin of serum A.P. in cholestasis. Increased liver A.P. synthesis together with interference with the membrane integrity of the liver cells, leads to an escape of liver A.P. from the liver cells into the serum and into the bile. Part of this A.P. is still associated with membrane fragments and can easily be detected in the serum.

Several attempts have been made to identify the organ source of serum alkaline phosphatase (A.P.) in health and disease. Paper, starch gel, cellulose acetate, Sephadex gel, polyacrylamide gel, and agar gel all have been found to be of value. We found agar gel electrophoresis to be a simple and satisfactory procedure for separating isozymes of liver, bone, placenta, and intestine (Fig. 1).

Separation is performed at 12 degrees centigrade, for 14 min at 20 v/cm in sodium veronal buffer of pH 8.4, in 1% Difco Special Agar Noble. Following electrophoresis the slabs are incubated for 2 hours at 37° centigrade using a gel overlay containing β-naphthylphosphate (Aldrich Chemical), Fast Blue RR (Difco), and magnesium sulphate in 1.5% agarose (Industrie Biologique Francaise) in borax borate buffer, pH9.2 (Wieme, 1965). After removal of the substrate layer, fixation is performed in 2% acetic acid for 2 hours. After drying, the slides are scanned photometrically (Quick-Scan, Helena Laboratories).

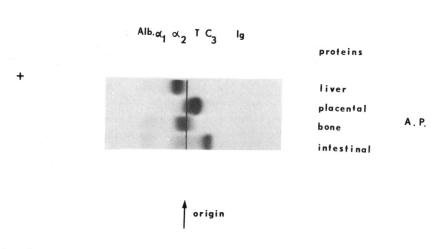

Fig. 1. The four typical A.P. isozymes as revealed by agar gel electrophoresis in human serum. At the location of serum albumin a trace of non-specific protein staining can be observed.

Using this technique we found the liver isozyme in the serum of 100 healthy adult blood donors with, in addition, in most cases, a small amount of bone A.P. In 48 individuals intestinal A.P. activity was also present in trace amounts.

Our experience with isozyme determinations on serum gathered from several thousands of patients, using agar gel electrophoresis, confirms the literature data as far as the bone and intestinal A.P. isozymes are concerned (Kaplan et al, 1969; Ohlen et al, 1970). This is not the case for liver disease. In this group of patients we have observed two different pat-

terns: in one an exclusive increase of the liver fraction
(L); in the second, there also appears a rapid fraction (R)
which is always associated with an increase of the L fraction.
This very typical combination we refer to as the doublet pattern (Fig. 2).

+

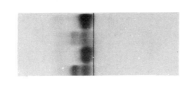

liver (L)

cholestatic doublet (R + L)

liver (L)

cholestatic doublet (R + L)

↑ origin

Fig. 2. Two different human sera with the R + L doublet,
contrasted with two other sera showing only an increase of the
L fraction.

In 208 of 235 patients with this isozyme doublet it was
possible, based on clinical radiological, surgical, and autop-
tic data to specify the diagnosis. As can be seen in Table 1,
the common denominator for all these clinical situations was
obstruction to bile flow or, in other terms, cholestasis in
its most general sense (Popper et al, 1970). Surgical release
of the obstruction or decongestion by medical treatment resulted in gradual disappearance of the R fraction. This R
fraction was never observed in the serum of normal controls.
It therefore seems justified to refer to the R + L doublet as
the cholestatic doublet pattern.

The activity of the R fraction expressed as percentage of
total activity was determined in 120 cases by photometrical
scanning. We found a mean value of 27% with a rather large
range of variation (4.8 - 68%). The determination itself can
be repeated with a coefficient of variation of only 2%.
Although, in most cases, the presence of the cholestatic
doublet was associated with higher than normal values for total
A.P., in other cases with normal total A.P. activity this iso-

TABLE 1
PATIENTS WITH THE CHOLESTATIC DOUBLET OF A.P.

Extrahepatic biliary obstruction
Cholelithiasis with or without cholecystitis	19
Bile duct tumor	16
pancreas tumor	8

Intrahepatic biliary obstruction
Hepatic metastases from different tumors	108
Hodgkin with liver invasion	2
Liver abcess	2
Cholangitis	2

Intrahepatic cholestasis with and without
morphological changes of hepatitis
Viral hepatitis	6
Drug induced hepatitis with cholestasis	2
Alcoholic hepatitis	1
Cirrhosis - postnecrotic	3
- Laennec	4
- primary biliary	1
Sepsis	12
Passive congestion	14
Postoperatively	8

T o t a l : 208

zyme doublet was still observed. The very important clinical value of this laboratory test has been reported elsewhere (Wieme et al, 1972).

One might wonder why this cholestatic doublet, so easily revealed by agar gel electrophoresis, has not been described by authors using other media for gel electrophoresis. Therefore we examined all sera that possessed the cholestatic doublet in agar gel using starch gel electrophoresis, and in addition a limited number (n = 20) with polyacrylamide gel.

Horizontal starch gel electrophoresis was performed according to Poulik (1957) (3 hours at 5 v/cm, using citrate buffer 76 mM) in a discontinuous buffer system. Incubation with the same gel substrate layer as for agar gel, was for 1 hour at 37°. Polyacrylamide 7.5% was used as described by Kaplan and Rogers (1969).

In all serum samples (235) with the isozyme doublet, starch gel electrophoresis revealed phosphatase activity mainly in the

classical liver position, with an inconstant minor zone of
activity in the lipoprotein region. However, there always
remains some activity at the origin (run 1 of Fig. 3). The
results with polyacrylamide gel were similar. It must, however, be stressed that in serum with only slightly elevated
total A.P. and with a weak R fraction, the origin activity can
be very low. In these media the R fraction, clearly detected
in agar gel, might easily be overlooked.

This observation is interesting and suggests the possibility
that the R fraction is of such high molecular weight, that it
does not penetrate into starch gel or polyacrylamide gel.
A more direct answer is given by two-dimensional electrophoresis. The first dimension of the run is agar gel, the second
is starch gel. Results are given in Fig. 3. Clearly, the R
fraction does not penetrate into the small pore gel medium.

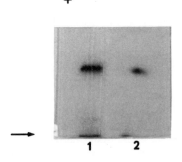

Fig. 3. Separation on starch gel of a human serum with the
R + L doublet. In 1 this serum was subjected to a conventional
one-dimensional run. In 2 the serum was first separated in
agar gel and part of the separation gel inserted in the starch
gel (the anode side of this separation is to the left). Note
that in both cases the R activity remained at the origin
(indicated by the arrow).

Other evidence was gained from ultracentrifugation experiments. Two cascaded preparative ultracentrifugations at rather
low speed (2 x 60 min at 36,000 g) separated a bottom layer
containing an A.P. that migrated as a R fraction in agar gel
and did not penetrate into a starch gel. In the supernatant
only the L fraction was found.

Haye and De Jong (1965), who also observed an R fraction in

agar gel electrophoresis, were of the opinion that the origin
of this fraction is bile regurgitation into the serum, and they
called it the bile fraction. Chiandussi, Green, and Sherlock
(1962) explained the appearance of an origin activity in starch
gel on the basis of regurgitation of bile into the circulation,
since the electrophoretic pattern of a limited number of bile
specimens in starch gel resembled, according to the authors,
the serum pattern of patients with extra- or intrahepatic obstruction. However, this statement may be an oversimplifica-
tion. We have studied bile A.P. in 14 patients, 3 of them
having the isozyme doublet in their serum. The other 11 had
a normal or increased total A.P. activity with a concomitant
rise in liver A.P. only. Bile was collected during surgical
intervention or in patients with T tube drainage of the common
bile duct after cholecystectomy. Several patterns have been
observed. In starch gel we always found activity at the
insertion slot and in the β-lipoprotein region; in a limited
number of the cases studied (5 out of 14) variable activity was
seen at the β-globulin or classical liver position. In agar
gel the predominant isozyme found in human bile had a rapid
migration in the region of albumin, minor activity was also
seen in the α_2, β-globulin position. Patients without the
doublet activity in the serum never had an A.P. of the R type
in their bile. But in the 3 patients with the cholestatic
doublet in their serum, R activity was seen in their bile.

This suggests the possibility that the R fraction originates
in the liver itself. However, A.P. activity of extracts of
normal human liver tissue is localized in the L position in
agar gel and in the β-globulin position in starch. Moreover,
in extract of liver tissue obtained from patients having the
isozyme doublet in their serum, A.P. activity is found only
at the classical L position in agar gel and in starch gel.
However, tissue extracts are generally obtained by n-butanol
treatment. Checking the possibility that n-butanol itself
could interfere, we submitted serum of patients with the
cholestatic doublet to treatment with an equal amount of
n-butanol. This resulted in the disappearance of the R fraction in agar gel and of the origin activity in starch gel.
The position of the L fraction in agar gel as well as in starch
gel remained unchanged, but some activation became manifest.
In the presence of increasing proportions of butanol (2:1,
3:1, 5:1, 10:1) a gradual shift towards the R fraction is
observed.

At a butanol:serum ratio of 1:50 the original position remains unaffected. It has to be noted that submitting sera of
patients with high concentration of only liver, only bone, or
only intestinal A.P. activity to similar treatment with butanol

does not influence the electrophoretic mobilities. Bile of
6 patients with high total A.P. activity in γ-globulin,
β-lipoprotein and occasionally in the β-globulin position, was
submitted to the same butanol treatment followed by starch gel
electrophoresis. This gave some activation of the A.P. but did
not induce a change in the electrophoretic mobilities. Yet
some decrease of the origin activity could be observed.

The conversion by butanol of the R fraction in agar gel and
of the origin activity in starch gel to a fraction with the
mobility of the L fraction, suggests basic similarity between
these two fractions, and this led us to study the behavior of
both fractions for different physico-chemical parameters.

Exposure of serum samples with the cholestatic doublet to
55°C for 15 minutes gives a parallel decrease of the A.P. activity of both isozymes. Both are completely inactivated at
65°C for 5 minutes. L-phenylalanine at a concentration of
5 mM in the substrate layer equally inhibits both fractions;
this inhibition is far less important than for the intestinal
component.

Incubation of serum (0.1 ml) with 0.4 ml of a neuraminidase
solution (from *Clostridium perfringens*, Sigma Chemical Co.
Ltd., type V, 1 mg/ml in Tris-buffer, pH 7.7) for 24 hours
changes the mobility in agar for both fractions, the L fraction
showing the greatest alteration in its electrophoretic position.
Neuraminidase does not influence the electrophoretic mobility
of the origin activity in starch gel but the mobility of A.P.
activity in the β-globulin region is greatly decreased and it
takes the position of the β-lipoproteins, which is typical for
liver A.P..

Levamisole and, more recently, its para-bromo derivative
have been proposed for the selective detection for intestinal
A.P. in the dog (Van Belle, 1972). Whereas all A.P. isozymes
are strongly inhibited except intestinal A.P., interesting
results were obtained when applied to our material. R as well
as L A.P. were equally inhibited (residual activity: 12%) in
contrast to intestinal A.P. (residual activity: 91%) at a
concentration of 1×10^{-4} M of the para-bromo derivative of
levamisole.

Incubation of serum (0.1 ml) with this isozyme doublet in a
medium containing 0.1 ml phospholipase solution (from *Clostridium welchii*, Sigma Chemical Co. Ltd., type 1, 1 mg/ml in Tris
buffer, pH 7.7) converts the doublet into a single fraction.
However, due to contaminating neuraminidase activity this
fraction has the well-known position of neuraminidase-treated
liver alkaline phosphatase.

It appears that the R and L fraction derive from the same
A.P. of liver, but the R fraction is a macromolecular complex

that can be stripped by n-butanol. The R fractions seem to correspond to fragments of the liver cell membrane loaded with L type A.P.

This interpretation fits well to recent histochemical and biochemical investigations. In normal human liver non-specific A.P. is localized mainly on the canalicular part of liver cell membrane.

But, experimental cholestasis is characterized by a great increase in A.P. activity of all parts of the liver cell membrane.

The studies of Kaplan and Righetti (1957) also seem to confirm our interpretation. These authors used polyacrylamide gel for the study of A.P. isozymes in the serum of rats submitted to experimental cholestasis. Within 12 hours after bile ligation the hepatic A.P. was increased 7 times and total serum A.P. activity 2 times. The elevation in the serum activity was entirely due to an increase in an isozyme that appears to originate in the microsomal and cell membrane fractions of the liver. After purification, the liver and serum isozyme seem to be identical. The inhibition of this increase with cycloheximide suggests the occurence of an intensified de-novo-synthesis of the liver isozyme. According to the authors, these data indicate that the elevated serum A.P. in obstructive jaundice should originate in the liver and these data should support the regurgitation theory.

Repeating some of Kaplan's experiments, but using starch gel instead of polyacrylamide gel, and in addition agar gel, we observed in the serum of rats with bile duct ligation an isozyme doublet very similar to that of man. Submitting both fractions to the same physico-chemical tests used for man, we found an identical response as far as temperature inhibition, L-phenylalanine sensitivity, and neuraminidase incubation is concerned.

CONCLUSIONS

1. Cholestasis in man and rat results in the appearance of a cholestatic doublet of A.P. (R + L fraction).
2. The R fraction is converted to the L fraction by butanol stripping and by phospholipase incubation.
3. The R fraction consists of L type A.P. associated with fragments of the liver cell membrane. This leads us to formulate a modified regurgitation theory for the origin of serum A.P. in cholestasis. The older theory postulates that bile regurgitates into the serum. We believe that increased A.P. synthesis, together with interference with membrane integrity, results in an escape of liver A.P. from the liver cells into

the blood stream and also into the bile. Part of this A.P. is still associated with membrane structures that can easily be detected in the serum.

LITERATURE CITED

Chiandussi, L., S. F. Greene, and S. Sherlock 1962. Serum Alkaline Phosphatase Fractions in Hepato-Biliary and Bone Diseases. *Clin. Sci.* 22: 425-434.

Haije, W. G. and M. De Jong 1963. Iso-enzyme Patterns of Serum Alkaline Phosphatase in Agar Gel Electrophoresis and their Clinical Significance. *Clin. Chim. Acta* 8: 620-623.

Kaplan, M. M. and A. Righetti 1970. Induction of Rat Liver Alkaline Phosphatase: the Mechanism of the Serum Elevation in Bile Duct Obstruction. *J. Clin. Invest.* 49: 508-516.

Kaplan, M. M. and L. Rogers 1969. Separation of Human Serum-Alkaline-Phosphatase Isoenzymes by Polyacrylamide Gel Electrophoresis. *Lancet* II: 1029-1031.

Ohlen, J., H. Pause, and J. Richter 1971. Alkaline Phosphatase: Diagnostic Value of Isoenzyme Determination. *Eur. J. Clin. Invest.* 1: 445-451.

Popper, H. and F. Schaffner 1970. Pathophysiology of Cholestasis. *Hum. Pathol.* 1: 1-24.

Poulik, M. D. 1957. Starch-gel Electrophoresis in a Discontinuous System of Buffers. *Nature* 180: 1477-1479.

Van Belle, H. 1972. Kinetics and Inhibition of Alkaline Phosphatases from Canine Tissues. *Biochim. Biophys. Acta* 289: 158-168.

Wieme, R. J. 1965. *Agar Gel Electrophoresis.* Elsevier Publishing Company, Amsterdam London New York, 163.

Wieme, R. J. and M. D. De Broe 1972. De Diagnostische Betekenis van een Snelle Fraktie in het Isoenzympatroon der Alkalische Fosfatase. *Tijdschr. v. Geneesk.* 28: 468-473.

THE MICROSOMAL MIXED FUNCTION OXIDASES
AND CHEMICAL CARCINOGENS

CHARLES R. SHAW
The University of Texas System Cancer Center
M.D. Anderson Hospital and Tumor Institute
Texas Medical Center, Houston, Texas 77025

ABSTRACT. The microsomal mixed function oxidases are a
group of related enzymes with similar activities and
structures. They are complexes composed of at least
two different polypeptide subunits bound to a phos-
phatidyl choline. The polypeptides are an NADPH dependent reductase and a P-cytochrome. The several forms of
the enzyme are differentially inducible by a number of
compounds, and the induction is thought to involve
mainly the cytochrome moiety. The enzymes have a wide
spectrum of activities, some of which are overlapping
in various degrees. Known substrates include a variety
of drugs, insecticides, steriod hormones, and many
polycyclic hydrocarbons including some of the chemical
carcinogens.
    The enzyme which metabolizes the hydrocarbon car-
cinogens is called aryl hydrocarbon hydroxylase (AHH).
It is thought to be responsible for converting the carcinogens to an active, oxygenated form, probably an
epoxide, which is then further metabolized to an inactive
hydroxide. Activity of AHH has been found to vary
genetically in different strains of mice. Recently we
have developed a method for assaying AHH in man, using
short term lymphocyte culture. We have demonstrated
genetic variation in man, probably polygenically controlled. Further, the amount of AHH activity and
inducibility appears to affect susceptibility to the
carcinogen of tobacco smoke, as there is a significant
correlation between occurrence of lung cancer and inducibility of AHH among human subjects. Possible correlation between this enzyme and other chemically induced
cancers in man is being investigated.

INTRODUCTION

For some years our laboratory has worked with a number
of isozyme systems using the well-known zymogram technique,
in which the multiple forms of an enzyme activity can be seen

by simple visual inspection. This report deals with a group
of isozymes whose demonstration and study depend upon a very
different kind of technique, namely differential induction
of the multiple forms by various inducing agents. This
method obviously has many disadvantages as compared to the
zymogram method, including technical difficulties and
especially the much greater amount of time and effort involved
in study of a single specimen. Moreover, minor differences
between individuals cannot be detected by the present somewhat inaccurate methods.

Efforts have been made, both in this laboratory and
others, to devise staining methods for the demonstration of
this group of enzymes in electrophoretic gels. These efforts
have failed, probably due mainly to the fact that the enzymes
are complex molecules, membrane bound and difficult to
solubilize in aqueous media; moreover, when separated into
their individual components, they lose activity and reconstitution is difficult. Some success has been reported in
electrophoretically separating one of the major components
of the molecule, namely the P cytochromes, but this requires
a considerable amount of purification accompanied by considerable loss of activity.

Our main interest in studying the mixed function oxidases
resides in the fact that they are involved in metabolism of
many of the chemical carcinogens. A variety of reports
during the past decade or so have implicated these enzymes
in susceptibility and/or resistance to certain chemically
induced tumors in experimental animals, with much speculation
as to their involvement in human carcinogenesis.

The mixed function oxidases are of interest also for
their metabolism of a large number of other compounds. They
appear to have an amazingly broad substrate specificity, and
metabolize not only various polycyclic hydrocarbons, but
also a wide variety of drugs, many of which appear to be
unrelated in structure, some steroid hormones, various
insecticides and other lipophilic endogenous substrates (see
review by Conney 1967).

The mixed function oxidases occur in the microsomal
fraction of the cell and are apparently bound to the microsomal membrane. Most are inducible enzymes, and can be
induced by a wide variety of substances including a number
of substrates as well as other compounds not known to
function as substrates.

The active molecule is a complex of at least three
different subunits: a CO-binding hemoprotein (cytochromes
P-448 or 450), an NADPH dependent reductase, and a lipid
recently identified as phosphatidylcholine. The basis for

the multiple molecular forms of the enzyme appears to reside in the cytochrome fractions, and recent electrophoretic focusing studies on partially purified cytochromes have indicated a relatively large number of such molecules, perhaps as many as 15 (A. Lu, personal communication). There is also evidence from fractionation studies that the reductase portion of the complex may be provided by two or more different proteins.

The early evidence for multiple forms of activity was based on the fact that different inducing agents showed varying degrees of induction in different species of organisms. Likewise, there is strain or species difference in the induction of activity on different substrates. For example, Cram et al., (1965) studied six different strains of rabbits. Phenobarbital injection caused large increase in the metabolism of hexobarbital and aminopyrene in one strain of rabbit and very little in another strain. Measuring benzpyrene in the same animals after phenobarbital induction, there was increase in the benzpyrene hydroxylation in certain strains, very little or none in others, with no relationship between the increase in hexobarbital and benzpyrene metabolism among the various strains. For some time it was considered that there were two major groups of enzymes: those induced by phenobarbital and related compounds, the second group by polycyclic hydrocarbons such as 3-methylcholanthrene. However, more recent evidence including some reported here from our laboratory, indicates that there are probably additional isozymic forms based on differential inducibility (Conney et al., 1973).

The work reported here is concerned mainly with the group of microsomal mixed-function oxidases designated aryl hydrocarbon hydroxylases (AHH), as they are involved in metabolism of the polycyclic hydrocarbon carcinogens. It should be emphasized that this group possesses the broad substrate specificity mentioned above, and the AHH's are involved to greater or lesser extent in metabolism of all the other classes of substrates listed above. It should be further noted that the physiological function of these enzymes is not at all well understood. The large variety of exogenous substances which they metabolize would not seem to account for the fact of their occurrence in evolution, as many of these agents were only recently introduced into the earth environment. Nor is it known to what extent they are involved in normal metabolism of the endogenous substances such as the steroid hormones. Clearly our ignorance regarding this important group of enzymes is vast.

ROLE IN CHEMICAL CARCINOGENESIS

The hydrocarbon carcinogens, in the forms in which they are ordinarily introduced into the body, are not actively carcinogenic. They undergo metabolic alteration into the active or so-called proximate carcinogen. While the active form of the molecule is still a matter of some controversy, a considerable body of evidence indicates that this is an epoxide (Huberman et al., 1972). There is no doubt that the K-region epoxides of the polycyclic hydrocarbons bind firmly to the cell nucleic acids (Grover and Sims, 1973). The epoxide is further metabolized by at least three different pathways, one by the enzyme epoxide hydrase which is itself a microsomal enzyme in vivo, another to a glutathione conjugate by the enzyme glutathione epoxide transferase, while the main pathway of epoxide metabolism appears to be spontaneous reduction to the phenol which is further conjugated and excreted (Grover, 1974).

Thus, AHH appears to metabolize the polycyclic hydrocarbons through the proximate carcinogenic form and on to the inactive form. The question arises as to whether the enzyme is good or bad for the organism in the sense that it may either produce the active molecule or it may remove the precarcinogen before it reaches the target cell (Gelboin and Wiebel, 1971; Wattenberg and Leong, 1970). Both views are held by various workers, and it would seem that both views would have validity depending upon the location of the enzyme system as related to the target cell. If the enzyme producing the active molecule is located in the target cell, then that cell would appear to be susceptible to the action of the carcinogen, since the molecule would be produced within the cell. If, on the other hand, the target cell is located beyond the AHH containing cell, then the carcinogen would be metabolized through the active form on to the inactive form before reaching the target cell. By far the largest amount of AHH is present in liver, and this organ may act as a detoxifying organ to remove the great majority of ingested or otherwise incorporated hydrocarbon carcinogens. On the other hand, the enzyme is present in measurable amount in nearly all tissues, and thus could theoretically provide the means for producing the active carcinogen in most tissues of the body. A prime target for such action in man would seem to be the lungs of cigarette smokers which are exposed to rather large doses of a well-known hydrocarbon carcinogen, benzpyrene. Moreover, the bronchiolar epithelium has been demonstrated to possess AHH activity.

Considering only the carcinogen metabolizing function
of this enzyme, a simplistic answer to the above issue would
seem to be that it would be better for the enzyme not to
exist. Thus, the precarcinogen would not be metabolized,
would be presumably excreted in its precarcinogenic form, and
the organism would thus be safe from cancer induction. But
the larger view must be taken, namely that the enzyme has a
very broad substrate specificity, as noted above, metabolizing
several classes of compounds including some endogenous
compound such as certain steroids as well as a number of
exogenous chemicals which are directly harmful. Thus, the
answer to the broad question of the relative selective advantage or disadvantage of the enzyme system cannot be answered
without the acquisition of a great deal more information.

Animal studies relating AHH activities to susceptibility
to chemically induced cancers have produced conflicting
results. For example, Kouri et al., (1973), using different
strains of mice with varying amounts of AHH activity, have
shown that the susceptibility to skin tumors induced by 3-
methylcholanthrene is directly related to the AHH inducibility
in the organism. Conversely, Nebert et al., (1972) found no
relation between AHH inducibility and susceptibility to
dimethylbenzanthracene induced skin tumors in mouse strains.

We have investigated this question in man, results of
which are here briefly reported.

HUMAN STUDIES

AHH activity has been demonstrated in a variety of human
tissues, mainly in casual biopsy material obtained at surgery.
As expected, by far the highest activity is in liver, with
much smaller amounts in most other tissues examined including
intestine, kidney, lung, skin, and placenta. Activity deter-
iorates rapidly postmortem, so that autopsy material has been
of no use in study of the enzyme in man. The first efforts
to look for individual variability were rather recent, and
included studies of fresh placental tissue (Welch et al., 1968;
Nebert et al., 1969) and fibroblasts cultured from newborn
foreskins (Levin et al., 1972). The sample numbers were
small, but did suggest individual variability. Indirect
evidence of variation, based on studies of drug metabolism in
man, was reported earlier (Vesell and Page, 1969).

The need for a readily accessible human material for
assay of the enzyme led to the development in our laboratory
of a method based on the use of short -term lymphocyte
culture (Busbee et al., 1972). Fresh blood cells, including
erythrocytes and all forms of leukocytes, contain no measur-

813

able activity. However, following three-day culture of the separated leukocytes, in which the majority of cells growing out in culture are lymphoblasts, there is significant and measurable AHH activity. Inducibility of the enzyme can also be measured in this system, by adding an inducing agent, such as 3-methylocholanthrene, at the end of the third day, then culturing for additional 24 hours to determine increase in activity over the noninduced sample. The amount of enzyme induction in cultured human lymphocytes ranges from about 1.5 to 5 times base levels, a range which approximates that found in mouse liver. Total activity in the human lymphoblasts is in the range of 0.1 - 0.4 pmoles 3-hydroxybenzpyrene per million cells.

The assay itself is based on a modification of the method of Nebert and Gelboin (1968). It utilizes the fact that the substrate benzpyrene and its 3-hydroxy derivative have markedly different fluorescence emmission spectra when activated by the same wave length. The results are best read on a very sensitive fluorometer - our laboratory uses the Aminco-Bowman spectrophotofluorometer with an ellipsoidal condensing system for increased sensitivity. Optimal excitation and emmission frequencies are 396 and 522 nanometers respectively.

The lymphoblast culture technique is at this stage difficult in the most experienced of hands, and reproducibility at times leaves much to be desired. Basis for some of the variation is difficult to assess. Like all tissue culture methods, rates of cell growth vary somewhat from sample to sample. Another difficulty lies in the counting of the lymphoblasts, as this requires some decision making based on cell size and appearance.

Despite the technical difficulties, the method is now in use in several laboratories and has provided sufficient reproducible data for several significant conclusions. First, there is considerable individual variability in AHH activity and inducibility in a normal human population (Kellermann et al., 1973a). Preliminary studies of a normal white population indicated that the variation in inducibility was hereditarily determined, and that the genetic pattern was that of two alleles and a single locus (Kellermann et al., 1973b). The population appeared to separate into three clusters, representing presumably homozygous low and high groups with the intermediate heterozygotes. This was at variance with the genetics described in different mouse strains, which presents a much more complex picture with controlled inducibility under probably more than two loci with multiple alleles (Thomas et al., 1972; Thomas and Hutton, 1973; Nebert and Gielen, 1972). More recent studies in our laboratory have failed to confirm the original finding of three clusters. Rather, the population shows a continuous distribution, with the median of inducibility at about 2.2 times base levels. Whether this finding is due to technical problems with diminished resolution of the method, or whether it represents the true picture, remains to be determined. Our present thinking is that the inheritance in man is polygenic.

Following the demonstration of AHH variation in the normal population, studies were undertaken on human cancer subjects. Bronchogenic carcinoma was selected as the initial cancer for investigation, since this is known to be induced by polycyclic hydrocarbons, mainly inhaled in cigarette smoke; moreover it is the most common cancer in the human male in this country, so that a significant sample is readily available. The study consisted of simply determining AHH activity and inducibility in the lung cancer subjects by the lymphocyte culture method, at the same time running two control groups, one group of normal subjects, another of an unselected group of patients with a variety of other tumors.

Preliminary findings on a group of 50 lung cancer subjects demonstrated a marked difference from both the normal control group and the other cancer control group with a higher mean inducibility in the lung cancers (Kellermann et al., 1973c). Whether the AHH differences in the lung cancer subjects were secondary to the occurrence of lung cancer, or whether they preceded the lung cancer, has not been determined. Studies are underway to evaluate this question, mainly by AHH measurements in families of the lung cancer subjects to see if they fit the genetically expected categories.

MULTI-ENZYME SYSTEMS IN CARCINOGEN METABOLISM

The mechanism of chemical carcinogenesis is not known. However, it is probably a form of mutagenesis, with binding of the carcinogen to the cell DNA followed by some alteration in the base sequence. Some evidence indicates that the phenomenon is a frameshift mutation. Whether a specific locus or loci are involved is not known. Ames and co-workers (1973 a and b) have developed a system for detecting the production of active mutagens from metabolism of carcinogens by a bacterial test system. The technique utilizes certain mutant strains of *Salmonella typhimurium* with frameshift mutations, back mutations of which are detectable in the system. The carcinogen to be tested is added to the medium together with the enzyme source, usually crude liver extract, and after a period of incubation, back mutations are counted.

We have employed a modification of the Ames technique to investigate the relationship between AHH activity and metabolism of the carcinogen acetylaminofluorine (AAF). We have employed six different strains of mice which are known to have varying activities of AHH. Liver extract from the mice was added to the bacterial culture together with AAF. Number of back mutations produced by the AAF metabolite correlated with AHH activity ($r = .88$). The experiment was repeated, after first injecting the animals with an inducer of AHH, in this case 3-methylcholanthrene. As expected, AHH activities increased in all the animals. AAF mutagenicity also increased, but not to a corresponding degree. In other words, the slope of the line in the second experiment was less than that in the first experiment. The interpretation is that other microsomal enzymes in addition to the AHH were induced, certain of which metabolize the AAF along pathways leading to less mutagenic or non-mutagenic metabolites. Use of phenobarbital as an inducer showed that there was no increase of AAF metabolism after induction, while AHH activity did increase (D. Stout, unpublished).

This experiment illustrates the complexity of interaction of metabolism of carcinogens, with multiple pathways leading to various products. The amount of active carcinogen present in the target cell depends on a variety of factors including the amounts of various enzymes.

CONCLUSION

Chemical carcinogenesis in man is a complex and as yet little understood phenomenon. It is clear, however, that most carcinogens must undergo metabolic alteration to their active forms. The concentration of the active, or proximate carcinogen within the target cell is presumably a major determinant of whether that cell undergoes maligant change. This concentration is in turn dependent upon a variety of factors including the amounts of a number of carcinogen metabolizing enzymes. It is further clear that at least some of these enzymes vary within the normal population and that this variation is at least partly under genetic control. Further study of the multiple forms of carcinogen metabolizing enzymes should lead to improved understanding of the mechanisms of carcinogenesis in man and the basis for the susceptibility and/or resistance to various cancers. Such understanding should in turn lead to the development of methods for cancer prevention.

ACKNOWLEDGEMENTS

I am grateful to my colleagues, James Baptist, Thomas Matney, Marilyn Rasco, Daniel Stout, and Toshio Yamauchi, for the use of various unpublished findings.

The work reported from this laboratory was supported in part by USPHS grants GM 15597, CA 15969, and CA 05831.

REFERENCES

Ames, B.N., F.D. Lee, and W.E. Durston 1973a. An improved bacterial test system for the detection and classification of mutagens and carcinogens. *Proc. Nat. Acad. Sci. U.S.A.* 70: 782-786.

Ames, B.N., W.E. Durston, E. Yamasaki, and F.D. Lee 1973b. Carcinogens are mutagens: a simple test system combining liver homogenates for activation and bacteria for detection. *Proc. Nat. Acad. Sci. U.S.A.* 70: 2281-2285.

Busbee, D.L., C.R. Shaw, E.T. Cantrell 1972. Aryl hydrocarbon hydroxylase induction in human leukocytes. *Science* 178: 315-316.

Conney, A.H. 1967. Pharmacological implications of microsomal enzyme induction. *Pharma. Rev.* 19: 317-366.

Conney, A.H., A.Y.H. Lu, W. Levin, A. Somogyi, S. West, M. Jacobson, D. Ryan, and R. Kuntzman 1973. Effect of enzyme inducers on substrate specificity of the cytochrome P-450's. *Drug Metabolism and Disposition* 1: 199-210.

Cram, R.L., M.R. Juchau, and J.R. Fouts 1965. Differences in hepatic drugs metabolism in various rabbit strains before and after pretreatment with phenobarbital. *Proc. Soc. Exp. Biol. Med.* 118: 872-875.

Gelboin, H.V. and F.J. Wiebel 1971. Studies on the mechanism of aryl hydrocarbon hydroxylase induction and its role in cytotoxicity and tumorigenicity. *Ann. N.Y. Acad. Sciences* 179: 529-547.

Grover, P.L. and P. Sims 1973. K-region epoxides of polycyclic hydrocarbons: reaction with nucleic acids and polyribonucleotides. *Biochem. Pharmacology* 22: 661.

Grover, Philip L. 1974. K-region epoxides of polycyclic hydrocarbon: formation and further metabolism by rat-lung preparation. *Chemical Pharmacology* 23: 333-343.

Huberman, E., T. Kuroki, H. Marquardt, J.K. Selkirk, C. Heidelberger, P.L. Grover, and P. Sims 1972. Transformation of hamster embryo cells by epoxides and other derivatives of polycyclic hydrocarbons. *Cancer Res.*

32: 1391-1396.

Kellermann, G., E. Cantrell, and C.R. Shaw 1973a. Variations in extent of aryl hydrocarbon hydroxylase induction in cultured human lymphocytes. *Cancer Res.* 33: 1654-1656.

Kellermann, G., M. Luyten-Kellermann, and C.R. Shaw 1973b. Genetic variation of aryl hydrocarbon hydroxylase in human lymphocytes. *Am. J. of Human Genetics* 25: 327-331.

Kellermann, G., C.R. Shaw, M. Luyten-Kellermann 1973c. Aryl hydrocarbon hydroxylase inducibility and bronchogenic carcinoma. *New England Journal of Medicine.* 289: 934-937.

Kouri, R.E., R.A. Salerno, and C.E. Whitmire 1973. Relationships between aryl hydrocarbon hydroxylase inducibility and sensitivity to chemically induced subcutaneous sarcomas in various strains of mice. *J. Nat. Can. Inst.* 50: 363-368.

Levin, W., A.H. Conney, A. Alvares, I. Merkatz, and A. Kappas 1972. Induction of benzo(α)pyrene hydroxylase in human skin. *Science* 176: 419-420.

Nebert, D.W. and H.V. Gelboin 1968. Substrate-inducible microsomal aryl hydroxylase in mammalian cell culture. *J. Biol. Chem.* 243: 6250-6261.

Nebert, D.W., J. Winker, and H.V. Gelboin 1969. Aryl hydrocarbon hydroxylase, epoxide hydrase, and 7, 12-dimethylbenz(α)anthracene-produced skin tumorigenesis in the mouse. *Molecular Pharma.* 8: 374-379.

Nebert, D., W.F. Benedict, and J.E. Gielen 1972. Aryl hydrocarbon hydroxylase, epoxide hydrase, and 7,12-dimethylbenz α anthracene-produced skin tumorigenesis in the mouse. *Molecular Pharma.* 8:374-379.

Nebert, D.W. and J.E Gielen 1972. Genetic regulation of aryl hydrocarbon hydroxylase induction in the mouse. *Fed. Proc.* 31: 1315-1325.

Thomas, P.E., R.E. Kouri, and J.J. Hutton 1972. The genetics of aryl hydrocarbon hydroxylase induction in mice: a single gene difference between C57BL/6J and DBA/2J. *Biochem. Genet.* 6: 157-168.

Thomas, P.E. and J.J. Hutton 1973. Genetics of aryl hydrocarbon hydroxylase induction in mice: additive inheritance in crosses between C3H/HeJ and DBA/2J. *Biochem. Genet.* 8: 249-257.

Vesell, E.S. and J.G. Page 1969. Genetic control of the phenobarbitol-induced shortening of plasma antipyrine half-lives in man. *The Journal of Clinical Investigation* 48: No. 12, 2202-2209, December.

Wattenberg, L.W. and J.L. Leong 1970. Inhibition of the carcinogenic action of benzo(α)pyrene by flavones. *Cancer
Research* 30: 1922-1925.

Welch, R.M., Y.E. Harrison, A.H. Conney, P.J. Poppers, and M.
Finster 1968. Cigarette smoking: stimulatory effect
on metabolism of 3, 4-benzypyrene by enzymes in human
placenta. *Science* 160: 541-542.

REGULATION OF ADENYLATE KINASES FROM RAT LIVER, MUSCLE, AND TUMOR TISSUES

WAYNE E. CRISS and TAPAS K. PRADHAN
Departments of Obstetrics and Gynecology
and Biochemistry
University of Florida College of Medicine
Gainesville, Florida 32610

ABSTRACT. The major enzymatic forms of adenylate kinase (EC 2.7.4.3) have been purified and compared from rat liver, skeletal muscle, and hepatoma (Morris tumor #3924A) tissues.

Citrate activates each enzyme such that Ka (citrate) was 0.09 mM, 0.26 mM, and 15.8 mM for liver, muscle, and tumor adenylate kinase, respectively. Apparent Michaelis constants ranged from 1.9 mM (ATP) with the liver enzyme to 38 mM (ATP) with the tumor enzyme. Each enzyme illustrated unique kinetics.

Physical protein analysis of each of the three enzymes showed the tumor enzyme to be composed of two polypeptide subunits of 13,000 daltons each. Liver adenylate kinase had three subunits, two at 13,000 and one at 11,000 daltons. The skeletal muscle enzyme was composed of four subunits of 13,000 daltons each. It is possible that the smaller (11,000) subunit found only with the liver adenylate kinase may be regulatory.

Adenylate kinase (ATP:AMP Phosphotransferase, EC 2.7.4.3) is a central component in the adenylate system (ATP, ADP, AMP, and adenylate kinase) of the living cell. The adenylate system may even be a central signal in the regulation of the metabolic pathways of glycolysis, gluconeogenesis, lipogenesis, and oxidative phosphorylation (Atkinson, 1965, 1966, 1968, 1969; Krebs, 1964; Liao and Atkinson, 1971a, 1971b).

We have identified four electrophoretic forms of adenylate kinase in normal and neoplastic rat tissues (Criss, 1970, 1971, 1973a, 1973b, 1974; Criss et al, 1970a, 1970b, 1974; Filler and Criss, 1971; Pradhan and Criss, 1974; Pradhan et al, 1974). The major liver form of adenylate kinase is subcellularly located in the outer mitochondrial compartment (Criss, 1970), shows quantitative activity which correlates with tissue respiration (Criss, 1971), is very low in fetal liver (Filler and Criss, 1971), responds to hormonal and dietary manipulations (Criss et al, 1970a), and is decreased about 10-fold in neoplasia (Criss et al, 1970a). We have measured tissue levels of ATP, ADP, and AMP and determined that adenylate kinase activity is near equilibrium (Q = 1.2) in normal liver but is

far from equilibrium (Q = 3.8) in fast-growing hepatomas
(Criss, 1973a, 1973b). Since a potential loss of the adenylate
control of cellular energy homeostasis in the hepatomas was
implied from these observations, we have purified the major
form of adenylate kinase from normal adult liver, skeletal
muscle, and a fast-growing tumor (Morris hepatoma #3924A). The
current report compares the kinetic and physical properties of
these three enzymes. Methods and techniques have been previously reported (Criss et al, 1970b, 1974; Pradhan and Criss,
1974; Pradhan et al, 1974).

General Kinetic Features

The final specific activities of liver, muscle, and tumor
adenylate kinases that were used in these studies were 700,
>1,000, and 400 units per mg protein, respectively. Each
enzyme was a single homogeneous entity according to analysis
by polyacrylamide disc gel electrophoresis and sedimentation
equilibrium ultracentrifugation. The substrates for each
enzyme were determined to be MgATP (or MgdATP):AMP, or MgADP:
ADP. The assay system routinely consisted of 50 mM MgATP and
22.5 mM AMP (or 30 mM ADP + 15 mM $MgCl_2$) in a final volume of
1 ml. Upon determination of the very high Km values with the
tumor enzyme, we employed 60 mM each of MgATP and AMP when
examining the kinetics of the tumor adenylate kinase. Ten to
100 μℓ of purified enzyme was added to start the reaction. The
assay proceeded for 15 min at 37°. The reaction was stopped
with 1 ml of 1.5 N perchloric acid and immersion into an ice
bath. 100 μℓ was directly spotted in Whatman filter paper.
Products and substrates were separated by high voltage electrophoresis and quantitated by spectrophotometry (Pradhan et al,
1974; Criss et al, 1974).

Effect of Citric Acid Cycle Intermediates

Most intermediates of the citric acid cycle increased the
activity of liver, muscle, and tumor adenylate kinases. The
general response was biphasic (Criss, 1974). Activation ranged
from 410 percent (at 0.1 mM citrate) with the liver enzyme to
only 20 percent (at 0.1 mM citrate) with the tumor enzyme
(Table 1).

Activation Constants for Citrate

Double reciprocal plots of final velocity minus initial
velocity divided by initial velocity versus final citrate concentration were linear and produced activation constants for

TABLE 1
ACTIVATION BY CITRATE

Citrate	LAK	TAK	MAK
0.01 mM	+230	+20	+30
0.10 mM	+410	+45	+80
1.00 mM	+320	+55	+70
10.00 mM	+160	+60	+60

Assay was performed by preincubation of nucleotides and intermediates in buffer for 15 min at 37°. Reaction was started by adding enzyme. Stimulation is expressed as ± % of control (minus citrate) activity.

---

liver, muscle, and tumor adenylate kinases of 0.09 mM, 0.26 mM, and 15.8 mM, respectively (Table 5). The in vivo concentration of citrate in normal rat liver is 0.2 mM. Therefore, it is possible that liver adenylate kinase may respond to citrate control in vivo. If such a form of regulation does occur in liver tissue, it probably does not occur in the liver tumor, since the Ka (citrate) with tumor adenylate kinase is several fold larger than in vivo citrate measurements. (Albeit, in vivo citrate concentrations have not been reported for this tumor, we assume they are in a range similar to that of liver tissue.)

Apparent Michaelis Constants

The apparent Michaelis constants of liver, muscle, and tumor adenylate kinases were determined in the presence and absence of 10 mM citrate (Table 2). The Km values ranged from 1.9 (AMP) for liver enzyme to 38 (ATP) for tumor enzyme. Km values for the liver adenylate kinase were lowest, while the corresponding Km values for the tumor enzyme were highest. Muscle adenylate kinase had Km values of intermediate range. Citrate decreased the Km (ATP) of liver enzyme; it decreased the Km (AMP) of tumor enzyme; it decreased both Kms (ATP and AMP) of muscle enzyme. Km (ADP) was not altered by citrate for any of the purified adenylate kinases.

Slope Values of Hill Plots

The degree of cooperative ligand-ligand interaction was analyzed with Hill plots (Table 3). Cooperative interaction was indicated for the liver enzyme with ATP as the variable

823

TABLE 2

COMPARISON MICHAELIS CONSTANTS

	ATP	
	- CITRATE	+ CITRATE
LAK	7.0	4.2
TAK	33.0	38.0
MAK	10.3	4.0

	AMP	
	- CITRATE	+ CITRATE
LAK	1.9	2.0
TAK	18.0	10.0
MAK	6.2	3.3

	ADP	
	- CITRATE	+ CITRATE
LAK	16.8	20.0
TAK	25.0	25.1
MAK	20.2	16.7

Assay was performed by preincubation of nucleotides and intermediates in buffer for 15 min at 37°. Reaction was started by adding enzyme. Final citrate concentration was 10 mM. Km values are expressed as mM. (Pradhan et al, 1974).

substrate ($\eta \simeq 2$). All other Hill plot slope values (for ATP) were near 1.5. The slopes for each enzyme with AMP were near 1.0, while the slopes for each enzyme with ADP were near 2.0.

Effect of Mercurial and Sulfhydryl Reagents

Mercurial and sulfhydryl reagents inhibited the activity of each enzyme. Concentrations of ethylmercurithiosalicylate, which had no effect on liver and muscle adenylate kinases, inhibited the tumor enzyme. Concentrations of dithiothreitol, which had no effect upon the activity of liver and tumor adenylate kinases, inhibited the muscle enzyme. ρ-hydroxymercuriphenylsulfonate (HMPS) and ρ-chloromercuribenzoate (CMB) showed varying degrees of inhibition with each enzyme. The concentration of mercurial reagents required to inhibit one-half of the activity of each enzyme (Ki) was determined in the presence and absence of citrate (Table 4). Ki values for HMPS deter-

824

TABLE 3
COOPERATIVE INTERACTION

	SLOPE OF HILL PLOTS FOR ATP	
	– CITRATE	+ CITRATE
LAK	2.1	1.4
TAK	1.4	1.5
MAK	1.4	1.5
	SLOPE OF HILL PLOTS FOR AMP	
	– CITRATE	+ CITRATE
LAK	0.5	0.7
TAK	1.0	1.2
MAK	1.4	1.1
	SLOPE OF HILL PLOTS FOR ADP	
	– CITRATE	+ CITRATE
LAK	2.3	2.3
TAK	2.0	2.3
MAK	1.9	2.2

Assay was performed by preincubation of nucleotides and intermediates in buffer for 15 min at 37°C. Reaction was started by adding enzyme. Final citrate concentration was 10 mM. (Pradhan et al, 1974).

mined in the absence of citrate were similar for each enzyme. However, addition of citrate decreased the K_i value (HMPS) for liver adenylate kinase. Upon addition of citrate, the K_i values for CMB with liver and muscle enzymes were increased over 4-fold, while the K_i value for CMB with the tumor enzyme remained consistently high.

Kinetic Comparison of the Liver, Muscle,
and Tumor Adenylate Kinases

Pertinent kinetic parameters for liver, muscle, and tumor adenylate kinases are tabulated in Table 5. The liver adenylate kinase is subcellularly located in the outer compartment of mitochondria. It has a K_m (ATP) and Hill plot slope value (ATP) which are decreased by citrate. It also shows a K_i (HMPS) which is decreased and a K_i (CMB) which is increased upon addition of citrate. The tumor adenylate kinase has a K_m (AMP) which is decreased by citrate. The muscle adenylate

TABLE 4
INHIBITOR CONSTANT (Ki) VALUES

ENZYMES	p-HMPS		p-CMB	
	– CITRATE	+ CITRATE	– CITRATE	+ CITRATE
LAK	30	11	0.62	>2.5
TAK	25	30	>2.50	>2.5
MAK	35	40	0.52	>2.5

Assay was performed by preincubation of nucleotides and additives in buffer for 15 min at 37°. Reaction was started by adding enzyme. Final concentration of citrate was 10 mM. Ki values are concentrations of inhibitor required to inhibit enzymatic activity by one-half. Ki values are expressed as mM. (p-HMPS = p-hydroxymercuriphenylsulfonate; p-CMB = p-chloromercuribenzoate). (Pradhan et al. 1974).

kinase has a Km (ATP) and Km (AMP) which are decreased by citrate. It also shows a Ki (CMB) which increases upon addition of citrate. Therefore, each of the three major forms of mammalian adenylate kinase have their own unique kinetic characteristics.

Molecular Weights

The fringe pattern analysis of molecular weights during a sedimentation-equilibrium ultracentrifugation run is given in Table 6. The weight average-molecular weights of liver and muscle adenylate kinases range from 39,000 to 41,000 and 49,000 to 59,000, respectively. The weight average-molecular weights of tumor enzyme range from 13,000 to 24,000. Calculated weight-average molecular weights of liver, muscle, and tumor adenylate kinases are 40,000, 52,000, and 17,500. Number average-molecular weight, weight average-molecular weight, and Z-average-molecular weight, all extrapolated to zero protein concentration, ranged from 33,000 to 39,000, 37,000 to 49,000, and 10,000 to 24,000 for liver, muscle, and tumor adenylate kinases, respectively. Therefore, using the minimum molecular weight of 13,000 for a single component subunit, liver adenylate kinase would be a trimer, muscle adenylate kinase would exist as a tetramer, and tumor adenylate kinase would be in a rapid equilibrium monomer-dimer formation.

TABLE 5

KINETIC STUDIES ON ADENYLATE KINASES

	LAK	TAK	MAK
SUBCELLULAR LOCATION	MITOCHONDRIA	CYTOPLASM	CYTOPLASM
Km (ATP) (+ CITRATE)	7.0 ↓	33.0↔	10.3 ↓
Km (AMP) (+ CITRATE)	1.9↔	18.0 ↓	6.2 ↓
Km (ADP) (+ CITRATE)	16.8↔	25.0↔	20.2↔
Ka (CITRATE)	0.09	15.8	0.26
HILL SLOPE (ATP) (+ CITRATE)	2.1 ↓	1.4↔	1.4↔
HILL SLOPE (AMP) (+ CITRATE)	0.5↔	1.0↔	1.4↔
HILL SLOPE (ADP) (+ CITRATE)	2.3↔	2.0↔	1.9↔
Ki (pHMPS) (+ CITRATE)	30 ↓	25 ↔	35 ↔
Ki (pCMB) (+ CITRATE)	0.62↑↑	>2.5	0.52↑↑
ETHYLMERCURITHIOSALICYLATE	NO EFFECT	INHIBITORY	NO EFFECT
DITHIOTHREITOL	NO EFFECT	NO EFFECT	INHIBITORY

This table summarizes data from several manuscripts (Criss, 1970, 1971, 1973a, 1973b; Criss et al, 1974; Pradhan et al, 1974). Arrows indicate change of representative value upon addition of citrate to assay. (↑ = increase; ↔ = no change; ↓ = decrease)

Polypeptide Mapping

Performic acid treated-tryptic digests of each of the three enzymes were analyzed by two dimensional paper chromatography-electrophoresis (Fig. 1). Muscle and tumor adenylate kinase patterns are very similar. However, several differences between the pattern of the liver enzyme and the patterns of the muscle and tumor enzymes are observed: 1) location of histidine containing peptide #1; 2) location of both peptides containing a reducing sulfur; 3) tyrosine containing peptide which is observed only in the liver pattern. Fourteen peptides are observed in the muscle and tumor patterns; 15 peptides are found in the pattern of the liver enzyme. A comparison of each pattern reveals that 11 out of 14 (or 15) peptides are common for all enzymes, while three or four peptides are variable.

Amino Acid Analysis

Analysis of the amino acid composition of liver, muscle, and tumor adenylate kinases is tabulated (Table 8). Using a

827

TABLE 6
FRINGE PATTERNS OF ADENYLATE KINASES

Net Fringe	WT AV MW		
	LAK	TAK	MAK
0.50000	0	0	0
1.00000	41,402	13,774	56,330
1.50000	39,992	12,917	49,692
2.00000	38,439	15,090	48,522
2.50000	39,001	15,786	49,018
3.00000	39,204	15,802	50,187
3.50000	39,411	15,577	52,010
4.00000	40,027	16,993	50,608
4.50000	39,831	17,928	51,675
5.00000	40,110	19,262	54,407
5.50000	41,369	19,917	55,145
6.00000	41,411	19,753	57,227
6.50000	42,076	21,370	59,301
7.00000	42,099	24,204	54,788
7.50000	43,157	23,264	55,110
8.00000	41,876	23,713	55,888
8.50000	41,222	22,039	59,448
9.00000	40,987	23,339	0
9.50000		0	0

Data calculated from a typical sedimentation-equilibrium ultracentrifuge run. (Criss et al, 1974).

TABLE 7
MOLECULAR WEIGHT OF ADENYLATE KINASES

	LAK	TAK	MAK
WT AV MW	40,072 ± 304	17,563 ± 533	52,300 ± 360
NU AV MW (C=0)	32,724 ± 841	9,721 ± 399	36,600 ± 803
WT AV MW (C=0)	39,304 ± 1,194	14,722 ± 880	48,700 ± 1,500
PT Z MW (C=0)	36,770 ± 4,887	24,274 ± 6,311	46,480 ± 8,670

Data calculated from sedimentation-equilibrium analysis of each enzyme. (Criss et al, 1974).

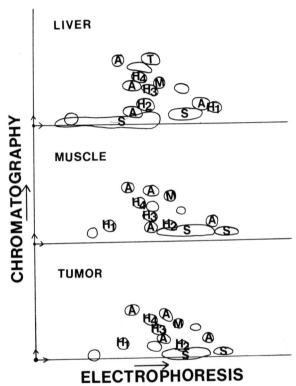

Fig. 1. Two-dimensional analysis of performic acid treated-tryptic digested enzymes. S = peptides containing reduced sulfur residues; M = peptides containing methionine residues; A = peptides containing arginine residues; T = peptides containing tyrosine residues.

minimum molecular-weight of 13,000, all calculated amino acid values for each enzyme are within a single integer, except alanine. Alanine residues are 11, 8, and 5 for liver, tumor, and muscle enzymes, respectively.

Subunit Analysis

Various purified preparations of the three adenylate kinases were digested in sodium-dodecyl-sulfate and electrophoresed on

TABLE 8
AMINO ACID ANALYSIS OF ADENYLATE KINASES

	LAK	TAK	MAK
	(Residues per 13,000)		
LYSINE	8	9	9
HISTIDINE	4	4	4
ARGININE	4	4	4
ASPARTIC ACID	8	9	9
THREONINE	5	5	5
SERINE	5	5	6
GLUTAMIC ACID	9	10	10
PROLINE	4	5	4
GLYCINE	8	8	8
ALANINE	11	8	5
HALF CYSTINE	1	-	-
VALINE	5	6	7
METHIONINE	1	1	1
ISOLEUCINE	4	4	5
LEUCINE	8	9	10
TYROSINE	2	1	1
PHENYLALANINE	3	3	3
CYSTEIC ACID	-	-	1

(Criss et al, 1974)

sodium-dodecyl-sulfate polyacrylamide disc gels (Fig. 2). A single protein is observed with both the muscle and tumor enzymes. This protein has a molecular weight of 13,000. Liver adenylate kinase showed two proteins. One protein has a molecular weight near 13,000; the other protein is near 11,000. The 11,000 molecular weight subunit is approximately 1/4 to 1/3 of the composition of the larger subunit in the structure of the liver enzyme.

Structural Models of Liver, Muscle, and Tumor Adenylate Kinases

Working structural models which are based exclusively upon molecular weights are illustrated in Fig. 3. Tumor adenylate kinase exists as a dimer with subunits of similar size. The liver enzyme is a trimer containing two subunits of similar size and a single smaller subunit. Muscle adenylate kinase is a quatramer of similar sized subunits. Since a major dif-

SDS Gels

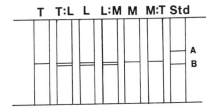

Fig. 2. Sodium dodecyl sulfate polyacrylamide disc gel analysis of adenylate kinase subunits. T = tumor AK; L = liver AK; M = muscle AK; A = growth hormone (50,000 daltons); B = insulin (11,500 daltons). The concentration of protein was 0.5 to 2 mg per gel.

TAK   LAK   MAK

Fig. 3. Molecular models, based exclusively upon molecular weight data, of the major native forms of adenylate kinases from tumor (Morris hepatoma #3924A), liver and skeletal muscle of the rat.

ference between the liver enzyme and the muscle and tumor enzymes is the presence of the smaller protein subunit, it is likely that this subunit is regulatory because: 1) liver enzyme has lower Km values for all substrates; 2) liver enzyme is activated by citrate to a much greater degree and at lower citrate concentrations (Ka = 0.09 mM); 3) liver enzyme is mitochondrial while muscle and tumor enzymes are cytoplasmic in subcellular location.

DISCUSSION

Intermediates of the citric acid cycle have been observed to activate several key pathway enzymes such as glycogen synthetase (Piras et al, 1970; Sevall and Kin, 1971), hexokinase (Kosow and Rose, 1971), phosphofructokinase (Pogson and Randle, 1966; Underwood and Newsholme, 1967), acetyl CoA carboxylase (Vagelos et al, 1963), and adenylate kinase (Pradhan and Criss, 1974). The role of the intermediates on all of the above-mentioned enzymes has in common a potential involvement of a metal (e.g., Mg^{2+}, Mn^{2+}, etc.) or a potential chelation phenomenon. Each enzyme system utilizes both ATP and a divalent metal either as a substrate or modulator of enzymatic activity. Citrate and ATP are both potent chelators of divalent cations. Since all of these above-listed enzymes play key roles in regulating their respective metabolic pathways, one must consider the possibility that in vivo concentrations of divalent cations, ATP, and citrate might effectively coordinate pathway flux.

Such a phenomenon might be called "chelation control." It is thus possible that chelation control might have evolved to allow an acute mode of regulation of pathway flux within various tissue types. The degree or type of such regulation would be dependent upon the enzymatic or isozymic form of the key pathway enzyme. Those tissues containing certain isozymic forms susceptible to regulation by citrate-Mg-ATP would have an acute form of regulation that tissues lacking susceptible isozymes would not have. In regard to adenylate kinase, acute chelation control is highly *likely* with the liver enzyme and highly *unlikely* with the liver tumor enzyme (Criss, 1974; Pradhan and Criss, 1974).

Two proteins which have similar enzymatic function but differing electrophoretic mobilities are isozymes only if more than one gene is involved. Allozyme is a term used to indicate two such enzymatic proteins which were produced from one gene but are modified postranslationally. The three adenylate kinase activities occur as uniquely charged proteins. We do not yet know if this is simply due to subunit aggregation, and a smaller subunit which is observed with only the liver enzyme, or whether the predominating subunits of identical size are different in primary sequence. We are currently examining the physical protein characteristics of these three adenylate kinases.

It is likely that the smaller subunit, identified in the liver adenylate kinase only, is regulatory. If this proves to be true, it would help to partially explain the influential role of adenylate kinase activity in maintaining energy homeostasis in liver tissues (Q = 1.4). Absence of the smaller polypeptide subunit in the tumor adenylate kinase and loss of the liver form of adenylate kinase in tumors would help to partially explain the decreased role of adenylate kinase activity in maintaining energy homeostasis in tumor tissues (Q = 3.8). Also, absence of the smaller subunit in the tumor enzyme can be interpreted as the loss of functioning of a specific gene which then results, not in the loss of overall enzymatic activity of an enzymatic system, but in a loss of acute metabolite regulatability of that specific enzymatic system.

ACKNOWLEDGMENTS

We wish to sincerely thank Dr. Sidney Weinhouse for his help and guidance throughout these many studies, and Dr. Harold P. Morris, for his very generous supply of tumors. We acknowledge the technical assistance of Dr. Paul Chun, Mrs. Caroline Easley, Mrs. Dorothy Fitzgerald, Mr. Willis Parker, and Mr. C. Ray Smith.

This work was supported by grants from the National Institutes of Health, CA-11818; Research Career Development Award, CA-70187 (WEC); and the Florida Division of the American Cancer Society, F73UF-4.

LITERATURE CITED

Atkinson, D. E. 1965. Biological Control at the Molecular Level. *Science* 150: 851-875.

Atkinson, D. E. 1966. Regulation of Enzyme Activity. *Ann. Rev. Biochem.* 35: 85-118.

Atkinson, D. E. 1968. The Energy Charge of the Adenylate Pool as a Regulatory Parameter. Interaction with Feedback Modifiers. *Biochem.* 70: 4030-4034.

Atkinson, D. E. 1969. Regulation of Enzyme Function. *Ann. Rev. Microbiol.* 23: 47-68.

Criss, W. E. 1970. Rat Liver Adenosine Triphosphate: Adenosine Monophosphate Phosphotransferase Activity II. Subcellular Localization of Adenylate Kinase Isozymes. *J. Biol. Chem.* 245: 6352-6356.

Criss, W. E. 1971. Relationship of ATP:AMP Phosphotransferase Isozymes to Tissue Respiration. *Arch. Biochem. Biophys.* 144: 138-142.

Criss, W. E. 1973a. Control of the Adenylate Charge in Morris "Minimal Deviation" Hepatomas. *Cancer Res.* 33: 51-56.

Criss, W. E. 1973b. Control of the Adenylate Charge in Novikoff Ascites Cells. *Cancer Res.* 33: 56-64.

Criss, W. E. 1974. Metabolite and Hormonal Control of Energy Metabolism in Experimental Hepatomas, in *Hormones and Cancer* (McKerns, ed). Academic Press, New York, VI: 169-202.

Criss, W. E., G. Litwack, H. P. Morris, and S. Weinhouse, 1970a. Adenosine Triphosphate: Adenosine Monophosphate Phosphotransferase Isozymes in Rat Liver and Hepatomas. *Cancer Res.* 30: 370-375.

Criss, W. E., T. K. Pradhan, and H. P. Morris 1974. Physical Protein Regulation of Adenylate Kinases from Muscle, Liver, and Hepatoma, submitted to *Cancer Res.*

Criss, W. E., V. Sapico, and G. Litwack 1970b. Rat Liver Adenosine Triphosphate:Adenosine Monophosphate Phosphotransferase Activity I. Purification and Physical and Kinetic Characterization of Adenylate Kinase III. *J. Biol. Chem.* 245: 6346-6351.

Filler, R. and W. E. Criss 1971. Development of Adenylate Kinase Isozymes in Rat Liver. *Biochem. J.* 122: 553-555.

Kemp, R. G. and E. G. Krebs 1967. Binding of Metabolites by Phosphofructokinase. *Biochem.* 6: 423-434.

Kosow, D. P. and I. A. Rose 1971. Activators of Yeast Hexokinase. *J. Biol. Chem.* 246: 2618-2625.

Krebs, H. A. 1964. Gluconeogenesis - The Croonian Lecture. *Proc. Royal Soc. London (B)* 159: 545-564.

Liao, C. L. and D. E. Atkinson 1971a. Regulation at the Phosphoenolpyruvate Branchpoint in Azobacter vinelandii: Phosphoenolpyruvate Carboxylase. *J. Bact.* 106: 31-36.

Liao, C. L. and D. E. Atkinson 1971b. Regulation at Phosphoenolpyruvate Branchpoint in Azobacter vinelandii: Pyruvate Kinase. *J. Bact.* 106: 37-44.

Piras, M. M., E. Bindstein, and R. Piras 1970. Regulation of Glycogen Metabolism in the Adrenal Gland I. Kinetic and Regulatory Properties of Glycogen Synthetase. *Arch. Biochem. Biophys.* 139: 121-129.

Pogson, C. I. and P. J. Randle 1966. Control of Rat Heart Phosphofructokinase by Citrate and Other Regulators. *Biochem. J.* 100: 683-693.

Pradhan, T. K. and W. E. Criss 1974. Modulation of Mitochondrial Adenylate Kinase by Citric-Acid-Cycle Intermediates, *Eur. J. Biochem.* in press.

Pradhan, T. K., W. E. Criss, and H. P. Morris 1974. Kinetic Regulation of Adenylate Kinases from Muscle, Liver, and Hepatoma, submitted to *Cancer Res.*

Sevall, J. S. and K. H. Kin 1971. Regulation of Hepatic Glycogen Synthetase of Rana catesbeianna. Significance of Citrate Activation with Reference to Insulin Activation. *J. Biol. Chem.* 246: 7250-7255.

Underwood, A. H. and E. A. Newsholme 1967. Some Properties of Phosphofructokinase from Kidney Cortex and Their Relation to Glucose Metabolism. *Biochem. J.* 104: 296-299.

Vagelos, P. R., A. W. Alberts, and D. B. Martin 1963. Studies on the Metabolism of Activation of Acetyl Coenzyme A Carboxylase by Citrate. *J. Biol. Chem.* 238: 533-540.

DEHYDROGENASE ISOZYMES IN THE HAMSTER AND HUMAN RENAL ADENOCARCINOMA

JONATHAN J. LI, SARA ANTONIA LI,
LESTER A. KLEIN, and CLAUDE A. VILLEE
Department of Biological Chemistry, Laboratory of Human
Reproduction and Reproductive Biology, and Department
of Urology, Beth Israel Hospital, Harvard Medical School,
Boston, Massachusetts   02115

ABSTRACT. Dehydrogenase isozymes (G6PD, M.E., LDH, MDH) were examined in extracts from estrogen dependent hamster renal adenocarcinoma, uterus, and whole embryo as well as in human renal cell cancer. Four isozymes of G6PD were demonstrated in the hamster renal tumor extracts compared to two in the untreated and estrogen-treated kidney. G6PD profiles similar to the renal tumor were observed in supernates of estrogen-stimulated uterus and whole embryo. The activity of the fastest migrating G6PD isozyme (G6PD-1) in the uterus is estrogen responsive. Two isozymes of M.E. were found in the hamster renal tumor, uterus, and whole embryo. Moderate to large shifts to the muscle type LDH isozymes (LDH-4, LDH-5) were observed in these tissues but no changes in MDH were noted. These data suggest that the hamster renal tumor may represent a reversion to a more embryonic state as well as a conversion to an estrogen-sensitive tissue like the uterus. Comparisons between renal cell cancer and adjacent normal cortex indicate the presence of an additional G6PD isozyme in the human tumor, a decrease in activity of the slowest migrating isozyme of M.E. (M.E.-4), and a concomitant increase in the most rapidly migrating isozyme (M.E.-1). Biochemical heterogeneity was observed in extracts from different areas of the same renal tumor despite apparent histological uniformity. Nevertheless, the overall pattern in the various tumor extracts was generally consistent with the extracts of renal tumors from other patients. The activities of the A_4 type LDH isozymes were substantially elevated in most renal cancers examined while a general decrease in MDH activity was found. These data suggest the possibility of distinguishing between the human clear cell and granular cell carcinoma. Alterations in the regulatory enzymes of lipogenesis may account in part for the high lipid content in both the hamster and human renal tumors.

INTRODUCTION

Multiple molecular forms of enzymes have provided information concerning the oncogenic process as well as the metabolic state of tumor cells compared to the corresponding normal tissue from which they are derived. The present study was undertaken to examine the similarities and differences between the estrogen-induced and dependent renal adenocarcinoma in the Syrian hamster and human renal cell cancer. In this regard, we have examined the isozymic patterns of four dehydrogenases: glucose-6-phosphate dehydrogenase (G6PD) and malic enzyme (ME), both regulatory enzymes which provide reducing equivalents for *de novo* fatty acid synthesis (the former also contributes pentose sugars for RNA synthesis) and two non-regulatory enzymes important in glycosis, lactate dehydrogenase (LDH) and malate dehydrogenase (MDH). It is interesting that both the hamster and human renal tumor apparently arise from the proximal convoluted tubule and contain considerable lipid in their cells (Bennington, 1973; Bloom, et al., 1963). Moreover, tumor incidence is greater in males than in females in both hamsters and humans; both tumors are responsive to hormonal therapy to varying extents (Bennington, 1973; Bloom et al., 1963). Richards and Hilf(1972) reported that G6PD activity was elevated in the estrogen-treated uterus and in certain mammary gland tumors as well as in the mammary gland during pregnancy and lactation. These studies indicated that one G6PD isozyme was responsive to treatment with estrogen. Dodge (1973) observed the activity of the slowest migrating G6PD isozyme was greater in the DES-dependent renal tumor of the hamster than in the nontumorous adjacent kidney tissue. Changes in LDH profiles, that is, an increase in the percentage of LDH-A subunits have been shown for a number of tumors including the human renal adenocarcinoma (Goldman et al., 1964; Macalalag and Prout, 1964). The data presented here have extended these observations in a more detailed study of the changes in isozyme profiles in renal carcinoma of the hamster and human.

MATERIALS AND METHODS

Mature intact and castrated male and ovariectomized female Syrian hamsters (LVG:LAK, outbred strain, Lakeview Hamster Colony, Newfield, N.J.) weighing 90-120 g were used. Tap water and Ralston Purina laboratory chow were available *ad lib*. Castrated male and female hamsters were retained at least two weeks prior to use. Estrogen pellets were implanted subpannicularly as described previously (Li et al., 1969). An ad-

ditional pellet was implanted every third month. Tumor induction time was 6-9 months. Estradiol benzoate dissolved in
sesame oil (40 μg/100 g body wt) was administered subcutaneously in a volume 0.1 ml daily for four days. Animals were either
exsanguinated under ether anesthesia and perfused through the
inferior vena cava with isotonic saline until the organs to be
excised were blanched or the tissues were immediately minced
and then washed twice in cold 0.25 M sucrose, 10 mM Tris (HCl),
1 mM dithiothreitol, pH 7.4 and blotted on filter paper. The
hamster uterus, whole embryo, and human renal tissues were
treated routinely with the latter procedure. Each tissue was
minced in 2.0-2.5 volumes of buffered sucrose. The hamster
kidneys were homogenized in chilled teflon Potter-Elvehjem
homogenizers and the homogenates, frozen-thawed three times,
were then centrifuged in a Spinco L2 ultracentrifuge for 60
min at 105,000 x g. Cytosols from uteri, whole embryo, and
human kidneys were prepared similarly except that tissues were
homogenized in a conical ground-glass homogenizer. After centrifugation, cytosols were filtered through a millipore (Millex,
0.45 μm) to remove residual lipid and cell debris. The clear
supernates were stored at -86° C until use.

Human renal tissues were obtained directly from surgery
and placed in glass beakers on ice. Fat, connective tissue,
necrotic and hemorrhagic areas were dissected away and the
remaining solid tumor was cut into pieces of approximately
1-2 g. Normal renal cortical tissue adjacent to the tumor was
also removed. The specimens were processed immediately according to the method described previously for hamster tissues.
A sample of each tissue was obtained for microscopic examination and on this basis the samples were categorized as clear
cell carcinoma, granular cell carcinoma, adenoma, or normal
kidney.

Acrylamide disc gel electrophoresis (Canalco) was performed by a technique similar to that described by Ornstein
(1964) and Davis (1964). This method employed a 7% separating gel 75 mm long and 18 mm spacer gel. All samples were
adjusted to contain equivalent amounts of protein (40-100 μl).
Electrophoresis was conducted at 4°C, using Tris-glycine buffer,
pH 8.3, 2.5 ma per gel for approximately 4 hrs. Protein concentrations of the supernates were determined by the method
of Lowry, et al. (1951) using bovine serum albumin as a standard.

Enzyme activity was visualized by incubating with tetrazolium in the dark at 37°C for 30-60 min. Glucose-6-phosphate
dehydrogenase (D-glucose-6-phosphate:NADP oxidoreductase, EC
1.1.1. 49), lactate dehydrogenase (L-lactate:NAD oxidoreductase,
EC 1.1.1. 27), and malate dehydrogenase (L-malate:NAD oxido-

reductase, EC 1.1.1. 37) were demonstrated by the method described by Smith (1968). Galactose-6-phosphate was substituted
for glucose-6-phosphate for visualizing hexose-6-phosphate
dehydrogenase (H6PD) activity. Malic enzyme (L-malate:NADP
oxidoreductase (decarboxylating), EC 1.1.1.40) was demonstrated
by the method described previously (Li, 1972). After incubation, the gels were immersed in 7.5% acetic acid to terminate
the enzyme reaction. The gels were scanned with a white light
using a Canalco model F microdensitometer.

RESULTS

G6PD AND ME ISOZYMES IN NORMAL AND TUMOR
TISSUES OF THE HAMSTER

Four G6PD isozymes were found in the estrogen-dependent
hamster renal adenocarcinoma and whole embryo compared to only
two G6PD isozymes in the kidneys of untreated or estrogenized
animals prior to tumor formation (Fig. 1). G6PD-2 and G6PD-3
are able to utilize galactose-6-phosphate as substrate and may
be considered hexose-6-phosphate dehydrogenases. G6PD-3 appears to utilize either substrate equally well (Fig. 2A). The
estrogenized hamster uterus H6PD patterns were similar to the
renal tumor. Fig. 2A illustrates the increased activity of
G6PD-1 in the uterus of the ovariectomized hamster following
estrogen treatment and the similarity of the G6PD pattern of
the estrogen-treated uterus to the renal tumor and whole embryo.
In contrast to G6PD, ME activity is not altered following estrogen treatment in the uterus (Fig. 2B). The other regulatory
dehydrogenase for lipogenesis, ME, may reflect more accurately
the state of *de novo* fatty acid synthesis in tumors since it
does not have the additional role of supporting RNA synthesis.
Fig. 3 dipicts a section of a renal tumor nodule stained for
ME activity. The intense reaction in the renal tumor is attributable to ME-1 (Fig. 4A, B). A marked increase in G6PD
staining reaction has also been reported in renal tumors (Murthy
and Russfield, 1968). It is interesting that the ME patterns
in the renal tumor are similar to the uterus and whole embryo
(Fig. 2B, 4B).

LDH AND MDH ISOZYMES IN NORMAL AND TUMOR TISSUES OF THE HAMSTER

A moderate but consistent increase in activity of LDH-4,
LDH-5 was observed in the renal tumor compared to the kidney
of untreated intact or castrate animals (Fig. 5A). It appears
also that the estrogen-treated uterus of ovariectomized hamsters shows an elevated LDH-5 activity compared to the untreated

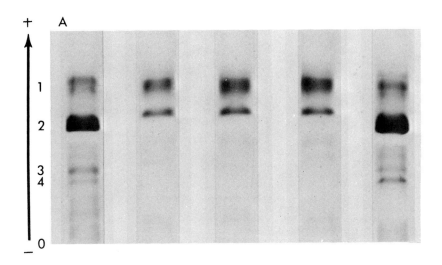

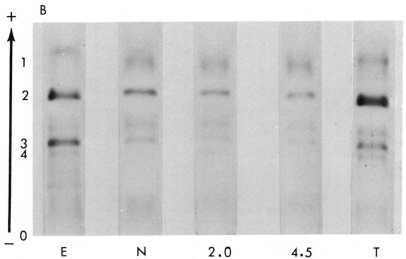

Fig. 1. Acrylamide gels of hamster tissues stained for A.
G6PD and B. H6PD. E, whole 14 day embryo; N, kidney of untreated control; 2.0, kidney of hamsters treated with DES for 2.0
months; 4.5, kidney of hamster treated with DES for 4.5 months;
T, primary renal tumor. Note the similarity of G6PD profiles
between the renal tumor and whole embryo. G6PD-2 and G6PD-3
are able to utilize galactose-6-phosphate as a substrate and
may be designated H6PD. Arrow indicates direction of migration
away from origin (O).

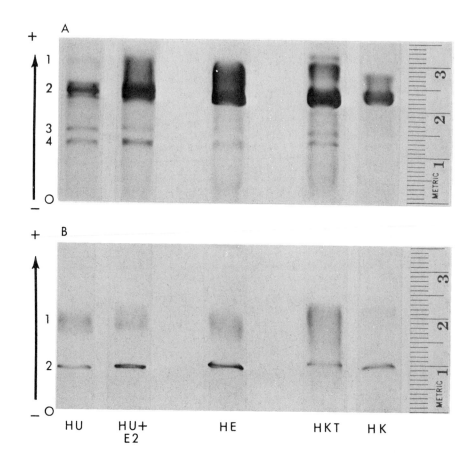

Fig. 2. Acrylamide gels of hamster tissues stained for A.
G6PD and B. ME. HU, uterus of 14 day ovariectomized animal;
HU + E$_2$, uterus of hamster ovariectomized 14 days and then
treated with estradiol benzoate (s.c.) for 4 days; HE, whole
embryo, 14 days; HKT, primary renal tumor; HK, kidney of untreated castrate hamster. Note the similarity of the G6PD and
ME patterns in the estrogenized uterus, renal tumor and whole
embryo. G6PD-1 is increased in the estrogenized hamster uterus.
control (Fig. 5B). Moreover, whole embryo extracts have very
low LDH-1 (B$_4$), LDH-2 (A$_1$B$_3$), LDH-3 (A$_2$B$_2$) activity and high
LDH-4 (A$_3$B$_1$), LDH-5 (A$_4$) activity. MDH activity appeared as a
single band migrating to the same region of the gel in all hamster tissues studied. In addition, MDH activity in the uterus

was not altered by estrogen treatment.

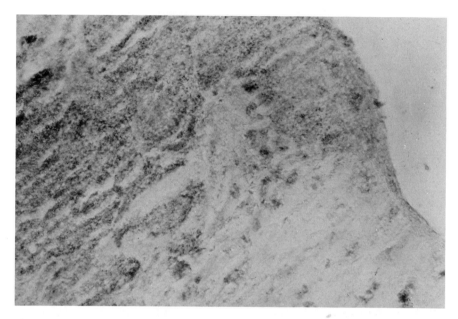

Fig. 3. Frozen section (12 µ) of hamster renal tumor nodule
stained specifically for malic enzyme activity. Section incubated without substrate served as control. Note the intense
staining reaction in the tumor nodule compared to only moderate reaction in adjacent nontumorous renal cortex. x 100.

G6PD AND ME ISOZYMES IN RENAL CELL CANCER AND ADJACENT
NON-TUMOROUS RENAL CORTEX OF THE HUMAN

Fig. 6 illustrates representative sections of renal clear
cell carcinoma, granular cell carcinoma, and borderline adenoma. Eight of the patients studied had clear cell carcinoma,
and one each had granular cell carcinoma and borderline adenoma. The clear cells are rich in lipids and glycogen but contain few cytoplasmic organelles whereas the granular cells contain little lipid or glycogen but are filled with numerous
mitochondria and other cytoplasmic organelles resulting in a
granular, eosinophilic cytoplasm. The borderline adenoma is
well differentiated and most resembles normal renal cortical
morphology.

Extracts of adjacent nontumorous renal cortex contain one
band of G6PD activity. An additional isozyme (G6PD-2) is present in the clear cells (Fig. 7A). No significant H6PD activity

was detectable in either isozyme. The granular cell carcinoma
and borderline adenoma showed essentially the same G6PD pattern as the clear cell carcinoma. The dehydrogenase isozyme
patterns in all cases studied in adjacent nontumorous renal
cortex is essentially similar to renal cortical extracts from a
patient whose kidney was removed becasue of a fistula. On this
basis, we concluded that the adjacent nontumorous renal cortex
represents normal renal cortical tissue and we shall now refer
to the former as normal renal cortex. Three isozymes of ME
were demonstrated in normal renal cortical extracts. The clear
cells show a marked reduction in ME-4 activity and a concomitant
increase in ME-1 activity (Fig. 7B). The tissue differences
could not be eliminated by dialyzing the renal extracts overnight against 0.05 M Tris (HCl) buffer, pH 7.4 according to
the procedure of Shaw and Koen (1968). Therefore, it appears
that ME-1 is not due to the migration of ME-4 as a result of
the latter's combination with a small molecule. Although the
borderline adenoma ME pattern was similar to the clear cell
carcinoma, the granular cell carcinoma did not show the change
in ME profile evident in the clear cells. Table I summarizes
the data on G6PD and ME isozyme patterns in the human renal
tumors studied. In all of the tumors examined G6PD-2 was present and six of these tumors demonstrated marked increases in
G6PD activity. Nine of ten tumors showed changes in ME profiles;
five of these had marked alterations. Fig. 8A and 8B depict
the considerable heterogeneity in the activities of individual
G6PD and ME isozymes of extracts made from different regions
of the same renal clear cell tumor despite apparent histological
homogeneity. It is interesting that the variability in G6PD
isozyme pattern is mainly due to the activity of G6PD-1. There
is considerable variability in the activity of both ME-1 and
ME-4.

LDH AND MDH ISOZYMES IN RENAL CELL CANCER AND ADJACENT NORMAL RENAL CORTEX OF THE HUMAN

The electrophoretic patterns of LDH isozymes in adjacent
normal renal cortex and corresponding clear cell tumor are
shown in Fig. 9A. Note the absence of LDH-1 in both normal
and tumor tissues. Renal tumor extracts showed a moderate increase in LDH-4 activity and a marked increase in LDH-5 activity
compared to normal cortex. There is a reduction in MDH activity
in many of the clear cell carcinomas which can be attributed
to the marked decrease in MDH-1 activity compared to the normal
renal cortex. Table I indicates that nine renal tumors showed
an increase in M-type subunits and that in five of these tumors

844

A

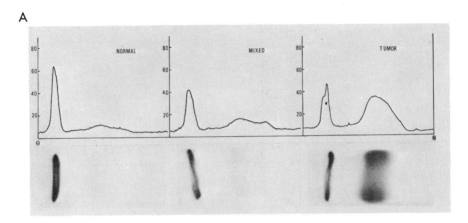

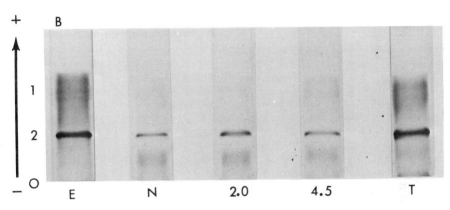

Fig. 4. Acrylamide gels of hamster tissues stained for ME.
A. Normal, untreated kidney; Mixed kidney of hamster treated
with DES for 8.0 months; Tumor, primary renal tumor. B. tube
designation as in Figure 1. Note the presence of ME-1 in the
renal tumor and whole embryo and its absence in untreated normal kidney.

the shifts were marked. Only six renal tumors showed a decrease
in MDH-1 activity and in four of these tumors the decreases
were marked. It is of interest that the granular cell carcinoma showed neither enhanced LDH-4, LDH-5 activity,nor decreased
MDH-1 activity demonstrated in most clear cell carcinomas whereas the borderline adenoma was similar to the clear cells in its
LDH and MDH profiles.

A

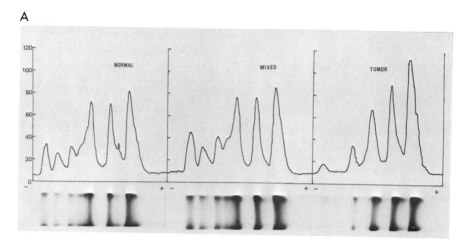

B

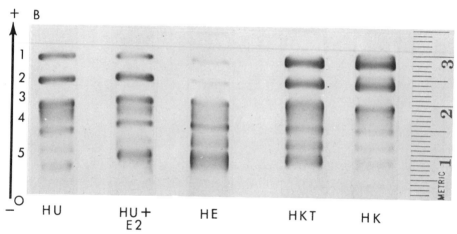

HU   HU+   HE   HKT   HK
       E2

Fig. 5. Acrylamide gels of hamster tissues stained for LDH.
A. tube designation as in Figure 4A. B. tube designation as
in Figure 2. Note the elevated activity of LDH-4, LDH-5 in
the estrogenized uterus, primary renal tumor, and whole embryo
compared to the kidney of the untreated castrate and the uterus
of the ovariectomized hamster.

DISCUSSION

In contrast to a previous report (Dodge, 1973), four isozymes of G6PD were demonstrated by disc gel electrophoresis in
primary estrogen dependent renal tumors of the hamster. Substantial amounts of G6PD-3 and G6PD-4 activities, not observed
in extracts of untreated and estrogenized kidneys, were present in estrogen-treated ovariectomized uterus and whole embryo.

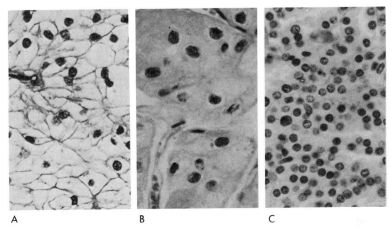

Fig. 6.   Sections of human renal tumors.   A. clear cell carcinoma.   B. granular cell carcinoma.   C.   borderline adenoma.

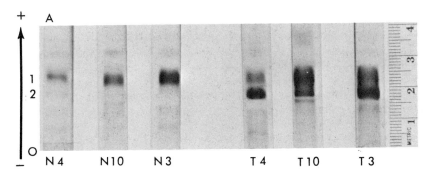

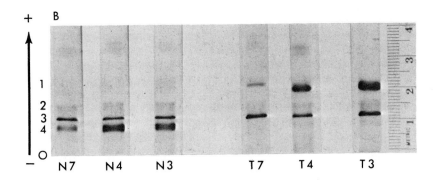

Fig. 7 (legend). Electrophoresis in acrylamide gels of human clear cell carcinoma (T) and adjacent nontumorous renal cortex (N) stained for A. G6PD and B. ME. Numbering refers to patients summarized in Table I. Note the presence of G6PD-2, ME-1 and absence of ME-4 in renal tumors.

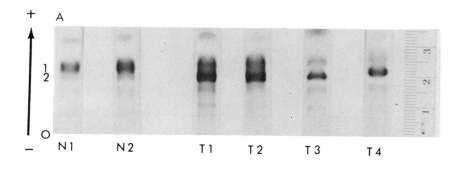

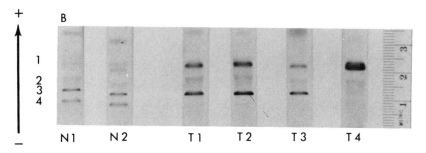

Fig. 8. Electrophoresis in acrylamide gels of human clear cell carcinoma (T) and adjacent nontumorous renal cortex (N) stained for A. G6PD and B. M.E. The tissues were taken from the kidney of a single patient. Note the heterogeneity of isozymic activity of different portions of the same tumor (T1-T4). In agreement with studies of G6PD isozymes of rat uterus and mammary gland (Richards and Hilf, 1972; Richards and Hilf, 1972a; Sartini et al., 1973), the activity of the most rapidly moving G6PD isozyme (G6PD-1) of the hamster uterus is considerably enhanced following estrogen treatment of ovariectomized females. Preliminary data indicate G6PD-1 activity in renal tumors is also enhanced by estrogen treatment in estrogen deprived tumor bearing animals. That estrogen-dependent renal tumors resemble uterine tissue in certain biochemical characteristics is not surprising since both tissues contain specific estrogen 5S

TABLE I

DEHYDROGENASE ISOZYMES IN HUMAN RENAL CELL CANCER

Patient	No.	Age	Sex	M.E.	G6PD	LDH	MDH	Histology
L.G.	1	60	M	+	++	+	++	Clear cell carcinoma nuclei typical
B.B.	2	42	M	++	+	++	++	Clear cell carcinoma nuclei moderately atypical
M.N.	3	56	F	++	++	++	-	Clear cell carcinoma nuclei markedly atypical
L.A.	4	59	F	++	++	+	-	Clear cell carcinoma, granular cell carcinoma minor component
S.S.	5	66	M	+	+	+	++	Clear cell carcinoma nuclei considerably atypical
R.H.	6	68	F	+	+	++	-	Clear cell carcinoma nuclei mainly atypical
A.G.	7	45	F	++	+	++	+	Clear cell carcinoma nuclei mainly atypical
A.G.	8	67	F	-	++	-	-	Granular cell carcinoma
H.M.	9	75	F	+	++	++	++	Clear cell carcinoma nuclei moderately atypical
A.C.	10	58	M	++	++	+	+	"Borderline" adenoma

nuclear receptors and 8S and 4S cytosol receptors (King et al.,
1970; Li et al., 1973). Although several of the renal tumor
G6PD isozymes appear to catalyze the dehydrogenation of galactose -6- phosphate to some extent only G6PD-3 is able to utilize both substrates equally well; hence H6PD-3. These studies
also demonstrate that ME-1 in the hamster uterus is not responsive to estrogen treatment, a finding consistent with similar
studies using extracts of treated and untreated rat uteri
(Eckstein and Villee, 1966). It is interesting that this ME
isozyme is also present in the renal tumor and whole embryo
but is absent in untreated and estrogenized kidneys. A moderate but consistent increase in LDH-4 and LDH-5 isozymes was

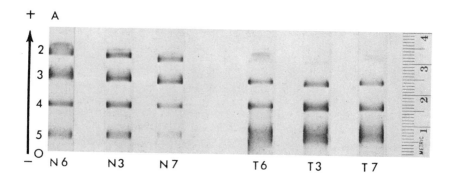

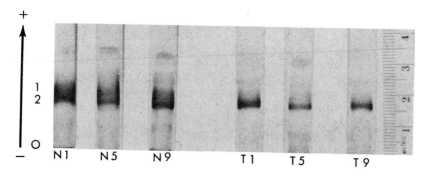

Fig. 9. Electrophoresis in acrylamide gels of human clear cell carcinoma (T) and adjacent nontumorous renal cortex (N) stained for A. LDH and B. MDH. Numbering refers to patients summarized in Table 1. Note the elevated activity of LDH-4, LDH-5 and decreased activity of MDH-1 in renal tumors.

observed in hamster tumor extracts which is at variance with results of similar investigations (Dodge, 1973). This shift to the A$_4$ type LDH isozymes also occurs in the uteri of ovariectomized hamsters after estrogen treatment and agrees with changes in estrogen-treated rat mammary gland and uteri (Farron et al., 1972; Richards and Hilf, 1972; Richards and Hilf, 1972a). The high activity of the slower migrating LDH isozymes is also evident in extracts of whole embryo and renal tumor. These data suggest that the renal tumor is a dedifferentiated tissue resembling fetal tissues and are in agreement with the results of receptor studies indicating conversion of the tumor to a uterine-like estrogen responsive tissue.

The presence of an additional G6PD isozyme in human renal cell carcinoma and concomitant alterations in ME profiles may explain in part the unusually high lipid content of these tumors.

Although the total ME activity was not significantly altered
in the clear cell carcinoma, unlike the hamster renal tumor,
the marked elevation of ME-1 activity in clear cells may perhaps affect lipogenesis by a shift of this isozyme, compared to
normal, to a cellular compartment which would lead to more
effective utilization of substrate or cofactor. It is interesting that the granular cell tumor which has considerably
less lipid in its cells, shows no increased ME-1 activity.
Additional studies are necessary to support this contention.
No correlation could be made between the degree of nuclear
pleomorphism and alterations in any of the isozyme profiles.

It is also apparent from these studies that extracts made
from various regions of the same tumor have different metabolic
activities despite histological homogeneity. Previous studies
of a few cases of human renal cancer have demonstrated the enhanced activity of LDH-4 and LDH-5 isozymes and concomitant
decreases in the fastest migrating LDH isozymes (Bloom et al.,
1963; Macalalag and Prout, 1964). The present study confirms
these findings. However, these reports do not distinguish
between clear cell and granular cell renal tumors as well as
renal adenomas. It is interesting that the borderline adenoma,
which resembles normal renal cortex histologically much more
than the carcinomas, has isozyme profiles similar to clear cell
tumors whereas the granular cell tumors resemble normal renal
cortex isozymic profiles (LDH, MDH, ME) more closely. Additional studies are needed to support these findings, but it seems
clear that biochemical characteristics of renal tumors can
serve as an important adjunct to histological characteristics
in efforts to elucidate the differentiation and metabilism of
these tumors.

ACKNOWLEDGEMENTS

We are especially grateful to Dr. Seymour Rosen, Department
of Pathology, Beth Israel Hospital, Harvard Medical School, for
the preparation of microscopic material and the evaluation and
grading of the human renal tumors. Estrogen (Sigma Chemical
Co.) pellets were prepared by Dr. George M. Krause, Copley
Pharmaceutical, Inc., Boston, Massachusetts. We gratefully
acknowledge the technical assistance of Miss Kathy Tomkinson in
these studies. We also appreciate the typing of the manuscript
by Ms. Kathleen Callinan.

This project was supported by Grant 5-TO1 HD 00006-14 from
the National Institutes of Health and the Charles A. King trust.

REFERENCES

Bennington, J.L.. 1973. Cancer of the kidney-etiology, epidemiology, and pathology. *Cancer* 32:1017-1029.

Bloom, H.J.G., C.E. Dukes, and B.C.V. Mitchley 1963. Hormone-dependent tumors of the kidney I. The estrogen-induced renal tumor of the Syrian hamster: Hormone treatment and possible relationship to carcinoma of the kidney in man. *Brit. J. Cancer* 17:611-664.

Criss, W.E. 1971. A review of isozymes in cancer. *Cancer Res.* 31:1523-1542.

Davis, B.J. 1964. Disc electrophoresis-II. Method and application to human serum proteins. *Ann. N.Y. Acad. Sci.* 121:404-427.

Dodge, A.H. 1973. Estrogen-dependent and independent renal tumors: G-6-PD and LDH isoenzyme analysis of estrogen-dependent and independent renal tumors of the Syrian hamster. *Oncology* 28:253-259.

Eckstein, B. and C.A. Villee 1966. Effect of estradiol on enzymes of carbohydrate metabolism in rat uterus. *Endocrinology* 78:409-411.

Farron, F., Howard H.T. Hsu, and W.E. Knox 1972. Fetal-type isoenzymes in hepatic and nonhepatic rat tumors. *Cancer Res.* 32:302-308.

Goldman, R.D., N.O. Kaplan, and T.C. Hall 1964. Lactic dehydrogenase in human neoplastic tissues. *Cancer Res.* 24:389-397.

King, R.J., J.A. Smith, and A.W. Steggles 1970. Estrogen binding and the hormone responsiveness of tumors. *Steroidologia* 1:73-88.

Kirkman, H. and R.L. Bacon 1952. Estrogen-induced tumors of the kidney. I. Incidence of renal tumors in intact and gonadectomized male golden hamsters treated with diethylstilbestrol. *J. Nat. Cancer Inst.* 13:745-755.

Kirkman, H. and M. Robbins 1959. Estrogen-induced tumors of the kidney. V. Histology and histogenesis in the Syrian hamster. In: Estrogen induced tumors of the kidney. *Natl. Cancer Inst. Monogr. No.*1:93-138.

Li, J.J., H. Kirkman, and R.L. Hunter 1969. Sex difference and gonadal hormone influence on Syrian hamster kidney esterase isozymes. *J. Histochem. Cytochem.* 17:386-393.

Li, Jonathan, J. 1972. NADP-malic enzyme. In-vitro interspecies hybridization of rat and hamster liver enzymes: Evidence for an isologous tetrameric structure. *Arch. Biochem. Biophys.* 150:812-814.

Li, J.J., D.J. Talley, S.A. Li, and C.A. Villee 1973. Estrogen binding in renal adenocarcinoma of the golden hamster: receptor specificity. *Federation Proc.* 32:453.

Lowry, O.H., N.J. Rosebrough, A.L. Farr, and R.J. Randall 1951. Protein measurements with the Folin phenol reagent. *J. Biol. Chem.* 193:265-276.

Macalalag, E.V. and G.R. Prout 1964. Confirmation of the source of elevated urinary lactic dehydrogenase in patients with renal tumor. *J. Urology* 92:416-423.

Murthy, A.S.K. and A.B. Russfield 1968. Dehydrogenases in diethylstilbesterol-induced kidney tumors of the Syrian hamster. *Experientia* 24:60-61.

Ornstein, L. 1964. Disc electrophoresis-I. Background and Theory. *Ann. N.Y. Acad. Sci.* 121:321-349.

Richards, A.H. and R. Hilf 1972. Influence of pregnancy, lactation and involution on glucose-6-phosphate dehydrogenase and lactate dehydrogenase isozymes in the rat mammary gland. *Endocrinology* 91:287-295.

Richards, A.H. and R. Hilf 1972a. Effect of estrogen administration on glucose-6-phosphate dehydrogenase and lactate dehydrogenase isoenzymes in rodent mammary tumors and normal mammary glands. *Cancer Res.* 32:611-616.

Sartini, J., D. Meadows, W.D. Rector, and R. Hilf 1973. Response of endometrial and myometrial glucose-6-phosphate dehydrogenase isoenzymes to estrogen. *Endocrinology* 93:990-993.

Shaw, C.R. and A.L. Kren 1968. Glucose-6-phosphate dehydrogenase and hexose-6-phosphate dehydrogenase of mammalian tissues. *Ann. N.Y. Acad. Sci.* 151:149-151.

Smith, I. 1968. Chromatographic and Electrophoretic Techniques. Zone electrophoresis. vol. II. *Interscience*, John Wiley and Sons. 343-351.

tRNA METHYLTRANSFERASES IN NORMAL AND NEOPLASTIC TISSUES

SYLVIA J. KERR

Department of Surgery
University of Colorado Medical Center
4200 East 9th Avenue
Denver, Colorado   80220

ABSTRACT: The tRNA methyltransferases are a complex family
of enzymes which modify the structure of preformed transfer
RNA by the insertion of methyl groups into specific positions
in the four major nucleotides comprising tRNA.   They are
ubiquitously distributed in procaryotes and eucaryotes.   The
enzymes are species, organ, base, and even site specific for
particular bases.

The activity of the enzymes can be modulated by a number
of naturally occurring compounds as well as by synthetic inhibitors.   The enzymes are also under hormonal regulation.
The mechanism of this regulation is complex, probably involving the induction or repression of the synthesis of the methyltransferases as well as affecting the levels of inhibitors.

In every neoplastic tissue examined alterations in the
methyltransferases have been detected.   These include both
qualitative and quantitative changes.

The enzymes have also been observed to undergo profound
alterations in a number of biological systems undergoing
changes in regulatory mechanisms.

Proper modification of the tRNA molecule has been implicated as a requirement for a number of the functions of tRNA.
The study of control mechanisms which regulate the activity
of the tRNA methyltransferases, then, is important for our
understanding of normal cellular controls.

INTRODUCTION

The tRNA methyltransferases are not isozymes in the
strict sense of multiple molecular forms of the same enzyme.
They do, however, comprise a family of enzymes which act on
the same general substrate, transfer RNA.   Transfer RNA consists of multiple amino-acid specific species and, thus, is a
heterogeneous substrate.   The tRNA methyltransferases in turn
are also a heterogeneous group within which exist base-specific,
sequence-specific and even conformation-specific enzymes.
That is not to say that isozymic forms of a specific tRNA
methyltransferase do not exist.   Due to difficulties in sepa-

855

ration and purification, experimental techniques are not yet refined enough to detect such isozymes.

The existence of enzymes which methylate tRNA was first demonstrated in extracts of *Escherichia coli* (Fleissner and Borek, 1962). Since then tRNA methyltransferases have been detected in every organism examined including yeast, insects, plants, and mammals (Borek and Srinivasan, 1966). The enzymes modify the structure of preformed tRNA by the addition of methyl groups at the macromolecular level. The methyl donor is the high energy compound S-adenosylmethionine.

The study of these enzymes which confer structural and conformational individuality on tRNA molecules is useful in understanding the complex interactions in which tRNA is involved at several levels of cellular control.

PROPERTIES

A. *MULTIPLICITY*

The tRNA methyltransferases are base, species, as well as organ specific (Srinivasan and Borek, 1964; Turkington and Riddle, 1970; Leboy and Piester, 1973).

In most methods of extraction of the methyltransferases, the homogenization of tissues in aqueous media is used. Under these conditions the enzymes are found in the high-speed supernatant fraction after centrifugation and thus were at first thought to be cytoplasmic components of the mammalian cell. However, Kahle et al. (1971) have reported that strictly non-aqueous extraction techniques indicate that the tRNA methyltransferases are located in the nuclei of rat liver cells.

Liau et al. (1972) have found a class of tRNA methyltransferases in the nucleoli of the Novikoff ascites cells. These enzymes formed a distinct subgroup of the total cellular methyltransferases since the pattern of methylated bases formed by the nucleolar enzymes differed from the pattern yielded by the soluble enzymes.

The first tRNA methyltransferases to be partially purified were those from *E. coli* (Fleissner and Borek, 1963; Hurwitz et al., 1964a). Hurwitz et al. (1964a, 1964b) separated six distinct methylating activities which yielded methylated derivatives of all four major bases.

Although only six enzyme activities were separated, this is undoubtedly only a lower limit of detection. The adenine-methylating fraction alone was capable of forming three different methylated adenine derivatives, which in view of the known specificity of these enzymes strongly suggests three separate enzymes. Taya and Nishimure (1973) have shown more recently that the cytosine-methyltransferase activity reported by Hurwitz et al. (1964a) is actually an enzyme whose product is 5-methylaminomethyl-2thiouridylate in *E. coli* glutamyl tRNA.

Evidence for the multiplicity of the tRNA methyltransferases also comes from fractionation of yeast enzymes. Bjork and Svensson (1969) separated 8 different methyltransferase activities from *Saccharomyces cerevisiae*, including three which all methylated uracil in the 5 position. Sequence analysis of the products of these enzymes revealed that they were distinctly different in sequence specificity (Svensson et al., 1969). Genetic studies on yeast have also provided evidence that the enzymes which monomethylate guanine at the N^2-position are distinct from those which dimethylate guanine in the same position (Phillips and Kjellin-Straby, 1967). This conclusion stems from finding mutants which lack the guanine dimethylating enzyme but still possess the monomethylating enzyme.

The instability of mammalian tRNA methyltransferases has been a hindrance in their isolation and purification. The first attempts to partially characterize mammalian tRNA methyltransferases were carried out by Rodeh et al. (1967) on rat liver and Simon et al. (1967) on rat brain. These workers found evidence for the existence of at least five different methyltransferases in rat liver and three to four different activities in rat brain.

Kuchino and Nishimura (1970) have separated three distinct guanine-specific methylating activities from rat liver using gradient elution from hydroxylapatite columns. They were able to distinguish the enzyme fractions by their elution profiles and on the basis of their reactivity toward individual purified tRNA species. Their results are shown in Table I where it can be seen that fractions I, II, and III have quite distinct specificities towards purified tRNA's. This was further confirmed by actual sequence analysis of the methylated bases in the purified tRNA's.

B. *SUBSTRATE SPECIFICITY*

All tRNA methyltransferases use S-adenosylmethionine as a methyl donor but their reactions with tRNA are governed by their species specificity. In the in vitro assay of the methyltransferases, tRNA from a heterologous source must be used as a methyl acceptor as the homologous tRNA has already been exposed in vivo to the indigenous methyltransferases and will no longer act as a substrate for them. The only exception is methyl deficient tRNA from *E. coli* which can be used as a substrate for the homologous methyltransferases.

TABLE I.

METHYL ACCEPTOR CAPACITIES OF INDIVIDUAL *E. COLI*
tRNA'S WITH LIVER METHYLASES IN VITRO[a]

	Extent of Methylation (cpm $^{14}CH_3$)		
tRNA	Methylase I	Methylase II	Methylase III
tRNA^{Asp}	210	290	150
tRNA$_2^{Glu}$	440	530	140
tRNA$_2^{Leu}$	1790	2340	170
tRNA^{fMet}	4580	4110	210
tRNA$_1^{Met}$	74	110	510
tRNA^{Phe}	33	240	770
tRNA$_1^{Ser}$	32	220	520
tRNA$_3^{Ser}$	2380	3270	110
tRNA^{Tyr}	220	480	210
tRNA^{Val}	64	350	710
E. Coli tRNA	1440	1280	460
Methyl deficient *E. coli* tRNA	1730	1600	540
Rat Liver tRNA	45	84	34
Yeast tRNA	230	240	70

[a]From Kuchino and Nishimura (1970).

The interaction with tRNA can be characterized in two
ways. The first is the standard enzymological method of determining rate (moles product/mg protein/time), which yields
the specific activity of a given enzyme preparation. The
second method takes advantage of the limited number of intro-
ducible methyl groups which are characteristic of each substrate and enzyme pair. Enzymes from a particular source
are able to introduce only a fixed number of methyl groups
into a unit amount of heterologous tRNA. This has been
called capacity or extent of methylation. It can be defined
as the absolute number of methyl groups per unit of tRNA
at infinite time and maximum protein concentration. This
number is a characteristic parameter of any given enzyme
source. It can be a sensitive probe for changes in enzyme
specificities.

The tRNA methyltransferases will not react with DNA, ribosomal RNA, viral RNA's, synthetic polynucleotides, or the
monomeric ribonucleotides.

The enzymes recognize both the sequence and conformation
of their substrate tRNA's. Baguley and Staehelin (1968) have
separated an adenine specific tRNA methyltransferase from
rat liver and spleen as well as from leukemic rat spleen.
The enzyme methylates adenine only in a particular sequence,
Ap1-MeApApUp. Tertiary structure of the tRNA plays an important role in the recognition of this sequence by the
enzyme. The 1-methyladenine methyltransferase derived from
leukemic rat spleen was able to methylate an adenine at the
19th position from the 3'-OH end of purified yeast $tRNA_1^{Ser}$
(Baguley et al.(1970). 1-Methyladenine already occurs in
this position normally in rat liver tRNA ^{Ser}. Furthermore,
other yeast tRNA species, $tRNA^{Tyr}$, $tRNA^{Phe}$ and $tRNA^{Val}$
also have a 1-methyladenine naturally occurring at that
position. Thus yeast must have a methyltransferase capable
of recognizing that sequence in the 19th position from the
3'-OH end in certain tRNA species but not in its own $tRNA_1^{Ser}$.
However, the rat spleen enzyme can react at that site with
the heterologous tRNA as well. Baguley et al. (1970) suggest that the yeast $tRNA_1^{Ser}$ exists in a closed conformation
unavailable to the yeast adenine methyltransferase for reaction and that the mammalian adenine methyltransferase can
induce a conformational change in the molecule or that the
high pH (9.25) required for in vitro methylation induces the
conformational change.

Kuchino and Nishimura (1970) have also demonstrated specificity of the enzymes both for sequence and conformation.
The sequence analysis of two purified tRNA's methylated by

enzyme fractions from rat liver is shown in Figure 1. The
sites of methylation are indicated by arrows. The methyltransferase in Fraction II methylated a guanine residue in
the 51st position from the 3'-OH of *E. coli* tRNA[fMet] but did
not act on tRNA[Val]. Methyltransferase III methylated a
guanine residue in tRNA[Val] but did not act on tRNA[fMet]. A
trinucleotide sequence alone was not sufficient for enzyme
recognition, as the same sequence occurred in other positions
in the same molecule and in other tRNA species but was not
methylated.

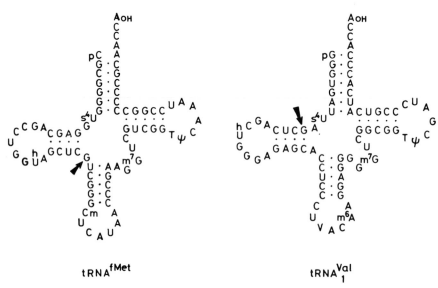

Fig. 1. Clover-leaf structure of *E. Coli* tRNA's, indicating
the site of methylation in tRNA[fMet] by methyltransferase II
and in tRNA[Val] by methyltransferase III. Enzyme II did not
accept valyl tRNA as a substrate while enzyme III would not
act on formyl-methionyl tRNA. V stands for uridin-5-oxy-
acetic acid, (Kuchino and Nishimura, 1970).

The importance of conformation was shown in experiments
with tRNA fragments (Kuchino et al. (1971). When the tRNA[fMet]
was split into two fragments comprising approximately three
quarters and one quarter of the molecule, the three quarter
fragment could be methylated by fraction II to a slight extent. However, the product was not the expected N^2-methyl-
guanine, but a 1-methyladenine at the 19th position from the

3'-OH end. When the fragments were recombined the methyltransferase fraction now methylated both adenine and guanine.

Thus it is clear that the conformation of the tRNA is essential for its recognition by tRNA methyltransferases and that alteration of the conformation can modify the specificity of recognition.

C. *pH OPTIMA AND IONIC STIMULATION*

The tRNA methyltransferases both from bacterial and mammalian sources exhibit a broad pH optimum from around pH 7.5 to pH 9.5. Individual tRNA methyltransferases have their own specific pH optimum within this range (Kerr and Borek, 1973).

The enzymes all require the presence of cations in some form for activity. Various workers have used Mg^{++}, polyamines or NH_4^+ to stimulate the reaction (Kerr and Borek, 1973).

The optimum requirements must be determined experimentally for each particular system or mixture of enzymes.

REGULATION

A. *INHIBITORS*

A number of inhibitors of the tRNA methyltransferases, both synthetic and natural, have been described.

Hurwitz et al. (1964b) showed that S-adenosylhomocysteine, which is the product derived from S-adenosylmethionine after methyl transfer, is a competitive inhibitor of the *E. coli* tRNA methyltransferases. They found adenosine to be a much weaker inhibitor.

Wainfan and Borek (1967) found that while adenosine did inhibit the *E. coli* guanine-specific methyltransferases, it was ineffective towards mammalian enzymes. Adenine was an inhibitor of the guanine-specific methyltransferases from both bacterial and mammalian sources. Wainfan also found that several analogs of these compounds such as 7-deaza-adenosine and a number of cytokinins-kinetin riboside, zeatin riboside, 6-benzyl-amino purine riboside and N^6 - (Δ^2-isopentenyl)adenosine - were active as inhibitors of the methyltransferases (Wainfan and Borek, 1967; Wainfan and Landsberg, 1971). These findings are particularly interesting as cytokinins are known to occur naturally as modifications of tRNA molecules (Armstrong et al. 1969).

Rodeh et al. (1967) showed that the tRNA methyltransferases from rat liver were inhibited by *M. lysodeikticus* DNA and by a number of synthetic ribopolynucleotides, including polyadenylic acid, polycytidylic acid, polyinosinic acid, polyuridylic acid, and polyguanylic acid. Double stranded copolymers such as poly UA did not inhibit the enzymes. Liau et al. (1973) have confirmed and extended the studies on inhibition of the tRNA methyltransferases by polyinosinate.

Moore and Smith (1969) investigated the effects of S-adenosylethionine and ethylthioadenosine on the methyl-transferases from *E. coli* and rat liver and observed competitive inhibition. Moore (1970) also showed that the inhibition by S-adenosylethionine was specific for the *E. coli* cytosine and adenosine specific tRNA methyltransferases.

Pegg (1971) has confirmed that the methyltransferases from rat liver and rat kidney are inhibited by S-adenosyleth-ionine as well as by S-adenosylhomocysteine. Further, he showed that these mammalian enzymes are also strongly inhibited by ethidium bromide, acridine orange, and proflavin.

Several investigators have reported elevated levels of the tRNA methyltransferases in organs of foetal and new-born animals compared to their adult counterparts (Hancock et al. 1967, Simon et al. 1967). Kerr (1970) has shown that this is in part due to the presence of inhibitors of the methyltransferases in extracts of adult tissues which are absent from foetal tissues and several tumor tissues, all of which exhibit elevated levels of methyltransferase activity. In the case of liver, kidney, and pancreas, the inhibitor has been identified as a competing enzyme system which methylates glycine to yield sarcosine (N-methyl-glycine) (Kerr, 1972). In tissue extracts of the above mentioned organs the specific activity of the glycine methyltransferase is orders of magnitude higher than the specific activity of the tRNA methyltransferases, so that in in vitro enzyme reactions using undialyzed extracts the glycine methyltransferase competes for the common substrate S-adenosylmethionine very effectively. Further examination of the system revealed another facet of this regulation which is a differential inhibition of the tRNA methyltransferases by S-adenosylhomocysteine. The tRNA methyltransferases are more sensitive by at least one order of magnitude to inhibition by S-adenosylhomocysteine than is the glycine methyl-transferase (Kerr, 1972).

In organs which lack the glycine N-methyltransferases there appear to be other competing enzyme systems which affect their tRNA methyltransferase activity.

Swiatek et al. (1971) have reported the appearance of an inhibitor of the tRNA methyltransferases soon after birth in pig brain. The system has not yet been characterized, but may well represent a competing enzyme.

In an organ which is under hormonal control - the uterus -

changes in the tRNA methyltransferases have been observed in response to estrogen (Sharma and Borek, 1970; Sharma et al., 1971). The enzyme level falls in the uterus of an ovariectomized animal but can be restored to normal levels by administration of physiological amounts of estradiol. These alterations in tRNA methyltransferase activity appear to be due to fluctuations in inhibitor levels in the uterus.

Thus, it would appear that just as there is an organ variation and specificity of the tRNA methyltransferases, there is an organ variation in the regulatory mechanisms controlling these enzymes.

B. *HORMONES*

Alterations in the levels and capacities of the tRNA methyltransferases have been observed in several hormonally regulated systems (Kerr and Borek; 1973).

Hacker (1969) found that in immature oviduct of chicks which had been maintained on a diet supplemented with diethylstilbesterol the tRNA methyltransferase levels rose several-fold.

Turkington (1969) showed that the tRNA methyltransferase levels were altered during the cell differentiation which occurs in mouse mammary gland during pregnancy. Similar alterations could be simulated by treatment of mammary gland explants in organ culture with the hormones insulin, hydrocortisone and prolactin.

As mentioned above, the tRNA methyltransferases in the uterus respond to physiological levels of estradiol. At these low levels of estrogen, no other organs except the target organ, uterus, exhibit any change in the enzymes.

If massive doses of estrogen are given to animals, changes in the tRNA methyltransferases are found in tissues other than the natural target organs (Sheid et al., 1970; Mays and Borek, 1971). Sheid et al. (1970) showed that high doses of mestranol caused a 60% enhancement of the tRNA methyltransferase capacity in rat liver. The uracil-specific methyltransferase appeared to be the most altered, showing a relative increase of 130%.

The synthesis of phosvitin, an egg yolk protein can be induced in the liver of the rooster by massive doses of estrogen (Greengard et al. 1964). The synthesis of this protein is normally restricted to hens. Mays and Borek (1971) have examined the tRNA methyltransferases in this system. They found that after administration of estradiol (10 mg/kg body weight) or diethystilbesterol (25 mg/kg body weight) the tRNA methyltransferase capacity of the rooster liver was decreased. Analysis of tRNA methylated in vitro by enzyme extracts showed that the different base specific enzymes were not uniformly affected. The enzyme activities producing N^2-methylguanine were higher in overall capacity after estrogen treatment, while those producing a number of other methylated bases were either constant or diminished after hormone treatment.

Hormonal influence on the tRNA methyltransferases has been shown in another system by Pillinger et al. (1971), who studied the enzymes in the giant bull frog, *Rana catesbeiana*, during thyroxine induced metamorphosis. Three days after the administration of physiological levels of thyroxine into the environment of the bull frog tadpole there was a diminution in methyltransferase capacity of extracts of both the liver and tail to 1/2 of that in the untreated animals. Four days after the administration, there was a beginning of return to normal in the capacity of the enzymes. Eight days after the administration of the hormone the enzyme capacity was almost that of the untreated animals.

Hormones, then, regulate the activity of the tRNA methyltransferases in a complex manner probably involving the induction or repression of their synthesis as well as modulating their activity through regulation of the levels of inhibitors of the methyltransferases.

TUMOR TISSUES

It has been found that in crude extracts of over 35 different neoplastic tissues the tRNA methyltransferases exhibit an abnormally high capacity, from a 2-fold to 10-fold increase, for methylation as compared to normal tissue counterparts (Borek and Kerr, 1972). This is in part due to a lack of the inhibitors of the methyltransferases found in normal adult tissue (Kerr, 1972) as well as to the appearance of enzymes with altered specificities (Mittelman et al., 1967; Sharma, 1973).

Mittelman et al. (1967) showed both qualitative and quantitative changes in the tRNA methyltransferases in an SV40 induced tumor of hamsters. They also demonstrated that the tumor enzymes could hypermethylate the host hamster homologous tRNA which normally does not serve as a substrate. This is further evidence of qualitative changes in the specificities of the enzymes from the tumor cells.

Sharma (1973) has compared the tRNA methyltransferases from normal rat liver and Novikoff hepatoma both in total extracts and in fractions separated according to the procedure of Kuchino and Nishimura (1970). Transfer RNA from a variety of sources was tested as a substrate as were individual tRNA species. The enzymes from Novikoff hepatoma had a greater specific activity and a higher capacity for methylation than the liver enzymes. They also showed qualitative differences when tested against purified tRNA species.

Kit et al. (1970) have studied the effect of the oncogenic virus SV40 on the tRNA methyltransferases during productive infection, abortive infection and transformation of cells. SV40 yields a productive infection with African green monkey kidney cells (CV-1). In this case neither the specific activity nor the capacity of tRNA methyltransferase appeared to be altered.

With primary mouse kidney cells, SV40 gives primarily an abortive infection and a small percentage of transformed cells. No changes in the methyltransferases were noted in the abortively infected cells. However, in mouse kidney cell lines transformed by SV40, 2-fold to 4-fold increases in both the specific activity and in the capacity of the enzymes were found.

Gallagher et al.(1971) have examined the tRNA methyltransferases after transformation of a rat embryo cell line by the oncogenic polyoma virus. They found a 3-fold to 6-fold increase in the capacity or extent of methylation in extracts from the transformed cell line, using $E. coli$ tRNA as a substrate. The differences were even more dramatic when yeast tRNA was used as a substrate.

Gantt et al. (1971) have found that a specific tRNA methyltransferase is located within the virion of the avian myeloblastosis virus. This would suggest that the enzyme is a requisite ancillary in the viral life cycle.

The regulatory factors or lack of them in tumor tissue which affect the tRNA methyltransferases are not completely clear, but some facts are emerging.

It has been established that growth rate,   is not the determinant in the abnormally high levels of enzyme activity in tumor tissues. Several lines of evidence bear on this. There is no correlation between the rate of growth of tumors and the level of tRNA methyltransferase activity found (Borek and Kerr, 1972).

Paired lines of cells, normal and neoplastic, in tissue culture with the same growth rate, still exhibit an elevated tRNA methyltransferase activity in the transformed cells (Gantt and Evans, 1969; Pillinger and Wilkinson, 1971).

A third line of evidence indicating that altered tRNA methyltransferase activity is not an obligatory concomitance of changes in growth rate comes from studies of regenerating liver by Rodeh et al. (1967). They found essentially no differences between normal adult and regenerating rat liver.

Another difference which has been discovered between normal and neoplastic tissues is the level of the competing glycine methyltransferase discussed earlier. This enzyme is present in high levels in normal adult liver, kidney and pancreas but is present at either very low levels or totally absent from foetal organs and several hepatomas (Kerr, 1972). Hepatomas have also been found to have low levels of S-adenosylmethionine synthesizing enzyme but normal tissue levels of S-adenosylmethionine itself(Lombardini and Tallalay, 1971). Thus, there is a difference between normal liver and hepatoma in the system governing the homeostasis between S-adenosylmethionine and S-adenosylhomocysteine and particularly in the availability of S-adenosylmethionine to the tRNA methyltransferases.

It is possible that the tRNA methyltransferases in transformed cells or in solid tumors represent a derepressed state or an expression of the total tRNA methyltransferase capacity of the species which is normally only expressed in embryonic tissue. Other similarities between embryonic and tumor tissues have been noted, such as the foetal antigens associated with some tumors (Gold, 1971) and, more pertinently, certain transfer RNA species found in tumor tissue, which have no counterparts in normal adult tissue but are present in extracts of embryonic tissue (Gonano et al., 1973).

BIOLOGICAL SIGNIFICANCE

Proper modification of the tRNA molecule has been implicated as a requirement for a number of the functions of tRNA. These include repression of transcription (Singer et al., 1972), reaction with the amino-acyl synthetases (Shugart et al., 1968), codon response (Capra and Peterkofsky, 1968), prevention of wobble (Yoshida et al., 1970), and ribosomal binding (Gefter and Russell, 1969).

The tRNA methyltransferases have been observed to undergo profound alterations, both qualitative and quantitative, in a number of biological systems undergoing changes in regulatory mechanisms (Table II.)

The changes in the tRNA modifying enzymes imply a change in the population of the tRNA molecules themselves, and this has been confirmed in many of the systems listed in Table II.

Specific tRNA species and tRNA, in general, have been implicated in regulation at a variety of levels in the cell. These include transcription (Singer et al. 1972), translation (Smith et al. 1966; Anderson and Gilbert, 1969; Wainwright, 1971) and expression of enzyme activity (Duda et al. 1968; Jacobson 1971).

TABLE II

MODULATIONS OF tRNA METHYLTRANSFERASES IN BIOLOGICAL SYSTEMS

System	References
Bacteriophage infection and induction	(Wainfan et al., 1965, 1966)
Insect metamorphosis	(Baliga et al., 1965)
Embryonic vs. neonatal tissue	(Hancock et al., 1967; Simon et al., 1967)
Colonizing slime mold	(Pillinger and Borek, 1969)
Differentiating lens tissue	(Kerr and Dische, 1970)
Mammary epithelial cell differentiation	(Turkington, 1969)
Ovariectomized uterus	(Sharma and Borek, 1970; Sharma et al., 1971)
Thyroxine-induced morphogenesis in the tadpole	(Pillinger et al., 1971)
Sea urchin embryogenesis	(Sharma et al., 1971)
Germination of spores	(Wong et al., 1971)
Hormone-induced phosvitin synthesis	(Mays and Borek, 1971)
Phytohemagglutinin induction of lymphocytes	(Riddick and Gallo, 1971)
Viral transformation	(Kit et al., 1970; Gallagher et al., 1971)
Vitamin A and D deficient bone	(Bradford et al., 1972)
Senescent tissues	(Wust and Rosen, 1972; Mays et al., 1973)
Neoplastic tissues	(Borek and Kerr, 1972)

Thus, the study of the control mechanisms which regulate the activity of the tRNA methyltransferases with their consequent influence on the structure and function of tRNA is important for our understanding of normal cellular controls, and in turn the derangement of these controls which results in neoplasia.

ACKNOWLEDGEMENTS

Studies from the author's laboratory were supported in part by Research Grant CA-12742 from the National Cancer Institute and National Institutes of Health Contract 71-2186.

The author is a Career Development Awardee of the National Cancer Institute.

REFERENCES

Anderson, W. F. and J. M. Gilbert 1969. tRNA-dependent translational control of in vitro hemoglobin synthesis. *Biochem. Biophys. Res. Commun.* 36: 456-462

Armstrong, D. J., W. J. Burrows, F. Skoog, K. L. Roy,and D. Söll. 1969. Cytokinins: Distribution in transfer RNA species of *Escherichia coli*. *Proc. Natl. Acad. Sci. U.S.A.* 63: 834-841.

Baguley, B. C. and M. Staehelin. 1968. Substrate specificity of adenine-specific transfer RNA methylase in normal and leukemic tissues. *Europ. J. Biochem.* 6: 1-7.

Baguley, B. C., W. Wehrli,and M. Staehelin. 1970. In vitro methylation of yeast serine transfer ribonucleic acid. *Biochemistry.* 9: 1645-1649.

Baliga, B. A., P. R. Srinivasan,and E. Borek. 1965. Changes in the tRNA methylating enzymes during insect metamorphosis. *Nature* 208:555-557.

Bjork, G. R. and I. Svensson. 1969. Studies on microbial RNA. Fractionation of tRNA methylases from *Saccharomyces cerevisiae*. *Europ. J. Biochem.* 9: 207-215.

Borek, E. and S. J. Kerr. 1972. Atypical transfer RNA's and their origin in neoplastic cells. *Adv. in Cancer Res.*15: 163-190.

Borek, E. and P. R. Srinivasan. 1966. The methylation of nucleic acids. *Ann. Rev. Biochem.* 35: 275-298.

Bradford, D. S., B. Hacker,and I. Clark. 1972. Transfer ribonucleic acid methylases of bone. Studies on vitamin A and D deficiency. *Biochem. J.* 126: 1057-1066.

Capra, J. D. and A. Peterkofsky. 1968. Effect of in vitro methylation on the chromatographic and coding properties of methyl-deficient leucine transfer RNA. *J. Mol. Biol.* 33: 591-607.

Duda, E.,M. Staub, P. Venetianer,and G. Denes, 1968. In-
    teraction between phenylalanine tRNA and the allosteric
    first enzyme of the aromatic amino acid biosynthetic
    pathway. *Biochem. Biophys. Res. Commun.* 32: 992-997.
Fleissner, E. and E. Borek. 1962. A new enzyme of RNA syn-
    thesis: RNA methylase. *Proc. Natl. Acad. Sci. U. S. A.*
    48: 1199-1203
Gallagher, R. E., R. C. Y. Ting,and R. C. Gallo. 1971. Trans-
    fer RNA methylase alterations in polyoma transformed rat
    embryo culture cells. *Proc. Soc. Exp. Biol. Med.* 136:
    819-823.
Gantt, R. R. and V. J. Evans. 1969. Comparison of soluble
    RNA methylase capacity in paired neoplastic and nonneo-
    plastic cell lines in vitro. *Cancer Res.* 29: 536-541.
Gantt, R. R., K. J. Stromberg,and F. Montes de Oca. 1971.
    Specific RNA methylase associated with avian myeloblastosis
    virus. *Nature.* 234: 35-37.
Gefter, M. L. and R. L. Russell. 1969. Role of modifications
    in tyrosine transfer RNA; a modified base affecting ribo-
    some binding. *J. Mol. Biol.* 39: 145-157.
Gold, P. 1971. Antigenic reversion in human cancer. *Ann.
    Rev. Med.* 22: 85-96.
Gonano, P., G. Pirro,and S. Silvetti. 1973. Foetal liver
    tRNA[Phe] in rat hepatoma. *Nature New Biol.* 242: 236-237.
Greengard, O., M. Gordon, M. A. Smith,and G. Acs. 1964.
    Studies on the mechanism of diethylstilbestrol-induced
    formation of phosphoprotein in male chickens. *J. Biol.
    Chem.* 239: 2079-2082.
Hacker, B. 1969. Estrogen-induced transfer RNA methylase
    activity in chick oviduct. *Biochim. Biophys. Acta* 186:
    214-216.
Hancock, R. L.,P. McFarland,and P. R. Fox. 1967. sRNA methy-
    lase of embryonic liver. *Experimentia.* 23: 806-809.
Hurwitz, J., M. Gold and M. Anders. 1964a. The enzymatic
    methylation of ribonucleic acid and deoxyribonucleic
    acid. III. Purification of soluble ribonucleic acid-
    methylating enzymes. *J. Biol. Chem.* 239: 3462-3473.
Hurwitz, J., M. Gold,and M. Anders. 1964b. The enzymatic
    methylation of ribonucleic acid and deoxyribonucleic acid.
    IV. The properties of the soluble ribonucleic acid-methy-
    lating enzymes. *J. Biol. Chem.* 239: 3474-3482.
Jacobson, K. B. 1971. Role of an isoacceptor transfer ribo-
    nucleic acid as an enzyme inhibitor: effect on tryptophan
    pyrrolase of *Drosophila. Nature New Biol.* 231: 17-19.

Kahle, P., P. Hoppe-Seyler, and H. Kroger. 1971. Transfer RNA methylases from rat liver nuclei. *Biochim. Biophys. Acta.* 240: 584-591.

Kerr, S. J. 1970. Natural inhibitors of the transfer ribonucleic acid methylases. *Biochemistry.* 9: 690-695.

Kerr, S. J. 1972. Competing methyltransferase systems. *J. Biol. Chem.* 247: 4248-4252.

Kerr, S. J. and E. Borek. 1973. Enzymic methylation of natural polynucleotides. In: *The Enzymes.* Vol. IX: 167-195.

Kerr, S. J. and Z. Dische. 1970. tRNA methylases in bovine lens. *Invest. Ophthalmol.* 9: 286-290.

Kit, S., K. Nakajima, and D. R. Dubbs. 1970. Transfer RNA methylase activities of cells and cells infected with animal viruses. *Cancer Res.* 30: 528-534.

Kuchino, Y. and S. Nishimura. 1970. Nucleotide sequence specificities of guanylate residue-specific tRNA methylases from rat liver. *Biochem. Biophys. Res. Commun.* 40: 306-313.

Kuchino, Y., T. Seno and S. Nishimura. 1971. Fragmented *Escherichia coli* methionine tRNA^f as methyl acceptor for rat liver tRNA methylase. *Biochem. Biophys. Res. Commun.* 43: 476-483.

Leboy, P. S. and P. Piester. 1973. Organ-specific differences in the methylation of transfer RNA in vitro. *Cancer Res.* 33: 2241-2246.

Liau, M. C., J. B. Hunt, D. W. Smith, and R. B. Hurlbert. 1973. Inhibition of transfer and ribosomal RNA methylases by polyinosinate. *Cancer Res.* 33: 323-331.

Liau, M. C., C. M. O'Rourke, and R. B. Hurlbert. 1972. Transfer ribonucleic acid methylases of nucleoli isolated from a rat tumor. *Biochemistry.* 11: 629-636.

Lombardini, J. B. and P. Talalay. 1971. Formation, functions and regulatory importance of S-adenosyl-L-methionine. *Adv. in Enz. Regul.* 9: 349-384.

Mays, L. L. and E. Borek, 1971. Transfer ribonucleic acid methyltransferases during hormone-induced synthesis of phosvitin. *Biochemistry.* 10: 4949-4954.

Mays, L. L., E. Borek, and C. E. Finch. 1973. Glycine N-methyltransferase: A regulatory enzyme which increases in aging animals. *Nature.* 243: 411-413.

Mittelman, A., R. H. Hall, D. S. Yohn, and J. T. Grace, Jr. 1967. The in vitro soluble RNA methylase activity of SV40-induced hamster tumors. *Cancer Res.* 27: 1409-1414.

Moore, B. G. 1970. Differential inhibition of bacterial tRNA methylases by S-adenosylethionine and other adenine derivatives. *Can. J. Biochem.* 48: 702-705.

Moore, B. G. and R. C. Smith. 1969. S-adenosylethionine as an inhibtor of tRNA methylation. *Can. J. Biochem.* 47: 561-565.

Pegg, A. E. 1971. Inhibitors of mammalian tRNA methylases. *FEBS Letters.* 16: 13-16.

Phillips, J. H. and K. Kjellin-Straby. 1967. Microbial ribonucleic acid. IV. Two mutants of *Saccharomyces cerevisiae* lacking N²-dimethylguanine in soluble ribonucleic acid. *J. Mol. Biol.* 26: 509-518.

Pillinger, D. J. and E. Borek. 1969. Transfer RNA methylases during morphogenesis in the cellular slime mold. *Proc. Nat. Acad. Sci. USA* 62: 1145-1150.

Pillinger, D. J., E. Borek and W. K. Paik. 1971. tRNA methylases during thyroxine-induced differentiation in bull frog tadpoles. *J. Endocrinol.* 49: 553-554.

Pillinger, D. J. and R. Wilkinson. 1971. Initiation of malignant transformation and the significance of changes in transfer RNA methylase enzymes. *Cancer Res.* 31: 630-632.

Riddick, D. H. and R. C. Gallo. 1971. Transfer RNA methylase activities of human lymphocytes. I. Induction by PHA in normal lymphocytes. *Blood.* 37: 282-292.

Rodeh, R., M. Feldman, and U. Z. Littauer. 1967. Properties of soluble ribonucleic acid methylases from rat liver. *Biochemistry.* 6: 451-460.

Sharma, O. K., 1973. Differences in the transfer RNA methyltransferases from normal rat liver and Novikoff hepatoma. *Biochim. Biophys. Acta.* 299: 415-427.

Sharma, O. K. and E. Borek. 1970. Hormonal effect on transfer ribonucleic acid methylases and on serine transfer ribonucleic acid. *Biochemistry.* 9: 2507-2513.

Sharma, O. K., S. J. Kerr, R. Lipshitz-Weiner and E. Borek. 1971. Regulation of the tRNA methylases. *Fed. Proc.* 30: 167-176.

Sharma, O. K., L. A. Loeb, and E. Borek. 1971. Transfer RNA methylases during sea urchin embryogenesis. *Biochim. Biophys. Acta.* 240: 558-563.

Sheid, B., E. Bilik, and L. Biempica. 1970. Estrogen mediated enhancement of rat liver sRNA methylase activity. *Arch. Biochem. Biophys.* 140: 437-442.

Shugart, L., G. D. Novelli, and M. P. Stulberg. 1968. Isolation and properties of undermethylated phenylalanine transfer RNA from a relaxed mutant of *Escherichia coli.* *Biochim. Biophys. Acta.* 157: 83-90.

Simon, L., A. J. Glasky, and T. H. Rejal. 1967. Enzymes in the central nervous system. I. RNA methylase. *Biochim. Biophys. Acta.* 142: 99-104.

Singer, C. E., G. R. Smith, R. Cortese, and B. N. Ames. 1972. Mutant tRNA[His] ineffective in repression and lacking two pseudouridine modifications. *Nature New Biol.* 238: 72-74.

Smith, J. D., J. N. Abelson, B. F. Clark, H. M. Goodman, and S. Brenner. 1966. Studies on amber suppressor tRNA. *Cold Spring Harbor Symp. Quant. Biol.* 31: 479-485.

Srinivasan, P. R. and E. Borek. 1964. Species variation of the RNA methylases. *Biochemistry.* 3: 616-619.

Svensson, I., G. R. Bjork, and P. Lundahl. 1969. Studies on microbial RNA. Properties of tRNA methylases from *Saccharomyces cerevisiae. Europ. J. Biochem.* 9: 216-221.

Taya, Y. and S. Nishimura. 1973. Biosynthesis of 5-methyl-aminomethyl-2-thiouridylate. I. Isolation of a new tRNA-methylase specific for 5-methylamino-methyl-2-thiouridylate. *Biochem. Biophys. Res. Commun.* 51: 1062-1068.

Turkington, R. W. 1969. Hormonal regulation of transfer ribonucleic acid-methylating enzymes during development of the mouse mammary gland. *J. Biol. Chem.* 244: 5140-5148.

Turkington, R. W. and M. Riddle. 1970. Transfer RNA-methylating enzymes in mammary carcinoma cells. *Cancer Res.* 30: 650-657.

Wainfan, E. and E. Borek. 1967. Differential inhibitors of tRNA methylases. *Mol. Pharmacol.* 3: 595-598.

Wainfan, E. and B. Landsberg. 1971. Cytokinins that inhibit transfer RNA methylating enzymes. *FEBS Letters.* 19: 144-148.

Wainfan, E., P. R. Srinivasan, and E. Borek. 1965. Alterations in the transfer ribonucleic acid methylases after bacteriophage infection or induction. *Biochemistry.* 4: 2845-2848.

Wainfan, E., P. R. Srinivasan, and E. Borek. 1966. Inhibition of tRNA methylases in lysogenic organisms after induction by ultraviolet light or heat. *J. Mol. Biol.* 22: 349-353.

Wainwright, S. D. 1971. Stimulation of hemoglobin synthesis in developing chick blastodisc blood islands by a minor alanine-specific transfer RNA. *Cancer Res.* 31: 694-696.

Wong, R.S.L., G. A. Scarborough, and E. Borek. 1971. Transfer ribonucleic acid methylases during germination of *Neurospora crassa. J. Bacteriol.* 108: 446-450.

Wust, C.J. and L. Rosen 1972. Aminoacylation and methylation of tRNA as a function of age in the rat. *Exp. Geront.* 7: 331-343.

Yoshida, M., K. Takeishi, and T. Ukita 1970. Anticodon structure of GAA-specific glutamic acid tRNA from yeast. *Biochem. Biophys. Res. Comm.* 39:852-857.

ISOZYME PATTERNS OF BRANCHED CHAIN AMINO ACID TRANSAMINASE DURING CELLULAR DIFFERENTIATION AND CARCINOGENESIS

AKIRA ICHIHARA, YASUHIDE YAMASAKI, HIROSHI MASUJI[*] and
JIRO SATO[*]
Institute for Enzyme Research,
School of Medicine
Tokushima University
Tokushima 770
and
Cancer Institute[*]
Okayama University Medical School
Okayama 700
JAPAN

ABSTRACT: Three isozymes (enzymes I-III) of the specific transaminase for valine, leucine, and isoleucine (branched chain amino acid transaminase EC 2.6.1.6) were isolated from various rat tissues. Enzyme I is found ubiquitously in the tissues, enzyme II only in liver, and enzyme III in brain, ovary and placenta. These three forms can be distinguished either chromatographically or immunochemically. Enzyme II is readily induced in rats by various hormonal or dietary treatments and is found in adult liver, but not in fetal liver, while enzyme I is not inducible and its activity decreases during development. Thus enzyme II could be considered as a differentiated type for hepatocytes and enzyme I as a proliferating type. This is supported by findings that hepatomas and cultured hepatocytes lost enzyme II, and transformed cells acquired enzyme III instead. Some slow growing Morris hepatomas have a similar isozyme pattern to that of adult liver, namely enzymes I and II, while others have all three isozymes. During transformation of cultured hepatocytes by chemical carcinogens enzyme III appeared in the cells. These findings indicate that change of the isozyme pattern of this transaminase in liver and hepatomas is not due to mere selection of a cell population, but to aberration of gene expression. Based on this conclusion the following scheme is proposed:

	Hepatomas		Fetal liver		Adult liver
		Decarcinogenesis?		Differentiation	
Expression of isozyme	I + III	$\xleftarrow{\hspace{1cm}} \dashrightarrow$	I	$\xleftarrow{\hspace{1cm}} \longrightarrow$	I + II
		Disdifferentiation		Dedifferentiation	

INTRODUCTION

The mechanisms involved in biological development and differentiation have been studied morphologically by biologists for many years, but in recent years this problem has attracted the attention of biochemists. Recent progress in cancer research has revealed that cancer cells tend to lose the specific characters of the cells from which they are derived and to show regression to an immature state. This process has been referred to as a state of retro-, de-, or dis-differentiation, disease of differentiation or blocked ontogeny (Walker et al., 1972, Knox, 1972). Therefore, studies on cellular differentiation are important, not only for the embryological point of view, but also for that of oncology. For studies on cellular differentiation accurate and quantitative parameters are required. Biochemical criteria and particularly isozymes are very useful for this purpose, because recent studies on isozymes in mammalian tissues have shown that tissues may have specific isozymes and that differentiation, in a biochemical sense, can be interpreted as acquisition of luxury and specific molecules by the cells (Criss, 1970).

ENZYMOLOGICAL PROPERTIES OF ISOZYMES OF BRANCHED CHAIN AMINO ACID TRANSAMINASE

In 1966 Ichihara et al. (1966) and Taylor et al. (1966) found a specific transaminase for branched chain amino acids (valine, leucine, and isoleucine) in hog heart and subsequent studies showed that there are three isozymes (enzymes I - III) in rat tissues (Aki et al., 1968; Ogawa et al., 1970). These isozymes can be separated either by DEAE cellulose column chromatography, as shown in Fig. 1 or by polyacrylamide disc gel electrophoresis. Table I shows their general properties. Enzymes I and III are very similar in their substrate specificities and their Km values, but their tissue distributions and chromatographic and immunochemical properties differ. Enzyme II is quite different from the other two isozymes, because it is found only in rodent liver (Ichihara et al., 1972) and is specific for leucine and its Km value is very high. Table II shows the tissue distributions of the three isozymes in rats. All the tissues examined contain enzyme I, while in addition liver has enzyme II and brain, ovary and placenta have enzyme III. It should be mentioned that the isozymes are found both in the supernatant and mitochondrial fractions and that their properties in the two fractions may differ (Aki et al., 1967). This paper is concerned only with the supernatant isozymes. Enzyme I has only been purified from hog heart (Taylor

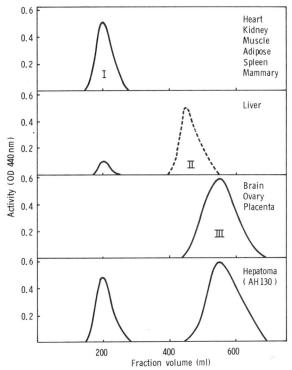

Fig. 1. Patterns of branched chain amino acid transaminase of various rat tissues on DEAE-cellulose column chromatography. Enzymes were eluted with a gradient of 0.005 M to 0.3 M phosphate buffer (pH 7.8).

et al., 1966) and enzyme III from hog brain and rat hepatoma (Aki et al., 1969; Ogawa et al., 1970). The molecular weights of enzymes I and III from hog tissues are 75,000 and 39,000, respectively and that of enzyme III from rat hepatoma is 42,000. Enzyme I of hog heart contains one mole of pyridoxal phosphate per mole of enzyme (Taylor et al., 1966) and the proteolytic susceptibilities and immunochemical properties of enzymes I and III from hog kidney suggest that the relation of these two isozymes cannot be explained on the basis of alleles, hybridization of subunits or aggregation-disaggregation (Aki et al., 1973). Therefore, it is very likely that the three forms of this transaminase are controlled by different genes.

PHYSIOLOGY OF THE ISOZYMES

It is known that leucine is ketogenic, valine is glycogenic

TABLE I

PROPERTIES OF ISOZYMES OF BRANCHED CHAIN
AMINO ACID TRANSAMINASE IN RATS

	ISOZYME		
	I	II	III
Concentration of phosphate buffer for elution from DEAE cellulose (M)	0.02	0.18	0.20
Mobility on polyacrylamide gel disc electrophoresis (Rm)	0.32	0.46	0.57
Inhibition by antiserum (%)			
against enzyme I	57	0	0
against enzyme III	0	0	94
K_m for substrate (mM)			
Valine	4.3		2.5
Leucine	0.8	25	0.6
Isoleucine	0.8		0.5
α-Ketoglutarate	1.0	0.07	0.5
Pyridoxal phosphate	0.03	0.004	
Optimal pH	8.2	8.7	8.4

TABLE II

DISTRIBUTION OF ISOZYMES IN VARIOUS RAT TISSUES

Tissue	Activity for leucine (nm/hr/g w.w.)		% Distribution of Isozymes in supernatant		
	Total	Supernatant	I	II	III
Liver	6.4	2.0	25	75	0
Kidney	139.2	46.8	100	0	0
Muscle	50.4	14.4	100	0	0
Brain	74.4	37.2	5	0	95
Heart	147.6	86.4	100	0	0
Lung	17.6	4.6	100	0	0
Spleen	21.8	7.3	100	0	0
Gut	7.4	3.0	100	0	0
Ovary	50.4	22.8	70	0	30
Placenta			45	0	55
Testis			100	0	0
Adipose tissue	7.8		100	0	0
Lactating mammary gland	100.0		100	0	0

and isoleucine has both properties. These amino acids are all essential for animal nutrition, but their metabolism in animals may be important as a source of energy when they are in excess or during severe starvation. We have shown that leucine metabolism is closely coupled with oxidative phosphorylation and that the transaminase is a rate limiting step for its metabolism in liver (Ichihara et al., 1973; Noda et al., 1974). We have also shown that these isozymes can be modulated by various treatments of rats. For instance, cortisol and a high protein diet increase the activity of enzyme II in liver within a few hours but do not change the activities of enzyme I in liver and kidney (Ichihara et al., 1967). Diabetes and hypophysectomy, on the contrary, induced enzyme I in kidney, but not enzyme II in liver. Daily treatment of rats with cortisol for a week also induced enzyme I in kidney. These results can be interpreted by supposing that both isozymes are under control of glucocorticoid, but that the turnover of enzyme II is more rapid than that of enzyme I and that enzyme II is more sensitive to induction by cortisol than enzyme I (Shirai et al., 1971). Ketogenesis from leucine is also controlled by enzyme I in lactating mammary gland (Ichihara, 1973). Therefore, it is evident that transaminase is important in regulating metabolism of branched chain amino acids. However, the significances of the respective isozymes are still obscure.

DEVELOPMENTAL ASPECTS OF THE ISOZYMES

Studies on changes in the isozymes during development of rats have shown that fetal liver contains only enzyme I and that its activity decreases as the fetus grows. After birth enzyme II appears and increases rapidly (Ichihara et al., 1968). Fetal brain contains enzyme III like adult brain. Similarly both fetal and adult kidney and heart have enzyme I only. Richter (1961) classified changes of enzyme activities during development into three types: enzyme I may be regarded as a growth type and enzymes II and III as maturation types. This is quite compatible with other results described in this paper. Other rapidly proliferating tissues, such as testis and spleen contain only enzyme I. Greengard discussed the mechanism of change of enzyme activities during development and emphasized that hormonal susceptibility changed during development (Greengard, 1970). Other papers suggest spontaneous changes of enzyme activities without the participation of hormones (Sereni et al., 1970; Rutter et al., 1973; Nakamura et al., 1973). We investigated the induction of these isozymes during development by circumfusion culture and found that glucocorticoid and glucagon are necessary to induce enzyme II in fetal liver (Ichihara et al., 1975). Therefore, hormones seem to play an important, if not essential,role in gene expression during development. It should be mentioned that during liver regeneration after partial hepatectomy there is no change in the isozyme pattern from that of adult liver.

CARCINOGENESIS AND THE ISOZYMES

Since Greenstein's proposal on the biochemical characteristics of cancer cells, many reports have suggested that cancer cells tend to lose differentiated characters (Knox, 1972). However, an interesting feature of cancer cells is that they acquire unusual properties of other normal tissues. Therefore, tissue specific isozymes are very useful markers for characterization of cancer cells (Criss, 1970). This ectopic expression of a phenotype in cancer cells is also seen with the isozymes of branched chain amino acid transaminase. Rat ascites Yoshida hepatoma (AH 130) has lost enzyme II and acquired enzyme III, which is normally found in brain (Ogawa et al., 1970). No difference could be found in the enzymological and immunochemical properties of the enzyme III's from brain and hepatoma. Five other strains of Yoshida hepatomas also showed similar deviation of isozyme patterns. Several Morris hepatomas had a variety of isozyme patterns, i.e. enzyme I only as in fetal liver, enzymes I and II as in adult liver, or all three isozymes (Ogawa et al., 1972). We also studied the isozymes in primary tumors induced in rat livers by administration of 3'-methyl DAB for different periods. The histological types of these tumors agreed well with their isozyme patterns. Namely, benign adenomas had the isozyme pattern found in normal adult liver, while hepatocellular carcinomas had a similar deviation pattern of isozymes to that of Yoshida ascites hepatomas.

From these results we proposed the following scheme to characterize the degree of differentiation and carcinogenesis of cells from the isozyme patterns of this transaminase (Ogawa et al., 1972).

	Rapidly growing hepatomas	Fetal immature liver cells	Adult mature liver cells

| Expression of isozymes | I + III | Decarcinogenesis? → ⟵ Disdifferentiation | I | Differentiation → ⟵ Dedifferentiation | I + II |

⟵ Carcinogenesis

The presence of enzyme II could be considered as an expression of the differentiated state of liver cells, and the

appearance of enzyme III in hepatocytes as a sign of transformation. If we add results on Morris hepatomas, it is possible
to draw a circular scheme of differentiation and carcinogenesis
as shown in Fig. 2 (Ichihara, 1973). In this scheme we can
visualize a continuous transition of isozyme patterns during
differentiation and transformation. This scheme also suggests
the possibility of direct transformation of differentiated
cells to cancer cells without passing through the immature
fetal stage. Further investigation is required on whether this
possibility is valid or not.

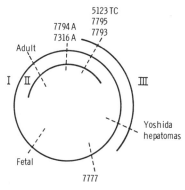

Fig. 2. Circular diagram of isozyme patterns of branched
chain amino acid transaminase in adult and fetal liver and
hepatomas.

ISOZYME PATTERNS IN CULTURED RAT HEPATOCYTES

There is still uncertainty about whether change in isozyme
patterns in tissues is due to alteration of gene expression in
the cells or change to a cell population with a different type
of isozyme pattern. This problem could be settled using a
procedure to identify hybrids of isozyme subunits, using fluorescent antibody to isozymes, or by culture of pure cloned cells.
We could not apply the two former methods, so we examined the
problem by culture of rat liver cells. We found that after
culture for a few months several cloned rat hepatocytes contained only enzyme I and no enzyme II and that treatment of
these cells with chemical carcinogens, such as DAB or 4-NQO
induced transformation and the cells acquired tumorigenicity.
Morphologically, the tumors induced by back-transplantation of
these transformed cells appeared to be hepatomas and their
isozyme patterns were just like those of Yoshida hepatomas

(Ogawa et al., 1973). These findings strongly suggest that acquisition of enzyme III is not due to selection of a cell population, but to aberration of gene expression. The finding that cells with very deviated chromosomal numbers express enzyme III supports this idea. The question of whether expression of enzyme III, deviation of chromosomal numbers, and acquisition of tumorigenicity occur in parallel was investigated using long-term cultures of hepatocytes, which tend to transform spontaneously as reported before (Evans et al., 1958; Katsuta et al., 1965; Sato et al., 1968; Borek, 1972; Oshiro et al., 1972; Breslow et al., 1973; Diamond et al., 1973). Table III shows that three strains of rat liver cells were cultivated and that two of them (RLN-B-2 and RLN-J-C-13) had fairly high proportions of diploid cells after short-term culture, but that their chromosomal numbers tended to become deviated to the triploid range during prolonged culture. RLN-J-C-13 acquired tumorigenicity after culture for about 580 days and RLN-B-2 after culture for 800 days. Until these times the cells contained only enzyme I, while after 700 to 1000 days slight, but significant, enzyme III was detected in these strains. RLN-8 contained only enzyme I even after 1300 days, although the cells became transformed around this time. Tumors and cells derived from tumors induced by cells transformed in vitro had a ratio of enzymes I and III of about one. These results suggest that transformation of hepatocytes causes expression of enzyme III as well as chromosomal deviation. It is also conceivable that the lower expression of enzyme III in transformed, long-term cultures than in cells from tumors is due to the much smaller population of transformed cells in cultures than in tumors, because on back-transplantation transformed cells would be selected, so that tumors should consist almost entirely of transformed cells. The results also indicate that back-transplantation *per se* is not the cause of expression of enzyme III, since transformed cells in long-term cultures could express enzyme III without back-transplantation. It is interesting that the total activity, namely that of enzyme I plus III, in hepatocytes in inversely correlated with the rate of expression of enzyme III, which is probably a function of transformation (Fig. 3). Thus in highly transformed cells, much enzyme III and low total activity could be expected. This conforms with the idea that catabolism of amino acids in general tends to decrease in cancer cells (Weber et al., 1966).

ISOZYMES OF CULTURED RAT HEPATOCYTES WITH DIPLOID CHROMOSOMES AND MORRIS HEPATOMA 7316A

There are many reports that cultured hepatocytes lose the characteristics of liver, such as secretion of albumin,

TABLE III

SPONTANEOUS TRANSFORMATION OF CULTURED HEPATOCYTES
AND THE APPEARANCE OF ENZYME III

The origins of cell lines RLN-B-2 and RLN-J-C-13 were described
in a previous report (Ogawa et al., 1973). The origin of RLN-8
has also been reported (Sato et al., 1968). For back-transplantation $5 \times 10^6 - 1 \times 10^7$ cells were injected intraperitoneally. None of these cell lines contained enzyme II.

Cells	Culture days	Specific activity (nm/min/mg protein)	% Distribution of isozymes		% Diploid chromosomes
			I	III	
RLN-8	1290	3.9	100	0	0
	1259 recultured a	2.7	60	40	0
RLN-J-C-13	174	14.0	100	0	58
	400				0
	646	15.3	100	0	
	720	3.0	90	10	
	579 recultured b	8.2	60	40	
	" tumor c	0.5	43	57	
RLN-B-2	278	23.1	100	0	82
	420	20.6	100	0	40
	685	3.6	100	0	0
	897	2.4	80	20	
	962				0
	1129	4.6	90	10	
	796 tumor d		40	60	

tyrosine transaminase, urea formation and gluconeogenesis.
However, recent reports suggest that it may be possible to
culture hepatocytes which retain liver-specific functions

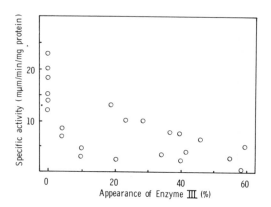

Fig. 3. Relation between total activities of isozymes of branched chain amino acid transaminase and degree of expression of enzyme III in cultured hepatocytes.

(Schapira et al., 1971; Lefert et al., 1972; Potter, 1972; Bissell et al., 1973).   It is also interesting that differentiated phenotypes are more stable in cell lines of some slow-growing hepatomas (van Rijn et al., 1974).   Use of multiple parameters to characterize cellular differentiation is necessary however (DeLuca et al., 1972; van Rijn et al., 1974).   We have found that hepatocytes cultured in one of our laboratories (J. S.) do not contain enzyme II, which can be considered as a specific character of liver cells.   However, strain RLN-B-2, with a high proportion of diploid chromosomes, has some characteristics of liver parenchymal cells (Table IV).   This

TABLE IV

DIFFERENTIATED CHARACTERS OF CULTURED RAT HEPATOCYTES

Tissue	Aldolase[1]	Glucose-ATP phosphotransferase[2]	Tyrosine transaminase[3]	Tryptophan pyrrolase[3]
Muscle	60	0.5	0	0
Liver	1.0	6.0	17.5	10.5
PC-2	50	0.5	0	–
RLN-B-2	14	1.4	1.2	3.3

1. Activity ratio for fructose 1,6-diphosphate and fructose-1-phosphate.
2. Activity ratio with high (1×10^{-1}M) and low (1×10^{-4}M) glucose concentrations.
3. Activities are expressed as   nmoles of product formed /min/mg protein.

cell line also secretes albumin (Namba, 1966). The spontaneously transformed cell line, PC-2, has no liver-type isozymes
of aldolase, glucokinase or tyrosine transaminase. It shows
rather a similar isozyme pattern to that of muscle or fetal
liver. We have previously shown that Morris hepatoma 7316A
contains the transaminase isozyme pattern of adult rat liver,
namely enzymes I and II (Ogawa et al., 1972). Cultured cells
of this hepatoma also had a similar pattern to adult liver
(Fig. 4). They also had a high level of tyrosine transaminase
(Table V). The modal chomosomal number of cultured hepatoma

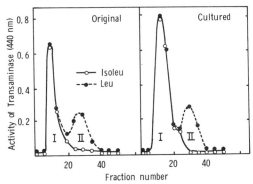

Fig. 4. Isozyme patterns of branched chain amino acid transaminase of Morris hepatoma 7316A and its cultured cells.

TABLE V

EXPRESSION OF DIFFERENTIATED PHENOTYPES
(LIVER SPECIFIC ENZYMES) IN CULTURED HEPATOCYTES

Cells	Culture days	Branched chain amino acid transaminase[1] II	Tyrosine transaminase[1]
RLN-B-2	420	0	1.1
" + Cortisol[2]		0	1.1
Morris 7316A	400	0.9	9.4
" + Cortisol[2]		2.5	52.7

[1] Specific activity = nmoles product /min/mg protein

[2] 1 x 10^{-5}M cortisol for 15 hr.

cells is 43, as reported by Nowell et al. (1969). Enzyme II and tyrosine transaminase can be induced by addition of cortisol to the medium. These two enzymes can also be induced by either cortisol or glucagon in a circumfusion culture of mouse fetal liver (Ichihara et al., 1975). Therefore, lack of enzyme II in cultured hepatocytes suggests that its expression may be suppressed by ordinary culture conditions. Therefore, studies on cultured cells may clarify the mechanism of differentiation and carcinogenesis and the isozymes of branched chain amino acid transaminase are very useful markers for this purpose.

ACKNOWLEDGEMENTS

This work was supported by research grants from the Ministry of Education to A. I. and J. S.

REFERENCES

Aki, K., K. Ogawa, and A. Ichihara 1968. Transaminase of branched chain amino acids IV. Purification and properties of two enzymes from rat liver. *Biochim. Biophys. Acta* 159: 276-284.

Aki, K., K. Ogawa, A. Shirai, and A. Ichihara 1967. Transaminase of branched chain amino acids III. Purification of the mitochondrial enzyme from hog heart and comparison with the supernatant enzyme. *J. Biochem.* 62: 610-617.

Aki, K., A. Yokojima, and A. Ichihara 1969. Transaminase of branched chain amino acids. VI. Purification and properties of the hog brain enzyme. *J. Biochem.* 65: 539-544.

Aki, K., T. Yoshimura, and A. Ichihara 1973. Transaminase of branched chain amino acids. IX. Conformational change of isozyme I *in vitro*. *J. Biochem.* 74: 779-784.

Bissell, D. M., L. E. Hammaker, and U. A. Meyer 1973. Parenchymal cells from adult rat liver in nonproliferating monolayer culture. I. Functional studies. *J. Cell Biol.* 59: 722-734.

Borek, C. 1972. Neoplastic transformation *in vitro* of a clone of adult liver epithelial cells into differentiated hepatoma-like cells under conditions of nutritional stress. *Proc. Natl. Acad. Sci. U. S. A.* 69: 956-959.

Breslow, J. L., H. R. Sloan, V. J. Ferrans, J. L. Anderson, and R. I. Levy.1973. Characterization of the mouse liver cell line FL83B. *Exptl. Cell Res.* 78: 441-453.

Criss, W. E. 1970. A review of isozymes in cancer. *Cancer Res.* 31: 1523-1542.

DeLuca, C., E. J. Massaro, and M. M. Cohen 1972. Biochemical and cytogenetic characterization of rat hepatoma cell lines *in vitro*. *Cancer Res.* 32: 2435-2440.

Diamond, L., R. McFall, Y. Tashiro, and D. Sabatini 1973. The WIRL-3 rat liver cell lines and their transformed derivatives. *Cancer Res.* 33: 2627-2636.

Evans, V. J., N. M. Hawkins, B. B. Westfall, and W. R. Earle 1958. Studies on culture lines derived from mouse liver parenchymatous cells grown in long-term tissue culture. *Cancer Res.* 18: 261-266.

Greengard, O. 1970. The developmental formation of enzymes in rat liver. In: *Biochemical Actions of Hormones*. Litwack, G., editor. Academic Press, New York. Vol. I, pp 53-87.

Ichihara, A. 1973. Cellular differentiation and isozymes of branched chain amino acid transaminase. *Enzyme* 15: 210-233.

Ichihara, A. and E. Koyama 1966. Transaminase of branched chain amino acids. I. Branched chain amino acids-α-ketoglutarate transaminase. *J. Biochem.* 59: 160-169.

Ichihara, A., J. Sato, and M. Kumegawa 1975. Isozyme patterns of branched chain amino acid transaminase in cultured rat liver cells; In: *Gene Expression and Carcinogenesis in Cultured Liver*. Gerschenson, L. E., and Thompson, E. B. editors. Academic Press, New York. (in press).

Ichihara, A. and H. Takahashi 1968. Transaminase of branched chain amino acids V. Activity change in developing and regenerating rat liver. *Biochim. Biophys. Acta* 167: 274-279.

Ichihara, A., C. Noda, and K. Ogawa 1973. Control of leucine metabolism with special reference to branched chain amino acid transaminase isozymes. *Adv. Enzyme Regul.* 11: 155-166.

Ichihara, A. and K. Ogawa 1972. Isozymes of branched chain amino acid transaminase in normal rat tissues and hepatomas. *Gann monograph* 13: 181-190.

Ichihara, A., H. Takahashi, K. Aki, and A. Shirai 1967. Transaminase of branched chain amino acids. II. Physiological changes in enzyme activity in rat liver and kidney. *Biochem. Biophys. Res. Commun.* 26: 674-678.

Katsuta, H., T. Takaoka, Y. Doida, and T. Kuroki 1965. Carcinogenesis in tissue culture. VII. Morphological transformation of rat liver cells in Nagisa culture. *Japan. J. Exp. Med.* 35: 513-544.

Knox, W. E. 1972. *Enzyme Patterns in Fetal, Adult and Neoplastic Rat Tissues*. Karger, Basel.

Lefert, H. L., and D. Paul 1972. Studies on primary cultures of differentiated fetal liver cells. *J. Cell Biol.* 52: 550-568.

Nakamura, T., and M. Kumegawa 1973. Induction of key glycolytic enzymes in mouse fetal liver cultured in circumfusion system by insulin. *Biochem. Biophys. Res. Commun.*

51: 474-479.

Namba, M. 1966. Function of the liver cells in short-term and long-term cultures. I. Albumin production of the liver cells *in vitro*. *Okayama Acta Med.* 20: 251-259.

Noda, C. and A. lchihara 1974. Control of ketogenesis from amino acids. II. Ketone bodies formation from α-ketoisocaproate, the keto-analogue of leucine, by rat liver mitochondria. *J. Biochem.* 76: 1123-1130.

Nowell, P. C., H. P. Morris, and V. R. Potter 1967. Chromosomes of "minimal deviation" hepatomas and some other transplantable rat tumors. *Cancer Res.* 27: 1565-1579.

Ogawa, K. and A. Ichihara 1972. Isozyme patterns of branched chain amino acid transaminase in various rat hepatomas. *Cancer Res.* 32: 1257-1263.

Ogawa, K., A. Ichihara, H. Masuji, and J. Sato 1973. Isozyme patterns of branched chain amino acid transaminase in various rat hepatomas. *Cancer Res.* 32: 1257-1263.

Ogawa, K., A. Ichihara, H. Masuji, and J. Sato 1973. Isozyme patterns of branched chain amino acid transaminase in cultured rat hepatocytes. *Cancer Res.* 33: 449-453.

Ogawa, K., A. Yokojima, and A. Ichihara 1970. Transaminase of branched chain amino acids. VII. Comparative studies on isozymes of ascites hepatoma and various normal tissues of rats. *J. Biochem.* 68: 901-911.

Oshiro, Y., L. E. Gerschenson, and J. A. Dipaolo 1972. Carcinomas from rat liver cells transformed spontaneously in culture. *Cancer Res.* 32: 877-879.

Potter, V. R. 1972. Workshop on liver cell culture. *Cancer Res.* 32: 1998-2000.

Richter, D. 1961. Enzymic activity during early development. *Brit. Med. Bull.* 17: 118-121.

Rutter, W. J., R. L. Pictet, and P. W. Morris 1973. Toward molecular mechanisms of developmental processes. *Ann. Rev. Biochem.* 42: 601-646.

Sato, J., M. Namba, K. Usiu, and D. Nagano 1968. Carcinogenesis in tissue culture VIII. Spontaneous malignant transformation of rat liver cells in long-term culture. *Japan. J. Exp. Med.* 38: 105-118.

Schapira, F., D. Delain, and Y. Lacroix 1971. Multiple molecular forms of aldolase in fetal liver cell cultures : action of dexamethasone. *Enzyme* 12: 545-552.

Sereni, F., and L. P. Sereni 1970. Spontaneous development of tyrosine aminotransferase activity in fetal liver cultures. *Adv. Enzyme Regul.* 8: 253-267.

Shirai, A. and A. Ichihara 1971. Transaminase of branched chain amino acids. VIII. Further studies on regulation of isozyme activities in rat liver and kidney. *J. Biochem.* 70: 741-748.

Taylor, R. T. and W. T. Jenkins 1966. Leucine aminotransferase II. Purification and characterization. *J. Biol. Chem.* 241: 4396-4405.

van Rijn, H., M. M. Bevers, R. van Wijk and W. D. Wicks 1974. Regulation of phosphoenolpyruvate carboxykinase and tyrosine transaminase in hepatoma cell cultures. III. Comparative studies in H35, HTC, MH, C, and RLC cells. *J. Cell Biol.* 60: 181-191.

Walker, P. R. and V. R. Potter 1972. Isozyme studies on adult, regenerating, precancerous and developing liver in relation to findings in hepatomas. *Adv. Enzyme Regul.* 10: 339-364.

Weber, G. and M. A. Lea 1966. The molecular correlation concept of neoplasia. *Adv. Enzyme Regul.* 4: 115-145.

CARCINOPLACENTAL ISOZYMES

WILLIAM H. FISHMAN
Tufts Cancer Research Center
Tufts University School of Medicine
136 Harrison Avenue, Boston, Massachusetts   02111

ABSTRACT. Isozyme studies in relation to carcinoplacental alkaline phosphatase (Regan isozyme) have gone through several phases of development. The use of starch gel electrophoresis is now restricted to phenotyping placental alkaline phosphatase whereas "microzone" cellulose acetate membrane electrophoresis is most frequently employed for detecting Regan isozyme. Polyacrylamide gel is now most popular for studying membrane-derived alkaline phosphatases, facilitated by the inclusion of non-ionic detergent in the gel. Isoelectric focusing in polyacrylamide gel rods is the most recent addition but the interpretation of the results is difficult. In all of these types of electrophoresis, identification of tissue origin of isozyme is aided by the use of organ-specific amino acid inhibitors, such as L-phenylalanine and L-homoarginine, by heat inactivation, and by employing specific antisera to the isozymes.

Studies carried out on ascitic fluids of patients with ovarian cancer have brought into view a heavy molecular weight fraction containing Regan isozyme, 5'-nucleotidase, and leucine aminopeptidase, "markers" of plasma membrane. Further, the coexpression with Regan isozyme of another placental protein, β-hCG (β-subunit of human chorionic gonadotropin) was observed as well as a correlation between Regan isozyme and the enzyme, histaminase. Reference is also made to the separate expression of alkaline phosphatase isozymes in two sublines of HeLa cells; TCRC-1 producing Regan isozyme and TCRC-2, non-Regan isozyme.

The significance of embryonic gene products appearing in cancer cells is viewed as one which may well include the mechanism for the initiation and perpetuation of the neoplastic process.

We have been working in the area of carcinoplacental isozymes (a carcinoplacental isozyme is one which is present in the placenta and absent in the fetus) for the past few years and the invitation to participate in the symposium gave us an opportunity to take stock of our efforts from the point of view of isozymes.

Our stress up to 1972 has been on the biochemical study of isozymes of alkaline phosphatase with the use of organ-specific

inhibitors. Thus, L-phenylalanine is an uncompetitive inhibitor of intestinal and placental alkaline phosphatases (Fishman, Green, Inglis, 1963; Fishman, Inglis, and Ghosh, 1968;
Ghosh and Fishman, 1966). On the other hand, L-homoarginine
(Fishman and Sie, 1970, 1971; Lin and Fishman, 1972) is an
uncompetitive inhibitor of liver and bone but not intestinal
and placental alkaline phosphatases. Automated techniques
were devised to fractionate the alkaline phosphatase isozymes
(Green, Anstiss, Fishman, 1971; Cantor et al, 1972).

With the availability of starch gel electrophoresis, we
were able to first separate the proteins on the gel slab and
then identify the protein zones by developing the gel in the
presence and absence of the appropriate amino acid inhibitor
reagents. The results of starch gel electrophoresis usually
produced diffuse zones of alkaline phosphatase activity unlike
the clear separation of the five bands of LDH so readily
obtained by others.

What will be discussed in the first part of this presentation is the history, current status, and projections into the
future of our work as it relates specifically to isozymes.
For reviews, the reader is referred to Fishman and Ghosh
(1967), Fishman (1973), and Fishman (1974).

Regan isozyme

This discussion is introduced by a brief word about the
carcinoplacental isozyme we have been interested in. This is
the Regan isozyme which is a placental form of alkaline phosphatase first discovered in a cancer patient named Regan
(Fishman et al, 1968).

We chose to identify this isozyme as Regan rather than as
placental because at that time there were so few properties
known of alkaline phosphatase isozymes which could serve as
a basis for their separate identification. These were heat-
inactivation, L-phenylalanine inhibition, neuraminidase cleavage, and electrophoresis. If it can be shown in every detail
that the Regan isozyme is the product of an embryonic gene
then we are dealing with the activation of embryonic genes in
human cancer. In this connection a recent effort by Greene
and Sussman (1973) shows similarity but not identity in the
peptide "fingerprint" patterns of Regan isozyme and placental
alkaline phosphatase. This problem demands that absolutely
pure isozymes be employed.

At that time, we made some progress with the help of Dr.
Ghosh and others in purifying placental alkaline phosphatase
(Ghosh and Fishman, 1968). A more convenient procedure has
recently given us a good product in a satisfactory yield
(Doellgast and Fishman, 1974).

Fig. 1. (With the permission of *Cancer Research.*) A starch
gel zymogram showing six Regan-positive serum samples three
of which demonstrated the slow-moving D variant and three
matching the F, FS or S of the placental phenotype standards
developed in slots 1, 5, and 9. Slot 2 displays the D as well
as F phenotype found in a normal pregnancy serum.

Electrophoretic techniques

The first isozyme electrophoresis we did was the classical
starch gel electrophoresis to which Norma Inglis contributed.
In Fig. 1, (Inglis et al, 1973) one can observe the location
of alkaline phosphatase as fluorescent spots resulting from
the hydrolysis of the fluorogenic substrate, α-naphthol phosphate. Our interest was to establish whether or not the
Regan (placental-type) alkaline phosphatase corresponded to
one or other of the known starch gel phenotypes reported
earlier for placental alkaline phosphatase (Boyer, 1961;
Robson and Harris, 1965; Beckman, Bjorling and Christodoulou,
1966). It was soon evident that the Regan isozymes matched
F, FS, and S phenotypes. However, what surprised us was the
fact that over 50% of the 36 Regan-positive sera we examined
exhibited the slow-moving rare D-phenotype. The latter was
first reported by Boyer (1961) as the D-variant and this was
subsequently confirmed by Beckman and Beckman (1968) in a
variety of populations as a rare phenotype. (The D-variant
phenotype is demonstrable so far on starch gel electrophoresis.) We also have noted the isozyme bands which either do
not leave the origin or move only a short distance away from
it. Unlike LDH which migrates readily alkaline phosphatase

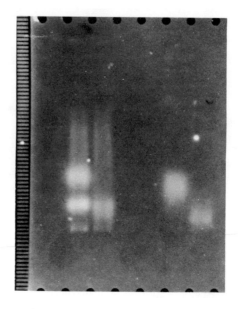

Fig. 2. 'Microzone electrophoresis membrane developed with Naphthol AS-MX phosphatase showing results of two ovarian fluid samples, heat-treated, and previously mixed with either saline for control or with diluted anitserum to placental alkaline phosphatase. The left side of membrane shows the development of two bands both of which were retarded and the right side showing the usual findings of one band which is also retarded.

---

in tissue extracts, in serum, in extracts of fetal tissue and HeLa cells does not move into the gel on electrophoresis. 1 Next we had to meet the demands of our co-workers and others for the identification of Regan isozyme in patients' sera, causing us to employ the "microzone" cellulose acetate membrane electrophoresis for this purpose. The test (Inglis et al, 1971) consists of heating the serum first for five min at 65° to inactivate the non-placental forms of alkaline phosphatase and to migrate the heat-stable isozyme in the absence and in the presence (mixed beforehand) of the antiserum to placental alkaline phosphatase. A positive test is shown by the retardation of the band. Definitive answers can be obtained on 30 to 40 specimens in one working day. Fig. 2 shows the results from a typical sample fluid on the right side of the membrane and an unusual two-banded pattern on the

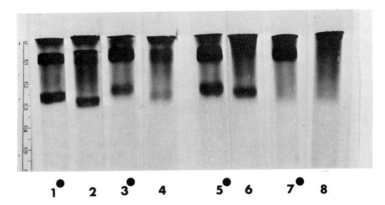

Fig. 3. Polyacrylamide gel electrophoresis of four samples run in the absence or presence of Triton X-100 incorporated in gel matrix. The two variants (A and B) of placental homogenates are shown in 1 and 2; a high Regan-rich fluid in 3 and 4; a bone tissue source in 5 and 6; and a bone serum source for 7 and 8 (with the permission of *Biochemical Medicine*). The black dots identify the Triton gels.

---

left.

In applying polyacrylamide gel electrophoresis to the separation of alkaline phosphatase isozymes (Green et al, 1972), the failure of a portion of the enzyme to enter the gel frustrated Lillian Fishman. She introduced Triton X-100 into the gel (Fishman, 1074) following the successful experience of Lin and Fishman, 1974 with this detergent in causing microsomal acid phosphatase to migrate into polyacrylamide. Fig. 3 illustrates the advantageous results with Triton X-100 which facilitated migration of variants A and B of placental alkaline phosphatase, bone tissue extract and bone alkaline phosphatase rich serum. These were purposely overloaded to demonstrate the extent of the effect of the Triton. This observation has had widespread consequences in our ability to obtain electrophoretic patterns for species of alkaline phosphatases which previously remained stubbornly at the origin. For example, in Fig. 4, fetal liver and intestinal alkaline phosphatases were coaxed into the gel migrating to identical positions in contrast to the corresponding faster adult forms of these isozymes.

In the next phase of our electrophoretic work, we thought

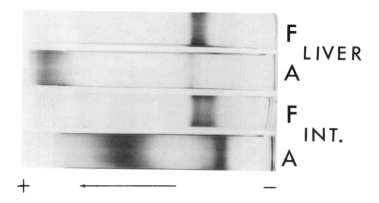

Fig. 4. The demonstration of alkaline phosphatase of F(fetal) and A(adult) liver and intestinal tissue extracts run in acrylamide gel (L. Fishman, 1974).

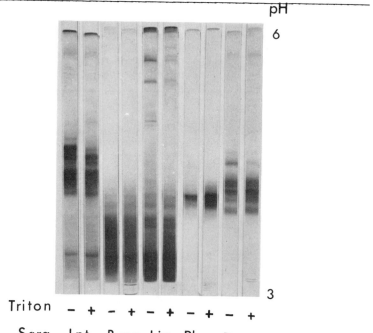

Fig. 5. Alkaline phosphatase patterns of isoelectric focusing in polyacrylamide gel rods of sources rich in intestine, bone, liver, placenta or Regan run in duplicate without or with Triton X-100 incorporated in the gel mixture.

we could be greatly rewarded by spending much effort on isoelectric focusing of alkaline phosphatases. However, the re-
wards proved to be much less than expected in terms of our
ability to interpret the results. In Fig. 5, (Angellis et
al, 1974) individual sera rich in liver, bone, intestine,
placenta and Regan isozymes of alkaline phosphatases (the
latter two were heated) were focused. One disappointment was
the observation that no distinction could be made in the iso-
focused bands of liver and bone isozyme-rich sera. In addition there was a multiplicity of bands ranging from 5 to 50
which plagued us. With regard to distinguishing phenotypic
forms alkaline phosphatase, we were likewise unrewarded.
Triton X-100 did facilitate the migration of isozymes into the
gel where they migrated sometimes to positions different from
the bands in gels lacking this detergent. It is possible that
the technique will still have utility in the physiochemical
study of homogeneous preparations of alkaline phosphatase.

Ovarian Carcinoma

We are now working extensively with ascitic fluids obtained
from patients with ovarian carcinoma. Why ovarian carcinoma?
It is the type of cancer with which Regan isozyme is most
frequently identified. Why ascitic fluids? These patients
when first seen are already in the late stages of cancer bearing peritoneal implants of tumor in the abdomen. These generate ascitic fluid which has to be removed for the comfort of
the patient. Thus, one, on centrifugation, may obtain from
these fluids the supernatant and in the pellet, populations
of cancer cells. The supernatant contains tumor products and
from the cells, we hope to culture individual lines of ovarian
cancer cells.

The chromatographic techniques which were introduced into
the laboratory by Dr. George Doellgast (see Doellgast and
Fishman, these proceedings) gave an interesting pattern. By
passing the ascitic fluid through a column of Sepharose-4B,
the void volume was found to contain the bulk of the heat
stable alkaline phosphatase and the phospholipid. In addition
5'-nucleotidase and leucine aminopeptidase, were detected and
measured by Dr. Lin and Norma Inglis. These are markers of
plasma membranes as is alkaline phosphatase. Thus, we have
reasons to believe that the fraction excluded in the void
volume of Sepharose-4B is enriched with membrane-derived material. On the other hand the β-glucuronidase, which is not
a plasma membrane constituent, is retained on the column.

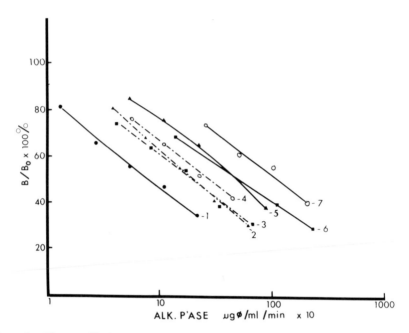

Fig. 6. The radioimmunoassay curves demonstrating similar slopes of three normal pregnancy sera, (2,3,4) as well as three Regan-isozyme rich fluid sources, (5,6,7) to the slope of the placental standard curve, plot 1 (Chang et al, 1974).

Radioimmunoassay

In evaluating certain enzyme components which remain at the origin after electrophoresis we are considering the possibility of enzyme antigen-antibody complexes. In this regard, we believe that radioimmunoassay techniques will be valuable in ascertaining the existence and nature of these complexes. With the help of Drs. Chang and Raam, (Chang, Raam, Fishman, 1974; Raam, Chang, Fishman, 1974) we have made progress in radioimmunoassay. Fig. 6 shows plots of the exchangeable antigen-antibody complex as a function of alkaline phosphatase concentration. The steep slope for placental alkaline phosphatase in contrast to the lines parallel to the X axis for intestine and liver are findings which attest to the specificity of the reagents for placental alkaline phosphatase. In unpublished work, we have observed slopes for the alkaline phosphatase of pregnancy serum and ovarian ascitic fluids which were closely parallel to the slope shown for placental alkaline phosphatase.

With the combination of the newer techniques of isolating a membrane-rich fraction coupled with the use of the radio-immunoassay and other techniques that we are developing, we hope we can approach the problem of detecting the antibodies produced in man in response to the appearance of Regan isozyme.

The question we have asked ourselves is "If one placental protein is being produced by this tumor, isn't it likely that there may be other placental proteins which are tumor products?".

Human Chorionic Gonadotrophin

Recently, there has been success achieved at the N.I.H. by Vaitukaitis, Braunstein, and Ross (1972) in developing a radioimmunoassay which is specific for the β-subunit to hCG (β-hCG) human chorionic gonadotrophin. This radioimmunoassay distinguishes very clearly between pituitary gonadotrophin and chorionic gonadotrophin. These workers have independently studied a large number of patients for the presence of serum β-hCG. We were able to interest them to assay a number of cancerous ascitic fluids for us. The first study (Fig. 7) was on 10 fluids of patients known to have Regan isozyme. Seven of these ten showed elevated values of human gonadotrophin. We could not resist the temptation to obtain measurements for CEA and α-fetoprotein. Four were strongly positive for CEA and none for α-fetoprotein. This information encouraged us to proceed with a systematic study of all the fluids whether they had Regan or not for the presence of β-hCG.

Correlation of Regan Isozyme with β-hCG

A point of some interest related to tumors in general can be gleaned from a comparison of statistics of Regan isozyme and β-hCG obtained in Boston (Stolbach, Krant, and Fishman, 1969; Nathanson and Fishman, 1971; unpublished data) and in Washington (Braunstein et al, 1973; Vaitukaitis, personal communication). Interestingly, the tumors in which hCG is most frequently found are testis, ovary and pancreas and the corresponding ones for Regan isozyme are ovary, testis, and pancreas. Here, the correlation is very evident.

In the actual study of the 60 ovarian ascitic fluids, 60% were Regan positive and there was correspondence in the expression or lack of it for both β-hCG and Regan isozyme in 55% of these 60 patients (Stolbach, Vaitukaitis, and Fishman, 1974).

We are not disturbed or disappointed in the lack of expression of other phenotypes or in the failure to observe a 100% correlation of placental proteins with cancer. What we are

Patient	Fluid Specimen	Site of Cancer	Carcinoplacental Phenotypes		Human Chorionic Gonadotropin	Carcinofetal Phenotypes	α-Fetoprotein
			Regan Isozyme Placental Units/100 ml Values	Ab test	Nanograms/ml	Carcino-embryonic Antigen Nanograms/ml	
P	Asc.	Ovary	0.77	+	1.46	3.6	–
C 1	Asc.	Ovary	2.8	+	2.20	2.6	–
C 2	Asc.	Ovary	304.0	+	4.70	< 2	–
O'C	Pl.	Ovary	239.0	+	1.24	< 2	–
B	Asc.	Cvary	11.5	+	1.05	14.5	–
M	Asc.	Ovary	0.74	+	0.66	< 2	–
Ch	Asc	Ovary	5.33	+	0.60	30	–
K	Asc.	Ovary	2.23	+	36.6	2	–
S	Asc.	Ovary	0.06	+	1.8	30	–
M	Pl.	Lung	200.0	+	25	89	–
					Max. value non-cancer = 1 ng/ml	Max. value non-cancer = 10 ng/ml	

Fig. 7. Comparison of 10 fluid samples from cancer patients producing various amounts of Regan isozyme with β-hCG, CEA and α-fetoprotein assays.

seeing is quite well known in spontaneous human cancer, namely phenotypic variability, a phenomenon not evident in experimental cancer which is manipulated to give uniformity of expression.

Another enzyme, histaminase, has been studied in our laboratory by Dr. Lin in relation to medullary thyroid carcinoma. In this tumor, the C-cell is neoplastic (Melvin and Tashjian, 1968) and produces calcitonin which is its identifying characteristic. We found that normal serum was usually free of histaminase in contrast to the markedly enriched pregnancy serum. This confirms findings which first appeared in the literature some 40 years ago, showing that term pregnancy serum is rich in histaminase as is the placenta. What is interesting is a concordance in the appearance and lack of appearance of histaminase in relation to the presence or absence of Regan isozyme is over twenty ovarian cancer ascitic fluids. These data are being prepared for publication (Lin et al, 1974).

Monophenotypic Sublines of HeLa Cells

This brings us to our newest interest in the sublines of HeLa cancer cells which have been discussed by Dr. R. M. Singer at this Conference two days ago. I should like to point out that the T C R C -l line of cells produces not only Regan isozyme but also fetal acidic isoferritins whereas the T C R C -2 line which produces non-Regan isozyme fails to synthesize fetal isoferritins (Singer, Drysdale, and Fishman, 1974). It would be interesting to find out whether there is a linkage between the genes expressing Regan isozyme and acidic isoferritins.

It is also suggestive from Singer and Fishman's data on the separate effects of antibody to placental and intestinal alkaline phosphatase that cancer cells (HEP-2) derived from cancer of the larynx, may be expressing a pair of ectopic phenotypes, i.e. intestinal and placental alkaline phosphatases. The amnion cells (Fl-amnion) appear to make these two phenotypes under the influence of prednisolone. Assuming that the HEP-2 cell line is homogeneous, there may be a coexpression of these two isozyme phenotypes which are normally not expressed together in adult tissues.

Significance of Embryonic Gene Products in Cancer?

What is the significance of the appearance of embryonic gene products in experimental and spontaneous tumors?

We have to look at the picture of fetal isozymes in a

different way from that of fetal antigens. In the latter one recognizes a fetal protein such as CEA, but we don't know its function. When a specific catalytic activity is attached to a particular protein and that catalytic activity is significant in a scheme of metabolism particularly in the glycolytic cycle, then one is able to construct an organized picture of the events which are taking place in the transformation of say the hepatocytes to fully undifferentiated hepatoma. This is being done in Dr. Weinhouse's and Dr. Schapira's laboratories (see Weinhouse, 1973, Schapira, 1973).

Briefly, these hepatomas produce fetal enzymes which are no longer under the control of host regulatory mechanisms. It is the loss of regulatory control which is at the basis of the cancer problem. If the organism were to lose control of CEA, there is no clue as to what the consequences would be because its function has not yet been established.

It is possible to imagine that the loss of control of the production of isozyme forms which are specifically geared to the more efficient utilization of metabolites and consequently the more efficient production of energy is a process which could be self-perpetuating. So these newer manifestations of cancer which have been recognized widely only within the past five years may possess the necessary ingredients for the initiation and perpetuation of the neoplastic process.

With this kind of a picture in front of us we are looking at the gene products in our studies and we are wondering what the consequences would be if they escaped host control. It will be worthwhile to determine whether these substances singly or in combination, may have effects on cell division and growth of cells. Certainly this is true of the polypeptide trophic hormones.

ACKNOWLEDGMENTS

Aided in part by grants-in-aid (CA12924, CA13332) from the National Cancer Institute. The author is the recipient of Career Research Award K6-CA-18543 of the National Cancer Institute. Grateful appreciation is expressed to Norma R. Inglis for her help in preparing the manuscript for publication.

REFERENCES

Angellis, D., N. R. Inglis, and W. H. Fishman 1974. Isoelectric focusing of alkaline phosphatase isoenzymes: effects of Triton X-100. Submitted for publication.

Beckman, L. and G. Beckman 1968. A genetic variant of placental alkaline phosphatase with unusual electrophoretic properties. *Acta Genet.-Basel* 18: 543-552.

Beckman, L., G. Bjorling, and C. Christodoulou 1966. Pregnancy enzymes and placental polymorphism 1. Alkaline phosphatase. *Acta Genet.-Basel* 16: 59.

Boyer, S. H. 1961. Alkaline phosphatases in human sera and placentae. *Science* 134: 1002-1004.

Braunstein, G. D., J. L. Vaitukaitis, P. P. Carbone, and G. T. Ross 1973. Ectopic production of human chorionic gonadotropin by neoplasms. *Ann. Intern. Med.* 78: 39-45.

Cantor, F., S. Green, L. L. Stolbach, and W. H. Fishman 1972. Quality control of an automated differential isoenzyme assay of alkaline phosphatase with the use of L-phenylalanine inhibition. *Clin. Chem.* 18: 391-392.

Chang, C. H., S. Raam, and W. H. Fishman 1974. One-step radioimmunoassay of human placental alkaline phosphatase by using solid-phase antibody polymer. Abstracts of the 60th Annual Meeting of the American Chemical Society. Atlantic City, September.

Doellgast, G. D. and W. H. Fishman 1974. Purification of human placental alkaline phosphatase: salt effects in affinity chromatography. *Biochem. J.* (in press).

Fishman, L. 1974. Acrylamide disc gel electrophoresis of alkaline phosphatase of human tissues, serum and ascites fluid using Triton X-100 in the sample and the gel matrix. *Biochem. Med.* 9: 309-315.

Fishman, W. H. 1973. Carcinoplacental isoenzyme antigens. *Advances in Enzyme Regulation II:* (Weber G. Ed) Oxford and New York, Pergamon Press, 293-322.

Fishman, W. H. 1974. Perspectives on alkaline phosphatase isoenzymes. *American J. Medicine,* May (in press).

Fishman, W. H. and N. K. Ghosh 1967. isoenzymes of human alkaline phosphatase. *Adv. Clin. Chem.* 10: 256-370.

Fishman, W. H., S. Green, and N. R. Inglis 1963. L-phenylalanine: an organ-specific stero-specific inhibitor of human intestinal alkaline phosphatase. *Nature* 198: 685-686.

Fishman, W. H., N. R. Inglis, and N. K. Ghosh 1968. Distinctions between intestinal and placental isoenzymes of alkaline phosphatase. *Clin. Chem Acta* 19: 71-79.

Fishman, W. H., N. R. Inglis, L. L. Stolbach, and M. J. Krant 1968. A serum alkaline phosphatase isoenzyme of human neoplastic cell origin. *Cancer Res.* 28: 150-154.

Fishman, W. H. and H. G. Sie 1970. L-homoarginine; an inhibitor of serum bone and liver alkaline phosphatase. *Clin. Chem. Acta* 29: 339-341.

Fishman, W. H. and H. G. Sie 1971. Organ-specific inhibition of human alkaline phosphatase isoenzymes of liver, bone, intestine and placenta; L-phenylalanine, L-tryptophan and L-homoarginine. *Enzymologia* 41: 141-167.

Ghosh, N. K. and W. H. Fishman 1966. On the mechanism of inhibition of intestinal alkaline phosphatase by L-phenylalanine. *J. Biol. Chem.* 241: 2516-2522.

Ghosh, N. K. and W. H. Fishman 1968. Purification and properties of molecular-weight variants of human placental alkaline phosphatase. *Biochem. J.* 108: 779-792.

Green, S., C. L. Anstiss and W. H. Fishman 1971. Automated differential isoenzyme analysis II. The fractionation of serum alkaline phosphatases into "liver," "intestinal" and "other" components. *Enzymologia* 41: 9-26.

Green, S., F. Cantor, N. R. Inglis, and W. H. Fishman 1972. Normal serum alkaline phosphatase isoenzymes examined by acrylamide and starch gel electrophoresis and by isoenzyme analysis using organ-specific inhibitors. *Am. J. Clin. Path.* 57: 52-64.

Greene, P. J. and H. H. Sussman 1973. Structural comparison of ectopic and normal placental alkaline phosphatase. *Proc. Natl. Acad. Sci. USA* 70: 2939-2942.

Inglis, N. R., D. Y. Guzek, S. Kirley, S. Green, and W. H. Fishman 1971. Rapid electrophoretic microzone membrane techniques for Regan isoenzyme (placental type alkaline phosphatase) using a fluorogenic substrate. *Clin. Chem. Acta* 33: 287-292.

Inglis, N. R., S. Kirley, L. L. Stolbach, and W. H. Fishman 1973. Phenotypes of the Regan isoenzyme and identity between the placental D-variant and the Nagao isoenzyme. *Cancer Res.* 33: 1657-1661.

Lin, C. W. and W. H. Fishman 1972. L-homoarginine: an organ-specific, uncompetitive inhibitor of human liver and bone alkaline phosphohydrolases. *J. Biol. Chem.* 247: 3082-3087.

Lin, C. W. and W. H. Fishman 1974. Microsomal and lysosomal acid phosphatase isoenzymes of mouse kidney, characterization and separation. *J. Histochem. and Cytochem.* 20: 487-498.

Lin, C. W., M. L. Orcutt, L. L. Stolbach, and W. H. Fishman 1974. Correlation of histaminase activity with Regan isoenzyme in ascitic fluids of ovarian cancer, unpublished data.

Melvin, K. E. W. and A. J. Tashjian 1968. The syndrome of excessive thyrocalcitonin produced by medullary carcinoma of the thyroid. *Proc. Natl. Acad. Sci. USA* 59: 1216-1222.

Nathanson, L. and W. H. Fishman 1971. New observations on the Regan isoenzyme of alkaline phosphatase in cancer patients. *Cancer* 27: 1388-1397.

Raam, S., C. H. Chang, and W. H. Fishman 1974. Dissociation of immobilized antigen-antibody complexes by freezing and thawing. Abstracts of 60th Annual Meeting of the American Chemical Society, Atlantic City, September.

Robson, E. G. and H. Harris 1965. Genetics of the alkaline phosphatase polymorphism of the human placenta. *Nature* 207: 1257-1259.

Schapira, F. 1973. Isozymes and Cancer. *Advances in Cancer Res.* 18: 77-153.

Singer, R. M., J. W. Drysdale, and W. H. Fishman 1974. Acidic isoferritins: presence in HeLa TCRC-1 and absence in HeLa TCRC-2, unpublished data.

Stolbach, L. L., M. J. Krant, and W. H. Fishman 1969. Ectopic production of an alkaline phosphatase isoenzyme in patients with cancer. *New Eng. J. Med.* 281: 757-762.

Stolbach, L. L., J. Vaitukaitis, and W. H. Fishman 1974. Correlation of Regan isoenzyme and HCG in serum and malignant effusion of patients with ovarian carcinoma. *Proc. Am. Assoc. for Cancer Research* 15: 309.

Vaitukaitis, J. L., G. D. Braunstein, and G. T. Ross 1972. A radioimmunoassay which specifically measures human chorionic gonadotropin in the presence of human luteinizing hormone. *Am. J. Obstet. Gynec.* 113: 751-758.

Weinhouse, S. 1973. Metabolism and isozyme alterations in experimental hepatomas. *Federation Proc.* 32 2162-2167.

GENETIC CONTROL OF δ-AMINOLEVULINATE DEHYDRATASE
IN ADULT LIVER, FETAL LIVER, AND A HEPATOMA OF INBRED MICE

DARRELL DOYLE and ROBERT MITCHELL
Department of Molecular Biology
Roswell Park Memorial Institute
Buffalo, New York   14203

ABSTRACT. δ-Aminolevulinate dehydratase isolated from fetal liver and a hepatoma has been compared to the enzyme from adult liver. The fetal liver enzyme is less stable to denaturation by heat and is catalytically more efficient than the enzyme from adult liver. The enzyme from a fast growing hepatoma has the same properties as the enzyme from fetal liver. The enzymes from all three tissue sources are similar in molecular size, electrophoretic mobility, apparent K_m for the substrate, δ-aminolevulinic acid, number of active sites, and pattern of peptides released by trypsin. These results indicate that the cellular mechanism controlling the structure of δ-aminolevulinate dehydratase in fetal liver is expressed again in the hepatoma and that a combined genetic-biochemical approach can provide insight into this mechanism of regulation.

A property of malignant tissues receiving considerable attention is the presence of isozymes, or other proteins, or antigens in the neoplasm that are characteristic not of the normal tissue of origin but of fetal tissues (Weinhouse, 1972; Criss, 1971; Knox, 1972; Schapira, 1970). This recurring pattern of fetal protein expression in malignant cells, suggests that one component in the complex program of events resulting in malignant growth is the reactivation of genes that become repressed during the normal course of embryonic development. In this report we describe our progress in the development of a system to examine the genetic basis for this reappearance of fetal function in neoplasia. The system is composed of an enzyme δ-aminolevulinate dehydratase and a series of mutations that affect the structure of this enzyme and its expression during the development of mouse liver.

δ-Aminolevulinate dehydratase catalyses the condensation of two molecules of δ-aminolevulinic acid to porphobilinogen, an intermediate in the pathway of porphyrin biosynthesis. The enzyme from adult mouse liver has been well characterized by Coleman (1966) and Doyle (1971). It is a 250,000 molecular weight protein composed of six apparently identical subunits. Thus, there should be one structural gene specifying the adult liver polypeptide. Mutations in this gene have been identified

907

by Coleman (1971) who showed that δ-aminolevulinate dehydratase from adult liver of the inbred mouse strains SM/J and C57BR/cdJ is differentially susceptible to inactivation by heat. This property of the enzyme presumably reflecting an altered primary structure is specified by a single genetic locus located on chromosome 4 (Hutton and Coleman, 1969).

The tissue concentration of δ-aminolevulinate dehydratase is also under genetic control. Most strains of inbred mice can be divided into three classes based on the activity of the enzyme in liver (Table 1). The difference in activity between

TABLE 1

Tissue Activities of δ-Aminolevulinate Dehydratase
in Different Strains of Inbred Mice

	δ-Aminolevulinate Dehydratase Activity Units/g Tissue				
Strain	Liver	Kidney	Spleen	Brain	Hepatoma
AKR/J	5.0	1.3	1.7	--	--
DBA/2J	4.2	1.5	1.6	0.52	--
C 57L/J	4.4	1.6	1.6	--	5.8
C 57BL/6J	1.3	0.50	0.40	0.06	--
SM/J	0.9	0.48	0.34	0.06	--

A Unit of activity is equal to one μ mole of porphobilinogen formed per hour. Conditions for assay and values for standard errors are given in Doyle and Schimke (1969).

the high activity strains AKR/J and DBA/2J and the low activity strain C57BL/6J is controlled by a single genetic locus denoted Levulinate (Lv) which also is located on chromosome 4 and which is closely linked or allelic to the structural gene locus (Russell and Coleman, 1963; Hutton and Coleman, 1969). Combined immunochemical and isotopic labelling methods have shown that the Lv locus controls the tissue concentration of δ-aminolevulinate dehydratase by regulating the rate at which this enzyme is synthesized in mouse liver (Doyle and Schimke, 1969). This locus does not affect the rate of degradation of the hepatic enzyme.

Thus far a variety of physical, chemical and enzymatic techniques have failed to detect a difference in primary structure of hepatic δ-aminolevulinate dehydratase from DBA/2J and C57BL/6J mice (Coleman, 1966; Doyle and Schimke, 1969; Doyle, 1971). Thus mutations at the levulinate locus affect the rate of δ-aminolevulinate dehydratase synthesis but do not appear to affect the structure of the dehydratase polypeptide. This locus is, however, probably very closely linked to the structural gene locus.

Although variants of δ-aminolevulinate dehydratase differing in stability to heat have been identified (Coleman, 1971) no variants differing in electrophoretic mobility were found in a screen of a large number of inbred *Mus musculus* lines or in a screen of the closely related *Mus castaneous* or in *Mus caroli* (Doyle, 1974). Also, all mouse tissues thus far examined including kidney, spleen, brain, and adult and fetal liver have only one electrophoretic form of the enzyme.

δ-Aminolevulinate dehydratase activity does show a characteristic pattern of change during development of the liver (Table 2). The enzymatic activity is high in fetal liver,

TABLE 2
Development of Liver δ-Aminolevulinate
Dehydratase in DBA/2J Mice

| Age | δ-Aminolevulinate Dehydratase Activity |
Days	Units/g Liver
−5	8.0
−1	4.2
Birth	2.3
+2	1.1
+7	1.8
+14	3.3
+21	4.1

The same pattern of development is shown in C57BL/6J mice except that the enzyme activity at the different ages is one-third to one-half that of the DBA/2J mice.

decreases markedly just before birth and then increases again to reach the adult level in about 21 days. The enzyme from fetal liver is probably confined to hematopoetic cells and differs in several important ways from the enzyme in hepatocytes of adult liver. The fetal enzyme is much more susceptible to inactivation by heat than is the adult enzyme (Table 3). Further, immunotitration of fetal and adult enzyme with an antibody monospecific for adult liver δ-aminolevulinate dehydratase indicates that the fetal enzyme is catalytically twice as efficient as the adult enzyme (Doyle and Schimke, 1969). Other properties of the enzymes are similar (Table 3).

TABLE 3

Some Properties of δ-Aminolevulinate Dehydratase
from Tissues of Different Inbred Mouse Strains

Source	Electrophoretic Mobility in Acrylamide gels Rf	K_m Moles	Stability to Heat Inactivation $t_{\frac{1}{2}}$ (min) 71°
DBA/2J Adult Liver	0.4	4×10^{-4}	120
DBA/2J Fetal Liver	0.4	4×10^{-4}	30
C57BL/6J Adult Liver	0.4	4×10^{-4}	125
SM/J Adult Liver	0.4	---	180
C57Br/cdJ Adult Liver	0.4	---	21
C57L/J Adult Liver	0.4	4×10^{-4}	120
C57L/J Fetal Liver	0.4	4×10^{-4}	30
C57L/J Hepatoma BW	0.4	4×10^{-4}	30

To determine electrophoretic mobility enzyme partially purified
from the tissue source was electrophoresed in gels as described
by Ornstein (1964) and Davis (1964). The gels were stained for
catalytic activity as described by Doyle (1969). Details for
the other experiments can also be found in the latter reference.

The differences between adult and fetal δ-aminolevulinate
dehydratase in catalytic efficiency and in stability to heat
could be due to some differential modification of the polypeptide after synthesis or to differences in primary amino
acid sequence. If the latter is correct, the levulinate locus
would have to be regulating the synthesis of two different polypeptides since this locus controls the concentration of aminolevulinate dehydratase in both fetal and adult liver (Table 2)
(Doyle and Schimke, 1969).

We determined which type of δ-aminolevulinate dehydratase
was present in a transplantable hepatoma, BW7756, which arose
spontaneously and is carried in the mouse inbred line C57L/J.
C57L/J is a high activity strain and the activity of δ-aminolevulinate dehydratase is slightly higher in the hepatoma than
in the host liver (Table 1). The hepatoma grows rapidly with
an average time between transplantations of 3 to 5 weeks.
δ-aminolevulinate dehydratase partially purified from the hepatoma is like the fetal enzyme in that it is heat labile
(Table 3) and catalytically twice as efficient as the normal
adult liver enzyme (Fig. 1).

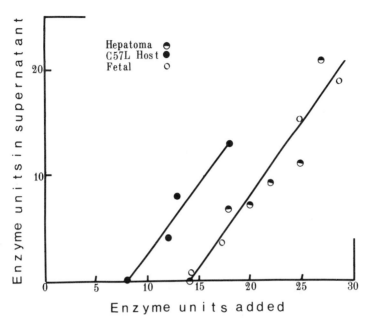

Fig. 1. Quantitative precipitin reactions of δ-aminolevulinate
dehydratase from fetal liver, adult liver, and hepatoma.

δ-Aminolevulinate dehydratase from either fetal liver (two
days before birth), adult liver, or hepatoma of the inbred
strain C57L/J was purified through the ammonium sulfate step
of Doyle and Schimke (1969). Increasing amounts of each extract containing the enzyme activity indicated was added to
1.0 ml of antiserum specific for adult liver δ-aminolevulinate
dehydratase. The mixtures were incubated at 37° for 30 min
and the immune precipitates were collected by centrifugation.
The precipitates were washed three times with cold 0.85% NaCl
and protein was determined. The supernatant fluids were
assayed for δ-aminolevulinate dehydratase. The amount of protein precipitated at the equivalence point (8 units for adult
enzyme and 15 units for hepatoma and fetal enzyme) of each
enzyme was identical, about 8 mg of protein. When the immune
precipitates were analysed by sodium dodecyl sulfate electrophoresis (Weber and Osborn, 1969), the amount of antigen in
each of the immune precipitates was also similar.

We have explored further the reasons for the apparent increased catalytic efficiency of δ-aminolevulinate dehydratase
from fetal liver and the hepatoma. Kinetic analysis shows
that the catalytic difference is due to a difference in V_{max}

and not K_m or apparent affinity of the enzyme for the substrate (Doyle and Schimke, 1969) (Table 3). A sucrose gradient analysis shown in Fig. 2, indicates that the enzyme from all three sources--hepatoma, fetal liver and adult liver--exists as a hexamer of 250,000 molecular weight.

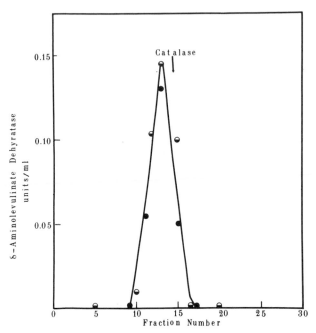

Fig. 2. Sucrose gradient analysis of δ-aminolevulinate dehydratase from C57L/J hepatoma ● and host liver ◒ .

The enzyme from each tissue was purified through the ammonium sulfate step of Doyle and Schimke (1969). Centrifugation through 5-20% sucrose gradients with beef liver catalase as a reference protein was performed as described by Martin and Ames (1961).

The mechanism by which this enzyme catalyses the condensation of two molecules of δ-aminolevulinic acid to porphobilinogen is known (Nandi and Shemin, 1968; Doyle, 1971). The enzyme catalyses an aldol condensation reaction involving the formation of an intermediate Schiff base with the substrate carbonyl group. This intermediate can be converted to a stable secondary amine derivative by reduction with $NaBH_4$. Thus, it is possible to titrate the number of active sites on the enzyme using radioactively labeled substrate and $NaBH_4$. As shown in Fig. 3, the amount of labeled substrate bound is propor-

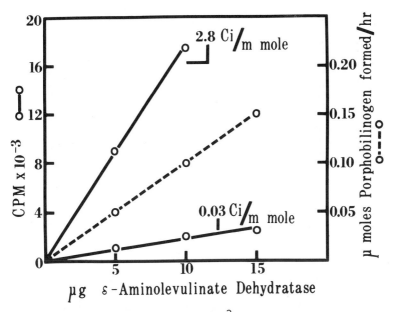

Fig. 3. Binding of amino- [3, 5 - ^{3}H] -levulinic acid to mouse liver α-aminolevulinate dehydratase.

Hepatic α-aminolevulinate dehydratase was purified to homogeneity from Swiss-Webster mice. The quantity of enzyme protein indicated was assayed using the standard assay procedure (Doyle and Schimke, 1969) and by coupling labeled substrate of different specific activity to the enzyme in the presence of NaBH$_4$ (see Doyle, 1971 for details).

tional to the amount of δ-aminolevulinate dehydratase protein present in the reaction mixture. Further, the amount of radioactivity bound to the enzyme is proportional to the specific radioactivity of the substrate. This coupling reaction is very specific. Even in crude extracts all of the bound substrate is coupled to δ-aminolevulinate dehydratase, since both the bound radioactivity and the enzymatic activity co-purify (Table 4). Further, SDS-polyacrylamide gel electrophoresis of the antibody precipitate from the final step of the purification shows that all of the labeled substrate is bound to polypeptide chains of molecular weight 40,000, which is the size of the dehydratase subunit (Doyle, 1971). Six moles of substrate are bound per mole of 250,000 molecular weight protein. When this coupling reaction was done in crude extracts of either fetal liver, hepatoma, or adult liver, the results shown in Table 5 were obtained. Each unit of δ-aminolevulinate dehydratase activity from adult liver can bind about twice as

TABLE 4

Purification of Hepatic δ-Aminolevulinate Dehydratase
Coupled to 3, 5-^{3}H - δ-Aminolevulinic Acid

C57L/J Adult Liver	Total Enzyme Activity (Units)	Yield (%)	Total Radioactivity (CPM x 10^{-3})	Yield (%)	$\dfrac{\text{Yield Enzyme Activity}}{\text{Yield Radioactivity}}$
Homogenate 10,000 x g	1.76	100	180	100	1.00
supernatant	1.54	88	169	94	0.94
105,000 x g supernatant	1.10	63	138	77	0.82
Heat (67°, 10 min)	0.97	55	90	50	1.1
Antibody Precipitate			86		

3, 5-^{3}H - δ-aminolevulinic acid was added to 25% (w/v) homogenate of C57L/J liver. The procedure for coupling the substrate to the enzyme is given in Doyle (1971). About 20% of the dehydratase enzymatic activity in the homogenate was inactivated by this procedure. Remaining enzymatic activity and trichloroacetic acid insoluble radioactivity were monitored through partial purification of the enzyme. Finally, the enzyme was precipitated with monospecific antiserum.

TABLE 5

Binding of ^{14}C-aminolevulinic Acid to
Aminolevulinate Dehydratase From Hepatoma,
Adult and Fetal Liver of C57L/J Mice

Tissue	^{14}C-aminolevulinic Acid Bound (DPM/Unit)	mg Enzyme Antigen/ Unit	^{14}C-aminolevulinic Acid Bound/mg
Adult Liver	22,000	0.100	220,000
Fetal Liver	9,000	0.050	180,000
Hepatoma	10,000	0.050	200,000

δ-Aminolevulinate dehydratase from each of the tissue sources
was partially purified through step 4 of Doyle and Schimke
(1969). One unit of enzymatic activity was incubated in a
reaction mixture containing 15 μ moles of 2 mercaptoethanol,
and 15 μ moles of potassium phosphate buffer, pH 6.5. δ-amino-
4-[14]C- levulinic acid, 50 μCi, specific activity 6 m Ci/m Mole,
was added simultaneously with a 2.5 μl portion of 1M NaBH$_4$.
Four additional 25 μl portions of NaBH$_4$ were added over the
course of 30 min at 37°. About 95% of the enzyme was inactivated. The extract was precipitated with 10% TCA which was 1
mM in unlabelled δ-aminolevulinic acid. The precipitates
were washed as described by Doyle (1971). Insoluble radioactivity was counted in a liquid scintillation spectrometer.

much labeled substrate as a unit of activity from fetal liver
or hepatoma. However, quantitative precipitin analyses (Fig.
1) indicate that a unit of adult enzyme activity corresponds
to about twice as much dehydratase protein (antigen) as a unit
of fetal dehydratase activity. We interpret these results to
mean that both adult and fetal δ-aminolevulinate dehydratase
contain the same number of active sites.

The question of whether the differences between fetal liver
or hepatoma and adult liver δ-aminolevulinate dehydratase result from a secondary modification or from a different primary
amino acid sequence has not yet been resolved. We have been
unable to obtain sufficient quantities of homogenous enzyme
from either fetal liver or hepatoma to do a rigorous structural
analysis. An indirect approach was used to show that the
enzymes from the different tissues are probably very similar in
primary sequence. Animals of the C57L/J strain containing the
hepatoma were injected with ^{3}H-lysine and ^{3}H-arginine. Host
C57L/J animals, not containing the hepatoma were injected with
[14]C lysine and ^{14}C arginine. One hour later, the animals were

killed; the livers and hepatomas were removed, mixed, and
homogenized together. δ-aminolevulinate dehydratase was partially purified from the mixed homogenates through the heat
step of Table 4. The enzyme was then precipitated with its
monospecific antiserum. The immunoprecipitate was washed,
oxidized with performic acid, hydrolysed with trypsin, and the
peptides then were chromatographed on Dowex with volatile buffer as described in detail in Doyle (1971). The pattern of
labeled peptides released by trypsin is shown in Fig. 4.

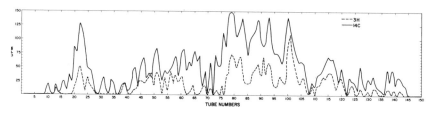

Fig. 4. Chromatography on Dowex 50 of the tryptic peptides of
δ-aminolevulinate dehydratase from C57L/J host liver and hepatoma.

Details for this experiment are given in the text and in
Doyle (1971). The ^{3}H counts represent the tryptic peptides
from the hepatoma enzyme; the ^{14}C counts represent the peptides
from host liver enzyme.

In this experiment the ^{3}H counts represent the tryptic peptides from δ-aminolevulinate dehydratase synthesized in the
hepatoma while the ^{14}C counts represent the peptides from the
enzyme synthesized in the host liver. Most of the peptides
are coincident. This method is too indirect and probably not
sensitive enough to differentiate small differences in primary
structure. Thus we only interpret the results to mean that
the enzymes from the two tissues are very similar in primary
amino acid sequence. We are currently using a genetic approach
to determine unequivocally whether fetal and adult δ-aminolevulinate dehydratase are specified by the same or a different
structural gene. This approach is based on the assumption
that the difference in stability to heat of the dehydratase
from strains SM/J and C57BR/cdJ is due to mutation resulting
in an amino acid substitution. Thus, we are presently finger-
printing the enzyme from these and other inbred mouse strains.
Our goal is to identify an altered tryptic peptide in the enzyme from adult liver and then to determine if the same mutation affects the primary structure of the enzyme in fetal
liver.

Regardless of the biochemical basis for the different forms of δ-aminolevulinate dehydratase, it is not unreasonable to conclude that a fetal gene, whether it be a different structural gene for the dehydratase polypeptide or a gene for a protein involved in modification of this enzyme, is expressed again in the hepatoma. Using the approach outlined here, it should be possible to further dissect the genetic-biochemical mechanisms involved in normal gene expression during development and to gain further insight into the alteration of this genetic program in neoplasia.

ACKNOWLEDGMENT

We acknowledge support for this work from the National Institutes of Health through grants HD 08410 and GM 19521.

REFERENCES

Coleman, D. L. 1966. Purification and properties of δ-aminolevulinate dehydratase from tissues of two strains of mice. *J. Biol. Chem.* 241: 5511-5517.

Coleman, D. L. 1971. Linkage of genes controlling the rate of synthesis and structure of aminolevulinate dehydratase. *Science* 173: 1245-1246.

Criss, W. E. 1971. A review of isozymes in cancer. *Cancer Res.* 31: 1523-1542.

Davis, B. J. 1964. Disc electrophoresis II. Method and application to human serum proteins. *Ann. N. Y. Acad. Sci.* 121: 404-427.

Doyle, Darrell 1971. Subunit structure of δ-aminolevulinate dehydratase from mouse liver. *J. Biol. Chem.* 246: 4965-4972.

Doyle, Darrell 1974. unpublished observation

Doyle, Darrell and R. T. Schimke 1969. The genetic and developmental regulation of hepatic δ-aminolevulinate dehydratase in mice. *J. Biol. Chem.* 244: 5449-5459.

Hutton, J. J. and D. L. Coleman 1969. Linkage analyses using biochemical variants in mice. II. Levulinate dehydratase and autosomal glucose-6-phosphate dehydrogenase. *Biochem. Genet.* 3: 517-523.

Knox, William E. 1972. Enzyme patterns in fetal, adult and neoplastic rat tissues. *Basal S., Karger AG.*

Marten, R. G. and B. N. Ames 1961. A method for determining the sedimentation behavior of enzymes: application to protein mixtures. *J. Biol. Chem.* 236: 1372-1379.

Nandi, D. L. and D. Shemin 1968. δ-Aminolevulinate dehydratase of *Rhodopseudomonas spheroides* III. Mechanism of porphobilinogen synthesis. *J. Biol. Chem.* 243: 1236-1242.

Ornstein, L. 1964. Disc electrophoresis I. Background and theory. *Ann. N. Y. Acad. Sci.* 121: 321-349.

Russell, R. L. and D. L. Coleman 1963. Genetic control of hepatic δ-aminolevulinate dehydratase in mice. *Genet.* 48: 1033-1039.

Schapira, F. 1970. Isozymes et cancer. *Pathol. Biol.* 18: 309-315.

Weber, K. and M. Osborn 1969. The reliability of molecular weight determinations by dodecyl sulfate-polyacrylamide gel electrophoresis. *J. Biol. Chem.* 244: 4406-4412.

Weinhouse, S. 1972. Glycolysis, respiration, and anomalous gene expression in experimental hepatomas. *Cancer Res.* 32: 2007-2016.

ISOZYME STUDIES OF HUMAN LIVER AND HEPATOMA
WITH PARTICULAR REFERENCE TO KINETIC PROPERTIES
OF HEXOKINASE AND PYRUVATE KINASE

DORIS BALINSKY, EFTIHIA CAYANIS, and I. BERSOHN
Enzyme Research Unit,
The South African Institute for Medical Research
P.O. Box 1038
Johannesburg, South Africa

ABSTRACT. Isozymes I, II, and III of hexokinase (EC 2.7.1.
1, HK) were partially purified from human tissues. The
limiting Michaelis constants of HK I, II, and III for MgATP
were 500, 360, and 440µM respectively, and for glucose 63,
140 and 16µM respectively, with inhibition of the latter
isozyme by glucose concentrations above 1mM. The apparent Michaelis constants for fructose were 12.4, 12.5,
and 6.1mM for HK I, II, and III respectively.

The L isozyme of pyruvate kinase (EC.2.7.1.40,PK) was
purified from human liver tissue and the M isozyme from
human hepatoma and fetal liver. The L isozyme was allosteric
with respect to phospho-enol-pyruvate, and could be
activated by fructose-1, 6-diphosphate (FDP) and to a
lesser extent by some hexose-monophosphates and triose
phosphates. It was allosterically inhibited by ATP and
alanine; this inhibition could be relieved by FDP. The
M isozymes did not show allosteric properties in the absence
of inhibitors, although the hepatoma M isozyme showed complex kinetics in double reciprocal plots. Alanine was an
equally potent inhibitor of all the isozymes; the inhibition
was relieved by FDP in all cases. It is suggested that the
L isozyme exists normally in an inactive conformation, but
is converted to an active one by FDP. The M isozymes exist
normally in the active conformation, but retain a binding
site for FDP.

Human primary malignant hepatocellular carcinoma occurs
with a high incidence in Southern Africa, expecially in some
areas of Mozambique, where an incidence of 1:1100 of the population occurs (Prates, 1961). Since the liver is an organ
active in carbohydrate metabolism, studies of the changes in
liver-specific enzymes and isozymes in primary liver cancer
are of interest.

HEXOKINASE

In rat hepatomas, the key glycolytic enzyme hexokinase
(ATP : D-hexose 6-phosphotransferase, EC 2.7.1.1) (HK) has

919

been shown to increase in level relative to the liver of the
tumor-bearing animal, the increase being correlated with
increasing growth rate (Weber, 1966). Four isozymes of hexokinase have been observed in rat tissues; these have been
named HK I, II, III, IV in order of increasing mobility towards
the anode (Katzen and Schimke, 1965). Decrease or disappearance
of the liver-specific HK IV, or glucokinase, has been observed
in rat hepatoma (Shatton et al., 1969; Sato et al., 1969).
Increases in the proportion of HK II in rat hepatomas were
observed by Sato et al. (1969) but not by Shatton et al.(1969).

In human hepatomas, hexokinase levels were either greater
than, equal to, or less than levels in adjacent, apparently
uninvolved "host" liver (Balinsky et al., 1973a). However,
the isozyme pattern was altered. As seen in Fig. 1A, normal
human liver, from a patient who had no liver pathology, showed
a predominance of HK I, and also HK III. Hepatoma tissue
showed a reduction of HK III and considerable activity of
HK II. "Host" tissue showed HK I and III, with a trace of
HK II, i.e. a pattern intermediate between that of hepatoma
and of liver from a normal subject. Some hepatomas showed
complete loss of HK III, while others showed little or no HK
II. A cell line has recently been established from one human
hepatoma (Prozesky et al., 1973). This was found to have
mainly HK II, with some HK I.

It seemed of interest to examine the kinetic properties
of these three hexokinases, in order to see the consequences,
in kinetic terms, of replacement of HK III by HK II. Initially,
HK II was partially purified from human hepatoma and some of
its kinetic properties studied (Balinsky et al., 1973b).
Subsequently, HK I was partially purified from human muscle,
and HK III from liver.

Double reciprocal plots of HK activity *vs* glucose or MgATP
concentration at various fixed concentrations of the other
substrate were made (Fig. 2). These show that:
(1) The apparent Michaelis constants of HK I and III for glucose and MgATP were independent of the concentration of the
second substrate, whereas the apparent Michaelis constant of
HK II for each substrate varied with the concentration of the
other substrate.
(2) HK I and II were not inhibited by high glucose concentrations - up to 66mM glucose had no inhibitory effect, whereas
HK III was inhibited by glucose concentrations above 1mM.

Fructose can also be utilized as substrate by HK. A
similar experiment was carried out in which fructose and MgATP
were varied. In this case the apparent Michaelis constants
for fructose and MgATP were found to be independent of the
concentration of the other substrate for all three HK isozymes.

Table I summarizes the kinetic parameters of human

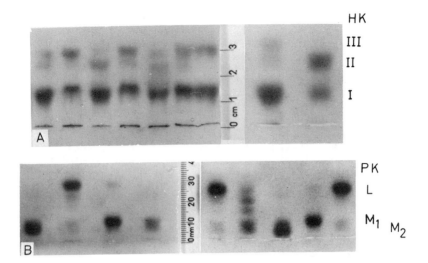

Fig. 1A. Starch gel electrophoresis of hexokinase - left to right: Cancer 1, host 1, cancer 2, host 2, cancer 3, host 3, normal liver, cancer 4, hepatoma cells (tissue culture). Electrophoresis and staining was according to Katzen and Schimke (1965).

Fig. 1B. Starch gel electrophoresis of pyruvate kinase - left to right: hepatoma cells (tissue culture), liver, muscle, malignant lymphocytes, host liver, hepatoma (tissue), hepatoma cells (tissue culture), muscle, normal liver. The electrode and gel buffers were 0.02M barbital (pH 8.6) - 0.25mM EDTA - 0.1mM DTT. The electrode buffer contained in addition 0.1mM FDP. Electrophoresis was carried out overnight at 1.5mA/cm gel width. Staining was according to Balinsky et al.(1973a).

TABLE I

MICHAELIS CONSTANTS FOR HUMAN HEXOKINASES

Isozyme	Michaelis constant (mM)			
	Glucose	MgATP with glucose	Fructose	MgATP with fructose
I	0.063	0.50	12.6	0.62
II	0.140	0.36	12.5	1.25
III	0.016	0.44	6.1	1.10

hexokinases. It will be noted that the K_m value for MgATP with

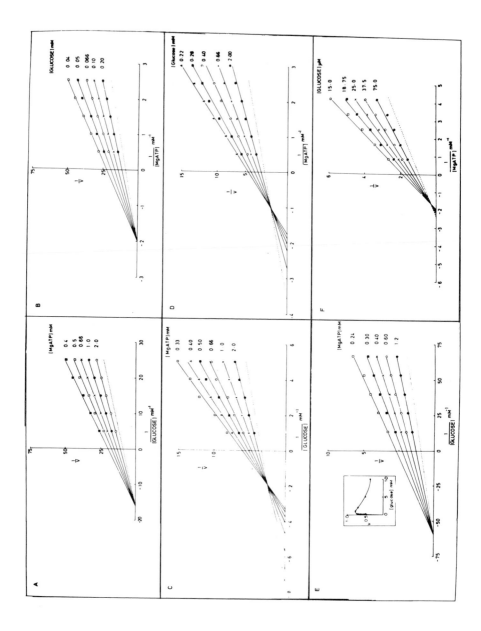

Fig. 2. Double reciprocal plots of human hexokinase, pH 8.0.
A,B: HK I; C,D: HK II; E,F: HK III. Assay conditions : 50mM
Tris-HCl buffer, pH 8.0, 4.5 U glucose-6-phosphate dehydrogenase, 0.36mM NADP, glucose, MgATP and enzyme. Activity was
assayed fluorimetrically. The lowest dotted line is that
extrapolated to infinite concentration of the second substrate.

---

fructose as sugar substrate was higher than that for glucose
in the case of both HK II and HK III. The K_m value of HK III
for glucose was considerably lower than that for the other two
isozymes, with HK I having an intermediate value. The values
obtained for glucose and MgATP are of the same order of
magnitude as those for rat tissues (cf. Grossbard and Schimke,
1966).

A very recent publication described partial purification
of human HK III from spleen and HK I from heart (Neumann et
al., 1974). The apparent Michaelis constants obtained for
ATP (1.4 - 1.6mM) and for glucose for HK I and III (0.1mM and
0.03 mM respectively) were somewhat higher than those in this
study.

Increase in tumors of HK II, which is unaffected by high
glucose concentrations, might be expected to compensate for
inhibition of HK III by glucose concentrations above 1mM.
However, the physiological significance of this is not clear
in the presence of a large excess of HK I, which is also
unaffected by high glucose concentrations. The explanation
may lie in the observation that rat HK II is subject to
hormonal regulation (Katzen and Schimke, 1965).

ISOZYMES OF PYRUVATE KINASE

Of 15 enzymes of carbohydrate metabolism assayed, only
pyruvate kinase (ATP : pyruvate 2-0-phosphotransferase, EC 2.
7.1.40)(PK) had consistently higher activity in all but one
case in hepatoma tissue compared to "host" tissue (Balinsky
et al., 1973a). This enzyme has previously been shown to
have raised activities in rat hepatomas, the increase in
activity correlating with increasing tumor growth rate (Weber,
1966). Starch gel electrophoresis of PK showed a change from
the typical pattern of normal human liver, with predominance
of the L band, to a pattern showing no L band in the hepatoma,
only the M band. Host tissue showed both bands (Balinsky et
al., 1973a). Imamura and Tanaka (1972) recently showed
separation of M_1 of muscle and M_2 of liver and hepatoma. We
subsequently succeeded in separating the M_1 isozyme of human
muscle and the M_2 isozyme of hepatoma and the hepatoma cell
line respectively by starch gel electrophoresis (Fig. 1B)

(K. D. Hammond and D. Balinsky, unpublished results). It should be noted that this particular hepatoma also had the L isozyme, as well as hybrids of L and M. A recent study of human placental and non-hepatic tumor PK has shown an isozyme with slower mobility than liver M-PK (Spellman and Fottrell, 1973).

Properties of pyruvate kinase isozymes. Detailed studies have been carried out on the L isozyme of rat liver and the M isozymes from rat muscle and hepatoma. The rat L isozyme has been shown to be allosteric with respect to the substrate phospho-enol-pyruvate (PEP). It is inhibited allosterically by ATP and alanine; this inhibition is relieved by fructose-1, 6-diphosphate (FDP) (Tanaka et al., 1967; Susor and Rutter, 1968). Rat muscle M_1-PK has normal Michaelis-Menten kinetics, is unaffected by FDP and less inhibited by ATP (Tanaka et al., 1967; Susor and Rutter, 1968). M_2-PK from Yoshida ascites hepatoma has recently been shown to have allosteric properties though less marked than those of the L isozyme (Imamura et al., 1972).

Human pyruvate kinases - kinetic properties. The L isozyme was purified from a normal adult liver, and the M isozyme from a human hepatoma and from fetal liver (Balinsky et al., 1973c). In human fetal liver, only 20-30% of the total PK activity is M.

Plots of reciprocal velocity *vs* reciprocal ADP concentration at various fixed concentrations of PEP were made (Fig. 3). The L isozyme showed substrate inhibition at high concentrations of ADP, especially at low levels of PEP. The Michaelis constant for ADP increased as the PEP concentration increased. Similar double reciprocal plots for the M isozyme from both hepatoma and fetal liver showed no inhibition, and a Michaelis constant for ATP independent of PEP concentration.

Replotting the data as reciprocal velocity *vs* reciprocal PEP concentration gave three different plots for the three isozymes. Lines curving upward were obtained for the liver L isozyme, since this enzyme is allosteric, as will be discussed below. The fetal liver M isozyme gave straight lines meeting on the abscissa, showing a Michaelis constant for PEP independent of ADP concentration. The hepatoma isozyme gave complex lines.

Effects of modulators. A detailed study of the effects of activators and inhibitors on the isozymes was carried out. Enzyme activity was plotted as a function of substrate, activator or inhibitor concentration. In addition, the data were analyzed using a Hill plot.

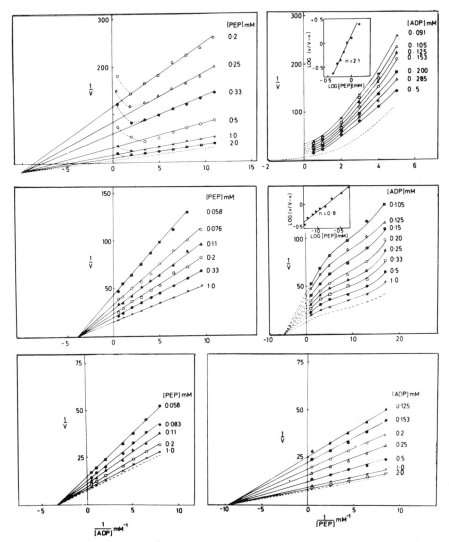

Fig. 3. Double reciprocal plots of human pyruvate kinase, pH 7.6 A,B: L-PK (liver); C,D: M-PK (hepatoma); E,F: M-PK (fetal liver). Assay conditions were as in Table II, except that PEP and ADP were varied. The lowest dotted line is that extrapolated to infinite concentration of the second substrate.

When enzyme activity was plotted as a function of PEP concentration, the liver L isozyme showed a sigmoid curve (Fig. 4.). The $K_{1/2}$ value for PEP derived from the Hill plot was 0.75mM, with a Hill coefficient of 2.1, representing the

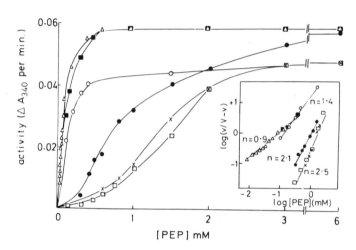

Figure 4. Activity of L-PK under various conditions ●-●
Control; Δ-Δ 0.1mM FDP; x-x 1mM ATP; o-o 1mM ATP + 0.1mM FDP;
□ - □ 4mM DL-alanine; ■ - ■ 4mM DL-alanine + 2 μM FDP. Inset:
Hill plot. Assay conditions were as in Table II.

binding of least 2 moles of PEP to the enzyme for activity.
FDP completely eliminated the allosteric effect, reduced the
$K_{1/2}$ value to 0.08mM and the slope of the Hill plot to 0.9,
i.e. no co-operativity and only 1 mole PEP bound per mole
enzyme. ATP inhibited the enzyme and increased the allosteric
effect - the slope of the Hill plot increased to 2.5, suggesting
that under these conditions at least 3 moles of PEP per mole
enzyme might be required for activity. FDP (0.1mM) partially
eliminated the allosteric effect. The Hill slope was reduced
to 1.4, but full activity was not regained. Alanine too
inhibited the enzyme allosterically; its effect was completely
eliminated by 2 μM FDP.

On a plot of velocity *vs* PEP concentration, the hepatoma
M enzyme showed no allosteric properties (Fig. 5). ATP
inhibited, but FDP had no effect on this inhibition. Alanine
inhibited allosterically - interestingly enough, this inhibition was completely eliminated by FDP. The fetal M isozyme
showed a similar plot.

A more detailed study of the individual modulators was
subsequently carried out.

Effects of sugar phosphates. FDP was by far the most potent
activator of liver L-PK. As seen in Table II, the concentration of FDP required for half-maximal activation, 0.28 μM, was
approximately 3000 times lower than that of fructose-6-P,

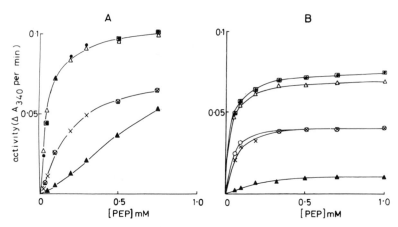

Fig. 5. Activity of M-PK from hepatoma (A) and fetal liver
(B) under various conditions. ●-● Control; □ - □ 0.1mM FDP;
x-x 1mM ATP; o-o 1mM ATP + 0.1mM FDP; ▲-▲ 4mM DL-alanine;
△-△ 4mM DL-alanine + 0.1mM FDP. Assay conditions were as in
Table II.

TABLE II

EFFECTS OF SUGARS AND SUGAR PHOSPHATES ON NORMAL HUMAN
ADULT LIVER TYPE L PYRUVATE KINASE ACTIVITY

Assay Conditions: 50mM triethanolamine-HCl Buffer, pH 7.4,
10mM $MgCl_2$, 5mM EDTA, 0.1M KCl, 0.5mM PEP, 1mM ADP, 0.15mM
NADH, 4U LDH and enzyme.

Hexose or hexose phosphate	$K_{\frac{1}{2}}$ (mM)	Amount of enzyme activation at $K_{\frac{1}{2}}$
None	–	1.0
FDP	0.32×10^{-3}	2.7
Glucose-6-phosphate	0.66	2.7
Fructose-6-phosphate	0.76	2.7
Fructose-1-phosphate	1.05	2.7
2-deoxy-glucose-6-phosphate	2.88	2.0
Glucose-1-phosphate	7.94	1.7
Galactose-6-phosphate	No effect	
Glucose	No effect	
Fructose	No effect	

fructose-1-P and glucose-6-P. It is interesting to note that
the maximal activation by these 4 sugar phosphates was

identical. 2-deoxy-glucose-6-phosphate and glucose-1-phosphate had even higher $K_{1/2}$ values, and did not activate as much even at saturating concentrations. Galactose-6-phosphate, glucose and fructose had no effect. There was also activation by some triose phosphates, notably dihydroxyacetone phosphate and, to a lesser extent, by glyceraldehyde-3-phosphate and 2-phospho-glycerate, whereas 3-phosphoglycerate and glycerol-1-phosphate had no effect (Balinsky et al., 1973d).

None of these compounds activated the hepatoma or fetal M isozymes.

Effects of ATP and alanine on FDP activation. Since both ATP and alanine inhibited the enzyme allosterically, it was considered of interest to test whether they bind at the same site or at different sites on the enzyme. If they bind at the same site, one would expect competition for this site in a mixture of the two inhibitors; if not, an additive effect would be expected.

Fig. 6. shows a plot of activity *vs* FDP concentration. In the absence of inhibitors, a hyperbolic plot was obtained, with $K_{1/2}$ of 0.28 μM and n_H value of 1.2. Addition of either 2mM ATP or 16mM alanine gave an allosteric plot, with n_H value raised to 1.5 and an increase in the apparent $K_{1/2}$ for FDP to 1μM. Addition of a mixture of ATP and alanine at these concentrations increased the slope of the Hill plot to 2.1, and raised the apparent $K_{1/2}$ for FDP still higher, to 2.8 M. This plot, and the one using 1mM ATP+8mM alanine, shows that inhibition by these two compounds probably occurs at different sites, each permitting relief by FDP.

Effect of different concentrations of ATP on the activities of the enzymes. Fig. 7 shows that inhibition of the liver isozyme by ATP was hyperbolic with respect to ATP concentration, with a Hill slope of 1.0. Addition of FDP relieved the inhibition partially, and caused the curve to become allosteric, with Hill slopes of 2.0 and 3.3 at 0.2 μM and 0.1mM FDP respectively. The activity of the hepatoma and fetal M-PK's at different ATP concentrations showed an allosteric plot with respect to ATP concentration, with Hill slope of 2.2. These plots and slopes were unaffected by FDP, though some slight relief of inhibition by 0.1mM FDP could be observed at high ATP concentrations for the fetal liver M-PK. The $K_{1/2}$ values of ATP for the liver L isozyme in the absence of FDP were considerably lower than those for the hepatoma and fetal liver type M enzymes. However, the apparent $K_{1/2}$ values of ATP for the liver L isozyme in the presence of FDP were comparable to the values for the M isozymes either with or without FDP. This

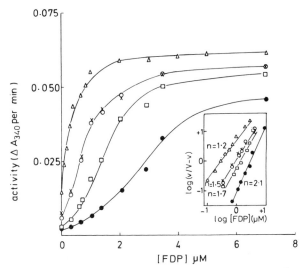

Fig. 6. Activity of L-PK at various concentrations of FDP.
Δ-Δ Control; x-x 2mM ATP; o-o 16mM DL-alanine; □ - □ 8mM DL-
alanine + 1mM ATP; ●-● 16mM DL-alanine + 2mM ATP. Assay
conditions were as in Table II.

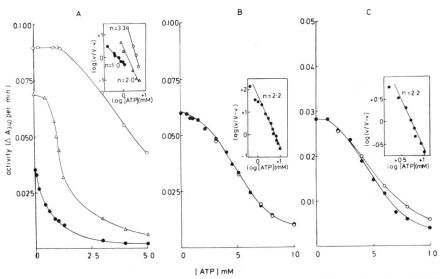

Fig. 7. Inhibition of PK by ATP. A : L-PK (liver); B : M-PK
(hepatoma; C : M-PK (fetal liver). ●-● Control; Δ-Δ 0.2 μM
FDP; o-o 0.1mM FDP. Assay conditions were as in Table II.

indicates that only 1 mole of ATP need bind per mole of enzyme
to give inhibition, but 2-4 moles are required in the presence

of increasing concentrations of FDP.

Effect of different concentrations of alanine on PK activity.
Alanine was found to be a very potent inhibitor of all three
isozyme preparations, with $K_{1/2}$ values of less than 1mM (Fig.
8). In all cases, as shown previously, FDP relieved the
inhibition. This is of great interest, as the other results
shown would appear to indicate that the M type isozymes are
not allosteric, yet the results with alanine point to the fact
that these isozymes too must have binding sites for the
activator as well as the inhibitors.

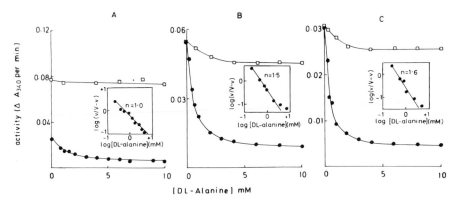

Fig. 8. Inhibition of PK by alanine. A. B. C as in Fig. 7.
●-● Control; □-□ 2μM FDP. Assay conditions were as in
Table II.

The $K_{1/2}$ and n_H values are summarized in Table III.

The allosteric properties of the human L isozyme are
comparable to those found for the rat L isozyme (Tanaka et al.,
1967; Susor and Rutter, 1968). The Michaelis-Menten kinetics
obtained for the fetal liver M isozyme are similar to those
of Jiménez de Asúa et al.(1971) for rat liver M-PK. The
present results for human tumor M-PK differ from those for
Yoshida ascites hepatoma M_2-PK (Imamura et al., 1972); this
isozyme was found to be allosteric, with reversal of the
allosteric effect by FDP. A similar effect was found for rat
liver M_2-PK by van Berkel et al. (1973), but only at pH 8.0:
at pH 7.5 the allosteric effect could barely be detected.
Michaelis-Menten kinetics were found for Morris hepatoma
PK (Taylor et al., 1969). Alanine inhibition and its reversal
by FDP was found for the M_2-PK of rat liver (van Berkel et
al., 1973), of Morris hepatoma (Taylor et al., 1969) and of
Ehrlich ascites tumor cells (Sparmann et al., 1973); relief
of alanine inhibition of Yoshida ascites hepatoma PK by
FDP was not observed by Imamura et al. (1972). These

TABLE III

KINETIC PARAMETERS FOR ATP AND ALANINE
INHIBITION OF PYRUVATE KINASE

Inhibitor	FDP (μM)	Liver L		Hepatoma M		Fetal liver M	
		$K_{\frac{1}{2}}$ (mM)	n_H	$K_{\frac{1}{2}}$ (mM)	n_H	$K_{\frac{1}{2}}$ (mM)	n_H
ATP	0	0.63	1.0	5.2	2.2	5.6	2.2
	0.2	1.5	2.0	–	–	–	--
	100	4.4	3.3	5.2	2.2	5.6	2.2
Alanine	0	0.63	1.0	0.52	1.5	0.40	1.6

workers have discussed the possibility that M_2-PK can exist in
interconvertible FDP-sensitive and FDP-insensitive forms. This
might possibly explain the differences in results of the
various studies, and the anomalous kinetics seen in Fig. 3d.
Anomalous plots for M_2-PK under certain conditions were also
found by Imamura et al. (1972).

From a consideration of the maximum slope of the Hill plots
obtained, *viz.* 3.3, one could expect approximately 4 binding
sites per mole enzyme for each modulator. Evidence of 4 subunits
for bovine liver L and muscle M pyruvate kinases has recently
been obtained (Cardenas and Dyson, 1973).

The present results are consistent with the Monod-Wyman-
Changeaux theory (Monod et al., 1969) that some enzymes can
exist in alternative conformations. The L isozyme probably
exists normally in the inactive form, in the absence of active
glycolysis. Increase of FDP produced by phosphofructokinase
during active glycolysis activates the enzyme. If ATP produced by the reaction is not rapidly utilized, the product
will allosterically inhibit the reaction. An excess of alanine
will inhibit the reaction, permitting PEP to be utilized by
enolase and thereby returning to the gluconeogenic pathway.
If glycolysis increases again, the FDP produced will reverse
the effect of alanine.

The M isozymes of hepatoma and liver appear to exist normally in the active conformation, but can be controlled by
alanine and FDP as discussed above.

ACKNOWLEDGEMENTS

This work was partly supported by the South African
Medical Research Council.

REFERENCES

Balinsky, D., E. Cayanis,E.W.Geddes, and I. Bersohn 1973a.

Activities and isoenzyme patterns of some enzymes of glucose metabolism in human primary malignant hepatoma. *Can. Res.* 33: 249-255.

Balinsky, D., E. Cayanis, and I. Bersohn 1973b. Properties of the altered hexokinase isoenzyme of human primary malignant hepatocellular carcinoma. In:*Liver*, Eds. S.J. Saunders and J. Terblanche, Pitman Medical, London 329-330.

Balinsky, D., E. Cayanis, and I. Bersohn 1973c. Comparative kinetic study of human pyruvate kinases isolated from adult and fetal livers and from hepatoma. *Biochemistry* 12: 863-870.

Balinsky, D., E. Cayanis, and I. Bersohn 1973d. The effects of various modulators on the activities of human pyruvate kinases isolated from normal adult and foetal liver and hepatoma tissue. *Int. J. Biochem.* 4: 489-501.

Cardenas, J. M. and R. D. Dyson 1973. Bovine pyruvate kinases II. Purification of the liver isozyme and its hybridization with skeletal muscle pyruvate kinase. *J. Biol. Chem.* 248: 6938-6944.

Grossbard, L. and R. T. Schimke 1966. Multiple hexokinases of rat tissues. Purification and comparison of soluble forms. *J. Biol. Chem.* 241: 3546-3560.

Imamura, K. and T. Tanaka 1972. Multimolecular forms of pyruvate kinase from rat and other mammalian tissues. I. Electrophoretic studies. *J. Biochem.* (Tokyo) 71: 1043-1051.

Imamura, K., K. Taniuchi, and T. Tanaka 1972. Multimolecular forms of pyruvate kinase. II. Purification of M_2-type pyruvate kinase from Yoshida ascites hepatoma 130 cells and comparative studies on enzymological and immunological properties of the three types of pyruvate kinases, L, M_1 and M_2. *J. Biochem.* (Tokyo) 72: 1001-1015.

Jiménez de Asúa, L., E. Rozengurt, J. J. Devalle, and H. Carminatti 1971. Some kinetic differences between the M isoenzymes of pyruvate kinase from liver and muscle. *Biochim. Biophys. Acta.* 235 : 326-334.

Katzen, H. M. and R. T. Schimke 1965. Multiple forms of hexokinase in the rat. Tissue distribution, age dependency and properties. *Proc. Natl. Acad. Sci. U.S.A.* 54:1218-1225.

Monod, J., J. Wyman, and J. P. Changeux 1965. Nature of allosteric transitions - a plausible model. *J. Mol. Biol.* 12: 88-118.

Neumann, S., F. Falkenberg, and G. Pfleiderer 1974. Purification and immunological characterization of the human hexokinase isoenzymes I and III (ATP-D-hexose 6-phosphotransferase EC 2.7.1.1). *Biochim. Biophys. Acta.* 334: 328-342.

932

Prates, M. D. 1961. Cancer and cirrhosis of the liver in the Portuguese East African with special reference to the specific age and sex rates in Lourenço Marques. *Acta Un. Int. Contr. Canc.* 17: 718-739.

Prozesky, O. W., C. Brits, and W. O. K. Grabow 1973. In vitro culture of cell lines from Australia antigen positive and negative hepatoma patients. In:*Liver*, eds. S.J. Saunders and J. Terblanche, Pitman Medical, London, 358-360.

Sato, S., T. Matsushima, and T. Sugimura 1969. Hexokinase isozyme patterns of experimental hepatomas of rats. *Can. Res.* 29: 1437-1446.

Shatton, J. B., H. P. Morris, and S. Weinhouse 1969. Kinetic, electrophoretic, and chromatographic studies on glucose-ATP phosphotransferases in rat hepatomas. *Can. Res.* 29: 1161-1172.

Sparmann, G., J. Schulz, and E. Hofmann 1973. Effects of L-alanine and fructose (1,6-diphosphate) on pyruvate kinase from Ehrlich ascites tumour cells. *FEBS Letters* 36: 305-308.

Spellman, C. M. and P. F. Fottrell 1973. Similarities between pyruvate kinase from human placenta and tumours. *FEBS Letters* 37: 281-284.

Susor, W. A. and W. J. Rutter 1968. Some distinctive properties of pyruvate kinase. *Biochem. Biophys. Res. Comm.* 30: 14-20.

Tanaka, T., Y. Harano, F. Sue, and H. Morimura 1967. Crystallization, characterization and metabolic regulation of two types of pyruvate kinase isolated from rat tissues. *J. Biochem.* (Tokyo) 62: 71-91.

Taylor, C. B., H. P. Morris, and G. Weber 1969. A comparison of the properties of pyruvate kinase from hepatoma 3924-A, normal liver and muscle. *Life Sci.* 8 (II) : 635-644.

van Berkel, Th. J. C., J. F. Koster, and W. C. Hülsmanns 1973. Some kinetic properties of the allosteric M-type pyruvate kinase from rat liver; influence of pH and the nature of the amino acid inhibition. *Biochim. Biophys. Acta* 321: 171-180.

Weber, G. 1966. The molecular correlation concept. Studies on the metabolic pattern of hepatomas. *Gann Monograph* 1: 99-115.

AN ALKALINE PHOSPHATASE VARIANT IN
HEPATOMA RESEMBLING THE PLACENTAL ENZYME

KAZUYA HIGASHINO, SHUNJIRO KUDO and YUICHI YAMAMURA
Third Department of Internal Medicine
Osaka University School of Medicine

ABSTRACT. An alkaline phosphatase having a unique electrophoretic mobility was found in hepatoma patients. The sera
and hepatic tissues of the non-hepatoma patients with hepatic or nonhepatic diseases failed to show this variant of
alkaline phosphatase.

The localization of this enzyme in cancer tissue and the
occurrence of the identical alkaline phosphatase in the
serum and tumor of other hepatoma patients, both male and
female, suggested that this enzyme is an actual product of
the hepatoma cells and not a degradation product of either
the liver or placental alkaline phosphatase.

The properties of this enzyme studied in comparison with
those of the liver and placental alkaline phosphatases revealed some unique characteristics. With respect to inhibition by L-phenylalanine, L-tryptophan and L-homoarginine,
inactivation by urea, susceptibility to neuraminidase and
molecular size, this variant enzyme was indistinguishable
from the placental enzyme (Regan isozyme). On the contrary,
with respect to pH optimum and sensitivity to inhibition by
phosphate, this enzyme was essentially the same as the liver
enzyme. The properties of our enzyme which distinguished
it from the other two phosphatases were electrophoretic
mobility, heat stability and susceptibility to $\underline{L}$ -leucine
and EDTA. Although cross reactivity of this enzyme with
anti-placental alkaline phosphatase antibody was recognized,
its immunochemical specificity seemed to differ in detail
from the placental alkaline phosphatase.

The alkaline phosphatases have been qualitatively analyzed
by many medical investigators since the time when the idea of
isozymes was first put forward by Markert and Møller (1959).
Knowledge of these isozymes has been of value in the diagnosis
of diseases, because when alkaline phosphatase isozymes associated with known disease states are identified in serum, the
diseased organ may be pinpointed (Fishman, 1967). In a series
of electrophoretic examinations of the serum for this purpose,
we found an abnormal alkaline phosphatase in a patient with
hepatocellular carcinoma (Higashino et al, 1972a). This
abnormal alkaline phosphatase had the fastest anodal migration amongst all the isozymes found in man (Fig. 1). For

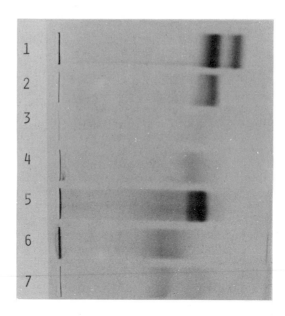

Fig. 1. Electropherogram of the alkaline phosphatase isozymes. Serum with variant alkaline phosphatase (1), liver alkaline phosphatase (2), serum with bone alkaline phosphatase (3), serum with Regan isozyme (4), crude placental alkaline phosphatase (5), crude lung alkaline phosphatase (6), and serum with intestinal alkaline phosphatase (7).

convenience's sake the enzyme was called the variant alkaline phosphatase.

In this report we describe first, the evidence for the occurrence of a variant alkaline phosphatase in different patients suggesting that the enzyme is not a partial degradation product, but an actual product of the hepatoma tissue, and second, the characteristics of this enzyme relative to the placental and liver alkaline phosphatases.

MATERIALS AND METHODS

Enzyme assay. Enzyme activity was measured at 37° C with phenyl phosphate as the substrate as previously described (Higashino et al, 1972a and Higashino et al, 1973).

Immunological studies. For the preparation of antibody against either the placental or the variant alkaline phosphatase, each of the purified antigens was mixed with an equal volume of complete Freund's adjuvant and the mixture was injected $S.C.$ into rabbits. A booster injection was given two to three weeks later (Higashino, Kudo, and Yamamura, in preparation). An antibody rich γ-globulin fraction was isolated from the antiserum by ethanol fractionation (Deutsch and Nichol, 1948). Immunodiffusion was carried out by Ouchterlony's technique (1949). Antigen-antibody complexes formed with the minute amounts of enzyme protein failed to produce visible precipitin lines, therefore the lines were detected by staining for enzyme activity.

Electrophoresis. Horizontal thin layer polyacrylamide gel electrophoresis was carried out as previously reported (Higashino et al, 1972a).

Serum and hepatic cancerous tissue. Serum and cancerous tissue were obtained from patients with hepatoma and various other diseases and from normal people.

Extraction of alkaline phosphatase from tissue. Extraction was performed with n-butanol and fractionation with acetone (Sussman, Small and Cotolove, 1968).

Purification of the alkaline phosphatase from serum or butanol extract. The methods were as described previously (Higashino et al, 1972a). The purification steps consisted of differential fractionation with ethanol (52-70%), DEAE-cellulose chromatography and gel filtration on Sephadex G-200.

OCCURRENCE OF THE VARIANT ALKALINE PHOSPHATASE IN HEPATOMA

In order to find the disease this enzyme occurred in, the serum and/or hepatic tissues of cases with various diseases were electrophoretically analyzed. Table I examines the occurrence of the variant alkaline phosphatase in serum and/or hepatic tissue of normal people and of patients with diseases, comprising hepatic and non-hepatic diseases. Although the number and variety of diseases in the examined patients was limited, only those patients with hepatoma showed the presence of the variant enzyme in the serum. Among thirteen hepatoma patients four had this enzyme. Table II shows the relation of the total alkaline phosphatase activity to the variant enzyme activity in the serum of these four patients. These values were variable case by case and during the course of the disease. The occurrence of this enzyme in the hepatic tissue was examined by using the acetone 30 to 60 percent fraction of the butanol extract as the enzyme. As shown in Table I, the hepatic specimens had no variant enzyme. However, among seven hepatoma specimens three were found to have this enzyme (Higashino et al, in preparation). These three

TABLE I

Diseases	Cases with variant AP in serum	Cases with variant AP in hepatic tissues
Normal control	0/100	
Patients with normal liver		0/23
Primary hepatoma	4/13	3/7
Metastatic carcinoma of the liver	0/9	0/3
Hemangioma of the liver	0/2	
Liver cirrhosis	0/12	0/13
Fatty liver	0/5	0/10
Acute viral hepatitis	0/11	
Acute alcoholic intoxication		0/1
Infantile hepatitis	0/1	
Intrahepatic cholestasis	0/4	
Diseases of biliary tract	0/7	0/1
Congenital biliary atresia	0/3	
Carcinoma of pancreas	0/4	
Pancreatitis	0/5	
Stomach carcinoma	0/5	
Renal tublar acidosis	0/3	
Carcinoma of the lung with bone metastasis	0/6	
Pregnancy (3M-6M)	0/10	
Cerebrovascular accidents		0/7
Cardiac diseases		0/31
Lobar pneumonia		0/5
Miscellaneous	0/15	0/3
TOTAL	4/215	3/104

Occurrence of the variant alkaline phosphatase in serum and/ or tissue of patients with various diseases.

were all from patients having this enzyme in their sera. The electropherogram of these hepatoma extracts is shown in Fig. 2. Two main bands of activity can be seen, that is, the one with faster electrophoretic mobility and the other with slower mobility. The latter corresponds to the liver alkaline phosphatase and the former to the variant alkaline phosphatase. The electrophoretic mobility of the tissue variant enzyme was identical with that of serum variant enzyme.

Table III shows the relation of the occurrence of the variant alkaline phosphatase to the macroscopic type of hepatoma.

TABLE II

No.	Name	AP activity in serum K.A.U.	Variant AP in serum K.A.U.
1	I K	81	17.5
2	T K	188	173.7
3	S S	116	12.8
4	K O	132	38.8

Relations between total and variant alkaline phosphatase activity in the serum of hepatoma patients with this enzyme.

The enzyme was found to arise from the massive type of hepatoma (Higashino et al, in preparation).

These results suggested that the enzyme arose from hepatoma. To further ascertain that the variant alkaline phosphatase is a product of the hepatoma tissues, the enzyme activity was examined in specimen from the central part and peripheral part of the hepatoma mass, and, also, from the uninvolved part of the same liver. n-Butanol extracts of these parts were subjected to electrophoresis, followed by densitometry of the zymogram.

One of these experiments is shown in Table IV. The variant enzyme was largely or absolutely localized in the hepatoma tissue (Higashino, Kudo and Yamamura, in preparation). Although little activity was present in the apparently intact

TABLE III

Type	Variant AP	
	(+)	(−)
	Type/Case	
Massive	4/4*	2/5
Diffuse	0	2/5
Nodular	0	1/5

*One is associated with nodular hepatoma.

Relationship between the occurrence of variant alkaline phosphatase and the macroscopic type of hepatoma.

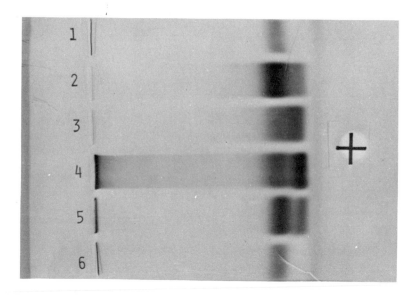

Fig. 2. Electropherogram of *n*-butanol extract from hepatoma tissues of the three patients with hepatocellular carcinoma having the variant alkaline phosphatase in their sera (2-4), and the liver alkaline phosphatase (1 and 6) and variant alkaline phosphatase in serum (5).

part of the liver, a separate experiment clearly confirmed the absence of the enzyme in the intact part. The presence of the enzyme in this case probably resulted from the contamination of the blood or cancer cells containing this enzyme in the specimen. The lower alkaline phosphatase activity in the central part of the tumor than that of the peripheral part may be ascribed to the necrosis of the hepatoma cells in the central part. The percentage of the variant enzyme in the serum was less than that in the cancerous tissue, but higher than that of the intact part. These facts may indicate that the tumor tissue produces the variant alkaline phosphatase and the enzyme passes into the blood stream together with the liver alkaline phosphatase derived mainly from the non-cancerous part of the liver.

The close association of the variant enzyme with hepatoma tissue led to the study of its enzymatic properties. But, as the variant enzyme occurred in the massive type of hepatoma where necrosis was usually present, a question arose as to whether this enzyme was a partially degraded product of either the placental or the liver alkaline phosphatases which were known to occur in other cancers and hepatomas, respectively.

TABLE IV
ACTIVITY OF THE VARIANT
ALKALINE PHOSPHATASE

Site of the Liver. from* which Specimen Obtained in a Hepatoma Patient	Total Activity** µmoles phenol/ min / mg protein	Percent Activity	
		Liver AP	Variant AP
Central part of a massive type carcinoma	0.39	23.5	76.5
Peripheral part of the same carcinoma	0.55	45.7	54.3
Macroscopically almost intact part of the liver	1.16	89.6	10.4
Serum	--	74.2	25.8

*Five grams of each specimen was extracted with n-butanol, and fractionated with acetone from 30 to 60 percent. The precipitate obtained by centrifugation, dissolved in 10 mM Tris-HCl buffer, pH 7.4, and the final volume of the fraction was made equal with the buffer after dialysis.

**Each extract contained both the variant and liver alkaline phosphatases.

In other words, the question was whether the variant alkaline phosphatase occurred generally in some of the hepatoma as a result of enzyme synthesis in the hepatoma cells. To elucidate this, cancerous tissue without necrosis was analyzed for the enzyme, and this revealed the presence of the enzyme in this tissue too. On the other hand, the enzymes obtained from different patients were examined for enzymatic properties. For this purpose, the enzyme was purified from the serum of patients T. K. and K. O. to almost the same extent, and from the hepatoma tissues of patient K. O. The specific activity of the variant enzyme from hepatoma tissue at the final step (13.63 µmoles phenol/min/mg protein) (Higashino, Kudo and Yamamura, in preparation) was more than twice that from the serum (5.77 µmoles phenol/min/mg protein) (Higashino et al, 1972a). As the materials were limited further purification was impossible. But the enzyme at this step was electrophoretically and chromatographically homogeneous.

Although the data is not shown, when enzymes from these three sources were tested, each of the amino acids, L-phenylalanine, L-tryptophan, L-leucine and L-homoarginine, caused the same grade of inhibition at each concentration tested, i.e., 1 and 5 mM. The sensitivity of inhibition by inorganic phosphate or EDTA and the inactivation by urea were also the same. The pH optimum and the molecular size as revealed by elution profile from the gel filtration were not different. The heat stability at 65° and 56° was almost alike. Finally, when tested with anti-placental alkaline phosphatase antibody, cross reactivity was observed with the enzyme from every source (Higashino, Kudo and Yamamura, in preparation).

EFFECTS OF AMINO ACIDS AND THEIR RELATED COMPOUNDS ON THE VARIANT ENZYME ACTIVITY

Since, as mentioned in the foregoing item, several amino acids were found to inhibit the variant alkaline phosphatase, the action of various other compounds was tested in order to examine certain structural characteristics.

L-α-amino acids with a hydrophobic ring or with a side chain, corresponding to C_5 and C_6 in the carbon chain length, such as L-tryptophan, L-phenylalanine, L-norleucine and L-leucine, were the best inhibitors. Among primary amines, n-hexyl amine was found to have a slight grade of inhibitory action. Esterification of a carboxyl radical of the inhibitory amino acid caused the decrease of inhibition. Monocarboxylic acids had no effect, and the blocking of the amino radical of L-leucine prevented the inhibitory action. The presence of ionizable groups at the ω-position hindered inhibition. These results suggested that both amino and carboxyl functions must be present at the α-position and that the function of the former is the more critical of the two. The inhibition by amino acids was pH dependent: maximum inhibition was observed at pH 9.3-9.5, and actually no inhibition was observed above pH 10.8.

The inhibition by amino acids was linearly uncompetitive in the absence of the product, but in the presence of constant concentration of inorganic phosphate it was linearly noncompetitive (Higashino et al, 1973).

COMPARISON OF THE VARIANT ALKALINE PHOSPHATASE WITH THE LIVER AND THE PLACENTAL ALKALINE PHOSPHATASES

It was previously reported (Fishman et al, 1968 and Nakayama, Yoshida and Kitamura, 1970) that the placental type alkaline phosphatase, such as the Regan or Nagao isozymes seemed to be produced by cancer cells. On the other hand, we have found that the increased alkaline phosphatase in the serum of many hepatoma patients was identical to the liver one at least electrophoretically. It seemed, therefore, of

942

interest to know how our variant alkaline phosphatase differed from those placental and liver alkaline phosphatases.

First, the effects of amino acids on these three enzymes were compared. L-phenylalanine, L-tryptophan and L-homoarginine were selected from among these inhibitors. The inhibition pattern by these amino acids for the variant alkaline phosphatase was quite similar to that of the placental enzyme and different from that of the liver enzyme, and the inhibition of the variant enzyme by L-phenylalanine and L-tryptophan was strong, but little or no inhibition was observed by L-homoarginine (Table V).

The effect of urea on the activity of the variant alkaline phosphatase was similar to that of the placental isozyme, but was different from the liver isozyme.

With regard to molecular size the liver enzyme seemed to be larger than the variant or the placental enzyme. The latter two enzymes did not appear to differ in their molecular weights.

In order to know the immunochemical properties, the three enzymes were reacted with an anti-placental alkaline phosphatase antibody or an anti-variant alkaline phosphatase antibody. The reactions were carried out at a suitable ratio of antigen to antibody.

Neither the anti-placental alkaline phosphatase antibody nor the anti-variant alkaline phosphatase antibody reacted with the liver alkaline phosphatase (Higashino, Kudo and Yamamura, in preparation).

As shown in Fig. 3, Ouchterlony double diffusion resulted in precipitin lines between the variant and placental alkaline phosphatases, and the anti-placental alkaline phosphatase antibody (Higashino, Kudo and Yamamura, in preparation). But the lines of the placental alkaline phosphatase spurred over that of the variant alkaline phosphatase. However, the precipitin lines of both antigens produced by anti-variant alkaline phosphatase antibody fused completely (Fig. 4). These results suggest two possibilities. One is that the variant alkaline phosphatase is immunochemically identical with one of the placental alkaline phosphatases, provided the latter enzyme fraction contains two alkaline phosphatases reacting with each anti-placental alkaline phosphatase antibody. Another possibility is that the variant alkaline phosphatase shares a part of its antigenic specificity with placental alkaline phosphatase. In other words, if there are two antigenic sites, i.e., one is common to the variant and the placental alkaline phosphatases and the other is specific to the placental alkaline phosphatase, then the problem is whether the placental alkaline phosphatase has the two antigenic sites in

TABLE V

COMPARATIVE PROPERTIES OF THREE
ALKALINE PHOSPHATASES

	VARIANT AP	PLACENTAL AP	PLACENTAL AP (REGAN)	LIVER AP
Inhibition by				
L-phenylalanine (5 mM)	78%	70%		10%
L-tryptophan (5 mM)	79%	72%		18%
L-homoarginine (5 mM)	18%	15%		78%
Inactivation by Urea (2 M)	33%	38%		81%
Molecular size Sephadex G-200	Almost the same as placental AP			Larger than placental AP
Cross reactivity to anti-placental AP antibody	+		+	–
Cross reactivity to anti-variant AP antibody	+		+	–
Effect of neuraminidase on electrophoretic mobility Rf with / Rf without	0.77		0.74	0.43
Inhibition by phosphate at 5 mM	6%		50%	3%
pH optimum	10.0--10.1		10.5--10.6	10.0--10.1

944

	fast α_2-globulin	α_2 to β-globulin	α_2-globulin
Inhibition by L-leucine (5 mM)	66%	28%	21%
Inhibition by EDTA			
0.1 mM	34%	10%	85%
0.5 mM	84%	33%	99%
Electrophoretic mobility corresponding site of serum protein	fast α_2-globulin	α_2 to β-globulin	α_2-globulin
Heat stability percent inactivation			
56°, 60 min	20%	1%	65%
65°, 3 min	62%	1%	98%
K_m value for phenyl phosphate at pH 9.9	1.1 mM	1.6 mM	1.9 mM

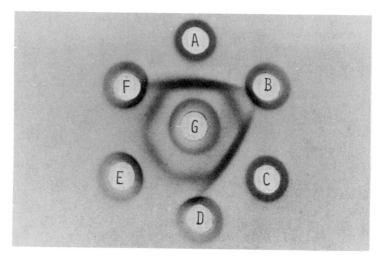

Fig. 3. Ouchterlony double diffusion in agar gel between anti-placental alkaline phosphatase antibody (G) and placental alkaline phosphatase (A,C), variant alkaline phosphatase from serum of patient K.O. (B,E), variant alkaline phosphatase from serum of patient T.K. (D), and variant alkaline phosphatase from hepatoma tissue of patient K.O. (F).

one enzyme molecule, or whether it actually consists of two enzymes having each proper antigenic site in each enzyme molecule.

The effect of neuraminidase treatment on the electrophoretic mobility was investigated. (Higashino, Hashinotsume and Yamamura, 1972b). The ratio of Rf values of the variant alkaline phosphatase with and without treatment was similar to that of the placental enzyme and different from that of the liver one.

With regard to the effect of the phosphate on the enzyme activity, the placental isozyme was most sensitive to this anion and the other two were unaffected at a 5 mM concentration.

The pH optimum for the variant and the liver enzyme was 10.1 to 10.0, while that of the placental alkaline phosphatase was about 10.5 to 10.6.

Inhibition by L-leucine was interesting. Because among amino acids L-leucine made it possible to distinguish the variant enzyme from the Regan isozyme. The variant enzyme was more sensitive than either the Regan or the liver alkaline phosphatase. The latter two showed an inhibition of about the same degree (Higashino et al, 1972a).

The grade of inhibition of the variant phosphatase by EDTA was found to fall between the liver and placental alkaline

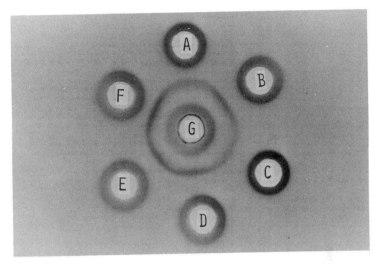

Fig. 4. Ouchterlony double diffusion in agar gel between anti-variant alkaline phosphatase antibody (G) and the placental alkaline phosphatase (A,C), and the variant alkaline phosphatase from different sources (B,D,E,F).
Samples in the peripheral wells are given in the legend to Fig. 3.

phosphatases.

The electrophoretic mobility of the variant enzyme was different from the liver and placental alkaline phosphatases. The mobility of the variant enzyme corresponds to fast α_2-globulin, but that of the placental or the liver alkaline phosphatase corresponds to the region between α_2 and β-globulin, or α_2-globulin.

The effect of heat treatment was compared among the three isozymes. The activity of the alkaline phosphatase isozymes remaining after incubation at 56° and 65° was examined as a function of time and the inhibition was expressed as the percentage obtained by comparison with the activity of the un-heated control. The variant enzyme was more labile than the placental enzyme, but was more stable than the liver enzyme (Higashino et al, 1972a).

The Michaelis constants for phenyl phosphate were not very different among the three enzymes.

DISCUSSION

The results presented in this paper bring up a question as to the structure of the variant alkaline phosphatase.

On the basis of its enzymatic properties, our variant alkaline phosphatase may be regarded as a variant of the placental alkaline phosphatase. What is the factor(s) responsible for the difference in properties between our variant and placental alkaline phosphatase? There are at least two possibilities. One is the difference in the carbohydrate moiety of the enzyme molecule, and another is whether the enzyme is a hybrid between the liver and the placental (Nagao) alkaline phosphatase.

As previous studies showed that the removal of sialic acid from the peptide hormone resulted in the loss of activity (Gottschalk, Whitten and Graham, 1960 and Goverde, Veenkamp and Homan, 1968) and as the alkaline phosphatase is a glycoprotein, the effect of neuraminic acid removal on the enzymatic properties was examined. Treatment of the enzyme with neuraminidase did not produce any changes in the enzymatic properties although the electrophoretic mobility was reduced (Higashino, Hashinotsume and Yamamura, 1972b). This does not rule out the possibility that some carbohydrate moiety other than neuraminic acid may be responsible for the differences in properties. The alkaline phosphatase from *E. coli* (Reid and Wilson, 1971) might serve as a supporting precedent for this hypothesis.

The possibility that the variant alkaline phosphatase is a hybrid seems interesting. If the alkaline phosphatases consist of subunits, situations analogous to aldolase isozyme in cancer cells (Sugimura et al, 1972) may obtain for this enzyme. These results suggest that in rapidly growing, poorly differentiated hepatoma, only prototype (Muscle-type) enzyme occurs, whereas in slowly growing, highly differentiated hepatoma, hybrids occur together with both the prototype and the differentiated form of the enzyme.

It is possible to regard the Regan or the Nagao as the prototype of the alkaline phosphatase. Is our variant enzyme, then, a hybrid between the liver and the placental alkaline phosphatase? A partial immunochemical identity of the variant alkaline phosphatase with the placental isozyme, and the enzymatic properties described may be consistent with this hypothesis, however, the electrophoretic patterns, the occurrence of only two molecular forms, i.e., the variant and liver enzymes, and the further lack of cross reactivity of the liver enzyme with anti-variant alkaline phosphatase isozymes are inconsistent with this idea.

The variant alkaline phosphatase is considered to be produced by the hepatoma cells as a result of altered gene expression. Therefore, the complete elucidation of the structure of the variant enzyme may provide a clue to the question whether hepatoma cells are in the state of disdifferentiation

(Sugimura et al, 1972) or dedifferentiation.

An abnormal alkaline phosphatase found by Warnock and Reisman (1969) in 1969 might be the same as our enzyme. This enzyme that occurred in hepatoma patients is inhibited by L-phenylalanine, and has an electrophoretic mobility corresponding to fast α_2-globulin on starch gel, and a molecular weight of about 158,000. The enzyme is relatively heat stable. These properties are similar to ours, but with respect to unique characteristics of our enzyme, such as immunochemical properties, pH optimum, sensitivity to the inhibition by L-leucine, inorganic phosphate, or EDTA, the similarity between these two enzymes is unknown yet.

ACKNOWLEDGMENTS

This report is the product of many co-workers. Their names are M. Hashinotsume, R. Ohtani, K. Kang, M. Fujioka, T. Hada, Y. Takahashi and T. Ohkochi. Their cooperation is sincerely appreciated. The authors, also, wish to express their thanks to Dr. G. W. Nace for his help and suggestions in preparing this manuscript.

REFERENCES

Deutsch, H. F. and J. C. Nichol 1948. Biophysical studies of blood plasma proteins. X. Fractionation studies of normal and immune horse serum. *J. Biol. Chem.* 176: 797-812.

Fishman, W. H. and N. K. Ghosh 1967. Isoenzymes of human alkaline phosphatase. *Adv. Clin. Chem.* 10: 255-370.

Fishman, W. H., N. I. Inglis, L. L. Stolbach, and M. J. Krant 1968. A serum alkaline phosphatase isoenzyme of human neoplastic cell origin. *Cancer Res.* 28: 150-154.

Gottschalk, A., W. K. Whitten, and E. R. B. Graham 1960. Inactivation of follicle-stimulating hormone by enzymic release of sialic acid. *Biochim. Biophys. Acta* 38: 183-184.

Goverde, B. C., F. J. N. Veenkamp, and J. D. H. Homan 1968. Studies on human chorionic gonadotrophin. II. Chemical composition and its relation to biological activity. *Acta Endocrin.* 59: 105-119.

Higashino, K., M. Hashinotsume, K. Kang, Y. Takahashi, and Y. Yamamura 1972a. Studies on a variant alkaline phosphatase in sera of patients with hepatocellular carcinoma. *Clin. Chim. Acta* 40: 67-81.

Higashino, K., M. Hashinotsume, and Y. Yamamura 1972b. Effect of neuraminic acid removal on the properties of a variant alkaline phosphatase in hepatoma. *Clin. Chim. Acta*

40: 305-307.

Higashino, K., M. Hashinotsume, Y. Yamamura, and M. Fujioka 1973. Kinetic studies of ahepatoma alkaline phosphatase. *Arch. Biochem. Biophys.* 158: 792-798.

Higashino, K., S. Kudo, and Y. Yamamura. Further studies on a variant of the placental alkaline phosphatase in human hepatic carcinoma. In preparation.

Higashino, K., R. Ohtani, M. Hashinotsume, S. Kudo, K. Kang, T. Hada, T. Ohkochi, Y. Takahashi, and Y. Yamamura. Hepatocellular carcinoma with reference to a variant of the placental alkaline phosphatase. In preparation.

Lowry, O. H., N. J. Rosebrough, A. L. Farr, and R. J. Randall 1951. Protein measurement with the Folin phenol reagent. *J. Biol. Chem.* 193: 265-275.

Markert, C. L. and F. Møller 1959. Multiple forms of enzymes: Tissue, ontogenetic, and species specific patterns. *Proc. Natl. Acad. Sci. Wash.* 45: 753-763.

Nakayama, T., M. Yoshida, and M. Kitamura 1970. L-Leucine sensitive, heat-stable alkaline-phosphatase isoenzyme detected in a patient with pleuritis carcinomatosa. *Clin. Chim. Acta* 30: 546-548.

Ouchterlony, Ö. 1949. Antigen-antibody reactions in gels. *Arkiv Kem. Mineral. Geol.* 26 B: 14, 1-9.

Reid, T. W. and I. B. Wilson 1971. In *The Enzymes* (P. D. Boyer, ed.), Vol. 4, 3rd ed., p. 387. Academic Press, New York and London.

Sugimura, T., T. Matsushima, T. Kawachi, K. Kogure, N. Tanaka, S. Miyake, M. Hozumi, S. Sato, and H. Sato 1972. Dis-differentiation and decarcinogenesis. *Gann Monograph on Cancer Res.* 13: 31-45.

Sussman, H. H., P. A. Small, Jr., and E. Cotlove 1968. Human alkaline phosphatase. Immunochemical identification of organ-specific isoenzymes. *J. Biol. Chem.* 243: 160-166.

Warnock, M. L. and R. Reisman 1969. Variant alkaline phosphatase in human hepatocellular cancers. *Clin. Chim. Acta* 24: 5-11.

ISOZYME PATTERNS OF GLYCOGEN PHOSPHORYLASE IN RAT HEPATOMAS

KIYOMI SATO, TSUYOSHI SATO,
HAROLD P. MORRIS AND SIDNEY WEINHOUSE
Fels Research Institute and Department of Biochemistry,
Temple University School of Medicine, Philadelphia;
Department of Biochemistry,
Hirosaki University School of Medicine, Hirosaki, Japan;
and the Department of Biochemistry,
Howard University School of Medicine, Washington, D.C.

ABSTRACT. A study of glycogen phosphorylase isozymes
in a series of transplantable rat hepatomas has confirmed
and extended our previous finding of profound alterations
of isozyme composition in a series of transplantable,
chemically-induced rat hepatomas. The general pattern
of isozyme alteration was similar to that found previously for the hexokinases, aldolases, and pyruvate kinases.
Using isoelectric focusing for separation and indenti-
fication, three distinct forms of phosphorylase b were
indentified. The form found in skeletal muscle focused
at pH 6.2, whereas the liver form gave a single peak
at pH 5.9. Fast growing, poorly differentiated hepatomas
had a single, third form, focusing at pH 5.6, whereas
less rapidly growing, well-differentiated hepatomas had
variable amounts of both the liver and hepatoma forms.
The three forms differed in kinetic properties and were
immunochemically distinct, with little or no cross-re-
activities. All of the hepatomas also contained the
phosphatases and cyclic-AMP requiring kinases for the
interconversion of the a and b forms.
Both the liver and tumor forms were present in fetal
liver, the tumor and muscle forms were present in fetal
skeletal muscle, but only the tumor form was present in
whole 14-day embryo. Adult spleen, kidney, testis, and
lung had both the liver and tumor forms; and brain and
heart had both fetal and muscle forms.
These data add further evidence for the occurrence
in tumors of an abnormality of gene expression manifested by the synthesis of protein inappropriate to the
adult tissue of origin. In the hepatomas, genes that
have been active in the fetal state but were inactivated
during normal embryonic development become re-activated
in cancer, while genes coding for isozymes under allosteric or hormonal regulation for normal hepatic function
become switched off.

INTRODUCTION

As part of a continuing study of the molecular background underlying the metabolism and growth of tumors, we have been studying the loss or retention of enzymes involved in specific hepatic functions in a series of rat liver neoplasms ranging widely in growth rate and degree of differentiation. Data from our own as well as other laboratories (Weinhouse,1972; Schapira, 1973; Ono and Weinhouse, 1972; Knox, 1973; Criss, 1971)has established for hepatomas a common pattern of isozyme alteration which supports and amplifies, and adds functional significance to many previous observations, particularly from the immunologic literature (Gold, 1969; Abelev, 1971; Lengerova, 1971; Baldwin, 1973; Alexander, 1972; Stonehill and Bendich, 1970; Anderson and Coggin, 1971), that point to a mis-programming of gene expression in cancer. Basically the pattern we and others have observed is represented in Figure 1.

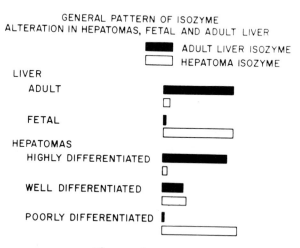

Figure 1

Pattern of isozyme alteration observed in Morris and other hepatomas. This figure is representative of data obtained by us and others for the glucose-ATP phosphotransferases, aldolases, adenylate kinases, and pyruvate kinases.

Of those enzymes which exist in multimolecular forms, the

adult rat liver has a sole or predominant isozyme, which is
under host control and is geared for hepatic functions, while
other isozymes are either low or absent in activity. In all
of the slow growing, highly differentiated hepatomas, which
resemble liver morphologically, cytologically and biochemically, there is a similar pattern, with virtually complete
retention of the liver-type isozyme. As dedifferentiation
proceeds, and growth rate increases, the hepatomas lose the
liver-type isozyme activity and other isozymes that are
normally low or may be undetectable in the adult liver begin
to appear. With progressive dedifferentiation to the poorly
differentiated state, associated with very rapid growth
rate, there is not only a complete or nearly complete disappearance of the liver type isozyme; but there is a replacement by a non-hepatic type isozyme, sometimes in extremely
high activity compared with the original total enzyme
activity in liver. It is notable that the isozyme patterns
of the poorly differentiated hepatomas resemble those of
fetal liver, where, like the poorly differentiated hepatomas,
the adult isozyme is low or absent. We have been struck by
the fact that loss of differentiation and increased growth
rates of hepatomas are associated with the loss of those
isozymes that are under host dietary or hormonal control, and
are geared for key function in hepatic metabolism; and
we feel that their replacement by other isozymes that are
not under host regulation may be an underlying molecular
foundation for the lack of control of cell proliferation that
characterizes the cancer cell.

In extending our studies, we have explored the isozyme
composition of glycogen phosphorylase. This enzyme is
highly appropriate for studies on loss of regulatory enzymes
in cancer, since it is highly regulated, it is involved in
the important hepatic function of glycogen mobilization, and
has been studied *in extenso* (Nigam and Cantero 1972; Ryman
and Whelan 1971; Brown and Cori 1960). Moreover as shown
in Figure 2, it participates, together with glycogen synthetase, in a synergistic cyclic activation and deactivation
catalyzed by kinases and phosphatases, in which cyclic AMP
plays a key role. This report will cover recent published
work supplemented by current unpublished work being conducted by Sato and Sato in Hirosaki.

METHODS

The experimental model for the most part was the Morris
hepatomas, a series of chemically induced, transplantable
rat hepatomas (Morris 1963, 1965, 1974; Morris and Wagner

SYNERGISTIC INTERCONVERSION OF
GLYCOGEN SYNTHETASES AND PHOSPHORYLASES

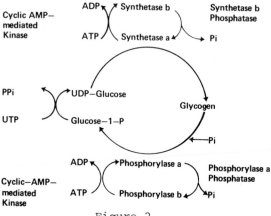

Figure 2

Diagram showing synergistic interconversions of glycogen synthetases and phosphorylases regulating glycogen synthesis and mobilization.

1968). Tumors were obtained as already described, and procedures for isoelectric focusing, assay and immunologic titration are published (Sato et al., 1973, 1973a). Disc gel electrophoresis was conducted according to Takeo et al., (1972).

RESULTS

Isoelectric focusing. Phosphorylase b isozymes are readily distinguished by isoelectric focusing. Figure 3 demonstrates the existence of three distinct molecular forms distinguished by their different isoelectric points. The muscle isozyme focuses at pH 6.2, and requires AMP for full activity. In contrast, the liver form, though exhibiting a somewhat broad base indicating some heterogeneity, gives a peak at pH 5.9, and is further distinguished from the muscle form in requiring, in addition to AMP, a divalent anion such as sulfate, for full activity. The third panel shows, however, that in the Novikoff hepatoma, a poorly differentiated, rapidly growing tumor which exemplifies other poorly

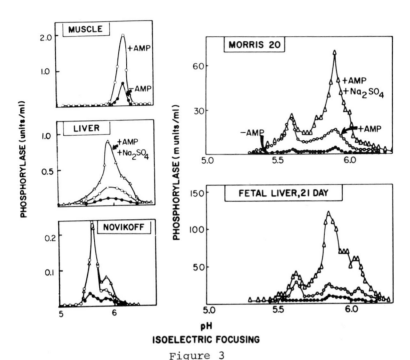

Figure 3

Isoelectric focusing patterns of phosphorylase b of normal
rat fetal and adult liver and rat hepatomas. Shaded circles
designate activities in absence of AMP, open circles with AMP,
and open triangles, with AMP and sodium sulfate.

differentiated hepatomas, the major form is distinctively
different from either the muscle or the liver form, focusing
at pH 5.6; and is further distinguished from the liver form
in not requiring sulfate ion for activation. It is accompanied, however, by a low activity of the liver form, which
focuses at pH 5.9 and requires sulfate ion. A well differentiated hepatoma, the Morris 20, has predominantly the liver
form, together with the tumor form, as does also the 21-day
fetal liver.

Isozyme properties. Table 1 summarizes the properties of the
hepatoma form of phosphorylase b and compares it with the

TABLE 1

A comparison of the Properties of Phosphorylases from Muscle, Liver and Novikoff Hepatoma (Fetal-type)

Properties	Muscle b-form	Liver inactive (b)-form	Novikoff b-form
AMP activation	+	±	+
Na$_2$SO$_4$ activation	−	+	−
cysteine activation	+	−	±
pI	6.15	5.90	5.60
Molecular weight	185,000	185,000	200,000
Tetramerization	+	−	−
Inhibition by antiserum to muscle phosphorylase b	+	−	−
liver "	−	+	±
Novikoff "	−	±	+

liver and muscle forms. All three forms are activated by AMP, but only the liver form requires sulfate ion for full activation. The hepatoma form is more like the liver than the muscle form in being only slightly activated by cysteine. Although it has a distinctive isoelectric point, it has about the same molecular weight as the other two forms. It also is similar to the liver form in that the b dimer form does not further dimerize to the tetramer on conversion to phosphorylase a by phosphorylase kinase and ATP. The last three lines of this table demonstrate that these three forms of phosphorylase b are immunologically distinctive. Antibody to the muscle phosphorylase has no inhibitory effect on either liver or hepatoma phosphorylase; and the antibodies to the liver and hepatoma phosphorylases have no effect on the muscle isozyme. However, the liver isozyme antibody does slightly inhibit the hepatoma isozyme, and the hepatoma isozyme antibody slightly inhibits the liver phosphorylase. This degree of cross-reactivity is in keeping with the presence of a minor activity of the liver isozyme in the Novikoff hepatoma, as seen in the electrofocusing pattern.

Isozyme composition. On the basis of the degree of immunologic cross-reactivity, coupled with the isozyme patterns obtained by electrofocusing, it was possible to quantitate approximately the relative activities of the three isozymes in

various normal and neoplastic cell types, and the results
are summarized in Figure 4.

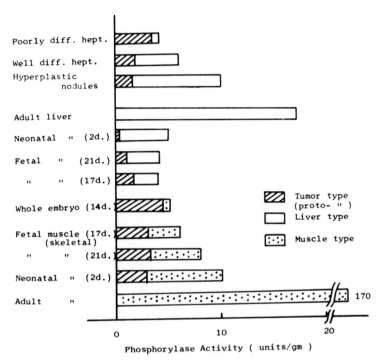

Changes of the phosphorylase isozymes in hepatomas and fetal
development

Figure 4
Isozyme composition of phosphorylases of various normal,
fetal, and neoplastic liver tissue. Activities are designated on the abscissa in units per g tissue, a unit being the
conversion of glucose-1-P to glycogen in μmoles per min.

Poorly differentiated hepatomas have predominantly the tumor
form, together with a low activity of the liver form. The
well differentiated hepatomas also contain the tumor form,
but a higher activity of the liver form. An early stage of
pre-neoplasia, the hyperplastic nodule, which develops
early in liver carcinogenesis and is still reversible on
discontinuance of carcinogen administration, resembles adult
liver in that it has a low activity of the tumor isozyme,

accompanied by a relatively high liver isozyme activity.

The normal adult liver has a high liver isozyme activity with no detectable tumor isozyme. However, the whole rat embryo, including the liver at 14 days has almost entirely the tumor isozyme, with only a very low muscle isozyme activity. Thus the embryonic form appears to be a true prototype isozyme, from which, with further embryonic development, the isozyme composition diverges. In the liver, the prototype isozyme decreases as the liver type appears, and by 2 days after birth the adult liver-type makes its appearance, and increases markedly, until by 2 days after birth the adult liver pattern is nearly established. In the skeletal muscle, however, the muscle type makes its appearance by 17 days and increases markedly until by 2 days after birth it is highly predominant, and becomes the sole form in the extremely high activity of adult skeletal muscle.

Application of disc gel electrophoresis. More recent work (Sato and Sato, unpublished) has shown that the three isozymes may be clearly distinguished by disc gel electrophoresis at pH 8.9, and staining of enzymatically synthesized glycogen by means of iodine. As shown in Figure 5, the fetal or prototype form migrates most rapidly, the muscle isozyme less rapidly, and the liver type least rapidly toward the anode. These electrophoretograms exhibited the expected requirements for AMP and sulfate ion, the muscle and fetal forms requiring only AMP and the liver form requiring both AMP and sulfate.

Figure 6 confirms the immunologic and electrofocusing data in demonstrating only prototype isozyme in whole 14-day embryo, both muscle and prototype isozymes in 21-day fetal muscle, and prototype and liver isozymes in 21-day fetal liver. Adult rat brain has predominantly the prototype isozyme, together with low activity of the muscle type.

Figure 7 summarizes additional preliminary data obtained by electrophoresis on the phosphorylase isozyme composition of other adult rat tissues. Spleen, kidney, testis, and lung have both the fetal type and a liver-like type which has a slightly lower electrophoretic mobility than the liver isozyme, but is otherwise similar immunochemically and kinetically, requiring sulfate for activation. Brain has predominantly the fetal type together with a low muscle type activity and heart muscle differs from adult skeletal muscle in having a low fetal type activity, together with the predominant muscle type. Thus far we have not detected any hybrid types, such as were reported for rabbit heart muscle by Davis et al., (1967).

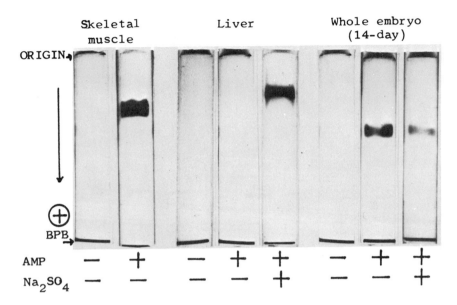

Figure 5
Polyacrylamide disc gel electrophoresis of phosphorylase
isozymes of liver, muscle and whole embryo according to the
procedure of Takeo et al., (1972)

Among the many individual strains of the Yoshida ascites
hepatoma which grow rapidly and are regarded generally to
be poorly differentiated, two strains, AH 130 and AH 66F,
differ in that the latter accumulates relatively large
amounts of glycogen, whereas the former does not (Sato and
Tsuiki 1972).

The wide variance among tumors and their capability for
glycogen storage have been puzzling features of tumor metabolism (Nigam and Cantero 1972; Sato and Tsuiki 1972).
Sato and Tsuiki (1972) found that the Yoshida hepatoma AH 130
which has a low glycogen content, had a glycogen synthetase
phosphatase that was inhibited by glycogen, whereas the
synthetase phosphatase of the Yoshida hepatoma AH 66F, which
has a high glycogen content, is relatively unaffected by
glycogen. The data of Figure 8 show that the liver form
phosphorylase predominates in the high glycogen storing AH
66F hepatoma but is the minor form in the low glycogen
storing AH 130 tumor. This quantitative difference in

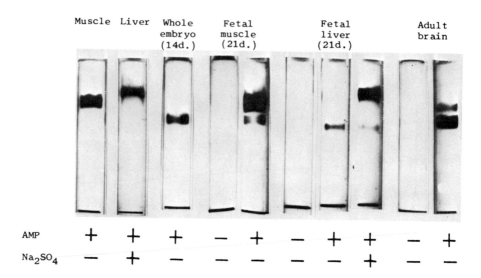

Figure 6

Polyacrylamide disc gel electrophoresis of phosphorylase isozymes in various fetal and adult tissues.

---

isozyme composition is borne out by titration against the liver phosphorylase antibody. As shown in Figure 9, there is the expected linear inhibition of the liver isozyme; the AH 66F hepatoma was inhibited about 80%, whereas the AH 130 phosphorylase was inhibited about 30%. These findings suggest that the high liver type phosphorylase activity in the AH 66F hepatoma may play a role in the high glycogen storing capability of this tumor. However, such reflections are speculative, since we still have no clear-cut understanding of why liver glycogen storage is high; nor can we yet state unequivocally that the Yoshida hepatoma isozyme is identical with the liver isozyme.

DISCUSSION

Emerging from these studies are a number of important inferences.

1. A single cell type of origin can give rise to tumors of great diversity, paralleled by diversity in the molecular structure of their enzymes.

2. Although our studies have been restricted largely to enzymes of carbohydrate metabolism, this phenomenon is

DISTRIBUTION OF PHOSPHORYLASE ISOZYMES IN RAT ADULT TISSUES

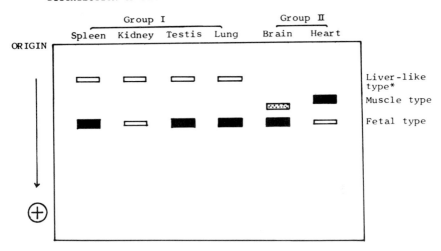

* Liver-like type migrates slightly slower than liver type.

Fig. 7. Diagrammatic representation of polyacrylamide disc gel electrophoretic patterns of various normal rat tissues.

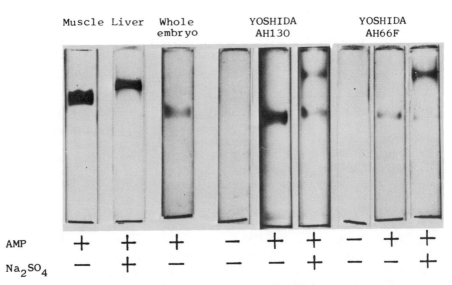

Fig. 8. Polyacrylamide disc gel electrophoretic patterns of phosphorylase isozymes of Yoshida hepatomas AH 130 and AH 66F.

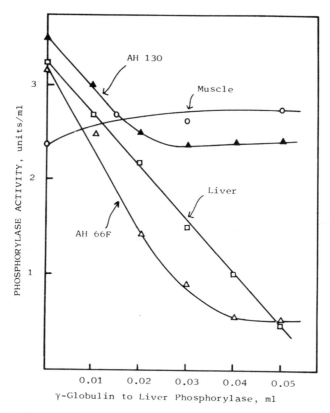

Fig. 9. Inhibition of phosphorylases of Yoshida hepatomas AH 130 and AH 66F by antiserum to rat liver phosphorylase.

probably a general one, extending into much of the cellular machinery. Similar isozyme alterations, with expression of fetal isozyme, have been reported with various aminotransferases (Ichihara and Ogawa 1972), with glutaminase (Katunuma et al., 1972), with thymidine kinase (Fujii et al., 1972) and no doubt many more surely remain to be discovered. An interesting example is the identification of the so-called Regan alkaline phosphatase isozyme as a placental form (Fishman et al., 1971).

3. It is noteworthy that the isozymes that are lost are those that are under host regulation and are geared to specific organ function; and those that are gained are not under host regulation but are geared for efficient utilization of substrates. This switch in isozyme pattern conceivably

may represent the molecular basis for the unbridled proliferation of cancer cells.

Alterations of phosphorylase isozyme composition with
decreased differentiation and increased growth rate of
hepatomas follow the same pattern of other enzymes and add
further substantiation for the generality of this phenomenon. Here again we observe the loss of a liver type
isozyme and its replacement by an isozyme that is normally
undectectable in adult normal liver, but is present in fetal
liver.

Perhaps the most striking feature of the isozyme studies
in hepatomas is their remarkable similarity to the alterations
that have been observed in the immunologic properties of
tumors. In cancer, antigens specific to the adult, differentiated cells of origin disappear, and tumor-specific or
tumor associated antigens appear, which, as with the isozymes,
are also present in fetal tissue (Abelev 1971; Baldwin 1973;
Alexander 1972; Anderson and Coggin 1971; Lengerova 1972;
Stonehill and Bendich 1970). Obviously these are opposite
sides of the same coin; a mis-programming of protein synthesis
in which genes that are expressed in embryonic or fetal
states but are inactivated during normal embryonic development become re-activated in cancer.

These embryonic manifestations in the antigen composition
of tumors have been described in such terms as derepressive
dedifferentiation (Gold 1971), retrogenic expression (Stonehill and Bendich 1970), retrodifferentiation (Anderson and
Coggin 1971), etc., implying a more or less systematic
reversion to an embryonic state; that is a reversal of
normal differentiation. Our isozyme data do indicate some
order in the observed alterations. For example, many well-
differentiated hepatomas have largely lost isozymes of the
differentiated liver cell but do not exhibit a resurgence of
the fetal isozyme; but when resurgence of the fetal does
occur, there is always a loss of normal hepatic isozyme
(Weinhouse 1972). These observations suggest that certain
genes may have to be switched off before others are switched
on. However, in general the alterations seem to be sporadic
and unpredictable, and therefore suggest a disordered rather
than a programmed mechanism of gene activation.

This conception of a disordered rather than a programmed
reversion is strongly bolstered by a large and ever-growing
body of clinical literature pointing to bizarre aberrations
of protein synthesis in certain tumors, resulting in the
ectopic production of polypeptide hormones by tumors of non-
endocrine origin (Eliel 1968; Goodall 1969; Omenn 1973; Roof
et al., 1971).

Whatever may be the fundamental nature of this impairment, it is not necessarily always expressed in cancer, since we see all degrees of divergence of isozyme expression; and the isozyme pattern of several highly differentiated hepatomas is indistinguishable from that of normal liver. What appears to happen, however, is that the normally rigid and highly selective mechanism that operates to determine which genes shall remain inactive and which can be expressed becomes "unlocked," so that genes that are normally inactive can become switched on, while other genes that are normally transcribed in the adult, differentiated cells can be switched off. It is conceivable that this impairment of gene control may be responsible for the familiar pattern of tumor progression which occurs unpredictably but inevitably as tumors grow or metastasize; involving changes in chromosome number and karyotype, loss or gain of antigens, alterations in surface properties, drug resistance, growth rate, etc. (Foulds 1969).

Despite all of the compelling evidence which points to errors of gene expression in cancer, it is important to ask whether these aberrations of protein synthesis are the cause or the effect of the neoplastic transformation. Neither isozyme nor antigen alterations are all-or-none phenomena; but rather are quantitative; many of the isozymes normally vary greatly in their activites, depending on dietary or hormonal conditions. Some antigens as well as isozymes appear in regenerating as well as in fetal liver; and many of the isozyme alterations seen in poorly differentiated hepatomas also appear in ostensibly normal liver cells when these are grown in vitro (Weinhouse 1972, 1973). Many more questions can be asked and one can speculate at great length in the light of these anomalies of gene expression; but we obviously need more imformation on the mechanisms of normal differentiation before we can understand these disorders of differentiation, which may lie at the heart of the neoplastic transformation.

ACKNOWLEDGMENT

The author gratefully acknowledges the support of the National Cancer Institute for grants CA-10729, CA-10916 and CA-12227; and the American Cancer Society for grant BC-74. The aid of Jennie B. Shatton, Albert Williams, Charity M. Jackson, Louise Lawson, and David Meranze is also acknowledged with appreciation.

Figure 3 reproduced through the courtesy of *Science*.

REFERENCES

Abelve. G.I. 1971. α-Fetoprotein in oncogenesis and its association with malignant tumors. *Cancer Res.* 14: 195-358.

Alexander, R. 1972. Foetal "Antigens" in cancer. *Nature.* 235: 137-140.

Anderson, N. and J.H. Coggin, Jr. 1971. Models of differentiation, retrogression and cancer. *Proceedings of the First Conference on Embryonic and Fetal Antigens in Cancer.* Oak Ridge National Laboratory, pp. 7-37.

Baldwin, R.W. 1973. Immunological aspects of chemical carcinogenesis. *Advan. Cancer Res.* 18: 1-75.

Brown, D.H. and C.F. Cori 1960. Animal and plant polysaccharide phosphorylase. In P.D. Boyer, H. Lardy, and K. Myrback (eds.), *The Enzymes,* Ed. 2, Vol 5, pp. 107-227, New York, Academic Press, Inc.

Criss, W.E. 1971. A review of isozymes in cancer. *Cancer Res.* 31: 1523-1542.

Davis, C.J., L.H. Schliselfied, D.P. Wolf, C.A. Leavitt, and E.G. Krebs 1967. Interrelationships among glycogen phosphorylase isozymes. *J. Biol. Chem.* 242: 4824-4833.

Eliel, L.P. 1968. Non endocrine secreting neoplasms: clinical manifestations. *Cancer Bull.* 20: 30-37.

Fishman, W.H., N.R. Inglis, and S. Green 1971. Regan isoenzyme: a carcinoplacental antigen[fn]. *Cancer Res.* 31: 1054-1057.

Foulds, L. 1969. *Neoplastic Development.* Vol. 1, Academic Press, London and New York.

Fujii, S., T. Hashimoto, T. Shiosaka, T. Arima, M. Masaka, and H. Okuda 1972. Thymidine kinases of neoplastic tissues, regenerating and embryonic liver, marrow cells, potato, and tetrahymena. *Gann Monograph.* 13: 107-119.

Gold, P. 1971. Embryonic origin of human tumor specific antigens. *Progr. Exptl. Tumor Res.* 14: 43-58.

Gold, P., and S.O. Freedman 1965. Specific carcino-embryonic antigens of human digestive tract. *J. Exptl. Med.* 122: 467-481.

Goodall, C.M. 1969. A review: on para-endocrine cancer syndromes. *Intern. J. Cancer.* 4: 1-10.

Ichihara, A., and K. Ogawa 1972. Isozymes of branched chain amino acid transaminase in normal rat tissues and hepatomas. *Gann Monograph.* 13: 181-190.

Katunuma, N., T. Katsunuma, I. Tomino, and Y. Natsuda 1968. Regulation of glutaminase activity and differentiation of the isozyme during development. *Advan. Enzyme Regulation.* 6: 117-242.

Katunuma, N., Y. Kuroda, T. Yoshida, Y. Sanada, and H.P. Morris 1972. Relationship between degree of differentiation of minimal deviation hepatomas and kidney cortex tumors studied with glutaminase isozymes. *Gann Monograph*. 13: 143-151.

Knox, W.E. 1972. *Enzyme Patterns in Fetal, Adult and Neoplastic Rat Tissues*. S. Karger AG., Basel.

Lengerova, L. 1972. Expression of normal histocompatability antigens in tumor cells. *Advan. Cancer Res*. 16: 235-272.

Morris, H.P. 1963. Some growth morphological and biochemical characteristics of hepatomas 5123 and other transplantable hepatomas. *Progr. Exptl. Tumor Res*. 3: 370-411.

Morris, H.P. 1965. Studies on the development, biochemistry and biology of experimental hepatomas. *Cancer Res*. 9: 227-302.

Morris, H.P. 1974 (in press). Biological and biochemical characteristics of transplantable hepatomas. In *Handbuch der Allgemeinen Pathologie*. E. Grumdmann (Ed.), Springer-Verlag, Berlin, Heidelberg, New York.

Morris, H.P., and B.P. Wagner 1968. Induction and transplantation of rat hepatomas with different growth rate (including minimal deviation hepatomas). In *H. Busch (ed.) Methods in Cancer Res*. 4: 125-152. Academic Press, London and New York.

Nigam, V.N., and A. Cantero 1972. Polysaccharides in cancer. *Advan. Cancer Res*. 16: 1-91.

Omenn, G.S. 1973. Pathobiology of ectopic hormone production by neoplasms in man. In *Pathobioloby Annual* (H.L. Ioachim, Ed.) Appleton, New York, pp. 177-216.

Ono, T., and S. Weinhouse (Eds.) 1972. Isozymes and enzyme regulation in cancer. *Gann Monograph*, 13, University of Tokyo Press.

Roof, B.S., B. Carpenter, D.J. Fink, and G.S. Gordon 1971. Some thoughts on the nature of ectopic parathyroid hormones. *Am. J. Med*. 50: 686-691.

Ryman, B.E., and W.J. Whelan 1971. New aspects of glycogen metabolism. *Advan. Enzymol*. 34: 285-443.

Sato, K., H.P. Morris, and S. Weinhouse 1972. Phosphorylase: a new isozyme in rat hepatic tumors and fetal liver. *Science*. 178: 879-881.

Sato, K., and S. Tsuiki 1972. Studies on metabolism of glycogen storage in ascites hepatomas. *Cancer Res*. 32: 1451-1452.

Sato, K., H.P. Morris, and S. Weinhouse 1973. Characterization of glycogen synthetases and phosphorylases in transplantable rat hepatoma. *Cancer Res*. 33: 724-733.

Sato, K., and S. Weinhouse 1973a. Purfication and character-

ization of the Novikoff hepatoma   glycogen phosphorylase
and its relations to a fetal form. *Arch. Biochem. Biophys.*
159:   151-159.

Schapira, F.   1973.   Isozymes in cancer.   *Cancer Res.*   18:
76-153.

Stonehill, E.H., and A. Bendich   1970.   The reappearance of
embryonal antigens in cancer cells. *Nature.*   228:   370-
371.

Takeo, K., and S. Nakamura   1972.   Dissociation constants of
glucan phosphorylases of rabbit tissues studied by polyacrylamide gel disc electrophoresis.   *Arch. Biochem. and
Biophys.*   153:   1-7

Weinhouse, S.   1972.   Glycolysis, respiration and anomalous
gene expression in experimental hepatomas: G.H.A. Clowes
Memorial Lecture.   *Cancer Res.*   32:   2007-2016.

Weinhouse, S.   1973.   Metabolism and isozyme alterations
in experimental hepatomas.   *Federation Proceedings.*   32:
2162-2167.

ALDEHYDE DEHYDROGENASE ISOZYMES IN CERTAIN HEPATOMAS

ROBERT N. FEINSTEIN
Division of Biological and Medical Research
Argonne National Laboratory
Argonne, Illinois 60439

ABSTRACT. The aldehyde dehydrogenase (Ald D) activity of autochthonous rat hepatomas induced by AAF averages a great deal higher than in normal rat liver. The hepatoma Ald D activity is more stable than that of normal liver to temperature, urea, and pH extremes. Its subcellular distribution is different, the hepatoma enzyme showing a greater proportion in the cytoplasm. The hepatoma enzyme appears to be aldehyde:NAD(P) oxidoreductase, EC 1.2.1.5, in contrast to the normal liver enzyme, which is aldehyde:NAD oxidoreductase, EC 1.2.1.3. The enzyme properties which we here designate as "hepatoma-like" are only slightly observed in the four minimal deviation (repeatedly retransplanted) hepatomas which we have examined, and are not at all observed in spontaneous hepatomas of the mouse or mastomy, nor in fetal or regenerating liver of the rat. Difficulties encountered due to the concomitant presence of the corresponding aldehyde oxidase and of "nothing dehydrogenase" are discussed.

It is convenient to divide this discussion into three sections: first, a general description of the phenomenon of increased and modified aldehyde dehydrogenase (Ald D) activity in certain rat hepatomas; second, a brief discussion of certain points of technique which have caused us considerable difficulty and which may be of interest to others using similar techniques in isozyme studies; and third, a more detailed description of the Ald D phenomenon, with particular emphasis on genetic aspects and its relationship to carcinogenesis per se.

I. The aldehyde dehydrogenase (Ald D) phenomenon

We began a survey of isozymes in tumors by looking at dehydrogenases in hepatomas induced in the Sprague-Dawley rat by 2-acetylaminofluorene (AAF). This was a convenient material because of the studies of these tumors being carried out at Argonne National Laboratory by Peraino, Fry, and Staffeldt (1971, 1973). The dehydrogenases represented a convenient starting point because techniques have been so well worked out for their detection on electrophoresis gels, and quantitative assay systems are available.

Although some minor points of interest were noted, nothing spectacular occurred until, in methodical progression, we began to test aldehydes as substrate. We are aware of no work other than our own (Feinstein and Cameron, 1972) on the Ald D activity of tumors, despite the extensive investigation of other members of the class of dehydrogenases. Preliminary work involved the use of polyacrylamide gel electrophoresis (PAGE), the location of the Ald D bands then being determined by incubation in a mixture cf NAD, PMS, NBT, and an aldehyde, most commonly benzaldehyde. Fig. 1 shows the striking differences between the appearance of Ald D in the hepatoma and in normal liver. We then turned to biochemical assays of the enzyme activity, using as the measure of activity the A_{340} of NADH produced from NAD. The A_{340} was corrected for non-enzymatic changes in A_{340} and, more importantly, for the increase in A_{340} in the absence of aldehyde. Table 1 summarizes the Ald D activity in one series of tumors and normal livers and confirms the PAGE gel findings of a greatly increased Ald D activity. (Diets are those of Peraino, Fry, and Staffeldt, 1971.)

TABLE 1

Aldehyde Dehydrogenase Activity in
Sprague-Dawley Rat Liver and Hepatoma

Tissue	Dietary Regimen	Number	Activity	Range
Normal liver	Normal	7	59 ± 5	51-67
	Phenobarbital	8	70 ± 18	39-88
Hepatoma	AAF, then normal	6	215 ± 106	74-355
	AAF, then pheno-barbital	10	239 ± 94	94-390
	Four cycles of AAF	33	187 ± 340	14-2010

Activities are mean and standard deviation from the mean.

The most likely explanation for the increased Ald D activity is somatic mutation or gene derepression. To distinguish between these two possibilities, the relevant questions are: (a) Does the oncogenic change represent merely an increase in the activity of the same isozyme(s), or are new molecular species produced? and (b) If new molecular species are produced, are they always the same, or is the change random from animal to animal? The answers are quite decisive: (a) New molecular species are produced, and (b) Within the limits of

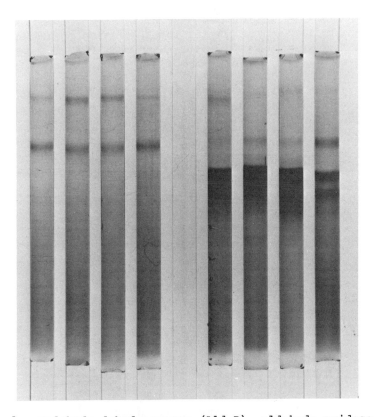

Fig. 1. Aldehyde dehydrogenase (Ald D), aldehyde oxidase (Ald Ox), and "nothing dehydrogenase" in normal liver and hepatoma extracts. Four PAGE gels on the left contained 10% normal liver extract; four gels on the right contained 10% hepatoma extract. Direction of migration is from cathode (top) to anode (bottom). As discussed in the text, the bands seen in the normal liver extract are not Ald D; rather, the slowest band is "nothing dehydrogenase," and the next is Ald Ox.

the techniques we employ, the new species are constant, not variable. The gene derepression mechanism is, therefore, most likely.

It must be noted that these remarks, and those that follow, are based on studies of the Ald D activity in crude extracts of normal rat liver or hepatoma. We do not yet have preparations of the pure isozymes, although an effort to prepare them is in progress. Molecular differences between the Ald D of hepatoma and of normal liver are suggested by the following evidence:

1. PAGE gels, as pictured in Fig. 1, indicate that the hepatoma enzyme has a much greater total activity than does the normal liver enzyme. The pattern shown in Fig. 1, however, cannot be accepted as evidence of a molecular variation. The slower moving bands are actually not Ald D. The true Ald D activity of normal liver is located on PAGE gels in approximately the same area as the hepatoma activity but can be detected only by the use of much more concentrated liver extracts than the 10% used in this sample. This is discussed later.

2. Fig. 2 shows the relative heat stability of the Ald D

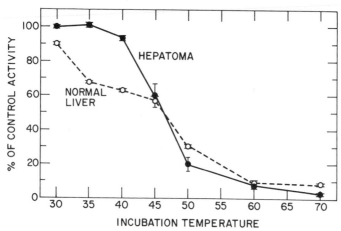

Fig. 2. Heat stability of Ald D activity of normal liver and hepatoma. Small volumes of normal liver or hepatoma extract were transferred into test tubes pre-warmed to the temperature shown, and maintained at that temperature for ten minutes before being plunged into an ice bath and assayed shortly thereafter.

activity in normal rat liver and in hepatoma. The difference in shape of the two curves suggests that different molecular species are represented. It is interesting to note that it is the normal form of the enzyme which is the more labile. It would generally be expected, from evolutionary considerations, that the normal form of an enzyme would be more stable than an uncommon, mutant form.

3. Fig. 3 shows curves of pH vs activity for the two sources of Ald D. The considerable difference in the shape of the two curves is probably due to stability differences rather than to activity optima, because a return to optimum pH before assay does not return the lost activity. If this is accepted, it is again apparent that the hepatoma enzyme has the greater stability.

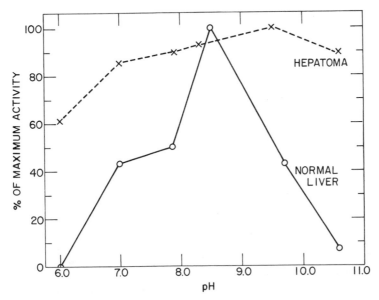

Fig. 3. pH vs Ald D activity. pH values were varied by the use of 0.1 M phosphate buffers, the pH of the mixture being determined at the end of the assay.

4. The same conclusion is reached if one investigates stability in the presence of five molar urea, as shown in Fig. 4. The curves showing incubation in the absence of urea are included to demonstrate that decreases are not due merely to the 25-26° room temperature at which the mixtures were incubated. The inactivation of the normal liver extract Ald D is due to the urea.

5. Fig. 5 shows that lower concentrations of urea will decrease normal liver Ald D activity while not harming that of hepatoma. In fact, Fig. 5 illustrates another interesting feature of the hepatoma activity. We have repeatedly, but not invariably, observed an *increased* Ald D activity upon incuba-

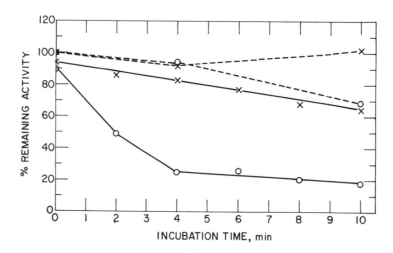

Fig. 4. Stability to urea of Ald D activity of normal liver and hepatoma. Extracts of normal liver or hepatoma were diluted with equal volumes of water or 10 M urea. The mixtures were maintained at room temperature for the time shown, at which time they were transferred directly into reaction mix- tures for assay, being diluted in the process by a factor of 60, at which dilution the urea is without further effect. O----O normal liver diluted with water; O——O normal liver diluted with urea; X----X hepatoma diluted with water; X——X hepatoma diluted with urea.

tion of hepatoma extracts with dilute urea, and occasionally upon incubation even in the absence of urea. We have never observed normal liver activity to increase; rather, it is found that incubation at 37° invariably causes a decrease in activity.

6. Further reason to suspect that the normal liver and hepatoma Ald D activities may represent different isozymes lies in the fact that the subcellular distribution of the activity is different (Table 2).

7. More conclusive evidence is obtained from electrofocusing the normal liver and hepatoma extracts. In this technique, which was carried out with a modification of the LKB Multiphor equipment, electrophoresis is carried out in polyacrylamide gel containing ampholytes, so that a pH gradient is formed in the gel, and proteins are concentrated at their isoelectric points (pI). This technique permits much finer and more subtle separation of molecular species, as is indicated in Fig. 6.

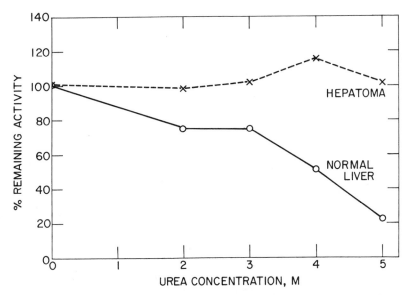

Fig. 5. Stability of Ald D of normal liver and of hepatoma to varied concentrations of urea. Tissue extracts were diluted with equal volumes of 10, 8, 6, or 4 M urea, or with water. The mixture was maintained at room temperature for 10 min, then transferred directly to a reaction mixture for assay.

TABLE 2
Subcellular Distribution of Aldehyde Dehydrogenase Activity

	Activity, Units		Activity, % of total	
Fraction	Normal liver	Hepatoma	Normal liver	Hepatoma
Nuclei	0	50	0.0	0.4
Mitochondria	640	1,530	12.9	11.5
Microsomes	2,150	1,670	43.2	12.6
Supernatant	2,190	10,000	44.0	75.5
Total	4,980	13,250	100.1	100.0

Total activity of each fraction is based on an arbitrary, but equal, weight of hepatoma or normal liver.

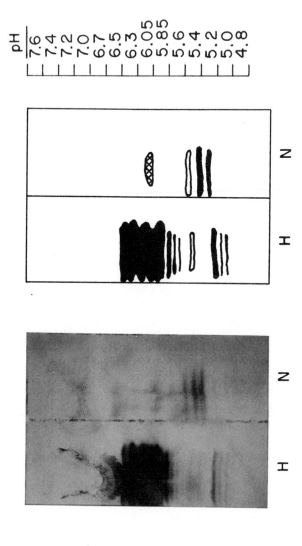

pH
7.6
7.4
7.2
7.0
6.7
6.5
6.3
6.05
5.85
5.6
5.4
5.2
5.0
4.8

N   H   N   H

Fig. 6. Electrofocusing of hepatoma and normal liver extracts in polyacrylamide gel, with NAD being added as coenzyme. Diagrams on right correspond to photographs on left.

(In the diagrams, the cathodal (upper) irregularities of the photograph have been eliminated; they are due to the initial sample deposition.) Although there is some daily variation in pI, presumably because of the difficulty of obtaining accurate pH measurements on the gel, this variation is never more than 0.1-0.3 pH units, and the pI of normal liver Ald D is approximately 5.4-5.5, whereas the great bulk of the hepatoma Ald D shows a pI in the range of 5.8-6.3, with apparently a very small amount of activity showing a pI of 5.2-5.3.

8. The most decisive evidence that the hepatoma Ald D activity represents new molecular species has to do with coenzyme specificity. Fig. 7 shows two more segments of the same

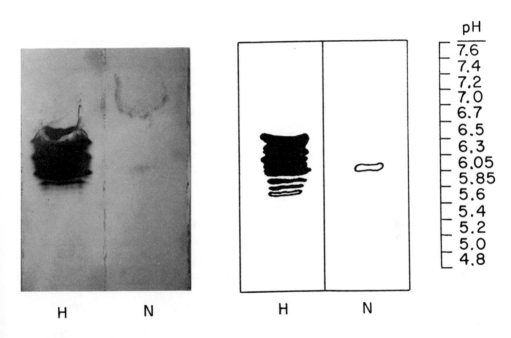

H   N   H   N

Fig. 7. Electrofocusing of hepatoma and normal liver extracts in polyacrylamide gel, with NADP being added as coenzyme.

Multiphor isoelectric focusing gel slab used in the last figure, with the same extracts being focused as on that figure. The only difference is that the gel segments, instead of being incubated in a mixture containing NAD, now are incubated in the presence of NADP. The results are striking. The great mass of hepatoma Ald D activity, with pI about 5.8-6.3, remains active. The minor hepatoma isozyme, pI 5.2-5.3, shows no activity, and most important, the Ald D of normal liver is also totally devoid of activity with NADP as coenzyme. (The faint band shown on the segment containing normal liver is not Ald D, but aldehyde oxidase, as is discussed later.)

These findings raise a question concerning the exact isozymic status of these two enzyme activities. Are the two reactions

(1) aldehyde + NAD $\rightleftharpoons$ acid + NADH, and

(2) aldehyde + NAD(P) $\rightleftharpoons$ acid + NAD(P)H

to be classed as isozymic? In the International Enzyme Commission nomenclature, these two enzymes are actually given separate numbers, namely 1.2.1.3 and 1.2.1.5, respectively. It is particularly interesting to note that the mammalian enzyme is generally NAD-specific. One exception that has come to our attention is the work of Ris and von Wartburg (1973); they found that rat brain Ald D is able to use NADP to a small extent. Generally, the NAD(P) enzyme is found only in microorganisms (Black, 1951; King and Cheldelin, 1956; Jakoby, 1958).

This coenzyme distinction does not apply to a biochemical assay system in solution. In such a system, NAD and NADP serve the normal liver enzyme equally well. This is illustrated in Table 3. Our explanation for this is that liver extract con-

TABLE 3

Coenzyme Specificity in Aldehyde Dehydrogenase Assay

Tissue	Coenzyme	Activity	Range
Normal liver	NAD	56 ± 5	50–60
	NADP	40 ± 12	28–57
Hepatoma	NAD	94 ± 14	80–110
	NADP	254 ± 53	216–330

tains a phosphatase which hydrolyzes NADP to NAD, which then serves as coenzyme. On electrophoresed gels, of course, such a phosphatase would be physically separated from the dehydrogenase system. Phosphatases capable of carrying out this reaction have been demonstrated in kidney, intestine, and other sources (Morton, 1955; Nakamoto and Vennesland, 1960), and there is no reason to doubt that liver also contains such an activity.

The second part of Table 3 perhaps constitutes further suggestive evidence for the non-identity of the Ald D of normal
liver and of hepatoma, because it is a consistent finding that
the hepatoma enzyme functions, in the assay system, several
times as effectively with NADP as coenzyme as it does with NAD.
It appears to be a valid generalization that the higher the
total Ald D activity of a hepatoma, the greater is the effectiveness ratio of NADP:NAD. In a series of 21 tumors, the
correlation of total NAD-based activity to ratio of NADP:NAD
effectiveness provided a coefficient of 0.622 ($p < 0.01$).

II. Pitfalls and artifacts

Let us turn briefly to a consideration of some of the factors complicating an investigation of the aldehyde dehydrogenases of normal rat liver and rat hepatoma.

The first complication arises from the fact that when a
gel is incubated with NAD, PMS, NBT, and an aldehyde, formazan
lines which are produced represent not only Ald D, but also
aldehyde oxidase (Feinstein and Lindahl, 1973) as well as the
so-called "nothing dehydrogenase" (Zimmerman and Pearse, 1959;
Decker and Rau, 1963; Shaw and Koen, 1965; Longo and Scandalios,
1968; Manwell and Baker, 1969). "Nothing dehydrogenase" is
the term applied to the enzyme activity which produces formazan bands on various electrophoresed gels even though no substrate has been added. To complicate matters further, we here
suggest that "nothing dehydrogenase" actually represents two
things: first, a true enzyme activity of unknown nature which
somehow brings about the ultimate reduction of NBT to formazan,
even in the absence of added substrate; and second, and possibly more common, an unintended but true enzyme reaction in
which substrate is actually present in trace concentrations
that are undetectable except by the very sensitive formazan
precipitation reaction.

Let me describe and illustrate this situation. In our
laboratory, gel cylinders after electrophoresis are transferred into small test tubes, approximately 9 x 75 mm. The
test tubes are nearly filled with the desired reaction mixture
and incubated at 37° C in covered metal containers about 7 x
11 x 13 cm. The inside of each container is a solid metal
block in which a series of holes has been bored, about 1.5 cm
apart from center to center. All reaction mixtures contain
buffer, NBT, and PMS. Some tubes also contain benzaldehyde
(but no NAD) to permit the demonstration of aldehyde oxidase
(Ald Ox), and some tubes also contain NAD (but no aldehyde)
to demonstrate the existence of nothing dehydrogenase. Finally,
some contain both NAD and benzaldehyde, needed for the demon-

stration of Ald D. The complete system causes formazan precipitation simultaneously for all three enzyme activities.

Because benzaldehyde boils at 180° C, and we were incubating at 37°, and because of the tall, thin nature of the test tubes, we considered possible cross-contamination as no problem, and we have routinely incubated all tubes in a single closed container. However, as results became more and more confused with the passage of time, we were finally led to investigate the extremely remote possibility of cross-contamination, with the results shown in Fig. 8. To obtain these results, a rat hepatoma extract was electrophoresed in replicate polyacrylamide gels, and the gel cylinders were placed into tubes that had been specially cleaned with sulfuric-dichromic cleaning mixture. Three of the metal incubator boxes were prepared in advance by heating them, with lids removed, at 37° overnight. The metal lids were well baked in a 100° oven. All eight gels shown in the figure were incubated in a mixture containing buffer, NAD, PMS, and NBT. The first pair of tubes, 1 and 2, were incubated with benzaldehyde added. Tubes 3 and 4 were incubated without added benzaldehyde, but in the same container as tubes 1 and 2. Tubes 5 and 6 were also incubated without benzaldehyde, entirely separately in a clean incubator box. Tubes 7 and 8 were incubated without benzaldehyde, but in another hole in the box was placed a tube containing only a solution of benzaldehyde in water. It seems quite clear first, that cross-contamination can indeed occur, and second, that the two slowest-moving bands are not Ald D; actually, we now consider the slowest moving band to be a true "nothing dehydrogenase," and the next band to be Ald Ox. This oxidase evidently has such a low K_m that our efforts to eliminate benzaldehyde were ineffective. Preliminary evidence suggests that the band now identified as Ald Ox may eventually prove to be a double band, a strong Ald Ox and a weak "nothing dehydrogenase."

This extraordinary sensitivity helps to explain why the "nothing dehydrogenase" so often appears to mimic whatever dehydrogenase an investigator is working with. I wish to emphasize that these remarks do *not* deny the existence of a true "nothing dehydrogenase." Apparently there *is* such an enzyme, much more active in some other tissues than in liver. I merely wish to emphasize that the formazan deposition can be extremely sensitive and that extreme care must be taken to guard against trace contamination.

A second complication, obvious but not always considered, is that with every improvement in gel technique, more molecular species become evident. An example may be noted in Fig. 6, where the great bulk of hepatoma Ald D activity appeared at a pI of 5.8-6.3, while a small, but real, activity appeared at a pI of 5.2-5.3. This result was obtained by electrofocus-

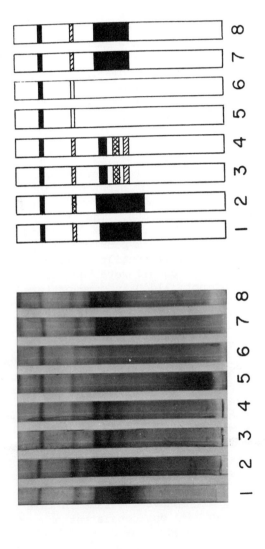

Fig. 8. Demonstration of substrate cross-contamination in incubating PAGE gel cylinders. Tubes 1 and 2 contained complete reagent system; tubes 3-8 lacked any aldehyde. Tubes 3 and 4 were incubated in the same box as tubes 1 and 2. Tubes 5 and 6 were incubated in a clean box with no other tubes present. Tubes 7 and 8 were incubated in a box which also held a tube containing a solution of benzaldehyde in water.

ing a 50% extract. In many earlier studies using 10% extracts, the pI 5.2-5.3 isozymes were unsuspected. Further subdivision of isozymes is, of course, also likely to occur with the use of a narrower range of ampholyte pH, spread out over a greater distance.

III. *Genetic and carcinogenic considerations*

The aldehyde dehydrogenase modification we have described is not a widespread phenomenon. We have examined several spontaneous hepatomas in mice or mastomys, and the level of Ald D activity is low, and no rat hepatoma-like isozyme pattern is evident on PAGE gels. We have also examined one intestinal, one lung, and two salivary gland tumors which appeared in Sprague-Dawley rats bearing AAF-induced hepatomas. The hepatomas showed the usual phenomena of high Ald D activity and complex isozyme patterns; the other tumors were low in activity and showed single bands (probably Ald Ox) or none.

The narrowness of the phenomena extends even further. We have examined four "minimal deviation" hepatomas kindly provided by Dr. H. P. Morris. These had been carried by repeated subcutaneous transplant in Buffalo rats. The original tumors had been induced by a variety of carcinogens, none of them AAF, and they represented a gamut from poorly differentiated, rapidly growing tumors to well differentiated, slowly growing neoplasms. The enzyme findings on all were the same: Although there was a faint, but distinct, Ald D isozyme pattern reminiscent of what we had seen in our AAF-induced hepatomas in Sprague-Dawley rats, the total Ald D activity was in the normal range.

We have tested the three obvious differences between the Morris hepatomas and our own, namely (a) autochthonous vs retransplanted tumors, (b) AAF vs other inducing agents, and (c) Buffalo rats vs Sprague-Dawley. The question is still moot. We suspect, however, that the critical matter is whether the tumor is autochthonous or the result of repeated serial transplant. The strain of rat is also apparently of some consequence. Hepatomas induced in Sprague-Dawley rats by DAB are at least as active as AAF-induced tumors and the isozyme pattern is the same. Also, Sprague-Dawley hepatomas induced by feeding AAF alone, or followed by phenobarbital (Peraino, Fry, and Staffeldt, 1971), all behave the same in this respect.

On the other hand, it has been our consistent observation that it is more difficult to induce hepatomas with AAF in Buffalo than in Sprague-Dawley rats. A longer time is required before the appearance of the first tumor, and at any given time the tumor masses are much smaller. In fact, we have not yet been able to induce hepatomas in Buffalo rats by a tech-

nique similar to that of Peraino, Fry, and Staffeldt (1971);
our rats were on an AAF-containing diet for one month, followed
indefinitely by a diet containing phenobarbital. We have,
however, produced hepatomas in Buffalo rats by the use of a
dietary regimen similar to that used by Reuber (1965): four
one-month cycles of AAF feeding, separated by single weeks on
a normal diet.

Hepatomas induced by this technique appear to be quite similar, from the point of view of Ald D, in the two strains of
rat. The consistently low Ald D activity of the Morris hepatomas we have tested may therefore be a consequence of the
repeated serial transplantation of those tumors. We have not
yet initiated a retransplantation series of a high-activity
hepatoma to check this point.

The hepatoma Ald D modification herein described differs
from most of the other neoplastic isozyme modifications described in the literature in yet another way. Potter's concept (1969) of "oncogeny as blocked ontogeny" has found repeated confirmation in the appearance of an isozyme pattern
in embryonic tissue similar to that in the corresponding neoplasm in the mature animal. In the case of Ald D, however,
16-20 day fetal liver shows an Ald D in the normal liver range,
and 10% extracts show no hepatoma-like isozymes. Samples of
regenerating liver taken 16 hours to five days after partial
hepatectomy in Sprague-Dawley rats are normal in these
respects.

Table 4 summarizes all these last findings. A noteworthy
feature of this table is the very wide range of Ald D activities seen in various AAF-induced hepatomas in contrast to
the relatively narrow activity range of the other materials.

The total activity, or the intensity and complexity of
isozyme banding, are in no way related to the histological
appearance of the various hepatomas. Many of the hepatomas
were examined histologically and rated as to extent of differentiation by one individual, and assayed enzymatically by
another person; when the results were merged, no correlation
between hepatoma histology and enzymatic activity could be
observed.

One last point merits comment. We have examined the livers
of Sprague-Dawley rats that had received AAF but in which
overt tumors had not yet appeared. In some instances, we
have examined biochemically segments of liver which were macro-
scopically normal while an overt hepatoma existed on another
lobe of the liver. In almost all these cases, we have seen a
hepatoma-like isozyme pattern even though the total Ald D
activity is normal. In other words, the biochemical changes
appear before the tumor can be identified histologically.

TABLE 4
Aldehyde Dehydrogenase Activity of Various Materials

Material		Number	Mean Activity	Range
S.-D. rat liver, normal		15	65	39–88
S.-D. rat hepatoma, AAF, 1 cycle		16	232	79–390
S.-D. rat hepatoma, AAF, 4 cycles		33	189	14–2010
S.-D. rat hepatoma, DAB		3	375	228–460
Buffalo rat liver, normal		3	68	62–73
Buffalo rat hepatoma, AAF, 4 cycles		40	129	0–633
Morris hepatoma 9618B		1	0	
7777		3	3	0–5
5123tc		3	11	2–26
7800		3	26	25–27
S.-D. rat lung tumor		1	9	
S.-D. rat intestinal tumor		1	2	
S.-D. rat salivary gland tumor		2	1	0–1
BCF_1 mouse liver, normal		3	19	5–42
BCF_1 mouse spontaneous hepatoma		3	13	1–32
Mastomy liver, normal		1	20	
Mastomy spontaneous hepatoma		1	0	
S.-D. rat liver, fetal, 16 day		3	6	2–12
20 day		3	6	5–8
S.-D. rat liver, regenerating,	16 hrs	1	51	
	2 days	1	50	
	5 days	1	57	

It is intriguing to speculate on the mechanism whereby these new Ald D isozymes are expressed with the progress of carcinogenesis. Because these isozyme changes are not detected in fetal liver; because their appearance is brought about by more than one carcinogen; because there appears to be a strain specificity to some extent; and because of the invariable nature of this change, we suggest the possibility that this isozyme modification represents a derepression of an "archeogene" (Anderson and Coggin, 1971), an activity which has been repressed for unknown ages and is now, under the influence of a variety of carcinogens, directly or indirectly derepressed and activated.

ACKNOWLEDGMENTS

I wish to express my appreciation to Miss Erma Cameron, who performed all of the assays and did most of the gel studies reported; to Drs. Carl Peraino and R. J. M. Fry for much of the early tumor material, for the histological examinations (R. J. M. F.), and for helpful discussions; and to Dr. H. P. Morris for the several hepatomas briefly mentioned above.

FOOTNOTE

*Abbreviations used in this paper include the following:

Ald D: Aldehyde dehydrogenase, aldehyde:NAD oxido-
        reductase, EC 1.2.1.3
Ald Ox: Aldehyde oxidase, aldehyde:O_2 oxidoreduc-
        tase, EC 1.2.3.1
AAF: 2-Acetylaminofluorene
DAB: Dimethylaminoazobenzene
PAGE: Polyacrylamide gel electrophoresis
NAD: Nicotinamide adenine dinucleotide
NADP: Nicotinamide adenine dinucleotide phosphate
PMS: Phenazine methosulfate
NBT: Nitroblue tetrazolium

LITERATURE CITED

Anderson, N. G., and J. H. Coggin, Jr. 1971. Models of differentiation, retrogression, and cancer. *Proc. First Conf. and Workshop on Embryonic and Fetal Antigens in Cancer*, 7–37.

Black, S. 1951. Yeast aldehyde dehydrogenase. *Arch. Biochem. Biophys*. 34: 86–97.

Decker, L. E., and E. M. Rau 1963. Multiple forms of glutamic-oxalacetic transaminase in tissues. *Proc. Soc. Exp. Biol. Med*. 112: 144–149.

Feinstein, R. N., and E. C. Cameron 1972. Aldehyde dehydrogenase activity in a rat hepatoma. *Biochem. Biophys. Res. Commun*. 48: 1140–1146.

Feinstein, R. N., and R. Lindahl 1973. Detection of oxidases on polyacrylamide gels. *Anal. Biochem*. 56: 353–360.

Jakoby, W. B. 1958. Aldehyde oxidation. I. Dehydrogenase from *Pseudomonas fluorescens*. *J. Biol. Chem*. 232: 75–87.

King, T. E., and V. H. Cheldelin 1956. Oxidation of acetaldehyde by *Acetobacter suboxydans*. *J. Biol. Chem*. 220: 177–191.

Longo, C. P., and J. G. Scandalios 1968. Specificity of the dehydrogenases of maize endosperm. *Biochem. Genet*. 2: 177–183.

Manwell, C., and C. M. A. Baker 1969. Hybrid proteins, heterosis, and the origin of species. I. Unusual variation of polychaete *Hyalinoecia* "nothing dehydrogenases" of quail *Coturnix* erythrocyte enzymes. *Comp. Biochem. Physiol.* 28: 1007-1028.

Morton, R. K. 1955. The action of purified alkaline phosphatases on di- and triphosphopyridine nucleotides. *Biochem. J.* 61: 240-244.

Nakamoto, T., and B. Vennesland 1960. The enzymatic transfer of hydrogen. VIII. The reactions catalyzed by glutamic and isocitric dehydrogenases. *J. Biol. Chem.* 235: 202-204.

Peraino, C., R. J. M. Fry, and E. Staffeldt 1971. Reduction and enhancement by phenobarbital of hepatocarcinogenesis induced in the rat by 2-acetylaminofluorene. *Cancer Res.* 31: 1506-1512.

Peraino, C., R. J. M. Fry, E. Staffeldt, and W. E. Kisieleski 1973. Effect of varying the exposure to phenobarbital on its enhancement of 2-acetylaminofluorene-induced hepatic tumorigenesis in the rat. *Cancer Res.* 33: 2701-2705.

Potter, V. R. 1969. Recent trends in cancer biochemistry: the importance of studies on fetal tissue. *Can. Cancer Conf.* 8:9.

Reuber, M. D. 1965. Development of preneoplastic and neoplastic lesions of the liver in male rats given 0.025 per cent N-2-fluorenyldiacetamide. *J. Natl. Cancer Inst.* 34: 697-709.

Ris, M. M., and J. -P. von Wartburg 1973. Heterogeneity of NADPH-dependent aldehyde reductase from human and rat brain. *Eur. J. Biochem.* 37: 69-77.

Shaw, C. R., and A. L. Koen 1965. On the identity of "nothing dehydrogenase". *J. Histochem. Cytochem.* 13: 431-433.

Zimmerman, H., and A. G. E. Pearse 1959. Limitations in the histochemical demonstration of pyridine nucleotide-linked dehydrogenases ("nothing dehydrogenase"). *J. Histochem. Cytochem.* 7: 271-275.

RESURGENCE OF SOME FETAL ISOZYMES IN HEPATOMA

F. SCHAPIRA, A. HATZFELD, AND A. WEBER
Institut de Pathologie Moléculaire[1]
Centre Hospitalo-Universitaire Cochin
24, rue du Faubourg St Jacques Paris 75014 - France

ABSTRACT. Investigation of isozymes provides a qualitative approach of enzyme modifications in cancer at the molecular level. Frequently isozymic modifications are a reversion towards an embryonic pattern. The adult molecular form tends to disappear while the fetal form increases or even appears while it was absent in the adult tissue.

We have studied several enzymes with multiple molecular forms: aldolase, pyruvate kinase, hexosaminidase and lactate dehydrogenase in fetal and regenerating liver, and in hepatomas with various rates of growth. We have found a rough parallelism between the extent of resurgence of fetal forms and the disappearance of the adult form. For example, in slow-growing hepatomas, there is a resurgence of aldolase A without resurgence of aldolase C, and aldolase B is only slightly decreased. On the contrary, in fast-growing hepatomas aldolase B almost completely disappears while both fetal aldolases A and C are present.

Isozymic modifications in regenerating liver are comparable to those occurring in slow-growing hepatomas, although the rate of cell multiplication is much greater. For example, M_2-type pyruvate kinase increases only slightly after hepatectomy or CCl_4 injection, while the L type is slightly decreased. This is in contrast with the pattern observed in fast-growing hepatomas, where the M_2-type is considerably increased and the L type almost completely disappears.

The mechanisms underlying these phenomena are still unclear. They may be due to the multiplication of stem cells with embryonic isozyme equipment. More likely, they represent an impairment of a control mechanism occurring at the transcriptional or post-transcriptional level.

---

[1] Université Paris V, Groupe U. 129 de l'Institut National de la Santé et de la Recherche Medicale, Laboratoire associe au Centre National de la Recherche Scientifique.

INTRODUCTION

The knowledge of isozymes has opened new perspectives in cancer enzymology, allowing us to find modifications at the molecular level. The Greenstein concept (Greenstein, 1954) that enzymic complements of tumors tend to resemble each other has been criticized: many examples of tumors with divergent enzymic patterns have been found. But the study of isozymes has renewed Greenstein's hypothesis on a new basis.

Our group was the first to find, in 1962, a modification of multiple forms of an enzyme, aldolase (EC 4.1.2.13 and 4.1.2.7), in human and experimental hepatomas, and to propose the hypothesis of the repression of the adult form with derepression of the fetal isozymes (Schapira et al., 1963; Schapira et al., 1962).

We recall first that there are three types of aldolases in higher animals, aldolase A, B, and C, which are tetrameric molecules . The hybridization of two types can give five isozymes, two pure tetramers and three hybrids: for example A_4, A_3B_1, A_2B_2, A_1B_3, and B_4. The three types differ by their relative activity towards two substrates: fructose-1-6-diphosphate (FDP) and fructose-1-phosphate (F1P) (Penhoët et al., 1966; Schapira, 1961). They differ also by their charge: type C is the most positively charged. Immunological methods allow a clear distinction to be made between the three types, but it must be noted that antisera against homo-polymers usually react also against the hybrid isozymes (Penhoët and Rutter, 1971).

We have shown that while type B is almost the only aldolase present in adult liver, aldolase A (muscle type) and aldolase C (brain type) are also present in fetal liver. Table 1 summarizes the properties and distribution of the three types in various rat tissues.

RESULTS

In our work, performed in collaboration with G. Schapira and J.C. Dreyfus, and then with Y. Nordmann, one of us has given evidence for the resurgence of both fetal aldolases A and C in hepatomas (Nordmann and Schapira, 1967; Schapira et al., 1971; Schapira et al., 1970; Weber and Schapira, 1972).

Fig. 1 schematizes the kinetic results with rat and human hepatomas and stresses the similarities between the properties of fetal liver aldolases and those of the aldolases of hepatomas, principally fast-growing ones.

TABLE 1

Properties and Distribution of Aldolase Isozymes
in Rat Tissues

	TYPES		
	A	B	C
Activity ratio FDP/F1P	≥ 50	1	7
Electrophoretic migration (in starch gel at pH 7.0)	Anodic slow	Cathodic	Anodic fast
Tissues for which this type is preponderant	Muscle	Liver	Brain

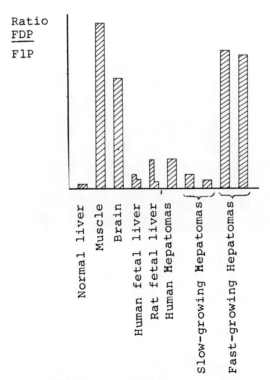

Fig. 1.   Aldolase activity ratios in various tissues.

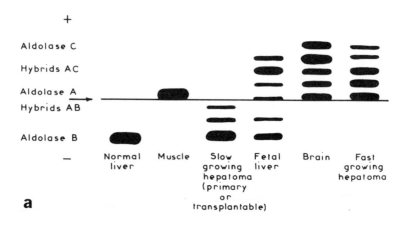

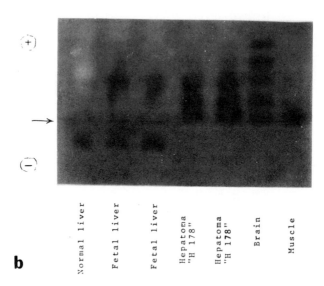

Fig. 2. Electrophoresis of aldolase isozymes in various rat tissues (starch gel). a. diagram; b. photograph.

While the substrate specificity of normal adult liver aldolase (aldolase B) is characterized by an FDP/F1P activity ratio constantly close to 1, in fetal liver this ratio is higher; in rat it decreases from 8 at the 15th day to 1 at birth.

In slow-growing hepatomas, primary or transplantable, this ratio is slightly abnormal between 1 and 2, but chiefly it is very high in fast-growing hepatomas (ascitic or solid), being about 30. Figure 1 shows also that this ratio is comparable to that found in muscle or in brain. In human hepatomas we have found that it was variable, but always abnormal, between 2 and 12 according to the stage of the disease.

Figure 2 schematizes our electrophoretic results (on starch gel at pH 7.0) and gives an example of a photograph. Adult rat liver aldolase is characterized by a single cathodic band, muscle aldolase by a slow anodic band, and aldolase of brain by five anodic isozymes, resulting from the hybridization of the tetrameric aldolases A and C.

In fetal liver, the three types are found, aldolase B (cathodic band) and aldolases A and C (3 to 5 anodic bands). In slow-growing hepatomas the aldolase A is hybridized with normal aldolase B. Principally it can be seen that the normal aldolase B has disappeared in fast-growing hepatomas; it is replaced by the two other types, A and C.

Immunological results (Table 2) confirm the presence of the two types, A and C, both in fetal liver and in fast-growing hepatomas. Antisera against each type were prepared by repeated injections of crystalline aldolases A, B, or C into chickens. They displayed a high specificity but no significant species specificity. Aldolases were isolated with modifications according to Penhoet et al., (1966). The action of antisera was tested by measuring aldolase activity in the supernatant after incubation of tissue extracts with each antiserum or with normal serum, followed by centrifugation. There is a good correlation between the degree of inhibition of the supernatant and the kinetic and electrophoretic results.

Fetal liver does contain the three types as early as the 15th day of fetal life. Hematopoietic tissue does not seem to contribute significantly to the results; there is no parallelism between its evolution and the isozymic evolution. Moreover, we have found that the aldolase activity ratio does not change when the hematopoietic cells regress under the influence of stress.

The extent of the resurgence of fetal aldolases (A alone or both A and C) with the disappearance of normal adult aldolase B corresponds roughly to the growth rate of the hepatomas studied.

TABLE 2
Inhibition of FDP Aldolase Activity
by the Antisera (in per cent)

RAT TISSUES	ANTI A	ANTI B	ANTI C
Normal liver	5 to 10	80 to 90	< 2
Muscle	90 to 95	< 2	< 5
Brain	85		95
Fetal liver			
15th day	30 to 35	60 to 70	20 to 50
Fast-growing			
hepatomas (H 178)	85 to 90	< 5	30 to 60
Slow-growing hepatomas	40 to 60	50 to 80	< 10

We may add that we have made analogous findings with human hepatomas, but normal adult human liver contains a small amount of aldolase C, contrary to the condition found in rat adult liver.

It must be stressed that the increase of aldolase A is not a characteristic of cancerous tissue. In organs where aldolase A is the preponderant isozyme of adult tissue (as for example muscle or spleen), we have found that aldolase A decreases after cancerization, while one of the other types increases (Schapira, 1966; Schapira et al., 1972).

The modifications of aldolase isozyme patterns represent at the present time the most striking example of the resurgence of fetal molecular forms in cancer, but similar findings are being made for an increasing number of isozymic systems.

We have also studied the isozymic modification of pyruvate kinase (2.7.1.40). The multiple molecular forms of this enzyme were discovered in 1965 by Tanaka et al., (1965) who showed that in ascitic hepatoma AH 130, the properties of pyruvate kinase were similar to those of the enzyme in muscle. Total activity was very high, as demonstrated by Lo et al., (1968). Taylor et al., (1969) demonstrated that pyruvate kinase of a fast-growing hepatoma was different, not only from the liver enzyme, but also from the muscle enzyme. Criss (1969), separating several forms by isoelectrofocusing, found a new form, predominant in the Novikoff hepatoma.

One of us, with C. Gregori (Schapira and Gregori, 1971), showed that pyruvate kinase of a poorly differentiated hepatoma (Reuber H 178) resembles placental pyruvate kinase from the point of view of its electrophoretic migration at pH 7.0 on starch gel. Moreover one of the three bands found in fetal liver (16th day) had the same migration as the tumor isozyme. The cancerous enzyme, like the placental enzyme, was activated by fructose-di-phosphate in the presence of a non-saturating concentration of substrate phosphoenol pyruvate, in contrast to muscle type.

With slow-growing hepatomas, we found two bands. The migration of one was identical to the main band of normal liver type; the other band, which is similar to the placental enzyme and is also found in normal liver, was increased.

Imamura and Tanaka (1972) demonstrated that there exist three types of pyruvate kinases in rat tissues: type L, predominant in liver; type M_1 in muscle, heart and brain; and type M_2 in many other tissues. The cancerous isozyme that we have described is identical to their M_2 type. The three isozymes that we found in fetal liver correspond to the L and M_2 types and to a fourth isozyme found also in erythrocytes, and which would be an hybrid (Whittel et al., 1973). In slow-growing hepatomas the two bands correspond to isozyme L and isozyme M_2. Fig. 3 schematizes these findings.

We have also worked on hexosaminidase (N-acetyl-β-glucosaminidase (2.2.1.30): the molecular forms of which have been more studied in human than in rat tissues. Our group has shown that in most rat tissues there are two bands, separable

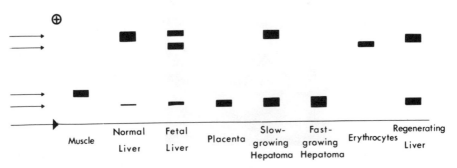

Fig. 3. Electrophoresis of pyruvate kinase isozymes in various rat tissues (starch gel).

by electrophoresis on cellulose acetate by using as substrate a methylumbelliferyl derivate. The predominant band has the slowest migration during electrophoresis in cellulose acetate and this cathodic band is thermostable and bound to lysosomes. The fastest band is weaker in normal liver; it is stronger in brain and strongest in fetal liver. This band is more

thermolabile and more easily extractible. We found that
in two types of transplantable fast-growing hepatomas ("LF"
of Frayssinet and "H 178" of Reuber) there is always a
diminution and sometimes even a disappearance of the slowest
band, with a considerable increase of the fastest. Principally,
as shown in Fig. 4, the cancerous pattern is similar to the
fetal pattern at the 16th day (Weber et al., 1973).

We have found a similar fetal-type with another enzyme:
lactate dehydrogenase (LDH) (Schapira and de Nechaud, 1968).
The modifications of LDH isozymes appear to be especially interesting, because it is known since the work of Goldman et
al. (1964) that generally the LDH-A subunits are increased
in cancerous tissues. We have found, on the contrary, an
increase of LDH-B subunits in hepatoma. Analogous findings
have been made by Kline and Clayton (1964) and by Johnson
and Kampschmidt (1965). We have pointed out that fetal
liver contains more LDH-B subunits than adult liver, while
many other fetal organs contain more LDH-A subunits.

The resurgence of fetal forms of enzymes in hepatoma is
now well demonstrated for an increasing number of enzymes
(Schapira, 1973). In order to try to explain the mechanism
of this resurgence of fetal enzymes in cancer, we have searched
for similar isozymic modifications in regenerating liver.

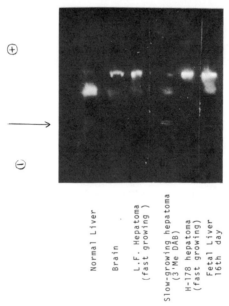

Fig. 4. Electrophoresis of hexosaminidase isozymes in various rat tissues   (cellulose acetate).

Rat liver regeneration was studied from 18 hours to 7 days after extensive hepatectomies according to the technique of Higgins and Anderson (1931), and also after injection of carbon tetrachloride (CCl_4). Comparison of livers after hepatectomies were made with livers after sham operation (Weber and Schapira, 1972).

Table 3 shows the aldolase activity ratio in regenerating liver at different times after hepatectomy. It is

TABLE 3
Aldolase Activites in Regenerating Liver

Time after surgery (*)	Aldolase activities (*) (in I.U. per g)		Activity Ratio $\frac{FDP}{F1P}$
	FDP	F1P	
18 h	17,4	15,9	1,13
24 h	16,4	13,4	1,20
48 h	16,8	14,6	1,10
3 - 4th day	20,8	18,1	1,14
5 - 6th day	18,5	16,9	1,10
7th day	16,7	14	1,20
Normal liver(**) (after sham operation)	23,1 $\pm$ 3,3	21,5 $\pm$ 3,2	1,065 $\pm$ 0,077

(*) Mean of 3 experiments
(**) Mean of 15 experiments $\pm$ SD

noticeable that the increase of this ratio is very slight (although significant). Its moderation is confirmed by the electrophoretic results, of which an example is given in Fig. 5. It must be noted that hybrids between A and B subunits of aldolase were not found in these experiments, contrary to the results obtained with slow-growing hepatomas as shown in the Fig. There is a moderate increase of aldolase A, confirmed by immunological results as shown in Table IV. There is no resurgence of aldolase C, unlike the pattern seen in fast-growing hepatomas and in fetal liver.

For hexosaminidase, results are comparable regardless of whether regeneration was induced by hepatectomy or by CCl_4, as shown in Fig. 6. The slight increase of the fastest band which we occasionally found was also sometimes present after sham operation.

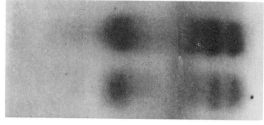

b

A₄

Hybrids
A-B

H 189

H 175

a

Muscle

Livers

Regenerating

Normal Liver

Brain

Fig. 5. Electrophoresis of aldolase isozymes (starch gel)
a) Normal and regenerating liver, muscle, and brain
b) Slow-growing hepatomas

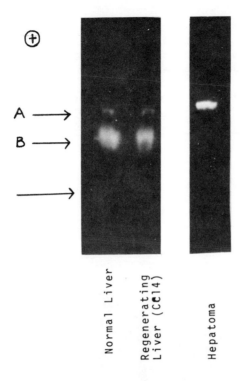

A ⟶
B ⟶
⟶

Normal Liver

Regenerating Liver (CCl4)

Hepatoma

Fig. 6. Electrophoresis of hexosaminidase isozymes in normal and regenerating liver and in fast-growing hepatoma.

Results for pyruvate kinase are similar. After extensive hepatectomy (from 18 hours to 7 days) the pattern remained almost normal. We only found a slight strengthening of the M_2 band, which was very similar to the pattern found in slow-growing hepatomas and very different from that in fast-growing hepatomas (Fig. 7). The action of anti M-pyruvate kinase antiserum allows the relative amount of "L" and "M_2" isozymes in normal and regenerating liver and in fast-growing hepatomas to be estimated. Table V shows that this ratio of L/M_2 is about 3 in normal liver. It is only slightly lower in regenerating liver (48 hours) while in fast-growing hepatomas it is very low, 0.1 or less. These findings are somewhat different from those of Suda et al., (1972), but comparable to the recent results of Bonney et al., (1973).

DISCUSSION

The similarity between the resurgence of fetal isozymes

TABLE 4
Action of Antiserum Anti Aldolase A on FDP Aldolase
Activity in Regenerating Liver

Time after hepatectomy	Inhibition percentage of aldolase activity of supernatant[1]
24 hours	13
48 hours	21
3 - 4 days	25
Normal liver	$\leqslant$ 15

[1] Mean of 3 experiments

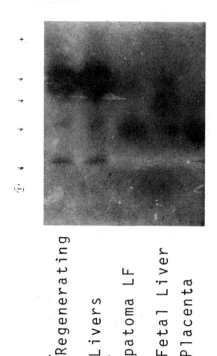

Fig. 7. Electrophoresis of pyruvate kinase isozymes in placenta, in fetal and regenerating liver, and in fast-growing hepatoma (starch gel).

TABLE 5
Pyruvate Kinase Activity Ratio $\dfrac{L}{M_2}$

Normal Liver	Regenerating Liver	Fast-Growing Hepatoma
	48 H	
2.96 ± 0.36	1.82 ± 0.35	≤ 0.1

in cancer and that of several fetal antigens, of which the
α-foeto-protein is the best known (Abelev, 1971), is obvious.
Moreover, these fetal enzymes have antigenic properties.
Some differences may be noted. While there is a remarkable
parallelism between the growth rate of tumors and the extent of isozyme resurgence, no parallelism can be found for
fetal antigens. We stress that in regenerating liver the
resurgence of α-foeto-protein and fetal isozymes remains very
moderate, and that both contrast with the considerable synthesis of fetal proteins in hepatoma. It seems that cell
multiplication alone is not sufficient to account for the
resurgence of fetal forms in hepatoma. The phenomenon is
not limited to cancerous liver and may be found in several
other tissues, spleen (Schapira, 1966), or muscle (Schapira
et al., 1972; Faron et al., 1972) for example. And it may
be added that, in some cases, the cancerous isozyme is found
not in the fetal organ but in another tissue of the same
species (Schapira, 1973), particularly in the placenta (Fishman et al., 1971) which, moreover, is an embryonic tissue.

In order to account for this general phenomenon, several
expressions have been used: "oncogeny as blocked ontogeny"
(Potter, 1969), "retro-differentiation" (Uriel, 1969), and
so on. The formulation of Sugimura et al., (1966) "dysdif-
ferentiation", seems more exact as also does the formulation
of Markert: "neoplasia is a disease of cell differentiation"
(1968). But at what level is the anomaly?

Is the anomaly located at the cellular level? In this
case, the same cells would not synthesize both adult and
fetal enzymic forms, and it must be supposed that during carcinogenesis stem cells or transitional cells would multiply
without complete maturation and synthesize fetal isozymes.
But we have found (Nordmann and Schapira, 1967) as did
Sugimura's group (1968) that hybrids are found between adult
aldolase B and fetal aldolase A. These hybrids show that synthesis in the same cells is very probable. The existence of
these hybrids in hepatoma and their absence in regenerating

liver is further evidence that the mechanisms of isozymic resurgence in regeneration and in cancer are different. Nevertheless, we shall search for more direct proof, using fluorescent antibodies against fetal and adult isozymes.

If the same cells synthezize both adult and fetal isozymes, is the anomaly a modification of the genome itself? The frequency of chromosomic abnormalities in cancer is well known, but, because the abnormal isozymes in tumors may always be found in other tissues of the same species, the chromosomic modifications probably would affect control mechanisms. We have suggested at the beginning of our work, that the control disturbance would be at the transcriptional level causing the repression of genes coding for the adult forms with derepression of genes coding for the fetal forms. But at the present time, there is no conclusive evidence for the existence of regulatory genes in higher animals (Dreyfus, 1972). A control impairment seems certain, but its mechanism remains unknown. Perhaps the non-histone proteins of chromatin are involved (Kruh, 1972), or perhaps the abnormality is at the post-transcriptional level.

A last point must be stressed. Although the phenomenon of a regression towards a fetal pattern is especially striking in hepatoma, and more generally in cancerous tissues, it is not really specific for carcinogenesis.

Our group and some others have demonstrated that, for instance, lactate dehydrogenase, aldolase, and creatine kinase take a fetal isozymic form in atrophied muscle (Schapira and Dreyfus, 1965; Schapira et al., 1968). It seems therefore that control impairment occurs in several pathological states.

We may conclude that the study of isozymes has given a new approach to the study of cell differentiation as well as to the cancer problem.

REFERENCES

Abelev, G.J. 1971. Alpha feto-protein in ontogenesis and its association with malignant tumors. *Advan. Cancer Res.* 14: 295-358.

Bonney, R.J., H.A. Hopkins, P. Roy Walker, and Van R. Potter 1973.Glycolytic isoenzymes and glycogen metabolism in regenerating liver from rats on controlled feeding schedules. *Biochem. J.* 136: 115-124.

Criss, W.E. 1969. A new pyruvate-kinase isozyme in hepatomas. *Biochem. Biophys. Res. Commun.* 35: 901-905.

Dreyfus, J.C. 1972. Bases moléculaires des maladies enzymatiques génétiques. *Biochimie* 54: 559-571.

Faron, F., H.T. Hsu, and W.E. Knox 1972. Fetal-type isozyme in hepatic and nonhepatic rat tumors. *Cancer Res.* 32: 302-308.

Fishman, W.H., N.R. Inglis, and S. Green 1971. Regan isoenzyme: a carcino-placental antigen. *Cancer Res.* 31: 1054-1057.

Goldman, R.O., N.O. Kaplan, and C. Hall 1964. Lactic de-hydrogenase in human neoplastic tissue. *Cancer Res.* 24: 389-399.

Greenstein, J.P. 1954. *Biochemistry of Cancer*, 2nd edition, Academic Press, New York.

Higgins, G.M. and R.M. Anderson 1931. Experimental pathology of the liver. Restoration of the liver of the white rat following partial surgical removal. *Arch. Pathol.* 12: 186-202.

Imamura, K. and T. Tanaka 1972. Multimolecular forms of pyruvate kinase from rat and other mammalian tissues. *J. Biochem.* 71: 1043-1051.

Johnson, H.L. and R.F. Kampschmidt 1965. Lactic dehydro-genase isozymes in rat tissues, tumors, and precancerous livers. *Proc. Soc. Exp. Biol.* 120: 557-561.

Kline, E.S. and G.C. Clayton 1964. Lactic dehydrogenase isozymes during development of Azo Dye tumors. *Proc. Soc. Exp. Biol.* 117: 891-894.

Kruh, J. 1972. RNA and the control of gene expression in animal cells. *Rev. Europ. Etud. Clin. Biol.* 17: 739-744.

Lo, C.H., F. Farina, H.P. Morris, and S. Weinhouse 1968. Glycolytic regulation in rat liver and hepatomas. *Advan. Enzyme Regul.* 6: 453-464.

Markert, C.L. 1968. Neoplasia: A disease of cell differenti-ation. *Cancer Res.* 28:1908-1914

Matsushima, T., S. Kawabe, M. Shibuya, and T. Sugimura 1968. Aldolase isozymes in rat tumor cells. *Biochem. Biophys. Res. Commun.* 30: 565-570.

Nordmann, Y. and F. Schapira 1967. Muscle type isoenzymes of liver aldolase in hepatomas. *European J. Cancer* 3: 247-250.

Penhoët, E.E., T.V. Rajkumar, and W.J. Rutter 1966. Multiple forms of fructose diphosphate aldolase in mammalian tissues. *Proc. Nat. Acad. Sci. U.S.A.* 56: 1275-1282.

Penhoët, E.E., U. Kochman, and W.J. Rutter 1966. Isolation of fructose diphosphate aldolase A, B, and C. *Biochem-istry* 8: 4391-4395.

Penhoët, E.E. and W.J. Rutter 1971. Catalytic and immuno-chemical properties of homomeric and heteromeric com-binations of aldolase subunits. *J. Biol. Chem.* 246: 318-323.

HUMAN PHOSPHOGLYCERATE MUTASE: ISOZYME MARKER FOR MUSCLE DIFFERENTIATION AND FOR NEOPLASIA

GILBERT S. OMENN, M.D., Ph.D. and
MARK A. HERMODSON, Ph.D.
Division of Medical Genetics, Department of Medicine,
University of Washington, Seattle, Washington 98195

ABSTRACT. Tissue-specific isozymes of the glycolytic enzyme phosphoglycerate mutase can be distinguished by electrophoresis, chromatography, heat stability, and amino
acid composition. Phosphoglycerate mutase (PGAM) from
skeletal muscle migrates slowly, while PGAM from brain and
other tissues migrates rapidly toward the anode in pH 7.5
Tris-citrate or pH 8.6 Tris-EDTA-borate horizontal starch
gels. Striking developmental and neoplastic transitions
for PGAM expression in man have been identified. Human
fetal skeletal muscle at 50-60 days gestation contains the
single-banded type B PGAM pattern, as in adult brain. The
appearance of a band of intermediate mobility signifies
production of type M PGAM subunits and formation of MB
hybrid enzyme molecules. By 166 days gestation and thereafter, the skeletal muscle contains almost exclusively the
type M PGAM phenotype. Fetal human brain has the same phenotype as adult brain for PGAM, as do relatively benign
brain tumors, such as meningioma or a grade II astrocytoma.
However, highly malignant grade III and IV astrocytomas and
a hemangioblastoma gave electrophoretic patterns with prominent hybrid and muscle type bands. PGAM promises to be
a useful isozyme marker for differentiation in vivo and
in cultured systems in vitro.

In the course of population screening for genetically-
determined variation of the glycolytic enzymes in human brain
(Cohen et al, 1973), we have found electrophoretic evidence
for tissue-specific isozymes of phosphoglycerate mutase (PGAM,
EC 2.7.5.3). PGAM reversibly converts 3-phosphoglycerate to
2-phosphoglycerate. Skeletal muscle PGAM migrates slowly,
while PGAM from brain and other tissues migrates rapidly toward
the anode in pH 7.5 Tris-citrate or pH 8.6 Tris-EDTA-borate
horizontal starch gels (Omenn and Cohen, 1971). Extracts of
heart muscle give a three-banded pattern of PGAM activity,
consistent with a dimer structure.

We have recently discovered striking developmental and neoplastic transitions in the phenotype of PGAM in certain human
tissues (Omenn and Cheung, 1974). In addition, we have made
some progress in characterizing the PGAM isozymes isolated
from muscle and brain.

MATERIALS AND METHODS

Thigh-muscle, brain, heart muscle, and other tissues from human fetuses were obtained from the Central Embryology Laboratory at the University of Washington (Dr. T. Shepard, Director). Fetal specimens ranged from approximately 50 to 166 days gestational age, estimated from crown-to-rump length; usually there was good agreement with estimates from the last menstrual period. Stillbirths and adult tissues were obtained from the autopsy service of the Department of Pathology, University Hospital, or from the Seattle-King County Medical Examiner's Office. Surgical specimens were provided by members of the Neurosurgery Department. Clinical records and histological diagnoses were reviewed carefully.

Animal tissues were obtained from the Primate Center and the Vivarium at the University of Washington. Transplantable tumors induced in Buffalo, Sprague-Dawley, and Wistar-Furth rats, together with normal tissues, were kindly provided by Dr. Ruth Shearer. These tumors included Morris hepatomas, hepatomas induced by azo dye and other carcinogens, and polyoma-induced kidney sarcomas.

Tissue extracts were prepared by homogenization with four volumes of 0.05M Tris-HCl pH 8.0 buffer. Horizontal starch gel electrophoresis was carried out overnight with pH 8.6 Tris-EDTA-borate and pH 7.5 Tris-citrate systems. Specific staining for PGAM activity employed 4 mM 3-phosphoglycerate, 1 mM 2,3-diphosphoglycerate, 20 mM $MgCl_2$, 3 mM ADP, 12 mg NADH and ten units each of enolase, pyruvate kinase, and lactate dehydrogenase in 10 ml pH 8.0 Tris-HCl buffer. The coupled system led to production of NAD (non-fluorescent) from fluorescent NADH. Quantitative enzyme assays utilized the same system, with measurement of the change in OD_{340} as a function of time in a Gilford model 2000 spectrophotometer. Specific staining for creatine phosphokinase (CPK, EC 2.7.3.2) activity employed 2 mM creatine phosphate, 3 mM ADP, 0.1 mM NADP, 5 mM glucose, 10 mM $MgCl_2$, ten units each of hexokinase and glucose-6-phosphate dehydrogenase, 5 mg MTT, 1 mg PMS, and 0.5% agar in 10 ml pH 7.0 Tris-HCl buffer. All reagents were purchased from Sigma Chemical Company.

RESULTS

Developmental Studies

As shown in Fig. 1, there is a striking transition in the PGAM pattern during development of human skeletal muscle. In the earliest specimens (approximately 50 days), fetal muscle

Fig. 1A

PGAM Activity
+
Origin
—
Adult brain
62
78
85
100
164
166
days
Fetal muscle
Adult muscle
164 days Fetal heart

Fig. 1B

PGAM Activity
+
Origin
—
Adult brain
62
78
85
100
164
166
days
Fetal muscle
Adult muscle
164 days Fetal heart

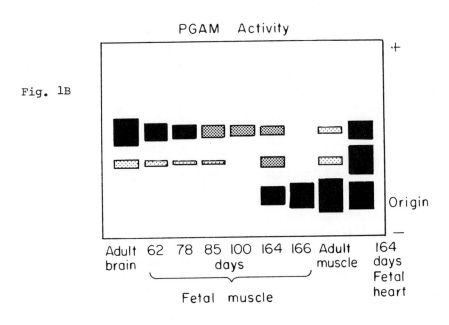

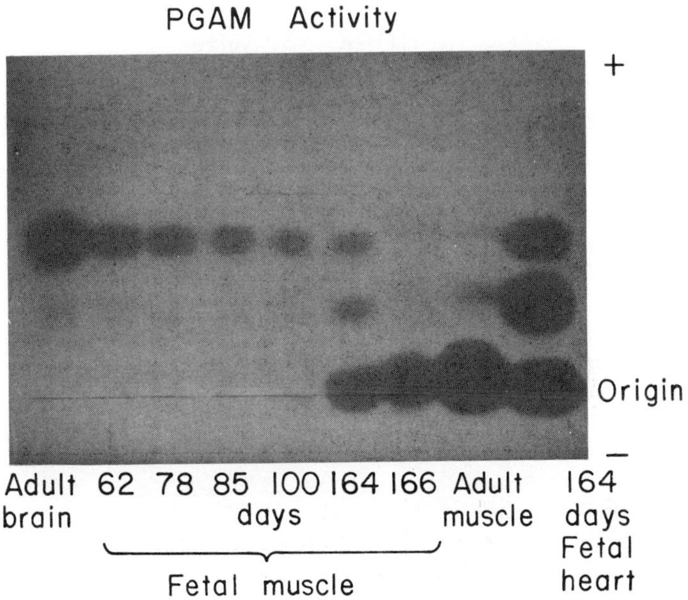

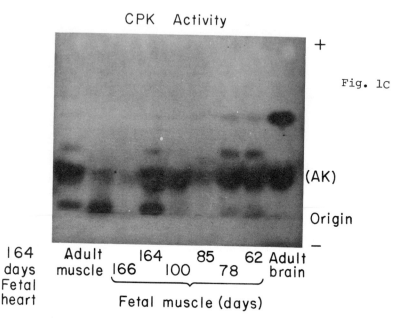

CPK   Activity

+

Fig. 1c

(AK)

Origin

−

| 164 days Fetal heart | Adult muscle | 164 166 | 85 100 78 | 62 | Adult brain |

Fetal muscle (days)

Fig. 1. Developmental transition in phosphoglycerate mutase (PGAM) in human fetal muscle. (A) pH 8.6 Tris-EDTA-borate starch gel stained specifically for PGAM activity in adult brain, adult muscle, fetal muscle, and fetal heart of gestational ages indicated. PGAM bands appear dark in fluorescent field; photographed under ultraviolet illumination. (B) Schematic diagram of frame 1A. (C) Developmental transition in creatine phosphokinase (CPK) in 'human fetal muscle. Top half of same gel stained for PGAM in Fig. 1A. The stain for CPK also demonstrates adenylate kinase activity (AK).

contains almost exclusively the single-banded type B PGAM pattern, as in adult brain. The appearance of the band of intermediate mobility (80-100 days) signifies type M PGAM production; at first type M subunits combine with type B subunits to make the MB hybrid dimer; then the three-banded pattern is recognized (115-164 days). In the specimen of 166 days gestation and on through term, the muscle-type pattern of the adult predominates. Only when gels are overstained can trace amounts of hybrid and type B PGAM be detected.

Creatine phosphokinase phenotypes were determined in the

same fetal muscle samples, since CPK is known to undergo a similar transition from a brain-type to a muscle-type pattern (Eppenberger et al, 1964). As shown in Fig. 1C, specimens of 62 and 78 days, which still show type B PGAM phenotype, already have definite expression of hybrid and muscle-type CPK. In fact, comparison of 19 specimens of 50 to 166 days gestation strongly suggests that the developmental transitions of PGAM and CPK are not synchronous, especially the first step of turning on the muscle-type genes in muscle (Omenn and Cheung, 1974).

Although we have not characterized the developmental process in other species we have tested, similarly distinguishable type M and type B isozymes of PGAM occur in the monkey, rat, mouse, and guinea pig. PGAM of chicken muscle and brain are not electrophoretically distinguishable in the Tris-citrate and Tris-EDTA-borate buffer systems we employ. Therefore, dark and light skeletal muscle of the chicken could not be compared. However, guinea pig skeletal muscle from thigh and calf was dissected by the method of Kark et al (1971) into fast and slow or red and white segments, both of which had the adult muscle-type of PGAM phenotype. Unlike the soleus, gastrocnemius, vastus lateralis, and vastus medialis, the guinea pig heart muscle gave a 3-banded PGAM phenotype with type B band more intense than the type M band.

Ectopic Production of Type M PGAM in Brain Tumors

A complementary transition in the electrophoretic phenotype and genetic expression of PGAM has been observed in human brain tumors (Fig. 2). Relatively benign tumors, such as meningioma or grade II astrocytomas, gave patterns identical with normal brain (overstained gels showed a trace of hybrid band). However, highly malignant grade III and IV astrocytomas (glioblastoma multiforme) and a recurrent cerebellar hemangioblastoma gave electrophoretic patterns with prominent hybrid and muscle-type bands. This phenomenon cannot be due to unrecognized contamination with blood, since erythrocyte PGAM is type B. No such neoplastic transformation to muscle-type enzyme occurred for CPK in these tumors. Thus far, we have analyzed 13 brain tumors; additional cases are being collected. In a small sampling of lung, liver, and intestinal carcinomas and lymphomatous malignancies we have found no shift from the type B PGAM phenotype. Finally, a few experimentally induced liver and kidney tumors in the rat were screened, but no shift from the PGAM phenotype of adult liver or kidney was detected.

PGAM Activity

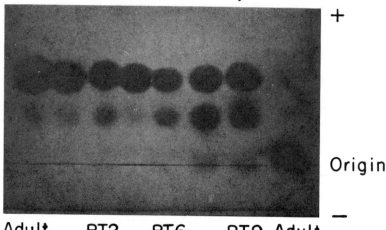

+

Origin

−

Adult brain BT1 BT2 BT4 BT6 BT7 BT9 Adult muscle

Fig. 2A

PGAM Activity

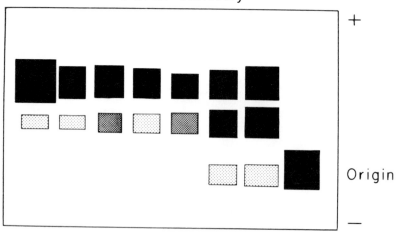

+

Origin

−

Adult brain BT1 BT2 BT4 BT6 BT7 BT9 Adult muscle

Fig. 2B

CPK Activity

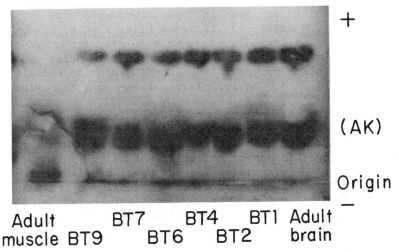

Fig. 2C

Fig. 2. Electrophoretic phenotype of PGAM in brain tumors
(BT). (A) BT 1 and BT 4, grade II astrocytomas, have only
minimal hybrid PGAM with an overstained brain band, as in normal brain control; BT 2, BT 6, and BT 7, astrocytomas of more
malignant appearance histologically, have excess hybrid band
and some muscle band, as does BT 9, a recurrent hemangioblastoma. Tris-EDTA-borate starch gel, pH 8.6. (B) Schematic
diagram of Fig. 2A. (C) The top half of the same gel stained
for CPK activity. All samples have electrophoretic phenotypes
identical with normal brain.

Characterization of the Isozymes

The phenotypes of PGAM in adult brain and muscle are readily
distinguished by DEAE-cellulose chromatography, amino acid
composition, and heat lability, as well as by electrophoresis.
Fig. 3 shows the greater lability to elevated temperature of
brain PGAM activity, compared with muscle PGAM activity, in
the presence and in the absence of the marked stabilizing
effect of 2,3-diphosphoglycerate.

With the expectation that muscle and brain PGAM are homologous gene products, we have initiated studies to determine the
amino acid sequences. Muscle-type PGAM has been prepared from
5 kg human skeletal muscle by acetone fractionation (35-60%

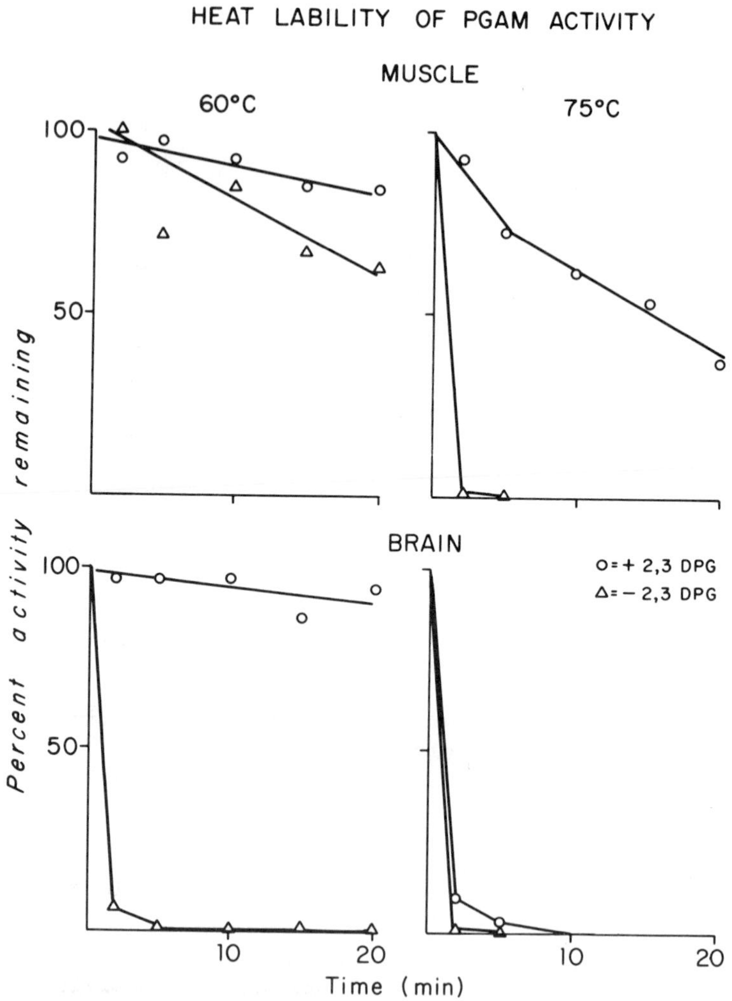

Fig. 3. Effect of elevated temperature (60° or 75° C) upon PGAM activity in adult muscle and adult brain extracts, with (o———o) or without (△———△) 2 mM 2,3-diphosphoglycerate. (From Omenn and Cheung, 1974.)

cut), suspension in buffer containing 2,3-diphosphoglycerate and then heating at 60° C for 10 min, DEAE-Sephadex A50 chromatography, and Sephadex G200 gel filtration. A homogeneous product, judging by aqueous and SDS-polyacrylamide electrophoresis and by chromatography, is obtained in over-all 25-30%

yield. A similar yield, starting from considerably less activity per gram of tissue, is obtained from human brain tissue, utilizing the same regimen, except for ammonium sulfate fractionation in the first step. The amino acid composition of the proteins is shown in Table 1, based upon 300 amino acid

TABLE 1

AMINO ACID ANALYSES

(Residues per 300 residues)

	Muscle PGAM	Brain PGAM
Asx	24.5	30.2
Thr	18.0	14.0
Ser	11.9	15.8
Glx	38.4	35.9
Pro	18.6	15.1
Gly	22.7	22.0
Ala	28.3	27.9
Val	15.9	21.0
Met	8.5	5.1
Ile	18.1	16.5
Leu	26.0	30.9
Tyr	8.8	9.0
Phe	7.9	11.1
Lys	27.0	21.9
His	9.0	7.9
Cys	3.0	4.7
Arg	20.7	16.6
Trp	n.d.	n.d.

These data represent time course analyses over a 96 hr period. Ser and Thr are extrapolated to zero time; Val and Ile are averages of the 96 hr samples. All others are averages of all samples. Calculated assuming 300 residues per monomer (30,000–32,000 monomer molecular weight).

residues per polypeptide chain, estimated from the dimeric molecular weight of about 60,000 reported for chicken, rabbit, and sheep skeletal muscle and porcine kidney (Torralba and Grisolia, 1966; Rose, 1970; James et al, 1971; Diederich et al, 1970). Note the very high number of acidic and basic residues, with the number of basic residues sufficiently higher in muscle-type PGAM to account for its slower electrophoretic mobility and for its lesser affinity for DEAE-cellulose upon chromatography.

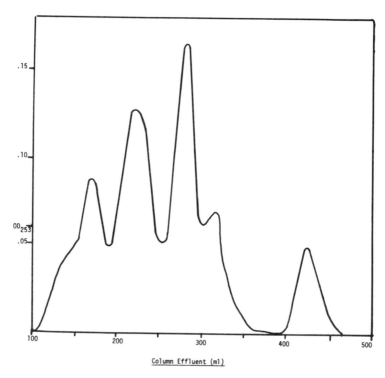

Fig. 4. Sephadex G75 separation of cyanogen bromide digest of reduced and alkylated muscle PGAM (80 mg portion). Lyophilized digest was passed over 2.5 x 100 cm column of G-75 (fine) in 9% formic acid at 20 ml/hr. Some peaks have one fragment, some two fragments, according to sequence data. The last peak contains no amino acids.

Cyanogen bromide fragments of muscle-type PGAM have been prepared. An 80 mg portion of the purified enzyme was digested with CNBr after reduction and alkylation. The lyophilized digest was passed over Sephadex G75 (Fig. 4), yielding at least three peptide-containing peaks, each of which on initial sequence analysis contains one or two fragments. The N-terminus of PGAM and of one of the fragments in peak two is blocked. Thus far, we have determined a 30-residue sequence of the one fragment and short runs in other fragments. A larger preparation of muscle PGAM has now been made, which should allow rapid progress in the sequence work.

Determinations of K_m for 3-phosphoglycerate, the physiological substrate, revealed no differences between the PGAM

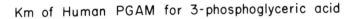

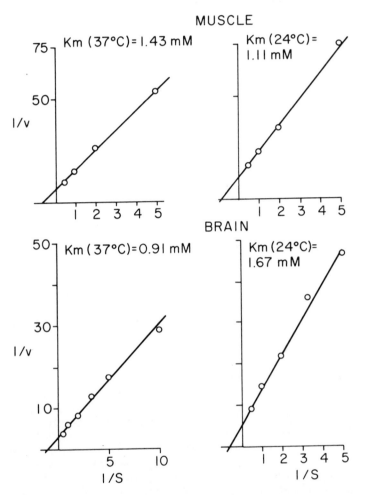

Fig. 5. Determination of Michaelis constants for the physiological substrate 3-phosphoglycerate. Muscle and brain PGAM
activity from adult human tissues; Tris-HCl buffer, pH 8.0,
at 24° or 37° C.

of skeletal muscle and of brain in the adult. At either 37° C
or 24° C, brain and muscle PGAM have K_m's of about 1.4 mM
(Fig. 5).

DISCUSSION

The developmental transition in the phenotype of phospho-glycerate mutase in human fetal skeletal muscle represents a two-step process, "turning on" or "turning up" the gene for muscle-type (type M) PGAM and "turning down" or "turning off" the gene for brain-type (type B) PGAM. The disappearance of the type B and hybrid bands on the gels depends also on the biological half-life of the enzyme. It may be objected that the transition represents not the maturation of muscle cells, but a shift in cell population from predominantly fibroblastic connective tissue to myoblastic cells. This explanation can be ruled out, however, by the asynchrony of the developmental transitions of PGAM and CPK.

The physiological stimulus for and consequences of this transition in PGAM phenotype are unknown. By contrast, for CPK, Eppenberger (this Conference) has provided evidence that the M-type CPK forms the M-line of myofibrils, while B-type CPK cannot do so.

PGAM, like CPK and aldolase (D.C. Turner, this Conference) can be detected in cultured muscle cells (Hauschka and Omenn, unpublished) and provides a good marker for investigation of differentiation of muscle in vitro. The transition during development of muscle detected here by change in the electro-phoretic phenotype also accounts for the developmental increase in sensitivity to mercuric ions reported for human and chicken skeletal muscle PGAM (Grisolia et al, 1970).

The transition in electrophoretic phenotype of PGAM in neo-plastic brain tissue appears to be correlated with the degree of malignancy of the tumors. A much larger series with various histological types of brain tumors will be necessary to con-firm this suggestion. We are gathering such material. In addition, examination of a variety of other malignancies will allow some conclusion about whether the transition is a par-ticular propensity of brain. Finally, the PGAM phenotype and other biochemical markers may prove useful in assessing the malignancy of human tumors and in following the induction of tumors experimentally.

It is likely that the neoplastic transformation somehow activates greater expression of the type M PGAM gene in brain cells; analogous "ectopic" production of polypeptides with distinctive hormonal, antigenic, and enzymatic properties has become a well-recognized clinical and biochemical phenomenon (Liddle et al, 1969; Omenn, 1970; Omenn, 1973). The associa-tion of particular sites or histology of tumors with particular types of ectopic polypeptide products is highly non-random. Since fetal brain does not express the type M PGAM, this transition resembles production of placental-type alkaline phosphatase by colon carcinomas (Stolbach et al, 1969) rather than the re-expression of fetal patterns as reported for aldolase or phosphorylase in neoplastic brain and liver (Schapira et al, 1970; Kumanishi et al, 1970; Schapira, this Conference; Weinhouse et al, this Conference).

Definitive assignment of these type B and type M isozymes of PGAM to different genes requires analysis of electrophoretic variants of the enzyme. No variants of PGAM were found in our survey of human, monkey, and mouse brain (Cohen et al, 1973). Two variants have been found recently after surveys of 3104 samples of human erythrocytes (Chen et al, 1974), but samples of other tissues were not available for comparison. In the absence of cases with electrophoretic variants of one isozyme without change in the other (but with change in position of the hybrid), definitive characterization of these isozymes may depend on the amino acid sequence determination now underway.

ACKNOWLEDGMENTS

Supported by grant GM 15253 and by a Research Career Development Award GM 43122 (G.S.O.) from the U.S. Public Health Service.

We gratefully acknowledge the technical assistance of Mrs. S. Cheung and the permission of the journal to publish figures 1A, 1C, 2A, and 3, which appeared in Omenn and Cheung (1974), © 1974 by the American Society of Human Genetics.

REFERENCES

Chen, S-H, J. Anderson, E. R. Giblett, and M. Lewis 1974. Phosphoglyceric acid mutase: rare genetic variants and tissue distribution. *Amer. J. Hum. Genet.* 26: 73-77.

Cohen, P. T. W., G. S. Omenn, A. G. Motulsky, S-H Chen, and E. R. Giblett 1973. Restricted variation in the glycolytic enzymes of human brain and erythrocytes. *Nature New Biol.* 241: 229-233.

Diederich, D., A. Khan, I. Santos, and S. Grisolia 1970. The effects of mercury and other reagents on phosphoglycerate mutase-2,3-diphosphoglycerate phosphatase from kidney, muscle and other tissues. *Biochim. Biophys. Acta* 212: 441-449.

Eppenberger, H. M., M. Eppenberger, R. Richterich, and H. Aebi 1964. The ontogeny of creatine kinase isoenzymes. *Develop. Biol.* 10: 1-16.

Grisolia, J., D. Diederich, and S. Grisolia 1970. Developmental increases in activity and sensitivity to mercury of human and chicken skeletal muscle phosphoglyceromutase. *Biochem. Biophys. Res. Commun.* 41: 1238-1243.

James, E., R. O. Hurst, and T. G. Flynn 1971. Purification and properties of phosphoglyceromutase from sheep muscle. *Can. J. Biochem.* 49: 1183-1194.

Kark, R. A. P., J. P. Blass, J. Avigan, and W. K. Engel 1971. The oxidation of β-hydroxybutyric acid by small quantities of type-pure red and white skeletal muscle. *J. Biol. Chem.* 246: 4560-4566.

Kumanishi, T., F. Ikuta, and T. Yamamoto 1970. Aldolase isoenzyme patterns of representative tumors in the human nervous system. *Acta Neuropathol.* 16: 220-225.

Liddle, G. W., W. E. Nicholson, D. P. Island, D. N. Orth, K. Abe, and S. C. Lowder 1969. Clinical and laboratory studies of ectopic humoral syndromes. *Recent Progr. Horm. Res.* 25: 283-314.

Omenn, G. S. 1970. Ectopic polypeptide hormone production by tumors. *Ann. Intern. Med.* 72: 136-138.

Omenn, G. S. 1973. Pathobiology of ectopic hormone production by neoplasms in man. *Pathobiology Ann.* 3: 177-216.

Omenn, G. S., and S. C-Y Cheung 1974. Phosphoglycerate mutase isozyme marker for tissue differentiation in man. *Amer. J. Hum. Genet.* 26: 393-399.

Omenn, G. S., and P. T. W. Cohen 1971. Electrophoretic methods for differentiating glycolytic enzymes of mouse and human origin. *In Vitro* 7: 132-139.

Rose, Z. B. 1970. Evidence for a phosphohistidine protein intermediate in the phosphoglycerate mutase reaction. *Arch. Biochem. Biophys.* 140: 508-513.

Schapira, F., M. D. Reuber, and A. Hatzfeld 1970. Resurgence of two fetal-type of aldolases (A and C) in some fast-growing hepatomas. *Biochem. Biophys. Res. Commun.* 40: 321-327.

Stolbach, L. L., M. J. Krant, and W. H. Fishman 1969. Ectopic production of an alkaline phosphatase isoenzyme in patients with cancer. *New Engl. J. Med.* 287: 757-761.

Torralba, A., and S. Grisolia 1966. The purification and properties of phosphoglycerate mutase from chicken breast muscle. *J. Biol. Chem.* 241: 1713-1718, 1966.

THE EXPRESSION OF HUMAN ALBUMIN IN A
MOUSE HEPATOMA/HUMAN LEUCOCYTE HYBRID

GRETCHEN J. DARLINGTON, HANS PETER BERNHARD,
AND FRANK H. RUDDLE
Yale University, Kline Biology Tower
New Haven, Connecticut   06520

ABSTRACT.   Cells derived from a murine hepatoma were
adapted to in vitro growth.   A substrain of the
hepatoma line, Hepa 1a, was selected for resistance to
6-thioguanine and has been shown to be deficient for
hypoxanthine: guanine phosphoribosyltransferase (HPRT).
Hepa 1a cells continue to synthesize and secrete
several serum proteins in vitro including albumin,
transferrin, ceruloplasmin, and α-fetoprotein.   In
addition, Hepa 1a expresses Es-2 esterase, an electrophoretic form of esterase found predominantly in
tissues of the liver and kidney.   The mouse line was
co-cultivated with normal human leucocytes in the
presence of inactivated Sendai virus.   Five colonies
appeared which were isolated and shown to be hybrid
by isozymic and karyotypic analysis.   Three of the
hybrids were tested for the production of the mouse
proteins transferrin and α-fetoprotein; they continued
to express these murine differentiated functions.   All
five hybrids produced both mouse serum albumin and Es-2
esterase.   In addition to the mouse proteins, two hybrid
populations were observed which produced human serum
albumin.   The activation of a non-hepatic human genome
to perform a liver function suggests the possibility
that the human structural loci for differentiated
phenotypes can be mapped in this somatic cell hybrid
system.

The use of somatic cell hybrids obtained from the
fusion of rodent and human lines to examine the linkage
relationships of human genes has been well established
(cf. Ruddle 1973).   Many of the gene products studied for
the purpose of mapping have been enzymes which are expressed
by cells in culture and whose electrophoretic mobilities are
different between man and mouse.   The segregation of human

chromosomes from the hybrid cells permits the investigator
to correlate the presence or absence of human enzymes with
the retention or segregation of human chromosomes. Human
genes responsible for particular phenotypes can be assigned
to specific human chromosomes. In addition, one can
correlate the expression of the various human isozymes with
each other, independent of the chromosomal analysis, and
establish linkage relationships between human phenotypes.

We were interested in examining hybrids between a line
of mouse hepatoma cells which continues to express several
hepatic properties in vitro and human diploid leucocytes.
In such hybrids, the interaction of the hepatoma genome
with sub-sets of the human complement could be studied.
The effect of the non-liver human genome on the expression
of murine liver traits was of particular interest.

In general, the observations of expression of organ
specific traits in hybrids between differentiated and non-
differentiated cells can be classified into three groups.

First, many specialized properties are extinguished
in hybrid cells, that is, the functions of the differentiated parent do not appear. Of the hepatic functions
examined, tyrosine aminotransferase inducibility (Schneider
and Weiss, 1972), the liver form of alcohol dehydrogenase
(Bertolotti and Weiss 1972a), and aldolase B (Bertolotti
and Weiss 1972b) are extinguished. It has been proposed
by Davidson (1966) that, extinction results from the
presence of a repressor-like product of the non-different-
iated genome which prevents the expression of the specialized
trait.

Other traits continue to be expressed in hybrid cells.
The production of the second component of complement
(Levisohn and Thompson, 1973) and the synthesis of serum
albumin as demonstrated by Peterson and Weiss (1972) and
by Darlington et al., (1974) are examples of this second
category of expression.

Finally, at least one characteristic (serum albumin)
not only continues to be expressed but the non-specialized
genome is activated to produce the homologous gene product.
The activation of a mouse fibroblast genome to produce
mouse albumin when combined with a rat hepatoma has been
observed by Peterson and Weiss (1972). The studies to be
presented here describe the activation of the human genome
to produce human serum albumin in mouse hepatoma/human
leucocyte hybrids.

The origin of the murine line is as follows. A transplantable hepatoma (BW7756) carried in C57 leaden mice was
excised, digested with trypsin, and plated in Falcon tissue

culture flasks. After a few days in culture, the cells were harvested, pelleted, and re-injected into adult, male C57L mice. Growth of a tumor occurred within 4 weeks. The process of alternate in vivo and in vitro passage was repeated six times, with each in vitro passage prolonged by several days, until a strain of cells adapted for growth outside the body was obtained. The mass population was called Hepa. This line was subsequently cloned to produce a subline, Hepa-1. Finally, Hepa-1 cells were treated with 6-thioguanine to select for a strain deficient in hypoxanthine: guanine phosphoribosyltransferase (HPRT). A colony lacking HPRT was isolated, designated Hepa la and became the parental population used in these hybrid studies. An examination of the Hepa la karyotypes revealed a modal chromosome number of 58 with 4 to 6 biarmed chromosomes. One long acrocentric chromosome was also present in the complement which served as a marker. When stained for heterochromatin, there were several interdigitated regions of darkly stained material in the long arms of this latter chromosome distinguishing it from all others.

Hepa la was also examined for a variety of hepatic phenotypes. Those traits which have been studied are listed in Table 1. As can be seen, many properties of liver are not found in the Hepa la line. It should be noted, however, that some of the traits which do not appear in Hepa la were expressed by the BW7756 tumor. The properties lost by the cells following adaptation to in vitro conditions were alcohol dehydrogenase, xanthine oxidase, and inducibility of tyrosine aminotransferase by dexamethasone.

Esterase-2 is one mouse esterase which is found primarily in tissues of the kidney and liver. This enzyme activity is measured by starch gel electrophoresis.

Several serum proteins have been identified in supernatant medium from Hepa la cultures. They are α-fetoprotein, transferrin, ceruloplasmin, and albumin. Figure 1 shows the pattern of migration of these proteins upon electrophoresis in a polyacrylamide gel. Albumin which is secreted in relatively large amounts compared to the other proteins was measured by immunoelectrophoretic procedures as well. Ceruloplasmin was identified in replicate gels using o-dianisidine as a substrate. Thus the mobility of this protein was established by a specific staining procedure and corresponded to that of the band seen in Fig. 1, stained for protein with Coomassie blue. Transferrin in Hepa la supernatant medium and amniotic fluid consistently moved less anodally than normal adult serum. As expected, α-fetoprotein is not seen in adult serum.

TABLE 1

Hepatic Phenotypes Examined in Hepa 1a

Absent	Present
aldolase B	esterase-2
alcohol dehydrogenase	α-fetoprotein
phenylalanine hydroxylase	transferrin
xanthine oxidase	ceruloplasmin
carbamyl phosphate synthetase	albumin
ornithine transcarbamylase	
tyrosine aminotransferase	
(TAT), inducibility by	
dexamenthasone	
esterase-1	

The Hepa 1a cells were combined with peripheral leu-cocytes from a normal male and treated with Sendai virus to promote fusion. The cell mixture was grown in a medium (HAT) containing hypoxanthine, aminopterin, and thymidine which selected against the Hepa 1a cells. The leucocytes did not attach to the growth surface and were washed away during medium exchanges. Several colonies appeared and five of those isolated grew to confluency. The clones were called Hal 3, Hal 5, Hal 6, Hal 7a, and Hal 7b. The latter two arose in the same culture flask and are not considered to be of independent origin.

Chromosomal and isozymic analysis of the colonies was performed to demonstrate that the populations were truly hybrid cells. The isozymes tested are listed in Table 2. Each clone with the exception of Hal 5, expressed at least one autosomally inherited enzyme. However, no hybrid had more than 3 human isozymes, in addition to the X-linked markers, HPRT, G6PD, and PGK. The observation that the hybrids expressed such a small number of human phenotypes suggested that few human chromosomes remained in the complement. The presence of these enzymes was used to predict which human chromosomes should be retained in the hybrids. All of these enzymatic phenotypes have been assigned to specific chromosomes. Glucose phosphate isomerase has been located on chromosome 19 (McMorris et al., 1973; Hamerton et al., 1973). The expression of this enzyme would suggest that chromosome 19 was present in Hal 3. However, upon cytological examination of metaphases

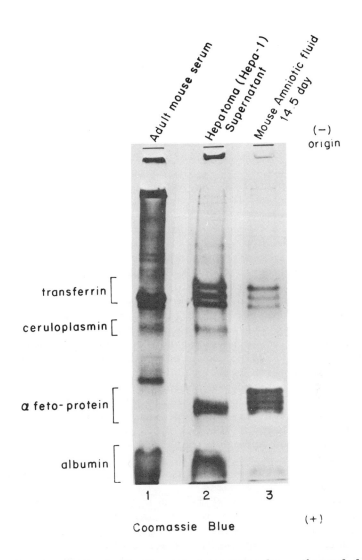

Fig. 1. Serum Proteins Secreted by Hepa 1a: Channel 1
contains adult mouse serum, 2 contains Hepa-1 supernatant
medium, concentrated 40 fold and 3 contains mouse amniotic
fluid also concentrated approximately 40 fold. Hepa-1 is
a clonal population of cells from which the drug resistant
Hepa 1a line was derived. The procedure for collecting
supernatant medium is to rinse a confluent monolayer of

Fig. 1., continued: cells twice with serum free medium, then allow the cells to remain in the absence of serum for 24 hours. The medium is then collected, spun free of cells, and concentrated by means of pressure filtration through PM-30 Amicon filters.

---

stained for Q and C banding, no chromosome 19 was observed. Other examples of a discrepancy between the observed chromosomal complement from that expected on the basis of the isozyme data is the absence of chromosome 21 in Hal 6 which expresses SOD-1 and the absence of chromosome 14 in Hal 7a and Hal 7b, both of which had activity for human nucleoside phosphorylase. The assignment of SOD-1 to chromosome 21 was reported by Tan et al., (1973) and Westerveld et al., (1973). Nucleoside phosphorylase has been assigned to chromosome 14 (Ricciuti and Ruddle, 1973).

With regard to the X-linked markers, all hybrids but Hal 7a expressed HPRT, G6PD, and PGK. Hal 7a had activity for the human form of HPRT. The loss of the other two phenotypes implied that the X chromosome was rearranged in Hal 7a. This prediction could not be confirmed cytologically as no human chromosomes, intact or rearranged, were found in Hal 7a. It is likely that a small portion of the X chromosome was present but that this segment could not be identified as being of human origin. Chromosomal rearrangement was also expected for chromosome 6, as the linkage assoication between SOD-2 and ME-1 was disrupted.

The modal chromosome numbers in Hal 3 and Hal 6 were similar to those of Hepa-1a, being 59 and 60 respectively. Hal 7a and Hal 7b were quite different in this respect with modal numbers of 106 and 110. Upon karyotypic analysis, it was apparent that these hybrids contained a double input from the hepatoma parent. A large proportion of the cells in these clones have 2 of the long acrocentric Hepa la marker chromosomes.

The Hal hybrids were tested for the presence of the hepatic traits expressed by Hepa la. All colonies were positive for Es-2 esterase as outlined in Table 3. Likewise all continued to express mouse serum albumin. Mouse α-fetoprotein and mouse transferrin were produced by the three hybrids tested. In Hal hybrids, therefore, all of the differentiated properties of the Hepa la parent were observed (Table 3). Extinction of these functions did not occur in this hybrid combination. It is to be remembered, that the number of human chromosomes is quite small. Possibly extinction would be observed if a greater human genetic

TABLE 2

Human Enxyme Phenotypes and Chromosomes in Hal Hybrids

Cell Strain	Human Isozymes Observed	Human Chromosomes Observed[1]	Human Chromosomes Expected
Hal 3	GPI		19
	HPRT	X	X
		(13/36)	
	GPD		
	PGK		
Hal 5	HPRT	not done	X
	GPD		
	PGK		
Hal 6	SOD-2	6	6
	ME-1	(17/47)	
	SOD-1		21
	HPRT	X	X
	GPD	(29/470	
	PGK		
		8	
		(29/47)	
Hal 7a	NP		14
	ME-1	none	rearranged 6
		(0/39)	
	HPRT		rearranged X
Hal 7b	NP		14
	HPRT	X	X
		(2/29)	
	GPD		
	PGK		

Human isozymes observed in one or more of the Hal hybrids are the following: glucose phosphate isomerase(GPI), hypoxanthine phosphoribosyl-transferase(HPRT), glucose 6 phosphate dehydrogenase(GPD), phosphoglycerate kinase(PGK), cytosol superoxide dismutase(SOD-2), malic enzyme(ME-1), mitochondrial superoxide dismutase(SOD-1), nucleoside phosphorylase(NP), Other enzymes tested, but which did not

Legend - Table 2, continued: appear in the hybrids were
adenylate kinase, adenosine deaminase, glutamic oxaloacetic
transaminase, lactate dehydrogenase A, lactate dehydrogenase
B, malate dehydrogenase, mannose phosphate isomerase,
peptidases A, B. C, and D, and xanthine oxidase.

[1] Numbers in parenthesis represent the number of cells in
which the chromosome was observed over the total number
of cells examined.

---

component were present.

Not only did the hybrids continue to express murine
serum proteins, but in Hal 7a and 7b, human serum albumin
was also produced. In these cells, the human genome was
activated to perform a function which is not normally
observed in leucocytes. Figure 2a depicts immunoelectrophoretic precipitin arcs of Hal 7a and Hal 7b supernatant
medium reacted with anti-human serum. Figure 2b illustrates
the identity of the protein in Hal 7a medium with human
serum albumin.

Activation of a non-hepatic genome was also observed
by Peterson and Weiss (1972) in hybrids between a rat
hepatoma line and mouse fibroblasts. Their observations as
well as those of Malawista and Weiss (1973) suggest that
mouse serum albumin is produced when appropriate gene
dosage relationships exist in the hybrids. An excess of
hepatoma chromosomes seems to promote activation of the
mouse genome. It may be significant that the Hal 7a and
7b hybrids in which the homologous human protein was
expressed contain a double hepatoma complement.

Activation of the non-hepatic human genome suggests a
model for mapping human loci governing specialized
functions. In appropriate hybrid combinations, it may be
possible to observe the production of tissue-specific
products by human leucocytes or fibroblasts. These tissues
are readily obtained whereas cellular material from organs
is much less accessible from adults and virtually unobtain-
able from fetuses. In our studies, it was impossible to
make an assignment for the human albumin structural locus
because of chromosomal rearrangement in the cells expressing
the trait. The potential for mapping this locus in a
larger series of hybrids is clearly present as is the
possibility of mapping any human loci which can be activated
in this hybrid system.

TABLE 3

Hepatic Phenotypes in Hal Hybrids

	Es-2	MSA	HSA	Mα-Fp	MTrf
Hal 3	+	+	-	+	+
Hal 5	+	+	-	n.d.	n.d.
Hal 6	+	+	-	+	+
Hal 7a	+	+	+	+	+
Hal 7b	+	+	+	n.d.	n.d.

Es-2 = esterase-2, MSA = mouse serum albumin, HSA = human serum albumin, Mα-Fp = mouse α-Fetoprotein, MTrf = mouse transferrin, n.d. = not done. Es-2 was analyzed by starch gel electrophoresis. Mouse serum albumin, mouse α-fetoprotein and mouse transferrin were examined on poly-acrylamide gel electrophoresis. In addition, mouse serum albumin was identified by immunologic procedures as was human serum albumin.

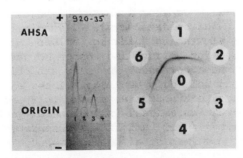

Figure 2a. Immunoelectrophoretic pattern of hybrid super-natant media. The conditions of immunoelectrophoresis have been described in a report by Bernhard et al., (1973). Well (1) contains supernatant medium of Hal 7a, concentrated 200 X

Figure 2a, continued: (2) supernatant medium of Hal 7b,
concentrated 200 x; (3) HSA, 250 ng.; (4) MSA, 250 ng.

Figure 2b. Double diffusion plate showing identity of
precipitin bands formed by Hal 7a supernatant medium and
human serum albumin with anti-human serum albumin. Well
(1) Hal 7a supernatant medium concentrated approximately
200 x; (2) Hepa 1a supernatant concentrated 200 x; (3) MSA,
Cohn Fraction V, 5 ng.; 4) Cell free growth medium; (5) HSA,
Cohn Fraction V, 5 ng.   (0) monospecific goat anti-HSA,
10 ng.

REFERENCES

Bernhard, H.P., G.J. Darlington, and F.H. Ruddle 1973. Expression of liver phenotypes in cultured mouse hepatoma cells. Synthesis and secretion of serum albumin. *Develop. Biol.* 35: 83-96.

Bertolotti, R. and M.C. Weiss 1972a. Expression of differentiated functions in hepatoma cell hybrids VI. Extinction and re-expression of liver alcohol dehydrogenase. *Biochemie* 54: 195-201.

Bertolotti, R. and M.C. Weiss 1972b. Expression of differentiated functions in hepatoma cell hybrids II. Aldolase. *J. Cell Physiol.* 79: 211-224.

Darlington, G.J., H.P. Bernhard, and F.H. Ruddle 1974. Human serum albumin phenotype activation in mouse hepatoma-human leucocyte cell hybrids. *Science* 185: 859-862.

Davidson, Richard L. 1971. Regulation of gene expression in somatic cell hybrids: a review. *In Vitro* 6: 411-426.

Hamerton, J.L., G.R. Douglas, P.A. Gee, and B.J. Richardson 1973. The association of glucose phosphate isomerase expression with human chromosome 19 using somatic cell hybrids. *Cytogenet. Cell Genet.* 12: 128.

Levisohn, S.R. and E. Thompson 1973. Contact inhibition and gene expression in HTC/L cell hybrid lines. *J. Cell Phys.* 81: 225-232.

Malawista, S.E. and M.C. Weiss 1974. Expression of differentiated functions in hepatoma cell hybrids: High frequency of induction of mouse albumin production in rat hepatoma-mouse lymphoblast hybrids. *Proc. Nat. Acad. Sci. USA* 71: 927-931.

McMorris, F.A., T.R. Chen, F. Ricciuti, J. Tischfield, R. Creagan, and F.H. Ruddle 1973. Chromosome assignments in man of the genes for two hexosephosphate isomerases. *Science* 179 1129-1131.

Peterson, J.A. and M.C. Weiss 1972. Expression of differentiated functions in hepatoma cell hybrids: induction of mouse albumin production in rat hepatoma-mouse fibroblast hybrids. *Proc. Nat. Acad. Sci. USA* 69: 571-575.

Ricciuti, F. and F.H. Ruddle 1973. Assignment of nucleoside phosphorylase to D-14 and localization of X-linked loci in man by somatic cell genetics. *Nature* 241: 180-182.

Ruddle, F.H. 1973. Linkage analysis in man by somatic cell genetics. *Nature* 242: 165-169.

Schneider, J.A. and M.C. Weiss 1971. Expression of differentiated functions in hepatoma cell hybrids I. tyrosine aminotransferase in hepatoma-fibroblast hybrids. *Proc. Nat. Acad. Sci. USA* 68: 127-131.

Tan, Y.H., J. Tischfield, and F.H. Ruddle 1973. The linkage of genes for the human interferon induced antiviral protein and indophenol oxidase-B traits to chromosome G-21. *J. Exp. Med.* 137: 317-330.

Westerveld, A., P.L. Pearson, H. van Someren, A. Johnsma, A. Hagemeijer,and D. Bootsma 1973. Assignment of genes to human chromosomes using man/Chinese hamster somatic cell hybrids. *Genetics* 74 (June Suppl.): 5295

Index